Photonics

Interface Transmission Tutorial Book Series

Series Editor
Leonard Dobrzyński

In this Series:
Phononics
Magnonics
Photonics

Interface Transmission Tutorial Book Series

Photonics

Leonard Dobrzyński

Abdellatif Akjouj

El Houssaine El Boudouti

Gaëtan Lévêque

Housni Al Wahsh

Yan Pennec

Cécile Ghouila-Houri

Abdelkrim Talbi

Bahram Djafari-Rouhani

Yabin Jin

Series Editor

Leonard Dobrzyński

ELSEVIER

Elsevier
Radarweg 29, PO Box 211, 1000 AE Amsterdam, Netherlands
The Boulevard, Langford Lane, Kidlington, Oxford OX5 1GB, United Kingdom
50 Hampshire Street, 5th Floor, Cambridge, MA 02139, United States

Library of Congress Cataloging-in-Publication Data
A catalog record for this book is available from the Library of Congress

British Library Cataloguing-in-Publication Data
A catalogue record for this book is available from the British Library

ISBN: 978-0-12-819388-4

Publisher: Matthew Deans
Acquisitions Editor: Kayla Dos Santos
Editorial Project Manager: Gabriela D. Capille
Production Project Manager: Vijayaraj Purushothaman
Designer: Matthew Limbert

Typeset by VTeX

Do simple things.

Jacques Friedel

Contents

Preface

The Interface Transmission Tutorial Book Series presents, at this stage, simple examples of Interface Transmission in Phononics, Magnonics, and Photonics. Other books are in preparation on Plasmonics, Electronics, Polaritonics, and other Mixed wave and particle interface sciences. This tutorial series is the original and tutorial fruit of rethought publications of a coherent scientific team. Its main aim is a new deeper physical understanding of nature out of the simplest possible linear models.

Such a deep rethinking produces original results. The first chapter of Magnonics and the seven first chapters of Photonics are completely new. All the other chapters of this Interface Transmission Tutorial Book Series have some original aspects and also contain many unpublished new results.

This opens broad new perspectives sketched at the end of the chapters and also in the whole last chapter of this book. We hope that some of these simple contributions may help those willing to try and use them. The Photonics book has three main parts: Paths, Circuits, and Materials. A path state is an original concept for confined states along linear paths. A confined state is strictly confined in its space subdomain.

The Photonic Path part of this book starts with the definition of loops. An open loop is a one-dimensional waveguide whose physical properties are independent of the loop curvature. A closed loop can be obtained by linear superposition of the free ends of an open loop. The main physical properties of loops are their states: eigenwavelengths (equivalently eigenfrequencies), eigenfunctions, and eigenfunction zeros (robust zeros).

Any such a zero is robust because its eigenstate cannot be activated by any action applied on it. This original concept of a robust zero is introduced here mostly to help and conceive new systems with confined states.

New final systems can be conceived out of reference systems, namely given numbers of independent open loops. The most interesting final systems are obtained by interconnecting these open loops through their robust zeros. The number of states of open loops is well known as well as the number of states of the reference systems. The number of eigenstates is invariant between any reference system and the final system out of which this system can be constructed. In order to know all the states of a system, it is essential to check this invariance.

The set of final system states has two main subsets: the path states and the shifted states. The shifted states are those for which eigenwavelengths, eigenfunctions and their zeros are modified by the construction process.

Original path states in the final systems are linear superpositions of the eigenstates of open and closed loops whose robust zeros, eigenwavelengths, and eigenfunctions remain unperturbed in the construction process. They are called path states because they are strictly confined in subdomains of the final system along some one-dimensional paths, without any decrease outside of their path.

Such path states are often degenerate. Degeneracy enables having different paths to go from one space point to another one, using different superpositions of degenerate linear eigenfunctions. This can enable going around defects and adds to the state robustness.

Original state theorems and general eigenfunction rules are defined. They enable finding, just by inspection, the path states of any final system. These inspection results are in agreement with the interface response theory exact calculations. The state phase shift enables determining all the final system states. This general approach is illustrated in this book by many exact tutorial models.

The path states are robust because they are confined within one-dimensional paths, and when they are degenerate, they can turn around defects.

Chapter 1 defines the basic properties of open loops. Chapter 2 introduces closed loops and their twin states. Chapter 3 gives state theorems and the path construction method. Simple examples follow within the next chapters for open loops in Chapter 4, closed loops in Chapter 5, and closed loops and stubs in Chapter 6. The general eigenfunction rules are demonstrated in Chapter 7. General prospectives about path and entangled path states are given in Chapter 8 of this book, including transpositions to other than photon particles.

The Photonic Circuits part of this Photonics book deals with some examples of fundamental photonic circuits. It is, in particular, essential to try and activate all the eigenstates of any circuit. This can be achieved by different theoretical and experimental approaches. The main approach used here employs interface transmission between different points of a circuit.

Chapter 9 reviews the theory and experiments related to Fano and electromagnetically induced transparency and absorption resonances. These effects are shown here to appear when the symmetry of bound in continuum path states is broken. Chapter 10 deals with demultiplexers based on photonic circuits. Chapters 11 and 12 are respectively about photonic monomode circuits made out of comb and loop structures. New results are reported about bound in continuum states. In particular, a flat band of bound in continuum states, shown to exist within a state continuum. Fibonacci photonic circuits, is the subject of Chapter 13. Microstub photonic device examples are given in Chapter 14. Optical absorption of silicon nanowires is reviewed in Chapter 15. Other experimental investigations of photonic circuits are reviewed in Chapter 16.

The Photonic Materials part of this Photonics book starts with a presentation of layered photonic materials in Chapter 17. Semiinfinite and finite one-dimensional photonic crystals are the subjects of Chapters 18 and 19, respectively. Omnidirectional bandgaps and selective transmission in photonic crystals is the matter of Chapter 20. An approach to one-dimensional photonic crystals with left-handed materials constitutes Chapter 21. Anisotropic one-dimensional photonic crystals are reviewed in Chapter 22. Chapter 23 deals with microelectromechanical systems and photonic materials.

We hope students, researchers, and professors may find within this book inspiration for their own research and lectures.

Leonard Dobrzyński
Lille, France
October 30, 2020

Acknowledgments

We dedicate this book to our families in partial compensation for taking so much of our time away from them.

It is a pleasure to thank Publishing Director Cathleen Sether, Acquisition Editor Kayla Dos Santos, Editorial Project Manager Gabriela Capille, Copyrights Coordinator Ashwathi Aravindakshan, Information Manager–Publishing Support Sajana Kanthan Payee, Cover Designer Matthew Limbert, Production Project Manager Vijayaraj Purushothaman, Publisher Matthew Deans and their Elsevier teams. They help produce today the best possible synthesis of some of the author journal publications and enable them to understand and share better the foundations of their contributions.

Leonard Dobrzyński
Lille, France
October 30, 2020

Part One

Photonic paths

Edited by Leonard Dobrzyński

Open loop

1

Leonard Dobrzyński[a], Housni Al Wahsh[b], Abdellatif Akjouj[a], El Houssaine El Boudouti[c], Cécile Ghouila-Houri[a], Abdelkrim Talbi[a], Gaëtan Lévêque[a], Bahram Djafari-Rouhani[a], Yan Pennec[a], and Yabin Jin[d]
[a]Univ. Lille, CNRS, Centrale Lille, Yncréa ISEN, Univ. Polytechnique Hauts-de-France, UMR 8520 - IEMN, Physics Department, Lille, France, [b]Benha University, Faculty of Engineering, Cairo, Egypt, [c]University Mohammed I, Faculty of Sciences, Laboratory of Physics of Matter and Radiation, Oujda, Morocco, [d]Tongji University, School of Aerospace Engineering and Applied Mechanics, Shanghai, China

1.1 Introduction

Networks can be constructed out of finite open loops. Open loops are guides such that their one-dimensional transmission properties may be considered independent of the open loop shape and radius. The wavelengths of the electromagnetic waves are assumed to be large compared to the lateral space dimensions of the open loops. Their response functions are derived from those of infinite open loops. A general definition of eigenstates is also given. The difference between resonant and force responses is made precise. These open loops have free boundary conditions at their two ends when they are used in network construction. The case of open loops with fixed end conditions is also considered.

The photonic analysis is done with the Maxwell electromagnetic propagation wave equation [1]. The photonic open loop properties are, from the mathematical point of view, similar to those of other waves, in particular to the magnon ones [2].

1.2 Infinite open loop

The Maxwell electromagnetic wave differential equation [1] for the electric field E, enabling to obtain electromagnetic waves in an infinite one-dimensional open loop without charges or electric currents is

$$\left(\frac{\partial^2}{\partial x^2} - \frac{1}{c^2}\frac{\partial^2}{\partial t^2}\right) E(x,t) = 0, \tag{1.1}$$

where x is the space position along the loop, t the time, and c the light speed within the loop.

Photonics. https://doi.org/10.1016/B978-0-12-819388-4.00010-1

The two times degenerate solutions of this differential equation are the following progressive waves:

$$E(x,t)=\cos\left[2\pi\left(\frac{x}{\lambda}-\frac{t}{\tau}\right)\right] \tag{1.2}$$

and

$$E(x,t)=\sin\left[2\pi\left(\frac{x}{\lambda}-\frac{t}{\tau}\right)\right], \tag{1.3}$$

where λ is the wavelength and τ the time period. In most of what follows, use is made of their standing wave form, when $t=0$ in the above two equations. These eigenfunctions are valid for a continuum of eigenvalues from $1/\lambda=0$ to $1/\lambda=\infty$.

Note also that when $t=0$ and for the eigenvalue $1/\lambda=0$, these eigenfunctions are equal to 1 for the invariance by translation state and to 0 for the ground state, respectively.

These Maxwell equation solutions provide the following relation:

$$\lambda=c\tau, \tag{1.4}$$

equivalent to $\omega=2\pi c/\lambda$, where ω is the angular frequency.

The loop lateral dimensions are supposed small compared to the loop length and to the photon wavelength λ.

With these solutions, the electromagnetic wave propagation equation may be rewritten as

$$\left(\frac{\partial^2}{\partial x^2}+\left(\frac{2\pi}{\lambda}\right)^2\right)E(x)=0. \tag{1.5}$$

Introduce now the Green's function **G** [3], also called the response function, by

$$\left(\frac{\partial^2}{\partial x^2}+\left(\frac{2\pi}{\lambda}\right)^2\right)\mathbf{G}(x-x')=\delta(x-x'), \tag{1.6}$$

where $\delta(x-x')$ is the Dirac delta function. This response function is

$$\mathbf{G}(x-x')=\frac{\lambda}{4j\pi}\exp\left(\frac{2j\pi\left|x-x'\right|}{\lambda}\right), \tag{1.7}$$

where $j^2=-1$.

Note that this response function diverges when the wavelength is infinite. This corresponds to the photon ground state.

The response function enables obtaining the field response $E(x)$ of the system when a force source term $F(x')$ is applied to it

$$E(x)=\mathbf{G}(x-x')F(x'). \tag{1.8}$$

The electromagnetic field propagates along the x axis, the electric field E is polarized along the y axis, and the magnetic field H is along the z axis.

The Maxwell equation for the magnetic field has the same form as the above equations for the electric field.

1.3 Free end semiinfinite open loop

In the interface response theory [4], the end of a semiinfinite loop is treated as a perturbation of the infinite loop. Within this approach, one defines first the surface cleavage operator [5] which is the perturbation

$$V(x) = \delta(x)\frac{\partial}{\partial x}, \tag{1.9}$$

where $\delta(x)$ is the Kronecker function indicating that this perturbation acts only at the point x, where the infinite loop is cleaved into two semiinfinite ones. This cleavage operator creates two semiinfinite loops with free ends.

Then one can define the surface response operator

$$\mathbf{A}_s(x, x') = V(x'')\mathbf{G}(x'' - x'), \tag{1.10}$$

for $x'' = x$. In this expression, x takes only the semiinfinite loop 0 end value.

Then one obtains

$$\mathbf{A}_s(0, 0) = -\frac{1}{2}. \tag{1.11}$$

The response function for each semiinfinite loop can then be found, using the corresponding truncated part of $\mathbf{G}$. For the semiinfinite loop such that $x \geq 0$, one finds

$$\mathbf{g}_s(x, x') = \frac{\lambda}{4j\pi}\left[\exp\left(\frac{2j\pi\,|x - x'|}{\lambda}\right) + \exp\left(\frac{2j\pi(x - x')}{\lambda}\right)\right]. \tag{1.12}$$

This cleavage procedure, creating two semiinfinite loops (final system) out of one infinite one (reference system), see Fig. 1.1, produces a phase shift η [5].

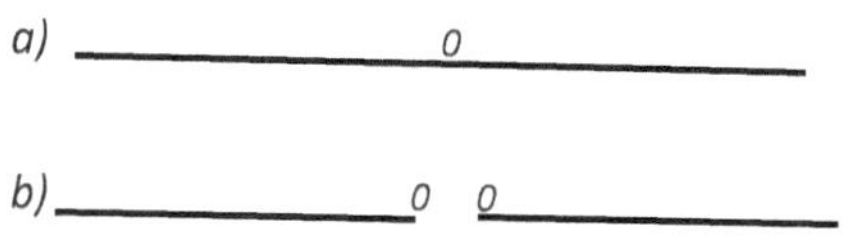

Figure 1.1 (a) The infinite loop is the reference system. (b) The two semiinfinite loops are the final system.

The derivative (divided by π) of η versus ω is the density of state variation between the final and reference systems. This phase shift cannot be calculated between one semiinfinite loop and the truncated corresponding part of the infinite loop, as such a truncated infinite loop has no well-defined eigenstates. That is why one has to get this phase shift between the two semiinfinite loops and the infinite one. It is

$$\eta(k) = \arg\left[\frac{[\mathbf{g}_s(0,0)]^2}{\mathbf{G}(0,0)}\right] = \arg\frac{\lambda}{2j\pi} = \frac{\pi}{2}, \tag{1.13}$$

where arg is the argument of the complex term which follows it, see Fig. 1.2. The argument of $1/\lambda$ shows that the two semiinfinite loops have one more state for $1/\lambda = 0$ than the infinite loop. Indeed, there are two ground states in the two semiinfinite loops and only one in the infinite loop. The argument of $1/j$ shows a $-\pi/2$ phase shift for $1/\lambda = 0$. So the phase shift is equal to $\pi/2$ all along the $1/\lambda$ axis. Note also that, because of the conservation of the state number rule, this phase shift has to drop back to zero when $1/\lambda$ is infinite. This $\pi/2$ phase shift is due to the fact that the antisymmetric states of the infinite loop shift and become symmetric states. So each semiinfinite loop has only symmetric states. This point is made clearer at the end of this section, where a finite loop L is cleaved into two loops $L/2$. Then the limit of infinite L shows this shift of the antisymmetric states into symmetric ones, with this $\pi/2$ phase shift.

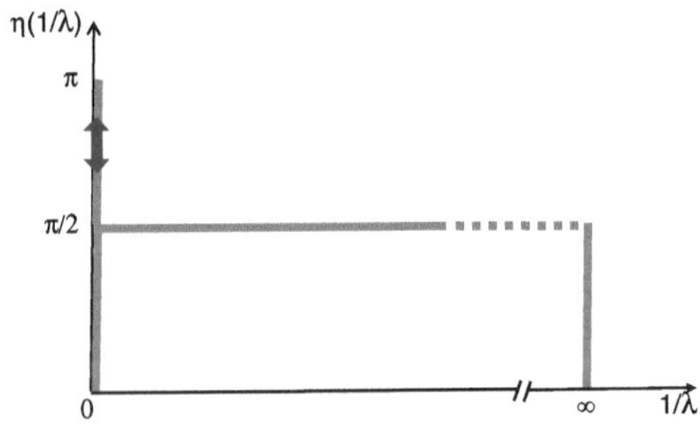

Figure 1.2 The phase shift $\eta(1/\lambda)$ between the two semiinfinite loops and the infinite loop.

This phase shift shows antisymmetric state collapse [6] and shift into symmetric ones.

1.4 Finite open loop

1.4.1 Matrix

In order to find the response function of an open loop (finite loop) of length L, it is possible to proceed as in the previous section. The free loop ends are treated as perturbations of the infinite loop within the response function approach.

The surface cleavage operator [5] acts now on the interface space M points $x = 0$ and $x = L$ in order to cleave, out of the infinite loop, a finite loop of length L with free ends.

The surface response operator is then obtained to be

$$\mathbf{A}_s(MM) = -\frac{1}{2}\begin{pmatrix} 1 & \exp\left(\frac{2j\pi L}{\lambda}\right) \\ \exp\left(\frac{2j\pi L}{\lambda}\right) & 1 \end{pmatrix}, \tag{1.14}$$

where $M = (0, L)$ is the two end loop interface space.

This enables obtaining the response function $\mathbf{g}$ of the finite loop and first

$$\Delta(MM) = \mathbf{I}(MM) + \mathbf{A}_s(MM). \tag{1.15}$$

Then

$$[\mathbf{g}(MM)]^{-1} = \Delta(MM)[\mathbf{G}(MM)]^{-1}, \tag{1.16}$$

where

$$\mathbf{G}(MM) = \frac{\lambda}{4j\pi}\begin{pmatrix} 1 & \exp\left(\frac{2j\pi L}{\lambda}\right) \\ \exp\left(\frac{2j\pi L}{\lambda}\right) & 1 \end{pmatrix}. \tag{1.17}$$

This provides

$$[\mathbf{g}(MM)]^{-1} = \frac{2\pi}{\lambda S}\begin{pmatrix} -C & 1 \\ 1 & -C \end{pmatrix}, \tag{1.18}$$

where

$$S = \sin\left(\frac{2\pi L}{\lambda}\right) \tag{1.19}$$

and

$$C = \cos\left(\frac{2\pi L}{\lambda}\right). \tag{1.20}$$

This is the interface space (2×2) matrix for the free ends of the finite loop.

1.4.2 Determinant

The determinant of the above matrix is equal to

$$\left|[\mathbf{g}(MM)]^{-1}\right| = -\left(\frac{2\pi}{\lambda}\right)^2. \tag{1.21}$$

1.4.3 State definition

In any linear system without damping, the eigenstates, called states, correspond to excitations, producing no source force F at any system point.

For the finite photonic networks considered here, this definition leads to the following equation:

$$F(M) = [\mathbf{g}(MM)]^{-1} E(M) = 0. \tag{1.22}$$

Here $g(x, x')$ is the network response function between points x and x', $E(x)$ the electric field of the photonic wave at point x. This means that the network eigenvalues are the solutions of

$$\left|[\mathbf{g}(MM)]^{-1}\right| = 0, \tag{1.23}$$

for all network eigenstates for which $E(M) \neq 0$, as there may be, for other systems than the one considered here, states for which $E(M) = 0$. To determine the eigenvalues when $E(M) = 0$, one has to look for solutions of Eq. (1.23) for other network points than those of a particular interface space.

1.4.4 State phase shift

The final system considered now is obtained by cleaving one finite loop and two semi-infinite loops out of an infinite one, see Fig. 1.3.

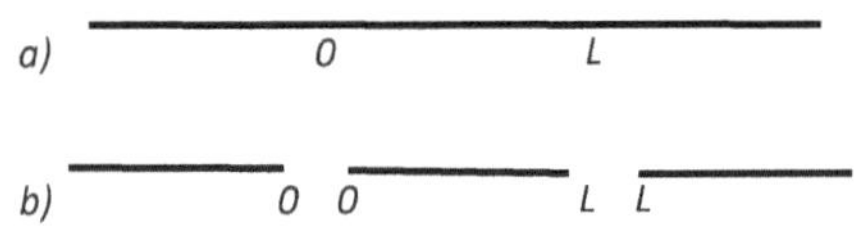

Figure 1.3 (a) The infinite loop is the reference system. (b) The two semiinfinite loops and the finite loop are the final system.

The eigenwavelengths and eigenfunctions are shifted compared to those of the infinite loop. This state phase shift [5] between the final and the reference system is obtained to be

$$\eta(\lambda) = \arg\left[\frac{g_s(0,0)g_s(L,L)\,|g(MM)|}{|G(MM)|}\right], \tag{1.24}$$

where $|g(MM)|$ is the determinant of $g(MM)$.

Using Eqs. (1.12), (1.17), and (1.21), one obtains

$$\eta(\lambda) = \arg\left[\frac{-j\lambda^2 \exp\left(-j\frac{2\pi L}{\lambda}\right)}{(2\pi)^2 S}\right] = \frac{3\pi}{2} - \frac{2\pi L}{\lambda} + \arg\left[\frac{1}{S}\right], \tag{1.25}$$

where account was taken of that the final system has two more states for infinite λ, than the reference one. This contributes a $+2\pi$ to this phase shift, see Fig. 1.4. The jump of η from 0 to $5\pi/2$ for infinite λ is due to the $\pi/2$ related to the creation of two semiinfinite loops, as in Fig. 1.2, added to the $+2\pi$ related to the creation of the finite loop.

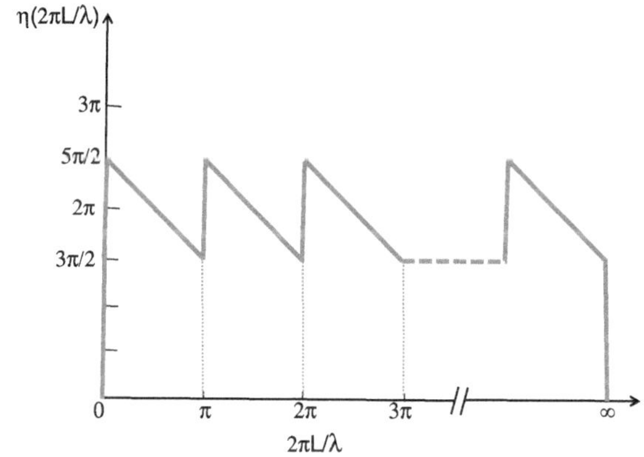

Figure 1.4 The phase shift $\eta(\frac{2\pi L}{\lambda})$ between the final system (one finite loop of length L with two semi-infinite loops) and the reference system (one infinite loop).

Note the appearance of the discrete $S = 0$ states of the finite loop of length L which replace the continuous shift $-\frac{2\pi L}{\lambda}$ of the bulk states. The finite loop eigenwavelengths have now discrete rather than continuous values.

Because of the state conservation rule, there is also a $-3\pi/2$ jump for $\lambda = 0$, related to the inverse jump for infinite λ.

1.4.5 Eigenvalues and interface eigenvectors

From the inverse of Eq. (1.18), one sees that the eigenvalues of this waveguide are the values of λ for which $S/\lambda = 0$.

In the limit of infinite λ, there are two degenerate eigenvalues. One is the ground state $1/\lambda = 0$ for which the eigenfunction is equal to 0 at all space points of the open loop. The other degenerate $1/\lambda = 0$ eigenvalue, contained within $S = 0$, is due to the invariance by translation state. The corresponding eigenfunction is equal to 1 at all space points of the open loop.

All the other open loop states are also given by $S = 0$. It is convenient to consider first the $C = 1$ states and then the $C = -1$ ones.

For

$$\begin{aligned} &C = 1, \\ &\lambda = \frac{L}{n}, \quad n = 0, 1, 2, 3, \ldots, \end{aligned} \tag{1.26}$$

where $n = 0$ recalls the invariance by translation state presented just before.

The corresponding loop end eigenvector is

$$\mathbf{E} = \frac{1}{\sqrt{2}} \begin{pmatrix} 1 \\ 1 \end{pmatrix}. \tag{1.27}$$

For

$$C = -1, \\ \lambda = \frac{2L}{1+2n}, \quad n = 0, 1, 2, 3, \ldots, \tag{1.28}$$

and the corresponding loop end eigenvector is

$$\mathbf{E} = \frac{1}{\sqrt{2}}\begin{pmatrix} 1 \\ -1 \end{pmatrix}. \tag{1.29}$$

1.4.6 Response function

The matrix inverse of $[\mathbf{g}(MM)]^{-1}$ is the interface response matrix

$$\mathbf{g}(MM) = \frac{\lambda}{2\pi S}\begin{pmatrix} C & 1 \\ 1 & C \end{pmatrix}. \tag{1.30}$$

It is also possible to obtain the response function between any two points x and x' belonging to the space D of this finite loop. This may be done with the help of the general equation [4]

$$\begin{aligned} \mathbf{g}(DD) = {} & \mathbf{G}(DD) - \mathbf{G}(DM)[\mathbf{G}(MM)]^{-1}\mathbf{G}(MD) \\ & + \mathbf{G}(DM)[\mathbf{G}(MM)]^{-1}\mathbf{g}(MM)[\mathbf{G}(MM)]^{-1}\mathbf{G}(MD), \end{aligned} \tag{1.31}$$

where $\mathbf{G}$ is the response function of the infinite loop given by Eq. (1.7), written here in its matrix form, while D is the space of the infinite loop.

The explicit form of the finite loop response function is

$$\mathbf{g}(x, x') = \frac{\lambda}{2\pi S}\cos\left[\frac{2\pi x}{\lambda}\right]\cos\left[\frac{2\pi(L-x')}{\lambda}\right], \quad \text{for } x \le x', \tag{1.32}$$

and

$$\mathbf{g}(x, x') = \frac{\lambda}{2\pi S}\cos\left[\frac{2\pi x'}{\lambda}\right]\cos\left[\frac{2\pi(L-x)}{\lambda}\right], \quad \text{for } x \ge x'. \tag{1.33}$$

It is interesting to stress that when one takes the response function limit as L becomes infinite, one recovers the response function in Eq. (1.12) of the semiinfinite loop, with the help of

$$\left[\frac{\cos\left[(\frac{2\pi}{\lambda} + j\epsilon)L\right]}{\sin\left[(\frac{2\pi}{\lambda} + j\epsilon)L\right]}\right]_{L\to\infty} = \frac{1}{j}, \tag{1.34}$$

where ϵ is an infinitely small positive number.

1.4.7 Complete eigenfunctions

With the help of Eq. (1.33), applying a unit excitation [5] on $x' = L$, one obtains the complete eigenfunctions, for $S = 0$, in agreement with its above corresponding loop end eigenvectors in Eqs. (1.27) and (1.29), to be

$$E(x) = \sqrt{\frac{2}{L}} \cos\left(\frac{n\pi x}{L}\right). \tag{1.35}$$

1.4.8 Resonant and forced responses

The system responds for all values of λ, once an action $\mathbf{F}$ is applied to it. However, the resonant responses for each of the eigenvalues, namely those for which $S = 0$, are the only ones for which the electric field diverges.

For values of λ for which there is no resonance, this loop may have forced responses.

1.4.9 Fixed end open loop

One may like to use such an open loop with other than free end boundary conditions. This may be done by adding source force terms F_1 and F_2 at the points 0 and L. This enables us to obtain the electric fields E_1 and E_2 due to these source forces as

$$\begin{pmatrix} E_1 \\ E_2 \end{pmatrix} = \frac{\lambda}{2\pi S} \begin{pmatrix} CF_1 + F_2 \\ F_1 + CF_2 \end{pmatrix}. \tag{1.36}$$

Note that this does not change the $(S/\lambda = 0)$ eigenvalues.

Consider the case of fixed ends ($E_1 = E_2 = 0$). For the eigenvalues given by $C = 1$, this leads to $F_1 = -F_2$. For the eigenvalues given by $C = -1$, this leads to $F_1 = F_2$. Although the eigenvalues are not shifted due to these new boundary conditions, the symmetric eigenfunctions $\cos(n\pi \frac{x}{L})$ are shifted by $\pi/2$ and become $\sin(n\pi \frac{x}{L})$, as both symmetric and antisymmetric eigenfunctions have now to be equal to zero at the open loop ends.

The source forces described here compensate exactly at the loop ends the inside source forces due to the derivatives of $E(x)$, in agreement with the general eigenfunction rules established in the last section of this chapter.

1.4.10 Breaking translational invariance

Break the translational invariance of an open loop of length $2L$ into two open loops of length L. With the help of Eq. (1.30), one finds for this cleavage the following phase shift:

$$\eta_L(\lambda) = \arg \frac{\lambda \cos\left(\frac{2\pi L}{\lambda}\right)}{2\pi \sin\left(\frac{2\pi L}{\lambda}\right)} = \arg \frac{\lambda}{2\pi} \frac{C}{S}. \tag{1.37}$$

In this cleavage, the initial antisymmetric states, $\lambda = 4L/(1 + 2n)$, $n = 0, 1, 2, 3, \ldots$, shift to the same wavelength, $\lambda = 2L/n$, $n = 0, 1, 2, 3, \ldots$, as the initial symmetric ones. The initial $\lambda = 4L$ state becomes one $\lambda \to \infty$ state. A second $\lambda \to \infty$ state appears also in the above phase shift. For the infinite wavelength, $\eta_L(\lambda)$ has then two $+\pi$ jumps. One of these jumps is the signature of the state [7,8], related to the breaking of translational symmetry invariance of any finite system. This state is the shifted antisymmetric $4L$ state of the open loop $2L$. Then the phase shift has a $-\pi$ jump for $\lambda = 4L$, followed by a $+\pi$ jump for $\lambda = 2L$ due to the shift of the antisymmetric $4L/3$ state, and so on. For $\lambda = 0$, it has a $-\pi$ jump from $+\pi$ to 0, in order to satisfy the state conservation rule. (See Fig. 1.5.)

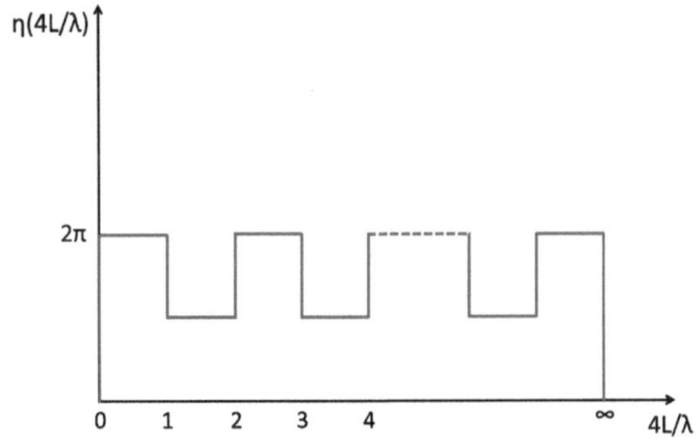

Figure 1.5 The state phase shift $\eta(4L/\lambda)$ between two finite loops of length L and a finite loop of length $2L$. The limit when L becomes infinite is the phase shift shown in Fig. 1.2. Note the 2π jump for $4L/\lambda = 0$. This shows the existence of two degenerate states when λ is infinite for any system finite and infinite. These ground and invariant by translation states are of fundamental importance in the physical understanding of the delta peaks at the edges of bulk bands. They play also a fundamental role in the invariance by translation breaks and in the surface and interface variations of the thermodynamic functions.

The initial system has the states of $\frac{1}{\lambda}SC = 0$. Using Eq. (1.37), the final system has then the states of

$$\left(\frac{1}{\lambda}SC\right)\left(\frac{1}{\lambda}\frac{S}{C}\right) = \left(\frac{1}{\lambda}\right)^2 S^2 = 0. \tag{1.38}$$

These are indeed the states of two finite open loops of length L.

Take now the $L \to \infty$ limit of the above $\eta_L(\lambda)$, with the help of

$$\left[\frac{\cos\left[\left(\frac{2\pi}{\lambda} + j\epsilon\right)L\right]}{\sin\left[\left(\frac{2\pi}{\lambda} + j\epsilon\right)L\right]}\right]_{L\to\infty} = \frac{1}{j}, \tag{1.39}$$

where ϵ is an infinitely small positive number.

One recovers the $\eta(\lambda) = \frac{\pi}{2}$ of Eq. (1.13) and Fig. 1.2.

In the cleavage of the finite loop, the antisymmetric states shift to symmetric ones. Each antisymmetric eigenwavelength reaches the same wavelength as its closest highest symmetric wavelength. At the $L \to \infty$ limit, the symmetric and antisymmetric states keep their own eigenfunctions, which differ only by a $\frac{\pi}{2}$ phase shift.

1.5 Perspectives

The open loop presented in this chapter is the elementary building element. It enables constructing most of the simple model systems presented in this book series. In Chapter 2 of this book, the closed loop is obtained just by linear superposition of the free ends of an open loop.

References

[1] J.C. Maxwell, A dynamical theory of the electromagnetic field, Philos. Trans. R. Soc. Lond. 155 (1865) 459–512.
[2] L. Dobrzyński, A. Akjouj, G. Lévêque, et al., Centered system magnons, in: Magnonics, in: Interface Transmission Tutorial Book Series, Elsevier, 2019.
[3] G. Green, An Essay. Applications to the Mathematical Analysis of the Theories of Elasticity and Magnetism, printed for the author by T. Wheelhouse, sold by Hamilton, Adams, Nottingham, and Deighton, Cambridge, 1828.
[4] L. Dobrzyński, Interface response theory of continuous composite systems, Surf. Sci. Rep. 11 (1990) 164.
[5] L. Dobrzyński, A. Akjouj, E. El Boudouti, et al., Interface response theory, in: Phononics, in: Interface Transmission Tutorial Book Series, Elsevier, 2017.
[6] W. Heisenberg, Über den anschaulichen Inhalt der quantentheoretischen Kinematik und Mechanik, Z. Phys. 43 (1927) 172.
[7] J. Goldstone, Field theories with "superconductor" solutions, Nuovo Cimento 19 (1961) 151.
[8] J. Goldstone, A. Salam, S. Weinberg, Broken symmetries, Phys. Rev. 127 (1962) 965.

Closed loop

2

Leonard Dobrzyński[a], Housni Al Wahsh[b], Abdellatif Akjouj[a], El Houssaine El Boudouti[c], Cécile Ghouila-Houri[a], Abdelkrim Talbi[a], Gaëtan Lévêque[a], and Bahram Djafari-Rouhani[a]
[a]Univ. Lille, CNRS, Centrale Lille, Yncréa ISEN, Univ. Polytechnique Hauts-de-France, UMR 8520 - IEMN, Physics Department, Lille, France, [b]Benha University, Faculty of Engineering, Cairo, Egypt, [c]University Mohammed I, Faculty of Sciences, Laboratory of Physics of Matter and Radiation, Oujda, Morocco

2.1 Introduction

A closed loop can be obtained by linear superposition of the two free ends of an open loop. Its main property is the double degeneracy of its states. The analytical expressions of its eigenvalues, eigenfunctions, and response function are presented. The case of two simultaneous identical measures at two different points of a closed loop is briefly addressed. The simultaneous activation of these two degenerate states is also considered.

2.2 Closing an open loop

The inverse of the response function of the open loop depicted in Fig. 2.1 is given (see the preceding chapter of this book) by

$$[\mathbf{g}(MM)]^{-1} = \frac{2\pi}{\lambda S}\begin{pmatrix} -C & 1 \\ 1 & -C \end{pmatrix}, \tag{2.1}$$

where $M = (0, L)$ is the interface space of the open loop,

$$S = \sin\left(\frac{2\pi L}{\lambda}\right), \tag{2.2}$$

and

$$C = \cos\left(\frac{2\pi L}{\lambda}\right), \tag{2.3}$$

where λ is the wavelength.

Photonics. https://doi.org/10.1016/B978-0-12-819388-4.00011-3

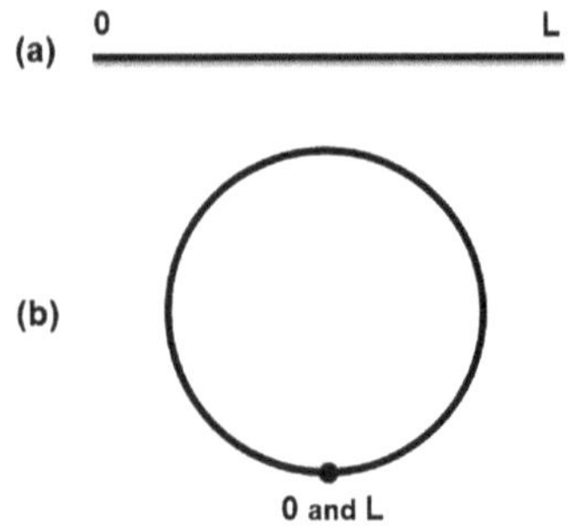

Figure 2.1 (a) The open loop of length L. (b) The closed loop of length L.

If we close this loop by superposing its ends 0 and L, the inverse of the interface response function on the connection point is

$$[g(0,0)]^{-1} = \frac{2\pi}{\lambda S}(-2C+2) = \frac{4\pi}{\lambda}\frac{\sin\left(\frac{\pi L}{\lambda}\right)}{\cos\left(\frac{\pi L}{\lambda}\right)}. \tag{2.4}$$

2.3 Eigenvalues and eigenfunctions

The eigenvalues of the reference system (open loop of length L) are given by $\frac{2\pi}{\lambda}S = 0$, see Eq. (1.30). Note also that $|[g(MM)]^{-1}| = -\left(\frac{2\pi}{\lambda}\right)^2$, see Eq. (1.21).

So taking into account the phase shift result and the state number conservation [1], the eigenvalues of the final system (closed loop of length L) are given by

$$\frac{4\pi}{\lambda}\sin\left(\frac{\pi L}{\lambda}\right)\cos\left(\frac{\pi L}{\lambda}\right)\frac{[g(00)]^{-1}}{|[\mathbf{g}(MM)]^{-1}|} = -\sin^2\left(\frac{\pi L}{\lambda}\right) = 0. \tag{2.5}$$

These eigenvalues are then two times degenerate. This two-times degeneracy is in agreement, when L becomes infinite, with the Born–von Karman cyclic boundary conditions.

The eigenwavelengths are

$$\lambda_m = \frac{L}{m}, \qquad m = 0, 1, 2, 3, \ldots \tag{2.6}$$

Note also that this double degeneracy is valid for the infinite λ states. One of these states is the ground state of the closed loop and the other is the translation invariance one.

It is also possible to understand the eigenwavelength shifts between the open and closed loops from the following physical explanation. Any open loop antisymmetric eigenfunction cannot be superposed at the loop ends because their values have the opposite signs at these ends and there must be continuity in the closed loop. So each antisymmetric eigenwavelength shifts to the same value as that of its closest and shorter symmetric eigenwavelength. Thanks to this eigenwavelength shift, the corresponding antisymmetric eigenfunction may keep its sine form as in the open loop. In the closed loop, its value becomes zero at the superposed ends.

So the antisymmetric eigenfunctions can remain, as in the open loop, to be

$$E(x) = \sin\left(2m\pi\frac{x}{L}\right), \tag{2.7}$$

and the symmetric ones

$$E(x) = \cos\left(2m\pi\frac{x}{L}\right), \tag{2.8}$$

where $m = 0, 1, 2, 3, \ldots$

We stress that when one of these two degenerate eigenfunctions is 0, the other takes the value $+1$ or -1.

The states of the closed loop are twice degenerate. Their eigenfunctions are the same as those of the infinite open loop, described in the preceding chapter. But their eigenvalues are discrete while those of the infinite open loop are continuous.

2.4 Response function

The response function $g(DD)$ for a closed loop can be derived using the general equation [1]:

$$\begin{aligned} g(DD) &= G(DD) - G(DM)[G(MM)]^{-1}G(MD) \\ &+ G(DM)[G(MM)]^{-1}g(MM)[G(MM)]^{-1}G(MD), \end{aligned} \tag{2.9}$$

where $G(DD)$ stands for the response function of the open loop of the same length L as that of the closed loop. Recall (see [1]) the following partial result, for $x \geq x'$:

$$\begin{aligned} G(DD) - G(DM)[G(MM)]^{-1}G(MD) &= \frac{-\lambda}{2\pi}\cos\left(\frac{2\pi x}{\lambda}\right)\sin\left(\frac{2\pi x'}{\lambda}\right) \\ &+ \frac{C\lambda \sin\left(\frac{2\pi x}{\lambda}\right)\sin\left(\frac{2\pi x'}{\lambda}\right)}{4\pi\cos\left(\frac{\pi L}{\lambda}\right)\sin\left(\frac{\pi L}{\lambda}\right)} \end{aligned} \tag{2.10}$$

and

$$\begin{aligned} &G(DM)[G(MM)]^{-1}g(MM)[G(MM)]^{-1}G(MD) = \\ &\frac{1}{S^2}\frac{\lambda\cos\left(\frac{\pi L}{\lambda}\right)}{4\pi\sin\left(\frac{\pi L}{\lambda}\right)}\begin{pmatrix} \sin\left[\frac{2\pi(L-x)}{\lambda}\right], & \sin\left(\frac{2\pi x}{\lambda}\right)\end{pmatrix} \\ &\times\begin{pmatrix} 1 & 1 \\ 1 & 1\end{pmatrix}\begin{pmatrix} \sin\left[\frac{2\pi(L-x')}{\lambda}\right] \\ \sin\left(\frac{2\pi x'}{\lambda}\right)\end{pmatrix}, \end{aligned} \tag{2.11}$$

where the four elements of $g(MM)$ have the same value because the ends 0 and L of the open loop are superposed and are one and have the same space point in the closed loop.

Therefore,

$$g(x,x') = \frac{\lambda}{4\pi}\frac{\cos\left[\frac{2\pi}{\lambda}\left(\frac{L}{2} - |x - x'|\right)\right]}{\sin\left(\frac{\pi L}{\lambda}\right)}. \tag{2.12}$$

2.5 Two simultaneous identical responses

Note that $g(x,x)$ is independent of the position x on the closed loop. So the closed loop response at x to a source force applied at x is the same all around the closed loop. Such identical responses, for all network space points, can be obtained also only along an infinite open loop. Here this is true for any value of the finite circumference length L.

Note that the responses at the positions x and x' are the same when identical source forces are applied for these two space positions. These identical responses are functions of only $|x - x'|$ and not x and x'. This is also only true for an infinite open loop.

2.6 Activation of the two states of closed loops

For the same wavelength λ, two degenerate states, $\left[\sin\left(\frac{2\pi}{\lambda}x\right)\right]$ and $\left[\cos\left(\frac{2\pi}{\lambda}x\right)\right]$, coexist within the same space domain. When one equals 1, the other equals 0. The only other case where this exists, to our knowledge, is the infinite open loop.

Simultaneous activation of these two degenerate states can be achieved for $\lambda = L$, for example, see Fig. 2.2. There $L/4$ is the distance between the nearest neighbor loop dots. The simultaneous inputs are I_1 and I_2. The simultaneous outputs are $O_1 = I_1$ and $O_2 = I_2$. This result is obtained here just by inspection of the robust zeros of each of the two degenerate loop states. It implies that the only interactions of the loop with its outside world can be achieved through these four points. The inputs are supposed to be brief laser shots with negligible influence on the original loop states.

Another activation method can be used, too. It uses four semiinfinite open loops for the inputs and outputs and two finite loops of length L' attached at each of these four points. This perturbs the loop states. However, when L' is incommensurable with L, it is possible to observe the same effects, but for $\lambda \approx L/2$, see, for example, Fig. 6.21, Chapter 6 of this Photonics book.

Such simple results pave the way for more efficient devices for wave and particle cross-transmission [2] and enhanced directional transmission [3].

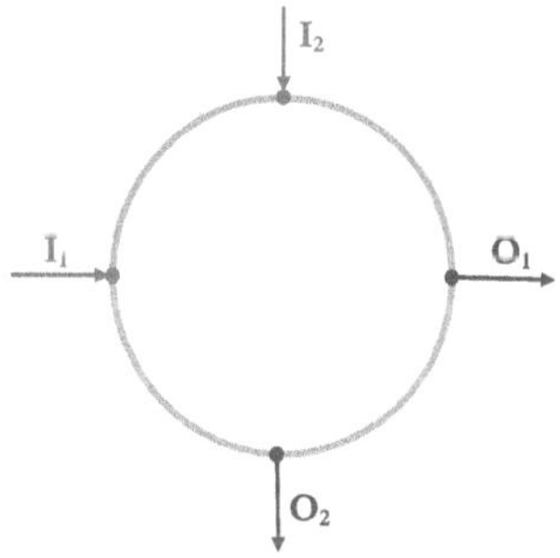

Figure 2.2 Simultaneous activation of the two degenerate loop states for $\lambda = L$, for example; L is the closed loop length, λ the eigenwavelength, and $L/4$ the distance between the nearest neighbor loop dots. The simultaneous inputs are I_1 and I_2. The simultaneous outputs are $O_1 = I_1$ and $O_2 = I_2$.

References

[1] L. Dobrzyński, A. Akjouj, G. Lévêque, et al., Centered system magnons, in: Magnonics, in: Interface Transmission Tutorial Book Series, Elsevier, 2019.
[2] L. Dobrzyński, A. Akjouj, Particle cross transfer, Surf. Sci. Rep. 63 (2008) 391.
[3] L. Dobrzyński, A. Akjouj, Enhanced directional optical transmission, Appl. Phys. Lett. 93 (2008) 043114.

Path states

3

Leonard Dobrzyński[a], Housni Al Wahsh[b], Abdellatif Akjouj[a], El Houssaine El Boudouti[c], Bahram Djafari-Rouhani[a], Cécile Ghouila-Houri[a], Abdelkrim Talbi[a], and Gaëtan Lévêque[a]
[a]Univ. Lille, CNRS, Centrale Lille, Yncréa ISEN, Univ. Polytechnique Hauts-de-France, UMR 8520 - IEMN, Physics Department, Lille, France, [b]Benha University, Faculty of Engineering, Cairo, Egypt, [c]University Mohammed I, Faculty of Sciences, Laboratory of Physics of Matter and Radiation, Oujda, Morocco

3.1 Introduction

Any open and closed loop eigenfunction has zero motion space points, called here robust zeros. Any such zero is robust because its eigenstate cannot be activated by any action applied on it. Indeed, such zero space positions remain unaffected when the loop is interfaced with its outside world only through one or several such robust zeros. The corresponding eigenwavelengths and eigenfunctions remain also unaffected by such interfacing.

When this interfacing introduces one or several new pathways whose lengths are commensurable with the unaffected eigenwavelength, the eigenstate becomes degenerate or more degenerate. Such states remain strictly confined within an ensemble of open or closed loop paths.

A confined state is strictly confined in its space subdomain. So path states belong to a subset of confined states. They are confined within combinations of open and closed loops.

In such an interfacing process, the other eigenvalues shift. They may have in general more complicated eigenfunctions. Similar considerations can be developed for any system state and lead to a confined state theorem. This enables also to propose a construction method of systems with confined states. This construction method is developed in what follows for path states. They are the simplest ones, as being confined along loop paths. When the path states are degenerate, their paths can go around defects.

The path states are robust because they are confined within one-dimensional paths, and when they are degenerate they can turn around defects.

This chapter presents also general eigenfunction rules. These rules are mainly eigenfunction continuities and outgoing source forces equal to zero, at each network point. The source forces are for electromagnetic waves, the outgoing derivatives of the wave electric and magnetic fields. For other types of waves, the source forces are, for

Photonics. https://doi.org/10.1016/B978-0-12-819388-4.00012-5

example, the outgoing stresses, currents, etc. These new tutorial results are illustrated by simple examples.

All this can lead to a construction procedure of robust transmission lines. The confined path states can pave the way for the future of information communication technology and contribute to many other sciences.

3.2 Path state properties

Information travels along a path within a communication network. This path goes through each physical component and material of the network. The used state eigenfunction gains to be confined within a one-dimensional path. It is important also to be able to turn around defects. When the path states are degenerate, they offer possibilities to turn around path defects. These are the main properties of path states.

New path states can be obtained with a confined state theorem and general eigenfunction rules. These rules are established in Chapter 7 of this Photonics book. It is shown there that at each space point of a network, for all eigenstates, the eigenfunctions are continuous and the sum of the outgoing source forces is equal to zero. For electromagnetic waves, the source forces are the derivatives of the fields.

The path states are different from usual surface [1], interface [2], edge and wedge [3], and adsorbed layer [4] states. The main difference is that the path states have no decaying character outside their space subdomain, in which they are strictly confined.

The edge states, first reported for electrons within the quantized Hall conductance [5], seem to be a particular kind of path states. Similar edge states were found also for photons [6] and other waves. These edge states can be path states when they are strictly confined within loops. If they present decays outside one-dimensional paths, they may be confined higher order states or decaying in space states.

Edge path states can be obtained by linear superposition of several degenerated bulk loop path states, as shown explicitly by many examples in Chapter 5 of this book. When the edge structure is modified as compared to the bulk, edge path states can exist as bound in continuum states or as confined states in gaps between the bulk bands. One considers that a bulk band exists when the dimensions of the system are big enough in order to get gathering of its discrete states within the limits of the bulk bands of the same infinite system.

3.3 State theorems

The theorems which follow are direct consequences of the properties of the response function elements $g(x, x')$ of any finite system [7]. The response function is related to the inverse of a Lagrangian, or a Hamiltonian. The determinant of any response function element, called also kernel, has the information about some eigenenergies (or wavelengths) of this finite system. The numerator of this response function elements contains the corresponding eigenfunctions, with their zeros.

Each state is defined by its eigenenergy, its eigenfunction, its eigenfunction domain or subdomain, and its robust zeros. This is particularly important for confined and degenerate states. For each such state, one has, for a degenerate eigenenergy, a different eigenfunction and a different robust zero set. Even if one part of such sets may be common between several degenerate states, the eigenfunction domain or subdomain may be the same for two different eigenfunctions, for example, for one closed loop. It is in general different. For more examples of such degenerate states, see Chapters 3, 4, and 5 of this Photonics book.

The response function enables finding the outgoing response at x' to an incoming action at x. This response diverges only for the discrete eigenenergies of the finite system and only when x and x' belong to one of its eigenfunction subdomains and are not its robust zeros.

Any state can be activated (excited). In order to achieve this, the system has to be connected to some device which perturbs the state parameters. As a simple example, it is possible to consider the additions of semiinfinite wires to the activation space points of a finite system. This introduces $2\pi j/\lambda$ diagonal terms in the inverse interface response matrix. This results in modifications of the response function: eigenwavelength degeneracy lifts and shifts, single state transformations into resonances, as well as eigenfunction modifications. So the state activation (measure) device has to be chosen to be as little perturbing as possible.

It can be easier to understand first these general response function properties on even simpler finite systems [8]. The generalization for infinite systems is then less tedious.

3.3.1 State number conservation theorem

The number of eigenstates is invariant between any reference system and the final system out of which this system can be constructed [7].

3.3.2 Confined state theorem

Any state interfaced only through some of its eigenfunction zeros with the other states of its system is a subdomain confined state. It has no interaction with its outside world.

3.3.3 Bound in continuum state theorem

Any state interfaced only through its robust zeros with a state continuum is a subdomain confined state. It is also a bound in a continuum state [9] when its eigenenergy is within a bulk band [8]. It can be also a flat band state when its eigenenergy is in a gap.

3.3.4 *State activation theorems*

3.3.4.1 *Two state active points*

Any state can be activated by one incoming action at one space point and one outgoing action at another point, provided that these two space points of its domain or subdomain are not zeros of its eigenfunction.

3.3.4.2 *One state active point and one system deformation point*

Any state can be activated just by one single action on one space point of its domain or subdomain, provided that this space point is not a zero of its eigenfunction and that another degenerate state does not coexist at this point.

This last condition can be fulfilled when the degeneracy is lifted by a small deformation at another space point of this state domain. Such a deformation shifts an original discrete state into a resonance, for example, that of Fano [10].

3.3.5 *Application to path states*

Any open or closed loop eigenfunction has zero motion space points, called robust zeros. Such zero space positions remain unaffected when the loop is interfaced with its outside world only through one or several such zeros. The corresponding loop eigenwavelengths and eigenfunctions remain also unaffected by such interfacing.

When this interfacing introduces one or several new confined pathways whose lengths are commensurable with the unaffected eigenwavelength, the eigenstate may become degenerate or more degenerate. Such states remain confined within open and (or) closed loop subdomains. They are called path states because they are confined strictly within space subdomains constituted out of open and (or) closed loops, without any decay away from their respective paths. In such an interfacing process, the other eigenvalues shift. They have in general more complicated eigenfunctions.

When a state is interfaced with a bulk continuum only through some of its robust zeros, it is a bulk in continuum state. Such examples appear in Chapter 9 of this Photonics book for simple path states. It is shown there that a specific detuning of some of the geometrical parameters of path state subdomains can partly deconfine such bound in continuum states [11]. Resonances, for example, that of Fano [10], then appear. Similar examples for flat band states appear in Chapters 5, 12, and 13.

3.4 General eigenfunction rules

3.4.1 *Rule 1*

The eigenfunctions are continuous.

3.4.2 Rule 2

The eigenfunction outgoing derivatives produce source forces. The sum of these forces vanishes at each space point for all eigenvalues. The sum of all such forces vanishes at all surface structure points, too [12].

3.5 Robust zeros and eigenvalues

Now the robust zeros of the basic eigenfunctions $\sin(\frac{2\pi x}{\lambda})$ and $\cos(\frac{2\pi x}{\lambda})$ are presented for loops and infinite lines.

3.5.1 Free end open loop

The eigenvalues for the open loop of length L depicted in Fig. 3.1(a) are given by $S = \frac{1}{\lambda}\sin(2\pi L/\lambda) = 0$, which provides the following eigenvalues:

$$\lambda_m = \frac{2L}{m}, \quad m = 0, 1, 2, 3, \ldots \tag{3.1}$$

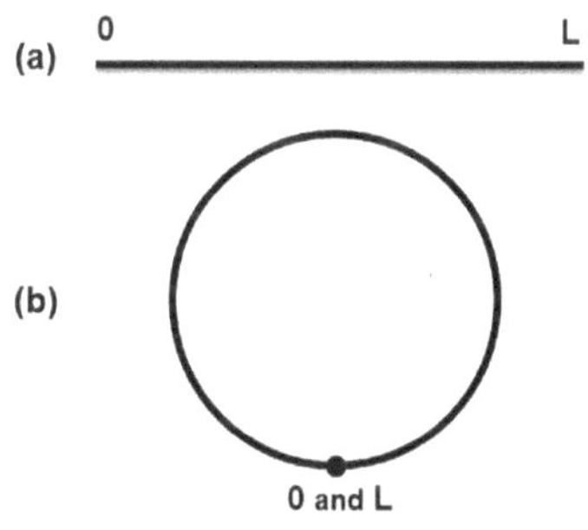

Figure 3.1 (a) The open loop of length L. (b) The closed loop of length L.

The $m = 0$ value provides the first two degenerate eigenvalues. It gives $\lambda = \infty$, which represents the ground state and the invariance by translation state of the whole loop. There is no eigenfunction zero associated to these states.

The eigenfunctions of this open loop are $E(x) = \cos(m\pi x/L)$.

So the corresponding eigenfunction zeros are

$$x_{m,n} = \left(\frac{2n+1}{2m}\right) L \le L, \quad m = 1, 2, 3, \ldots, \quad n = 0, 1, 2, 3, \ldots \tag{3.2}$$

In order to be more explicit, here are the robust zeros for $m = 1, 2, 3, 4$:

$$\begin{aligned}
\lambda_1 &= 2L, & m &= 1, & x_{m,n} &= \frac{L}{2}, \\
\lambda_2 &= L, & m &= 2, & x_{m,n} &= \frac{L}{4}, \frac{3L}{4}, \\
\lambda_3 &= \frac{2L}{3}, & m &= 3, & x_{m,n} &= \frac{L}{6}, \frac{L}{2}, \frac{5L}{6}, \\
\lambda_4 &= \frac{L}{2}, & m &= 4, & x_{m,n} &= \frac{L}{8}, \frac{3L}{8}, \frac{5L}{8}, \frac{7L}{8}.
\end{aligned} \tag{3.3}$$

Note that once a zero is robust for a given eigenvalue λ_m, it is also robust for the following sequence of eigenvalues:

$$\lambda_p = \left(\frac{1}{2p+1}\right)\frac{2L}{m}, \quad m = 1, 2, 3, \ldots, \quad p = 0, 1, 2, 3, \ldots, \tag{3.4}$$

where m is equal, for this free end open loop, to the number of robust zeros associated with the eigenwavelength $\lambda_m = \frac{2L}{m}$.

Equivalently, the λ_p are the roots of

$$\cos\left(\frac{\pi L}{m\lambda_p}\right) = 0. \tag{3.5}$$

Note that $L/(2m)$ is the distance between $x = 0$ and the first robust zero associated to the wavelength λ_m.

It is also interesting to notice, using Eq. (1.32) of Chapter 1 and Eq. (3.2), that for open loops $g(x, x') = 0$, when x and x' are robust zeros.

3.5.2 Fixed end open loop

The eigenvalues, for a fixed end open loop of length L, are given by $S = 0$, i.e., $\sin(2\pi L/\lambda) = 0$, which provides the following eigenvalues:

$$\lambda_m = \frac{2L}{m}, \quad m = 0, 1, 2, 3, \ldots \tag{3.6}$$

Note that $L/(2m)$ is also the distance between two nearest neighbor robust zeros associated with the wavelength λ_m.

The $m = 0$ value provides the first eigenvalue. It gives $\lambda = \infty$, which is the loop ground state. There is no eigenfunction zero associated with this state.

The corresponding eigenfunctions are $E(x) = \sin(m\pi x/L)$. So the corresponding eigenfunction zeros are

$$x_{mn} = \left(\frac{n}{m}\right)L \leq L. \tag{3.7}$$

In order to be more explicit, here are the robust zeros for $m = 1, 2, 3, 4$:

$$\begin{aligned}
\lambda_1 &= 2L, & m &= 1, & x_{m,n} &= 0, L, \\
\lambda_2 &= L, & m &= 2, & x_{m,n} &= 0, \frac{L}{2}, L, \\
\lambda_3 &= \frac{2L}{3}, & m &= 3, & x_{m,n} &= 0, \frac{L}{3}, \frac{2L}{3}, L, \\
\lambda_4 &= \frac{L}{2}, & m &= 4, & x_{m,n} &= 0, \frac{L}{4}, \frac{L}{2}, \frac{3L}{4}, L.
\end{aligned} \tag{3.8}$$

Note that once a zero is robust for a given eigenvalue λ_m, it is also robust for the following sequence of eigenvalues:

$$\lambda_p = \left(\frac{1}{p}\right)\frac{2L}{m}, \quad m = 1, 2, 3, \ldots, \quad p = 1, 2, 3, \ldots, \tag{3.9}$$

where $m + 1$ is equal, for this fixed end open loop, to the number of robust zeros associated with the eigenwavelength $\lambda_m = \frac{2L}{m}$.

Equivalently, the λ_p are the roots of

$$\sin\left(\frac{2\pi L}{m\lambda_p}\right) = 0. \tag{3.10}$$

Note that L/m is the distance between two nearest neighbor robust zeros associated with the wavelength $\lambda_m = 2L/m$.

3.5.3 Closed loop

A closed loop of length L is depicted in Fig. 3.1(b).

3.5.3.1 For the first degenerate state

The path eigenvalues for a closed loop of length L are given by $\frac{1}{\lambda}\sin\left(\frac{\pi L}{\lambda}\right) = 0$, which provides the following eigenvalues:

$$\lambda_m = \frac{L}{m}, \quad m = 0, 1, 2, 3, \ldots \tag{3.11}$$

The $m = 0$ value provides the first degenerate eigenvalue. It gives $\lambda = \infty$, which represents the loop ground state. There is no eigenfunction zero associated with this state.

Consider $E(x) = \sin(2m\pi x/L)$ as the first of the two degenerate eigenfunctions. So the corresponding eigenfunction zeros are:

$$x_{mn} = \left(\frac{n}{2m}\right)L \leq L, \quad n = 0, 1, 2, 3, \ldots \tag{3.12}$$

In order to be more explicit, here are the robust zeros for the $m = 1, 2, 3, 4$:

$$\begin{aligned}
&\lambda_1 = L, && m = 1, && x_{m,n} = 0, \frac{L}{2}, \\
&\lambda_2 = \frac{L}{2}, && m = 2, && x_{m,n} = 0, \frac{L}{4}, \frac{L}{2}, \frac{3L}{4}, \\
&\lambda_3 = \frac{L}{3}, && m = 3, && x_{m,n} = 0, \frac{L}{6}, \frac{L}{3}, \frac{L}{2}, \frac{2L}{3}, \frac{5L}{6}, \\
&\lambda_4 = \frac{L}{4}, && m = 4, && x_{m,n} = 0, \frac{L}{8}, \frac{L}{4}, \frac{3L}{8}, \frac{L}{2}, \frac{5L}{8}, \frac{3L}{4}, \frac{7L}{8}.
\end{aligned} \tag{3.13}$$

Note that once a zero is robust for a given eigenvalue λ_m, it is also robust for the following sequence of eigenvalues:

$$\lambda_p = \left(\frac{1}{p}\right)\frac{L}{m}, \quad m = 1, 2, 3, \ldots, \quad p = 0, 1, 2, 3, \ldots, \tag{3.14}$$

where m is equal, for this closed loop, to half the even number of robust zeros associated with the eigenwavelength $\lambda_m = \frac{L}{m}$.

Equivalently, the λ_p are the roots of

$$\sin\left(\frac{\pi L}{m\lambda_p}\right) = 0. \tag{3.15}$$

Note that $L/(2m)$ is also the distance between two nearest neighbor robust zeros associated with the wavelength λ_m.

3.5.3.2 For the second degenerate state

Consider $E(x) = \cos(2m\pi x/L)$ as the second of the two degenerate eigenfunctions. So the corresponding eigenfunction zeros are:

$$x_{mn} = \left(\frac{1+2n}{2m}\right)\frac{L}{2} \leq L. \tag{3.16}$$

In order to be more explicit, here are the robust zeros for $m = 1, 2, 3, 4$:

$$\begin{aligned}
&\lambda_1 = L, && m = 1, && x_{m,n} = \frac{L}{4}, \frac{3L}{4}, \\
&\lambda_2 = \frac{L}{2}, && m = 2, && x_{m,n} = \frac{L}{8}, \frac{3L}{8}, \frac{5L}{8}, \frac{7L}{8}, \\
&\lambda_3 = \frac{L}{3}, && m = 3, && x_{m,n} = \frac{L}{12}, \frac{L}{4}, \frac{5L}{12}, \frac{7L}{12}, \frac{3L}{4}, \frac{11L}{12}, \\
&\lambda_4 = \frac{L}{4}, && m = 4, && x_{m,n} = \frac{L}{16}, \frac{3L}{16}, \frac{5L}{16}, \frac{7L}{16}, \frac{9L}{16}, \frac{11L}{16}, \frac{13L}{16}, \frac{15L}{16}.
\end{aligned} \tag{3.17}$$

Note that the present zeros cannot be robust together with those obtained just before for the first degenerate eigenfunction. As explained in the above subsection where these two degenerate eigenfunctions were obtained, when one has the value 0, the other has the value ± 1. It is also interesting to notice, using Eqs. (2.12), (3.12), and (3.16), that for closed loops, $g(x, x')$ does not always vanish when x and x' are robust zero sites.

3.5.4 Infinite open loop

The infinite line has two degenerate states for each of his continuous eigenwavelength λ. The corresponding robust zeros are for each of these states as follows.

3.5.4.1 For the first degenerate state

Consider $\sin(2\pi x/\lambda)$ as the first eigenfunction. So the corresponding eigenfunction zeros are:

$$x_m = m\frac{\lambda}{2}, \quad m = 0, 1, 2, 3, \ldots, \tag{3.18}$$

where x is positive on both sides of the origin 0, in order to remain consistent with the convention used before for the finite loops. So the distance between the two nearest neighbor robust zeros is $\frac{\lambda}{2}$.

3.5.4.2 For the second degenerate state

Consider $\cos(2\pi x/\lambda)$ as the second eigenfunction. So the corresponding eigenfunction zeros are:

$$x_m = (1 + 2m)\frac{\lambda}{4}, \quad m = 0, 1, 2, 3, \ldots \tag{3.19}$$

Note that the present zeros cannot be robust together with those obtained just before for the first degenerate eigenfunction. As explained in the above subsection where these two degenerate eigenfunctions were obtained, when one has the value 0, the other has the value ± 1.

It is also interesting to notice, using Eqs. (1.7), (3.18), and (3.19), that for the infinite line, $g(x, x')$ does not vanish when x and x' are robust zero sites.

The first connection with the infinite line interacts with the $\cos(2\pi x/\lambda)$ eigenfunction. The connection point defines the 0 for x. The $\sin(2\pi x/\lambda)$ remains a path eigenfunction. Many other connections are possible.

3.6 Path state construction

Once the robust zeros are defined, it is possible to create confined path state using some of these zeros and keeping them unperturbed, as well as the states associated with them. This procedure defines, for a given choice of robust zeros, which states remain robust.

The simple examples given in the next three chapters are aimed to try and better understand the above general creation strategy.

3.7 Some perspectives

This chapter can be better understood when one knows some of the simple examples given in Chapters 4, 5, and 6 of this book. It tries and generalizes the results of these examples. It tries and foresees many new research directions, from the solutions of simple exact models. For the sake of simplicity and better physical understanding, this chapter invites to address first simple finite structures. It invites also to conceive novel structures constructed with the help of the robust eigenfunction zeros. One of the advantages of such an approach is the possibility to try and obtain all the eigenstates of a final system from a reference system for which all the eigenstates are known. Once all the states of such finite systems are obtained, it is easy to understand all its states when its dimensions become infinite. In particular, the existence of eventual bound in continuum states and flat state bands appears in a clear and logical way.

It is also possible to start from a finite or infinite system and try to find some new unknown states and new fundamental and applied properties of a system. This is the main approach used in the Circuits and Materials part of this book. This research direction uses here mainly theoretical and experimental methods of interface transmission. This research direction is widely developed elsewhere for edge states, for example, since its pioneering papers [5] and [6]. It uses mainly small geometrical deformation methods. Are all states of such materials known? This remains often an open question.

Another research direction is worth to be explored. Each final system can be decomposed into a given number of elementary pieces for which all the eigenstates are known. Then with the rule of conservation of the number of states and the state phase shift [7], all the states of the final system can be obtained.

All such research directions are complementary and enable our progress.

Acknowledgments

One of us (L.D.) thanks Ludwik Dobrzynski, emeritus research professor at CNRS and CERN, for stimulating scientific exchanges.

References

[1] L. Rayleigh, On waves propagated along the plane surface of an elastic solid, Proc. Lond. Math. Soc. 17 (1887) 4.

[2] R. Stoneley, Elastic waves at the surface of separation of two solids, Proc. R. Soc. Lond. A 106 (738) (1924) 416.
[3] L. Dobrzyński, A.A. Maradudin, Electrostatic edge modes in a dielectric wedge, Phys. Rev. B 6 (1972) 3810.
[4] L. Dobrzyński, D.L. Mills, Vibrational properties of an adsorbed surface layer on a simple model crystal, J. Phys. Chem. Solids 30 (1969) 1043.
[5] B.I. Halperin, Quantized Hall conductance, current-carrying edge states, and the existence of extended states in a two dimensional disordered potential, Phys. Rev. B 25 (1982) 2185.
[6] F.D.M. Haldane, S. Raghu, Possible realization of directional optical waveguides in photonic crystals with broken time-reversal symmetry, Phys. Rev. Lett. 100 (2008) 013904.
[7] L. Dobrzynski, A. Akjouj, E.H. El Boudouti, et al., Interface response theory, in: Phononics, in: Interface Transmission Tutorial Book Series, Elsevier, 2017, Chapter 1.
[8] L. Dobrzyński, A. Akjouj, G. Lévêque, et al., Centered system magnons, in: Magnonics, in: Interface Transmission Tutorial Book Series, Elsevier, 2019.
[9] J. Von Neuman, E. Wigner, Über merkwürdige diskrete Eigenwerte, Phys. Z. 30 (1929) 465.
[10] U. Fano, Effects of configuration interaction on intensities and phase shifts, Phys. Rev. 124 (1961) 1866.
[11] E.H. El Boudouti, A. Akjouj, B. Djafari-Rouhani, et al., Electromagnetic induced transparency, induced absorption and Fano resonances in photonic circuits, in: Photonics, in: Interface Transmission Tutorial Book Series, Elsevier, 2021, Chapter 9.
[12] L. Dobrzyński, H. Al Wahsh, A. Akjouj, et al., Eigenfunction rules, in: Photonics, in: Interface Transmission Tutorial Book Series, Elsevier, 2021, Chapter 7.

Open loop examples

4

Housni Al Wahsh[a], Leonard Dobrzyński[b], Abdellatif Akjouj[b], and El Houssaine El Boudouti[c]
[a]Benha University, Faculty of Engineering, Cairo, Egypt, [b]Univ. Lille, CNRS, Centrale Lille, Yncréa ISEN, Univ. Polytechnique Hauts-de-France, UMR 8520 - IEMN, Physics Department, Lille, France, [c]University Mohammed I, Faculty of Sciences, Laboratory of Physics of Matter and Radiation, Oujda, Morocco

4.1 Introduction

Free-end open loop examples of robust zeros and of path states may be found already in [1]. The simplest general robust zero is the middle of an open loop. When the middles of several free-end open loops of the same length are superposed, their antisymmetric eigenfunctions remain unperturbed and the final network has several degenerate states. This is published for magnons and phonons in [1]. The transposition of these results for photons is straightforward. The first two simple examples, which follow here, confirm these results. The next three examples take into account spatial symmetry breaks. It is assumed for simplicity that all loops are made out of the same dielectric material.

A robust eigenfunction zero is robust because its eigenstate cannot be activated by any action applied on it. The path states are robust because they are confined within one-dimensional paths and when they are degenerate they can turn around defects, see Chapter 3 of this book.

4.2 T network

The present simple T network, see Fig. 4.1, can be considered as a broken symmetry case of the four interacting objects structure published in the preceding book of this series [1] for magnons and for phonons. Indeed, the effects of changing one of the three constituting loops of this network are considered here. This is a simple example of one robust zero connected to another loop.

As a reference system, one takes here two loops of length L and another loop of length L', all made of the same material. The final system is obtained by superposing the 0 ends of these three loops, see Fig. 4.1. The x coordinates vary from 0 to L on the first two loops and from 0 to L' on the third loop.

Photonics. https://doi.org/10.1016/B978-0-12-819388-4.00013-7

4.2.1 Path states by inspection

The 0 of the two connected loops of length L is a robust zero. The confined state theorem, see Chapter 3 of this book, says that this zero remains unperturbed by the connection with the third loop, as do its eigenvalues $C=0$, namely

$$\lambda_p = \frac{4L}{2p+1}, \quad p=0,1,2,3,\ldots, \tag{4.1}$$

and its antisymmetric eigenfunctions

$$E(x) = \pm \sin\left[\pi(2p+1)\frac{x}{2L}\right], \quad p=0,1,2,3,\ldots, \tag{4.2}$$

with the "+" sign when x is within one L loop and the "−" sign when x is within the other L loop.

This state is the simplest example of a path state as it is strictly confined within the loop $2L$ without any decay away from its localization space.

Each of these antisymmetric path eigenvalues, λ_p, becomes two times degenerate in the final network, for some specific values of L'/L derived below. The symmetric states of the two connected loops of length L and the states of the loop of length L' collapse in the final system. New states replace them and are fully derived in what follows.

All the eigenvalues of the final system are found from two approaches: the state phase shift and the kernels of the response functions. Both analyses confirm the path eigenvalues obtained above by inspection. The eigenfunctions of the new states are obtained from the numerators of the response functions.

4.2.2 The interface response inverse matrix

The inverse of the interface response matrix of the network depicted in Fig. 4.1 is obtained by superposition of the inverses of the interface response matrices, see Eq. (1.18), of the three reference systems and is

$$[g(MM)]^{-1} = \frac{2\pi}{\lambda}\begin{pmatrix} -\frac{2C}{S}-\frac{C'}{S'} & \frac{1}{S} & \frac{1}{S} & \frac{1}{S'} \\ \frac{1}{S} & -\frac{C}{S} & 0 & 0 \\ \frac{1}{S} & 0 & -\frac{C}{S} & 0 \\ \frac{1}{S'} & 0 & 0 & -\frac{C'}{S'} \end{pmatrix}, \tag{4.3}$$

where $S=\sin\left(\frac{2\pi L}{\lambda}\right)$, $S'=\sin\left(\frac{2\pi L'}{\lambda}\right)$, $C=\cos\left(\frac{2\pi L}{\lambda}\right)$, and $C'=\cos\left(\frac{2\pi L'}{\lambda}\right)$.

Its determinant is

$$|[g(MM)]^{-1}| = -\left(\frac{2\pi}{\lambda}\right)^4 \frac{C(2SC'+S'C)}{S^2S'}. \tag{4.4}$$

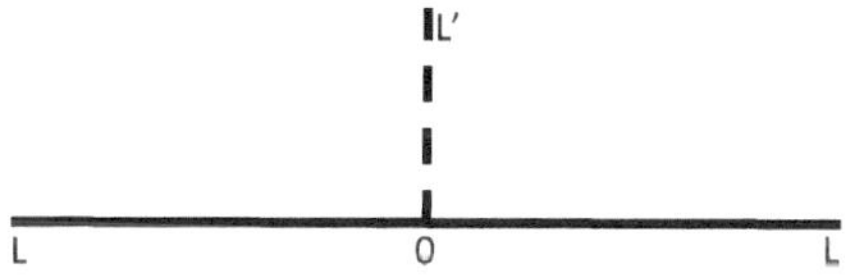

Figure 4.1 The T network. One robust zero of an open loop of length $2L$ connected with another free-end open loop of length L'.

4.2.3 All eigenvalues from the state phase shift

The eigenvalues of an open loop of length L are given by $\left(\frac{2\pi}{\lambda}S\right) = 0$, see Eq. (1.30). Note also that its determinant is $|[g(MM)]^{-1}| = -\left(\frac{2\pi}{\lambda}\right)^2$, see Eq. (1.21). So the reference system eigenvalues are given by

$$\left(\frac{2\pi}{\lambda}\right)^3 S^2 S' = 0. \tag{4.5}$$

The state phase shift [1] between the final and the reference system is obtained, as for the preceding examples, to be

$$\eta(\lambda) = -\arg \frac{\lambda^2 C(2SC' + S'C)}{S^2 S'}. \tag{4.6}$$

So the eigenvalues of this final system are given by

$$\frac{C(2SC' + S'C)}{\lambda} = 0. \tag{4.7}$$

This result confirms that $C = 0$ provides path eigenvalues as obtained from the confined state theorem. The relation $(2SC' + S'C) = 0$ provides the new eigenvalues of the final system. Note also the two times degenerate states for infinite λ. These results can also be found from the kernels of the full response functions derived in what follows.

4.2.4 Complete response function

By inverting the above matrix, one obtains the interface space response matrix

$$g(MM) = \frac{\lambda}{2\pi C(2SC' + S'C)} \times \begin{pmatrix} C^2C' & CC' & CC' & C^2 \\ CC' & -[CSS' + C'(1 - 2C^2)] & C' & C \\ CC' & C' & -[CSS' + C'(1 - 2C^2)] & C \\ C^2 & C & C & C(CC' - 2SS') \end{pmatrix}. \tag{4.8}$$

The response function between any two space points of the final network can be obtained with the general Eq. (2.9) as is done above for the response function of the closed loop. Here one has to distinguish the following four cases.

4.2.4.1 *x and x′ in the same wire L*

In this case, one gets $g(x,x')$ for $x \leq x'$ to be

$$g(x,x') = \frac{\lambda}{2\pi}\cos\left(\frac{2\pi(L-x')}{\lambda}\right) \times \left[-\frac{1}{C}\sin\left(\frac{2\pi x}{\lambda}\right) + \frac{C'}{C(2SC'+S'C)}\cos\left(\frac{2\pi(L-x)}{\lambda}\right)\right]. \tag{4.9}$$

4.2.4.2 *x and x′ in the same wire L′*

In this case, one gets $g(x,x')$ for $x \leq x'$ to be

$$g(x,x') = \frac{\lambda}{2\pi(2SC'+S'C)}\cos\left(\frac{2\pi(L'-x')}{\lambda}\right) \times \left[C\cos\left(\frac{2\pi x}{\lambda}\right) - 2S\sin\left(\frac{2\pi x}{\lambda}\right)\right]. \tag{4.10}$$

4.2.4.3 *x in the wire L and x′ in the wire L′*

After calculation one obtains

$$g(x,x') = \frac{\lambda}{2\pi(2SC'+S'C)}\cos\left(\frac{2\pi(L'-x')}{\lambda}\right)\cos\left(\frac{2\pi(L-x)}{\lambda}\right). \tag{4.11}$$

4.2.4.4 *x in one wire L and x′ in the other wire L*

After calculation one obtains

$$g(x,x') = \frac{\lambda C'}{2\pi C(2SC'+S'C)}\cos\left(\frac{2\pi(L-x)}{\lambda}\right)\cos\left(\frac{2\pi(L-x')}{\lambda}\right). \tag{4.12}$$

4.2.5 All eigenvalues and eigenfunctions from the response function

From the numerators of the above response function elements, one obtains their eigenfunction for each eigenvalue.

4.2.5.1 *The ground state*

The ground state eigenvalue is $1/\lambda$. The corresponding eigenvector is the 0 displacement of all system points.

4.2.5.2 For $C = 0$

When $C = 0$, $\lambda_p = 4L/(2p+1)$, $p = 0, 1, 2, 3, \ldots$ are path eigenvalues. The eigenfunctions associated to them are confined within the two loops of length L. Their explicit expression is

$$E(x) = \pm \sin\left(\frac{(1+2p)\pi x}{2L}\right), \tag{4.13}$$

with the "+" sign when x is within one L loop and the "−" sign when x is within the other L loop.

These states remain confined in the final system as expected from the confined state theorem.

4.2.5.3 For $2SC' + S'C = 0$

The eigenfunctions associated to these eigenvalues are localized within the three loops. Keep in mind that in what follows λ has the discrete values found from $2SC' + S'C = 0$. The explicit expressions of the eigenfunctions, associated to these eigenvalues, are

$$E(x) = C' \cos\left[\frac{2\pi(L-x)}{\lambda}\right], \tag{4.14}$$

within the loops L and

$$E(x) = C \cos\left[\frac{2\pi(L'-x)}{\lambda}\right], \tag{4.15}$$

within the loop L'. The factors C and C' in the above eigenfunctions assure their continuity at $x = 0$.

Note that the eigenvalue $1/\lambda = 0$ corresponds here to the invariant by translation displacement of the whole structure.

One has $C = C' = 0$ when $\frac{L}{L'} = \frac{1+2n}{1+2n'}$ with $n = 0, 1, 2, 3, \ldots$ and $n' = 0, 1, 2, 3, \ldots$ This particular case provides other path eigenfunctions confined within one loop L and the loop L'. Their path eigenfunctions are as those given just above for the $C = 0$ states. When this happens, there is also another state localized within the three loops whose eigenfunction expression reads as those just above.

4.2.6 A possible application: a path bifurcation

Consider the path state described above, excited in the two identical branches L of the T network. Changing the length of one of these branches to L', one obtains a shift of the initial path state into a new one confined within the two branches L'. It is a path bifurcation. It is also possible to call it a pathway switch.

4.3 Asymmetric cross

The present asymmetric cross network, see Fig. 4.2, can be considered as a broken symmetry case of the five interacting objects structure published in the preceding book of this series [1] for magnons and phonons. Indeed, here the effects of changing the lengths of two of the four constituting loops of this network are considered. The analysis of the mathematical results, from the point of view of the path states, done here can also be transposed to magnons, phonons, and other waves. This is an example of two superposed robust zeros.

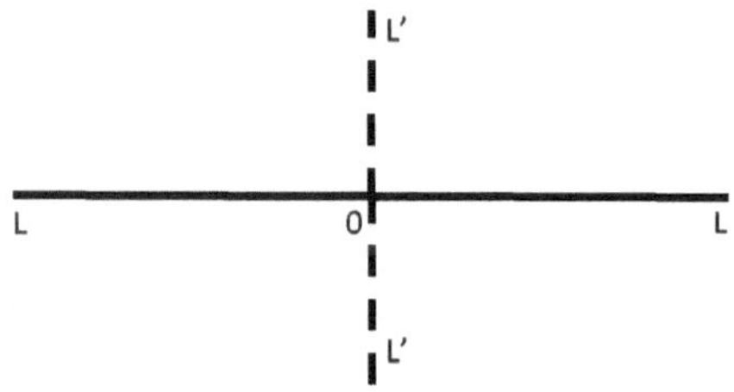

Figure 4.2 The asymmetric cross. Two superposed different robust zeros of open loops of length $2L$ and $2L'$.

As a reference system, one takes here two loops of length L and two loops of length L', all made out of the same material. The final network is obtained by superposing the 0 ends of these four loops, see Fig. 4.2. The x_i coordinates vary from 0 to L on the first two loops and from 0 to L' on the two other loops.

4.3.1 Path states by inspection

The middle zero of the loop of length $2L$ and the middle zero of the loop $2L'$ remain robust zeros, as stated by the confined state theorem.

The eigenvalues and confined antisymmetric eigenfunctions of the loop $2L$ are

$$\lambda_p = \frac{4L}{2p+1}, \quad p = 0, 1, 2, 3, \ldots \tag{4.16}$$

and

$$E(x) = \pm \sin\left[\pi(2p+1)\frac{x}{2L}\right], \quad p = 0, 1, 2, 3, \ldots, \tag{4.17}$$

with the "+" sign when x is within one L loop and the "−" sign when x is within the other L loop.

The eigenvalues and eigenfunctions confined within the loop $2L'$, once L is replaced by L', have the same expressions as those just given above.

These states are simple examples of path states as they are strictly confined within the loop $2L$ or $2L'$ without any decay away from their localization space.

Each of these path antisymmetric states becomes three times degenerate in the final network, for some specific values of L'/L derived hereafter. The symmetric states of the two connected loops of length L and the two connected loops of length L' collapse in the final system. New states replace them.

4.3.2 The interface response inverse matrix

The inverse of the response function of the network depicted in Fig. 4.2 is obtained by superposition of the inverses of the response functions, see Eq. (1.18), of the four reference system loops and is given by

$$[g(MM)]^{-1} = \frac{2\pi}{\lambda}\begin{pmatrix} -2(\frac{C}{S}+\frac{C'}{S'}) & \frac{1}{S} & \frac{1}{S'} & \frac{1}{S} & \frac{1}{S'} \\ \frac{1}{S} & -\frac{C}{S} & 0 & 0 & 0 \\ \frac{1}{S'} & 0 & -\frac{C'}{S'} & 0 & 0 \\ \frac{1}{S} & 0 & 0 & -\frac{C}{S} & 0 \\ \frac{1}{S'} & 0 & 0 & 0 & -\frac{C'}{S'} \end{pmatrix}, \tag{4.18}$$

where $S = \sin\left(\frac{2\pi L}{\lambda}\right)$, $S' = \sin\left(\frac{2\pi L'}{\lambda}\right)$, $C = \cos\left(\frac{2\pi L}{\lambda}\right)$, and $C' = \cos\left(\frac{2\pi L'}{\lambda}\right)$.

Its determinant is

$$|[g(MM)]^{-1}| = -\left(\frac{2\pi}{\lambda}\right)^5 \frac{CC'}{S^2 S'^2}\sin\left[\frac{2\pi(L+L')}{\lambda}\right]. \tag{4.19}$$

4.3.3 All eigenvalues from the state phase shift

Note also that for each reference loop one has $|[g(MM)]^{-1}| = -\left(\frac{2\pi}{\lambda}\right)^2$, see Eq. (1.21). The state phase shift [1] between the final and the reference system is then

$$\eta(k) = -\arg\left(\frac{\lambda^3 CC'}{S^2 S'^2}\sin\left[\frac{2\pi(L+L')}{\lambda}\right]\right). \tag{4.20}$$

The eigenvalues of an open loop of length L are given by $(S/\lambda = 0)$, see Eq. (1.30). So the reference system eigenvalues are given by $S^2 S'^2/\lambda^4 = 0$.

Thus the eigenvalues of this final system are given by

$$\frac{CC'}{\lambda}\sin\left[\frac{2\pi(L+L')}{\lambda}\right] = 0. \tag{4.21}$$

This result confirms that $C = 0$ and $C' = 0$ provide path eigenvalues as obtained from the confined state theorem. The relation $\sin\left[\frac{2\pi(L+L')}{\lambda}\right] = 0$ provides the new eigenvalues of the final system. Note also the two degenerate states for infinite λ.

These results can be found also from the kernels of the full response functions derived hereafter.

4.3.4 All eigenvalues and interface space eigenvectors

In what follows, for each eigenvalue λ found from the above equation, its corresponding interface space eigenvector(s) are given.

4.3.4.1 The ground state

The ground state eigenvalue is $1/\lambda$. The corresponding eigenvector is the 0 displacement of all system points.

4.3.4.2 For $C = 0$

When $C = 0$, $\lambda = 4L/(1+2n)$, $n = 0, 1, 2, 3, \ldots$ eigenvalues correspond to the path antisymmetric states confined in the two loops L. Their explicit interface eigenvectors are

$$E(M) = \begin{pmatrix} 0 \\ 1 \\ 0 \\ -1 \\ 0 \end{pmatrix}. \tag{4.22}$$

These states remain confined in the final system as found from the confined state theorem.

4.3.4.3 For $C' = 0$

When $C' = 0$, $\lambda = 4L'/(1+2n)$, $n = 0, 1, 2, 3, \ldots$ eigenvalues correspond to the path antisymmetric states confined in the L' loops. Their explicit interface eigenvectors are

$$E(M) = \begin{pmatrix} 0 \\ 0 \\ 1 \\ 0 \\ -1 \end{pmatrix}. \tag{4.23}$$

These states remain confined in the final system as found from the confined state theorem.

4.3.4.4 For $C = C' = 0$

When $C = C' = 0$, $\lambda = 4L/(1+2n)$, $n = 0, 1, 2, 3, \ldots$ and $\lambda = 4L'/(1+2n)$, $n' = 0, 1, 2, 3, \ldots$, eigenvalues correspond to the superposition of three robust zeros. The first one is that of the two connected loops of length L, the second belongs to the two loops of length L', and the third is that of an open loop of length $L + L'$ with $\frac{L'}{L} = \frac{2n'+1}{2n+1}$. The path antisymmetric states of these open loops of length $2L$, $2L'$, and $L + L'$ remain confined in this final network and provide for each eigenvalue $\lambda = \frac{4L}{2n+1} = \frac{4L'}{2n'+1}$ the following three degenerate interface eigenvectors:

$$E(M) = \begin{pmatrix} 0 \\ 1 \\ 0 \\ -1 \\ 0 \end{pmatrix}, \begin{pmatrix} 0 \\ 0 \\ 1 \\ 0 \\ -1 \end{pmatrix}, \begin{pmatrix} 0 \\ 1 \\ -1 \\ 0 \\ 0 \end{pmatrix} \quad \text{when} \quad S' = S, \tag{4.24}$$

or

$$E(M) = \begin{pmatrix} 0 \\ 1 \\ 0 \\ -1 \\ 0 \end{pmatrix}, \begin{pmatrix} 0 \\ 0 \\ 1 \\ 0 \\ -1 \end{pmatrix}, \begin{pmatrix} 0 \\ 1 \\ 1 \\ 0 \\ 0 \end{pmatrix} \quad \text{when} \quad S' = -S. \tag{4.25}$$

4.3.4.5 For $C = \pm C' \neq 0$

The $C = \pm C' \neq 0$ eigenvalues correspond to $\lambda = \frac{2(L+L')}{m}$, where $m = 0, 1, 2, 3, \ldots$ Their explicit interface eigenvectors are

$$E(M) = \begin{pmatrix} C \\ 1 \\ 1 \\ 1 \\ 1 \end{pmatrix} \quad \text{when} \quad C' = C \tag{4.26}$$

and

$$E(M) = \begin{pmatrix} -C \\ -1 \\ 1 \\ -1 \\ 1 \end{pmatrix} \quad \text{when} \quad C' = -C. \tag{4.27}$$

Note that the $1/\lambda = 0$ eigenvalue corresponds here to the invariant by translation displacement of the whole structure.

4.3.5 The interface response matrix

By inverting the above interface response inverse matrix, one obtains the interface response matrix

$$g(MM) = \frac{\lambda}{4\pi CC' \sin[\frac{2\pi(L+L')}{\lambda}]} \times \begin{pmatrix} (CC')^2 & CC'^2 & C^2C' & CC'^2 & C^2C' \\ CC'^2 & C'\cos[\frac{2\pi(2L+L')}{\lambda}] & CC' & C'^2 & CC' \\ C^2C' & CC' & C\cos[\frac{2\pi(2L'+L)}{\lambda}] & CC' & C^2 \\ CC'^2 & C'^2 & CC' & C'\cos[\frac{2\pi(2L+L')}{\lambda}] & CC' \\ C^2C' & CC' & C^2 & CC' & C\cos[\frac{2\pi(2L'+L)}{\lambda}] \end{pmatrix}. \tag{4.28}$$

The kernel of the above interface matrix provides the eigenvalues of this network given by

$$\frac{CC'}{\lambda}\sin\left[\frac{2\pi(L+L')}{\lambda}\right] = 0, \tag{4.29}$$

as obtained from the state phase shift.

We remark that the eigenvalues $\sin\left[\frac{2\pi(L+L')}{\lambda}\right] = (C'S + CS') = 0$ are equivalent to those given by $C = \pm C'$. Among these eigenvalues, first those for which $C = \pm C' = 0$ are considered below.

4.3.6 The complete response functions

The complete response function expressions are obtained from the general equation of the interface response theory, as explained for the previous example to be as follows.

4.3.6.1 When x and x' are in the same wire L

In this case, one gets $g(x, x')$ for $x \leq x'$ to be

$$g(x, x') = \frac{\lambda}{4\pi}\cos\left(\frac{2\pi(L - x')}{\lambda}\right) \times \left[\frac{1}{\sin\left[\frac{2\pi(L+L')}{\lambda}\right]}\cos\left(\frac{2\pi(L' + x)}{\lambda}\right) - \frac{1}{C}\sin\left(\frac{2\pi x}{\lambda}\right)\right]. \tag{4.30}$$

4.3.6.2 When x and x' are in the same wire L'

In this case, one gets $g(x,x')$ for $x \leq x'$ to be

$$g(x,x') = \frac{\lambda}{4\pi}\cos\left(\frac{2\pi(L'-x')}{\lambda}\right) \times \left[\frac{1}{\sin\left[\frac{2\pi(L+L')}{\lambda}\right]}\cos\left(\frac{2\pi(L+x)}{\lambda}\right) - \frac{1}{C'}\sin\left(\frac{2\pi x}{\lambda}\right)\right]. \tag{4.31}$$

4.3.6.3 When x is in the wire L and x' in the other wire L

In this case, one has

$$g(x,x') = \frac{\lambda}{4\pi}\cos\left(\frac{2\pi(L-x')}{\lambda}\right) \times \left[\frac{1}{\sin\left[\frac{2\pi(L+L')}{\lambda}\right]}\cos\left(\frac{2\pi(L'+x)}{\lambda}\right) + \frac{1}{C}\sin\left(\frac{2\pi x}{\lambda}\right)\right]. \tag{4.32}$$

4.3.6.4 When x is in one wire L and x' in one wire L'

This time one obtains

$$g(x,x') = \frac{\lambda}{4\pi\sin\left[\frac{2\pi(L+L')}{\lambda}\right]}\cos\left(\frac{2\pi(L'-x')}{\lambda}\right)\cos\left(\frac{2\pi(L-x)}{\lambda}\right). \tag{4.33}$$

4.3.7 All eigenvalues and eigenfunctions from the full response function

The eigenfunctions can be obtained from the numerators of the above response functions. Their explicit expressions are given below.

4.3.7.1 The ground state

The ground state eigenvalue is $1/\lambda$. The corresponding eigenvector is the 0 displacement of all system points.

4.3.7.2 For the $C = 0$ path states

The eigenfunctions are confined within the two L loops. Their explicit expression is

$$E(x) = \pm\sin\left[\frac{(2n+1)\pi x}{2L}\right], \quad 0 \leq x < L. \tag{4.34}$$

The sign in the above equation changes when one goes from one wire L to the other.

4.3.7.3 For the $C' = 0$ path states

The eigenfunctions are confined within the two L' loops. Their explicit expression is

$$E(x) = \pm \sin\left[\frac{(2n'+1)\pi x}{2L'}\right], \quad 0 \le x < L'. \tag{4.35}$$

The sign in the above equation changes when one goes from one wire L' to the other.

4.3.7.4 For the $C = C' = 0$ path states

Their eigenfunctions are confined within one of the two L loops and one of the two L' loops. Their explicit expressions are the same as those given just above, but with a sign change when one goes from wire L to wire L' for $S' = S$ and without a sign change when going from wire L to wire L' for $S' = -S$.

4.3.7.5 For the new states

For the new states, with C and C' nonzero, confined within the four loops, one gets $\lambda = 2(L + L')/m$, with $m = 1, 2, 3, \ldots$

The eigenfunctions corresponding to these states are as follows:

Within the wire L,

$$E(x) = C' \cos\left[\frac{2\pi(L-x)}{\lambda}\right], \quad 0 \le x < L. \tag{4.36}$$

Within the wire L',

$$E(x) = C \cos\left[\frac{2\pi(L'-x)}{\lambda}\right], \quad 0 \le x < L'. \tag{4.37}$$

We remark that the sum of the outgoing derivatives of these eigenfunctions, at the $x = 0$ node point, is equal to $-2(C'S + CS') = 0$. This equation gives the eigenwavelengths. This property is a general eigenfunction rule, as explained already in Chapter 3 of this book.

Remember also that for these states one has $C' = C$ or $C' = -C$. So these eigenfunctions may also be written as follows:

$$\begin{aligned} &\text{within the wire } L, \quad E(x) = \cos\left[\tfrac{2\pi(L-x)}{\lambda}\right], \\ &\text{within the wire } L', \quad E(x) = \cos\left[\tfrac{2\pi(L'-x)}{\lambda}\right], \end{aligned} \tag{4.38}$$

when $C' = C$, and

$$\begin{aligned} &\text{within the wire } L, \quad E(x) = -\cos\left[\frac{2\pi(L-x)}{\lambda}\right], \\ &\text{within the wire } L', \quad E(x) = \cos\left[\frac{2\pi(L'-x)}{\lambda}\right], \end{aligned} \tag{4.39}$$

when $C' = -C$.

Note that the $1/\lambda = 0$ eigenvalue corresponds here, when $C' = C$, to the invariant by translation displacement of the whole structure.

4.4 Step ladder

This H network is called a one-step ladder, see Fig. 4.3. It is the simplest case of ladders with N steps, $N = 1, 2, 3, \ldots$ Such ladders are expected to show interesting results as far as path states are concerned. The fact that they have, at both ladder ends, two free-end loops enables foreseeing the existence of path states within their two longest pieces. It is also possible, thanks to the confined state theorem, to foresee the existence of path states in such N step ladders.

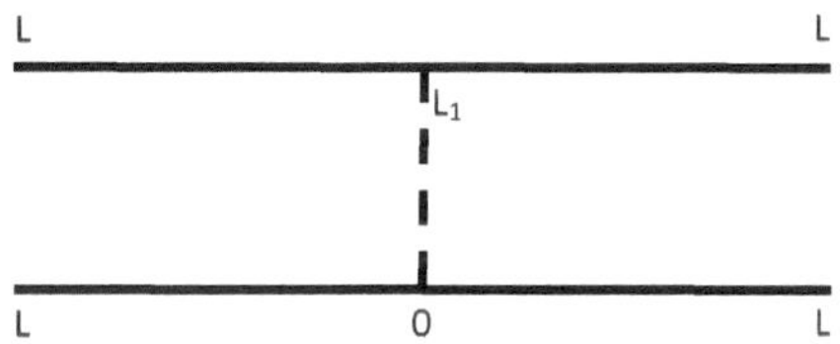

Figure 4.3 One-step ladder. Two different robust zeros of open loops of length $2L$ connected by an open loop of length L_1.

As a reference system, one can use here four loops of length L and one loop of length L_1, all made of the same material. The final system, called one step ladder, is obtained by superposing the 0 ends of two of the L loops with one of the 0 ends of the L_1 loop and the 0 ends of the other two L loops with the other L_1 end of the L_1 loop, see Fig. 4.3. The x coordinates vary from 0 to L on the four loops L and from 0 to L_1 on the loop L_1.

4.4.1 *Path states by inspection*

Each loop $2L$ has a robust zero in its middle. According to the confined state theorem, these zeros remain unperturbed by their connection with the ends of the L_1 loop. The explicit expressions of the path eigenvalues and eigenfunctions, associated with these zeros, are the same as those given by Eqs. (4.1) and (4.2).

Note also that the ends of the L_1 loop are not robust zeros.

The symmetric states of the two connected loops of length L and the states of the connected loops of length L_1 collapse in the final system. They combine together and shift to new states.

4.4.2 The interface space response inverse matrix

The inverse of the interface response matrix of the structure depicted in Fig. 4.3 is given by

$$[g(MM)]^{-1} = \frac{2\pi}{\lambda}\begin{pmatrix} -\frac{C_1}{S_1}+\frac{2S}{C} & \frac{1}{S_1} \\ \frac{1}{S_1} & -\frac{C_1}{S_1}+\frac{2S}{C} \end{pmatrix}, \tag{4.40}$$

where $S=\sin\left(\frac{2\pi L}{\lambda}\right)$, $S_1=\sin\left(\frac{2\pi L_1}{\lambda}\right)$, $C=\cos\left(\frac{2\pi L}{\lambda}\right)$, and $C_1=\cos\left(\frac{2\pi L_1}{\lambda}\right)$.

The determinant of this matrix is

$$\left|[g(MM)]^{-1}\right| = \left(\frac{2\pi}{\lambda}\right)^2 \frac{(-5C^2S_1+4S_1-4C_1CS)}{C^2S_1}. \tag{4.41}$$

4.4.3 All eigenvalues from the state phase shift

The phase shift η [1] between the final and reference systems is

$$\eta(k) = -\arg\left[\frac{\lambda^4C^2(-5C^2S_1+4S_1-4C_1CS)}{S^4S_1}\right], \tag{4.42}$$

where account is taken of the fact that the four L loops bring a contribution $\arg\left(\frac{S}{\lambda C}\right)^4$ as only one of their two ends is used in this interface space.

The reference system has the states $S_1/\lambda=0$ and $(S/\lambda)^4=0$. They collapse and are replaced by new ones. So the eigenvalues of this final system are given by

$$C^2(-5C^2S_1+4S_1-4C_1CS)/\lambda=0. \tag{4.43}$$

This result confirms that $C^2=0$ provides path eigenvalues as obtained from the confined state theorem. The $-5C^2S_1+4S_1-4C_1CS=0$ provides the new eigenvalues of the final system. Note again the two degenerate states for infinite λ.

These results can also be found from the kernels of the full response functions derived hereafter.

4.4.4 The complete response function

The inverse of the above matrix is given by

$$g(MM) = \frac{\lambda C}{2\pi(-5C^2S_1+4S_1-4C_1CS)}\begin{pmatrix} -CC_1+2SS_1 & -C \\ -C & -CC_1+2SS_1 \end{pmatrix}. \tag{4.44}$$

Note that in the final system all the four elements of $g(MM)$ vanish, when $C=0$.

The explicit response function expressions between any two points x and x' of this network are found from the elements of the response function, see Eqs. (1.33) and (1.32), of an open loop to be as shown here after.

4.4.4.1 *For x within one of the L loops and x' within the adjacent L loop*

In such a case, one has

$$\begin{aligned} g(x,x') &= \frac{\lambda C(-CC_1+2SS_1)}{2\pi(-5C^2S_1+4S_1-4C_1CS)} \\ &\times \left[\cos\left(\frac{2\pi x}{\lambda}\right)\cos\left(\frac{2\pi x'}{\lambda}\right)+\frac{S}{C}\sin\left(\frac{2\pi(x+x')}{\lambda}\right)\right. \\ &\left.+\frac{S^2}{C^2}\sin\left(\frac{2\pi x}{\lambda}\right)\sin\left(\frac{2\pi x'}{\lambda}\right)\right]. \end{aligned} \tag{4.45}$$

Note that $C = 0$ remains in its kernel for the eigenstates confined within each of the two $2L$ loops.

4.4.4.2 *For x within one of the L loops and x' within one of the nonadjacent L loops*

In this case, we obtain

$$\begin{aligned} g(x,x') &= \frac{-\lambda C^2}{2\pi(-5C^2S_1+4S_1-4C_1CS)} \\ &\times \left[\cos\left(\frac{2\pi x}{\lambda}\right)\cos\left(\frac{2\pi x'}{\lambda}\right)+\frac{S}{C}\sin\left(\frac{2\pi(x+x')}{\lambda}\right)\right. \\ &\left.+\frac{S^2}{C^2}\sin\left(\frac{2\pi x}{\lambda}\right)\sin\left(\frac{2\pi x'}{\lambda}\right)\right]. \end{aligned} \tag{4.46}$$

Note that the $C = 0$ eigenstates do not allow communication between the two $2L$ loops. A forced communication is, however, possible between these loops at these $C = 0$ wavelengths.

4.4.4.3 *For x and x' within one and the same L loop*

This time one has

$$\begin{aligned} g(x,x') &= -\frac{\lambda}{2\pi}\sin\left(\frac{2\pi x}{\lambda}\right)\left[\cos\left(\frac{2\pi x'}{\lambda}\right)+\frac{S}{C}\sin\left(\frac{2\pi x'}{\lambda}\right)\right] \\ &+\frac{\lambda C(-CC_1+2SS_1)}{2\pi(-5C^2S_1+4S_1-4C_1CS)} \\ &\times \left[\cos\left(\frac{2\pi x}{\lambda}\right)\cos\left(\frac{2\pi x'}{\lambda}\right)+\frac{S}{C}\sin\left(\frac{2\pi(x+x')}{\lambda}\right)\right. \\ &\left.+\frac{S^2}{C^2}\sin\left(\frac{2\pi x}{\lambda}\right)\sin\left(\frac{2\pi x'}{\lambda}\right)\right]. \end{aligned} \tag{4.47}$$

Note that $C = 0$ remains in the kernel of this response function for the localization of these states within the L loop.

4.4.4.4 For x within one of the L loops and x' within the L_1 loop

In this case, one has

$$g(x,x') = \frac{\lambda}{2\pi} \frac{C\cos\left(\frac{2\pi x}{\lambda}\right) + S\sin\left(\frac{2\pi x}{\lambda}\right)}{(-5C^2S_1 + 4S_1 - 4C_1CS)} \times \left[(-CC_1 + 2SS_1)\cos\left(\frac{2\pi x'}{\lambda}\right) - (2SC_1 + S_1)\sin\left(\frac{2\pi x'}{\lambda}\right)\right]. \tag{4.48}$$

4.4.4.5 For x and x' within the L_1 loop

This time one gets

$$\begin{aligned} g(x,x') = {} & \frac{\lambda}{2\pi(-5C^2S_1 + 4S_1 - 4CC_1S)} \\ & \times \left[(-6C^2S_1 + 4S_1 - 6CC_1S)\sin\left(\frac{2\pi x}{\lambda}\right)\cos\left(\frac{2\pi x'}{\lambda}\right)\right. \\ & + C(2SS_1 - CC_1)\cos\left(\frac{2\pi x}{\lambda}\right)\cos\left(\frac{2\pi x'}{\lambda}\right) \\ & - (-5C^2C_1 + 4C_1 + 2CSS_1 + CC_1)\sin\left(\frac{2\pi x}{\lambda}\right)\sin\left(\frac{2\pi x'}{\lambda}\right) \\ & \left. - C(2SC_1 + S_1)\cos\left(\frac{2\pi x}{\lambda}\right)\sin\left(\frac{2\pi x'}{\lambda}\right)\right], \end{aligned} \tag{4.49}$$

where $x \le x'$.

4.5 Perspectives

There are many simple extensions of the three open loop examples presented above. Let us cite just a few of them.

The case of the asymmetric cross presents two different sets of path states confined within two loops $2L$ and $2L'$. Obvious extensions of this asymmetric network are those with more than two such crossed loops. Varying the lengths of several such loops enables selecting more path states.

The physical results given here for the step ladder are for incommensurable L and L_1. When these loop lengths are commensurable, it is possible to use the above mathematical results. This then provides specific results for the eigenvalues and their eigenfunction localizations. Also remember the case of ladders with N steps, $N = 1, 2, 3, \ldots$, introduced at the beginning of this example.

It is also possible to consider these $N = 1, 2, 3, \ldots$ step ladders. When the lengths L and L_1 are incommensurable and in the limit of infinite N, it is possible to foresee a flat band of path eigenvalues confined within the N loops $2L$.

Any eigenstate can be activated within its subdomain without any subsystem deformation, see the activation state theorem in Chapter 3 of this Photonics book. This paves the way to the use of the degenerate path states for many applications.

Although all this is at a tutorial analytical level, there is no doubt that experiments with coaxial cables and optical fibers can be done, too, as well as numerical simulations with application-aimed prototypes in the micrometric and even nanometric domains.

References

[1] L. Dobrzyński, A. Akjouj, G. Lévêque, et al., Centered system magnons, in: Magnonics, in: Interface Transmission Tutorial Book Series, Elsevier, 2019.

Closed loop examples

5

Housni Al Wahsh[a], *Leonard Dobrzyński*[b], *Abdellatif Akjouj*[b], *and El Houssaine El Boudouti*[c]
[a]Benha University, Faculty of Engineering, Cairo, Egypt, [b]Univ. Lille, CNRS, Centrale Lille, Yncréa ISEN, Univ. Polytechnique Hauts-de-France, UMR 8520 - IEMN, Physics Department, Lille, France, [c]University Mohammed I, Faculty of Sciences, Laboratory of Physics of Matter and Radiation, Oujda, Morocco

5.1 Introduction

The specific properties of closed loop eigenfunctions deserve to be better understood. This chapter underlines that responses in closed loops differ from the responses in open loops. Within the following simple examples, constructed on closed loop robust zeros, many path states are found.

A robust eigenfunction zero is robust because its eigenstate cannot be activated by any action applied on it. The path states are robust because they are confined within one dimensional paths, and when they are degenerate they can turn around defects, see Chapter 3 of this book.

5.2 Circle and diameter

In this first simple example, as a reference system, one takes two open loops of length $L/2$ and one open loop of length L_1, all made of the same material, see Fig. 5.1.

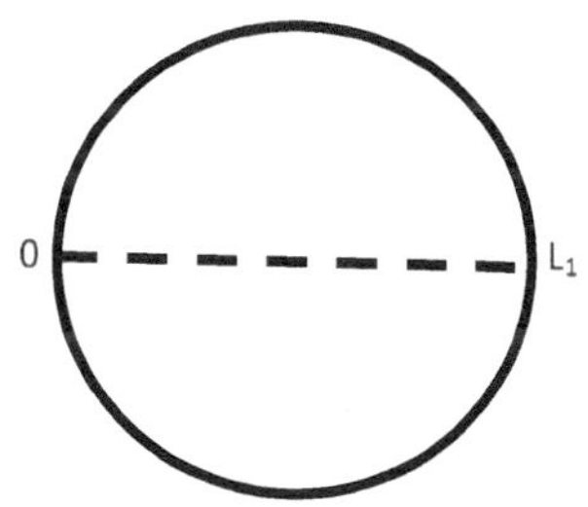

Figure 5.1 Closed loop and diameter. One closed loop of length L with an open loop of length L_1 fixed as a diameter.

The final system, called circle and diameter, is obtained by superposing at one interface point the 0 ends of these three loops and at another interface point their two $L/2$ and one L_1 ends. The x coordinates vary from 0 to $L/2$ on the first two loops and from 0 to L_1 on the third loop.

Photonics. https://doi.org/10.1016/B978-0-12-819388-4.00014-9

5.2.1 Path states by inspection

Once connected, the loops of length $L/2$ form a circle whose circumference length is $L = \pi L_1$. The two connection points are robust zeros of the closed loop L but are not robust zeros of the open loop of length L_1, as shown above in Chapter 3 of the Photonics book where the robust zeros are defined.

In agreement with the confined state theorem, see Chapter 3 of this book, the present robust zeros remain unperturbed by their connections with the open loop of length L_1, as well as their associated eigenvalues ($S = 0$) and eigenfunctions confined within the closed loop L. The eigenfunctions associated with these two robust zeros are those of one of the two times degenerate eigenstates of the closed loop.

The path eigenvalues confined within the loop L are

$$\lambda_p = \frac{L}{p}, \quad p = 1, 2, 3, \ldots \tag{5.1}$$

Their eigenfunctions are

$$E(x) = \sin\left[2\pi p \frac{x}{L}\right], \quad p = 1, 2, 3, \ldots \tag{5.2}$$

These path eigenfunctions are confined on the external surface of the network. They do not decrease towards the inside of the structure. They differ from the surface [1], edge [2], and adsorbed layer [3] states which decay towards the bulk of the medium. They are simple examples of path states for a closed loop.

The other eigenfunctions of the closed loop interact with the eigenfunctions of the open loop L_1. This interaction leads to their collapses. New states appear in the final system. They are localized within the whole network space.

5.2.2 The interface response inverse matrix

The inverse of the interface response matrix for a circle with a fixed open loop as its diameter, see Fig. 5.1, is

$$[g(MM)]^{-1} = \frac{2\pi}{\lambda} \begin{pmatrix} \frac{-2C}{S} - \frac{C_1}{S_1} & \frac{2}{S} + \frac{1}{S_1} \\ \frac{2}{S} + \frac{1}{S_1} & \frac{-2C}{S} - \frac{C_1}{S_1} \end{pmatrix}, \tag{5.3}$$

where $S = \sin\left(\frac{\pi L}{\lambda}\right)$, $S_1 = \sin\left(\frac{2\pi L_1}{\lambda}\right)$, $C = \cos\left(\frac{\pi L}{\lambda}\right)$, and $C_1 = \cos\left(\frac{2\pi L_1}{\lambda}\right)$.

The determinant of the above matrix is

$$\left|[g(MM)]^{-1}\right| = \left(\frac{2\pi}{\lambda}\right)^2 \frac{(-5SS_1 + 4C_1C - 4)}{SS_1}. \tag{5.4}$$

5.2.3 All eigenvalues from the state phase shift

The phase shift η [4] between the final and reference systems is

$$\eta(k) = -\arg\left[\frac{\lambda^4(-5SS_1 + 4C_1C - 4)}{SS_1}\right]. \tag{5.5}$$

The original system has the states given by $S_1/\lambda = 0$ and $(S/\lambda)^2 = 0$. So the eigenvalues of this final system are given by

$$\lambda S(-5SS_1 + 4C_1C - 4) = 0. \tag{5.6}$$

The final system has then the states $\lambda S = 0$ and $(-5SS_1 + 4C_1C - 4) = 0$. The path $S = 0$ states correspond to one of the two degenerate states of the closed loop L. The $(-5SS_1 + 4C_1C - 4) = 0$ states are the new states due to the interaction between the $S_1 = 0$ states with the other degenerate state $S = 0$ of the closed loop. Note again the two degenerate states for infinite λ appearing in the $(-5SS_1 + 4C_1C - 4) = 0$ term.

5.2.4 The complete response function

The interface response matrix is

$$g(MM) = \frac{\lambda}{2\pi[5SS_1 + 4(1 - CC_1)]}\begin{pmatrix} 2CS_1 + C_1S & 2S_1 + S \\ 2S_1 + S & 2CS_1 + C_1S \end{pmatrix}. \tag{5.7}$$

The complete interface function elements between any points x and x' of this network are as follows.

5.2.4.1 When x is on one half of the circle and x' is on the other half

In such a case, we have

$$g(x, x') = \frac{\lambda}{2\pi[5SS_1 + 4(1 - CC_1)]}\Big[(2CS_1 + C_1S)\cos\left(\frac{2\pi(x + x')}{\lambda}\right) + (1 + 2SS_1 - CC_1)\sin\left(\frac{2\pi(x + x')}{\lambda}\right) + \frac{2}{S}(C_1 - C)\sin\left(\frac{2\pi x}{\lambda}\right)\sin\left(\frac{2\pi x'}{\lambda}\right)\Big]. \tag{5.8}$$

Note the S in the kernel of this response function. This confirms the localization of the $S = 0$ states.

5.2.4.2 When x is on one half of the circle and x' is on the line

In this case, one has

$$g(x,x') = \frac{\lambda}{2\pi[5SS_1 + 4(1-CC_1)]}\left[(2CS_1 + C_1S)\cos\left(\frac{2\pi x}{\lambda}\right)\cos\left(\frac{2\pi x'}{\lambda}\right)\right.$$
$$-(2SC_1 + CS_1)\sin\left(\frac{2\pi x}{\lambda}\right)\sin\left(\frac{2\pi x'}{\lambda}\right)$$
$$+(2SS_1 - CC_1 + 1)\sin\left(\frac{2\pi x}{\lambda}\right)\cos\left(\frac{2\pi x'}{\lambda}\right)+$$
$$\left.(-2CC_1 + SS_1 + 2)\cos\left(\frac{2\pi x}{\lambda}\right)\sin\left(\frac{2\pi x'}{\lambda}\right)\right]. \tag{5.9}$$

Note that $S = 0$ and S_1 are not in the kernel of this $g(x,x')$. So the states $S = 0$ are confined within the circle. They are path eigenvalues.

5.2.4.3 When x and x' are in the same half of the circle

This time, we have

$$g(x,x') = \frac{-\lambda}{2\pi}\sin\left(\frac{2\pi x}{\lambda}\right)\cos\left(\frac{2\pi x'}{\lambda}\right)$$
$$+\frac{2\lambda(C - C_1)}{2\pi S[5SS_1 + 4(1-CC_1)]}\sin\left(\frac{2\pi x}{\lambda}\right)\sin\left(\frac{2\pi x'}{\lambda}\right)$$
$$+\frac{\lambda}{2\pi[5SS_1 + 4(1-CC_1)]}\left[(5CS_1 + 4C_1S)\sin\left(\frac{2\pi x}{\lambda}\right)\sin\left(\frac{2\pi x'}{\lambda}\right)\right.$$
$$\left.+(SC_1 + 2CS_1)\cos\left(\frac{2\pi(x+x')}{\lambda}\right) + (2SS_1 - CC_1 + 1)\sin\left(\frac{2\pi(x+x')}{\lambda}\right)\right], \tag{5.10}$$

where $x \leq x'$. Note that the states $S = 0$ are path eigenstates of the whole circle.

Their eigenfunctions are the $\sin[2p\pi\frac{x}{L}]$, $p = 1, 2, 3, \ldots$ These eigenfunctions vanish for $x = 0$ and $x = L/2$. They are given here by the confined state theorem and appear in the numerators of the above response functions.

5.2.4.4 When x and x' are within the line

In such a case, one gets

$$g(x,x') = \frac{-\lambda}{2\pi}\sin\left(\frac{2\pi x}{\lambda}\right)\cos\left(\frac{2\pi x'}{\lambda}\right)$$
$$+\frac{\lambda}{2\pi[5SS_1 + 4(1-CC_1)]}\left[(5SC_1 + 4CS_1)\sin\left(\frac{2\pi x}{\lambda}\right)\sin\left(\frac{2\pi x'}{\lambda}\right)\right.$$

$$+ (2CS_1 + C_1 S)\cos\left(\frac{2\pi(x+x')}{\lambda}\right) + (SS_1 - 2CC_1 + 2)\sin\left(\frac{2\pi(x+x')}{\lambda}\right)\Big], \tag{5.11}$$

where $x \leq x'$.

Note that the $S_1 = 0$ states collapse in the final system, as predicted by the state phase shift.

5.3 Circle and cross

As a reference system, one takes here four loops of length L' and four loops of length L, all made of the same material. The final system, called circle and cross, is obtained by superposing the ends of these eight loops in order to obtain a circle of a $4L'$ circumference and a cross structure made out of four loops of length L, see Fig. 5.2.

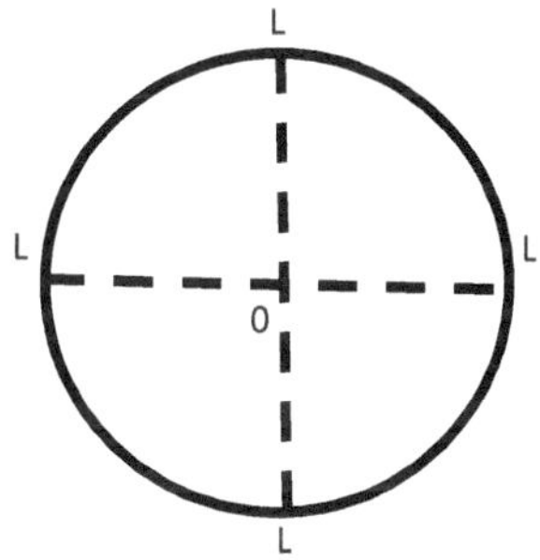

Figure 5.2 Circle and cross. One closed loop of length $4L'$ is connected with a cross whose four segments each have length L.

5.3.1 Path states by inspection

Circle and cross are connected through four equidistant robust zeros of the circle, as shown in the section where the robust zeros are defined. The ends of the cross structure are not robust zeros. The space position x coordinates vary from 0 to L' on the first four loops and from 0 to L on the four other loops. Once connected, the loops L' constitute a circle whose circumference length is $4L' = 2\pi L$. In agreement with the confined state theorem, the present robust zeros remain unperturbed by their connection with the cross, as well as their associated eigenvalues $S'/\lambda = 0$ and eigenfunctions confined within the closed loop $4L'$. The path eigenfunctions associated with these four robust zeros are those of one of the two times degenerate eigenstates of the closed loop.

The path eigenvalues are then

$$\lambda_p = \frac{2L'}{p}, \quad p = 1, 2, 3, \ldots, \tag{5.12}$$

and the eigenfunctions confined within each quarter L' of the circle circumference $4L'$ are then

$$E(x) = \sin\left(\pi p \frac{x}{L'}\right), \quad p = 1, 2, 3, \ldots \tag{5.13}$$

These path eigenfunctions are equal to 0 on all the four robust zeros.

Such path states are another example of surface path states.

The other eigenfunctions of these two times degenerate closed loop states interact with the eigenfunctions of the cross. This interaction leads to the collapse of these states and to the appearance of new states in the final system, localized within the whole network space.

Notice also that the independent cross network has two superposed robust zeros in its middle. But as its connection to the closed loop is through its nonrobust space points, this middle point is no longer robust in the final system, as well as the path eigenstates associated with it in the independent cross network.

5.3.2 The interface response inverse matrix

The inverse of the response function of the network depicted in Fig. 5.2 is given by

$$[g(MM)]^{-1} = \frac{2\pi}{\lambda} \times \begin{pmatrix} \frac{-4C}{S} & \frac{1}{S} & \frac{1}{S} & \frac{1}{S} & \frac{1}{S} \\ \frac{1}{S} & \frac{-C}{S} - 2\frac{C'}{S'} & \frac{1}{S'} & 0 & \frac{1}{S'} \\ \frac{1}{S} & \frac{1}{S'} & \frac{-C}{S} - 2\frac{C'}{S'} & \frac{1}{S'} & 0 \\ \frac{1}{S} & 0 & \frac{1}{S'} & \frac{-C}{S} - 2\frac{C'}{S'} & \frac{1}{S'} \\ \frac{1}{S} & \frac{1}{S'} & 0 & \frac{1}{S'} & \frac{-C}{S} - 2\frac{C'}{S'} \end{pmatrix},$$

where $S = \sin\left(\frac{2\pi L}{\lambda}\right)$, $S' = \sin\left(\frac{2\pi L'}{\lambda}\right)$, $C = \cos\left(\frac{2\pi L}{\lambda}\right)$, and $C' = \cos\left(\frac{2\pi L'}{\lambda}\right)$.

The determinant of the above matrix is

$$\begin{aligned} \left|[g(MM)]^{-1}\right| &= -4\left(\frac{2\pi}{\lambda}\right)^5 \frac{1}{(SS')^4}(CS' + 2C'S)^2 \\ &\quad \times [CS' + 2S(C' + 1)][-SS' + 2C(C' - 1)] \\ &= -\left(\frac{2\pi}{\lambda}\right)^5 \frac{1}{(S)^4(S')^3}\left[8 + \cos\left[\frac{2\pi(L' - 2L)}{\lambda}\right]\right. \\ &\quad \left. - 9\cos\left[\frac{2\pi(L' + 2L)}{\lambda}\right]\right] \\ &\quad \times \left[3\sin\left[\frac{2\pi(L' + L)}{\lambda}\right] - \sin\left[\frac{2\pi(L' - L)}{\lambda}\right]\right]^2. \end{aligned} \tag{5.14}$$

5.3.2.1 *All eigenvalues from the state phase shift*

The state phase shift η [4] between the final and reference systems is

$$\eta(\lambda) = -\arg\left[8+\cos\left[\frac{2\pi(L'-2L)}{\lambda}\right]-9\cos\left[\frac{2\pi(L'+2L)}{\lambda}\right]\right]$$
$$-\left[3\sin\left[\frac{2\pi(L'+L)}{\lambda}\right]-\sin\left[\frac{2\pi(L'-L)}{\lambda}\right]\right]^2-\arg\frac{\lambda^{11}}{S^4 S'^3}. \tag{5.15}$$

The original system has the states $(S/\lambda)^4 = 0$ and $(S'/\lambda)^4 = 0$. So the eigenvalues of this final system are given by

$$\lambda^3 S'\left[8+\cos\left[\frac{2\pi(L'-2L)}{\lambda}\right]-9\cos\left[\frac{2\pi(L'+2L)}{\lambda}\right]\right]$$
$$\times\left[3\sin\left[\frac{2\pi(L'+L)}{\lambda}\right]-\sin\left[\frac{2\pi(L'-L)}{\lambda}\right]\right]^2=0. \tag{5.16}$$

This result can also be rewritten as

$$\lambda^3 S'[2-2C'(2C^2-1)+5SS'C][S'C+2C'S]^2=0. \tag{5.17}$$

Taking the limit of small $1/\lambda$, one sees that again that there are two degenerate states for infinite λ.

This state phase shift also confirms that $S'/\lambda = 0$ states remain path states of the final system and that all the other states, namely those given by $(S/\lambda)^4 = 0$ and $(S'/\lambda)^3 = 0$, collapse and are replaced by new ones.

5.3.3 *Perspectives*

The above general result for a circle and a cross is valid for any values of L and L', even when they are incommensurable. It may also be used to find the eigenvalues when L and L' are commensurable. As a simple example, when $L' = 2L$, one obtains that the eigenvalues are

$$\lambda^3 S^5 C^3 [3C^2-1]^2 = 0. \tag{5.18}$$

5.4 Four squares

As a reference system, one takes here 12 loops of length L all made out of the same material. The final system, called four squares, is obtained by superposing the ends of these loops in order to obtain the four-square network of Fig. 5.3.

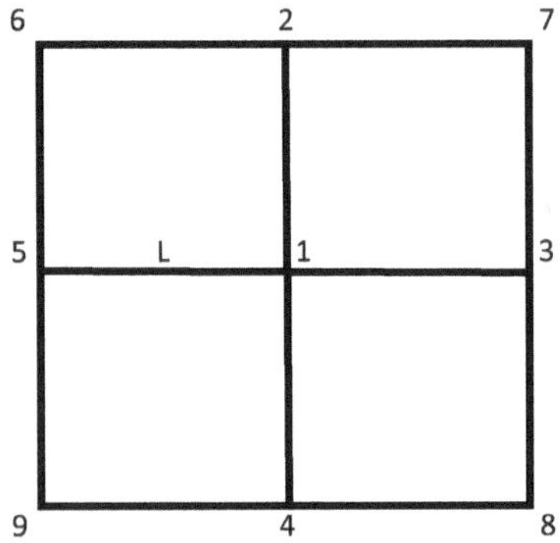

Figure 5.3 Four squares built out of 12 open loops of length L. The numbers refer to the interface points.

The difference between the present structure and that depicted by Fig. 5.2 is that here L' and L are commensurate, as one has $L' = 2L$. In both networks one has four internal closed loops.

5.4.1 Path states by inspection

The surface loop of circumference $8L$ has four equidistant robust zeros connected with the cross structure. The confined state theorem enables to associate this loop with the following path eigenvalues:

$$\lambda_p = \frac{4L}{p}, \quad p = 1, 2, 3, \ldots, \tag{5.19}$$

and path eigenfunctions are

$$E(x) = \pm \sin[\pi p \frac{x}{2L}], \quad p = 1, 2, 3, \ldots \tag{5.20}$$

The $\pm$ signs have to be chosen so that they are opposite on both sides of a robust zero. Also x varies from 0 at one robust zero site till it comes back to this zero after a complete closed loop revolution of $4L$, $6L$, or $8L$ depending on the eigenvalue, the eigenfunction, and the way of excitation. Each path eigenvalue λ_p is degenerate. So some structure defects may affect just one of their degenerate eigenfunctions and not the other(s). It is possible to find by inspection all these degenerate eigenfunctions. The elementary degenerate interface eigenvectors are given below, after determination of all the eigenvalues of this network.

5.4.2 The interface response inverse matrix

The interface response inverse matrix of the simple square network depicted in Fig. 5.3 is

$$[g(MM)]^{-1} = \frac{2\pi}{\lambda S}$$
$$\times \begin{pmatrix} -4C & 1 & 1 & 1 & 1 & 0 & 0 & 0 & 0 \\ 1 & -3C & 0 & 0 & 0 & 1 & 1 & 0 & 0 \\ 1 & 0 & -3C & 0 & 0 & 0 & 1 & 1 & 0 \\ 1 & 0 & 0 & -3C & 0 & 0 & 0 & 1 & 1 \\ 1 & 0 & 0 & 0 & -3C & 1 & 0 & 0 & 1 \\ 0 & 1 & 0 & 0 & 1 & -2C & 0 & 0 & 0 \\ 0 & 1 & 1 & 0 & 0 & 0 & -2C & 0 & 0 \\ 0 & 0 & 1 & 1 & 0 & 0 & 0 & -2C & 0 \\ 0 & 0 & 0 & 1 & 1 & 0 & 0 & 0 & -2C \end{pmatrix}. \tag{5.21}$$

The determinant of the above matrix is

$$|[g(MM)]^{-1}| = \frac{576(2\pi)^9 C^3(1-3C^2)^2}{\lambda^9 S^7}. \tag{5.22}$$

5.4.3 All eigenvalues from the state phase shift

The phase shift η [4] between the final and reference systems is

$$\eta(\lambda) = -\arg \frac{\lambda^{15} C^3(1-3C^2)^2}{S^7}. \tag{5.23}$$

The reference system (12 finite loops of length L) has the eigenvalues of $(S/\lambda)^{12} = 0$. The determinant $|[g(MM)]^{-1}|$ of each finite loop of length L is equal to $-\left(\frac{2\pi}{\lambda}\right)^2$. So the eigenvalues of the final system are those satisfying

$$\lambda^3 S^5 C^3(1-3C^2)^2 = 0. \tag{5.24}$$

This confirms the result obtained above in Eq. (5.18). Notice again the two degenerate states for infinite λ.

5.4.4 The interface eigenvectors

5.4.4.1 For the $S = 0$ eigenvalues

Figs. 5.4, 5.5, 5.6, and 5.7 provide illustrations within the interface space of all the elementary eigenvectors associated to the eigenvalues of the four square network.

Fig. 5.4 shows five out of the six eigenvectors for $C = 1$, or equivalently $\lambda = L/n$, $n = 1, 2, 3, \ldots$ The trivial ground state eigenvector of the ground state, with 0 on all sites is not represented on this figure. The first one (1) shown here is for all values of n, even $n = 0$, which is for the invariant by translation eigenfunction. The other four illustrate for $\lambda = L$ the path elementary eigenvectors within each of the four internal loops. For $n = 2, 3, \ldots$, there are just more interface points. Note that the superposition of these four eigenvectors provides the surface path eigenvector around this four-square network. These four path eigenvectors agree with those found above from the confined state theorem.

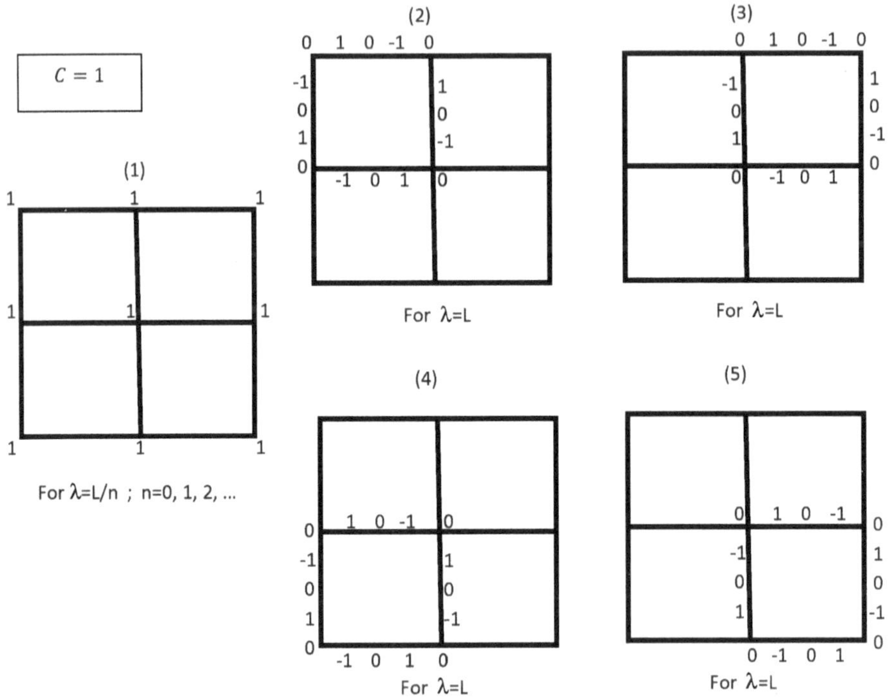

Figure 5.4 Four squares. Five of the six eigenvectors for the eigenvalues ($C = 1$), where $C = \cos(2\pi L/\lambda)$, or equivalently for $\lambda = L/n$, $n = 1, 2, 3, \ldots$, are shown here. The trivial ground state eigenvector with 0 on all sites is not depicted here. Figure (1) is for all values of n. Figures (2), (3), (4), and (5) illustrate for $\lambda = L$ four elementary loop path eigenvectors. In these figures the zero interface values of the eigenvectors on the loops, where they are not confined, are not shown. Their superposition provides one surface path eigenvector.

Fig. 5.5 illustrates the five eigenvectors for $C = -1$ or $\lambda = 2L$. The first one is a bulk eigenvector. The last four illustrate the path elementary eigenvectors within each of the four internal loops. Note that the superposition of these four eigenvectors provides the surface path eigenvector around this four-square network. These four path eigenvectors agree with those found above from the confined state theorem.

5.4.4.2 *For the $C = 0$ eigenvalues*

Fig. 5.6 illustrates the three eigenvectors for $C = 0$, or equivalently for $\lambda = 4L/(1 + 2n)$, $n = 0, 1, 2, 3, \ldots$

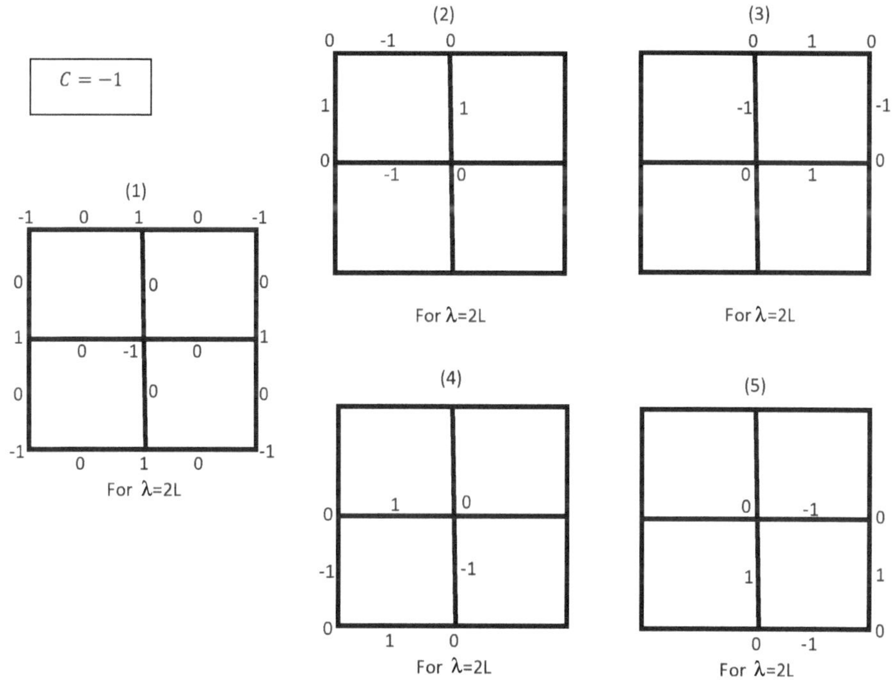

Figure 5.5 Four squares. This figure illustrates the five eigenvectors for $\lambda = 2L$, one of the eigenvalues given by $C = -1$. The first is a bulk eigenvector. The last four illustrate the path elementary eigenvectors confined within each of the four internal loops. In these figures, the zero interface values of the eigenvectors on the loops, where they are not confined, are not shown. Note that the superposition of these four eigenvectors provides the surface path eigenvector around this four square network.

The first two are path eigenvectors as they do not decay away, from their localization in two diagonal closed loops $4L$, because their four connections with the rest of the network are through robust zeros. They agree with those found above from the confined state theorem. The superposition of these two eigenvectors provides the surface path eigenvectors confined around this four-square network, as found before from the confined state theorem. Also we remark that each of these first two eigenvectors is also a wedge state localized in two opposite wedges of this network. Each of them is also a path eigenvector as it does not show any decay away from its localization within two diagonal closed loops $4L$. The eigenvector (c) is a surface state decaying towards the bulk. It is not a path state as its amplitude is maximum on the robust zeros of the closed loop $8L$.

5.4.4.3 *For the $C = \pm 1/\sqrt{3}$ eigenvalues*

Fig. 5.7 illustrates the two eigenvectors for $C = 1/\sqrt{3}$ and the two eigenvectors for $C = -1/\sqrt{3}$. We remark that each of these four eigenvectors is a wedge state localized in two opposite wedges of this network.

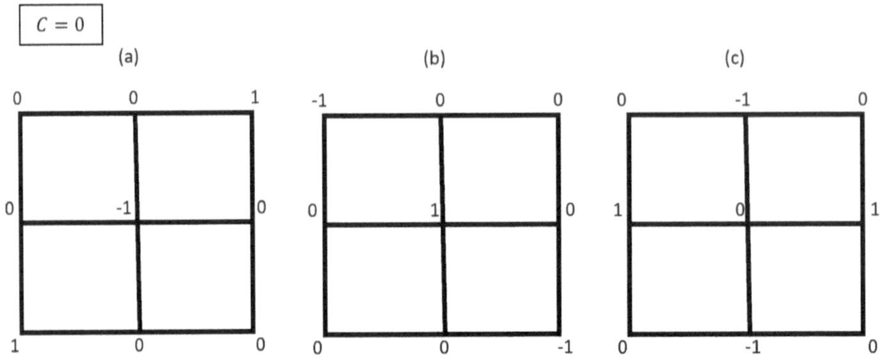

Figure 5.6 Four squares. This figure illustrates the three eigenvectors for $C = 0$, or equivalently for $\lambda = 4L/(1+2n)$, $n = 0, 1, 2, 3, \ldots$ The first two are path eigenvectors as they do not decay away from their localization in two diagonal closed loops $4L$, because their four connections with the rest of the network is through robust zeros. The superposition of these two eigenvectors provides the surface path eigenvectors confined around this four-square network. Note also that each of this first two eigenvectors is also a wedge state localized in two opposite wedges of this network. Each of them is also a path eigenvector as it does not show any decay away from its localization within two closed loops $4L$. The eigenvector (c) is a surface state decaying towards the bulk.

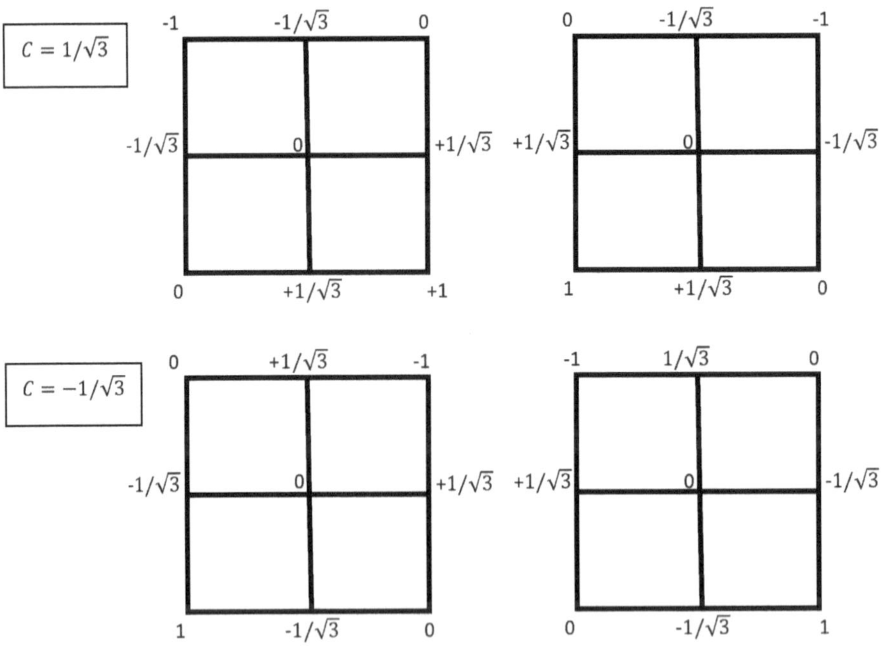

Figure 5.7 Four squares. This figure illustrates the two eigenvectors for $C = 1/\sqrt{3}$ and the two eigenvectors for $C = -1/\sqrt{3}$. Note that each of these four eigenvectors is a wedge state localized in two opposite wedges of this network.

5.5 Nine squares

As a reference system, one takes here 24 loops of length L all made of the same material. The final system, called nine squares, is obtained by superposing the ends of these loops in order to obtain the nine-square network of Fig. 5.8. The difference between the present structure and that depicted by Fig. 5.3 is that here the network has nine rather than four internal closed loops of perimeter $4L$.

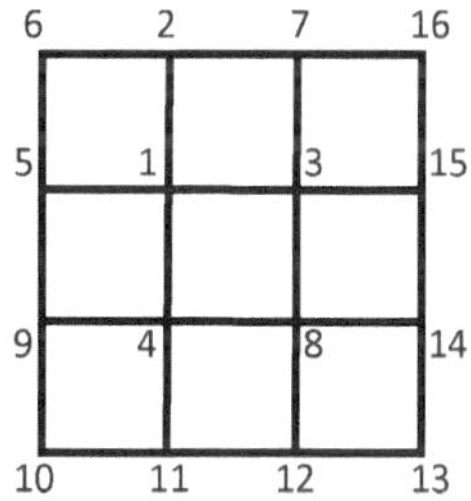

Figure 5.8 Nine squares. As a reference system, one takes here 24 open loops of length L. The final system is obtained by superposing the ends of these loops in order to obtain the nine-square network. The difference between the present structure and that depicted by Fig. 5.3 is that here the network has 9 rather than 4 internal closed loops of perimeter $4L$. The numbers refer to the interface points.

5.5.1 Path states by inspection

The surface loop of perimeter $12L$ has 12 equidistant robust zeros, eight of them are connected with the bulk structure. Each of the nine internal loops of circumference $4L$ has four robust zeros used in its connection with the other internal loops, see Fig. 5.8. The confined state theorem enables finding the following path eigenvalues in each of these nine internal loops:

$$\lambda_p = \frac{2L}{p}, \quad p = 1, 2, 3, \ldots, \tag{5.25}$$

and path eigenfunctions

$$E(x) = \pm \sin\left[\pi p \frac{x}{L}\right], \quad p = 1, 2, 3, \ldots, \tag{5.26}$$

where x varies from 0 at one robust zero to $4L$ after one revolution within each internal closed loop. The $\pm$ signs have to be chosen so that they are opposite for two adjacent loops. These path eigenvalues are nine times degenerate. A superposition of 2 to 9 of them enables obtaining path eigenfunctions around 2 to 9 internal closed loops. In particular, the superposition of the 9 degenerate eigenfunctions provides the path eigenfunction confined on the external surface path of the network without any decrease towards the inside of the structure. Some structure defects may affect just one of their degenerate eigenfunctions and not the others. So these path eigenfunctions can turn around defects.

5.5.2 The interface response inverse matrix

The interface response inverse matrix of the network depicted in Fig. 5.8 can be obtained using the same procedure shown explicitly for the previous example. This matrix has dimension 16 × 16, and the value of its determinant is given by

$$|[g(MM)]^{-1}| = \frac{256}{S^{16}}\left(\frac{2\pi}{\lambda}\right)^{16} C^4(-1+C)(1+C)(-1+2C)(1+2C) \\ \times (-1+3C)^2(1+3C)^2(-11+18C^2)^2. \quad (5.27)$$

5.5.3 All eigenvalues from the state phase shift

The phase shift η [4] between the final and reference systems is

$$\eta(\lambda) = -\arg\left(\frac{\lambda^{32}}{S^{16}}C^4(-1+C)(1+C)(-1+2C)(1+2C)\right) \\ -\arg(-1+3C)^2(1+3C)^2(-11+18C^2)^2. \quad (5.28)$$

The reference system (24 finite loops of length L) has the eigenvalues of $(S/\lambda)^{24} = 0$. The determinant $|[g(MM)]^{-1}|$ of each finite loop of length L equals to $-\left(\frac{2\pi}{\lambda}\right)^2$. So the eigenvalues of the final system are those of

$$\lambda^8 S^{10} C^4(-1+2C)(1+2C)(-1+3C)^2(1+3C)^2(-11+18C^2)^2 = 0. \quad (5.29)$$

Note again the two degenerate states for infinite λ.

5.5.4 The interface eigenvectors

The following figures illustrate within the interface space all the eigenvectors associated with the eigenvalues of the nine-square network.

5.5.4.1 For the $S = 0$ eigenvalues

Fig. 5.9 illustrates 10 out of 11 eigenvectors for $C = 1$, or equivalently $\lambda = L/n$, $n = 0, 1, 2, 3, \ldots$ The trivial ground state eigenvector, with 0 on all sites, is not shown in this figure. The first one shown, a bulk eigenvector (a), is for all values of n. The two other illustrate for $\lambda = L$ the path eigenvectors within two of the nine internal loops. For $n = 2, 3, \ldots$ there are just more interface points. Note that the superposition of these nine eigenvectors provides the surface path eigenvector around this nine-square network. Other superpositions of these nine path eigenvectors can be excited, too. These eigenvectors are also robust against defects.

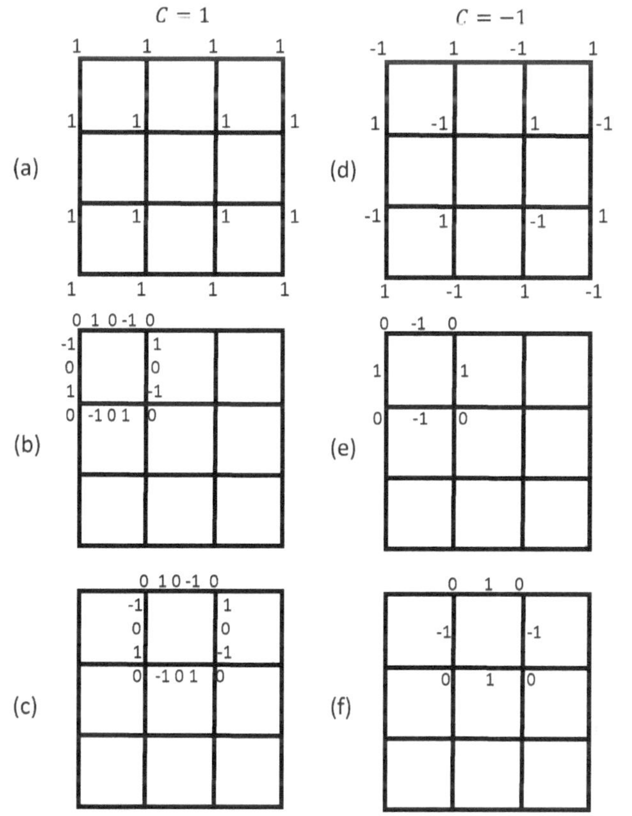

Figure 5.9 Nine squares. The figures illustrate 10 out of the 11 eigenvectors for $C = 1$, or equivalently $\lambda = L/n$, $n = 0, 1, 2, 3, \ldots$ The trivial ground state eigenvector with 0 on all sites is not depicted here. The first one shown here, a bulk eigenvector (a), is for all values of n. The two other ((b) and (c)) illustrate for $\lambda = L$ the path eigenvectors within two of the nine internal loops. In these figures only the nonzero interface values of the eigenvectors are shown. Note that the superposition of these nine eigenvectors provides the surface path eigenvector around this nine-square network. This figure illustrates also the ten eigenvectors for $C = -1$ ($\lambda = 2L$). The first is a bulk eigenvector. The last two ((e) and (f)) illustrate the nine path eigenvectors within each of the nine internal loops. In these figures only the nonzero interface values of the eigenvectors are shown. Remember that the superposition of these nine eigenvectors provides the surface path eigenvector around this nine-square network.

Fig. 5.9 also illustrates the ten eigenvectors of $C = -1$ for $\lambda = 2L$. The first one is a bulk eigenvector. The last two illustrate the nine path eigenvectors within each of the nine internal loops. Remember that the superposition of these nine eigenvectors provides the surface path eigenvector around this nine-square network. Other superpositions of these nine path eigenvectors can be excited, too. These eigenvectors are also robust against defects.

5.5.4.2 *For the $C = 0$ path eigenvalues*

In Fig. 5.10, (a) and (d) illustrate two path eigenvectors for $C = 0$, or equivalently for $\lambda = 4L/(1+2n)$, $n = 0, 1, 2, 3, \ldots$

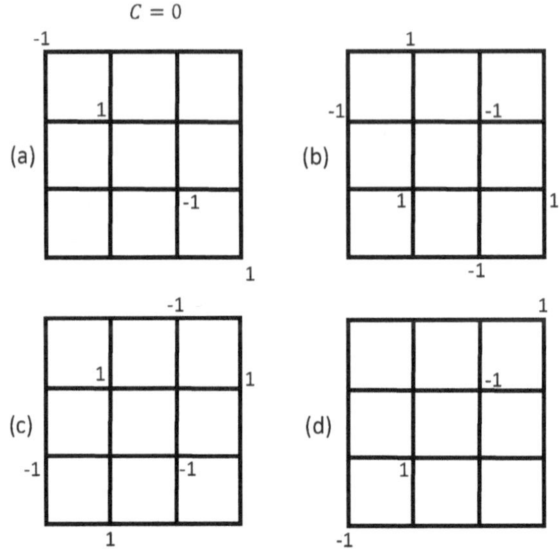

Figure 5.10 Nine squares. Eigenvectors for eigenvalues $C = 0$, or equivalently for $\lambda = 4L/(1+2n)$, $n = 0, 1, 2, 3, \ldots$ In these figures only the nonzero interface values of the eigenvectors are shown. Figures (a) and (d) illustrate two path eigenvectors confined within three diagonal closed loops (4L). They are path eigenvectors as they do not decrease away from their localization, because their six connections with the rest of the network are through robust zeros. The superposition of (a) and (d) is also a path eigenvector. The (b) and (c) eigenvectors are confined in seven of the nine squares. The eigenvectors (b)–(d) and (a)–(c) are surface states confined on the external 12 loops. They decrease towards the network bulk through 6 out of their 12 connections with the network bulk.

The eigenvectors (a) and (d) are confined within three diagonal closed loops $4L$. They are path eigenvectors as they do not decrease away from their localization, because their six connections with the rest of the network are through robust zeros. The eigenvector (a)+(d) is also a path eigenvector.

5.5.4.3 *For the other $C = 0$ eigenvalues*

In Fig. 5.10, the (b) and (c) eigenvectors are localized in seven of the nine squares. Each of them leaves two wedge loops unexcited in two opposite corners of the network. Their value is zero along one diagonal of the nine-square network. Note also the zeros on the lines parallel to this diagonal. These eigenvectors are localized in two opposite wedges of the network.

The eigenvectors (b)–(d) and (a)–(c) are surface states localized on the external $12L$ loop. They decrease towards the network bulk through 6 out of their 12 connections with the network bulk.

5.5.4.4 For the $C = \pm 1/2$ eigenvalues

In Fig. 5.11, (a) and (b) illustrate the eigenvectors for $C = 1/2$ and $C = -1/2$. They are bulk eigenvectors with a slow decrease from the surface towards the network bulk.

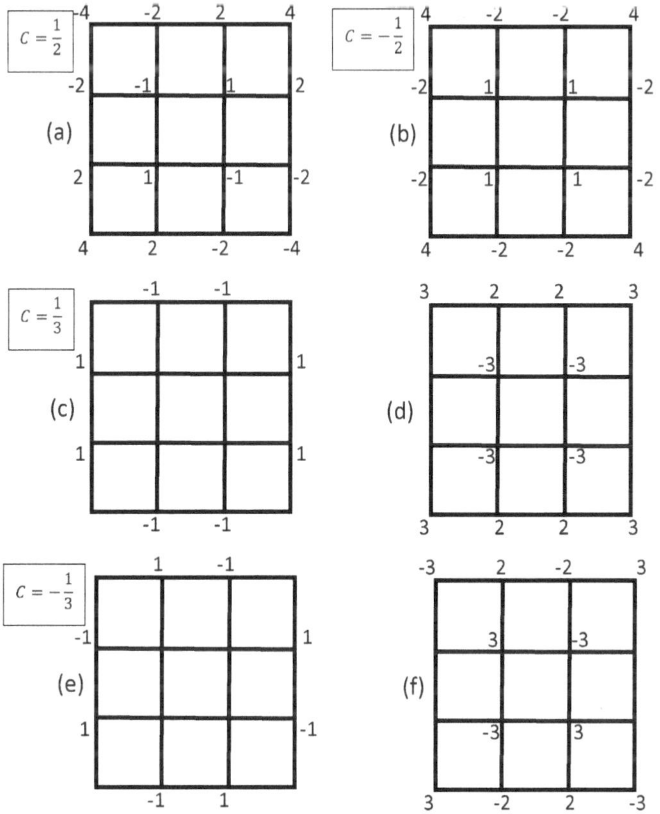

Figure 5.11 Nine squares. Figures (a) and (b) illustrate the eigenvectors for $C = 1/2$ and $C = -1/2$. They are bulk eigenvectors with a slow decrease from the surface towards the network bulk. Figures (c), (d), (e), and (f) illustrate the eigenvectors for $C = 1/3$ and $C = -1/3$. The eigenvectors (c) and (e) are surface eigenvectors with a decrease depth by L. The eigenvectors (d) and (f) are bulk eigenvectors.

5.5.4.5 For the $C = \pm 1/3$ eigenvalues

In Fig. 5.11, (c), (d), (e) and (f) illustrate the eigenvectors for $C = 1/3$ and $C = -1/3$. The eigenvectors (c) and (e) are surface eigenvectors with a decreased depth of L. The eigenvectors (d) and (f) are bulk eigenvectors.

5.5.4.6 For the $C = \pm\sqrt{11/18}$ eigenvalues

Fig. 5.12 illustrates the eigenvectors for $C = \sqrt{11/18}$ and $C = -\sqrt{11/18}$. The eigenvectors (a) and (d) are bulk ones. The eigenvectors (b) and(d) are bulk-wedge eigenvectors.

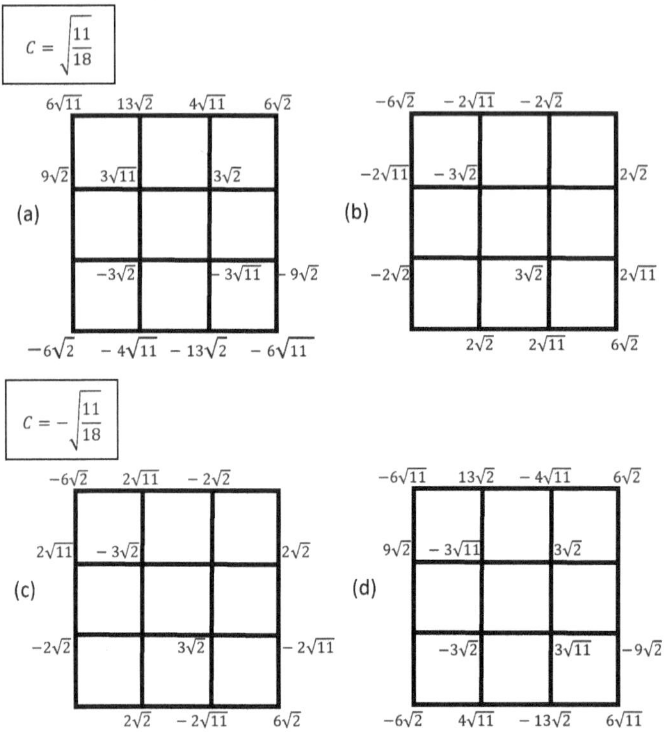

Figure 5.12 Nine squares. The figures illustrate the eigenvectors for $C = \sqrt{11/18}$ and $C = -\sqrt{11/18}$. The eigenvectors (a) and (d) are bulk ones. The eigenvectors (b) and (c) are bulk-wedge eigenvectors.

5.6 Sixteen squares

As the reference system, one takes here 40 loops of length L all made of the same material. The final system, called 16 squares, is obtained by superposing, as in the previous two examples, the ends of these loops in order to obtain the 16-square network.

5.6.1 Path states by inspection

In the present case, each of the 16 internal loops of circumference $4L$ has robust zeros used in its connection with the other internal loops. These zeros are used in the connections of each internal loop with the other neighbor internal loops. So the confined state theorem enables, as in the previous two examples, finding the following path eigenvalues in each of these 16 internal loops:

$$\lambda_p = \frac{2L}{p}, \quad p = 1, 2, 3, \ldots, \tag{5.30}$$

and path eigenfunctions

$$E(x) = \pm \sin\left[\pi p \frac{x}{L}\right], \quad p = 1, 2, 3, \ldots, \tag{5.31}$$

where x varies from 0 at one robust zero to $4L$ after one revolution within each internal closed loop. The $\pm$ signs have to be chosen so that they are opposite for two adjacent loops. These path eigenvalues are 16 times degenerate. A superposition of 2 to 16 of them enables obtaining path eigenfunctions around 2 to 16 internal closed loops. In particular, the superposition of the 16 degenerate eigenfunctions provides the path eigenfunctions confined on the external surface path of the network without any decrease towards the inside of the structure. Some defects may affect just one of their degenerate eigenfunctions and not the others. So these path eigenfunctions can turn around defects.

5.6.2 The interface response inverse matrix

The interface response inverse matrix of this network can be obtained by using the same procedure shown explicitly for the previous examples. This matrix is of dimension 25×25, and the value of its determinant is given by

$$\begin{aligned}|[g(MM)]^{-1}| = \frac{18432}{S^{25}}\left(\frac{2\pi}{\lambda}\right)^{25} C^5 S^2(-1+2C^2)(-11+36C^2)^2 \\ \times (-1+30C^2-201C^4+216C^6)^2.\end{aligned} \tag{5.32}$$

5.6.3 All eigenvalues from the state phase shift

The state phase shift η between the final and reference systems is

$$\begin{aligned}\eta(\lambda) = -\arg\left(\frac{\lambda^{55}}{S^{25}} C^5 S^2(-1+2C^2)(-11+36C^2)^2\right) \\ -\arg(-1+30C^2-201C^4+216C^6)^2.\end{aligned} \tag{5.33}$$

The reference system, 40 finite loops of length L, has the eigenvalues of $(S/\lambda)^{40} = 0$. The determinant $|[g(MM)]^{-1}|$ of each finite loop of length L equals to $(-\left(\frac{2\pi}{\lambda}\right)^2)$. So the eigenvalues of the final system are those of

$$\lambda^{15} S^{17} C^5(-1+2C^2)(-11+36C^2)^2(-1+30C^2-201C^4+216C^6)^2 = 0. \tag{5.34}$$

Again this system has two degenerate states for infinite λ.

5.6.4 The interface eigenvectors

The following figures provide an illustration within the interface space of all the eigenvectors associated to the eigenvalues of the 16-square network.

5.6.4.1 For the $S = 0$ eigenvalues

In Fig. 5.13, (a), (b), and (c) illustrate the 17 eigenvectors for $C = 1$, or equivalently $\lambda = L/n$, $n = 0, 1, 2, 3, \ldots$ The ground state eigenfunction with 0 on all structure points is not shown in this figure. The first bulk eigenfunction (a) is for all values of n and represents for $n = 0$ the invariant by translation eigenfunction. The two other illustrate for $n = 1$ the 16 degenerate path eigenvectors within 2 of the 16 internal loops. For $n = 2, 3, \ldots$, there are just more interface points. Remember that the superposition of these 16 eigenvectors provides the surface path eigenvector around this 16-square network. These eigenvectors are also robust against defects.

In Fig. 5.13, (d), (e), and (f) illustrate the 17 eigenvectors for $C = -1$, when $\lambda = 2L$. The last two illustrate the 16 path eigenvectors within each of the 16 internal loops. Remember that the superposition of these 16 eigenvectors provides the surface path eigenvector around this 16-square network. These eigenvectors are also robust against defects.

5.6.4.2 For the $C = 0$ path eigenvalues

In Fig. 5.14, (a) and (d) illustrate two path eigenvectors for $C = 0$, or equivalently for $\lambda = 4L/(1 + 2n)$, $n = 0, 1, 2, 3, \ldots$

These eigenvectors are confined within four diagonal internal loops $4L$. They are path eigenvectors as they do not decrease away from their localization, because their eight connections with the rest of the network are through robust zeros. The eigenvector (a)+(d) is also a path eigenvector. It is a four path crossing state. Such a state can be used for most efficient signal crossing devices. One can also construct systems with such path crossing states for other number of paths than four.

5.6.4.3 For the other $C = 0$ eigenvalues

The eigenvectors of Fig. 5.14, (b) and (e), are localized in 14 of the 16 squares. Each of them leaves two wedge loops unexcited in two opposite corners of the network. Their value is zero along one diagonal of the 16-square network. Note also the zeros on the lines parallel to this diagonal. They are localized states between two opposite wedges of the network.

The eigenvector (c) is localized in 32 out of the 40 pieces L of this structure. The four two corner pieces remain unactivated by this state motions.

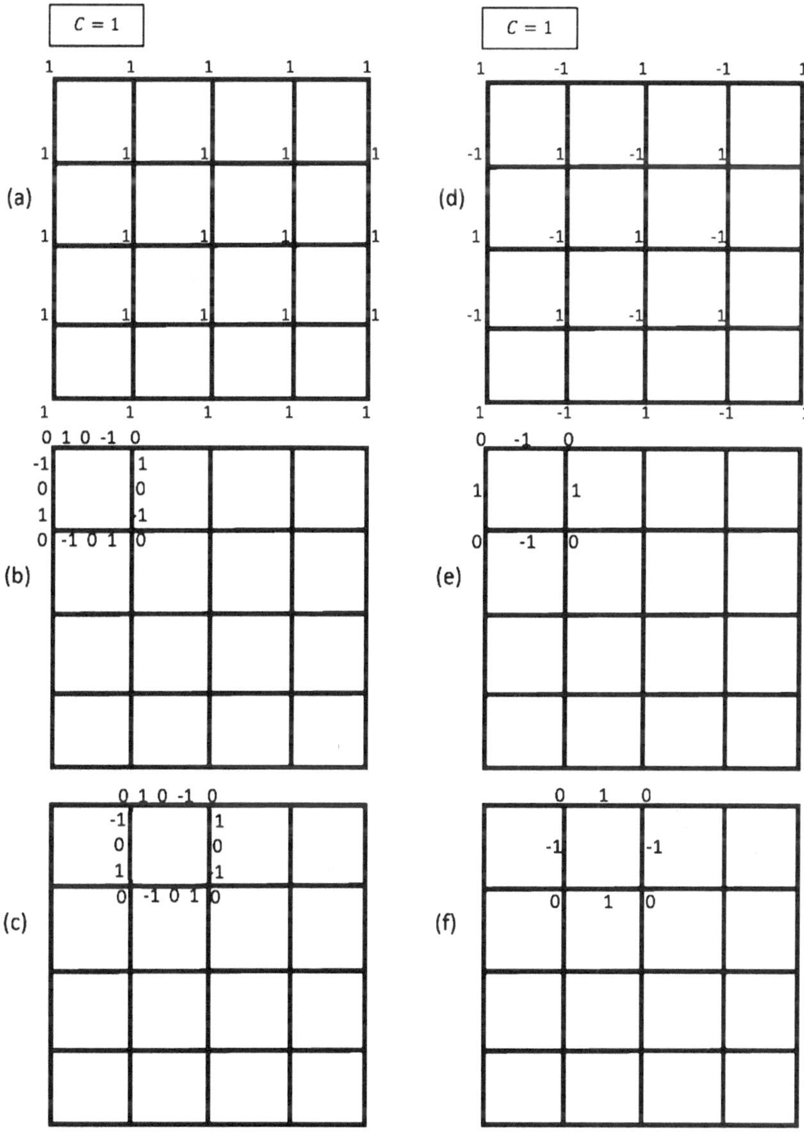

Figure 5.13 16 Squares. Figures (a), (b), and (c) illustrate 17 out of the 18 eigenvectors for $C = 1$, or equivalently $\lambda = L/n$, $n = 0, 1, 2, 3, \ldots$ The trivial ground state eigenvector with 0 on all sites is not depicted here. The first shown, a bulk one (a), is for all values of n. The two other illustrate for $n = 1$ the 16 degenerate path eigenvectors within two of the 16 internal loops. In these figures the zero interface values of the eigenvectors on the loops, where they are not confined, are not shown. Remember that the superposition of these 16 eigenvectors provides the surface path eigenvector around this 16 square network. Figures (d), (e), and (f) illustrate the 17 eigenvectors (of $C = -1$) for $\lambda = 2L$. The last two illustrate the 16 path eigenvectors within each of the 16 internal loops. In these figures the zero interface values of the eigenvectors on the loops, where they are not confined, are not shown. Remember that the superposition of these 16 eigenvectors provides the surface path eigenvector around this 16 square network.

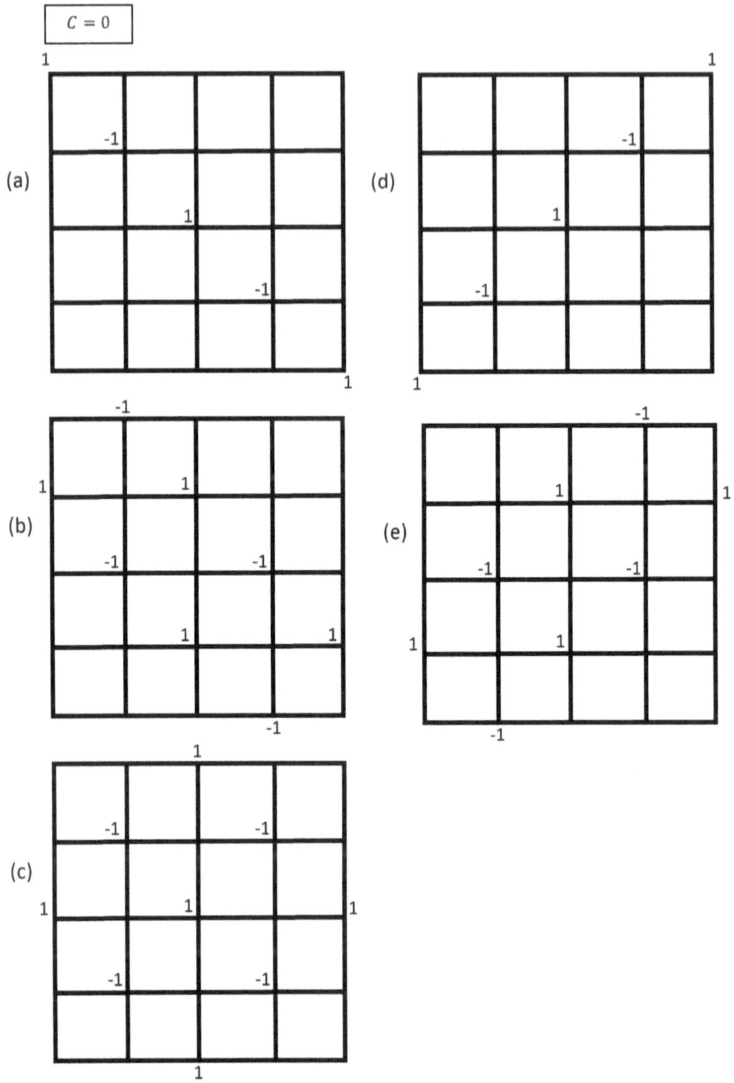

Figure 5.14 16 Squares. Eigenvectors for eigenvalues $C = 0$, or equivalently for $\lambda = 4L/(1+2n)$, $n = 0, 1, 2, 3, \ldots$ In these figures only the nonzero interface values of the eigenvectors are shown. Figures (a) and (d) illustrate two path eigenvectors. These eigenvectors are confined within four diagonal internal $4L$ loops. They are path eigenvectors as they do not decrease away from their localization, because their eight connections with the rest of the network are through robust zeros. The eigenvector (a)+(d) is also a path eigenvector. It is a four path crossing state. Such a state can be used for most efficient signal crossing devices. The eigenvectors (b) and (e) are confined in 14 of the 16 squares. Each of them leaves two wedge loops unexcited in two opposite corners of the network. Their space values are zero along one diagonal of the 16 square network. Note also the zeros on the lines parallel to this diagonal. They are confined states between two opposite wedges of the network. The eigenvector (c) is confined in 32 out of the 40 pieces L of this structure. The four two corner pieces remain unactivated by this state motions. Its space values are zero along lines parallel to the diagonal along which it is confined. The eigenvector (b)–(e) is surface state localized on the external $16L$ loops. Their decrease length towards the bulk is L.

The eigenvectors (b)–(e) are surface states localized on the external $16L$ loop. They have decreased length by L towards the network bulk through 8 out of their 12 connections with the network interior nodes.

5.6.4.4 *For the $C = \pm 1/\sqrt{2}$ eigenvalues*

Fig. 5.15 illustrates the eigenvectors for $C = -1/\sqrt{2}$ and $C = 1/\sqrt{2}$. They are wedge eigenvectors localized in the four wedges of this network. Their decrease of length towards the bulk is $2L$.

5.6.4.5 *For the $C = \pm 1/\sqrt{11/36}$ eigenvalues*

In Fig. 5.15, (c) and (d) illustrate the eigenvectors for $C = \sqrt{11/36}$ and $C = -\sqrt{11/36}$. Eigenvectors (d) and (e) are surface eigenvectors. Their decrease of length towards the bulk is $2L$. Eigenvectors (c) and (f) are bulk eigenvectors.

5.6.4.6 *For the $(-1 + 30C^2 - 201C^4 + 216C^6)^2 = 0$ eigenvalues*

Figs. 5.16, 5.17, and 5.18 illustrate the eigenvectors for $(-1 + 30C^2 - 201C^4 + 216C^6)^2 = 0$. This equation provides six eigenvalues for C, and from the matrix $[g(MM)]^{-1}$ one finds two degenerate eigenvectors for each of these eigenvalues. The results of these numerical calculations are presented in these figures for each of the six eigenvalues as a function of 12 parameters y_i, $i = 1, 2, \ldots, 12$ given in Table 5.1. The most striking in all these eigenvectors is the 0 in the middle of the network. The eigenvectors reflect also the symmetries of the network. All these eigenvectors are surface-wedge eigenvectors with more or less fast decrease towards the central 0.

Table 5.1 Values of y_i, $i = 1, 2, \ldots, 12$ for different eigenvalues.

C	**±0.868**	**±0.218942**	**±0.357745**
y_1	0.35	0.019	0.212
y_2	0.30	0.215	0.035
y_3	0.18	0.214	0.178
y_4	0.05	0.165	0.281
y_5	0.02	0.435	0.008
y_6	0.31	0.201	0.057
y_7	0.25	0.001	0.367
y_8	0.12	0.179	0.199
y_9	0.01	0.044	0.014
y_{10}	0.09	0.146	0.284
y_{11}	0.20	0.234	0.192
y_{12}	0.13	0.196	0.215

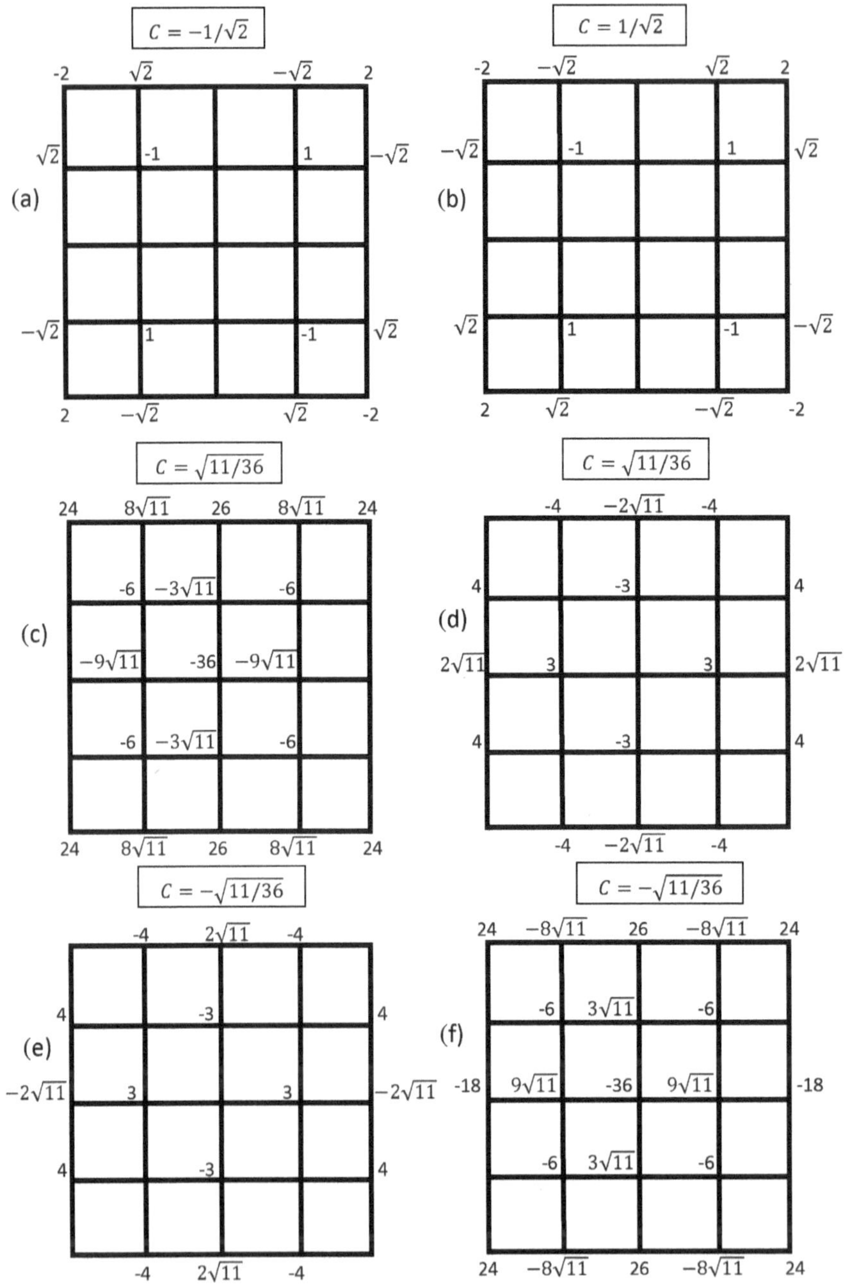

Figure 5.15 16 Squares. Eigenvectors for eigenvalues $C = \pm\sqrt{1/2}$ and $C = \pm\sqrt{11/36}$. In these figures only the nonzero interface values of the eigenvectors are shown. These figures illustrate the eigenvectors for $C = -1/\sqrt{2}$ and $C = 1/\sqrt{2}$. They are wedge eigenvectors localized in the four wedges of this network. Their decrease of length towards the bulk is $2L$.

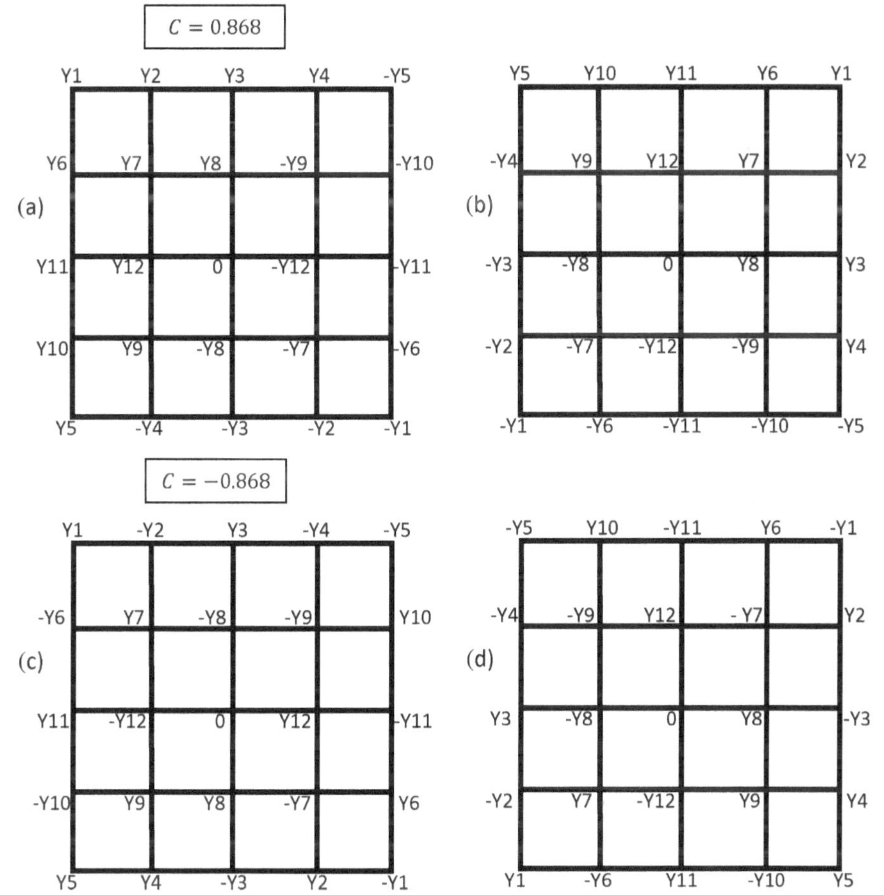

Figure 5.16 16 Squares. Two times degenerate eigenvectors for eigenvalues $C = \pm 0.868$, solutions of $(-1+30C^2-201C^4+216C^6)^2 = 0$. The results of numerical calculations are presented in these figures as a function of 12 parameters y_i, $i = 1, 2, \ldots, 12$ given in Table 5.1. The most striking in all these eigenvectors is the 0 in the middle of the network. The eigenvectors also reflect the symmetries of the network. All these eigenvectors are surface-wedge eigenvectors with more or less fast decrease towards the central 0.

5.7 Cube

As a reference system, one takes here 12 loops of length L all made of the same material. The final network, called cube, is obtained by superposing the ends of these loops in order to obtain the cube network of Fig. 5.19.

5.7.1 Path states by inspection

This cube has six faces (six closed loops $4L$) connected to their neighbors through their four robust zeros. So the confined state theorem provides again the following

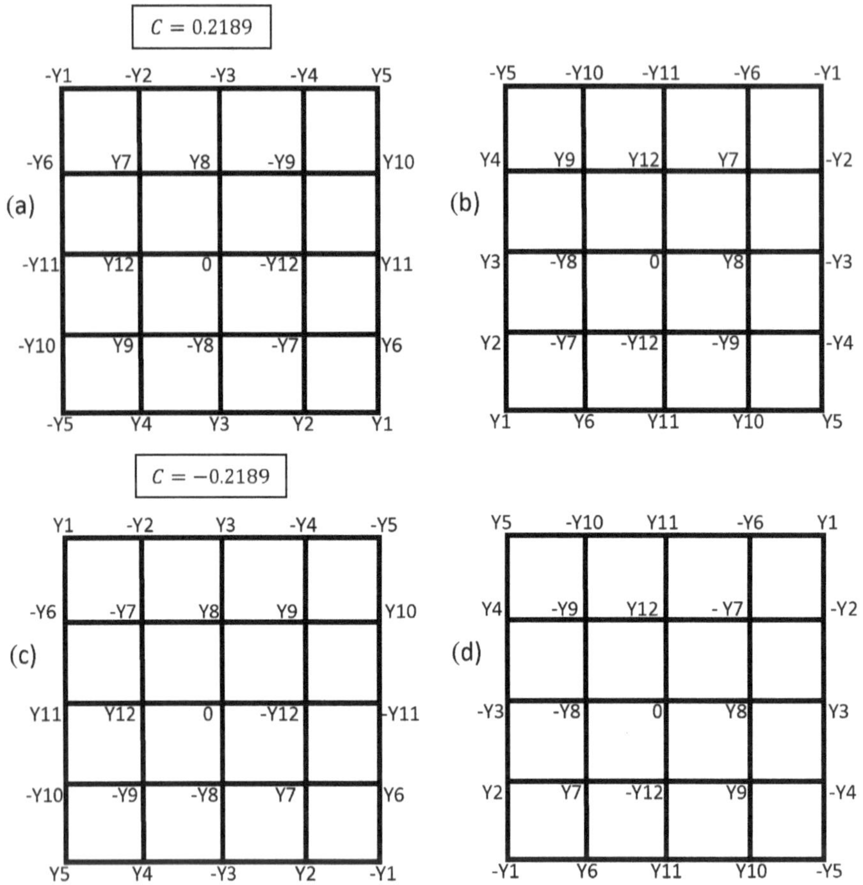

Figure 5.17 16 Squares. Eigenvectors for eigenvalues $C = \pm 0.2189$, solutions of $(-1+30C^2-201C^4+216C^6)^2 = 0$. The results of numerical calculations are presented in these figures as a function of 12 parameters y_i, $i = 1, 2, \ldots, 12$ given in Table 5.1. The most striking in all these eigenvectors is the 0 in the middle of the network. The eigenvectors also reflect the symmetries of the network. All these eigenvectors are surface-wedge eigenvectors with more or less fast decrease towards the central 0.

path eigenvalues:

$$\lambda_p = \frac{2L}{p}, \quad p = 1, 2, 3, \ldots, \tag{5.35}$$

and five elementary path eigenfunctions within each of the 5 cube faces (no matter which five out of six are chosen):

$$E_{1,n}(x) = \pm \sin\left[\pi p \frac{x}{L}\right], \quad n = 1, 2, 3, 4, 5, \quad p = 1, 2, 3, \ldots, \tag{5.36}$$

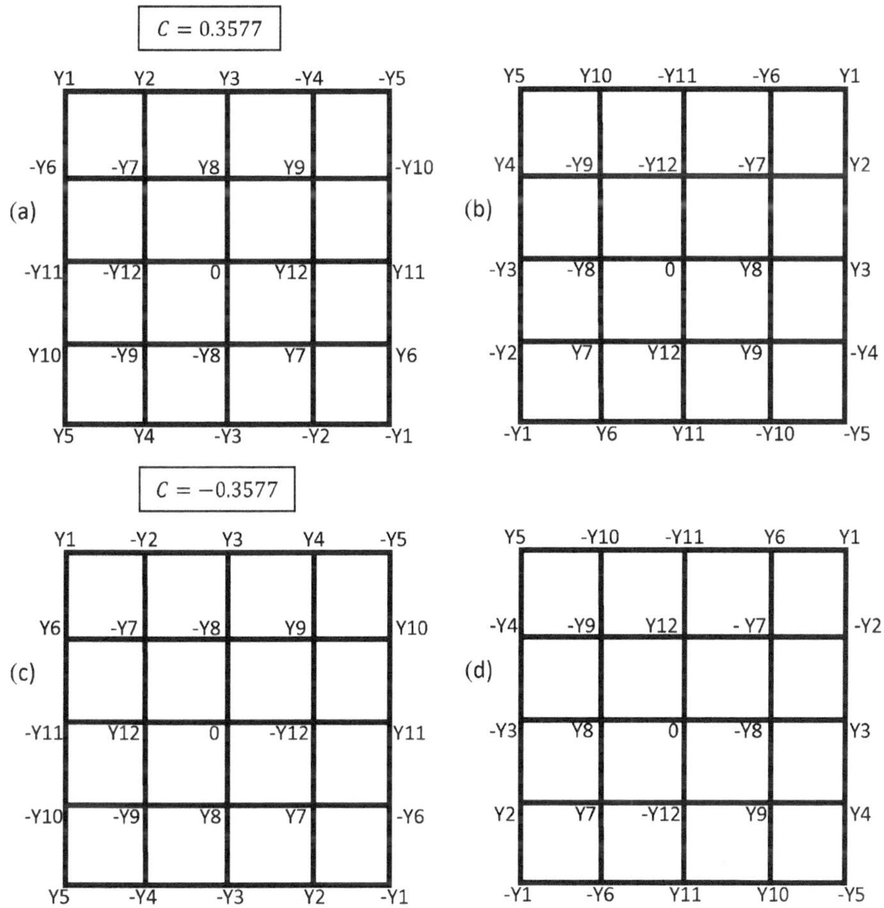

Figure 5.18 16 Squares. Eigenvectors for eigenvalues $C = \pm 0.3577$, solutions of $(-1 + 30C^2 - 201C^4 + 216C^6)^2 = 0$. The results of numerical calculations are presented in these figures as a function of 12 parameters y_i, $i = 1, 2, \ldots, 12$ given in Table 5.1. The most striking in all these eigenvectors is the 0 in the middle of the network. The eigenvectors reflect also the symmetries of the network. All these eigenvectors are surface-wedge eigenvectors with more or less fast decrease towards the central 0.

where x varies from 0 at one robust zero to $4L$ after one revolution within each internal closed loop. The $\pm$ signs have to be chosen so that they are opposite for two adjacent cube faces. These path eigenvalues are 5 times degenerate. The superposition of these five elementary eigenvectors provides the eigenvectors of the face 6. The superposition of four of these adjacent eigenvectors provides the eigenvectors of the two other loops $4L$ or of a loop $6L$. The superposition of three of these adjacent eigenvectors provides the eigenvectors of a loop $8L$. The superposition of two of these adjacent eigenvectors provides the eigenvectors of a loop $6L$.

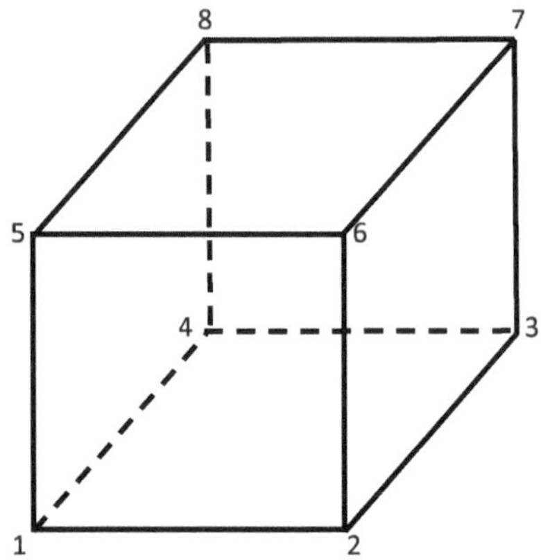

Figure 5.19 Cube network. As a reference system, one has 12 open loops of length L. The final network, called cube, is obtained by superposing the ends of these loops in order to obtain the cube. The interface numbers in this figure are the same as in the interface response matrices.

5.7.2 The interface response inverse matrix

The interface response inverse matrix of the cube network can be obtained using the same procedure shown explicitly for the previous examples. This matrix of dimension 8×8 is

$$[g(MM)]^{-1} = \frac{2\pi}{\lambda S}\begin{pmatrix} -3C & 1 & 0 & 1 & 1 & 0 & 0 & 0 \\ 1 & -3C & 1 & 0 & 0 & 1 & 0 & 0 \\ 0 & 1 & -3C & 1 & 0 & 0 & 1 & 0 \\ 1 & 0 & 1 & -3C & 0 & 0 & 0 & 1 \\ 1 & 0 & 0 & 0 & -3C & 1 & 0 & 1 \\ 0 & 1 & 0 & 0 & 1 & -3C & 1 & 0 \\ 0 & 0 & 1 & 0 & 0 & 1 & -3C & 1 \\ 0 & 0 & 0 & 1 & 1 & 0 & 1 & -3C \end{pmatrix}. \tag{5.37}$$

The determinant of the above matrix is

$$|[g(MM)]^{-1}| = -\frac{9(2\pi)^8 S^2(-1+3C)^3(1+3C)^3}{\lambda^8 S^8}. \tag{5.38}$$

5.7.3 All eigenvalues from the state phase shift

The phase shift η [4] between the final and reference systems is

$$\eta(\lambda) = -\arg\left(\frac{\lambda^{16}(-1+3C)^3(1+3C)^3}{S^6}\right). \tag{5.39}$$

The reference system (12 finite loops of length L) has the eigenvalues of $(S/\lambda)^{12} = 0$. The determinant $|[g(MM)]^{-1}|$ of each finite loop of length L equals to $-\left(\frac{2\pi}{\lambda}\right)^2$. So the eigenvalues of the final system are those of

$$\lambda^4 S^6(-1+3C)^3(1+3C)^3=0. \tag{5.40}$$

Again this final network has two degenerate states for infinite λ.

5.7.4 The interface eigenvectors

5.7.4.1 For the two times degenerate $1/\lambda=0$ eigenvalue

For the two times degenerate $1/\lambda=0$ eigenvalue, the two eigenfunctions are the 0 displacement and the invariance of translation of all structure points.

Their representations within the interface space, see Fig. 5.19, are respectively

$$E_{0,1}=\begin{pmatrix} 0 & 0 & 0 & 0 & 0 & 0 & 0 & 0 \end{pmatrix} \tag{5.41}$$

and

$$E_{0,2}=\begin{pmatrix} 1 & 1 & 1 & 1 & 1 & 1 & 1 & 1 \end{pmatrix}. \tag{5.42}$$

5.7.4.2 For the $S=0$ eigenvalues

The confined state theorem provides already 5 out of the 6 eigenvectors when $S=0$. These path eigenvectors are not rewritten here.

The sixth $S=0$ eigenvector is in the interface space, Fig. 5.19, namely

$$E_{1,6a}=\begin{pmatrix} 1 & 1 & 1 & 1 & 1 & 1 & 1 & 1 \end{pmatrix}, \tag{5.43}$$

for $C=1$, while

$$E_{1,6b}=\begin{pmatrix} 1 & -1 & 1 & -1 & 1 & -1 & 1 & -1 \end{pmatrix}, \tag{5.44}$$

for $C=-1$.

5.7.4.3 For the $S=1/3$ eigenvalues

For the $S=1/3$ eigenvalues, the three degenerate eigenvectors, see Fig. 5.19, are:

$$E_{2,1}=\begin{pmatrix} 0 & 0 & -1 & -1 & 1 & 1 & 0 & 0 \end{pmatrix}, \tag{5.45}$$

$$E_{2,2}=\begin{pmatrix} -1 & 0 & 1 & 0 & -1 & 0 & 1 & 0 \end{pmatrix}, \tag{5.46}$$

and

$$E_{2,3}=\begin{pmatrix} 0 & -1 & -1 & 0 & 1 & 0 & 0 & 1 \end{pmatrix}. \tag{5.47}$$

5.7.4.4 For the $S = -1/3$ eigenvalues

For the $S = -1/3$ eigenvalues, the three degenerate eigenvectors, see Fig. 5.19, are:

$$E_{3,1} = \begin{pmatrix} 0 & 0 & -1 & 1 & -1 & 1 & 0 & 0 \end{pmatrix}, \tag{5.48}$$

$$E_{3,2} = \begin{pmatrix} 1 & 0 & -1 & 0 & -1 & 0 & 1 & 0 \end{pmatrix}, \tag{5.49}$$

and

$$E_{3,3} = \begin{pmatrix} 0 & 1 & -1 & 0 & -1 & 0 & 0 & 1 \end{pmatrix}. \tag{5.50}$$

5.8 Two cubes

As a reference system, one takes here 20 loops of length L, all made of the same material. The final system, called two cubes, is obtained by superposing the ends of these loops in order to obtain the two-cubes network of Fig. 5.20.

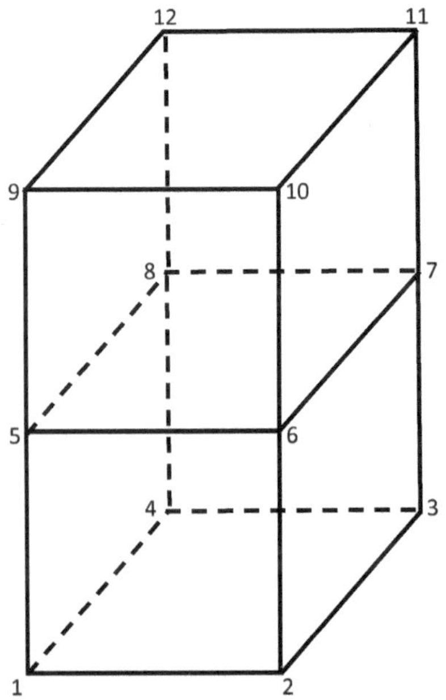

Figure 5.20 Two-cube network. The two cubes are build out of 20 open loops of length L. The interface numbers in this figure are the same as in the interface response matrices.

5.8.1 Path states by inspection

This two-cube network has 11 faces, 10 external and 1 internal (the 5678 one).

It has 11 closed loops $4L$ connected to their neighbors through their robust zeros. So the confined state theorem provides again the following path eigenvalues:

$$\lambda_p = \frac{2L}{p}, \quad p = 1, 2, 3, \ldots, \tag{5.51}$$

and nine elementary path eigenfunctions within each of nine cube external faces, excluding, for example, the 1234 external face. The corresponding eigenvectors are

$$E_{1,n}(x) = \pm \sin\left[\pi p \frac{x}{L}\right], \quad n = 1, 2, \ldots, 9, \quad p = 1, 2, 3, \ldots, \tag{5.52}$$

where x varies from 0 at one robust zero to $4L$ after one revolution within each internal closed loop. The $\pm$ signs have to be chosen so that they are opposite for two adjacent cube faces. These path eigenvalues are nine times degenerate. The superposition of these nine elementary eigenvectors provides the eigenvector of the face 1234. The superposition of the upper five elementary eigenvectors provides the eigenvector of the internal face 5678, see Fig. 5.20.

5.8.2 The interface response inverse matrix

The interface space response inverse matrix of the two-cube network can be obtained using the same procedure shown explicitly for the previous examples. This matrix has dimension 12×12 and then

$$[g(MM)]^{-1} = \frac{2\pi}{\lambda S} \times \begin{pmatrix} -3C & 1 & 0 & 1 & 1 & 0 & 0 & 0 & 0 & 0 & 0 & 0 \\ 1 & -3C & 1 & 0 & 0 & 1 & 0 & 0 & 0 & 0 & 0 & 0 \\ 0 & 1 & -3C & 1 & 0 & 0 & 1 & 0 & 0 & 0 & 0 & 0 \\ 1 & 0 & 1 & -3C & 0 & 0 & 0 & 1 & 0 & 0 & 0 & 0 \\ 1 & 0 & 0 & 0 & -4C & 1 & 0 & 1 & 1 & 0 & 0 & 0 \\ 0 & 1 & 0 & 0 & 1 & -4C & 1 & 0 & 0 & 1 & 0 & 0 \\ 0 & 0 & 1 & 0 & 0 & 1 & -4C & 1 & 0 & 0 & 1 & 0 \\ 0 & 0 & 0 & 1 & 1 & 0 & 1 & -4C & 0 & 0 & 0 & 1 \\ 0 & 0 & 0 & 0 & 1 & 0 & 0 & 0 & -3C & 1 & 0 & 1 \\ 0 & 0 & 0 & 0 & 0 & 1 & 0 & 0 & 1 & -3C & 1 & 0 \\ 0 & 0 & 0 & 0 & 0 & 0 & 1 & 0 & 0 & 1 & -3C & 1 \\ 0 & 0 & 0 & 0 & 0 & 0 & 0 & 1 & 1 & 0 & 1 & -3C \end{pmatrix}. \tag{5.53}$$

The determinant of the above matrix is

$$|[g(MM)]^{-1}| = -\frac{144}{S^{12}} \left(\frac{2\pi}{\lambda}\right)^{12} C^2 S^2 (-2 + 3C)(2 + 3C) \times (-1 + 6C)(1 + 6C)(-1 + 6C^2)^2. \tag{5.54}$$

5.8.3 All eigenvalues from the state phase shift

The phase shift η [4] between the final and reference systems is

$$\eta(\lambda) = -\arg\left[\frac{\lambda^{28}}{S^{10}} C^2(-2+3C)(2+3C)(-1+6C)(1+6C)(-1+6C^2)^2\right]. \tag{5.55}$$

The reference system (20 finite loops of length L) has the eigenvalues of $(S/\lambda)^{20} = 0$. The determinant $|[g(MM)]^{-1}|$ of each finite loop of length L equals to $(-\left(\frac{2\pi}{\lambda}\right)^2)$. So the eigenvalues of the final system are those of

$$\lambda^8 S^{10} C^2(-2+3C)(2+3C)(-1+6C)(1+6C)(-1+6C^2)^2 = 0. \tag{5.56}$$

Again two degenerate states for infinite λ.

5.8.4 The interface eigenvectors

5.8.4.1 For the two times degenerate $1/\lambda = 0$ eigenvalue

For the two times degenerate $1/\lambda = 0$ eigenvalue, the two eigenfunctions are the 0 displacement and the invariance of translation of all structure points.

Their representations within the interface space, see Fig. 5.20, are respectively

$$E_{0,1} = \begin{pmatrix} 0 & 0 & 0 & 0 & 0 & 0 & 0 & 0 & 0 & 0 & 0 & 0 \end{pmatrix} \tag{5.57}$$

and

$$E_{0,2} = \begin{pmatrix} 1 & 1 & 1 & 1 & 1 & 1 & 1 & 1 & 1 & 1 & 1 & 1 \end{pmatrix}. \tag{5.58}$$

5.8.4.2 For the $S = 0$ eigenvalues

The confined state theorem provides already 9 out of the 10 eigenvectors for $S = 0$. These path eigenvectors are not rewritten here.

The tenth $S = 0$ eigenvector is in the interface space, see Fig. 5.20, namely

$$E_{1,10a} = \begin{pmatrix} 1 & 1 & 1 & 1 & 1 & 1 & 1 & 1 & 1 & 1 & 1 & 1 \end{pmatrix}, \tag{5.59}$$

for $C = 1$, and

$$E_{1,10b} = \begin{pmatrix} 1 & -1 & 1 & -1 & 1 & -1 & 1 & -1 & 1 & -1 & 1 & -1 \end{pmatrix} \tag{5.60}$$

for $C = -1$.

5.8.4.3 For the $C = 0$ eigenvalues

For the $C = 0$ eigenvalues, the two degenerate (see the two-cube figure) wedge eigenvectors are

$$E_{2,1} = \begin{pmatrix} 1 & 0 & -1 & 0 & 0 & 0 & 0 & 0 & -1 & 0 & 1 & 0 \end{pmatrix} \quad (5.61)$$

and

$$E_{2,2} = \begin{pmatrix} 0 & 1 & 0 & -1 & 0 & 0 & 0 & 0 & 0 & -1 & 0 & 1 \end{pmatrix}. \quad (5.62)$$

5.8.4.4 For the $C = 2/3$ eigenvalues

For the $C = 2/3$ eigenvalues, the eigenvector is

$$E_3 = \begin{pmatrix} -1 & -1 & -1 & -1 & 0 & 0 & 0 & 0 & 1 & 1 & 1 & 1 \end{pmatrix}. \quad (5.63)$$

It is a breathing mode of the two more distant faces of this two cube network.

5.8.4.5 For the $C = -2/3$ eigenvalues

For the $C = -2/3$ eigenvalues, the eigenvector is

$$E_4 = \begin{pmatrix} 1 & -1 & 1 & -1 & 0 & 0 & 0 & 0 & -1 & 1 & -1 & 1 \end{pmatrix}. \quad (5.64)$$

It is an asymmetric breathing mode of the two more distant faces of this two-cube network.

5.8.4.6 For the $C = 1/6$ eigenvalues

For the $C = 1/6$ eigenvalues, the eigenvector is

$$E_5 = \begin{pmatrix} 1 & 1 & 1 & 1 & -\frac{3}{2} & -\frac{3}{2} & -\frac{3}{2} & -\frac{3}{2} & 1 & 1 & 1 & 1 \end{pmatrix}. \quad (5.65)$$

It is a different breathing mode but for the three parallel faces of this network.

5.8.4.7 For the $C = -1/6$ eigenvalues

For the $C = -1/6$ eigenvalues, the eigenvector is

$$E_6 = \begin{pmatrix} -1 & 1 & -1 & 1 & -\frac{3}{2} & \frac{3}{2} & -\frac{3}{2} & \frac{3}{2} & -1 & 1 & -1 & 1 \end{pmatrix}. \quad (5.66)$$

This mode is similar to the preceding one but with an additional asymmetry.

5.8.4.8 For the $C = 1/\sqrt{6}$ eigenvalues

For the $C = 1/\sqrt{6}$ eigenvalues, the two degenerate eigenvectors are

$$E_{7a} = \begin{pmatrix} -1 & 0 & 1 & 0 & -\sqrt{\frac{3}{2}} & 0 & \sqrt{\frac{3}{2}} & 0 & -1 & 0 & 1 & 0 \end{pmatrix} \tag{5.67}$$

and

$$E_{7b} = \begin{pmatrix} 0 & -1 & 0 & 1 & 0 & -\sqrt{\frac{3}{2}} & 0 & \sqrt{\frac{3}{2}} & 0 & -1 & 0 & 1 \end{pmatrix}. \tag{5.68}$$

These two modes are wedge modes affecting the two pairs of opposite edges of length $2L$.

5.8.4.9 For the $C = -1/\sqrt{6}$ eigenvalues

For the $C = -1/\sqrt{6}$ eigenvalues, the two degenerate eigenvectors are

$$E_{8a} = \begin{pmatrix} -1 & 0 & 1 & 0 & \sqrt{\frac{3}{2}} & 0 & -\sqrt{\frac{3}{2}} & 0 & -1 & 0 & 1 & 0 \end{pmatrix} \tag{5.69}$$

and

$$E_{8b} = \begin{pmatrix} 0 & -1 & 0 & 1 & 0 & \sqrt{\frac{3}{2}} & 0 & -\sqrt{\frac{3}{2}} & 0 & -1 & 0 & 1 \end{pmatrix}. \tag{5.70}$$

These two modes are similar to the above two wedge modes but with a sign change for the sites being on the middle face of this network.

5.9 Hexagon

As a reference system, one takes here 24 loops of length L, all made of the same material. The final system, called hexagon, is obtained by superposing the ends of these loops in order to obtain the hexagon network of Fig. 5.21.

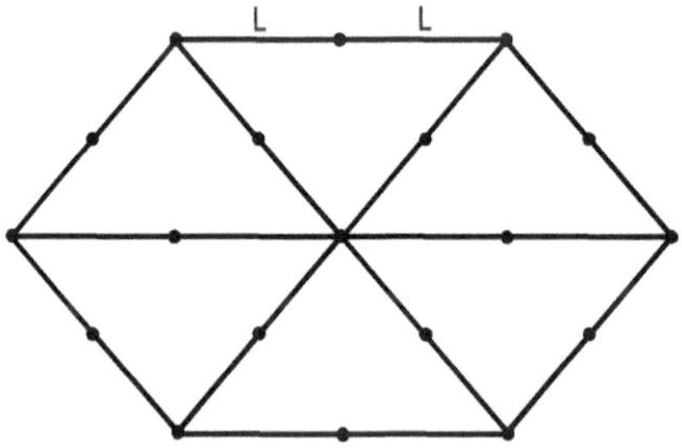

Figure 5.21 Hexagon interface space. This hexagon is built out of 24 open loops of length L. The dots in this figure indicate the interface points.

5.9.1 Path states by inspection

This hexagon network has one surface closed loop $12L$ and six robust zeros.

According to the confined state theorem, this network has the following path eigenvalues:

$$\lambda_p = \frac{4L}{p}, \quad p = 1, 2, 3, \ldots, \tag{5.71}$$

and eigenfunctions

$$E_1(x) = \pm \sin\left[\pi p \frac{x}{2L}\right], \quad p = 1, 2, 3, \ldots, \tag{5.72}$$

where x varies from 0 at one robust zero till the end of all its possible closed loop revolutions.

The $\pm$ signs are opposite on each side of all the robust zeros that are on the paths of all the eigenfunctions. These path eigenvalues are degenerate. So when there is one defect, the eigenfunction may change its path and remain unperturbed.

5.9.2 The interface response inverse matrix

The interface response inverse matrix can be obtained using the same procedure shown explicitly for the previous examples. This matrix is of dimension 19×19. Its determinant is

$$|[g(MM)]^{-1}| = -\frac{12288(2\pi)^{19}(C^5)(-1+6C^2)(2-3C^2)^2(-1+3C^2)^3}{\lambda^{19}S^{17}}. \tag{5.73}$$

5.9.3 All eigenvalues from the state phase shift

The phase shift η [4] between the final and reference systems is

$$\eta(\lambda) = -\arg\left[\frac{\lambda^{29}}{S^{17}}(C^5)(-1+6C^2)(2-3C^2)^2(-1+3C^2)^3\right]. \tag{5.74}$$

The reference system (24 finite loops of length L) has the eigenvalues of $(S/\lambda)^{24} = 0$. The determinant $|[g(MM)]^{-1}|$ of each finite loop of length L equals to $-\left(\frac{2\pi}{\lambda}\right)^2$. So the eigenvalues of the final system are those of

$$\lambda^5 S^7 C^5(-1+6C^2)(2-3C^2)^2(-1+3C^2)^3 = 0. \tag{5.75}$$

Again two degenerate states for infinite λ. Fig. 5.22 enables checking the state conservation around the trigonometric circle.

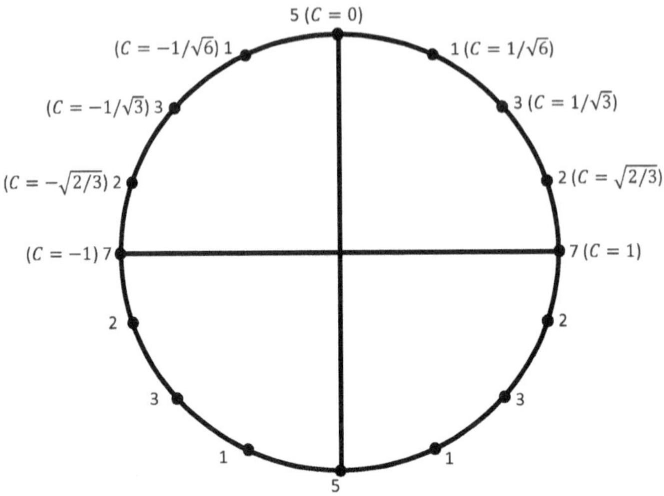

Figure 5.22 Hexagon eigenstate number conservation. This figure enables checking the state conservation around the trigonometric circle.

5.9.4 The interface eigenvectors

The eigenvectors within the hexagon interface space are as follows.

5.9.4.1 For the $S = 0$ eigenvalues

Fig. 5.23 illustrates 7 out of the 8 eigenvectors for $C = 1$, or equivalently $\lambda = 2L/n$, $n = 0, 1, 2, 3, \ldots$ The trivial ground state eigenvector, with 0 on all sites, is not shown in this figure. The first eigenvector shown in this figure, (a), is a bulk eigenvector for all values of n, including the invariant by translation eigenvector for $n = 0$. The six other illustrate for $\lambda = 2L$ the path eigenvectors within the six internal loops. For $n = 2, 3, \ldots$, there are just more interface points. Note that the superposition of these 6 eigenvectors provides the surface path eigenvector around this hexagon. Remember that these eigenvectors are also robust against defects.

5.9.4.2 For the $C = 0$ eigenvalues

Fig. 5.24 illustrates now the 5 eigenvectors of $C = 0$ for $\lambda = 4L$. They are confined within two interior loops. Notice that the superpositions of the eigenvectors (a)+(b)+(c) and (c)+(d)+(e) provide the surface path eigenvector around this hexagon network. Remember that these eigenvectors are also robust against defects.

5.9.4.3 For the $C = \pm 1/\sqrt{3}$ eigenvalues

Fig. 5.25 illustrates the eigenvectors for $C = 1/\sqrt{3}$ and $C = -1/\sqrt{3}$. The eigenvectors (a) and (d) are bulk-surface eigenvectors. Their decrease of length from surface towards the bulk is $2L$. The four others are bulk eigenvectors.

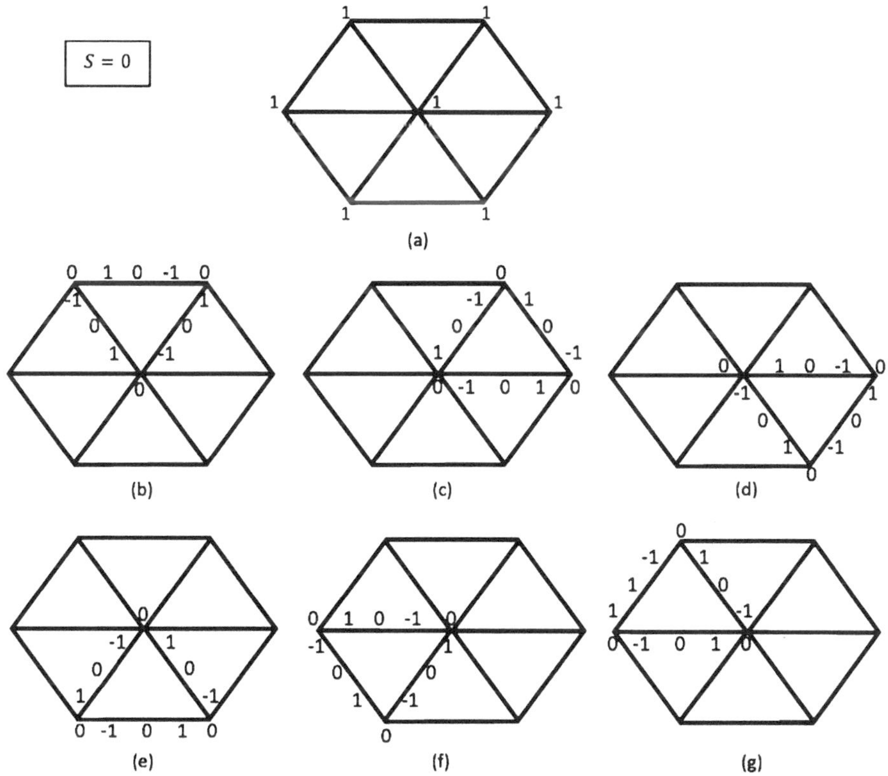

Figure 5.23 Hexagon eigenvectors for eigenvalues $S = 0$. These figures illustrate 7 out of the 8 eigenvectors for $C = 1$, or equivalently $\lambda = 2L/n$, $n = 0, 1, 2, 3, \ldots$ The trivial ground state eigenvector, with 0 on all sites, is not shown in this figure. The first eigenvector (a), shown on this figure, is a bulk eigenvector for all values of n. The six other illustrate for $\lambda = 2L$ the path eigenvectors within the six internal loops. In these figures the zero interface values of the eigenvectors on the loops, where they are not localized, are not shown. Notice that the superposition of these 6 eigenvectors provides the surface path eigenvector around this hexagon.

5.9.4.4 For the $C = \sqrt{2/3}$ eigenvalues

Fig. 5.26 illustrates the eigenvectors for $C = \sqrt{2/3}$. They are surface eigenvectors. Their decrease of length towards the bulk is $2L$.

5.9.4.5 For the $C = -\sqrt{2/3}$ eigenvalues

Fig. 5.27 illustrates the eigenvectors for $C = -\sqrt{2/3}$. They are bulk-surface eigenvectors. Their decrease of length towards the bulk is $2L$.

5.9.4.6 For the $C = \pm\sqrt{1/6}$ eigenvalues

Fig. 5.28 illustrates the eigenvectors for $C = \pm\sqrt{1/6}$. They are bulk-surface eigenvectors. Their decrease of length towards the bulk is $2L$.

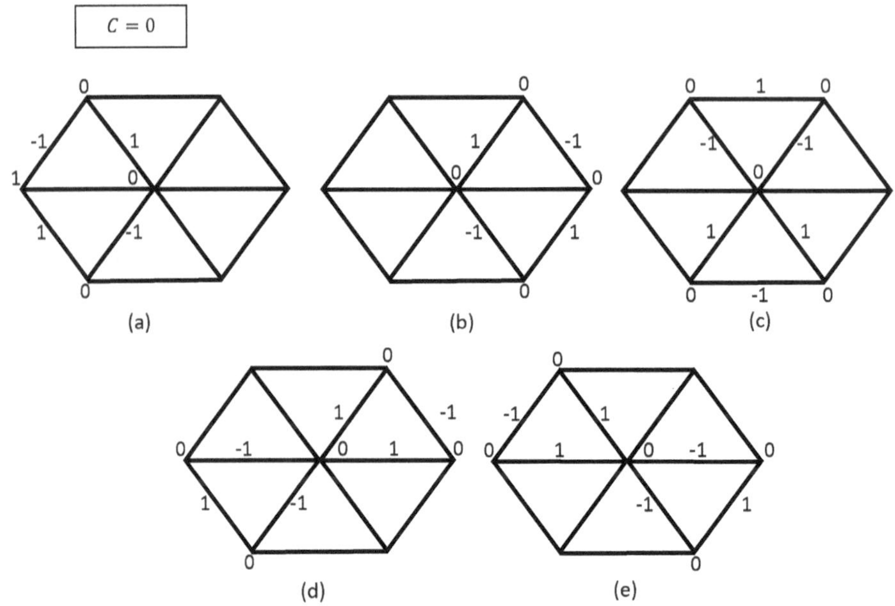

Figure 5.24 Hexagon eigenvectors for eigenvalues $C = 0$. These figures illustrate now the 5 eigenvectors for $\lambda = 4L$. They are confined within two interior loops. Notice that the superpositions of the eigenvectors (a)+(b)+(c) and (c)+(d)+(e) provide the surface path eigenvector around this hexagon network.

5.10 Loop chains

Closed loop chains are present in all periodic structures. Usually, they interact with their neighbors. However, for some states, their robust zeros isolate them from the rest of the structure to which they belong. Examples of such loop states appear already in the square networks investigated above.

It is interesting to try and understand the main physical properties of such isolated closed loop chains.

5.10.1 Two-loop chain

As a reference system, one takes here two open loops of length L, both made out of the same material. The final system, called two-loop chain, is obtained by superposing at $x = 0$ the ends of these two loops in order to obtain the network of Fig. 5.29.

5.10.1.1 Path states by inspection

Each loop of circumference L has one robust zero connected with the other loop robust zero. The confined state theorem enables associating to this loop the following path

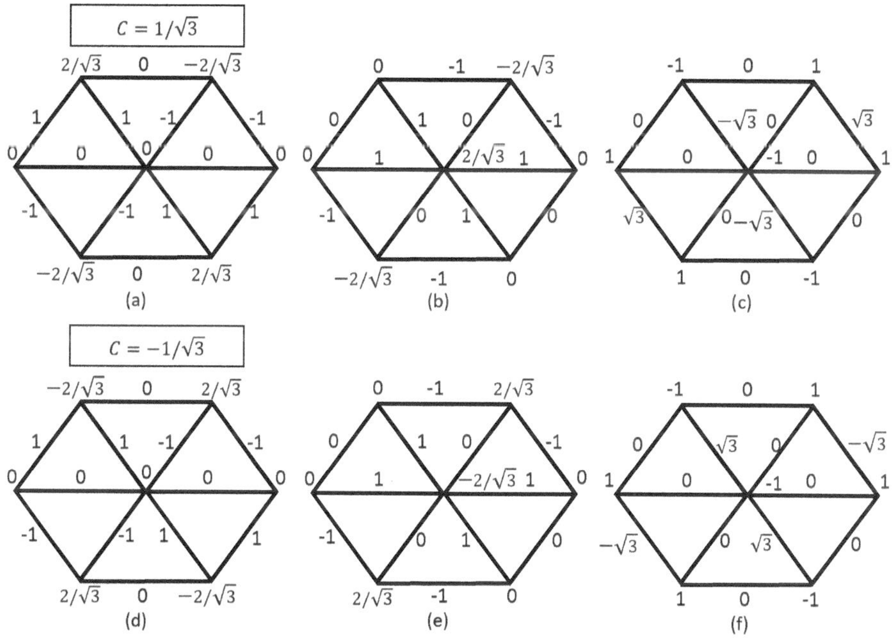

Figure 5.25 Hexagon. These figures illustrate the eigenvectors for $C = 1/\sqrt{3}$ and $C = -1/\sqrt{3}$. The eigenvectors (a) and (d) are bulk-surface eigenvectors. Their decrease of length from surface towards the bulk is $2L$. The four others are bulk eigenvectors.

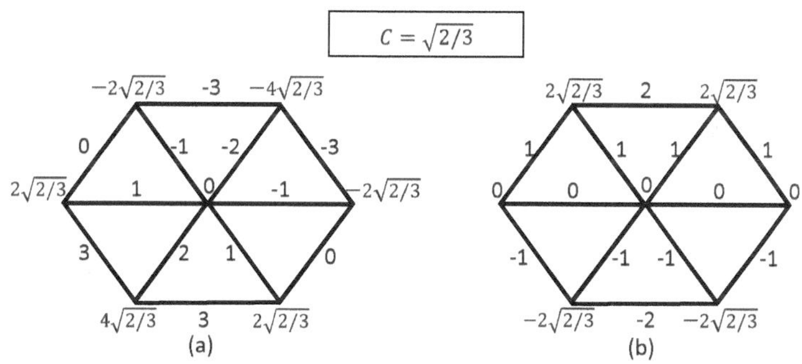

Figure 5.26 Hexagon. These figures illustrate the eigenvectors for $C = \sqrt{2/3}$. They are surface eigenvectors. Their decrease of length towards the bulk is $2L$.

eigenvalues:

$$\lambda_p = \frac{L}{p}, \quad p = 1, 2, 3, \ldots \tag{5.76}$$

Their first eigenfunctions are

$$E(x) = \sin\left(2p\pi \frac{x}{L}\right), \quad p = 1, 2, 3, \ldots, \tag{5.77}$$

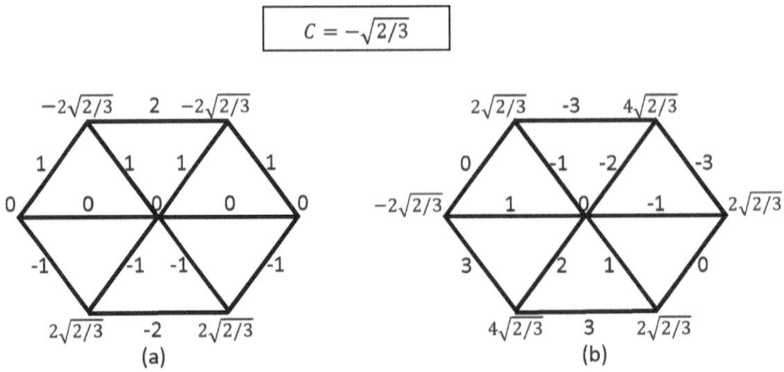

Figure 5.27 Hexagon. These figures illustrate the eigenvectors for $C = -\sqrt{2/3}$. They are surface eigenvectors. Their decrease of length towards the bulk is $2L$.

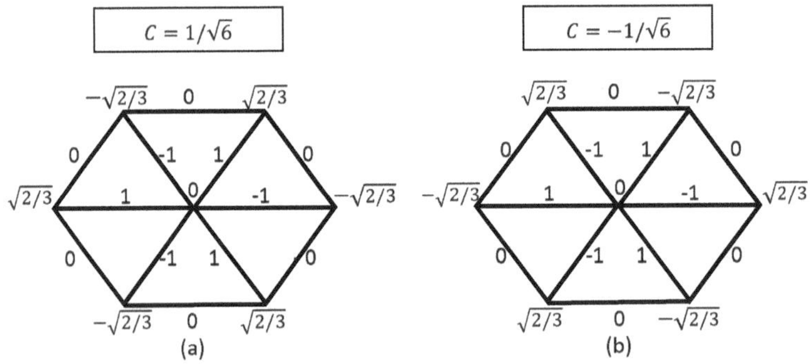

Figure 5.28 Hexagon. These figures illustrate the eigenvectors for $C = \pm\sqrt{1/6}$. They are bulk-surface eigenvectors. Their decrease of length towards the bulk is $2L$.

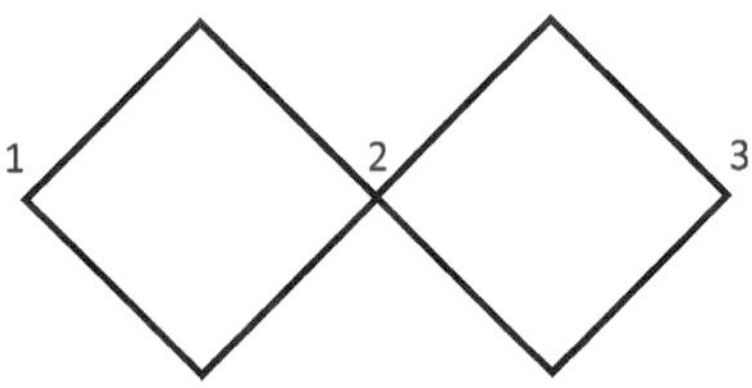

Figure 5.29 Two closed loop chain. As a reference system, one takes here two open loops of length L, both made out of the same material. The final system, called two-loop chain, is obtained by superposing at $x = 0$ the ends of these 2 loops in order to obtain such a network. The numbers in this figure indicate another interface space enabling to obtain the same results.

where x starts from 0 at the connection point and runs all around one loop of this network.

Their second eigenfunctions are

$$E(x) = \sin\left(2p\pi\frac{x}{L}\right), \quad p = 1, 2, 3, \ldots, \tag{5.78}$$

where x starts from 0 at the connection point and runs all around the "other" loop.

Their third eigenfunctions are

$$E(x) = \cos\left(2p\pi\frac{x}{L}\right), \quad p = 1, 2, 3, \ldots, \tag{5.79}$$

where x starts from 1 at the connection point and runs all around this two loop chain.

Note that when at the same space point one of these three degenerate eigenfunctions is 0, the other takes the value ±1. These eigenfunctions are presented in Fig. 5.30(a), in which just their ±1 values are indicated. This means that in the middle between these values, the eigenvalues go through robust zeros. So a source force acting on the ±1 points of the first eigenfunction excites only this one and not the other two.

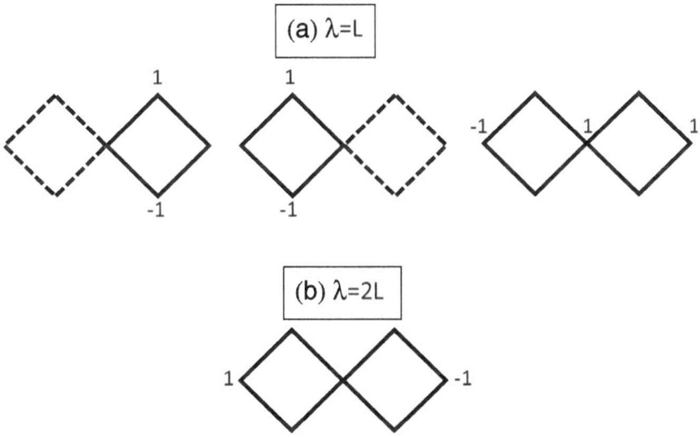

Figure 5.30 Two closed loop chain eigenfunctions are presented in these figures just by their ±1 values. This means that at half the distance between these values, the eigenvalues go through robust zeros. So a source force acting on the ±1 points of the first eigenfunction excites only this one and not the other two. Figure (a) presents for $\lambda = L$ three degenerate eigenfunctions. The first two are surface path eigenfunctions. Note that the third path eigenfunction is created by the connection between the two loops L. It is a bulk path eigenfunction of this network, as it does not decay away from his path. Notice that when one of these three degenerate eigenfunctions is 0, another takes the value ±1. Figure (b) presents, for $\lambda = 2L$, one new eigenfunction localized within the two loops, with a 0 between the two loops. It is also a bulk path one.

We remark that the third path eigenfunction is created by the connection between the two loops L. It is a bulk path eigenfunction of this network, as it does not decay away from its path. It is shown in the next examples that it remains a path eigenfunction within more elaborated networks. Each path eigenvalue λ_p is degenerate. So some structure defects may affect just one of their degenerate eigenfunction and not the other(s).

This connection creates also the following new path eigenvalues:

$$\lambda_m = \frac{2L}{1+2p}, \qquad p = 0, 1, 2, 3, \ldots, \tag{5.80}$$

and eigenfunctions

$$E(x) = \sin\left((1+2p)\pi\frac{x}{L}\right), \qquad p = 0, 1, 2, 3, \ldots, \tag{5.81}$$

where x starts from 0 at the connection point and does complete λ_m revolution(s) all around this two loop chain, see Fig. 5.30(b). It is also a path bulk state of this network. These new states are path states as they do not decay out of their path. It is also possible to understand this eigenfunction from the following physical point of view. One open loop antisymmetric eigenfunction can now be continuous without an eigenvalue shift when the open loop ends are superposed (at 0 and L), as now four loops join together at $x = 0$.

5.10.1.2 Closing two open loops with one interface point

Close two open loops by superposing all their four ends. The inverse of the interface response function on the connection point of the two closed loop is

$$[g(0,0)]^{-1} = \frac{2\pi}{\lambda S}(-4C+4) = \frac{8\pi}{\lambda}\frac{\sin\left(\frac{\pi L}{\lambda}\right)}{\cos\left(\frac{\pi L}{\lambda}\right)}. \tag{5.82}$$

5.10.1.3 All states from the state phase shift

The phase shift η [4] between the final and reference systems is

$$\eta(\lambda) = -\arg\left[\frac{\lambda^3 \sin\left(\frac{\pi L}{\lambda}\right)}{\cos\left(\frac{\pi L}{\lambda}\right)}\right]. \tag{5.83}$$

The eigenvalues of the reference system (two open loops of length L) are given by $(S/\lambda)^4 = 0$, see Eq. (2.5). Note also that the determinant is $|[g(MM)]^{-1}| = -\left(\frac{2\pi}{\lambda}\right)^2$. So all the eigenvalues of the two loop chain are given by

$$\lambda \sin^3\left(\frac{\pi L}{\lambda}\right)\cos\left(\frac{\pi L}{\lambda}\right) = 0. \tag{5.84}$$

This confirms the eigenvalues and eigenfunctions obtained above from the confined state theorem.

Note the two degenerate states for infinite λ.

5.10.1.4 Response function for the two loop chain

The response function $g(x, x')$ for the two closed loop chain can be derived using the general equation [5] as done before for one closed loop, see Eq. (2.11). The explicit form of this response function is as follows.

When x is in one loop and x' in the other,

$$g(x,x') = \frac{\lambda}{8\pi}\left[\frac{\sin\left(\frac{\pi L}{\lambda}\right)}{\cos\left(\frac{\pi L}{\lambda}\right)}\sin\left(\frac{2\pi x}{\lambda}\right)\sin\left(\frac{2\pi x'}{\lambda}\right)\right]$$
$$+\frac{\lambda}{8\pi}\left[\frac{\cos\left(\frac{\pi L}{\lambda}\right)}{\sin\left(\frac{\pi L}{\lambda}\right)}\cos\left(\frac{2\pi x}{\lambda}\right)\cos\left(\frac{2\pi x'}{\lambda}\right)+\sin\left(\frac{2\pi(x+x')}{\lambda}\right)\right], \tag{5.85}$$

where x and x' start from $x = 0$ and are positive in each loop.

When x and x' are in the same loop and $x \leq x'$,

$$g(x,x') = \frac{\lambda}{8\pi}\left(3\frac{\cos\left(\frac{\pi L}{\lambda}\right)}{\sin\left(\frac{\pi L}{\lambda}\right)} - \frac{1}{\sin\left(\frac{\pi L}{\lambda}\right)\cos\left(\frac{\pi L}{\lambda}\right)}\right)\sin\left(\frac{2\pi x}{\lambda}\right)\sin\left(\frac{2\pi x'}{\lambda}\right)$$
$$+\frac{\lambda}{8\pi}\left[\frac{\sin\left(\frac{\pi L}{\lambda}\right)}{\cos\left(\frac{\pi L}{\lambda}\right)}\cos\left(\frac{2\pi x}{\lambda}\right)\cos\left(\frac{2\pi x'}{\lambda}\right)+\sin\left(\frac{2\pi(x+x')}{\lambda}\right)\right], \quad (5.86)$$

where x and x' start from $x = 0$ and are positive in each loop.

The numerators of these response functions confirm the eigenfunctions found above from the confined state theorem. The kernels of these response functions confirm the eigenvalues. The whole explicit expressions of these response functions enable finding these network responses to any action.

5.10.2 Three-loop chain

As a reference system, one takes here six open loops of length $L/2$, all made of the same material. The final system, called three-loop chain, is obtained by superposing the ends of these loops in order to obtain the three-loop network.

5.10.2.1 Path states by inspection

Each loop of circumference L has (one or two) robust zero(s) connected with the other loop robust zeros. The confined state theorem enables associating to this loop the following path eigenvalues:

$$\lambda_p = \frac{L}{p}, \quad p = 1, 2, 3, \ldots, \quad (5.87)$$

and the four degenerate path eigenfunctions which follow.

The first three sets of eigenfunctions are confined in each of the three loops L and their explicit expression is

$$E(x) = \sin\left(2p\pi\frac{x}{L}\right), \quad p = 1, 2, 3, \ldots, \quad (5.88)$$

where x starts from 0 at one robust zero common to two loops L and runs all around one loop of the three-loop chain.

The fourth is confined within the whole network and is

$$E(x) = \cos\left(2p\pi\frac{x}{L}\right), \quad p = 1, 2, 3, \ldots, \quad (5.89)$$

where x starts from 0 at one robust zero common to two loops L and runs all around the three-loop chain.

Note that when at the same space point one of these three degenerate eigenfunctions is 0, the other takes the value ± 1. These eigenfunctions have the same properties as those sketched before. In particular, a source force acting on the ± 1 points of one eigenfunction excites only this one and not the three others.

Each path eigenvalue λ_p is degenerate. So some structure defects may affect just one of their degenerate eigenfunctions and not the other(s).

This connection creates also new path eigenvalues. The path $\lambda_p = \frac{2L}{p}$ eigenvalues of the two closed loop chain cannot exist in this three-loop chain, as their eigenfunctions cannot achieve complete λ_p revolutions in closed loop chains when the number of closed loops is odd.

5.10.2.2 All states from the state phase shift

Following the same approach as above for the two closed loop chain, one finds the eigenvalues of the three-loop chain to be given by

$$\lambda^2 \sin^4\left(\frac{\pi L}{\lambda}\right)\left[4\cos^2\left(\frac{\pi L}{\lambda}\right) - 1\right] = 0. \tag{5.90}$$

This confirms the path eigenvalues and eigenfunctions obtained above from the confined state theorem.

For the new eigenvalues $\cos\left(\frac{\pi L}{\lambda}\right) = \frac{1}{2}$, the eigenvector is found to be a bulk eigenvector, namely

$$E = \begin{pmatrix} 2 & 1 & -1 & -2 \end{pmatrix}. \tag{5.91}$$

For the new eigenvalues $\cos\left(\frac{\pi L}{\lambda}\right) = -\frac{1}{2}$, the eigenvector is found to be

$$E = \begin{pmatrix} 2 & -1 & -1 & 2 \end{pmatrix}. \tag{5.92}$$

5.10.3 Four-loop chain

Repeating the same analysis as above, one obtains similar results for a four-loop chain as for the two-loop chain.

In particular, the eigenvalues are now given by

$$\lambda^3 \sin^5\left(\frac{\pi L}{\lambda}\right)\cos\left(\frac{\pi L}{\lambda}\right)\left[-2\cos^2\left(\frac{\pi L}{\lambda}\right) + 1\right] = 0. \tag{5.93}$$

This confirms the path eigenvalues and eigenfunctions obtained here from the confined state theorem. More precisely, one has now 5 times degenerate, $\lambda_p = \frac{L}{p}$, $p = 1, 2, 3, \ldots$, eigenvalues. Their first four eigenvectors are the path ones, confined in each one of the four loops. The fifth is the path bulk one confined in the whole network.

For the new eigenvalues $\cos\left(\frac{\pi L}{\lambda}\right) = \frac{1}{\sqrt{2}}$, the eigenvectors are found to be the bulk eigenvectors

$$E = \begin{pmatrix} \sqrt{2} & 1 & 0 & -1 & -\sqrt{2} \end{pmatrix}. \tag{5.94}$$

For the new eigenvalues $\cos\left(\frac{\pi L}{\lambda}\right) = -\frac{1}{\sqrt{2}}$, the bulk eigenvectors are found to be

$$E = \begin{pmatrix} \sqrt{2} & -1 & 0 & 1 & -\sqrt{2} \end{pmatrix}. \tag{5.95}$$

5.10.4 N-loop chain

Consider now a chain of N closed loops, with $N = 1, 2, 3, \ldots$

A recurrent analysis enables obtaining the path eigenvalues for N closed loop chains to be

$$\begin{aligned} &\lambda^{N-1} \sin^{N+1}\left(\tfrac{\pi L}{\lambda}\right) = 0, && \text{for } N \text{ odd,} \\ &\lambda^{N-1} \sin^{N+1}\left(\tfrac{\pi L}{\lambda}\right) \cos\left(\tfrac{\pi L}{\lambda}\right) = 0, && \text{for } N \text{ even.} \end{aligned} \tag{5.96}$$

Their eigenfunctions are also simple generalizations of those obtained for the examples derived here above.

This network has, for each value of p, one path state $\lambda_p = \frac{L}{p}$, $p = 1, 2, 3, \ldots$ with eigenfunction $\sin(2p\pi\frac{x}{L})$ confined in one single loop. The $N + 1$ state with eigenfunctions $\cos(2p\pi\frac{x}{L})$ is localized within all the N-loop structure. This network has also, only when N is odd, the path state $\lambda_p = 2L/(1 + 2p)$, $p = 0, 1, 2, 3, \ldots$, with eigenfunction $\sin((1 + 2p)\pi\frac{x}{L})$.

In the limit of infinite loop chains, the bulk path states become flat bulk passing bands. But each of the N path states corresponding to $\sin(2p\pi\frac{x}{L})$ remains confined in its loop.

Note that symmetry breaking enables getting confined $\sin(2p\pi\frac{x}{L})$ states within any number of loops smaller than N. When N is infinite, such confined loop states are bulk in continuum states within their corresponding flat bands. This may be obtained also for the $\cos(2p\pi\frac{x}{L})$ states with any even number of loop pairs.

This network has also other more complex states. Examples of such states are given here up to the four-loop chain.

5.11 Degenerate confined state activation

Activation of the three degenerate states for $\lambda = L$ is presented in Fig. 5.31. Here L is the closed loop length, λ the eigenwavelength, and $L/4$ the distance between the nearest neighbor loop dots. The inputs are I_1, I_2, and I_3. The outputs are $O_1 = I_1$, $O_2 = I_2$, and $O_3 = I_2$. Each input can be activated alone or simultaneously with one or the two others.

This result is obtained just by inspection of the robust zeros of each of the two degenerate loop states. It implies that the only interactions of the loop with its outside world can be achieved through these four points. The inputs are supposed to be brief laser shots with negligible influence on the original loop states.

Another activation method can be used, too. It uses four semiinfinite open loops for the inputs and outputs and two finite loops of length L' attached at each of these four

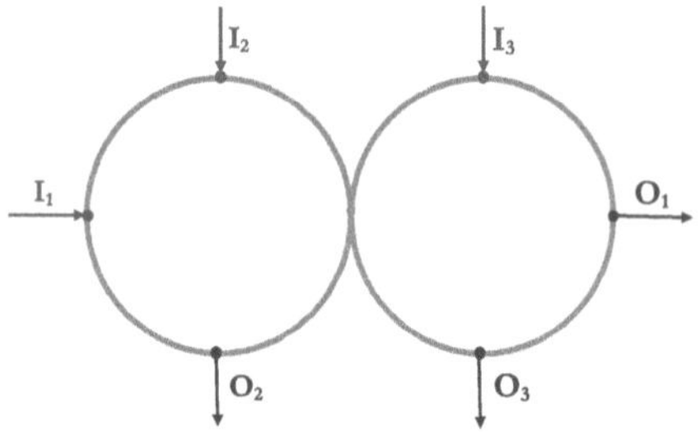

Figure 5.31 Activation of the three degenerate states for $\lambda = L$ presented in Fig. 5.30. Here L is the closed loop length, λ the eigenwavelength, and $L/4$ the distance between the nearest neighbor loop dots. The inputs are I_1, I_2, and I_3. The outputs are $O_1 = I_1$, $O_2 = I_2$, and $O_3 = I_2$. Each input can be activated alone or simultaneously with one or the two others.

points. This perturbs the loop states. However, when L' is incommensurable with L, it is possible to observe the same effects, but for $\lambda \approx L/2$, see, for example, Fig. 6.21, Chapter 6 of this Photonics book.

It is straightforward to generalize this activation procedure to any number of such confined degenerate states. This applies also to the cases of similar states existing in the square and cube structures appearing above in this chapter.

The linear superpositions of such confined degenerate states can be also activated like shown in the above simple example.

Such simple results provide possible devices for wave and particle cross-transmission [6] and enhanced directional transmission [7].

5.12 Perspectives

From the above tutorial closed loop examples, it is obvious that path closed loop states are omnipresent. Their multiple degeneracy enables the eigenfunctions to go around defects. They are confined in all possible single and double loops of the networks. They form also confined paths within loop chains which seem to be the most direct paths between two network points. These path states are simple as made out of half-wavelength sinusoidal oscillations fitted inside all the network possible closed loops. They are simple compared to the other states of the same network. They represent about half the total number of states of a finite network. The other half comprises the more complex (bulk, surface, wedge-edge, etc.) ones.

This is the point of view which comes out of the above simple examples using small finite networks. This point of view is complementary to the now classical point of view of the condensed matter physicist, obtained by starting and analyzing first infinite lattices.

The understanding of the respective roles of the path states and of the more complex ones deserves much more investigations. Nevertheless, it is clear that information communication technology can benefit from the preferential use of the path states.

Any eigenstate can be activated within its subdomain without any subsystem deformation, see the activation state theorem in Chapter 3 of this Photonics book. This paves the way to the use of the degenerate path states for many applications.

References

[1] L. Rayleigh, On waves propagated along the plane surface of an elastic solid, Proc. Lond. Math. Soc. 17 (1887) 4.
[2] L. Dobrzynski, A.A. Maradudin, Electrostatic edge modes in a dielectric wedge, Phys. Rev. B 6 (1972) 3810.
[3] L. Dobrzynski, D.L. Mills, Vibrational properties of an adsorbed surface layer on a simple model crystal, J. Phys. Chem. Solids 30 (1969) 1043.
[4] L. Dobrzyński, A. Akjouj, E. El Boudouti, et al., Interface response theory, in: Phononics, in: Interface Transmission Tutorial Book Series, Elsevier, 2017.
[5] L. Dobrzyński, A. Akjouj, G. Lévêque, et al., Centered system magnons, in: Magnonics, in: Interface Transmission Tutorial Book Series, Elsevier, 2019.
[6] L. Dobrzyński, A. Akjouj, Particle cross transfer, Surf. Sci. Rep. 63 (2008) 391.
[7] L. Dobrzyński, A. Akjouj, Enhanced directional optical transmission, Appl. Phys. Lett. 93 (2008) 043114.

Closed loop and stubs

6

Leonard Dobrzyński[a], *Housni Al Wahsh*[b], *Abdellatif Akjouj*[a], *Cécile Ghouila-Houri*[a], *Abdelkrim Talbi*[a], *Gaëtan Lévêque*[a], and *El Houssaine El Boudouti*[c]
[a]Univ. Lille, CNRS, Centrale Lille, Yncréa ISEN, Univ. Polytechnique Hauts-de-France, UMR 8520 - IEMN, Physics Department, Lille, France, [b]Benha University, Faculty of Engineering, Cairo, Egypt, [c]Laboratory of Physics of Matter and Radiation, University Mohammed I, Physics Department, Faculty of Sciences, Oujda, Morocco

6.1 Introduction

Closed loops with finite open loops attached to the robust zeros of one of their two twin states are the systems investigated in this chapter. Such attached finite open loops are called stubs. For simplicity, here only the cases where all the stubs have the same length are considered. This length is chosen to be commensurable with the closed loop length. So the interface points of such path systems remain robust zeros for many eigenfunctions of the final systems. As a consequence, the corresponding eigenwavelengths and eigenfunction paths are robust. All the states of such systems are obtained with particular attention to the path ones.

An eigenfunction zero is robust because its eigenstate cannot be activated by any action applied on it. The path states are robust because they are confined within one-dimensional paths, and when they are degenerate they can turn around defects, see Chapter 3 of this book.

Multiplexer applications of such systems are also presented. It is shown that some path states can be transferred from one infinite line to another without any perturbation. Such single wavelength transfers are of infinite quality. It is possible to transform them into finite-width resonances just by tuning the stub lengths. In the next four sections we consider simple examples of one, two, three, and four robust zeros of one closed loop and N stubs are attached to these loop points. In these sections the emphasis is on the robustness of the path states of such devices. The last two sections present two applications. The first is for multiplexers which can be tuned up to infinite quality for such systems when all losses are compensated by appropriate amplifications. The second application is for systems with wide reflection bands, as well for loops and for cubes with stubs.

6.2 Closed loop with N stubs

One closed loop of length $2L$, with $N\ (= 1, 2, 3, \ldots)$ stubs of length L attached to one space point of the closed loop, is the system of interest here, see Fig. 6.1.

Photonics. https://doi.org/10.1016/B978-0-12-819388-4.00015-0

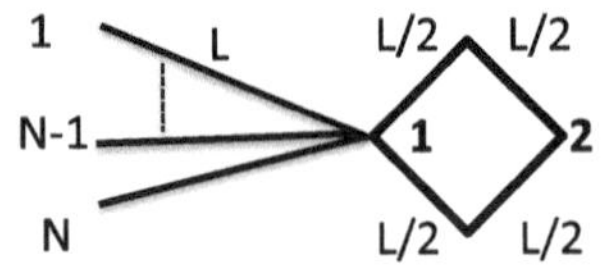

Figure 6.1 One closed loop of length $2L$, with N $(= 1, 2, 3, \ldots)$ stubs of length L.

6.2.1 Reference system

In order to build the system described by Fig. 6.1, it is possible to use as reference system $N+2$ loops of length L, $N = 1, 2, 3, \ldots$, and an interface space $M \equiv (1, 2)$.

Define

$$\alpha = \frac{2\pi}{\lambda}, \tag{6.1}$$

where λ is the wavelength, $S = \sin(\alpha L)$, and $C = \cos(\alpha L)$.

The reference system states are given by

$$(\alpha S)^{N+2} = 0. \tag{6.2}$$

The determinant of the inverse of the corresponding $g_R(MM)$ is [1]

$$|[g_R(MM)]^{-1}| = (-\alpha^2)^2 (\frac{\alpha S}{C})^N. \tag{6.3}$$

6.2.2 Path states by inspection

In the reference system, the interface point 1 is a robust zero for N independent open loops of length $2L$. The confined state theorem says that this zero remains unperturbed, as well as its N times degenerate eigenvalues for $C = 0$, equivalently the eigenwavelengths are as follows:

$$\lambda_p = \frac{4L}{2p+1}, \quad p = 0, 1, 2, 3, \ldots, \tag{6.4}$$

and their antisymmetric eigenfunctions are

$$E(x) = \pm \sin\left[\pi(2p+1)\frac{x}{2L}\right], \quad p = 0, 1, 2, 3, \ldots, \tag{6.5}$$

with the "+" sign to be used when x is within one L loop and the "−" sign when x is within the other L loop.

The interface point 1 is also a robust point, but for the closed loop of length $2L$. The confined state theorem enables finding by inspection two times degenerate states $S = 0$, equivalently

$$\lambda_p = \frac{2L}{p}, \quad p = 0, 1, 2, 3, \ldots, \tag{6.6}$$

and their symmetric eigenfunctions

$$E(x) = \sin\left[2\pi p \frac{x}{L}\right], \quad p = 1, 2, 3, \ldots \tag{6.7}$$

6.2.3 All states from the interface response theory

In order to obtain the final system depicted by Fig. 6.1, one superposes one end of all the independent $N+2$ loops at the same interface point 1 and two other ends at the interface point 2. This enables obtaining by linear superposition

$$[g_f(MM)]^{-1} = \alpha \begin{pmatrix} -\frac{2C}{S} + \frac{NS}{C} & \frac{2}{S} \\ \frac{2}{S} & -\frac{2C}{S} \end{pmatrix}. \tag{6.8}$$

The determinant of the above matrix is

$$|[g_f(MM)]^{-1}| = 2\alpha^2(N+2). \tag{6.9}$$

With the above results and the conservation of the number of states between the reference and the final system, one obtains that the final system states are given by

$$S^2 C^N = 0, \tag{6.10}$$

as obtained above from the confined state theorem.

6.2.4 Eigenvectors for $N = 2$

In what follows, the eigenvectors for $N = 2$ stubs of length L attached to the same point of a closed loop of length $2L$ are presented. These eigenvectors cannot be zero at the stub ends because in this chapter this system is constructed out of open loops, unless we consider the cases of dead stub branches. Note also that the sum of the outgoing derivatives at all system points has to vanish. These remarks help us find the path eigenvectors just by fitting a sine function within the structure.

It is also possible to construct the same kind of system with closed stub free ends. This is left as tutorial exercises.

6.2.4.1 Eigenvectors for $S^2 = 0$

The eigenvectors for $S^2 = 0$, or equivalently for

$$\lambda = \frac{2L}{n}, \quad n = 0, 1, 2, 3, \ldots, \tag{6.11}$$

are given in Fig. 6.2. These states are two times degenerate. They correspond to path states.

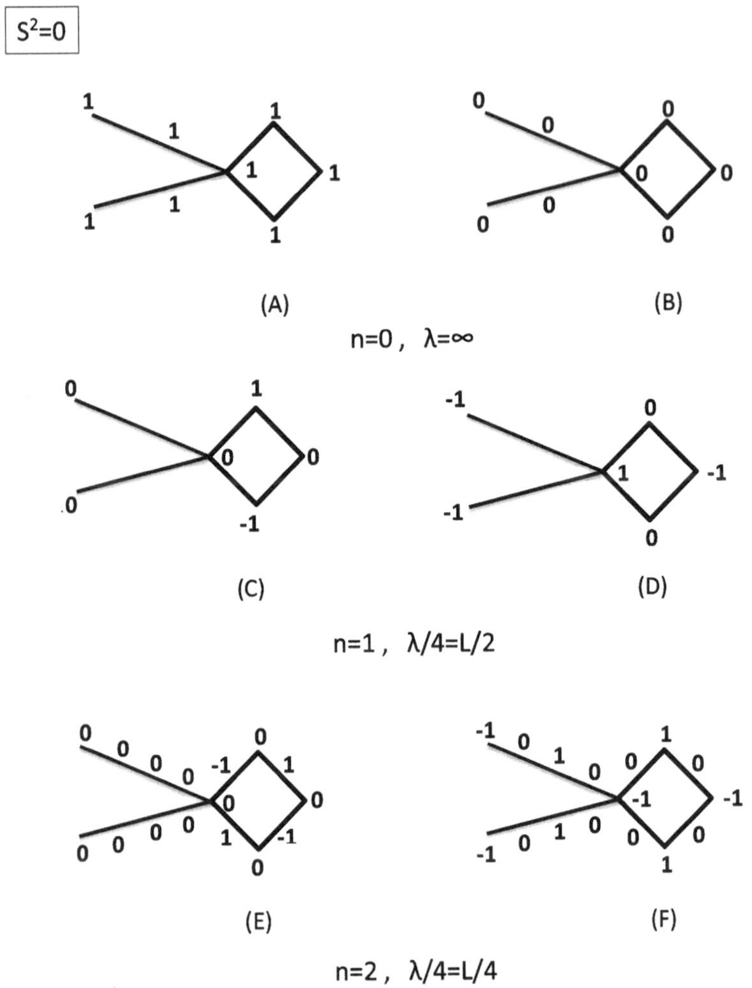

Figure 6.2 One closed loop of length $2L$ with two stubs of length L attached to the same point of the closed loop. Eigenvectors for $S^2 = 0$, or equivalently for $\lambda = 2L/n$, $n = 0, 1, 2, 3, \ldots$

6.2.4.2 Eigenvectors for $C^2 = 0$

The eigenvectors for $C^2 = 0$, or equivalently for

$$\lambda = \frac{4L}{1+2n}, \qquad n = 0, 1, 2, 3, \ldots, \tag{6.12}$$

are given in Fig. 6.3. These states are two times degenerate. They also correspond to path states.

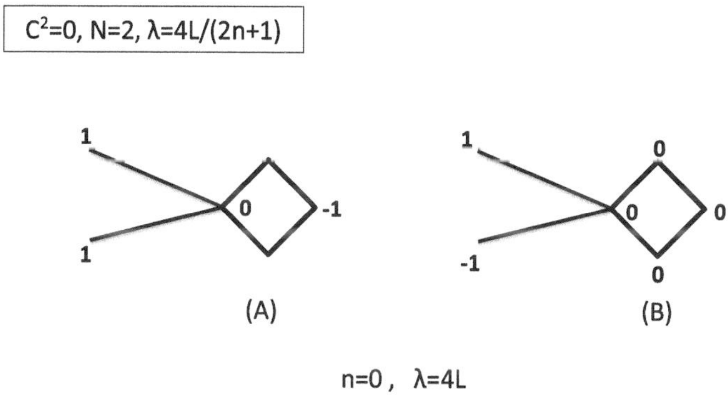

Figure 6.3 One closed loop of length $2L$ with two stubs of length L attached to the same point of the closed loop. Eigenvectors for $C^2 = 0$, or equivalently for $\lambda = 4L/(1+2n)$, $n = 0, 1, 2, 3, \ldots$

6.3 Closed loop with $2N$ stubs

Consider one closed loop of length $2L$, with $2N$ ($N = 1, 2, 3, \ldots$) stubs of length L. Half of these stubs are attached to one robust zero of one loop state. The other half are attached to another robust zero of the same loop eigenfunctions, see Fig. 6.4.

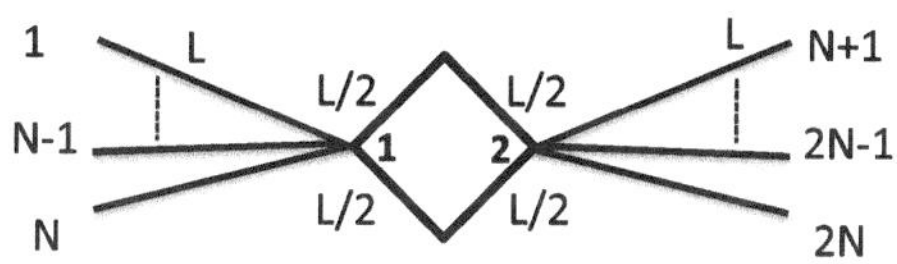

Figure 6.4 One closed loop of length $2L$, with $2N$ ($N = 1, 2, 3, \ldots$) stubs of length L, attached to two robust zeros of the same eigenfunctions.

6.3.1 Reference system

In order to build the system described by Fig. 6.4, it is possible to use as a reference system $2(N+1)$ open loops of length L, $N = 1, 2, 3, \ldots$, and an interface space $M \equiv (1, 2)$.

With λ as the wavelength, $\alpha = 2\pi/\lambda$, $S = \sin(\alpha L)$, and $C = \cos(\alpha L)$, the reference system states are given by

$$(\alpha S)^{2(N+1)} = 0. \tag{6.13}$$

The determinant of the inverse of the corresponding $g_R(MM)$ is

$$|[g_R(MM)]^{-1}| = (-\alpha^2)^2 (\frac{\alpha S}{C})^{2N}. \tag{6.14}$$

6.3.2 Path states by inspection

In the reference system, the interface points 1 and 2 are robust zeros for $2(N-1)$ independent open loops of length $2L$. The confined state theorem says that this zero remains unperturbed, as well as its N times degenerate eigenvalues for $C=0$, equivalently the eigenwavelengths are as follows:

$$\lambda_p = \frac{4L}{2p+1}, \quad p=0,1,2,3,\ldots, \tag{6.15}$$

and their antisymmetric eigenfunctions are

$$E(x) = \pm\sin\left[\pi(2p+1)\frac{x}{2L}\right], \quad p=0,1,2,3,\ldots, \tag{6.16}$$

with the "+" sign to be used when x is within one L loop and the "−" sign when x is within the other L loop.

The interface point 1 is also a robust point, but for the closed loop of length $2L$. The confined state theorem enables finding by inspection two times degenerate states $S=0$, equivalently

$$\lambda_p = \frac{2L}{p}, \quad p=0,1,2,3,\ldots, \tag{6.17}$$

and their symmetric eigenfunctions

$$E(x) = \sin\left[2\pi p\frac{x}{L}\right], \quad p=1,2,3,\ldots \tag{6.18}$$

6.3.3 All states from the interface response theory

In order to obtain the final system depicted by Fig. 6.4, we proceed as follows: superpose one end of the N open loops at the same interface point 1, then do the same with N other open loops at the interface point 2. Then we superpose one end of two other open loops at point 1 and their other end at point 2. Thus we obtain [1]

$$[g_f(MM)]^{-1} = \alpha\begin{pmatrix} -\frac{2C}{S}+\frac{NS}{C} & \frac{2}{S} \\ \frac{2}{S} & -\frac{2C}{S}+\frac{NS}{C}\end{pmatrix}. \tag{6.19}$$

The determinant of the above matrix is

$$|[g_f(MM)]^{-1}| = \frac{\alpha^2}{S^2C^2}[N-(N+2)C^2+2C][N-(N+2)C^2-2C]. \tag{6.20}$$

With the above results and the conservation of the number of states between the reference and the final system, one obtains that the final system states are given by

$$S^2C^{2(N-1)}[(N+2)C+N][(N+2)C-N]=0. \tag{6.21}$$

The $S^2C^{2(N-1)}=0$ states were obtained above from the confined state theorem.

6.3.4 Eigenvectors for $S = 0$ and $C = 0$

In what follows, examples of eigenvectors for the $S = 0$ and $C = 0$ eigenvalues are presented. These eigenvectors cannot be zero at the stub ends because in this chapter this system is constructed out of open loops, unless we treat the cases of dead stub branches. Note also that the sum of the outgoing derivatives at all system points has to vanish. These remarks help us find the path eigenvectors just by fitting a sine function within the structure.

It is also possible to construct the same kind of system with closed stub free ends. This is left as tutorial exercises.

6.3.4.1 Eigenvectors for $N = 2$

The eigenvectors for $N = 2$ and $S^2 = 0$, or equivalently for

$$\lambda = \frac{2L}{n}, \qquad n = 0, 1, 2, 3, \ldots, \tag{6.22}$$

are given in Fig. 6.5.

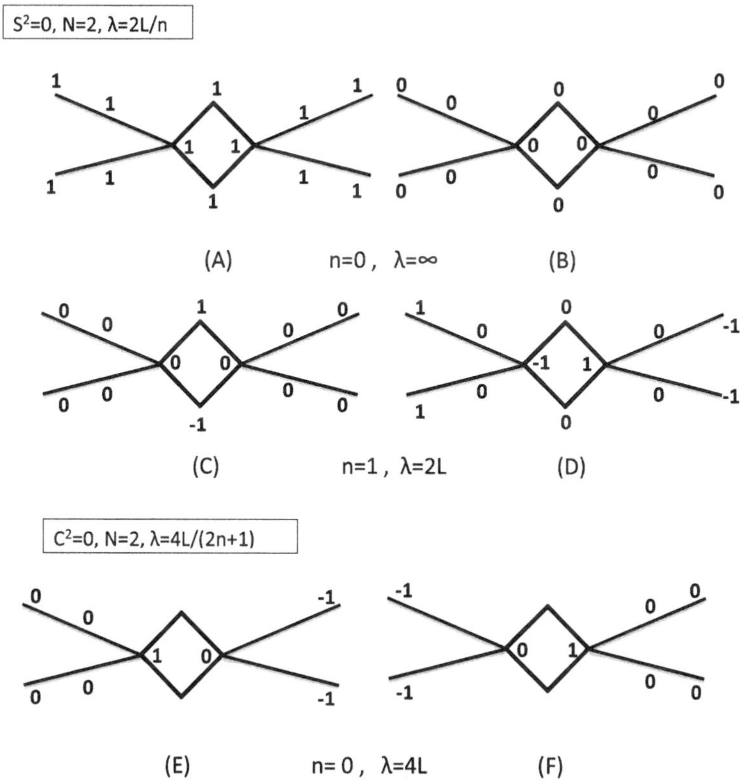

Figure 6.5 One closed loop of length $2L$, with $2N$ ($N = 2$) stubs of length L. Eigenvectors for $S^2 = 0$ and $C^2 = 0$.

The eigenvectors for $C^2 = 0$, or equivalently for

$$\lambda = \frac{4L}{1+2n}, \qquad n = 0, 1, 2, 3, \ldots, \tag{6.23}$$

are also given in Fig. 6.5.

All these states are two times degenerate. They correspond to path states.

6.3.4.2 *Eigenvectors for $N = 3$*

For the same kind of eigenvectors as above but for $N = 3$, see Fig. 6.6.

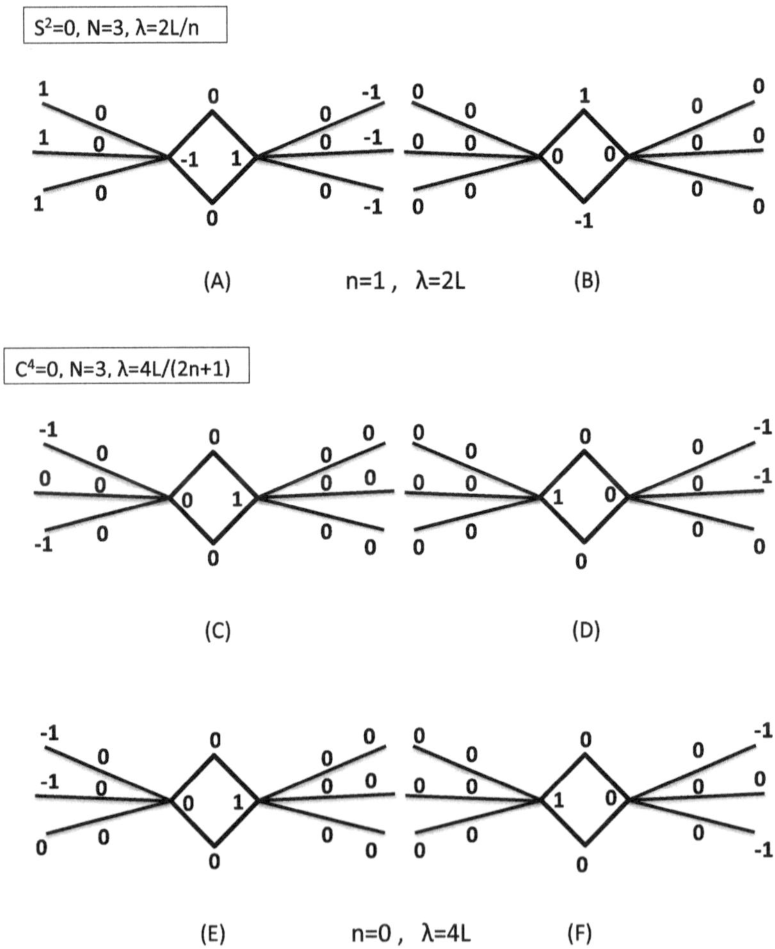

Figure 6.6 One closed loop of length $2L$, with $2N$ ($N = 3$) stubs of length L. Eigenvectors for $S^2 = 0$ and $C^2 = 0$.

6.3.5 Eigenvectors for $C = 1/2$ and $C = -1/2$ with $N = 2$

For $N = 2$ and $C = 1/2$, these eigenwavelengths are

$$\lambda = \frac{6L}{1+6n}, \qquad n = 0, 1, 2, 3, \ldots, \tag{6.24}$$

and

$$\lambda = \frac{6L}{5+2n}, \qquad n = 0, 1, 2, 3, \ldots \tag{6.25}$$

For $N = 2$ and $C = -1/2$, the eigenwavelengths are

$$\lambda = \frac{6L}{1+3n}, \qquad n = 0, 1, 2, 3, \ldots, \tag{6.26}$$

and

$$\lambda = \frac{6L}{2+n}, \qquad n = 0, 1, 2, 3, \ldots \tag{6.27}$$

Examples of eigenvectors for $C = 1/2$ and $C = -1/2$ when $N = 2$ are shown in Fig. 6.7. Their eigenfunctions are robust cosine type, although the expressions of their eigenvalues are slightly different.

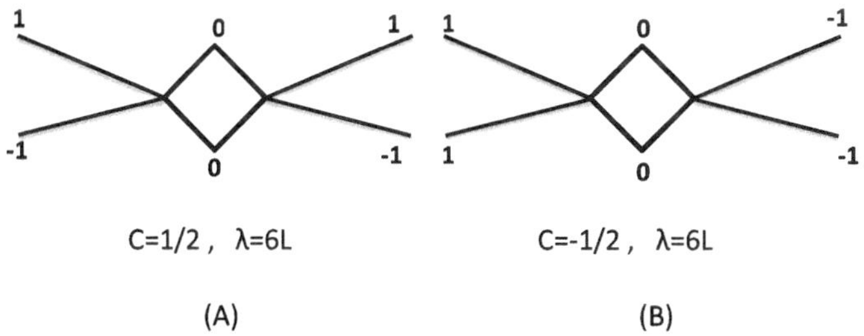

Figure 6.7 One closed loop of length $4L$, with $2N$ $(N = 2)$ stubs of length L. Eigenvectors for $N = 2$ and $C = 1/2$, as well as $C = -1/2$.

6.4 Closed loop with $3N$ stubs

Consider one closed loop of length $4L$, with $3N$ $(N = 1, 2, 3, \ldots)$ stubs of length L. These N stubs are attached to one robust zero of one loop state. The second group of

N stubs are attached to another robust zero of the same loop state. The third set of N stubs is attached to a third robust zero of the same loop state. See Fig. 6.8.

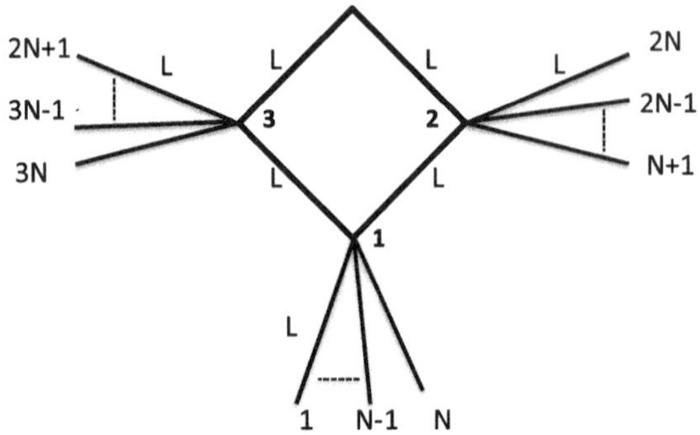

Figure 6.8 One closed loop of length $4L$, with $3N$ $(N = 1, 2, 3, \ldots)$ stubs of length L, attached to three robust zeros of the same eigenfunctions.

6.4.1 Reference system

In order to build the system described by Fig. 6.8, it is possible to use as a reference system $3N + 2$ open loops of length L, $N = 1, 2, 3, \ldots$, one open loop of length $2L$, and an interface space $M \equiv (1, 2, 3)$.

Define λ as the wavelength, $\alpha = 2\pi/\lambda$, $S = \sin(\alpha L)$, and $C = \cos(\alpha L)$.

The reference system states are given by

$$(\alpha S)^{3N+3} C = 0. \tag{6.28}$$

The determinant of the inverse of the corresponding $g_R(MM)$ is

$$|[g_R(MM)]^{-1}| = (-\alpha^2)^3 (\frac{\alpha S}{C})^{3N}. \tag{6.29}$$

6.4.2 Path states by inspection

As for the two preceding examples, it is possible to find the path state eigenwavelengths to be given here by

$$S^2 C^{3N-2} = 0. \tag{6.30}$$

The eigenfunctions are made out of the usual sinusoidal functions.

6.4.3 All states from the interface response theory

In order to obtain the final system depicted by Fig. 6.8, one proceeds as for the two preceding cases. The inverse of the interface space response function here is [1]

$$[g_f(MM)]^{-1} = \alpha \begin{pmatrix} -\frac{2C}{S} + \frac{NS}{C} & \frac{1}{S} & \frac{1}{S} \\ \frac{1}{S} & -\frac{C}{S} + \frac{NS}{C} - \frac{\cos(2\alpha L)}{\sin(2\alpha L)} & \frac{1}{\sin(2\alpha L)} \\ \frac{1}{S} & \frac{1}{\sin(2\alpha L)} & -\frac{C}{S} + \frac{NS}{C} - \frac{\cos(2\alpha L)}{\sin(2\alpha L)} \end{pmatrix}$$

$$= \frac{\alpha}{2SC} \begin{pmatrix} A & 2C & 2C \\ 2C & B & 1 \\ 2C & 1 & B \end{pmatrix}, \qquad (6.31)$$

where $A = 2[N - (N+2)C^2]$ and $B = 2N + 1 - 2(N+2)C^2$.

The determinant of the above matrix is

$$|[g_f(MM)]^{-1}| = \frac{\alpha^3}{SC^3}[(N+2)C^2 - N][(N+2)^2C^2 - N(N+1)]. \qquad (6.32)$$

With the above results and the conservation of the number of states between the reference and the final system, one obtains that the final system states are given by

$$S^2C^{3N-2}[(N+2)C^2 - N][(N+2)^2C^2 - N(N+1)] = 0. \qquad (6.33)$$

The $S^2C^{(3N-2)} = 0$ states were obtained above from the confined state theorem.

6.4.4 Eigenvectors of the final system for $N = 2$

A few examples of eigenvectors for $N = 2$ are given in what follows.

6.4.4.1 Eigenvectors for $S = 0$

The eigenvectors for $S^2 = 0$, or equivalently for

$$\lambda = \frac{2L}{n}, \qquad n = 0, 1, 2, 3, \ldots, \qquad (6.34)$$

are given in Fig. 6.9.

All these states are two times degenerate. They correspond to path states.

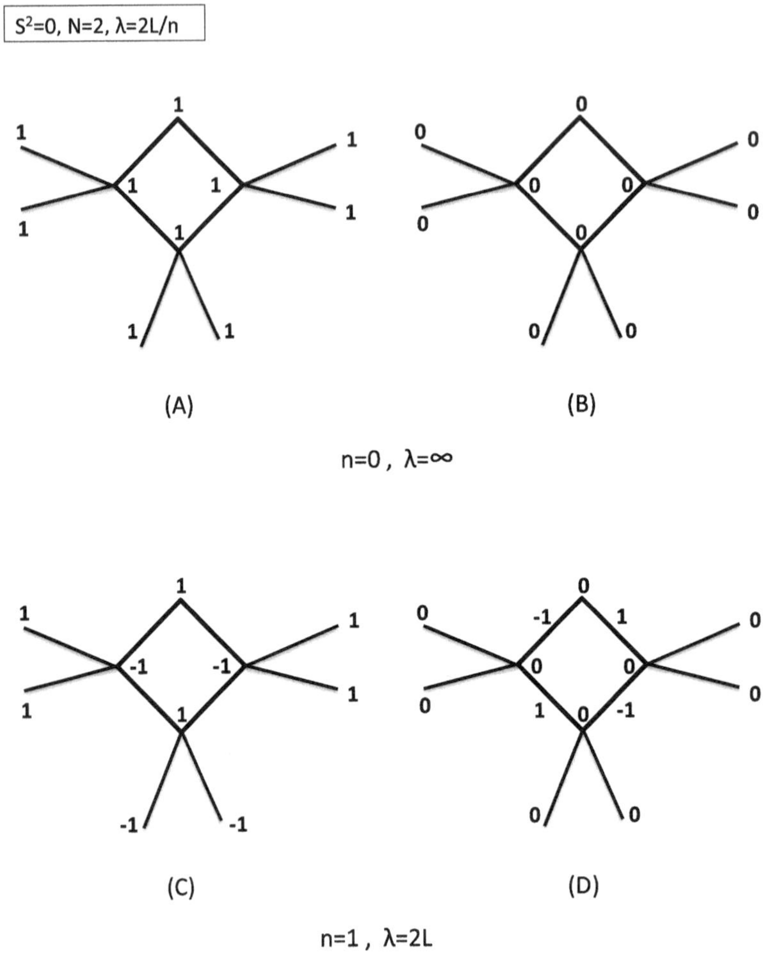

Figure 6.9 One closed loop of length $4L$, with $3N$ $(N=2)$ stubs of length L. Eigenvectors for $S^2=0$.

6.4.4.2 *Eigenvectors for* $C=0$

The eigenvectors for $C^4=0$, or equivalently for

$$\lambda = \frac{4L}{1+2n}, \qquad n=0,1,2,3,\ldots, \tag{6.35}$$

are given in Fig. 6.10.

These states are four times degenerate. They correspond to path states.

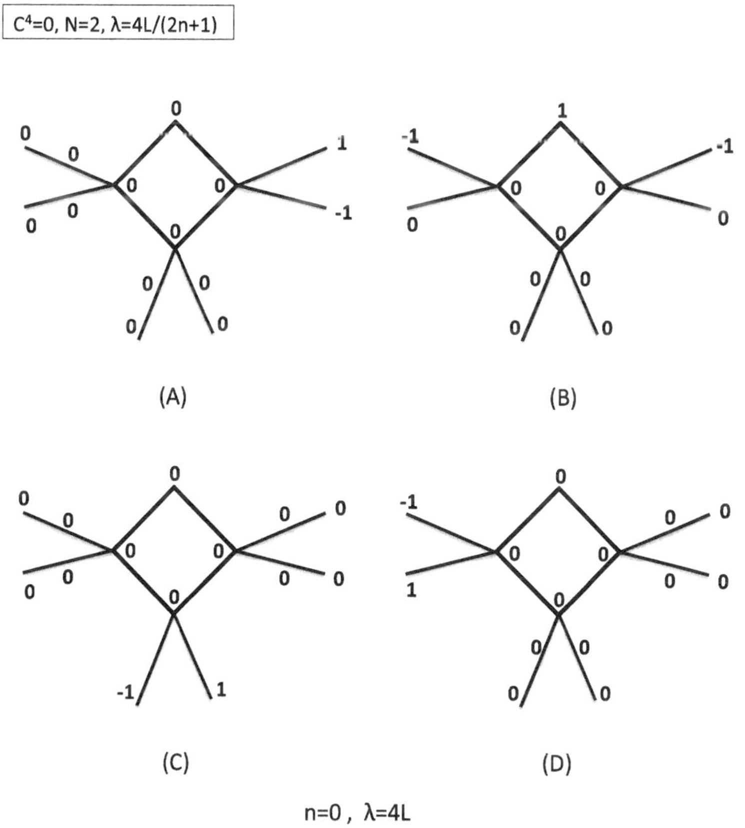

Figure 6.10 One closed loop of length $4L$, with $3N$ ($N = 2$) stubs of length L. Eigenvectors for $C^4 = 0$.

6.4.4.3 Eigenvectors for $C = 1/\sqrt{2}$, $C = -1/\sqrt{2}$, $C = \sqrt{3/8}$, and $C = -\sqrt{3/8}$

The eigenvectors for $C = 1/\sqrt{2}$, $C = -1/\sqrt{2}$, $C = \sqrt{3/8}$, and $C = -\sqrt{3/8}$ are shown in Fig. 6.11. In order to obtain them, the best way is to rewrite the matrix $[g_f(MM)]^{-1}$ within the 12 point interface space, in which use is made of all two ends of the 12 open loops of length L.

6.5 Closed loop with $4N$ stubs

Consider one closed loop of length $8L$, with $4N$ ($N = 1, 2, 3, \ldots$) stubs of length L. Here N stubs are attached to one robust zero of one loop state. The second set of N stubs are attached to another robust zero of the same loop state. The third set of N stubs is attached to a third robust zero of the same loop state. Finally, the fourth set of N stubs is attached to a fourth robust zero of the same loop state. See Fig. 6.12.

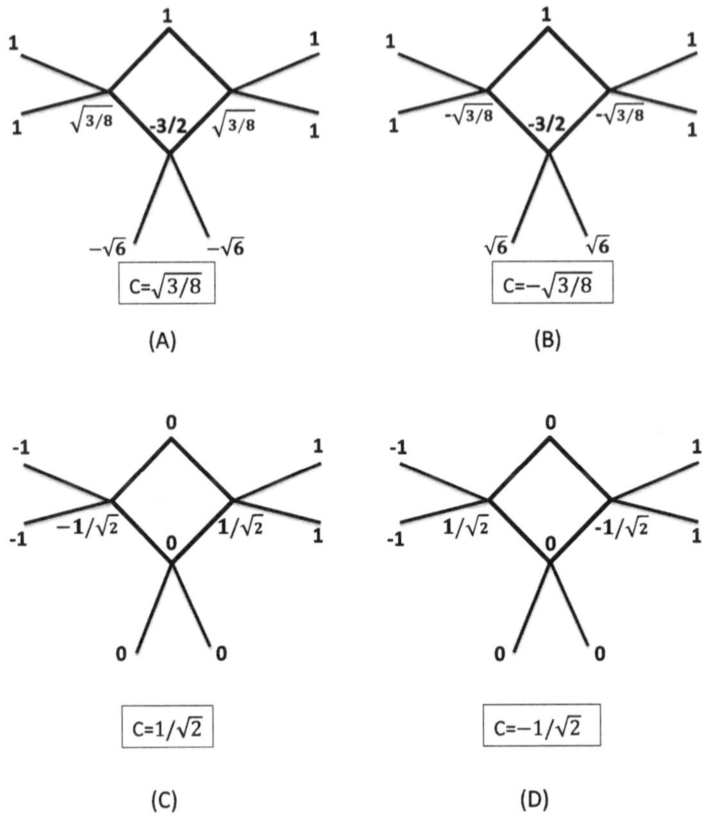

Figure 6.11 One closed loop of length $4L$, with $3N$ $(N = 2)$ stubs of length L. Eigenvectors for $C^2 = 3/8$ and $C^2 = 1/2$.

6.5.1 Reference system

In order to build the system described by Fig. 6.12, it is possible to use as a reference system $4N$ open loops of length L whose states are $(\alpha S)^{4N}$, $N = 1, 2, 3, \ldots, 4$ an open loop of length $2L$ whose states are $[2\alpha SC]^4$, and an interface space $M \equiv (1, 2, 3, 4)$.

Define λ as the wavelength, $\alpha = 2\pi/\lambda$, $S = \sin(\alpha L)$, and $C = \cos(\alpha L)$.

The reference system states are given by

$$2^4(\alpha S)^{4(N+1)}C^4 = 0, \tag{6.36}$$

and the determinant of the inverse of the corresponding $g_R(MM)$ is

$$|[g_R(MM)]^{-1}| = (-\alpha^2)^4(\frac{\alpha S}{C})^{4N}. \tag{6.37}$$

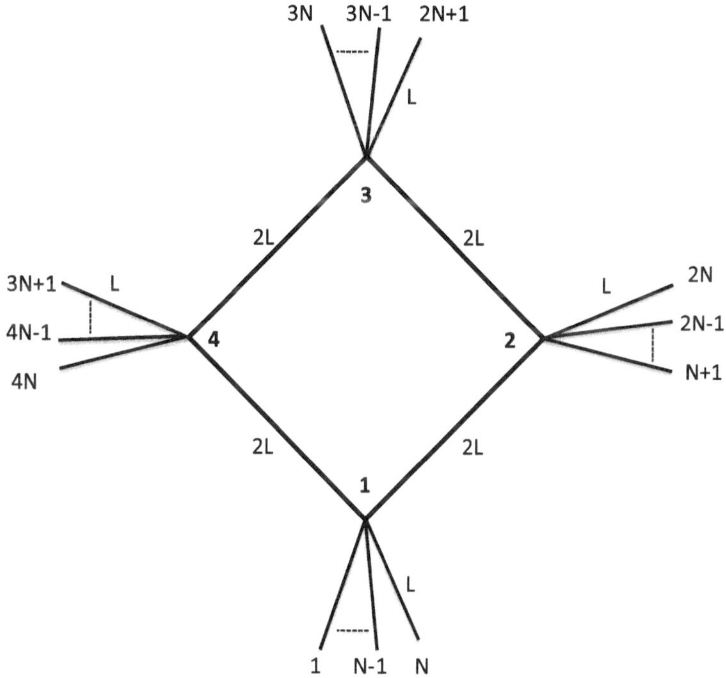

Figure 6.12 One closed loop of length $8L$, with $4N$ ($N = 1, 2, 3, \ldots$) stubs of length L, attached to four robust zeros of the same eigenfunctions.

6.5.2 Path states by inspection

As for the preceding examples, it is possible to find the path state eigenwavelengths to be given here by

$$S^2C^{4N} = 0. \tag{6.38}$$

The eigenfunctions are made out of the usual sine functions.

6.5.3 All states from the interface response theory

In order to obtain the final system depicted by Fig. 6.12, one proceeds as for the two preceding cases. The inverse of the interface space response function here is [1]

$$[g(MM)]^{-1} = \frac{\alpha}{2SC}\begin{pmatrix} A & 1 & 0 & 1 \\ 1 & A & 1 & 0 \\ 0 & 1 & A & 1 \\ 1 & 0 & 1 & A \end{pmatrix}, \tag{6.39}$$

where

$$A = 2SC\left[N\frac{S}{C} - 2\frac{\cos(2\alpha L)}{\sin(2\alpha L)}\right] = 2[1 + N - (2 + N)C^2]. \tag{6.40}$$

The determinant of the above matrix is

$$\begin{aligned} |[g(MM)]^{-1}| &= 4(\frac{\alpha}{2SC})^4 A^2(A^2 - 4) \\ &= 4(\frac{\alpha}{2SC})^4 \left[1 + N - (2 + N)C^2\right]^2 \\ &\quad \left[2N - 2(2 + N)C^2\right]\left[4 + 2N - 2(2 + N)C^2\right]. \end{aligned} \tag{6.41}$$

With the above results and the conservation of the number of states between the reference and the final system, one obtains that the final system states are given by

$$\begin{aligned} &2^4\alpha^{4(N+1)}S^{4(N+1)}C^4 \times 4(\frac{\alpha}{2SC})^4\left[1 + N - (2 + N)C^2\right]^2 \\ &\quad \left[2N - 2(2 + N)C^2\right]\left[4 + 2N - 2(2 + N)C^2\right] \times (-\alpha^2)^{-4}(\frac{\alpha S}{C})^{-4N} \\ &\quad = 16(2 + N)S^2C^{4N}\left[1 + N - (2 + N)C^2\right]^2\left[N - (2 + N)C^2\right] = 0. \end{aligned} \tag{6.42}$$

The $S^2C^{4N} = 0$ states were obtained above from the confined state theorem.

6.5.4 Some general eigenvectors of the final system

It is interesting here and for what follows to outline the following eigenvalues and eigenvectors.

6.5.4.1 Eigenvectors for $A = 0$

The two times degenerate eigenvectors given by $A^2 = 0$ are

$$E = \begin{pmatrix} 0 & -1 & 0 & 1 \end{pmatrix} \tag{6.43}$$

and

$$E = \begin{pmatrix} -1 & 0 & 1 & 0 \end{pmatrix}. \tag{6.44}$$

These eigenvectors show that, for the corresponding eigenvalues, only the opposite interface points, namely 2 and 4, on the one hand, and 1 and 3, on the other hand, can be activated together.

6.5.4.2 Eigenvector for $A = 2$

The eigenvector given by $A = 2$ is

$$E = \begin{pmatrix} -1 & 1 & -1 & 1 \end{pmatrix}. \tag{6.45}$$

6.5.4.3 Eigenvector for $A = -2$

The eigenvector given by $A = -2$ is

$$E = \begin{pmatrix} 1 & 1 & 1 & 1 \end{pmatrix}. \tag{6.46}$$

These eigenvectors show that, for the corresponding eigenvalues, all the four interface points are activated together.

6.5.5 Eigenvectors of the final system for $N = 2$

Here, for $N = 2$, the eigenvectors of this system are presented explicitly.

6.5.5.1 Eigenvectors for $S = 0$

The eigenvectors for $S^2 = 0$, or equivalently for

$$\lambda = \frac{2L}{n}, \qquad n = 0, 1, 2, 3, \ldots, \tag{6.47}$$

are given in Fig. 6.13.

All these states are two times degenerate. They correspond to path states.

6.5.5.2 Eigenvectors for $C = 0$

The eigenvectors for $C^8 = 0$, or equivalently for

$$\lambda = \frac{4L}{1 + 2n}, \qquad n = 0, 1, 2, 3, \ldots, \tag{6.48}$$

are given in Figs. 6.14 and 6.15.

These states are eight times degenerate. They correspond to path states.

6.5.5.3 Eigenvectors for $C = 1/\sqrt{2}$

The eigenvalues for $C = 1/\sqrt{2}$ are provided by

$$\lambda = \frac{8L}{8n + 1}, \qquad n = 0, 1, 2, 3, \ldots \tag{6.49}$$

and

$$\lambda = \frac{8L}{8n + 7}, \qquad n = 0, 1, 2, 3, \ldots \tag{6.50}$$

An example of their eigenvectors is shown in Fig. 6.16.

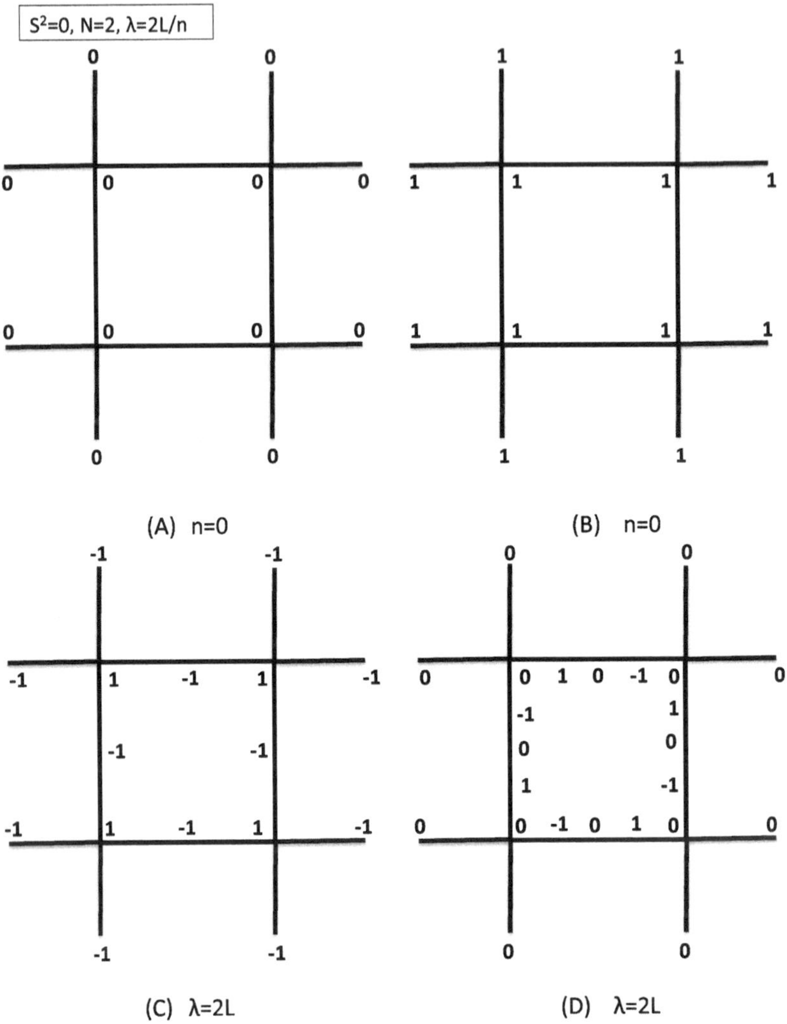

Figure 6.13 One closed loop of length $8L$, with $4N$ $(N=2)$ stubs of length L. Eigenvectors for $S^2=0$.

6.5.5.4 *Eigenvectors for* $C=-1/\sqrt{2}$

The eigenvalues for $C=-1/\sqrt{2}$ are given by

$$\lambda=\frac{8L}{8n+3}, \qquad n=0,1,2,3,\ldots, \tag{6.51}$$

and

$$\lambda=\frac{8L}{8n+5}, \qquad n=0,1,2,3,\ldots \tag{6.52}$$

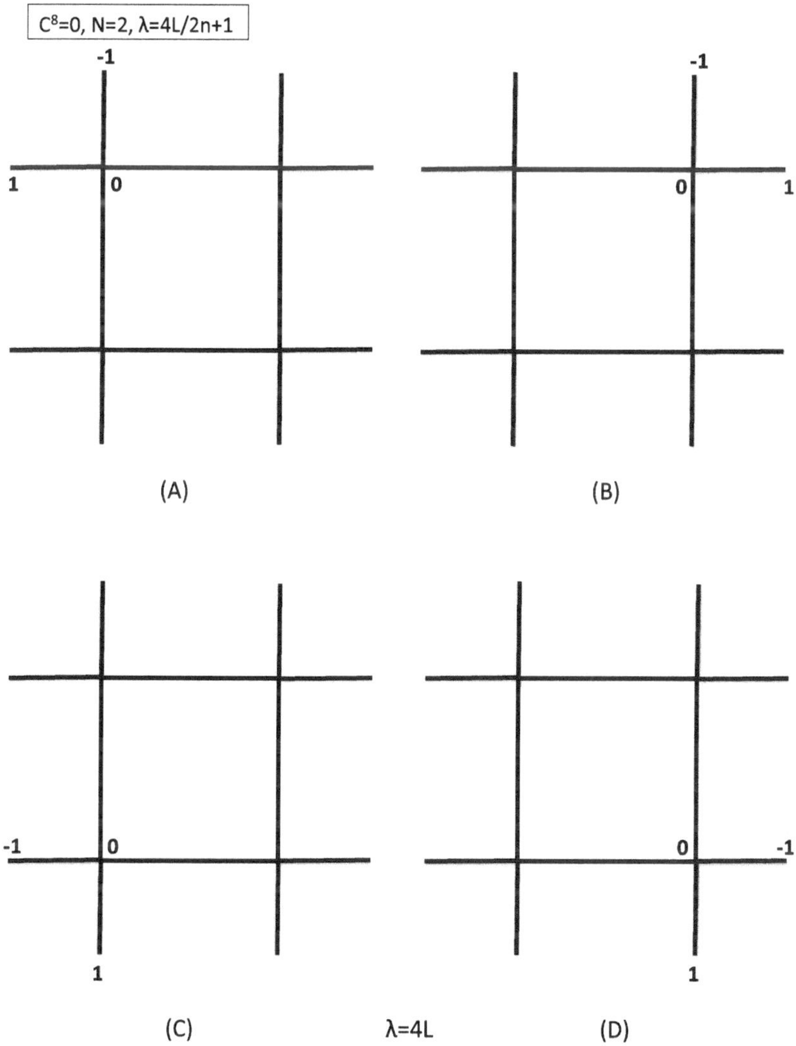

Figure 6.14 One closed loop of length $8L$, with $4N$ ($N=2$) stubs of length L. Four eigenvectors for $C^8=0$.

Their eigenvectors are similar to that shown in Fig. 6.16.

6.5.5.5 *Eigenvectors for* $C=\sqrt{3}/2$

The eigenvalues for $C=\sqrt{3}/2$ are

$$\lambda=\frac{12L}{12n+1}, \qquad n=0,1,2,3,\ldots, \tag{6.53}$$

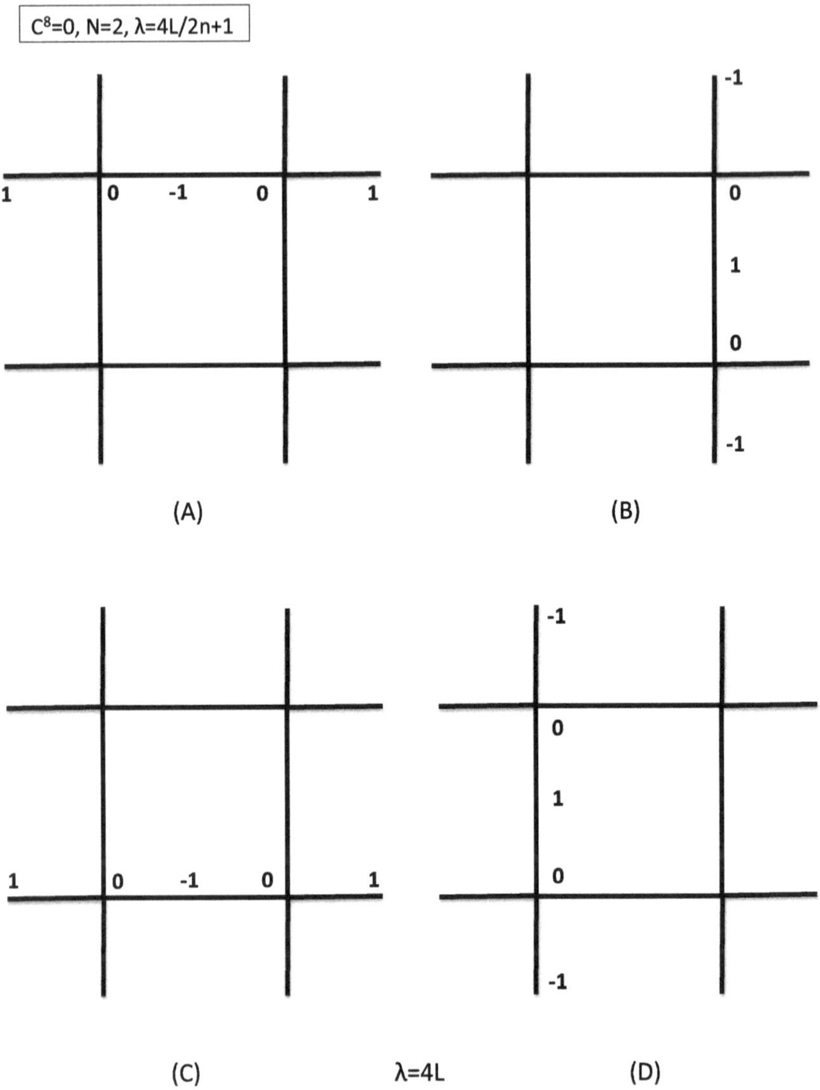

Figure 6.15 One closed loop of length $8L$, with $4N$ $(N = 2)$ stubs of length L. The other four eigenvectors for $C^8 = 0$.

and

$$\lambda = \frac{12L}{12n + 11}, \qquad n = 0, 1, 2, 3, \ldots \tag{6.54}$$

An example of their eigenvectors is shown in Fig. 6.17.

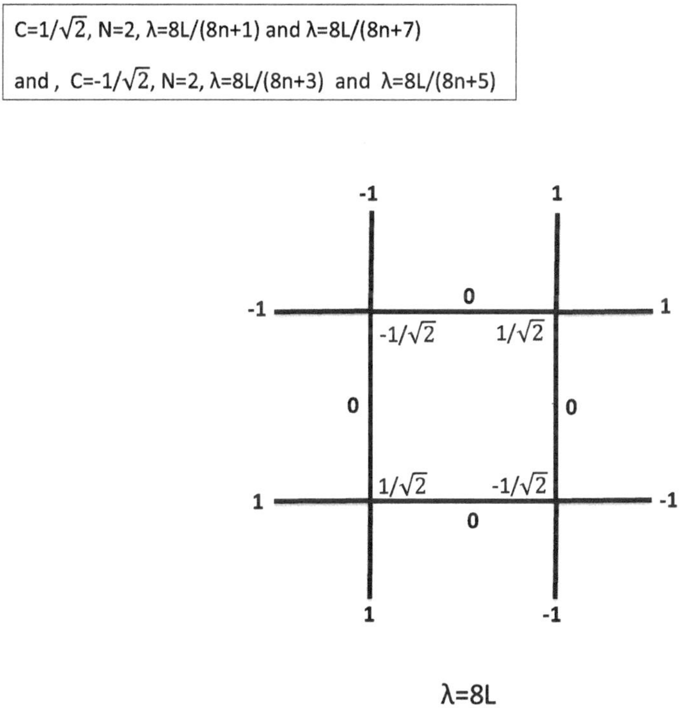

Figure 6.16 One closed loop of length $8L$, with $4N$ ($N = 2$) stubs of length L. The eigenvectors for $C^2 = 1/2$.

6.5.5.6 *Eigenvectors for $C = -\sqrt{3}/2$*

The eigenvalues for $C = -\sqrt{3}/2$ are

$$\lambda = \frac{12L}{12n+5}, \qquad n = 0, 1, 2, 3, \ldots, \tag{6.55}$$

and

$$\lambda = \frac{12L}{12n+7}, \qquad n = 0, 1, 2, 3, \ldots \tag{6.56}$$

An example of their eigenvectors is shown in Fig. 6.18.

6.5.6 *Mode number conservation*

Fig. 6.19 enables checking the conservation of the number of states between the reference and the final systems.

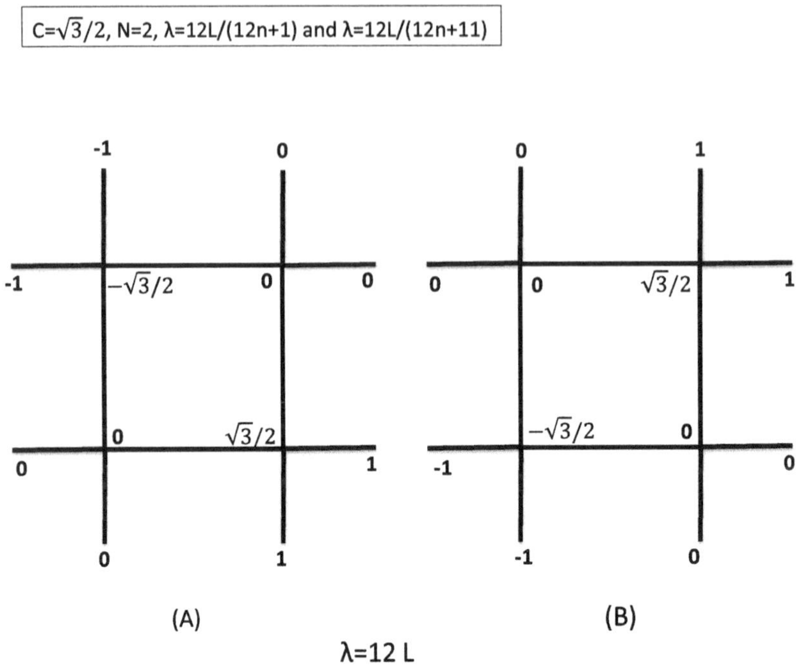

Figure 6.17 One closed loop of length $8L$, with $4N$ $(N = 2)$ stubs of length L. The eigenvectors for $C = \sqrt{3}/2$.

6.6 Applications: multiplexers and wide reflection bands

In the preceding sections of this chapter, simple models of a closed loop with stubs grafted to its robust zeros were presented. The emphasis in those sections was on the robust states of the final systems.

There are many other such systems one can build and there are many practical applications such systems may provide. The aim here is not to be exhaustive but tutorial. Nevertheless, let us have a quick look at two of the many possible applications of these systems, namely the multiplexers and wide reflection bands. Use is taken of a generalization of the system depicted in the last section above. This generalization is just made to enable and tune the lengths of all stubs to L' rather than L. Another generalization is done by replacing the loop by a cube.

6.6.1 *The finite system*

In order to obtain the final system depicted by Fig. 6.12, but now with L' rather than L for the length of all stubs, one proceeds as above. The inverse of the interface space

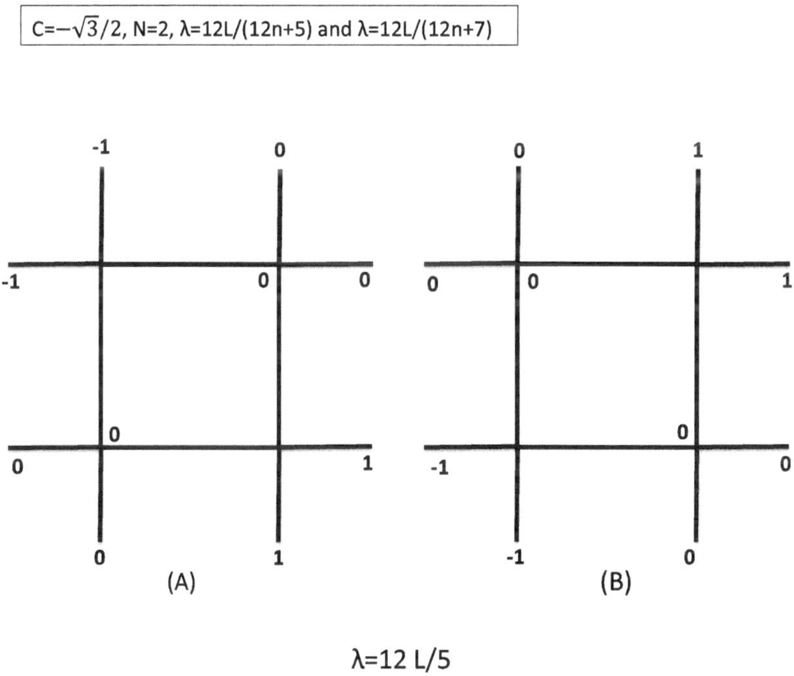

Figure 6.18 One closed loop of length $8L$, with $4N$ $(N = 2)$ stubs of length L. The eigenvectors for $C = -\sqrt{3}/2$.

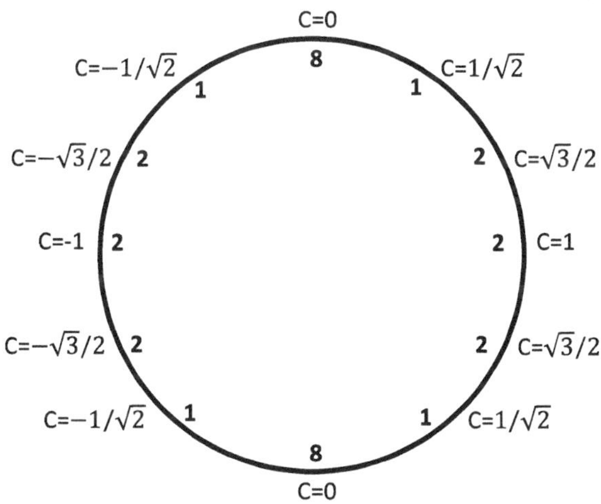

Figure 6.19 One closed loop of length $8L$, with $4N$ $(N = 2)$ stubs of length L. Conservation of the number of states between the reference and final systems.

response function here is [1]

$$[g(MM)]^{-1} = \frac{\alpha}{2SCC'} \begin{pmatrix} A_0 & C' & 0 & C' \\ C' & A_0 & C' & 0 \\ 0 & C' & A_0 & C' \\ C' & 0 & C' & A_0 \end{pmatrix}, \tag{6.57}$$

where $C' = \cos(\alpha L')$, $S' = \sin(\alpha L')$, and

$$A_0 = 2\big[(1 - 2C^2)C' + NSCS'\big]. \tag{6.58}$$

The determinant of the above matrix is

$$|[g(MM)]^{-1}| = (\frac{\alpha}{2SCC'})^4 A_0^2 (A_0^2 - 4C'^2). \tag{6.59}$$

With the above results and the conservation of the number of states between the reference and the final system, one obtains that the final system states are given by

$$SCC'^{4(N-1)}\big[(1 - 2C^2)C' + NSCS'\big]^2 (2SC' + NCS')(-2CC' + NSS') = 0. \tag{6.60}$$

6.6.2 The finite system with four semi-infinite wave guides

Add now to the above finite system four semiinfinite waveguides attached to the four interface points. See Fig. 6.20.

In the above $[g(MM)]^{-1}$, one has just to replace A_0 by

$$A = A_0 + 2jSCC'. \tag{6.61}$$

6.6.2.1 The interface response function elements

By inversion of the above $[g(MM)]^{-1}$, one obtains the interface response function elements

$$g(MM) = \frac{2SCC'}{\alpha A(A^2 - 4C'^2)} \times \begin{pmatrix} A^2 - 2C'^2 & -C'A & 2C'^2 & -C'A \\ -C'A & A^2 - 2C'^2 & -C'A & 2C'^2 \\ 2C'^2 & -C'A & A^2 - 2C'^2 & -C'A \\ -C'A & 2C'^2 & -C'A & A^2 - 2C'^2 \end{pmatrix}. \tag{6.62}$$

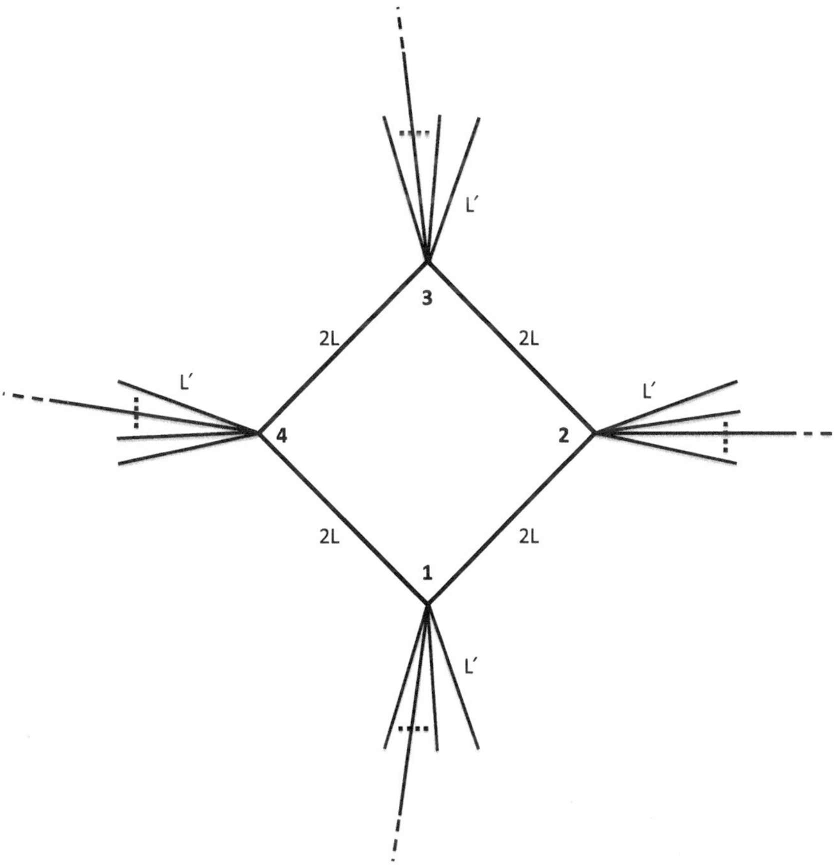

Figure 6.20 One closed loop of length $8L$, with 4 times one semiinfinite guide and N $(= 1, 2, 3, \ldots)$ stubs of length L, attached to four robust zeros of the same eigenfunctions.

For any value of L', it is helpful to write more explicitly the elements of this response matrix by using

$$A = 2\big[(1 - 2C^2)C' + NSCS'\big] + 2jSCC', \tag{6.63}$$

$$(A^2 - 4C'^2) = 4SC(2SC' + NCS' + jCC')(-2CC' + NSS' + jSC'), \tag{6.64}$$

and

$$\begin{aligned}(A^2 - 2C'^2) = &\big[(2 + \sqrt{2} - 4C^2)C' + 2NSCS' + 2jSCC'\big] \\ &\times \big[(2 - \sqrt{2} - 4C^2)C' + 2NSCS' + 2jSCC'\big].\end{aligned} \tag{6.65}$$

For $L' = L$, these expressions are

$$A = 2C\left[N + 1 - (2+N)C^2 + jSC\right], \tag{6.66}$$

$$(A^2 - 4C^2) = 4SC^2\left[(2+N)S + jC\right]\left[N - (2+N)C^2 + jSC\right], \tag{6.67}$$

and

$$\begin{aligned}(A^2 - 2C^2) = C^2&\left[2 + \sqrt{2} + 2N - 2C^2(2+N) + 2jSC\right] \\ &\times\left[2 - \sqrt{2} + 2N - 2C^2(2+N) + 2jSC\right].\end{aligned} \tag{6.68}$$

6.6.2.2 Transmissions and reflection general expressions

With an input unity at port 1 of the multiplexer, the transmissions to the other three ports and the reflection are obtained from [1]

$$T_{1i} = |2j\alpha g_{1i}|^2, \quad i = 2, 3, 4, \tag{6.69}$$

$$R = |1 - 2j\alpha g_{11}|^2. \tag{6.70}$$

6.6.2.3 Multiplexers for L' different from L

In the above finite model with four robust points, it is remarkable that the diagonal elements $A_0 = 0$ of the inverse interface response matrix provide path states whose eigenfunction elements are nonzero only between two opposite interface points. This is indeed one of the properties a robust multiplexer is expected to have. On the other hand, by inspection of the above response function matrix, one sees that $T_{13} = 0$, when $C' = C = 0$.

Nevertheless, let us inspect what happens to the response function elements when the diagonal element A_0, see Eq. (6.63), of the finite system vanish, while C' is different from C. The diagonal element A of the system with the four semiinfinite waveguides is then just $A = 2jSCC'$, when $A_0 = 0$. From the above general expressions, one obtains in this case

$$T_{13} = \frac{1}{(S^2C^2 + 1)^2}, \tag{6.71}$$

$$T_{12} = \frac{(SC)^2}{(S^2C^2 + 1)^2}, \tag{6.72}$$

and

$$R = \frac{(SC)^4}{(S^2C^2 + 1)^2}. \tag{6.73}$$

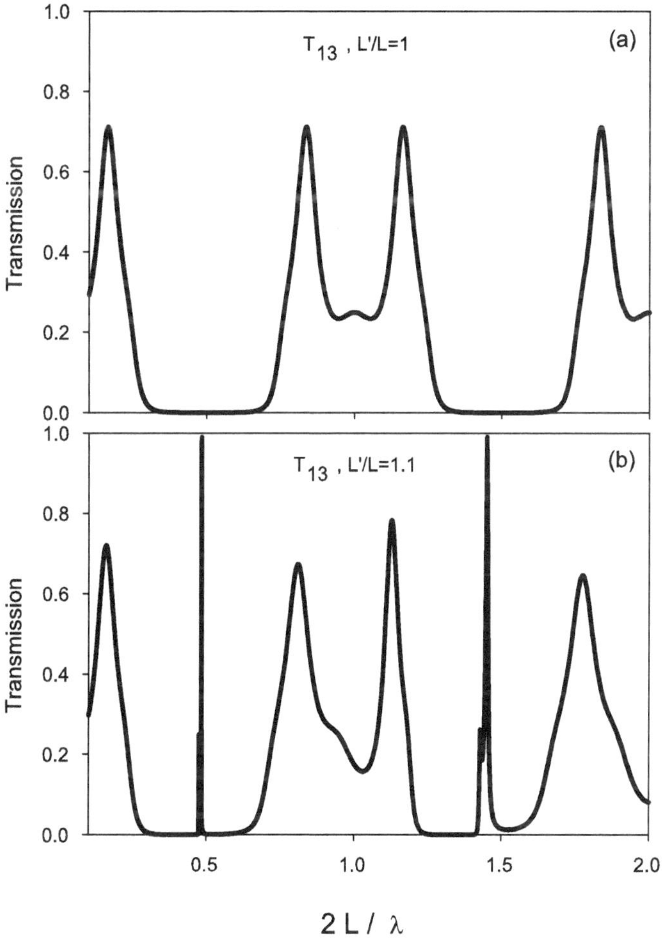

Figure 6.21 The multiplexer, for $N = 2$, with $L' = L$ (top curve) and $L'/L = 1.1$ (bottom curve). The transmission T_{13} between ports 1 and 3.

So this quick inspection shows that, for C' different from C, one can get $T_{13} = 1$ when $SC = 0$. Going back to the explicit expressions of A and A_0, one obtains, for this to happen, two conditions

$$SC = 0 \tag{6.74}$$

and

$$C' = 0. \tag{6.75}$$

These conditions provide

$$\lambda = \frac{4L}{n}, \qquad n = 1, 2, 3, \ldots, \tag{6.76}$$

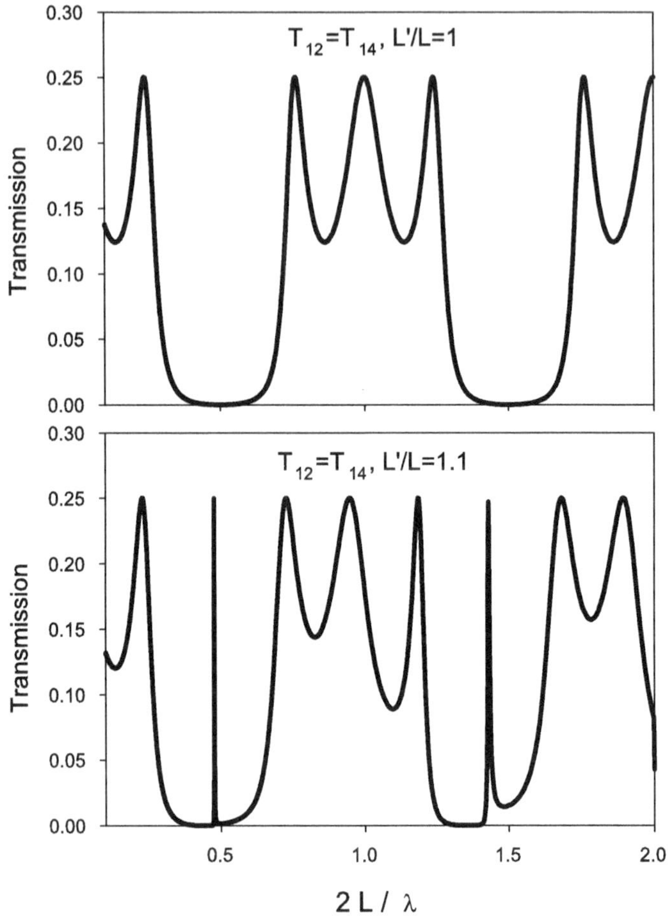

Figure 6.22 The multiplexer, for $N = 2$, with $L' = L$ (top curve) and $L'/L = 1.1$ (bottom curve). The transmissions $T_{12} = T_{14}$ between ports 1 to 2 and 1 to 4.

$$\lambda = \frac{4L'}{1 + 2n'}, \qquad n' = 0, 1, 2, 3, \ldots, \tag{6.77}$$

and

$$\frac{L'}{L} = \frac{1 + 2n'}{n}, \qquad n = 1, 2, 3, \ldots, \qquad n' = 0, 1, 2, 3, \ldots \tag{6.78}$$

Note also that when $SC = 0$ the imaginary part of A vanishes. So when the two conditions expressed by Eqs. (6.74) and (6.75) are fulfilled, the real and imaginary parts of A vanish. When this happens for the same wavelengths, this is the signature of the existence of bound in continuum states, not interacting with the multiplexer transmission coefficients. So in what follows, one uses the above three expressions in order to get the values of L'/L for which these bound in continuum states exist and

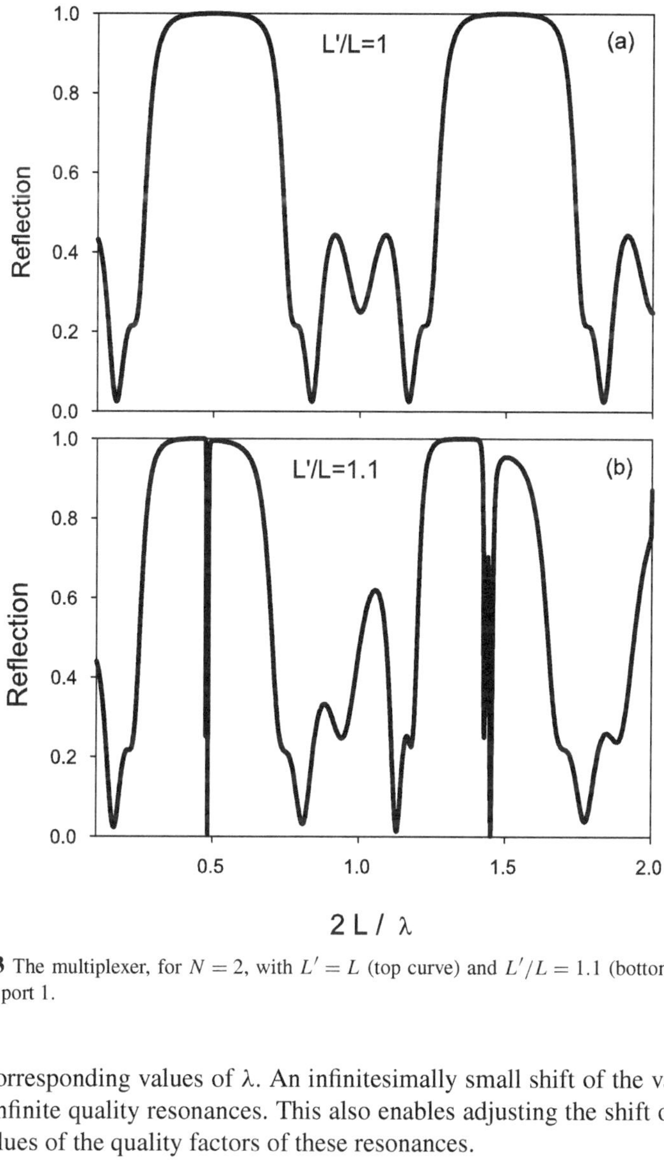

Figure 6.23 The multiplexer, for $N = 2$, with $L' = L$ (top curve) and $L'/L = 1.1$ (bottom curve). The reflection by port 1.

then the corresponding values of λ. An infinitesimally small shift of the values of L' provides infinite quality resonances. This also enables adjusting the shift of L' to the desired values of the quality factors of these resonances.

In what follows, Fig. 6.21 shows an example of very high quality multiplexing resonance peaks in T_{13} obtained for $L'/L = 1.1$ and appearing in the middle of the very large gaps situated around $2L/\lambda = (1 + 2n)/2$, where $n = 0, 1, 2, 3, \ldots$ The wavelengths of these peaks come from $A_0 = 0$, namely

$$(1 - 2C^2)C' + NSCS' = 0. \tag{6.79}$$

Note also that close to these high quality peaks also high quality peaks appear, but of amplitude 0.25. These peak wavelengths are related with the vanishing, when the same conditions given by Eqs. (6.74) and (6.75) are fulfilled, of the real and imaginary parts of two other factors appearing in the denominator of the response function elements, see Eq. (6.64), namely

$$(2SC' + NCS' + jCC') = 0 \tag{6.80}$$

and

$$(-2CC' + NSS' + jSC') = 0. \tag{6.81}$$

See also the transmission factors $T_{12} = 0.25$ and $T_{14} = 0.25$ given in Fig. 6.22. These results agree with the eigenvectors obtained above for the corresponding eigenstates, $A_0^2 - 4C'^2 = 0$, of the finite system.

6.6.2.4 Wide reflection bands for L' equal to L

The above expressions show that around the wavelengths for which $C = 0$, large nonpassing bands exist, see Fig. 6.23, for inputs in the system by any of the four ports.

The widths of these nonpassing bands increase when the number N of stubs increases, see Fig. 6.24, where the number of stubs at each port is taken to be 10 rather than 2.

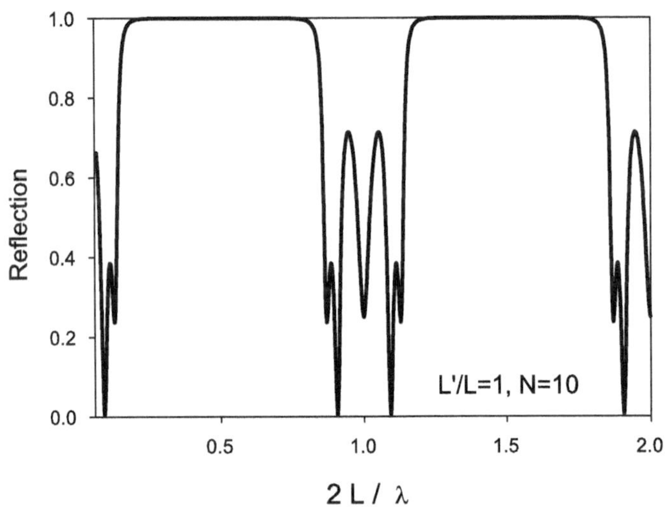

Figure 6.24 The multiplexer, for $N = 10$ with $L' = L$. The reflection by port 1.

Remember that here use is made only of four robust points. There is no doubt that by increasing the number of these input points, it is possible to obtain a larger part of the closed loop to be covered by such wide reflection bands. The presented model can be extended from a loop to a three-dimensional surface covered by multiple robust zeros with many stubs attached to them. This is illustrated here by using the cube structure of Chapter 5 of this book, see Fig. 5.19. In order to compare easier with the present loop and stubs structure, the edges of the cube are taken to be of length $2L$. At each of the 8 corners of the cube, one attaches, as for the loop, N stubs of length L and one semiinfinite guide. Then as for the loop and stub structure, one activates this cube and stub structure through port 1. Fig. 6.25 presents the reflection results for $N = 2$ and 10. This shows that wide reflection bands exist also in three dimensions for such structures.

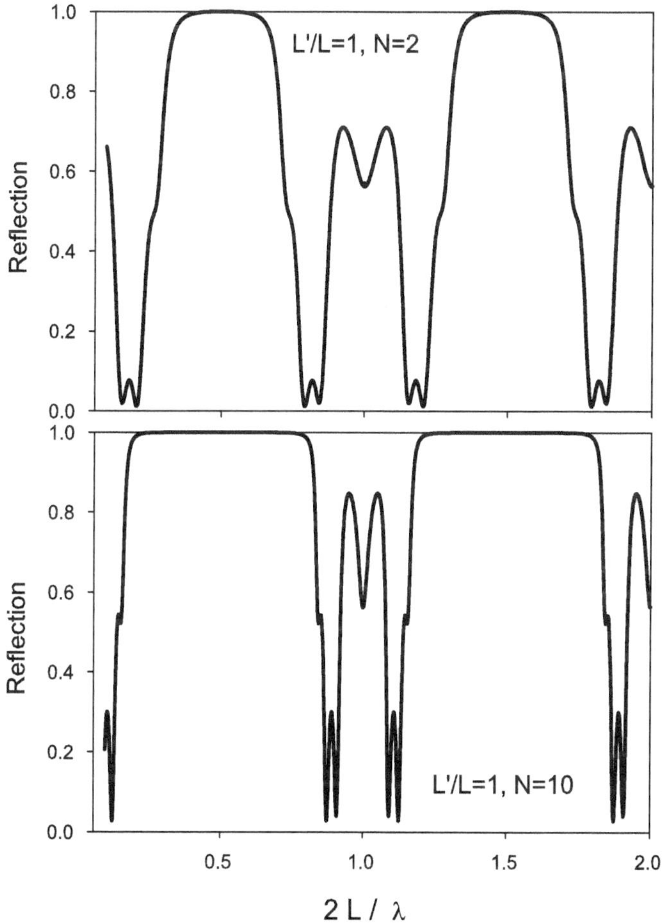

Figure 6.25 The cube with edges $2L$ and stubs L. Top curve for $N = 2$ and bottom curve for $N = 10$. The reflection coefficient.

Fig. 6.26 shows also, for $N = 2$, the three different transmission coefficients $T_{12} = T_{14} = T_{15}$, $T_{13} = T_{16} = T_{18}$, and T_{17}. This enables, together with the reflection coefficient, checking that the sum of all transmissions and the reflection is equal to 1 for all wavelengths.

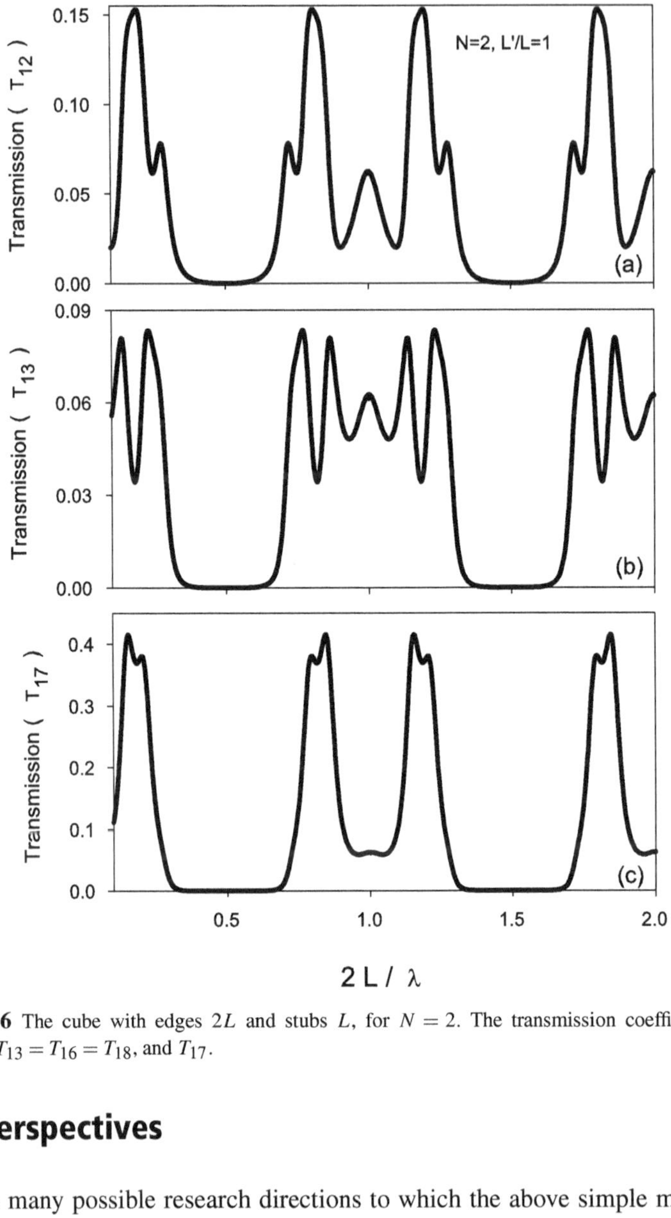

Figure 6.26 The cube with edges $2L$ and stubs L, for $N = 2$. The transmission coefficients $T_{12} = T_{14} = T_{15}$, $T_{13} = T_{16} = T_{18}$, and T_{17}.

6.7 Perspectives

There are many possible research directions to which the above simple models may lead. Let us just stress the three-dimensional models, for example, spheres with stubs. The case of the cube structure with stubs, used here just above, can be considered as a simple model of a sphere with stubs. This example is worth to be expanded by using more robust points on the sphere and more elaborated stub structures. This can enable designing novel path states. Such structures may be highly protective to what can be inside such spheres with stubs. Although robust and highly protective, such structures have some weaker features. For example, it is important to learn more about their nonrobust states. Indeed, by exciting very large amplitude eigenfunctions, it can be possible to induce nonlinear effects and irreversible mutations of the whole structure of such spheres with stubs.

It is worth also to stress that any eigenstate can be activated within its subdomain without any subsystem deformation, see the activation state theorem in Chapter 3 of this Photonics book. This paves the way to the use of the degenerate path states for many applications.

References

[1] L. Dobrzyński, A. Akjouj, G. Lévêque, et al., Centered system magnons, in: Magnonics, in: Interface Transmission Tutorial Book Series, Elsevier, 2019.

Eigenfunction rules

7

Leonard Dobrzyński[a], Housni Al Wahsh[b], Abdellatif Akjouj[a], and El Houssaine El Boudouti[c]
[a]Univ. Lille, CNRS, Centrale Lille, Yncréa ISEN, Univ. Polytechnique Hauts-de-France, UMR 8520 - IEMN, Lille, France, [b]Benha University, Faculty of Engineering, Cairo, Egypt, [c]University Mohammed I, Faculty of Sciences, Laboratory of Physics of Matter and Radiation, Oujda, Morocco

7.1 Introduction

It is well known that eigenfunctions have to be continuous at each space point of the structure in which they are localized. This is not the only rule they have to fulfill. Rules involving eigenfunction derivatives are also used in the bulk and at the boundary of all physical systems.

The aim of this section is to try and reach a general formulation of the eigenfunction rules. Let us start from a simple finite star network, Fig. 7.1.

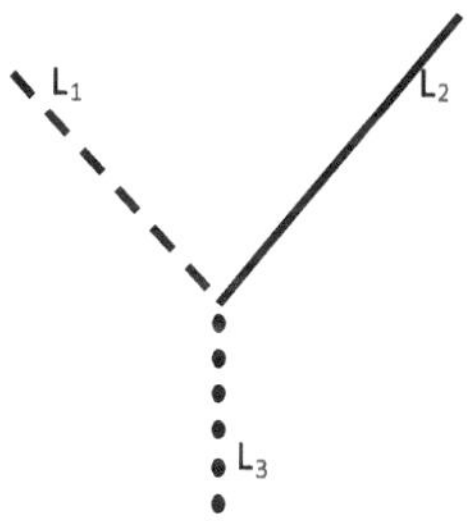

Figure 7.1 Sketch of the geometry of a simple star network build out of three one-dimensional wave open loops. The open loops may have different dielectric constants and different lengths. The point of connexion of these loops is a node.

At each space point of this network, it is found that the eigenfunctions are continuous and the sum of their outgoing first derivatives is equal to zero. These rules fix the values of the discrete eigenvalues. In order to show that these rules are general, one proceeds by iteration. First, one goes one step further than the starting model and finds that the above rules remain true. Then one assumes that these general rules hold for any kind of network formed out of N finite loops. Then one goes one step further to an $(N + 1)$-loop new network and shows that these rules are general.

It is shown also that when one adds to a finite network some infinite and semiinfinite loops, the first two general eigenfunction rules (continuity of the field and of its outgoing derivative at each space point of the network) apply. The semiinfinite and infinite loops add continua of eigenstates to the discrete set of eigenstates of the initial finite network. So the initial discrete states become peaks in the final density of

Photonics. https://doi.org/10.1016/B978-0-12-819388-4.00016-2

states, with their maxima near their initial frequency positions. The first two general rules provide, in this case, the electric field transmission and reflection coefficients at each network space point. The electromagnetic wave intensity is also shown to be conserved.

7.2 Star network

In this system, the network node is formed by the connection of one free end of all the N loops. It is assumed that the wavelengths of the electromagnetic standing waves of such a network are large compared to the spatial dimensions of all the nodes. This enables neglecting the loop-bending effects on the electrical field polarization. Let us remind that these wavelengths are also considered to be large compared to the lateral dimensions of the loops.

7.2.1 The node element of the response function

The sum of the inverses of the $g_i(0,0)$ element, see Eq. (1.18), of the finite loops of lengths L_i and wavelengths λ_i, $i = 1, 2, \ldots, N$, enables obtaining [1] the inverse of the $g_N(0,0)$ element of the response function of the node system build out of N finite loops to be

$$[\mathbf{g}_N(0,0)]^{-1} = \sum_{i=1}^{N} \frac{2\pi S_i}{\lambda_i C_i}, \tag{7.1}$$

where

$$S_i = \sin\left(\frac{2\pi L_i}{\lambda_i}\right) \tag{7.2}$$

and

$$C_i = \cos\left(\frac{2\pi L_i}{\lambda_i}\right). \tag{7.3}$$

7.2.2 The complete response function

The complete response function of such a node network can be obtained from the general equation of the interface response theory where the reference response function $\mathbf{g}_N(D,D)$ is now formed by the $\mathbf{g}_i(x_i, x_i')$ given by Eq. (1.33) for each independent loop of length L_i of this network. Let us proceed step by step starting with a single loop.

7.2.2.1 One-loop network

Let us rewrite the response function obtained before, Eqs. (1.32) and (1.33), in the present form

$$\mathbf{g}_1(x_1, x_1') = \frac{\mathbf{g}_1(0,0)}{C_1} \cos\left(\frac{2\pi x_1}{\lambda_1}\right) \cos\left(\frac{2\pi (L_1 - x_1')}{\lambda_1}\right), \quad x_1 \le x_1', \tag{7.4}$$

and

$$\mathbf{g}_1(x_1, x_1') = \frac{\mathbf{g}_1(0,0)}{C_1} \cos\left(\frac{2\pi x_1'}{\lambda_1}\right) \cos\left(\frac{2\pi (L_1 - x_1)}{\lambda_1}\right), \quad x_1 \ge x_1'. \tag{7.5}$$

7.2.2.2 Two-loop network

The network is build out of two different finite loops of length L_1 and L_2. Its response function elements are as follows.

For x_1 within the guide 1 and x_2 within the loop 2,

$$\mathbf{g}_2(x_1, x_2) = \mathbf{g}_2(0,0) \frac{\cos\left(\frac{2\pi x_1}{\lambda_1}\right)}{C_1} \frac{\cos\left(\frac{2\pi (L_2 - x_2)}{\lambda_2}\right)}{C_2}. \tag{7.6}$$

For x_1 and x_1' within the loop 1, with $x_1 \le x_1'$,

$$\mathbf{g}_2(x_1, x_1') = \frac{\cos\left(\frac{2\pi (L_1 - x_1')}{\lambda_1}\right)}{C_1} \left[\mathbf{g}_2(0,0) \frac{\cos\left(\frac{2\pi (L_1 - x_1)}{\lambda_1}\right)}{C_1} - \frac{\lambda_1}{2\pi} \sin\left(\frac{2\pi x_1}{\lambda_1}\right) \right]. \tag{7.7}$$

For x_2 and x_2' within the loop 2, with $x_2 \le x_2'$,

$$\mathbf{g}_2(x_2, x_2') = \frac{\cos\left(\frac{2\pi (L_2 - x_2')}{\lambda_2}\right)}{C_2} \left[\mathbf{g}_2(0,0) \frac{\cos\left(\frac{2\pi (L_2 - x_2)}{\lambda_2}\right)}{C_2} - \frac{\lambda_2}{2\pi} \sin\left(\frac{2\pi x_2}{\lambda_2}\right) \right]. \tag{7.8}$$

When $x_i \ge x_i'$, $i = 1$ or 2, just interchange x_i and x_i' in the second term of the above two equations.

7.2.2.3 Three-loop node network

The network is build out of three different finite loops of length L_1, L_2, and L_3, respectively. Its response function elements are as follows.

For x_i within the loop i and x_j within the loop j with $i \ne j$,

$$\mathbf{g}_3(x_i, x_j) = \mathbf{g}_3(0,0) \frac{\cos\left(\frac{2\pi (L_i - x_i)}{\lambda_i}\right)}{C_i} \frac{\cos\left(\frac{2\pi (L_j - x_j)}{\lambda_j}\right)}{C_j}. \tag{7.9}$$

For x_i and x_i' within the loop $i(=1,2,3)$, with $x_i \leq x_i'$,

$$\mathbf{g}_3(x_1,x_1') = \frac{\cos\left(\frac{2\pi(L_1-x_1')}{\lambda_1}\right)}{C_1}\left[\mathbf{g}_3(0,0)\frac{\cos\left(\frac{2\pi(L_1-x_1)}{\lambda_1}\right)}{C_1} - \frac{\lambda_1}{2\pi}\sin\left(\frac{2\pi x_1}{\lambda_1}\right)\right], \tag{7.10}$$

$$\mathbf{g}_3(x_2,x_2') = \frac{\cos\left(\frac{2\pi(L_2-x_2')}{\lambda_2}\right)}{C_2}\left[\mathbf{g}_3(0,0)\frac{\cos\left(\frac{2\pi(L_2-x_2)}{\lambda_2}\right)}{C_2} - \frac{\lambda_2}{2\pi}\sin\left(\frac{2\pi x_2}{\lambda_2}\right)\right], \tag{7.11}$$

$$\mathbf{g}_3(x_3,x_3') = \frac{\cos\left(\frac{2\pi(L_3-x_3')}{\lambda_3}\right)}{C_3}\left[\mathbf{g}_3(0,0)\frac{\cos\left(\frac{2\pi(L_3-x_3)}{\lambda_3}\right)}{C_3} - \frac{\lambda_3}{2\pi}\sin\left(\frac{2\pi x_3}{\lambda_3}\right)\right]. \tag{7.12}$$

When $x_i \geq x_i'$, $i = 1, 2, 3$, just interchange x_i and x_i' in the second term of the above three equations.

7.2.2.4 N-loop node network

This star network is build out of N different finite loops of length L_i, $i = 1, 2, \ldots, N$. Its response function elements are as follows.

For x_i within the loop i and x_j within the loop j with $i \neq j$,

$$\mathbf{g}_N(x_i,x_j) = \mathbf{g}_N(0,0)\frac{\cos\left(\frac{2\pi(L_i-x_i)}{\lambda_i}\right)}{C_i}\frac{\cos\left(\frac{2\pi(L_j-x_j)}{\lambda_j}\right)}{C_j}. \tag{7.13}$$

For x_i and x_i' within the loop $(i = 1, 2, 3, \ldots, N)$, with $x_i \leq x_i'$,

$$\mathbf{g}_N(x_i,x_i') = \frac{\cos\left(\frac{2\pi(L_i-x_i')}{\lambda_i}\right)}{C_i}\left[\mathbf{g}_N(0,0)\frac{\cos\left(\frac{2\pi(L_i-x_i)}{\lambda_i}\right)}{C_i} - \frac{\lambda_i}{2\pi}\sin\left(\frac{2\pi x_i}{\lambda_i}\right)\right]. \tag{7.14}$$

When $i = 1, 2, \ldots, N$, with $x_i \geq x_i'$, just interchange x_i and x_i' in the second term of the above last equation.

7.2.3 The eigenfunctions

From the above expressions of the response functions, one obtains the following expression for the eigenfunctions within each of the loops of the above star network:

$$E(x_i) = \frac{\cos\left(\frac{2\pi(L_i - x_i)}{\lambda_i}\right)}{C_i}. \tag{7.15}$$

The normalization of this eigenfunction is such that $|E(x_i)| = 1$ at the node point $x_i = 0$. This assumes the continuity of all these eigenfunctions at this node.

Note also that

$$\frac{\partial E(x_i)}{\partial x_i} = \frac{2\pi \sin\left(\frac{2\pi(L_i - x_i)}{\lambda_i}\right)}{\lambda_i C_i}. \tag{7.16}$$

In particular, this partial derivative at the node point is

$$E_i'(0) = \frac{2\pi S_i}{\lambda_i C_i}. \tag{7.17}$$

The partial derivatives at all free ends of the star network are

$$E_i'(L_i) = 0. \tag{7.18}$$

At each of these ends, there is only one outgoing first derivative, as these ends are assumed to be perfect reflectors in this finite system.

At all the other points x_i of this network, but when $x = 0$, there are two outgoing first partial derivatives of the eigenfunctions whose sum is equal to zero.

Looking now at $[\mathbf{g}_i(0,0)]^{-1}$, one sees that

$$[\mathbf{g}_i(0,0)]^{-1} = \sum_{i=1}^{N} E_i'(0). \tag{7.19}$$

The network eigenvalues can also be obtained by equating to zero, at the star network node, the sum of all outgoing first partial derivatives of the eigenfunctions.

7.2.4 Summary

At each space point of this network, the eigenfunctions are continuous. The corresponding kernel of the response function is the sum of the eigenfunction outgoing first partial derivatives. The zero values of this kernel determine all the values of the discrete eigenvalues, but those for which the eigenfunction is zero at this space point.

7.3 One step further

In order to demonstrate that the above conclusions remain true for any network, let us use an iterative procedure.

Consider first the above star network made out of three loops. Add to the L_1 end of loop 1, called node 0_1, a fourth different loop of length L_4, see Fig. 7.2. These choices of only three loops in the star network and only one additional fourth loop are made for simplicity.

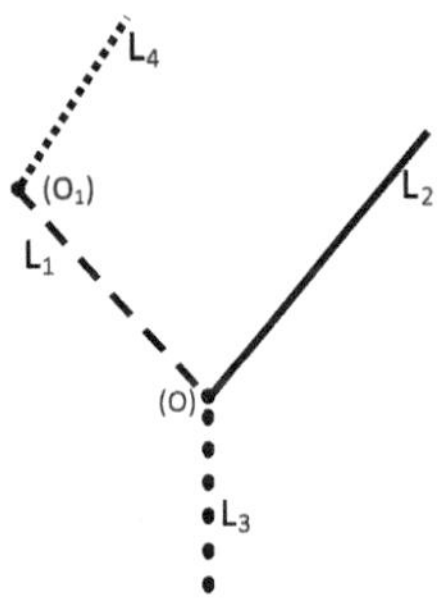

Figure 7.2 The star network of Fig. 7.1 with a fourth different loop of length L_4.

In order to determine the eigenfunctions within this new network, it is convenient to work out first some elements of its response function $\mathbf{g}_4$. By linear superposition, one obtains

$$[\mathbf{g}_4(0_1, 0_1)]^{-1} = [\mathbf{g}_3(L_1, L_1)]^{-1} + \frac{2\pi S_4}{\lambda_4 C_4}. \tag{7.20}$$

Use then the general equation of the interface response theory. Here the response function of the reference system is made out of the disconnected $\mathbf{g}_3$ of the star network and $\mathbf{g}_4$ of the single loop 4. The other needed elements of the response function of the final system are:

$$\mathbf{g}_4(x_2, x_1) = \mathbf{g}_3(x_2, L_1)[\mathbf{g}_3(L_1, L_1)]^{-1}\mathbf{g}_4(0_1, 0_1)[\mathbf{g}_3(L_1, L_1)]^{-1}\mathbf{g}_3(L_1, x_1), \tag{7.21}$$

$$\mathbf{g}_4(x_3, x_1) = \mathbf{g}_3(x_3, L_1)[\mathbf{g}_3(L_1, L_1)]^{-1}\mathbf{g}_4(0_1, 0_1)[\mathbf{g}_3(L_1, L_1)]^{-1}\mathbf{g}_3(L_1, x_1), \tag{7.22}$$

and

$$\mathbf{g}_4(x_1, x_4) = \mathbf{g}_3(x_1, L_1)[\mathbf{g}_3(L_1, L_1)]^{-1}\mathbf{g}_4(0_1, 0_1)[\mathbf{g}_1(0_1, 0_1)]^{-1}\mathbf{g}_1(0_1, x_4). \tag{7.23}$$

One then obtains

$$[\mathbf{g}_3(0, 0)]^{-1} = \sum_{i=1}^{3} \frac{2\pi S_i}{\lambda_i C_i}, \tag{7.24}$$

$$[\mathbf{g}_1(0_1, 0_1)]^{-1} = \frac{2\pi S_4}{\lambda_4 C_4}, \tag{7.25}$$

$$\mathbf{g}_1(0_1, x_4) = \frac{\lambda_4 \cos\left[\frac{2\pi(L_4 - x_4)}{\lambda_4}\right]}{2\pi S_4}, \tag{7.26}$$

$$\mathbf{g}_3(x_2, L_1) = \mathbf{g}_3(0,0)\frac{\cos\left[\frac{2\pi(L_2 - x_2)}{\lambda_2}\right]}{C_2}, \tag{7.27}$$

$$\mathbf{g}_3(x_3, L_1) = \mathbf{g}_3(0,0)\frac{\cos\left[\frac{2\pi(L_3 - x_3)}{\lambda_3}\right]}{C_3}, \tag{7.28}$$

and

$$\mathbf{g}_3(x_1, L_1) = \frac{1}{C_1}\left[\mathbf{g}_3(0,0)\frac{\cos\left[\frac{2\pi(L_1 - x_1)}{\lambda_1}\right]}{C_1} - \frac{\lambda_1 \sin\left(\frac{2\pi x_1}{\lambda_1}\right)}{2\pi}\right]. \tag{7.29}$$

So finally,

$$\begin{aligned}\mathbf{g}_4(x_2, x_1) = {} & \frac{\cos\left[\frac{2\pi(L_2 - x_2)}{\lambda_2}\right]}{C_2}\frac{[\mathbf{g}_3(L_1, L_1)]^{-2}}{[\mathbf{g}_3(0,0)]^{-2}}\mathbf{g}_4(0_1, 0_1) \\ & \times\left[\frac{\cos\left[\frac{2\pi(L_1 - x_1)}{\lambda_1}\right]}{C_1} - [\mathbf{g}_3(0,0)]^{-1}\frac{\lambda_1 \sin\left(\frac{2\pi x_1}{\lambda_1}\right)}{2\pi}\right],\end{aligned} \tag{7.30}$$

$$\begin{aligned}\mathbf{g}_4(x_3, x_1) = {} & \frac{\cos\left[\frac{2\pi(L_3 - x_3)}{\lambda_3}\right]}{C_3}\frac{[\mathbf{g}_3(L_1, L_1)]^{-2}}{[\mathbf{g}_3(0,0)]^{-2}}\mathbf{g}_4(0_1, 0_1) \\ & \times\left[\frac{\cos\left[\frac{2\pi(L_1 - x_1)}{\lambda_1}\right]}{C_1} - [\mathbf{g}_3(0,0)]^{-1}\frac{\lambda_1 \sin\left(\frac{2\pi x_1}{\lambda_1}\right)}{2\pi}\right],\end{aligned} \tag{7.31}$$

and

$$\begin{aligned}\mathbf{g}_4(x_1, x_4) = {} & \left[\frac{\cos\left[\frac{2\pi(L_1 - x_1)}{\lambda_1}\right]}{C_1} - [\mathbf{g}_3(0,0)]^{-1}\frac{\lambda_1 \sin\left(\frac{2\pi x_1}{\lambda_1}\right)}{2\pi}\right] \\ & \times\frac{[\mathbf{g}_3(L_1, L_1)]^{-2}}{[\mathbf{g}_3(0,0)]^{-1}}\mathbf{g}_4(0_1, 0_1)\frac{\lambda_4 \cos\left[\frac{2\pi(L_4 - x_4)}{\lambda_4}\right]}{2\pi S_4}.\end{aligned} \tag{7.32}$$

These results provide the eigenfunctions within the four loops of this network, which are

$$E_2(x_2) = \frac{\cos\left[\frac{2\pi(L_2-x_2)}{\lambda_2}\right]}{C_2}, \tag{7.33}$$

$$E_3(x_3) = \frac{\cos\left[\frac{2\pi(L_3-x_3)}{\lambda_3}\right]}{C_3}, \tag{7.34}$$

$$E_1(x_1) = \frac{\cos\left[\frac{2\pi(L_1-x_1)}{\lambda_1}\right]}{C_1} - [\mathbf{g}_3(0,0)]^{-1}\frac{\lambda_1 \sin\left(\frac{2\pi x_1}{\lambda_1}\right)}{2\pi}, \tag{7.35}$$

and

$$E_4(x_4) = E_1(L_1)\frac{\cos\left[\frac{2\pi(L_4-x_4)}{\lambda_4}\right]}{C_4}, \tag{7.36}$$

where

$$E_1(L_1) = \frac{1}{C_1} - \frac{\lambda_1 S_1[\mathbf{g}_3(0,0)]^{-1}}{2\pi}. \tag{7.37}$$

These eigenfunctions satisfy the continuity conditions at the nodes 0 and 0_1. The sum of the outgoing derivatives at node 0 is

$$E_1'(0) + E_2'(0) + E_3'(0) = \frac{2\pi S_1}{\lambda_1 C_1} + \frac{2\pi S_2}{\lambda_2 C_2} + \frac{2\pi S_3}{\lambda_3 C_3} - [\mathbf{g}_3(0,0)]^{-1} = 0. \tag{7.38}$$

The sum of the outgoing derivatives of the eigenfunctions at node 0_1 is

$$E_4'(0_1) - E_1'(L_1) = E_1(L_1)\left[\frac{2\pi S_4}{\lambda_4 C_4} + [\mathbf{g}_3(L_1, L_1)]^{-1}\right] = 0. \tag{7.39}$$

In order to obtain this last result, one needs Eq. (1.23).

So the general conclusions obtained before for the star network are the same for this new network. It is possible now to finish this iteration demonstration.

7.4 The last iteration step

Suppose now that the general results we obtained above for the eigenfunctions hold for any finite network a. Let g_a be its response function. Create a new network by superposing, on any of the space points, called here 0_a of the previous network, one end of another finite loop b. Let g_b be the response function of this finite loop and g the response function of this new network.

As the a network is assumed to posses all the general eigenfunction properties, this means in particular that for any of its eigenfunctions E_a, not having a zero at the node 0_a, one has

$$[\mathbf{g}_a(0_a, 0_a)]^{-1} = \sum_{i=1}^{n_a} E'(0_a), \tag{7.40}$$

where n_a is the number of derivatives outgoing from the point 0_a.

The above defined superposition enables us also to obtain the equation providing the new eigenvalues

$$[\mathbf{g}(0_a, 0_a)]^{-1} = [\mathbf{g}_a(0_a, 0_a)]^{-1} + \frac{2\pi S_b}{\lambda_b C_b}. \tag{7.41}$$

The general equation of the interface response theory provides also, with $x_b \leq x_b'$,

$$\mathbf{g}(x_b, x_b') = \frac{\cos\left[\frac{2\pi(L_b - x_b')}{\lambda_b}\right]}{C_b}\left[\mathbf{g}_a(0_a, 0_a)\frac{\cos\left[\frac{2\pi(L_b - x_b)}{\lambda_b}\right]}{C_b} - \frac{\lambda_1 \sin\left(\frac{2\pi x_b}{\lambda_b}\right)}{2\pi}\right]. \tag{7.42}$$

This enables taking for the new eigenfunctions within the b loop

$$E(x_b) = \frac{\cos\left[\frac{2\pi(L_b - x_b)}{\lambda_b}\right]}{C_b}. \tag{7.43}$$

The partial derivative at the free end of the finite loop b is

$$E_b'(L_b) = 0. \tag{7.44}$$

The sum of all outgoing first derivatives at the node 0_a is then

$$E'(0_a) + \sum_{i=1}^{n_a} E'(0_a) = \frac{2\pi S_b}{\lambda_b C_b} + [\mathbf{g}(0_a, 0_a)]^{-1} = 0 \tag{7.45}$$

for the eigenfunctions.

So the general eigenfunction rules remain valid. If one wishes to check that they remain valid for all points of this final network, one may redo from this b end the demonstration done above.

7.5 Infinite networks

Start from a general finite network sketched in Fig. 7.3, to which three (for simplicity) semiinfinite loops are added at a node 0. In all these loops, the space positions are positive and start from 0 at this node point.

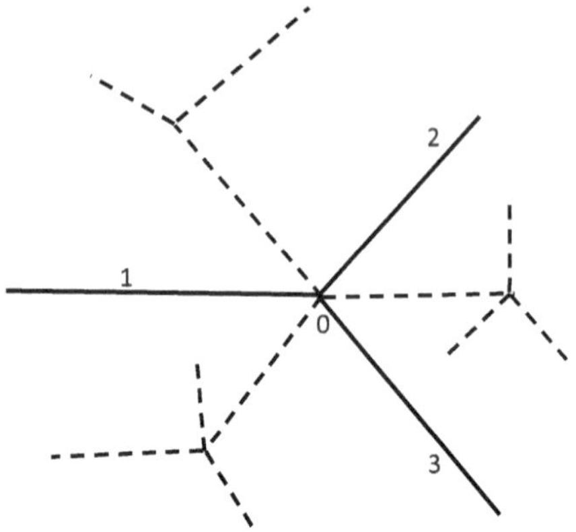

Figure 7.3 A general infinite network made out of a finite one (in dotted lines) to which three (for simplicity) semiinfinite loops (in full lines) are added at a node 0.

Let $g(0,0)$ be the response function at this node point, so by linear superposition

$$[g(0,0)]^{-1} = j\left(\frac{2\pi}{\lambda_1} + \frac{2\pi}{\lambda_2} + \frac{2\pi}{\lambda_3}\right) + \Delta, \tag{7.46}$$

where the $\lambda_1, \lambda_2, \lambda_3$ are the wavelengths in the corresponding semiinfinite loops and Δ is the sum of outgoing (from 0) derivatives of the initial finite network. Remember, from what was said above, that the eigenstates of the finite network are obtained from $\Delta = 0$.

A progressive wave $e^{-2\pi j x_1/\lambda_1}$ comes to the node 0. It creates reflected $re^{2\pi j x_1/\lambda_1}$, transmitted $te^{2\pi j x_2/\lambda_2}$, $te^{j2\pi j x_3/\lambda_3}$, and stationary waves, such as, for example, $t\cos[2\pi(x_i - L_i)/\lambda_i]$ in the initial finite network. The same value of the transmission coefficient t has to be in the transmitted waves in order to satisfy the electric field continuity at the node 0. The reflection r and transmission t coefficients can then be obtained from the following equation of the interface response theory:

$$\begin{aligned}\mathbf{u}(D) = {} & \mathbf{U}(D) - \mathbf{G}(D,0)[\mathbf{G}(0,0)]^{-1}\mathbf{U}(0) \\ & + \mathbf{G}(D,0)[\mathbf{G}(0,0)]^{-1}\mathbf{g}(0,0)[\mathbf{G}(0,0)]^{-1}\mathbf{U}(0),\end{aligned} \tag{7.47}$$

where $\mathbf{U}(D)$ and $\mathbf{u}(D)$ are the reference and final network eigenfunctions, D stands for the space position, and $\mathbf{G}(D, D)$ is the contribution of the semiinfinite chains to the response function of the reference system. It is constituted of disconnected truncated parts within each semiinfinite space of the corresponding bulk response functions, see Eq. (1.7).

The above equation enables obtaining

$$t = \frac{4\pi j \frac{1}{\lambda_1}}{2\pi j(\frac{1}{\lambda_1} + \frac{1}{\lambda_2} + \frac{1}{\lambda_3}) + \Delta} \tag{7.48}$$

and

$$r = \frac{2\pi j(\frac{1}{\lambda_1} - \frac{1}{\lambda_2} - \frac{1}{\lambda_3}) - \Delta}{2\pi j(\frac{1}{\lambda_1} + \frac{1}{\lambda_2} + \frac{1}{\lambda_3}) + \Delta}. \tag{7.49}$$

These results may be also obtained from the eigenfunction rules. The continuity at the node 0 of all the above defined eigenfunctions gives

$$r = t - 1. \tag{7.50}$$

The vanishing of the sum of the eigenfunction outgoing derivatives at this node provides

$$-2\pi j \frac{1}{\lambda_1}(1-r) + [2\pi j(\frac{1}{\lambda_2} + \frac{1}{\lambda_3}) + \Delta]t = 0. \tag{7.51}$$

These last two results yield the values of t and r, obtained before from the interface response approach.

This demonstrates that the general rules implying, at each network point, the continuity of the eigenfunctions and the vanishing of their outgoing derivatives hold as well for progressive and for standing waves.

Let us now check the conservation of the electromagnetic intensity at this node. This intensity I_i may be obtained in any dielectric loop i from

$$I_i = \frac{c_i \epsilon_i E_i^2}{2}, \tag{7.52}$$

where E_i is the amplitude of the electric field.

The reflected intensity in our case is

$$I_R = |r|^2 I_I, \tag{7.53}$$

where I_I is the incident intensity.

Let us now introduce the index n_i with $\omega n_1 = 2\pi c/\lambda_1$ and $\epsilon_i = n_i^2$. This enables writing

$$\frac{c_i \epsilon_i}{c_1 \epsilon_1} = \frac{n_i}{n_1}. \tag{7.54}$$

The transmitted intensities are then

$$I_{T2} = \frac{n_2}{n_1} |t|^2 I_I \tag{7.55}$$

and

$$I_{T3} = \frac{n_3}{n_1} |t|^2 I_I. \tag{7.56}$$

The sum of the reflected and transmitted intensities is then found to be

$$I_R + I_{T2} + I_{T3} = \frac{(n_1 - n_2 - n_3)^2 + 4n_1 n_2 + 4n_1 n_3 + (\frac{c\Delta}{\omega})^2}{(n_1 + n_2 + n_3)^2 + (\frac{c\Delta}{\omega})^2} I_I = I_I. \quad (7.57)$$

So in a network without any loss, there is conservation of the incident intensity at any point.

7.6 General rules

7.6.1 Rule 1

The eigenfunctions are continuous.

7.6.2 Rule 2

The eigenfunction derivatives produce source forces. The sum of these forces vanishes at each space point. The sum of all such surface forces vanishes.

7.6.3 Rule 3

For finite networks, the kernel of the response function is the sum of the outgoing source force terms. The zero value of this kernel determines all the discrete eigenvalues, but those for which the eigenfunction is zero at this space point. In the latter case, such eigenvalues are obtained from another network point.

For infinite networks, the kernel of the response function has a real part which is the sum of the outgoing source force terms of the finite parts of the network and of an imaginary part related to the semiinfinite and infinite parts of the network. The zero value of this kernel real part determines the position of peaks in the network density of states. The imaginary part of this kernel determines the width of these peaks.

7.7 Perspectives

In the above demonstration of the general eigenfunction rules, it was assumed that there is no damping and the eigenfunction wavelengths are much larger than the loop lateral dimensions and the node space dimensions. The investigations of such effects remain an open problem. It is expected that the damping effects should mostly enlarge the resonance frequency peaks. Taking into account the lateral loop dimensions should mostly shift the eigenstate frequency positions.

Let us stress also that the couplings between different waves was not considered above. As an example of such couplings, let us cite longitudinal with transverse elastic waves and polaritons.

The investigations of these effects are important, but more complicated and specific to each special case. However, the above general rules may remain valid.

References

[1] L. Dobrzyński, Interface response theory of continuous composite systems, Surf. Sci. Rep. 11 (1990) 164.

General wave perspectives

8

Leonard Dobrzyński[a], *Housni Al Wahsh*[b], *Abdellatif Akjouj*[a], *and El Houssaine El Boudouti*[c]
[a]Univ. Lille, CNRS, Centrale Lille, Yncréa ISEN, Univ. Polytechnique Hauts-de-France, UMR 8520 - IEMN, Physics Department, Lille, France, [b]Benha University, Faculty of Engineering, Cairo, Egypt, [c]University Mohammed I, Faculty of Sciences, Laboratory of Physics of Matter and Radiation, Oujda, Morocco

8.1 Introduction

Many other waves may be investigated with a second order differential equation, function of time and space position, similar to that of Maxwell used in this Photonics book. In many cases, these equations differ one from another only by some coefficients. Without trying to be exhaustive, a few other waves are cited, for which the results obtained in this book may be easily transposed.

Using the simple example of path states within one closed loop, it is possible to conjecture the existence of entangled linear photons. This example may be transposed to other kind of waves.

8.2 Phonons

Let us cite first the transverse elastic waves in an infinite string whose propagation equation is

$$\left(c_{44}\frac{\partial^2}{\partial x^2}-\rho\frac{\partial^2}{\partial t^2}\right)u_t(x,t)=0, \tag{8.1}$$

where x is the space position along the string, t the time, c_{44} the transverse stress coefficient, ρ the mass density by length unit, and $u_t(x,t)$ the transverse motion displacement.

All what is in this book may be easily transferred to these transverse elastic waves. The only things which have to be changed are as follows: the square of the speed of light (c^2) becomes c_{44}/ρ and the speed of light becomes the speed of sound. The continuity of the electric field E becomes the continuity of the displacement u_t, and the vanishing at each space point of the outgoing derivatives of the electric field E becomes the vanishing at each space point of the outgoing stresses $(c_{44}\partial u_t(x)/\partial x)$.

Photonics. https://doi.org/10.1016/B978-0-12-819388-4.00017-4

For longitudinal elastic waves, one needs just to replace c_{44} by c_{11} and the transverse speed of sound c_t by the longitudinal c_l.

8.3 Electrons

Free electron propagation along a wire may be investigated with the Schrödinger equation

$$\left(-\frac{\hbar^2}{2m^*}\frac{\partial^2}{\partial x^2} - j\hbar\frac{\partial}{\partial t}\right)\psi(x,t) = 0, \tag{8.2}$$

where $\hbar$ is the Planck constant divided by 2π, m^* is the free electron effective mass, and $\psi(x,t)$ the electron wave function. It is possible to transfer what's in this book from electromagnetic waves to electrons. The main points are the continuity at any point x of such an electronic network of the wave function $\psi(x)$ and the vanishing of the outgoing $\frac{\hbar^2}{2m^*}\frac{\partial}{\partial x}\psi(x)$, which when multiplied by the electron charge e are the outgoing electronic currents. This last result is in agreement with the Kirchhoff [1] node current rule.

8.4 Magnons

For long wavelength magnons, the transfer of the results of this book is similar to that for phonons and electrons: at each space point, one requires continuity of the spin deviations from equilibrium and vanishing of their outgoing derivatives multiplied by the appropriate factor existing in its source force term, must be fulfilled. In the long wavelength limit of the Heisenberg Hamiltonian, for example, this factor is proportional to the exchange constant J and to the spin value S within each material.

8.5 Entangled linear photons

An entangled system state is defined to be one whose state eigenfunction cannot be factorized as a product of its local reference eigenfunctions. In an entangled eigenfunction, a local reference constituent cannot be described without considering the other(s) local reference one(s). Entanglement is the cause of the correlation between spatially separated particles. Von Neuman and Wigner [2] discover bound in continuum states. Einstein, Podolsky, and Rosen [3] invoke incompleteness in quantum mechanical description of physical reality. Indeed, to our knowledge, the fact that some entangled particle states can be bound in continuum ones remains to be investigated. The main point about spatially separated entangled particles is their instantaneous correlation. This can be checked with the help of the response functions used in this book. Consider the entangled twin states of one closed loop [4], see Fig. 8.1.

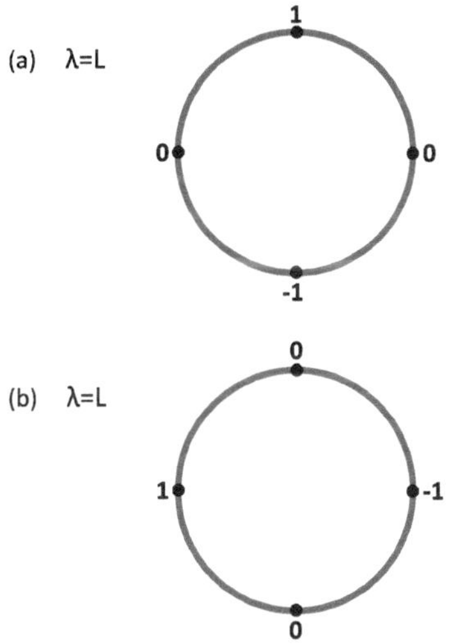

Figure 8.1 The two twin states of the closed loop of length L for $\lambda = L$.

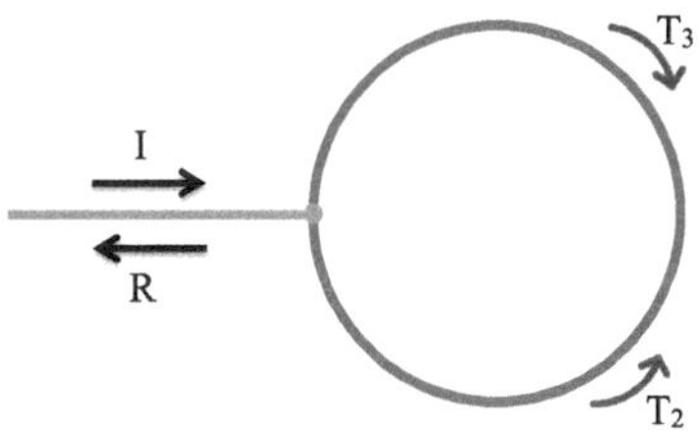

Figure 8.2 Possible activation of two entangled photons. I is for the incident photons, R the reflected ones. The incoming photons have the same probability to take the clockwise as the anticlockwise path along the closed loop.

Activate one of them like depicted in Fig. 8.2, taking due account of the right boundary conditions [5].

Active both of them by an adequate stimulus, for example $\delta(\lambda - L)$ applied at one single origin point $x = 0$ of the closed loop. This can be achieved provided the right boundary conditions [5] at the activation point remain fulfilled during the activation, for both eigenfunctions. Such activation generates the ($\lambda = L$) two times degenerate photons within the closed loop. This activation method is depicted by Fig. 8.3.

The structure is made out of an infinite open loop tangent at $x = 0$ to a closed loop of length L. I is the incident wave, R the reflected one, and T_1, T_2, and T_3 are the transmitted ones. Both infinite open loop and finite closed loop share the same twin states.

The incompleteness predicted by Einstein et al. may be due to the fact that the current quantum mechanical approaches to the entanglement questions use incomplete

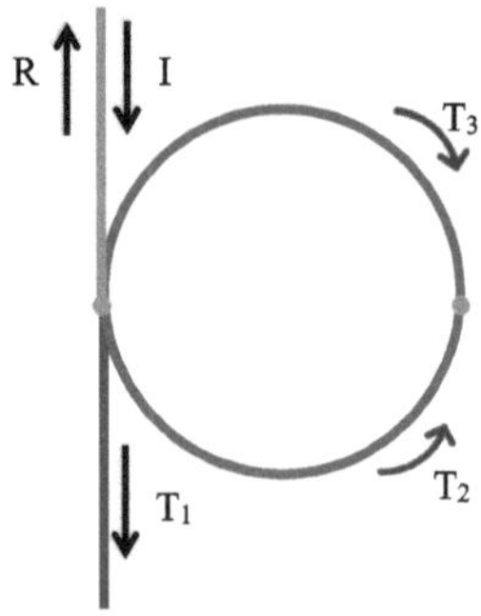

Figure 8.3 Possible activation of two times two entangled photons. The structure is made out of an infinite open loop tangent at $x = 0$ to a closed loop of length L. I is for the incident photons, R the reflected ones, and T_1, T_2, and T_3 for the transmitted ones.

reference systems. They do not check the state conservation number between reference and final state. In the simple photon entangled system sketched here above this verification is done and shows that Von Neuman and Wigner bound in continuum state play a crucial role.

8.6 General perspectives

This book defines path states as a subset of all eigenstates of a given system.

The determination of these path states can be done just by inspection of the final system with the help of a confined state theorem and general eigenfunction rules, both defined in this book.

The robustness of these states is even reinforced when they are degenerate. Their degeneracy gives them healing possibilities as they may go around defects, when the defects do not affect all the degenerate states.

These states can exist in all possible space localizations of the final system: bulk, surface, interface, wedge, edge, adsorbed layers, etc.

Any eigenstate can be activated within its subdomain without any subsystem deformation, see the activation state theorem in Chapter 3 of this Photonics book. This paves the way to the use of the degenerate path states for many applications.

Any elaborated system has also nonrobust states. By exciting very large amplitude eigenfunctions, it is possible to induce nonlinear effects and irreversible mutations of any structure. It is important also to learn more about these properties.

Nevertheless, it is possible to foresee the construction, based on the robust zeros, of new and more performing networks and communication devices. These path states may pave the way for the future of information communication technology and contribute to many other sciences.

Acknowledgments

One of us (L.D.) acknowledge Ludwik Dobrzynski, emeritus research professor at CNRS, CERN and Palaiseau Polytechnic school, for stimulating scientific exchanges about particles. He benefits also from the critical comments of Jean-Claude Garreau, research professor at CNRS and Lille University, about the herein proposed research path on linear entangled photons.

References

[1] G. Kirchhoff, Über die Auflösung der Gleichungen auf welche man bei der Untersuchung der linearen Vertheilung galvanischer Ströme geführt wird, Ann. Phys. Chem. 72 (1847) 497–508.

[2] J. Von Neuman, E. Wigner, Über merkwürdige diskrete eigenwerte, Phys. Z. 30 (1929) 465.

[3] A. Einstein, B. Podolsky, N. Rosen, Can quantum-mechanical description of physical reality be considered complete?, Phys. Rev. 47 (1935) 777.

[4] H. Al Wahsh, L. Dobrzyński, A. Akjouj, E.H. El Boudouti, Closed loop examples, in: L. Dobrzyński (Ed.), Photonics, in: Interface Transmission Tutorial Book Series, Elsevier, 2021, Part I, Chapter 5.

[5] L. Dobrzyński, H. Al Wahsh, A. Akjouj, E.H. El Boudouti, Eigenfunction rules, in: L. Dobrzyński (Ed.), Photonics, in: Interface Transmission Tutorial Book Series, Elsevier, 2021, Part I, Chapter 7.

Part Two

Photonic circuits

Edited by El Houssaine El Boudouti

Electromagnetic induced transparency, induced absorption, and Fano resonances in photonic circuits

9

El Houssaine El Boudouti[a], Abdellatif Akjouj[b], Bahram Djafari-Rouhani[b], Abdelkrim Talbi[c], and Leonard Dobrzyński[b]
[a]Laboratoire de Physique de la Matière et de Rayonnement, Département de Physique, Faculté des Sciences, Université Mohamed I, Oujda, Morocco, [b]Institute of Electronic, Microelectronics and Nanotechnology (IEMN), UMR CNRS 8520, Department of Physics, Faculty of Sciences and Technologies, Lille University, Villeneuve d'Ascq Cedex, France, [c]Institut d'Electronique, de Microélectronique et de Nanotechnologie (IEMN), UMR CNRS 8520, Ecole Centrale de Lille, Villeneuve d'Ascq, France

9.1 Introduction

Often the scattering of waves in composite materials involves propagation along different paths and, as a consequence, results in interference phenomena, where constructive interference corresponds to resonant enhancement and destructive interference to resonant suppression of the transmission. This kind of resonance followed by antiresonance is called Fano resonance [1] and is characterized by an asymmetric line profile. The asymmetry originates from a close coexistence of resonant transmission and resonant reflection, and can be reduced to the interaction of a discrete localized mode with a continuum of propagation modes [2]. Fano was the first to describe such resonances when he was attracted by unusual sharp peaks in the absorption spectra of noble gases observed by Beutler [3]. The asymmetric line shapes have been reported in the electronic transport in mesoscopic systems using the Aharonov–Bohm systems [4–9]. The subject of these studies was mainly to use these interferometric systems to show the conditions for the existence and the collapse of Fano resonances as a function of the applied current voltage and magnetic flux. These studies are also related to the investigation of the electronic states of quantum dots [4–9], as well as to the understanding [10–13] of the transmission phase jumps by π between two adjacent resonances in relation with the experiments of Yacoby et al. [14] and Schuster et al. [15]. The analogy between scattering properties of electrons and other elementary excitations suggests that such a feature can also appear for other types of waves such as phonons, photons, magnons, etc. [16].

When the resonance falls between two antiresonances, Fano resonance behaves like an electromagnetic induced transparency (EIT) resonance. EIT [17] is the phenomenon in which a sharp transparent window associated with steep dispersion is

Photonics. https://doi.org/10.1016/B978-0-12-819388-4.00019-8

induced into opaque atomic media. It is due to the destructive quantum interference between different excitation pathways of the excited states [18], leading to a dip in the absorption within a frequency window. These systems have shown potential applications for slow light and optical data storage [17–21]. Several works have demonstrated that Fano and EIT-like behaviors are not restricted to quantum systems and can be extended to classical systems, such as coupled microresonators [22–30], photonic crystal waveguides coupled to cavities [31–38], acoustic slender tube waveguides [39–43], periodic arrays of square rods [44], plasmonic nanostructures, and metamaterials [45–51]. With regard to optical resonances, microring, microdisk, and microsphere resonators [22–30] coupled to a straight waveguide have shown EIT and Fano resonances with high quality factors. The resonances can be tuned by changing the diameters of the resonators, their separation (coupling) to the straight waveguide, and the distance between the resonators. It is worth mentioning that the EIT resonance can be realized by two main mechanisms: the first uses two coupled resonators with high (dark resonator) and low (radiative resonator) Q factors resonating at the same frequency [52,53], while the second suggests two radiative coupled resonators with closely-spaced frequencies decaying to the same continuum [54–56].

In addition to EIT resonances, one can also change the phase of the interaction between the resonators to realize constructive interference and, hence, achieve an enhancement of absorbance, the so-called electromagnetic induced absorption (EIA) in atomic physics. A few works on classical analogue of EIA have been reported recently in planar metamaterials [57] and plasmonic materials [58]. The transition between EIT and EIA can be obtained in microwave metamaterials by adjusting the dissipative loss of the dark and bright resonators or the coupling strength between them [57]. While the EIT phenomenon gives rise to enhanced transmission and a dip in the absorption, EIA phenomenon shows a narrow peak of enhanced absorption larger than the background absorption of the radiative element [57,58].

Besides Fano, EIT, and EIA resonances, bound in continuum (BIC) states or trapped modes have found an increasing interest in recent years [59,60]. These states can be found when Fano and EIT resonances collapse in the transmission and reflection spectra. Therefore, they remain confined in some parts of the system even though they coexist with a continuous spectral range of radiating waves that can transport energy away. This is a counterintuitive idea which was originally proposed several decades before in quantum mechanics [60], but found a high amount of interest in the recent literature [59]. These states may exhibit potential applications as they are characterized by a high quality factor when they are revealed by a small amount of perturbation in the structure. Several types of perturbations have been discussed in the literature to reveal such hidden modes [59]. Although these states were first proposed in quantum mechanics, it was demonstrated that such modes without leakage can be found also in classical systems as discrete modes in the band structure of a periodic system, or as a consequence of the transformation of Fano and EIT resonances into hidden resonances with practically infinite lifetimes when their line-widths are reduced to zero [59,61–73]. In this chapter, we show several examples of how EIT and Fano resonances collapse into a BIC state, and conversely how a BIC state can reveal itself in the shape of Fano or EIT resonances when introducing a small detuning in the geometrical parameters.

It is worth noting that one-dimensional (1D) photonic crystals made of coaxial cables have shown several interesting properties like in their 2D and 3D counterpart systems. Among these properties, the existence of band gaps with or without defect modes in periodic and quasiperiodic structures [74–82], add/drop filters [83], slow and fast light, the so-called superluminal and subluminal, in periodic structures [79–81], and Mach–Zehnder-like interferometers [84,85]. Also, recent works [86–89] have shown that several properties of quantum graphs could be successfully simulated by microwave networks made of coaxial cables. These structures are attractive since their production is more feasible at any wavelength scale and they require only simple analytical and numerical calculations.

In this chapter, we consider two different simple structures that may support BIC, Fano, EIT, and EIA phenomena: (i) U-like structure made of two stubs of lengths d_1 and d_2 placed at two different positions and separated by a segment of length d_3 [90] (Fig. 9.1). We show both theoretically and experimentally the possibility of existence of BIC, Fano, and EIT resonances in the transmission spectra of such a photonic circuit made of coaxial waveguides in the radio-frequency domain; and (ii) cross-like structure constituted of two stubs of lengths d_1 and d_2 placed at the same position [91] (Fig. 9.10). We give the theoretical and experimental results of the transmission, reflection, and absorption coefficients and discuss the conditions for the existence of the EIA resonance, showing the effect of loss and detuning $\delta = d_2 - d_1$ on such resonances. In particular, we shall give closed-form expressions (Taylor expansion) of the transmission, reflection, and absorption around Fano, EIT, and EIA resonances. These will enable us to deduce implicitly the position, width, and Fano shape of all these resonances [92]. Also, we give for both structures a comparison between the phase of the determinant of the scattering matrix, the so-called Friedel phase, and the phase of the transmission amplitude. Such a comparison has been only studied theoretically [10,11] in mesoscopic systems because of the difficulties in the measurement of the reflection coefficient. The possible realization and performance of such a system in the optical domain is also discussed.

9.2 Method of theoretical calculation of U-structure

Our theoretical analysis is performed with the help of the interface response theory [93] of continuous media, which allows the calculation of the Green's function of any composite material and then total and local densities of states, as well as transmission and reflection coefficients. In what follows, we give first the expressions of the Green function elements for the building blocks (finite and semiinfinite wires), as well as of the U-structure made of two separated stubs of lengths d_1 and d_2 separated by a segment of length d_3 (Fig. 9.1).

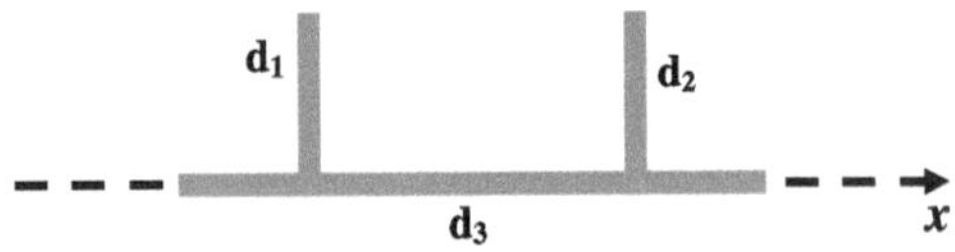

Figure 9.1 Schematic illustration of the one-dimensional electromagnetic waveguide of length d_3 with dangling resonators on both sides. The whole structure is inserted between two semiinfinite waveguides. The lengths of the dangling resonators are d_1 and d_2.

9.2.1 Inverse surface Green's functions of the elementary constituents

We consider an infinite homogeneous isotropic dielectric wire i characterized by its characteristic impedance Z_i and the relative permittivity ε_i. An electromagnetic field is propagating through the cable in the dielectric-filled space between the center conductor and shield. In the limit where the widths of the wires are much smaller than their lengths, the telegrapher's equation describing microwave networks, the so-called TEM waves, is given by

$$\left(\frac{\partial^2}{\partial x^2} + \omega^2 \varepsilon_i / c^2\right) E(x) = 0, \tag{9.1}$$

where c is the speed of light and ω is the angular frequency of the wave. Then, the corresponding Green's function between two points x and x' of this wire is defined by

$$\left(\frac{\partial^2}{\partial x^2} + \omega^2 \varepsilon_i / c^2\right) G(x, x') = \delta(x - x'), \tag{9.2}$$

whose solution is given by

$$G_i(x, x') = -\frac{e^{j\alpha_i |x - x'|}}{2F_i} \tag{9.3}$$

where $F_i = -j\omega / Z_i$, $\alpha_i = \omega\sqrt{\varepsilon_i}/c$, and $j = \sqrt{-1}$; see [16]. Let us recall that $\sqrt{\varepsilon_i} = n_i$ defines the index of refraction of the coaxial cable. Its inverse, called the velocity factor, is given by $V_F = \dfrac{1}{\sqrt{L_0 C}}$, where L_0 and C are the inductance and capacitance per unit length of the cable, respectively [94]. Also L_0 and C are frequency independent.

Before addressing the problem of the U-structure presented in this chapter (see Fig. 9.1), it is helpful to know the surface elements of its elementary constituents, namely, the Green function of a finite wire of length d_i, $i = 1, 2, 3$, and of a semiinfinite wire. The boundary conditions at the ends of all the wires are $H = 0$ (vanishing magnetic field). The wire of length d_3 is bounded by two surfaces located at $x = 0$ and $x = d_3$ (Fig. 9.1). These surface elements can be written in the form of a (2×2) matrix $g_3(M, M)$, within the interface space $M = \{0, +d_3\}$. The inverse of this matrix takes the following form [16]:

$$g_3^{-1}(M,M) = \begin{pmatrix} -\frac{\omega C_3}{Z_3 S_3} & \frac{\omega}{Z_3 S_3} \\ \frac{\omega}{Z_3 S_3} & -\frac{\omega C_3}{Z_3 S_3} \end{pmatrix}. \tag{9.4}$$

The inverse of the surface Green's functions of the dangling resonators grafted at the sites $\{0\}$ and $\{d_3\}$ is given by $g_1^{-1}(0,0) = -\omega S_1/Z_1 C_1$ and $g_2^{-1}(d_3,d_3) = -\omega S_2/Z_2 C_2$, where $C_i = \cos(\frac{\omega d_i \sqrt{\varepsilon}}{c})$, $S_i = \sin(\frac{\omega d_i \sqrt{\varepsilon}}{c})$, $i = 1, 2, 3$. The inverse of the surface Green's functions of the two semiinfinite waveguides surrounding the whole structure is given by $g_s^{-1}(0,0) = g_s^{-1}(d_2,d_2) = j\omega/Z_s$, where Z_s is the characteristic impedance of the wire s. In what follows, we suppose that all the wires are standard coaxial cables with the same characteristic impedances and permittivities (i.e., $Z_1 = Z_2 = Z_3 = Z_s = Z \approx 50\ \Omega$ and $\varepsilon_1 = \varepsilon_2 = \varepsilon_3 = \varepsilon_s = \varepsilon = 2.3$). We report on results of calculated transmission coefficients and phase or phase time as a function of frequency. Using the Green's function method [16], the expression giving the inverse of the Green's function of the whole system given in Fig. 9.1 can be obtained from a linear superposition of the above inverse Green's functions of the constituents, namely

$$g^{-1}(M,M) = \frac{\omega}{Z} \begin{pmatrix} \frac{C_3}{S_3} - \frac{S_1}{C_1} - j & \frac{1}{S_3} \\ \frac{1}{S_3} & \frac{C_3}{S_3} - \frac{S_2}{C_2} - j \end{pmatrix}, \tag{9.5}$$

where $C_i = \cos(\alpha d_i)$, $S_i = \sin(\alpha d_i)$, and $\alpha = \omega\sqrt{\varepsilon}/c$ $(i = 1, 2, 3)$.

9.2.2 *Transmission coefficient*

Let us consider an incident wave $U(x) = e^{-j\alpha x}$ launched in the left semiinfinite waveguide (Fig. 9.1). With the help of Eqs. (9.3) and (9.5), one easily finds the transmission wave in the right semiinfinite waveguide (Fig. 9.1), namely [16] $t = (2j\omega/Z)g(0,d_3)$, or equivalently

$$t = \frac{2C_1C_2}{\chi_1 - j\chi_2}, \tag{9.6}$$

where

$$\chi_1 = 2C_1C_3C_2 - S_3S, \tag{9.7}$$

$$\chi_2 = 2C_1C_2S_3 + C_3S - S_1S_3S_2 \tag{9.8}$$

and $S = \sin(\alpha d)$ (with $d = d_1 + d_2$). In the same way, the reflection coefficient is given by $r = -1 + (2j\omega/Z)g(0,0)$, or equivalently

$$r = \frac{\chi_1' + j\chi_2'}{\chi_1 - j\chi_2}, \tag{9.9}$$

where

$$\chi_1' = S_3(S_1C_2 - S_2C_1) \tag{9.10}$$

and

$$\chi_2' = C_3S - S_1S_3S_2. \tag{9.11}$$

From the expression of t (Eq. (9.6)), one can deduce the transmission coefficient

$$T = \frac{4C_1^2C_2^2}{\chi_1^2 + \chi_2^2}, \tag{9.12}$$

as well as the phase

$$\varphi = \arctan(\chi_2/\chi_1) + \pi\Theta[C_1C_2] \tag{9.13}$$

where Θ means the Heaviside function. From Eq. (9.12), one can notice that the transmission zeros are induced by the side branches (i.e., $C_1 = 0$ or $C_2 = 0$). These two equations give the eigenmodes of the two stubs with $H = 0$ boundary conditions on both sides. When the expression C_1C_2 changes sign at some frequencies denoted by ω_n, then the phase (Eq. (9.13)) exhibits a jump of π. Another interesting quantity is the first derivative of φ with respect to the pulsation ω, which is related to the delay time taken by the photons to traverse the structure. This quantity, called delay time, is defined by [95]

$$\tau_\varphi = \frac{d\varphi}{d\omega} \tag{9.14}$$

and can be written as

$$\tau_\varphi = \frac{d}{d\omega}\arctan(\chi_2/\chi_1) + \pi\sum_n \text{sgn}\left[\frac{d}{d\omega}(C_1C_2)_{\omega=\omega_n}\right]\delta(\omega - \omega_n) \tag{9.15}$$

where sgn means the sign function. Furthermore, the density of states (DOS) of the present composite system from which we have subtracted the DOS of the semiinfinite cable is given by [95]

$$\Delta n(\omega) = \frac{1}{\pi}\frac{d}{d\omega}\left[\arctan(\chi_2/\chi_1)\right]. \tag{9.16}$$

Because of the second term in the right-hand side of Eq. (9.15), one can deduce that $\tau_\varphi \neq \pi\Delta n(\omega)$ as τ_φ (Eq. (9.14)) may exhibit δ functions at the transmission zeros that do not exist in the variation of the DOS (Eq. (9.16)). However, if the system does not exhibit transmission zeros, then $\Theta[C_1C_2] = 0$ or 1 and $\tau_\varphi = \pi\Delta n(\omega)$.

From the expression of r (Eq. (9.9)), one can deduce the reflection coefficient

$$R = \frac{\chi_1'^2 + \chi_2'^2}{\chi_1^2 + \chi_2^2}. \tag{9.17}$$

From Eqs. (9.12) and (9.17), one can show easily the conservation energy, namely

$$R + T = 1. \tag{9.18}$$

9.2.3 Bound in continuum modes

It is well known that the poles of the transmission and reflection coefficients (Eqs. (9.6) and (9.9)), corresponding also to the poles of the Green function, giving the eigenstates of the system, namely

$$\chi_1 - j\chi_2 = 0. \tag{9.19}$$

This latter equation is a complex quantity. Its real part gives the position of the resonance in the transmission coefficient, whereas its imaginary part is related to the width of the resonance. In general, it is not easy to simultaneously cancel real and imaginary parts of Eq. (9.19) at the same frequency. However, when this happens, it will correspond to a bound state falling in a continuum of states. Indeed, if the lengths d_1, d_2, and d_3 are chosen commensurate, then Eq. (9.19) can be factorized by a real term whose zeros give the BIC states. The factorization can take place if

$$d_1 = \frac{(2m_1 + 1)d}{2(m_1 + m_2 + 1)}, \tag{9.20}$$

$$d_2 = \frac{(2m_2 + 1)d}{2(m_1 + m_2 + 1)}, \tag{9.21}$$

and

$$d_3 = \frac{2m_3 d}{2(m_1 + m_2 + 1)}, \tag{9.22}$$

where m_1, m_2, and m_3 are integers. Then we obtain $C_1 = C_0(-1)^{m_1}U_{2m_1}(S_0)$, $C_2 = C_0(-1)^{m_2}U_{2m_2}(S_0)$, and $S = C_0(-1)^{m_3-1}U_{2m_3-1}(S_0)$ where we set $C_0 = \cos(\frac{\alpha d}{2(m_1+m_2+1)})$, $S_0 = \sin(\frac{\alpha d}{2(m_1+m_2+1)})$. Also $U_m(S_0)$ denotes Chebyshev polynomial of the second kind of degree m. Therefore, Eq. (9.19) can be factored by the term C_0. Then, if $C_0 = 0$, one obtains the frequencies of the BIC states given by

$$\alpha d = (2n + 1)(m_1 + m_2 + 1)\pi, \tag{9.23}$$

where n is an integer.

It should be pointed out that the validity of our results is subject to the requirement that the cross-section of the waveguide is negligible compared to their length and to the propagation wavelength. The assumption of mono-mode propagation is then satisfied.

9.3 EIT-like resonance in U-structure

9.3.1 Theoretical results

Eq. (9.12) clearly shows that the transmission zeros are due only to the dangling resonators, these occur when $C_1 = 0$ or $C_2 = 0$ (i.e., at $\omega d_1\sqrt{\varepsilon}/c = (2m_1 + 1)\pi/2$ or $\omega d_2\sqrt{\varepsilon}/c = (2m_2 + 1)\pi/2$). Also, one can expect the resonance (i.e., transmission unity) induced by the segment of length d_3 around $S_3 = 0$ (i.e., at $\omega d_3\sqrt{\varepsilon}/c = m\pi$) as it can be easily shown from Eqs. (9.6)–(9.12) (m, m_1, and m_2 are positive integers). Therefore, in order to show the possibility of existence of symmetric Fano resonance which behaves like EIT-resonance, i.e., a resonance squeezed between two transmission zeros, one choice would be to take d_1 and d_2 slightly different from $0.5d_3$. Indeed, at $\Omega = \omega d_3\sqrt{\varepsilon}/c = m\pi$, $m = 1, 2, \ldots$, the expression of the transmission function (Eq. (9.6)) becomes

$$t = \pm\frac{2C_1C_2}{2C_1C_2 + j\sin[\alpha(d_1 + d_2)]}. \tag{9.24}$$

In particular, if $\alpha(d_1 + d_2) = m\pi$, $\alpha d_1 \neq (m_1 + 0.5)\pi$, and $\alpha d_2 \neq (m_2 + 0.5)\pi$, one obtains a resonance that reaches unity (i.e., $T = 1$). An example corresponding to this situation is given in Fig. 9.2(a) where $d_1 = 0.46d_3$ and $d_2 = 0.54d_3$ (with $d_1 + d_2 = d_3$). All the curves are given with the dimensionless frequency Ω/π in order to discuss the Fano and EIT resonances around $\Omega/\pi = 1$. One can notice that the resonance at $\Omega = \pi$ is squeezed between two zeros (indicated by solid circles on the abscissa of Fig. 9.2(a)) induced by the dangling resonators as it is also illustrated in the plot describing the variation of the phase (Fig. 9.2(c)). The width of this resonance increases as far as d_1 and d_2 deviate from $0.5d_3$ (see below). In the particular case where $\alpha d_1 = (m_1 + 0.5)\pi$ and $\alpha d_2 = (m_2 + 0.5)\pi$, the numerator and denominator of t (Eq. (9.24)) vanish altogether. In this case, the resonance and the two zeros induced by the resonators fall at the same position, then the resonance collapses, giving rise to a BIC state (Eq. (9.23)) with $m_1 = m_2 = 0$, $m_3 = 1$, and $n = 0$. Then, the transmission coefficient vanishes and the phase drops by π as will be shown in Section 9.4.

The resonance in Fig. 9.2(a) shows the same characteristics as a Fano resonance, but with two zeros of transmission around the resonance instead of one as is usually the case [1,2]. Such resonances behave like EIT resonances [17]. Indeed, we can obtain an approximate analytical expression for the transmission function (Eq. (9.6)) in the vicinity of the resonance. A Taylor expansion around $\Omega = \pi$ (i.e., $\Omega = \pi + \varepsilon$ with $\varepsilon/\pi \ll 1$) enables us to obtain

$$t = -\frac{\zeta\zeta'}{\varepsilon^2/2 + \zeta\zeta' + j\varepsilon(1 - \zeta\zeta'/2)}, \tag{9.25}$$

where $\zeta = \Delta + \varepsilon(\frac{1}{2} + \Delta)$, $\zeta' = -\Delta + \varepsilon(\frac{1}{2} - \Delta)$ and Δ is the detuning of d_1 and d_2 from $0.5d_3$ (i.e., $\Delta = \pi(0.5 - d_1/d_3)$).

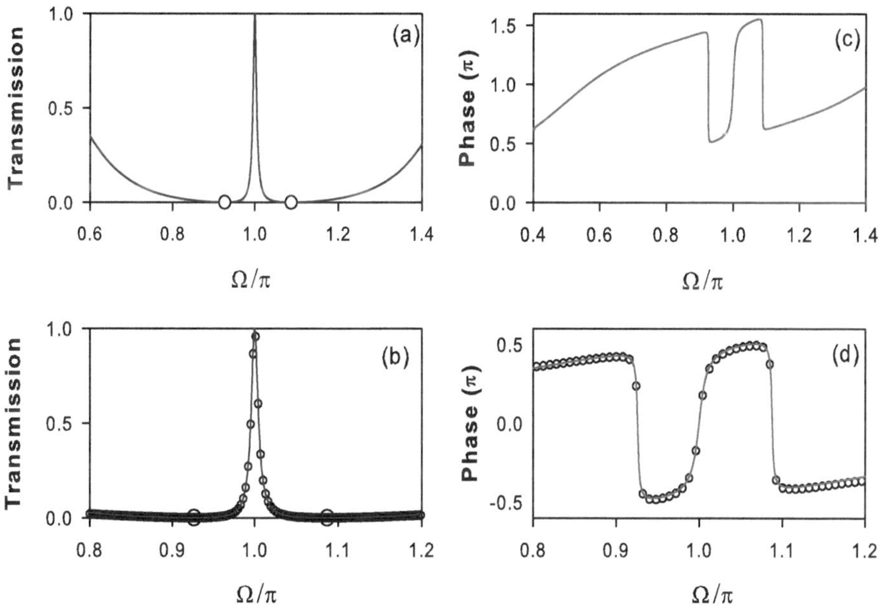

Figure 9.2 (a) Theoretical (solid line) variations of the transmission coefficient versus the dimensionless frequency Ω for the structure depicted in Fig. 9.1. The lengths of the resonators are taken such that $d_1 = 0.46d_3$ and $d_2 = 0.54d_3$. White circles on the abscissa indicate the positions of the transmission zeros induced by the dangling resonators on both sides of the resonance. (c) Same as in (a) but for the variation of the phase. (b) and (d) compare the approximate results obtained by Taylor expansion (open circles) around the resonance with the exact calculations (solid lines).

From Eq. (9.25), one can show that the transmission coefficient T can be written (following the symmetric Fano lineshape or EIT-like resonance) in the form [39]

$$T = A\frac{(\varepsilon + q_1\Gamma)^2(\varepsilon - q_2\Gamma)^2}{\varepsilon^2 + \Gamma^2}, \tag{9.26}$$

where $A = \dfrac{(\frac{1}{4} - \frac{\Delta^2}{\pi^2})^2}{(1 + \frac{\Delta^2}{2})^2}$. Here $\Gamma = \dfrac{\Delta^2}{1 + \Delta^2/2}$ characterizes the width of the resonance falling at $\varepsilon = 0$ (i.e., $\Omega = \pi$), $q_1 = \dfrac{1 + \frac{\Delta^2}{2}}{\Delta(\frac{1}{2} + \frac{\Delta}{\pi})}$ and $q_2 = \dfrac{1 + \frac{\Delta^2}{2}}{\Delta(\frac{1}{2} - \frac{\Delta}{\pi})}$ are the coupling parameters; they give qualitatively the strength of the interference between the bound states and the propagating continuum states [1,2]. One can notice that when increasing Δ, Γ increases, and q_1 and q_2 decrease.

The results of the approximate expression (Eq. (9.26)) are shown in Fig. 9.2(b) by open circles. These results are in accordance with the exact ones (solid lines) and clearly show that the resonance is a Fano symmetric or EIT-like resonance with $q_1 = 14.85$, $q_2 = 17.43$, and width $2\Gamma = 0.031\pi$. As a matter of comparison, in periodic photonic crystals, a perturbation is often introduced to the system in order to

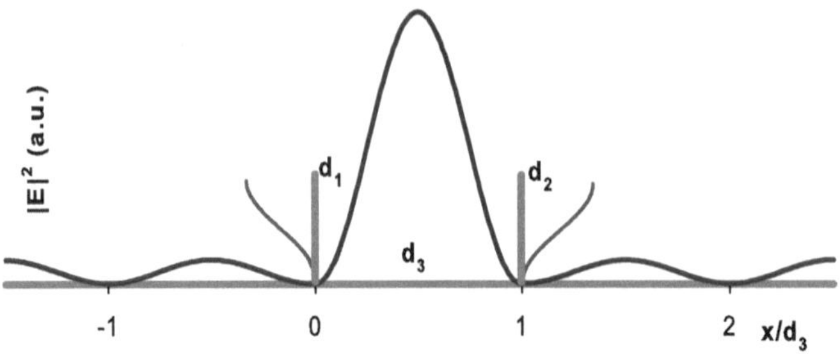

Figure 9.3 The square modulus of the electric field $| E |^2$ (in arbitrary units) as a function of the space positions along the horizontal waveguide x/d_3 (red curve, dark gray in print version) and vertical waveguide y/d_3 (blue curves, mid gray in print version), respectively, for the EIT resonance in Fig. 9.2(a).

create the resonance state in a bandgap [77–79]. However, the above calculation shows that, without introducing any perturbation in the structure, one can find a well-defined symmetric Fano resonance with a width 2Γ and coupling parameters q_1 and q_2 that can be adjusted by tailoring the lengths of the resonators (i.e., Δ, see below). Eq. (9.25) enables us also to deduce an approximate expression for the phase as

$$\varphi = \pi\Theta(\zeta) + \pi\Theta(\zeta') - \arctan\left[\varepsilon(1-\zeta\zeta'/2)/(\varepsilon^2/2+\zeta\zeta')\right]. \tag{9.27}$$

This function is plotted by open circles in Fig. 9.2(d) and clearly shows two abrupt phase drops of π at $\zeta = 0$ and $\zeta' = 0$ (i.e., $\varepsilon_1 = q_1\Gamma$ and $\varepsilon_2 = q_2\Gamma$) in accordance with the exact results (solid line).

The Green's function approach enables one also to deduce the spatial localization of square modulus of the electric field $| E |^2$. The details of this calculation are given in [16]. An analysis of $| E |^2$ as a function of the space position (Fig. 9.3) clearly shows that the EIT resonance in Fig. 9.2(a) is confined both inside the horizontal cable and stubs. In particular, the electric field is maximal in the middle of the finite horizontal waveguide of length d_2 and vanishes at its extremities, while it is maximal at the end of the stubs as the boundary conditions at these points are $H = 0$. Therefore, these resonances could be classified as local resonances.

9.3.2 Experimental results

In what follows, we give an experimental evidence of the analytical and numerical results in Fig. 9.2. The coaxial cable circuit studied consists of segments of RG-58/U coaxial cable of different lengths with characteristic impedance $Z = 50\ \Omega$. The cables were filled with polyethylene ($\varepsilon = 2.3$), corresponding to a nominal propagation speed of $0.66c$. The segments of coaxial cables were connected by standard BNC T connectors. The upper ends of the stubs in Fig. 9.3 are open circuit (i.e., $H = 0$ boundary condition). The scattering matrix S_n of the 1D comb-like photonic crystal was measured in the frequency range of 1–200 MHz with a ZVA8 Rohde–Schwarz 4 ports vector network analyzer. The experimental results include the amplitude and the phase of the transmission (S_{21}) and reflection (S_{11}) coefficients. The attenuation inside the coaxial cables was simulated by introducing a complex dielectric constant ε ($\varepsilon = \varepsilon' - j\varepsilon''$). The attenuation coefficient α'' can be expressed as $\alpha'' = \varepsilon''\omega/c$. On the other hand, the attenuation specification data supplied by the manufacturer of the coaxial cables in the frequency range of 10 to 200 MHz can be approximately fitted with the expression $\ln(\alpha'') = \gamma + \delta \ln(\omega)$, where γ and δ are two constants. From this fitting procedure, a useful expression for ε'' as a function of frequency can be obtained under the form $\varepsilon'' = 0.017 f^{-0.5}$ where the frequency f is expressed in hertz.

Fig. 9.4(a) gives the theoretical (solid line) and experimental (open circles) variations of the transmission coefficient versus the frequency (in MHz) for the structure depicted in Fig. 9.2 (the dimensionless frequency $\Omega = \omega d_3\sqrt{\varepsilon}/c$ is indicated at the top of the figure). The lengths of the stubs are chosen slightly different from those in Fig. 9.2 (i.e., $d_1 = 0.4$ m, $d_3 = 0.6$ m, and $d_3 = 1$ m). The experimental results (open circles) are very well fitted by the 1D model (solid line) using the Green's function method. One can notice that the dimensionless frequency domain $0 < \Omega < 2\pi$ in Fig. 9.4 corresponds to the radio-frequency range f such that $0 < f < 200$ MHz. The attenuation inside the cables induces transmission depletion in Fig. 9.4(a), especially at high frequencies, that is why the transmission does not reach unity.

One can notice that the resonance in Fig. 9.4(a) is well squeezed between two transmission zeros as it is illustrated in the phase (Fig. 9.4(b)), and the delay time (Fig. 9.4(c)). The EIT resonance is very well confined inside the stubs and the segment lying between them (see Fig. 9.3) giving rise to a large delay time in Fig. 9.4(c). An interesting result that can be deduced from the delay time τ (Fig. 9.4(c)) is the group velocity v_g defined as the first derivative of the angular frequency ω versus the wave vector $k = \omega n(\omega)/c$ where $n(\omega)$ is the effective refractive index. Indeed, the group velocity of a pulse is the velocity at which the peak of its envelope propagates and is related to $n(\omega)$ by [96]

$$v_g = \frac{c}{n(\omega) + \omega \dfrac{dn}{d\omega}}. \tag{9.28}$$

By regarding the system in Fig. 9.1 as a 1D metamaterial, the refractive index $n(\omega)$ is related to the phase φ of the transmitted wave by the following relation:

$$n(\omega) = \frac{\varphi c}{\omega L}, \tag{9.29}$$

where L is the total effective length of the finite structure (here $L \simeq 1.5d_3$). Substituting Eq. (9.29) into Eq. (9.28), the group velocity as a function of the effective length L and delay time $\tau = d\varphi/d\omega$ can be obtained as

$$v_g = L/\tau. \tag{9.30}$$

Eqs. (9.28), (9.29), and (9.30) clearly show that in frequency regions of normal dispersion (i.e., $\frac{dn}{d\omega} > 0$), v_g can lead to a very small group velocity $v_g \ll c$ (*slow light* or subluminal group velocity), whereas in the case of anomalous dispersion (i.e.,

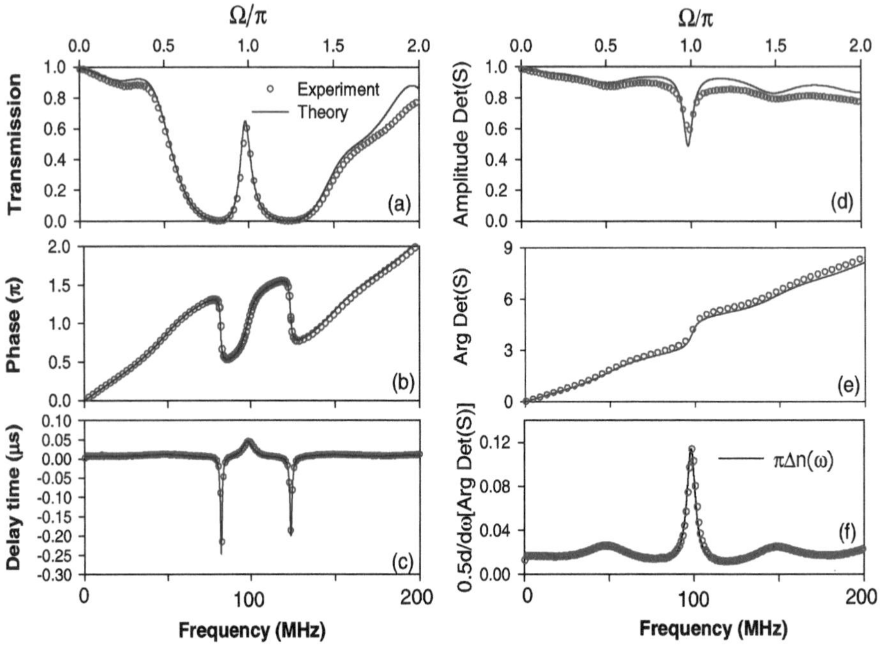

Figure 9.4 (a) Theoretical (solid line) and experimental (open circles) variations of the transmission coefficient versus the dimensionless frequency $\Omega = \omega d_3\sqrt{\varepsilon}/c$ for the structure depicted in Fig. 9.1. The lengths of the stubs are chosen such that $d_1 = 0.4d_3$ and $d_2 = 0.6d_3$. (b) Same as in (a) but for the phase. (c) Same as in (a) but for the delay time. (d)–(e) Variation of the amplitude and the phase (in units of π) of the determinant of the scattering matrix S versus the dimensionless frequency Ω/π. Open circles and solid lines correspond to experimental and theoretical results. (f) Variation of $\frac{d}{d\omega}$[Arg det(S)] (in units of μs) obtained from (e) (open circles) and the variation of the DOS ($\Delta n(\omega)$) obtained from Eq. (9.16) (black curve).

$\frac{dn}{d\omega} < 0$ or $\frac{d\varphi}{d\omega} < 0$) the group velocity can be larger than the speed of light in vacuum (superluminal group velocity) or even negative (pulse tunneling). The latter cases are known as *fast light*.

The theoretical and experimental phase function $\varphi(\omega)$ is used in Eqs. (9.29) and (9.30) to obtain the interferometer's effective index $n(\omega)$ and the expected group velocity $v_g(\omega)$. In the example of Fig. 9.4(c), negative group delay around -0.2 μs is reached giving rise to a value of v_g as small as $-0.02c$ at the peaks situated around $\Omega = \pi$. The idea of negative group velocity amounts to several decades with the challenge of realizing pulse advancement comparable to pulse width, with a low level of pulse distortion. This phenomenon has been found first in media made of atomic vapor cells in the presence of gain lines [97]. In all these works, it was clearly pointed out that such superluminal behavior is not at odds with either causality or Einstein's theory of special relativity, but it exclusively results from interference between the different frequency components of the pulse in an anomalous dispersion region [98]. Negative group velocities have been demonstrated in a linear and passive asymmetric loop made of two or three coaxial cables [79,84,85] where constructive and destructive interferences of the waves behave like in a Mach–Zehnder interferometer. The advantage of the structure presented here lies in its simplicity, as well as in the fact that the Fano resonances can be tuned by varying the boundary conditions at the ends of the resonators, which is not the case for the loop structures.

The above results show that the group velocity at the resonance ($\Omega = \pi$) is very slow ($v_g \approx 0.06c$), whereas it becomes very fast around the transmission zeros. Away from the resonance and antiresonance, the group velocity v_g is equal to the phase velocity $v_\varphi \approx c/n = 0.65c$ where $n \approx 1.51$ is the refractive index of the homogeneous coaxial cables. Therefore, all these results confirm the *slow* and *fast* light behaviors of the EIT resonance observed in this system.

9.3.3 Friedel phase and the phase of the transmission

In addition to the information that can be deduced from the amplitude and the phase of the transmission coefficient, one can also use the scattering matrix defined by

$$S = \begin{pmatrix} r & t' \\ t & r' \end{pmatrix}, \tag{9.31}$$

where t and r are the transmission and reflection coefficients (Eqs. (9.6) and (9.9)) for incident waves coming from the left of the double stubs in Fig. 9.1; t' and r' are the transmission and reflection coefficients for incident waves coming from the right side of the double stubs. Their expressions are obtained respectively from Eqs. (9.6) and (9.9) by permuting the subscripts 1 and 3 (i.e., interchanging the stubs 1 and 3). One can remark here that $t = t'$.

The Friedel sum rule relates the density of states (DOS) to the scattering matrix such as [99]

$$\frac{d}{d\omega}\theta_F(\omega) = \frac{1}{2}\frac{d}{d\omega}\left[\text{Arg } \det(S)\right] = \pi\,\Delta n(\omega) \tag{9.32}$$

where θ_F is called the Friedel phase and $\Delta n(\omega)$ is the variation of the DOS or the difference in the DOS due to the presence of scatterer. Several works devoted to electronic transport in mesoscopic systems have shown theoretically [10,11] the difference between the Friedel phase and phase of transmission amplitude in quasi-one-dimensional quantum wires in relation with the experiments of Yacoby et al. [14] on quantum dots in the so-called Bohm–Aharonov interferometers. However, to our knowledge, such a study has not been performed yet in photonic crystals because of the difficulties in measuring both reflection and transmission coefficients. Such measurements are, however, very easy in coaxial cables [89].

Figs. 9.4(c) and 9.4(d) give the amplitude and the phase of $\det(S)$ for the structure depicted in Fig. 9.3. The experimental and theoretical results are plotted by open circles and solid lines, respectively. Because of the absorption in the cables, the amplitude of $\det(S)$ is different from unity and reflects the strength of the absorption in the system, in particular around the Fano resonance at $\Omega = \pi$ as $A = 1 - |\det(S)|$. Contrary to the phase of the transmission amplitude (Fig. 9.4(b)) where phase drops exist at the transmission zeros, the phase of $\det(S)$ increases monotonically (Fig. 9.4(d)) with a fast slope around the resonance frequency, therefore evidencing a positive resonance in its derivative versus ω (Fig. 9.4(e)). In Fig. 9.4(e), we have also plotted (black curve) the variation of the DOS ($\pi\Delta n(\omega)$) given by Eq. (9.16). A very good agreement between $\pi\Delta n(\omega)$ and $\frac{1}{2}\frac{d}{d\omega}[\text{Arg}\det(S)]$ is obtained (Eq. (9.32)). These results clearly show how the scattering matrix measurement enables extracting experimentally both the DOS and absorption. In addition, these results are in accordance with those found in single channel quantum scattering systems [10,11,100] and give an experimental demonstration of Eqs. (9.15) and (9.16), namely the variation of the DOS (Fig. 9.4(e)) is the same as the transmission delay time (Fig. 9.4(c)) except at the transmission zeros where the delay time exhibits additional negative delta peaks. It is worth noticing a very recent paper giving an extensive study about the comparison between DOS and scatetring parameters in photonic crystals based on coaxial cables [101].

9.4 Asymmetric Fano-like resonance in U-structure

The structure studied in this work (Fig. 9.1) can exhibit also an asymmetric Fano resonance by adjusting the transmission zeros on only one side of the resonance, this can be obtained by considering a structure where the resonators are supposed to be identical with lengths slightly different from $0.5d_3$. This is shown in Fig. 9.5(a) for $d_1 = d_2 = 0.56d_3$. Indeed, an analytical Taylor expansion around $\Omega = \pi$ enables us to write the transmission function (Eq. (9.6)) as

$$t = \frac{-2\eta^2}{(j+\eta)(\varepsilon + 2\eta - j\eta\varepsilon)}, \tag{9.33}$$

where $\eta = \Delta + \varepsilon(\frac{1}{2} + \frac{\Delta}{\pi})$ and Δ is the detuning of the lengths of the two resonators from $0.5d_3$ (i.e., $\Delta = \pi(d_1/d_3 - 0.5)$).

From the expression of t (Eq. (9.33)), one can deduce the following Fano line shape transmission coefficient:

$$T = B\frac{1}{1+\eta^2}\frac{(\varepsilon - \varepsilon_R + q\Gamma)^4}{(\varepsilon - \varepsilon_R)^2 + \Gamma^2} \simeq B\frac{(\varepsilon - \varepsilon_R + q\Gamma)^4}{(\varepsilon - \varepsilon_R)^2 + \Gamma^2}, \tag{9.34}$$

where $B = \dfrac{(\frac{1}{2} + \frac{\Delta}{\pi})^4}{(1 + \frac{\Delta}{\pi})^2}$. $\Gamma = \dfrac{1}{4}\dfrac{\Delta^2(1 - \frac{\Delta}{\pi})}{(1 + \frac{\Delta}{\pi})^2}$ and $\varepsilon_R = \dfrac{-\Delta}{1 + \frac{\Delta}{\pi}}$ characterize the width and the shift of the resonance, respectively, whereas $q = \dfrac{2(1 + \frac{\Delta}{\pi})}{\Delta(\frac{1}{2} + \frac{\Delta}{\pi})(1 - \frac{\Delta}{\pi})}$ is the Fano parameter.

One can notice that the resonance shifts slightly from $\Omega = \pi$ and its width is small as compared to the preceding case (Fig. 9.2(a)). Also q increases when Δ decreases and tends to infinity when Δ vanishes. In this case the resonance falls at $\varepsilon_R = 0$ and, as expected, its width 2Γ reduces to zero, giving rise to a BIC mode (see below). The results of the approximate expression (Eq. (9.34)) are sketched (open circles)

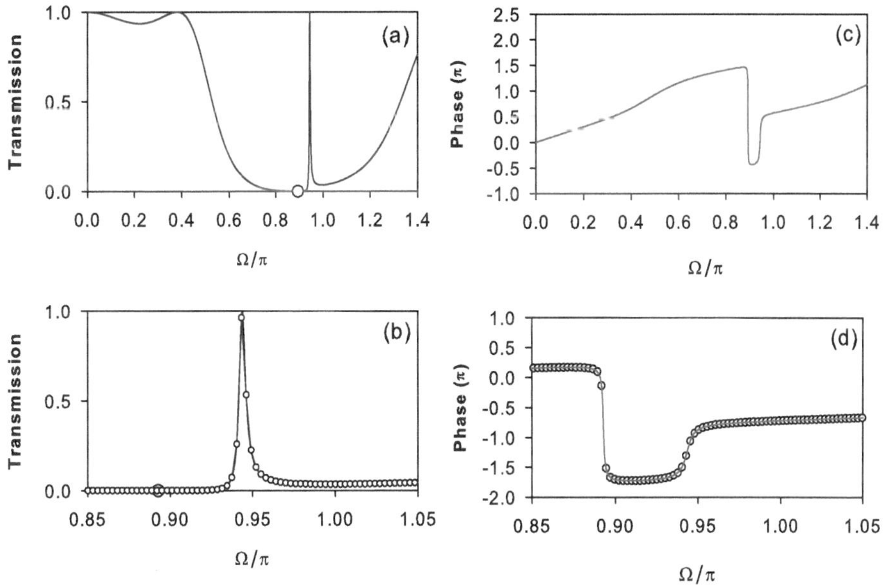

Figure 9.5 (a) Theoretical (solid line) variations of the transmission coefficient versus the dimensionless frequency Ω for the structure depicted in Fig. 9.1. The lengths of the resonators are taken such that $d_1 = d_2 = 0.56d_3$. White circles on the abscissa indicate the positions of the transmission zeros induced by the dangling resonators on both sides of the resonance. (c) Same as in (a) but for the variation of the phase. (b) and (d) compare the approximate results obtained by Taylor expansion (open circles) around the resonance with the exact calculations (solid lines).

in Fig. 9.5(b) for $\Delta = \pi(d_1/d_3 - 0.5) = 0.06\pi$ (i.e., $d_1/d_3 = 0.56$). These results are in accordance with the exact ones (solid lines) and clearly show that the resonance presents an asymmetrical Fano shape with $|q| \simeq 21.38$ and width $2\Gamma \simeq 0.015$. Concerning the evolution of the phase of the transmitted waves in this structure, one can notice from Eq. (9.6) that the numerator of the transmission function t vanishes when $C_1 = C_2 = 0$ (or equivalently $\eta = 0$ in the approximate result (Eq. (9.33))) at $\Omega = 0.893\pi$ indicated by a filled circle on the abscissa of Fig. 9.5(a). The transmission zeros induced by the two identical resonators fall at the same frequency, therefore the phase (Fig. 9.5(c)–(d)) shows a phase drop of 2π at these frequencies. Indeed, as the phase is defined modulo 2π, the 2π phase change can be observed if we take into account the absorption in the system which is the case for the coaxial cables studied here.

An experimental evidence of these theoretical results is given in Fig. 9.6, where $d_1 \simeq d_2 \simeq 0.6$ m (i.e., $d_1 = 0.605d_3$ and $d_2 = 0.625d_3$, the small difference between the two cables is due to the experimental precision). The experimental results (open circles) are in agreement with the theoretical calculation (solid lines). One can notice here also that the intensity of the resonance (Fig. 9.6(a)) does not reach unity because of the attenuation inside the cables. However, the main features of the asymmetric Fano resonance can be observed. The two resonators give rise to a phase drop of 2π as it is illustrated in Fig. 9.6(b). However, because of the small difference between the

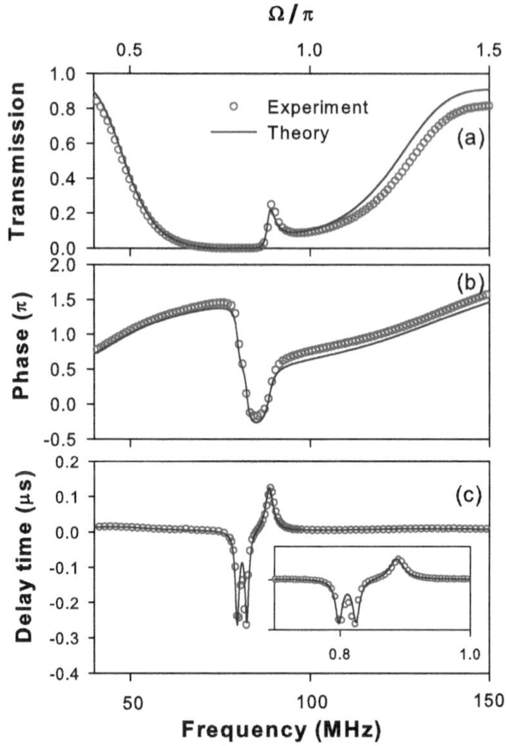

Figure 9.6 (a)–(c) Same as in Fig. 9.4(a)–(c) but the lengths of the resonators are taken such that $d_1 = 0.605d_3$ and $d_2 = 0.625d_3$ (i.e., $d_1 \simeq d_2$). The inset in (c) gives a highlight of the delay time in the frequency domain $0.75 < \Omega/\pi < 1$.

lengths of d_1 and d_2, the two transmission zeros induced by the two resonators do not fall exactly at the same frequency as it can be seen in Fig. 9.6(c) (see also the inset) where two negative delta peaks appear in the delay time around the transmission zeros at $\Omega = 0.846\pi$ and $\Omega = 0.85\pi$.

In the particular case where $d_1 = d_3 = 0.5d_2 = 0.5$ m, the frequency of the Fano resonance coincides exactly with the transmission zeros induced by the two stubs at $\Omega = \pi$ (see Eqs. (9.26) and (9.34)). The width of the resonance Γ vanishes and therefore the resonance collapses, giving rise to a BIC mode. This result is clearly demonstrated in Fig. 9.7(a) where a well-defined dip is observed around $\Omega = \pi$. However, because of the existence of two antiresonances and one resonance at $\Omega = \pi$, the phase exhibits a jump of π (Fig. 9.7(b)), giving rise to a negative delta peak in the delay time (Fig. 9.7(c)). This result gives a clear signature of the existence of the so-called ghost-Fano [102,103] resonance in the transmission amplitude (Fig. 9.7(a)). To our knowledge, this is a first experimental evidence showing the existence of such resonances or BIC modes characterized by an infinite lifetime [90].

In order to give a better insight about the profile of the Fano resonances as a function of the parameter Δ (or equivalently d_1/d_3), Fig. 9.8 gives theoretical results of

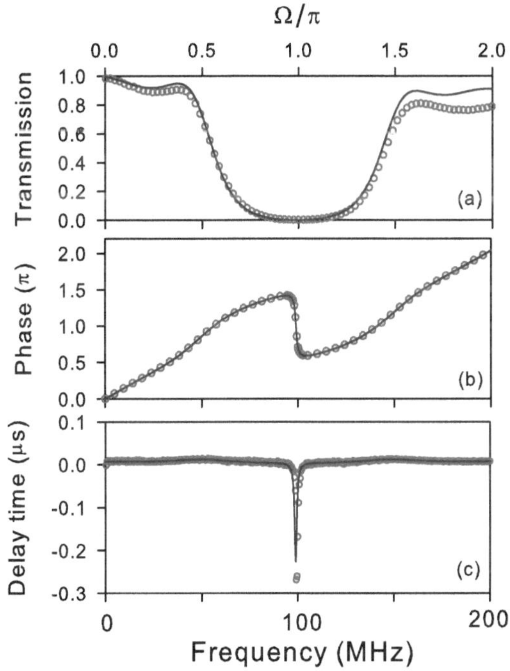

Figure 9.7 (a) Theoretical (solid line) and experimental (open circles) variations of the transmission coefficient versus the dimensionless frequency $\Omega = \omega d_3 \sqrt{\varepsilon}/c$ for the structure depicted in Fig. 9.1. The thicknesses of the stubs are chosen such that $d_1 = d_2 = 0.5 d_3 = 0.5$ m. (b) Same as in (a) but for the phase. (c) Same as in (a) but for the delay time.

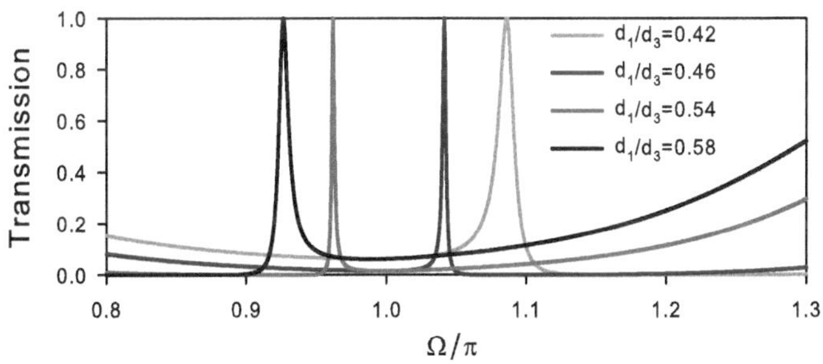

Figure 9.8 Theoretical (solid line) variations of the transmission coefficient versus the dimensionless frequency Ω for different values of d_1/d_3 in the structure depicted in Fig. 9.1 with $d_1 = d_2$.

the characteristic features of the resonances around $d_1/d_3 = 0.5$ for an asymmetric resonance. One can notice that the frequency position of the resonance decreases as function of d_1/d_3, its asymmetric Fano profile becomes symmetric and changes sign for $d_1/d_3 \simeq 0.5$. In other words, the parameter q responsible for the asymmetric Fano profile of the resonance diverges and changes sign around $d_1/d_3 = 0.5$. The width of the resonance decreases when d_1/d_3 tends to 0.5 and vanishes when d_1/d_3 is ex-

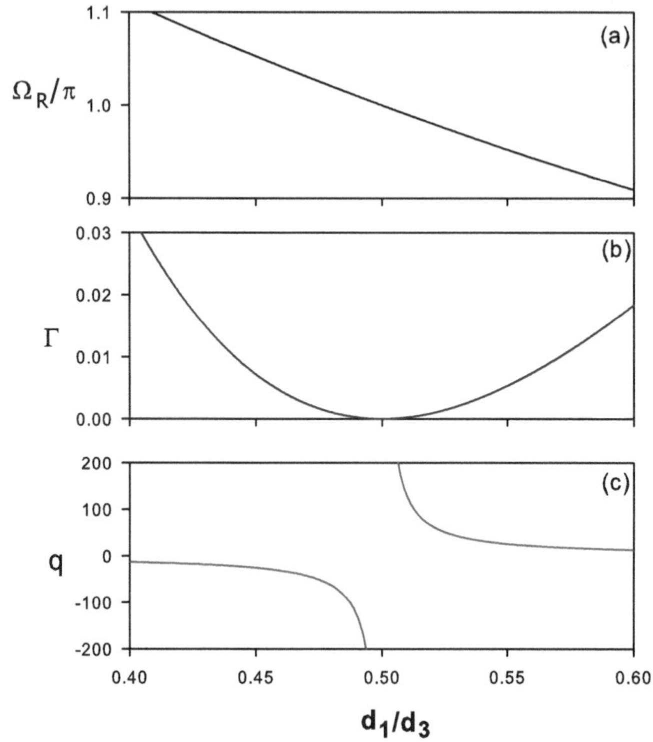

Figure 9.9 Theoretical variations of the quantities ε_R (a), Γ (b), and q (c) as functions of d_1/d_3 around $d_1/d_3 = 0.5$. The corresponding expressions are given after Eq. (9.34). Let us notice that the approximate expression of the position of the resonance in (a) is given by $\Omega_R = \pi + \varepsilon_R$.

actly equal to 0.5, giving rise to the collapse of the resonance and therefore a BIC mode (Fig. 9.8). These results are well illustrated by the plots of the approximate expressions of ε_R (Fig. 9.9(a)) (or equivalently $\Omega_R = \pi + \varepsilon_R$), Γ (Fig. 9.9(b)) and q (Fig. 9.9(c)) around $d_1/d_3 = 0.5$. Let us notice that the inversion of the asymmetry of the line-shape of Fano resonances has been discussed in photonic, phononic and mesoscopic systems by tuning the geometric parameters, or the angle of incidence [34,35,71–73].

9.5 EIA-like resonance in a cross structure

9.5.1 Theoretical and experimental demonstration

In this section, we consider a cross structure made of two stubs of lengths d_1 and d_2 grafted at the same position (Fig. 9.10(a)). Our goal is to show the possibility to realize the classical analogue of electromagnetic induced absorption. Indeed, in the study developed in Section 9.4, we have discussed only the behavior of the amplitude and the phase of the transmission coefficient to qualify the resonances as Fano or EIT. The same study can be performed also for the cross structure studied here. However, considering only the amplitude and the phase of the transmission is not enough to ensure that the resonance is an EIT-like resonance, and a study of the absorption is essential to guarantee the EIT or EIA behavior of the resonance. The intrinsic absorption in the system is introduced by means of the imaginary part ε'' of the permittivity ε of the coaxial cables. The EIA-like resonance in a double stub structure can be demonstrated by means of the determination of transmission and reflection coefficients through such structures. However, in order to understand the origin of the EIA resonance studied here, we shall need to develop also the results for each stub taken alone. Also, the two stubs can be treated as two coupled resonators to an incident wave in the waveguide. Therefore, in the presence of loss, the contribution of each stub to the absorption should be proportional to the square modulus of the wave amplitude in this stub. All these points will be developed theoretically and supported by experimental data. The experimental setup used in the measurement of the amplitude and phase of the different elements of the scattering matrix is depicted in Fig. 9.10(b).

Let us consider an incident wave $E_i(x) = e^{-jk_0x}$ launched in the left semiinfinite waveguide (Fig. 9.1(a)). Here $j = \sqrt{-1}$ and $k_0 = \omega\sqrt{\varepsilon'}/c$ is the incident wavevector (ε' is the real part of the permittivity ε), ω is the angular frequency, and c the speed of light. The transmission amplitude in the right semiinfinite waveguide (Fig. 9.10(a)) can be obtained easily from the one obtained in Section 9.2 by just eliminating the segment of length d_3 between the two stubs of lengths d_1 and d_2, i.e., by taking $d_3 = 0$ in the transmission and reflection coefficients (Eqs. (9.6) and (9.9)), respectively. The transmission coefficient can be written in the following form:

$$t = \frac{2k_0C_1C_2}{2k_0C_1C_2 - jkS} = \frac{Z_1Z_2}{(Z_1 - Z_c/2)(Z_2 - Z_c/2) - (Z_c/2)^2}, \tag{9.35}$$

where $C_i = \cos(kd_i)$, $S_i = \sin(kd_i)$ $(i = 1, 2)$, $S = \sin(kd)$, and $d = d_1 + d_2$ is the total length of the resonators. Also $k = k_0 + jk''$ where $k'' = \frac{\omega\varepsilon''}{2c\sqrt{\varepsilon'}}$ is the imaginary part of the wavevector in the stubs; $Z_i = -j\frac{Z_c}{1+j\varepsilon''/2\varepsilon'}\frac{C_i}{S_i}$ is the impedance of the resonator i reduced at the bottom surface when the boundary condition at the top surface is $H = 0$.

In the same way, the reflection amplitude in the left semiinfinite waveguide (Fig. 9.10(a)) is given by

$$r = \frac{jkS}{2k_0C_1C_2 - jkS} = \frac{Z_c(Z_1 + Z_2)/2}{(Z_1 - Z_c/2)(Z_2 - Z_c/2) - (Z_c/2)^2}. \tag{9.36}$$

The amplitudes of the waves in the two stubs can be written as

$$E_1(x) = a_1\cos(k(x - d_1)) \text{ and } E_2(x) = a_2\cos(k(x - d_2)), \tag{9.37}$$

where

$$a_1 = \frac{2k_0C_2}{2k_0C_1C_2 - jkS} \text{ and } a_2 = \frac{2k_0C_1}{2k_0C_1C_2 - jkS}. \tag{9.38}$$

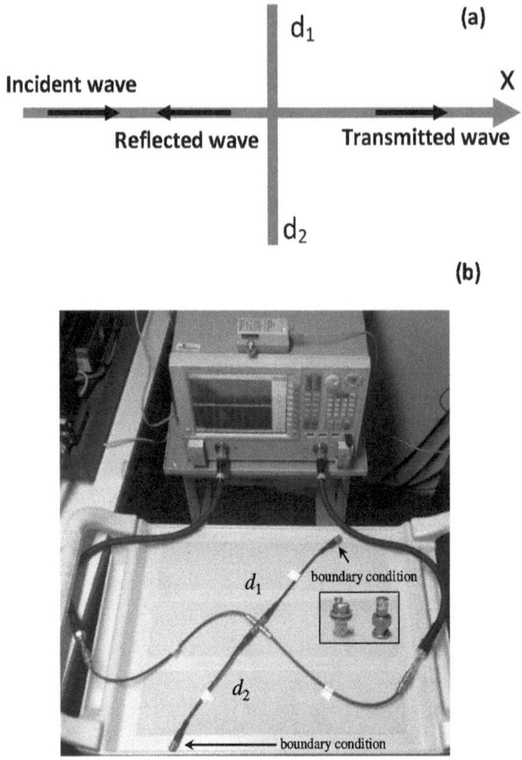

Figure 9.10 (a) A schematic representation of the studied structure consisting of one-dimensional electromagnetic waveguide (coaxial cable) with two dangling resonators of lengths d_1 and d_2 grafted at the same site on the waveguide. (b) Experimental setup considering open loads or $H = 0$ as boundary conditions. The inset shows the other BNC termination (short-circuit or $E = 0$) used as boundary conditions (to be connected to the output arm of the dangling resonators).

From the expressions of t (Eq. (9.35)) and r (Eq. (9.36)), one can respectively deduce the transmission and reflection coefficients $T = |t|^2$ and $R = |r|^2$.

Also, in order to understand the EIT and EIA phenomena in the double stub structure (Fig. 9.10), it is interesting to relate the transmission and reflection coefficients of the whole structure to those of its elementary constituents, namely, each stub taken separately. From Eqs. (9.35) and (9.36), one can deduce the transmission and reflection coefficients of one attached resonator stub as follows:

$$t_1 = \frac{k_0 C_1}{k_0 C_1 - jkS_1/2} = \frac{Z_1}{Z_1 - Z_c/2} \quad \text{and} \quad r_1 = \frac{ikS_1/2}{k_0 C_1 - jkS_1/2} = \frac{Z_c/2}{Z_1 - Z_c/2}. \tag{9.39}$$

One can notice from Eqs. (9.35), (9.36), and (9.39) that the reflection and transmission coefficients of the whole structure can be written as functions of those of the individual resonators (Eq. (9.39)):

$$t = \frac{t_1 t_2}{1 - r_1 r_2} \quad \text{and} \quad r = \frac{r_1 + r_2 + 2r_1 r_2}{1 - r_1 r_2}. \tag{9.40}$$

Eq. (9.40) clearly shows that the transmission zeros (total reflection) of the double stub structure are exactly the same as those of the stubs taken individually, whereas the total reflection is the sum of the reflection of each stub alone plus a third term representing the interference between the two stubs. In addition, if we call f_1 and f_2 two closed frequencies for which the transmission vanishes (f_1 and f_2 are induced by the stubs 1 and 2, respectively), then one can easily show that the reflection coefficient always vanishes (total transmission) for a given frequency f_3 lying between f_1 and f_2 (see below).

The possibility of existence of EIA resonances requires the presence of loss in the materials as it is the case for the coaxial cables studied here. However, before discussing such resonances, we shall first show theoretically the existence of EIT resonances in lossless materials (i.e., $\varepsilon'' = 0$). In this case, one can easily deduce from Eqs. (9.35) and (9.36) the conservation energy, namely $R + T = 1$. Also we will show below how the lengths of the two stubs should be chosen in order to realize the EIT or EIA resonances.

Eq. (9.35) clearly shows that the transmission zeros are due only to the dangling resonators, these occur when $C_1 = 0$ or $C_2 = 0$ (i.e., $\omega d_1 \sqrt{\varepsilon'}/c = (m_1 + 0.5)\pi$ or $\omega d_2 \sqrt{\varepsilon'}/c = (m_2 + 0.5)\pi$). These frequencies correspond respectively to the eigenmodes of the stubs 1 and 2, with $H = 0$ boundary condition on one side and $E = 0$ boundary condition on the other side [16]. Now, if $\omega(d_1 + d_2)\sqrt{\varepsilon'}/c = m\pi$, (i.e., $S = 0$), but $\omega d_1 \sqrt{\varepsilon'}/c \neq (m_1 + 0.5)\pi$ and $\omega d_2 \sqrt{\varepsilon'}/c \neq (m_2 + 0.5)\pi$ (i.e., $C_1 \neq 0$ and $C_2 \neq 0$), then one obtains a resonance that reaches unity (i.e., $T = 1$).

In the particular case where $\omega d_1 \sqrt{\varepsilon'}/c = (m_1 + 0.5)\pi$ and $\omega d_2 \sqrt{\varepsilon'}/c = (m_2 + 0.5)\pi$, we get $\omega(d_1 + d_2)\sqrt{\varepsilon'}/c = (m_1 + m_2 + 1)\pi$ and the numerator and denominator of t (Eq. (9.35)) vanish altogether. In this case, the resonance and the two zeros induced by the resonators fall at the same position, thus the resonance collapses and the transmission coefficient vanishes, giving rise to BIC modes (see below). These modes coincide with those of the long stub of length $d = d_2 + d_1$ with $H = 0$ boundary condition on both sides as $S = 0$. The above conditions are satisfied if d_1 and d_2 are chosen such that

$$d_1^0 = \frac{(2m_1 + 1)d}{2(m_1 + m_2) + 2}, \tag{9.41}$$

and

$$d_2^0 = \frac{(2m_2 + 1)d}{2(m_1 + m_2) + 2}. \tag{9.42}$$

These equations are the same as in Eqs. (9.20), (9.21), and (9.22), where we should take $m_3 = 0$ in Eq. (9.22) as $d_3 = 0$. The above expressions show how the lengths

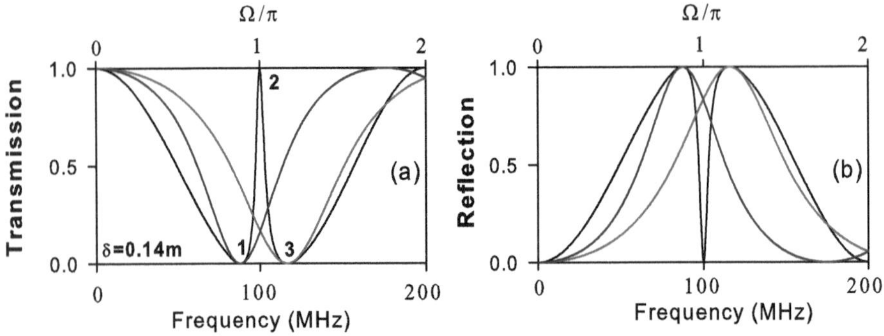

Figure 9.11 Theoretical transmission (a) and reflection (b) versus the dimensionless frequency $\Omega = \omega d\sqrt{\varepsilon'}/c$ in the absence of loss for $\delta = d_2 - d_1 = 0.14$ m where d_1 and d_2 are the lengths of the two stubs and $d = d_1 + d_2 = 1$ m. The blue (dark gray in print version) and red (light gray in print version) curves in (a) and (b) correspond respectively to each stub alone with $d_2 = 0.57$ m and $d_1 = 0.43$ m, whereas the black curves correspond to the two stubs present together.

of the two stubs should be chosen appropriately in order to realize the BIC mode before detuning the two lengths d_1 and d_2 in order to realize EIT resonance. In what follows, we consider the first modes of the two stubs given by $m_1 = m_2 = 0$ and keep the total length d constant in order to fix the resonance at the same frequency. In this case, the above lengths of the two stubs become $d_1^0 = d_2^0 = d/2$. In what follows, we shall use both the radio-frequency (in MHz) as well as the dimensionless frequency $\Omega = \omega d\sqrt{\varepsilon'}/c$ for plotting the different curves. The BIC mode (Eq. (9.23)) falls at $\Omega = \pi$.

In order to show the possibility of existence of EIT-resonance, i.e., a resonance squeezed between two transmission zeros, we have to take d_1 and d_2 slightly different from $d/2$ such that $d_1 = d/2 - \delta/2$ and $d_2 = d/2 + \delta/2$ (where $\delta = d_2 - d_1$). An example corresponding to this situation is given in Fig. 9.11(a) for $\delta = 0.14$ m and $d = 1$ m. One can notice that the resonance at $\Omega = \pi$ is squeezed between two transmission zeros induced by the two stubs. As predicted, in the absence of loss, the intensity of this resonance reaches unity (Fig. 9.11(a)) while the reflection vanishes (Fig. 9.11(b)). As a matter of comparison, we have also plotted the transmission amplitudes for each stub attached alone using Eq. (9.39) (see the red [light gray in print version] and blue [dark gray in print version] curves in Figs. 9.11(a)–(b)). As mentioned above (Eq. (9.40)), one can notice that the reflection coefficient of the whole structure is the same as the reflection coefficient of each resonator around the stub resonances falling at $\Omega = 0.88\pi$ and $\Omega = 1.16\pi$ (i.e., $C_1 = 0$ and $C_2 = 0$), respectively, whereas at $\Omega = \pi$ (i.e., $S = 0$) the stubs interfere destructively, giving rise to a zero reflection (or total transmission). These results are in accordance with Eq. (9.40) and the discussion given above where total transmission induced by the two stubs taken together always occurs between two transmission zeros induced by the two stubs. In the absence of loss, this phenomenon can be qualified as EIT resonance.

In the presence of loss, a part of energy is lost in the stubs as $A = 1 - R - T$. Recent works [57,58] have treated the EIA phenomenon in the framework of two

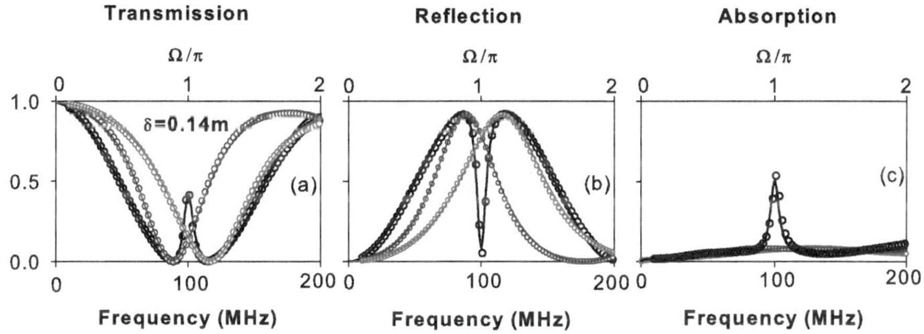

Figure 9.12 (a)–(b) Same as in Fig. 9.11(a)–(b) but in the presence of loss. (c) Absorption coefficient. Open circles show the experimental data.

coupled resonators model where the absorption can be related to the amplitudes of the waves in the two resonators. In order to demonstrate this relation in our case, we use a microscopic model where the absorption can be deduced as follows [104]:

$$A = \frac{\varepsilon''\omega}{c\sqrt{\varepsilon'}}\left(\int_0^{d_1} |E_1(x)|^2 dx + \int_0^{d_2} |E_2(x)|^2 dx\right) \tag{9.43}$$

where $E_1(x)$ and $E_2(x)$ are the electric fields of the waves in the two stubs given by Eq. (9.37). After some straightforward calculation, one can show that the absorption A can be related to the square modulus of the amplitudes a_1 and a_2 in the two stubs (Eq. (9.38)) by

$$\begin{aligned} A = \frac{\varepsilon''\omega}{c\sqrt{\varepsilon'}}\Bigg(&|a_1|^2\left[\frac{\sin(2k_0 d_1)}{2k_0} + \frac{\sinh(2k'' d_1)}{2k''}\right] \\ &+ |a_2|^2\left[\frac{\sin(2k_0 d_2)}{2k_0} + \frac{\sinh(2k'' d_2)}{2k''}\right]\Bigg). \end{aligned} \tag{9.44}$$

Figs. 9.12(a)–(c) give the transmission, reflection, and absorption for $\delta = 0.14$ m. One can notice that the intensity of the resonance does not reach unity because of the loss in the cables. The reflection reaches almost zero and the absorption presents a sharp peak on a shallow background with almost the same intensity as in the transmission. The experimental results are represented by open circles. They are very well fitted by the 1D model (solid lines) using the Green's function method. One can notice that the dimensionless frequency domain $0 < \Omega < 2\pi$ in Fig. 9.2 corresponds to the radio-frequency range f such that $0 < f < 200$ MHz. The attenuation inside the cables induces transmission and reflection depletions in Figs. 9.12(a) and 9.12(b) especially at high frequencies. The experimentally observed peak in the absorption (Fig. 9.12(c)) is 50% compared to 10% in the uncoupled case (red and blue curves; light gray and dark gray in print version, respectively), which corresponds to a five-fold enhancement. This phenomenon is equivalent to that found by Tassin et al. [57] in

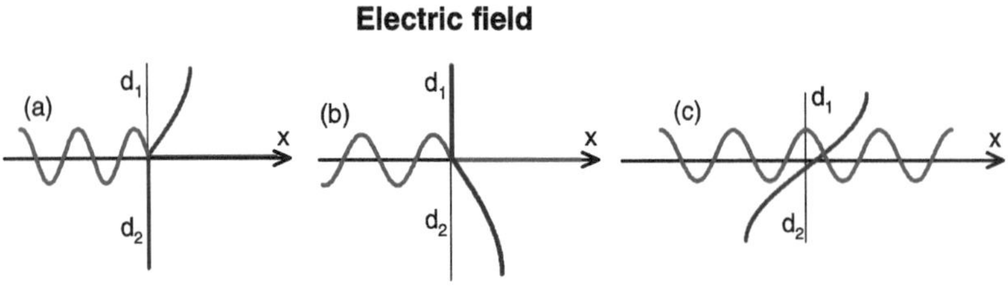

Figure 9.13 (a)–(c) Theoretical variation of the electric field (in arbitrary units) as a function of the space position along the waveguide (red curves, light gray in print version) and the stubs (blue curves, dark gray in print version) for the modes labeled 1, 2, 3 in Fig. 9.11(a), respectively.

microwave metamaterials and by Taubdert et al. [58] in plasmonics and can be qualified as EIA resonance. One can clearly see that the EIA resonance is a consequence of the effect of the interference between the waves in the two stubs in the presence of loss.

The behavior of the electric field in the structure is given in Fig. 9.13 for different values of Ω. Figs. 9.13(a) and 9.13(b) give the electric fields for the modes labeled 1 and 3 in Fig. 9.11(a) and corresponding to $\Omega = 0.88\pi$ and $\Omega = 1.16\pi$, respectively. As predicted, one can notice that the electric field is localized in the corresponding stubs and does not propagate in the right-hand side waveguide. In this case, each stub behaves as a quarter wavelength wire. However, for the mode labeled 2 in Fig. 9.11(a) and corresponding to $\Omega = \pi$ (i.e., $S = 0$), the electric field is localized in the two stubs and propagates along the whole waveguide (Fig. 9.13(c)). In this case, the two stubs taken together behave like a half-wavelength wire.

As mentioned above, for $\delta = 0$ (i.e., $d_1 = d_2 = 0.5$ m) the resonance collapses in the transmission spectra, giving rise to a BIC mode. This mode is given by Eq. (9.23) for $m_1 = m_2 = n = 0$ (i.e., $\Omega = \alpha d = \pi$). In this case, the system behaves like an opaque medium around $\Omega = \pi$ as it is shown in Fig. 9.14(e). The reflection reaches almost unity and the absorption shows a weak value around 0.1. Even if the BIC state is not well defined in the presence of loss, Fig. 9.14(e) shows that even for a lossy system, a small detuning of the geometrical parameters can highlight this state as an EIA peak, simultaneously with a reflection dip and a small transmission peak. Indeed, the width of the resonance and its intensity increase in the transmission as δ increases (Fig. 9.14, from bottom to the top). The reflection decreases and reaches almost zero at $\delta = 0.3$ m, whereas the absorption increases for small values of δ reaches a maximum around $\delta = 0.07$ m and then decreases for large values of δ as it is depicted in Fig. 9.15(a) at the resonance $\Omega = \pi$. The maximum absorption ($\simeq 0.5$) occurs when the transmission and reflection amplitudes cross each other ($R = T \simeq 0.25$). The experimental results (open circles) are very well fitted by the theoretical curves (solid lines). These results show that in order to avoid the loss effect on the transmission, the detuning between the two resonators should be chosen larger, but at the price of the quality factor which becomes smaller.

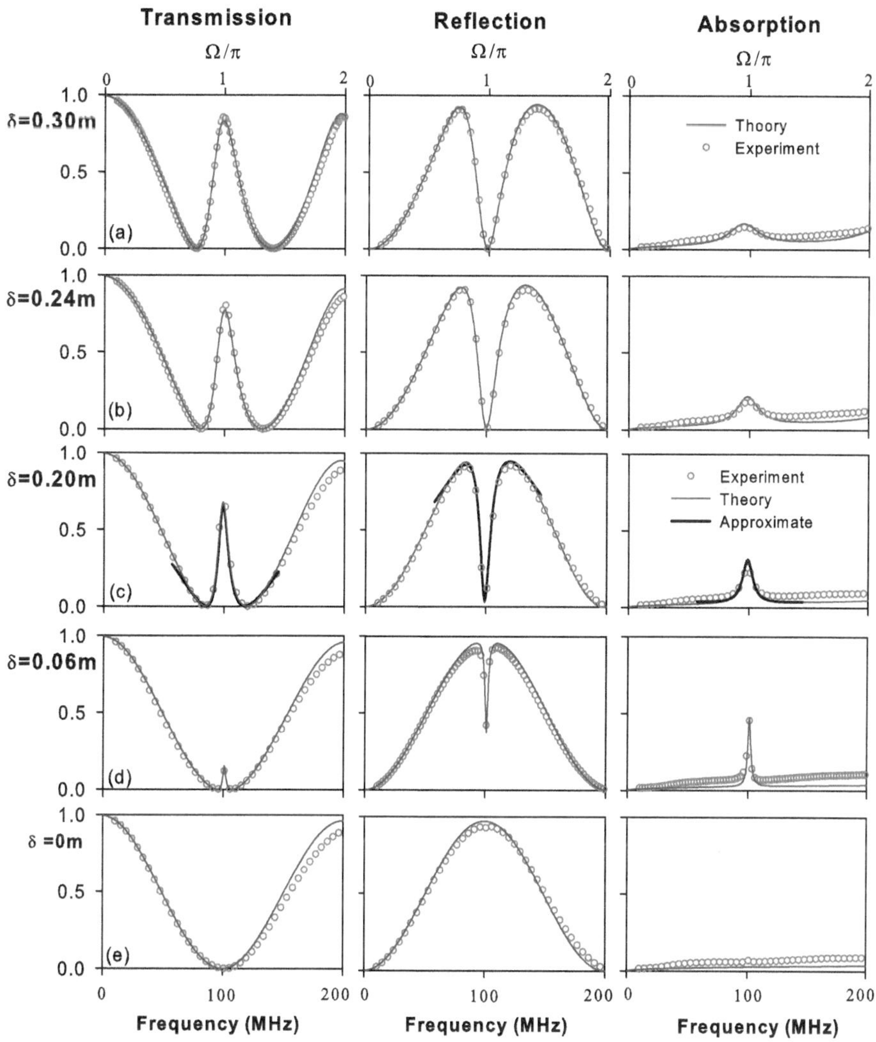

Figure 9.14 (a) Theoretical (blue solid line, dark gray in print version) and experimental (red open circles, light gray in print version) transmission (left panel), reflection (middle panel), and absorption (right panel) versus the frequency (in MHz) or the dimensionless frequency $\Omega = \omega d\sqrt{\varepsilon'}/c$ for different values of the detuning $\delta = d_2 - d_1$. The black curves in (c) correspond to the approximate results obtained by Taylor expansion (see below).

9.5.2 Analytical results around the EIA resonance

In order to give a quantitative study of the shape, width, and intensity of the EIA resonance in the presence of loss, we shall give in this section approximate expressions of the different quantities studied in Section 9.5.1 around the EIA resonance. First, in order to give a clear evidence about the coupling of the two resonators with the

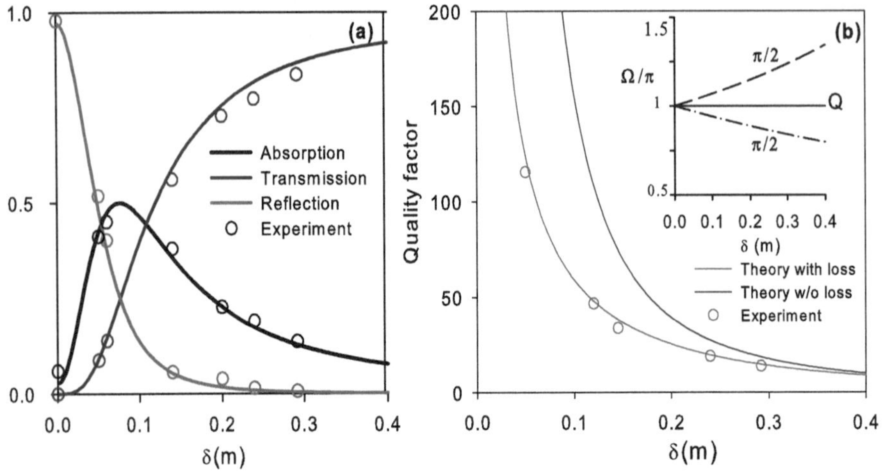

Figure 9.15 (a) Theoretical variation of the transmission (blue line, dark gray in print version), reflection (red line, light gray in print version), and absorption (black line) versus δ at the resonance $\Omega = \pi$ (i.e., $f \simeq 100$ MHz) in Fig. 9.14. Open circles correspond to experimental data. (b) Theoretical variation of the quality factor of the transmission resonance around $\Omega = \pi$ as a function of δ with (red line, light gray in print version) and without (blue line, dark gray in print version) loss. Red (light gray in print version) open circles correspond to experimental data. The insert gives the variation of the frequencies of the two resonators (dashed and dashed–dotted lines) versus δ. The quality factors of each stub alone is about $\pi/2$ as indicated in the insert.

incident wave, it is interesting to write the amplitude of the waves inside the two stubs following a two coupled resonators model [41,45,57,105]. Indeed, a Taylor expansion around $\Omega = \pi$ (i.e., $\Omega = \pi + \varepsilon$) enables us to write the amplitudes in Eq. (9.38) as follows:

$$a_1 = -j\frac{\gamma_1\left(j(\varepsilon-\varepsilon_2)-\gamma'\right)}{(j(\varepsilon-\varepsilon_1)-(\frac{\gamma_1}{2}+\gamma'))(j(\varepsilon-\varepsilon_2)-(\frac{\gamma_2}{2}+\gamma'))-\frac{\kappa^2}{4}} \tag{9.45}$$

and

$$a_2 = -j\frac{\gamma_2\left(j(\varepsilon-\varepsilon_1)-\gamma'\right)}{(j(\varepsilon-\varepsilon_1)-(\frac{\gamma_1}{2}+\gamma'))(j(\varepsilon-\varepsilon_2)-(\frac{\gamma_2}{2}+\gamma'))-\frac{\kappa^2}{4}}, \tag{9.46}$$

where $\varepsilon_1 = -\pi\frac{\delta}{2d_1}$ and $\varepsilon_2 = \pi\frac{\delta}{2d_2}$ are the resonance frequencies of the modes of each resonator 1 and 2; $\gamma_i = \frac{d}{d_i}$ $(i = 1, 2)$ and $\gamma' = \frac{\omega d\varepsilon''}{2c\sqrt{\varepsilon'}}$ are the linewidths that originate from the coupling to the semiinfinite waveguides, and the last one is due to intrinsic cavity losses; $\kappa = \sqrt{\gamma_1\gamma_2}$ is the coupling parameter between the two resonators. Eqs. (9.45) and (9.46) qualitatively describe the coupling of both radiative resonators with the incident wave as in the V-type atom [41]. In a Γ-type atom, only one resonator (the radiative) couples strongly to the incident wave, the other resonator (the dark) weakly couples to the incident wave [41,45,57,105].

In the same way, a Taylor expansion of the transmission amplitude (Eq. (9.35)) around $\Omega = \pi$ enables us to write T following the Fano line-shape or EIT-like resonance as

$$T = B\frac{(\varepsilon + q_1\Gamma)^2(\varepsilon + q_2\Gamma)^2}{\varepsilon(\varepsilon + \beta) + \Gamma^2}, \tag{9.47}$$

where $B = \frac{(\frac{1}{4}-\frac{\Delta^2}{\pi^2})^2}{\frac{1}{4}+\frac{4\Delta^4}{\pi^2}+\frac{\gamma'}{2}}$, $\beta = \frac{\frac{4\Delta^2}{\pi}(\Delta^2+\gamma')}{\frac{1}{4}+\frac{4\Delta^4}{\pi^2}+\frac{\gamma'}{2}}$ and $\Delta = \pi\delta/2d$. The full width at half maximum of the EIT resonance falling at $\varepsilon = 0$ (i.e., $\Omega = \pi$) is given by

$$\Gamma = \sqrt{\frac{\Delta^4 + \Delta^2\gamma' + (\frac{\gamma'}{2})^2}{\frac{1}{4} + \frac{4\Delta^4}{\pi^2} + \frac{\gamma'}{2}}}, \tag{9.48}$$

$q_1 = \frac{\Delta}{\Gamma(\frac{1}{2}+\frac{\Delta}{\pi})}$ and $q_2 = \frac{\Delta}{\Gamma(\frac{1}{2}-\frac{\Delta}{\pi})}$ are the Fano parameters; they qualitatively give the strength of the interference between the bound state and the propagating continuum states [1,27,29]. One can notice that when increasing Δ (or δ), Γ increases and q decreases (see below). Eq. (9.47) shows that in the presence of loss, the transmission coefficient does not reach unity around $\Omega = \pi$ (i.e., $\varepsilon = 0$). Indeed, the maximum value of T around $\Omega = \pi$ is given by

$$T_{\max} = \frac{\Delta^4}{\Delta^4 + \Delta^2\gamma' + (\frac{\gamma'}{2})^2}. \tag{9.49}$$

As predicted, $T_{\max}$ reaches unity when $\gamma' = 0$ and decreases when Δ increases (see Fig. 9.15(a)). An example of the results of the approximate expression (Eq. (9.47)) is shown by the black line in Fig. 9.14(c) for $\delta = 0.2$ m and clearly shows that the resonance is Fano-like with $q_1 = -3.77$, $q_2 = 2.52$, and width $2\Gamma = 0.41\pi$. As a matter of comparison, in the periodic photonic crystals, a perturbation is often introduced to the system in order to create the resonance state in the gap [79]. However, the above calculation shows that, without introducing any perturbation in the structure, one can find a well-defined EIT resonance with a width 2Γ and coupling parameters q_1 and q_2 that can be adjusted by tailoring the lengths of the resonators (i.e., δ).

Similarly a Taylor expansion of the reflection amplitude (Eq. (9.36)) around $\Omega = \pi$ enables us to write R as follows:

$$R = C\frac{\frac{\varepsilon^2}{4} + \frac{\gamma'^2}{4}}{\varepsilon(\varepsilon + \beta) + \Gamma^2}, \tag{9.50}$$

where $C = \frac{1}{\frac{1}{4}+\frac{4\Delta^4}{\pi^2}+\frac{\gamma'}{2}}$. Eq. (9.50) shows that in the presence of loss, the reflection coefficient does not reach zero around $\Omega = \pi$. Indeed, the minimum value of R

at $\Omega = \pi$ is given by

$$R_{\min} = \frac{(\frac{\gamma'}{2})^2}{\Delta^4 + \Delta^2\gamma' + (\frac{\gamma'}{2})^2}. \tag{9.51}$$

As predicted, $R_{\min}$ vanishes when $\gamma' = 0$. Also, one can notice that $R_{\min}$ increases as far as Δ decreases, in accordance with the results in Fig. 9.14. An example of the approximate curves of the reflection (Eq. (9.50)) is given in Fig. 9.14(c) by the black line for $\delta = 0.2$ m. The experimental results are well fitted by the approximate expression (Eq. (9.50)). The absorption can be extracted from the reflection and transmission as $A = 1 - R - T$. From Eqs. (9.47) and (9.50), one can deduce

$$A = C\frac{\gamma'(\Delta^2 + \frac{\varepsilon^2}{4} + \frac{\gamma'^2}{4})}{\varepsilon(\varepsilon + \beta) + \Gamma^2}. \tag{9.52}$$

As predicted, one can notice that A vanishes when γ' (the intrinsic loss) approaches zero. An example of the approximate curves of the absorption (Eq. (9.52)) is given in Fig. 9.14(c) by the black line for $\delta = 0.2$ m which fits well the experimental results (open circles). The absorption at $\Omega = \pi$ (i.e., $\varepsilon = 0$) becomes

$$A = \gamma'\frac{\Delta^2 + (\frac{\gamma'}{2})^2}{\Delta^4 + \Delta^2\gamma' + (\frac{\gamma'}{2})^2}. \tag{9.53}$$

This expression shows that when Δ vanishes, A becomes as small as $A = \gamma'^2/2$, and for large values of Δ, A goes to zero. For intermediate values of Δ, one can show easily the existence of a threshold value of Δ (or δ) such that $\delta_{th} \simeq \frac{d}{\pi}\sqrt{2\gamma'}$ where A goes through a maximum $A_{\max} \simeq 1/2$. These results are in accordance with those in Fig. 9.15(a) and show that the peak in the absorption can be controlled by detuning the two stubs. Also, δ_{th} decreases with the attenuation γ' (or equivalently ε'') for a given value of d.

In Fig. 9.15(b), we have displayed the variation of the quality factor Q of the EIA resonance defined by $Q = \frac{\pi}{\Gamma}$ as a function of δ. The width Γ of the resonance is given by Eq. (9.48). The curves are sketched in both cases with and without loss. When δ increases, the Q factor decreases and goes to an asymptotic limit ($\simeq 4$) for large values of δ. The effect of the loss on Q can be well appreciated for small values of δ. In the insert of Fig. 9.15(b), we have plotted by dashed and dashed–dotted lines the variation of the frequency positions of the two resonators versus δ. The EIA resonance induced by the whole stub of length d falls at the same frequency $\Omega = \pi$ for all values of δ. One can show easily that the Q factor of each resonator in presence of loss is about $\pi/2$, whereas the Q factor of the EIT resonance in Fig. 9.15(b) can reach a value as large as 120.

9.5.3 Effect of boundary conditions on the EIA resonance

Consider now the case where the boundary conditions at the end of the resonators are $E = 0$ on the stub of length d_2 and $H = 0$ on the stub of length d_1. In this case the transmission and reflection coefficients are given respectively by

$$t = \frac{2k_0 C_1 S_2}{2k_0 C_1 S_2 - jkC} \tag{9.54}$$

and

$$r = \frac{jkC}{2k_0 C_1 S_2 - jkC}, \tag{9.55}$$

where $C = \cos(kd)$ and $d = d_1 + d_2$. Eq. (9.54) clearly shows that the transmission zeros are due only to the dangling resonators, these occur when $C_1 = 0$ or $S_2 = 0$ (i.e., $\omega d_1\sqrt{\varepsilon}/c = (m_1 + 0.5)\pi$ or $\omega d_2\sqrt{\varepsilon}/c = m_2\pi$). If $\omega(d_1 + d_2)\sqrt{\varepsilon}/c = (m + 0.5)\pi$, (i.e., $C = 0$), but $\omega d_1\sqrt{\varepsilon}/c \neq (m_1 + 0.5)\pi$ and $\omega d_2\sqrt{\varepsilon}/c \neq m_2\pi$ (i.e., $C_1 \neq 0$ and $S_2 \neq 0$), one obtains a resonance that reaches unity (i.e., $T = 1$). In the particular case where $\omega d_1\sqrt{\varepsilon}/c = (m_1 + 0.5)\pi$ and $\omega d_2\sqrt{\varepsilon}/c = m_2\pi$, we get $\omega(d_1 + d_2)\sqrt{\varepsilon}/c = (m_1 + m_2 + 0.5)\pi$ and the numerator and denominator of t (Eq. (9.54)) vanish altogether. In this case, the resonance and the two zeros induced by the resonators fall at the same position, thus the resonance collapses and the transmission coefficient vanishes. The above conditions are satisfied if d_1 and d_2 are chosen such that

$$d_1^0 = \frac{(2m_1 + 1)d}{2(m_1 + m_2) + 1} \tag{9.56}$$

and

$$d_2^0 = \frac{2m_2 d}{2(m_1 + m_2) + 1}. \tag{9.57}$$

In what follows we shall focus on the first modes of the two stubs given by $m_1 = 0$ and $m_2 = 1$ (i.e., $d_2^0 = 2d_1^0$). These modes coincide with the first mode of the long stub of length $d = 3d_1^0$ with $H = 0$ boundary condition on one side and $E = 0$ boundary condition on the other side. These modes are given by $C = 0$ (i.e., $\omega d\sqrt{\varepsilon}/c = 3\pi/2$).

In order to show the possibility of existence of EIA-resonance, i.e., a resonance squeezed between two transmission zeros, we have to take d_1 and d_2 slightly different from $d/3$ and $2d/3$, respectively, such that $d_1 = d/3 - \delta/2$ and $d_2 = 2d/3 + \delta/2$ (where $\delta = d_2 - d_1 - d/3$). Some examples of the transmission, reflection, and absorption coefficients are given in Fig. 9.16 for different values of δ and fixed $d = 1.5$ m. The results are similar to those presented in Fig. 9.14, but the resonance shows an asymmetrical profile for large values of δ due to the asymmetry in the length of the stubs and the boundary conditions. The behavior of the transmission, reflection, and absorption at the resonance as a function of δ is sketched in Fig. 9.17. This figure

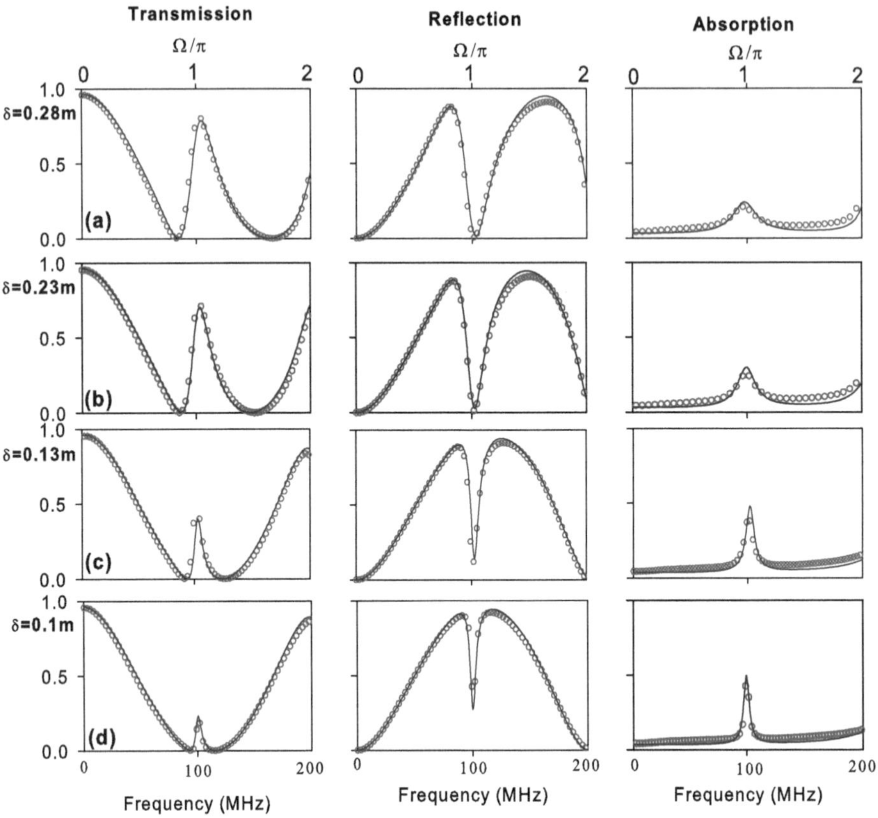

Figure 9.16 Theoretical (blue line, dark gray in print version) and experimental (red open circles, light gray in print version) transmission (left panel), reflection (middle panel), and absorption (right panel) versus the dimensionless frequency $\Omega = \omega d\sqrt{\varepsilon'}/c$ for different values of the detuning $\delta = d_2 - d_1 - d/3$. The boundary conditions are $E = 0$ on the stub of length d_2 and $H = 0$ on the stub of length d_1.

exhibits the same behavior as in the case where boundary conditions at the end of the stubs are $H = 0$ (Fig. 9.14). The case where the boundary conditions at the ends of the resonators are of $E = 0$ type on both sides was also studied. The results (not shown here) are similar to those obtained in the case of $H = 0$ boundary condition on both sides. However, the resonance falls around $\Omega = 2\pi$ instead of π. The theoretical expressions are exactly the same as in Section 9.5.1 where we have to replace C_1 by S_1 and C_2 by S_2, and vice versa.

9.6 Summary and conclusion

In this chapter we have demonstrated both theoretically and experimentally the possibility to realize the classical analogue of Fano, EIT, and EIA in two simple pho-

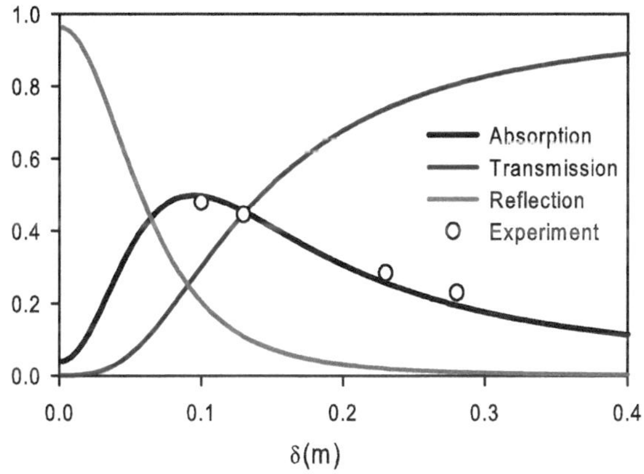

Figure 9.17 Theoretical variation of the transmission (blue line, dark gray in print version), reflection (red line, light gray in print version), and absorption (black line) versus δ at the resonance $\Omega = 3\pi/2$ in Fig. 9.8. The boundary conditions are $E = 0$ on the stub of length d_2 and $H = 0$ on the stub of length d_1. Open circles correspond to experimental data.

tonic structures made of coaxial cables. Two structures are considered with U-shape (Fig. 9.1) and cross-shape (Fig. 9.10).

For the U-structure, we have shown, through an analysis of the amplitude and phase of the transmission coefficient, different kinds of resonances like EIT and asymmetric Fano resonances. These resonances lie at the vicinity of a transmission zero or are squeezed between two transmission zeros and collapse when they coincide with the transmission zeros. The Fano resonances are obtained by tailoring the lengths of the different branches constituting the structure. A study of the phase of the transmission function enables us to deduce several properties on the wave propagation through such structures as the delay times (or the group velocities) and the density of states. For the system studied here, the phase of the amplitude transmission exhibits phase drops at the transmission zeros, whereas the Friedel phase (or equivalently the DOS) increases monotonically with a fast slope around the resonance frequency. The resonant modes give rise to well-defined peaks in the delay time and the DOS. The theoretical results are confirmed experimentally by using standard coaxial cables connected with T connectors.

For the cross-structure, we have demonstrated the existence of EIA resonances which appear as a consequence of the coupling between the two stubs where a sharp peak can exist at the same time inside the transmission and the absorption. Depending on the detuning δ between the two stubs, the absorption may reach a maximum value around 0.5 for a threshold value δ_{th} which depends on the attenuation in the system. At $\delta = \delta_{th}$, the transmission and the reflection become equal ($\simeq 0.25$). In order to understand the behavior of all the scattering parameters around the EIA resonance, we have performed in addition to numerical calculations, closed-form analytical calculations of the transmission, reflection, and absorption coefficients. In particular, we have shown that the transmission can be written following the Fano-like form, which enables us to deduce the position, width, as well as the Fano parameters of the resonance as functions of δ.

The experimental results are in very good agreement with the theoretical results obtained using the Green's function method. It is worth noting that all the results presented here for the amplitude and the phase of the transmission remain valid in microwave and optical ranges by taking properly the thicknesses of the two finite waveguides in Figs. 9.1 and 9.10. By inserting more than two stubs chosen appropriately at the same point, one can realize multi-Fano, or EIT, or EIA resonances. Also, the results presented here may be of interest for undergraduate students to understand easily a classical analogue of Fano, EIT, and EIA phenomena that take place in atomic systems [105]. Finally, it is worth mentioning that EIT and Fano resonances can be used to realize a perfect Y-shape demultiplexer composed of one input line and two output lines, where each output line contains appropriate stubs [106]. This demultiplexer enables to filter a given frequency in one line without perturbing the other line. This study will be the subject of the next chapter.

Acknowledgments

The authors would like to thank Abdelkader Mouadili, Ali Soltani, and Kamal Haddadi for their fruitful contribution in the original papers from which this chapter is inspired.

References

[1] U. Fano, Effects of configuration interaction on intensities and phase shifts, Phys. Rev. 124 (1961) 1866.
[2] For a review, see A.E. Miroshnichenko, S. Flach, Y.S. Kivshar, Fano resonances in nanoscale structures, Rev. Mod. Phys. 82 (2010) 2257.
[3] H. Beutler, Über Absorptionsserien von Argon, Krypton und Xenon zu Termen zwischen den beiden Ionisierungsgrenzen 2P32/0 und 2P12/0, Z. Phys. A 93 (1935) 177.
[4] D. Goldhaber-Gordon, H. Shtrikman, D. Mahalu, D. Abush-Magder, U. Meirav, M.A. Kastner, Kondo effect in a single-electron transistor, Nature (London) 391 (1988) 156.
[5] J. Gores, D. Goldhaber-Gordon, S. Heemeyer, M.A. Kastner, H. Shtrikman, D. Mahalu, U. Meirav, Fano resonances in electronic transport through a single-electron transistor, Phys. Rev. B 62 (2000) 2188.
[6] Y. Ji, M. Heiblum, H. Shtrikman, Transmission phase of a quantum dot with Kondo correlation near the unitary limit, Phys. Rev. Lett. 88 (2002) 076601.
[7] K. Kobayashi, H. Aikawa, S. Katsumoto, Y. Iye, Tuning of the Fano effect through a quantum dot in an Aharonov–Bohm interferometer, Phys. Rev. Lett. 88 (2002) 256806.
[8] J. Kim, J.-R. Kim, Jeong-O. Lee, J.W. Park, H.M. So, N. Kim, K. Kang, K.-H. Yoo, J.-J. Kim, Fano resonance in crossed carbon nanotubes, Phys. Rev. Lett. 90 (2003) 166403.
[9] B. Babic, C. Schönenberger, Observation of Fano resonances in single-wall carbon nanotubes, Phys. Rev. B 70 (2004) 195408.
[10] T. Taniguchi, M. Buttiker, Friedel phases and phases of transmission amplitudes in quantum scattering systems, Phys. Rev. B 60 (1999) 13814.

[11] H.W. Lee, Generic transmission zeros and in-phase resonances in time-reversal symmetric single channel transport, Phys. Rev. Lett. 82 (1999) 2358.
[12] J. Wu, B.L. Gu, H. Chen, W. Duan, Y. Kawazoe, Resonant tunneling in an Aharonov–Bohm ring with a quantum dot, Phys. Rev. Lett. 80 (1998) 1952.
[13] H. Al-Wahsh, E.H. El Boudouti, B. Djafari-Rouhani, A. Akjouj, L. Dobrzynski, Transmission gaps and sharp resonant states in the electronic transport through a simple mesoscopic device, Phys. Rev. B 75 (2007) 125313.
[14] A. Yacoby, M. Heiblum, D. Mahalu, H. Shtrikman, Coherence and phase sensitive measurements in a quantum dot, Phys. Rev. Lett. 74 (1995) 4047.
[15] R. Schuster, E. Buks, M. Heiblum, D. Mahalu, V. Umansky, H. Shtrikman, Phase measurement in a quantum dot via a double-slit interference experiment, Nature (London) 385 (1997) 417.
[16] J.O. Vasseur, A. Akjouj, L. Dobrzynski, B. Djafari- Rouhani, E.H. El Boudouti, Photon, electron, magnon, phonon and plasmon mono-mode circuits, Surf. Sci. Rep. 54 (2004) 1.
[17] For a review, see M. Fleischhauer, A. Imamoglu, J.P. Marangos, Electromagnetically induced transparency: optics in coherent media, Rev. Mod. Phys. 77 (2005) 633.
[18] S.E. Harris, Electromagnetically induced transparency, Phys. Today 50 (1997) 36.
[19] L.V. Hau, S.E. Harris, Z. Dutton, C.H. Behroozi, Light speed reduction to 17 metres per second in an ultracold atomic gas, Nature 397 (1999) 594.
[20] C. Liu, Z. Dutton, C.H. Behroozi, L.V. Hau, Observation of coherent optical information storage in an atomic medium using halted light pulses, Nature 409 (2001) 490.
[21] I. Novikova, R.L. Walsworth, Y. Xiao, Electromagnetically induced transparency-based slow and stored light in warm atoms, Laser Photonics Rev. 6 (2012) 333.
[22] L. Maleki, A.B. Matsko, A.A. Savchenkov, V.S. Ilchenko, Tunable delay line with interacting whispering-gallery-mode resonators, Opt. Lett. 29 (2004) 626.
[23] D.D. Smith, H. Chang, K.A. Fuller, A.T. Rosenberger, R.W. Boyd, Coupled-resonator-induced transparency, Phys. Rev. A 69 (2004) 063804.
[24] Q. Xu, S. Sandhu, M.L. Povinelli, J. Shakya, S. Fan, M. Lipson, Experimental realization of an on-chip all-optical analogue to electromagnetically induced transparency, Phys. Rev. Lett. 96 (2006) 123901.
[25] K. Totsuka, N. Kobayashi, M. Tomita, Slow light in coupled-resonator-induced transparency, Phys. Rev. Lett. 98 (2007) 213904.
[26] M. Tomita, K. Totsuka, R. Hanamura, T. Matsumoto, Tunable Fano interference effect in coupled-microsphere resonator-induced transparency, J. Opt. Soc. Am. B 26 (2009) 813.
[27] C.H. Raymond Ooi, C.H. Kam, Controlling quantum resonances in photonic crystals and thin films with electromagnetically induced transparency, Phys. Rev. B 81 (2010) 195119.
[28] Y.F. Xiao, M. Li, Y.C. Liu, Y. Li, X. Sun, Q. Gong, Asymmetric Fano resonance analysis in indirectly coupled microresonators, Phys. Rev. A 82 (2010) 065804.
[29] B.B. Li, Y.F. Xiao, C.L. Zou, X.F. Jiang, Y.C. Liu, F.W. Sun, Y. Li, Q. Gong, Experimental controlling of Fano resonance in indirectly coupled whispering-gallery microresonators, Appl. Phys. Lett. 100 (2012) 021108.
[30] W. Ding, B. Luḱyanchuk, C.-W. Qiu, Ultrahigh-contrast-ratio silicon Fano diode, Phys. Rev. A 85 (2012) 025806.
[31] S. Fan, J.D. Joannopoulos, Analysis of guided resonances in photonic crystal slabs, Phys. Rev. B 65 (2002) 235112.
[32] M.F. Yanik, W. Suh, Z. Wang, S. Fan, Stopping light in a waveguide with an all-optical analog of electromagnetically induced transparency, Phys. Rev. Lett. 93 (2004) 233903.

[33] X. Yang, M. Yu, D.-L. Kwong, C.W. Wong, All-optical analog to electromagnetically induced transparency in multiple coupled photonic crystal cavities, Phys. Rev. Lett. 102 (2009) 173902.
[34] M. Galli, S.L. Portalupi, M. Belotti, L.C. Andreani, L. ÓFaolain, T.F. Krauss, Light scattering and Fano resonances in high-photonic crystal nanocavities, Appl. Phys. Lett. 94 (2009) 071101.
[35] Lj. Babíc, M.J.A. de Dood, Interpretation of Fano lineshape reversal in the reflectivity spectra of photonic crystal slabs, Opt. Express 18 (2010) 26569.
[36] J. Zhou, Da Mu, J. Yang, W. Han, Xu Di, Coupled-resonator-induced transparency in photonic crystal waveguide resonator systems, Opt. Express 19 (2011) 4856.
[37] Y. Huo, S. Sandhu, J. Pan, N. Stuhrmann, M.L. Povinelli, J.M. Kahn, J.S. Harris, M.M. Fejer, S. Fan, Experimental demonstration of two methods for controlling the group delay in a system with photonic-crystal resonators coupled to a waveguide, Opt. Lett. 36 (2011) 1482.
[38] Y. Sato, Y. Tanaka, J. Upham, Y. Takahashi, T. Asano, S. Noda, Strong coupling between distant photonic nanocavities and its dynamic control, Nat. Photonics 6 (2012) 56.
[39] E.H. El Boudouti, T. Mrabti, H. Al-Wahsh, B. Djafari-Rouhani, A. Akjouj, L. Dobrzynski, Transmission gaps and Fano resonances in an acoustic waveguide: analytical model, J. Phys. Condens. Matter 20 (2008) 255212.
[40] Arturo Santillán, S.I. Bozhevolnyi, Acoustic transparency and slow sound using detuned acoustic resonators, Phys. Rev. B 84 (2011) 064304.
[41] W. Tan, C.Z. Yang, H.S. Liu, Z.G. Wang, H.Q. Lin, H. Chen, Manipulating classical waves with an analogue of quantum interference in a V-type atom, Eur. Phys. J. 97 (2012) 24003.
[42] S. Hein, W. Koch, L. Nannen, Fano resonances in acoustics, J. Fluid Mech. 664 (2010) 238;
S. Hein, W. Koch, L. Nannen, Trapped modes and fano resonances in two-dimensional acoustical duct-cavity systems, J. Fluid Mech. 692 (2012) 257.
[43] J. Fey, W.M. Robertson, Compact acoustic bandgap material based on a subwavelength collection of detuned Helmholtz resonators, J. Appl. Phys. 109 (2011) 114903.
[44] F. Liu, M. Ke, A. Zhang, W. Wen, J. Shi, Z. Liu, P. Sheng, Acoustic analog of electromagnetically induced transparency in periodic arrays of square rods, Phys. Rev. E 82 (2010) 026601.
[45] N. Liu, L. Langguth, T. Weiss, J. Kästel, M. Fleischhauer, T. Pfau, H. Giessen, Plasmonic analogue of electromagnetically induced transparency at the Drude damping limit, Nat. Mater. 8 (2009) 758.
[46] B. Luḱyanchuk, N.I. Zheludev, S.A. Maier, N.J. Halas, P. Nordlander, H. Giessen, C.T. Chong, The Fano resonance in plasmonic nanostructures and metamaterials, Nat. Mater. 9 (2010) 707.
[47] J. Wu, B. Jin, J. Wan, L. Liang, Y. Zhang, T. Jia, C. Cao, L. Kang, W. Xu, J. Chen, P. Wu, Superconducting terahertz metamaterials mimicking electromagnetically induced transparency, Appl. Phys. Lett. 99 (2011) 161113.
[48] R. Singh, I.A.I. Al-Naib, Y. Yang, D.R. Chowdhury, W. Cao, C. Rockstuhl, T. Ozaki, R. Morandotti, W. Zhang, Observing metamaterial induced transparency in individual Fano resonators with broken symmetry, Appl. Phys. Lett. 99 (2011) 201107.
[49] X. Piao, S. Yu, S. Koo, K. Lee, N. Park, Fano-type spectral asymmetry and its control for plasmonic metal–insulator–metal stub structures, Opt. Express 19 (2011) 10907.
[50] X. Piao, S. Yu, N. Park, Control of Fano asymmetry in plasmon induced transparency and its application to plasmonic waveguide modulator, Opt. Express 20 (2012) 18994.

[51] G. Wang, H. Lu, X. Liu, Dispersionless slow light in MIM waveguide based on a plasmonic analogue of electromagnetically induced transparency, Opt. Express 20 (2012) 20902.
[52] C. Kurter, P. Tassin, L. Zhang, T. Koschny, A.V. Ustinov, S.M. Anlage, C.M. Soukoulis, Classical analogue of electromagnetically induced transparency with a metal-superconductor hybrid metamaterial, Phys. Rev. Lett. 107 (2011) 043901.
[53] X. Liu, J. Gu, R. Singh, Y. Ma, J. Zhu, Z. Tian, M. He, J. Han, W. Zhang, Electromagnetically induced transparency in terahertz plasmonic metamaterials via dual excitation pathways of the dark mode, Appl. Phys. Lett. 100 (2012) 131101.
[54] K.J. Boller, A. Imamoglu, S.E. Harris, Observation of electromagnetically induced transparency, Phys. Rev. Lett. 66 (1991) 2593.
[55] S.Y. Zhu, M.O. Scully, Spectral line elimination and spontaneous emission cancellation via quantum interference, Phys. Rev. Lett. 76 (1996) 388.
[56] P. Zhou, S. Swain, Quantum interference in probe absorption: narrow resonances, transparency, and gain without population inversion, Phys. Rev. Lett. 78 (1997) 832.
[57] P. Tassin, L. Zhang, R. Zhao, A. Jain, T. Koschny, C.M. Soukoulis, Electromagnetically induced transparency and absorption in metamaterials: the radiating two-oscillator model and its experimental confirmation, Phys. Rev. Lett. 109 (2012) 187401.
[58] R. Taubert, M. Hentshel, J. Kastel, H. Giessen, Classical analog of electromagnetically induced absorption in plasmonics, Nano Lett. 12 (2012) 1367.
[59] C.W. Hsu, B. Zhen, A.D. Stone, J.D. Joannopoulos, M. Soljacic, Bound states in the continuum, Nat. Rev. Mater. 1 (2016) 16048.
[60] J. Von Neuman, E. Wigner, Über merkwürdige diskrete Eigenwerte, Phys. Z. 30 (1929) 465.
[61] Y. Plotnik, O. Peleg, F. Dreisow, M. Heinrich, S. Nolte, A. Szameit, M. Segev, Experimental observation of optical bound states in the continuum, Phys. Rev. Lett. 107 (2011) 183901.
[62] E.N. Bulgakov, A.F. Sadreev, Bound states in the continuum in photonic waveguides inspired by defects, Phys. Rev. B 78 (2008) 075105.
[63] M.I. Molina, A.E. Miroshnichenko, Y.S. Kivshar, Surface bound states in the continuum, Phys. Rev. Lett. 108 (2012) 070401.
[64] D.C. Marinica, A.G. Borisov, S.V. Shabanov, Bound states in the continuum in photonics, Phys. Rev. Lett. 100 (2008) 183902.
[65] S. Weimann, Y. Xu, R. Keil, A.E. Miroshnichenko, A. Tünnermann, S. Nolte, A.A. Sukhorukov, A. Szameit, Y.S. Kivshar, Compact surface Fano states embedded in the continuum of waveguide arrays, Phys. Rev. Lett. 111 (2013) 240403.
[66] G. Corrielli, G. Della Valle, A. Crespi, R. Osellame, S. Longhi, Observation of surface states with algebraic localization, Phys. Rev. Lett. 111 (2013) 220403.
[67] J. Mur-Petit, R.A. Molina, Chiral bound states in the continuum, Phys. Rev. B 90 (2014) 035434.
[68] R.A. Vicencio, C. Cantillano, L. Morales-Inostroza, B. Real, C. Mejía-Cortés, S. Weimann, A. Szameit, M.I. Molina, Observation of localized states in Lieb photonic lattices, Phys. Rev. Lett. 114 (2015) 245503.
[69] C.W. Hsu, Bo Zhen, J. Lee, S.-L. Chua, S.G. Johnson, J.D. Joannopoulos, M. Soljačić, Observation of trapped light within the radiation continuum, Nature (London) 499 (2013) 188.
[70] L.-S. Li, H.-C. Yin, Bound states in the continuum in double layer structures, Sci. Rep. 6 (2016) 26988.

[71] I. Quotane, E.H. El Boudouti, B. Djafari-Rouhani, Trapped-mode-induced Fano resonance and acoustical transparency in a one-dimensional solid–fluid phononic crystal, Phys. Rev. B 97 (2018) 024304.
[72] T. Mrabti, Z. Labdouti, O. El Abouti, E.H. El Boudouti, F. Fethi, B. Djafari-Rouhani, Transmission gaps, trapped modes and Fano resonances in Aharonov-Bohm connected mesoscopic loops, Phys. Lett. A 382 (2018) 613.
[73] T. Mrabti, Z. Labdouti, A. Mouadili, E.H. El Boudouti, B. Djafari-Rouhani, Aharonov-Bohm-effect induced transparency and reflection in mesoscopic rings side coupled to a quantum wire, Physica E 116 (2020) 113770.
[74] Y. El Hassouani, H. Aynaou, E.H. El Boudouti, B. Djafari-Rouhani, A. Akjouj, V.R. Velasco, Surface electromagnetic waves in Fibonacci superlattices: theoretical and experimental results, Phys. Rev. B 74 (2006) 035314.
[75] B. Djafari-Rouhani, E.H. El Boudouti, A. Akjouj, L. Dobrzynski, J.O. Vasseur, A. Mir, N. Fettouhi, J. Zemmouri, Surface states in one-dimensional photonic band gap structures, Vacuum 63 (2001) 177.
[76] L. Poirier, A. Haché, Nonlinear coaxial photonic crystal, Appl. Phys. Lett. 78 (2001) 2626.
[77] G.J. Schneider, S. Hanna, J.L. Davis, G.H. Watson, Defect modes in coaxial photonic crystals, J. Appl. Phys. 90 (2001) 2642.
[78] J.O. Vasseur, B. Djafari-Rouhani, L. Dobrzynski, A. Akjouj, J. Zemmouri, Defect modes in one-dimensional comblike photonic waveguides, Phys. Rev. B 59 (1999) 13446.
[79] E.H. El Boudouti, N. Fettouhi, A. Akjouj, B. Djafari-Rouhani, A. Mir, J.O. Vasseur, L. Dobrzynski, J. Zemmouri, Experimental and theoretical evidence for the existence of photonic bandgaps and selective transmissions in serial loop structures, J. Appl. Phys. 95 (2004) 1102.
[80] A. Haché, L. Poirier, Anomalous dispersion and superluminal group velocity in a coaxial photonic crystal: theory and experiment, Phys. Rev. E 65 (2002) 036608;
J.N. Munday, W.M. Robertson, Negative group velocity pulse tunneling through a coaxial photonic crystal, Appl. Phys. Lett. 80 (2002) 518.
[81] H. Aynaou, E.H. El Boudouti, Y. El Hassouani, A. Akjouj, B. Djafari-Rouhani, J. Vasseur, A. Benomar, V.R. Velasco, Propagation and localization of electromagnetic waves in quasiperiodic serial loop structures, Phys. Rev. E 72 (2005) 056601.
[82] E.H. El Boudouti, Y. El Hassouani, B. Djafari-Rouhani, H. Aynaou, Two types of modes in finite size one-dimensional coaxial photonic crystals: general rules and experimental evidence, Phys. Rev. E 76 (2007) 026607.
[83] L. Dobrzynski, B. Djafari-Rouhani, A. Akjouj, J.O. Vasseur, J. Zemmouri, Resonant tunneling between two continua, Phys. Rev. B 60 (1999) 10628.
[84] M.M. Sánchez-López, A. Sánchez-Meroño, J. Arias, J.A. Davis, I. Moreno, Observation of superluminal and negative group velocities in a Mach–Zehnder interferometer, Appl. Phys. Lett. 93 (2008) 074102.
[85] J. Arias, A. Sánchez-Meroño, M.M. Sánchez-López, I. Moreno, Slow and fast light in three-beam interferometers: theory and experiment, Phys. Rev. A 85 (2012) 033815.
[86] O. Hul, S. Bauch, P. Pakonski, N. Savytskyy, K. Zyczkowski, L. Sirko, Experimental simulation of quantum graphs by microwave networks, Phys. Rev. E 69 (2004) 056205.
[87] M. Ławniczak, O. Hul, S. Bauch, P. Seba, L. Sirko, Experimental and numerical investigation of the reflection coefficient and the distributions of Wigner's reaction matrix for irregular graphs with absorption, Phys. Rev. E 77 (2008) 056210.
[88] M. Ławniczak, S. Bauch, O. Hul, L. Sirko, Experimental investigation of the enhancement factor for microwave irregular networks with preserved and broken time reversal symmetry in the presence of absorption, Phys. Rev. E 81 (2010) 046204.

[89] O. Hul, M. Ławniczak, S. Bauch, A. Sawicki, M. Kuś, L. Sirko, Are scattering properties of graphs uniquely connected to their shapes?, Phys. Rev. Lett. 109 (2012) 040402.

[90] A. Mouadili, E. El Boudouti, A. Soltani, A. Talbi, A. Akjouj, B. Djafari-Rouhani, Theoretical and experimental evidence of Fano-like resonances in simple monomode photonic circuits, J. Appl. Phys. 113 (2013) 164101.

[91] A. Mouadili, E. El Boudouti, A. Soltani, A. Talbi, B. Djafari-Rouhani, A. Akjouj, K. Haddadi, Electromagnetically induced absorption in detuned stub waveguides: a simple analytical and experimental model, J. Phys. Condens. Matter 26 (2014) 505901.

[92] H. Al-Wahsh, E.H. El Boudouti, B. Djafari-Rouhani, A. Akjouj, T. Mrabti, L. Dobrzynski, Evidence of Fano-like resonances in mono-mode magnetic circuits, Phys. Rev. B 78 (2008) 075401.

[93] L. Dobrzynski, Interface response theory of continuous composite systems, Surf. Sci. Rep. 11 (1990) 139.

[94] C.R. Paul, Analysis of Multiconductor Transmission Lines, Wiley, 1994.

[95] M.L.H. Lahlaouti, A. Akjouj, B. Djafari-Rouhani, L. Dobrzynski, M. Hammouchi, E.H. El Boudouti, A. Nougaoui, B. Kharbouch, Theoretical analysis of the density of states and phase times: application to resonant electromagnetic modes in finite superlattices, Phys. Rev. B 63 (2001) 035312.

[96] L. Brillouin, Wave Propagation and Group Velocity, Academic, New York, 1960.

[97] L.J. Wang, A. Kusmich, A. Dogarin, Gain-assisted superluminal light propagation, Nature (London) 406 (2000) 277.

[98] B. Macke, B. Segard, Propagation of light-pulses at a negative group-velocity, Eur. Phys. J. D 23 (2003) 125.

[99] J. Friedel, The distribution of electrons round impurities in monovalent metals, Philos. Mag. 43 (1952) 153.

[100] S. Souma, A. Suzuki, Local density of states and scattering matrix in quasi-one-dimensional systems, Phys. Rev. B 65 (2002) 115307.

[101] S. Khattou, M. Amrani, A. Mouadili, E.H. El Boudouti, A. Talbi, A. Akjouj, B. Djafari-Rouhani, Comparison of density of states and scattering parameters in coaxial photonic crystals: theory and experiment, Phys. Rev. B 102 (2020) 165310.

[102] M.L. Ladrón de Guevara, F. Claro, Pedro A. Orellana, Ghost Fano resonance in a double quantum dot molecule attached to leads, Phys. Rev. B 67 (2003) 195335.

[103] H. Barkay, E. Narevicius, N. Moiseyev, Non-Hermitian scattering theory: resonant tunneling probability amplitude in a quantum dot, Phys. Rev. B 67 (2003) 045322.

[104] F. Richter, M. Florian, K. Henneberger, Poynting's theorem and energy conservation in the propagation of light in bounded media, Europhys. Lett. 81 (2008) 67005.

[105] L.G. Alzar, M.A.G. Martinez, P. Nussenzveig, Classical analog of electromagnetically induced transparency, Am. J. Phys. 70 (2002) 37.

[106] A. Mouadili, E.H. El Boudouti, A. Soltani, A. Talbi, K. Haddadi, A. Akjouj, B. Djafari-Rouhani, Photonic demultiplexer based on electromagnetically induced transparency resonances, J. Phys. D 52 (2019) 075101.

Photonic demultiplexers based on Fano and induced transparency resonances

10

El Houssaine El Boudouti[a], *Abdellatif Akjouj*[b], *Bahram Djafari-Rouhani*[b], *Abdelkrim Talbi*[c], *and Leonard Dobrzyński*[b]
[a]Laboratoire de Physique de la Matière et de Rayonnement, Département de Physique, Faculté des Sciences, Université Mohamed I, Oujda, Morocco, [b]Institute of Electronic, Microelectronics and Nanotechnology (IEMN), UMR CNRS 8520, Department of Physics, Faculty of Sciences and Technologies, Lille University, Villeneuve d'Ascq Cedex, France, [c]Institut d'Electronique, de Microélectronique et de Nanotechnologie (IEMN), UMR CNRS 8520, Ecole Centrale de Lille, Villeneuve d'Ascq, France

10.1 Introduction

A wavelength demultiplexer is a very important device for improving the capacity of optical communications to transmit several pieces of information in the form of electromagnetic signals simultaneously in a single optical fiber with a low cross-talk ratio between the output channels [1,2]. This requires the use of multiplexers (MUX) and demultiplexers (DEMUX) [3]. Both techniques are often based on either the spatial division multiplexing (SDM) or the wavelength division multiplexing (WDM). The first technique is used to divide the fiber into multiple independent channels, and to transmit a single signal on each channel, while the second technique is used to operate the bandwidth to get several different wavelength signals on the same fiber. Let us quote also the technique called dense wavelength division multiplexed optical networks [4].

The use of a WDM communication system allows for better use of the resources of the bandwidth provided by the system. Along these lines, the proposed WDM systems use various technologies, such as planar circuits based on grilled waveguide arrays [5,6] and fiber networks [7]. Much attention has been paid for designing demultiplexers, trying to find an adapted solution for their characteristics. Decreasing the size of the structure, increasing the number of channels, and decreasing cross-talk between channels and spacing between adjacent channels are the perspectives of new designs [2]. The conventional wavelength demultiplexers in modern optical devices and communications perform excellent wavelength filtering function. But these WDMs have large size and cannot be utilized in highly integrated optical circuits due to the diffraction limit. For this reason, WDM devices based on photonic crystals (PhCs) have been proposed [8,9].

Photonics. https://doi.org/10.1016/B978-0-12-819388-4.00020-4

PhCs are promising as a platform for compact photonic integrated circuits because of certain characteristics such as their ability to manipulate light at wavelength scale, substantial size reduction, low loss, electromagnetic wave emission controllability, high rate of performance speed, long life period, and the feasibility of their integration in optical circuits [10]. Various optical devices have been realized based on PhCs, such as optical switches [11,12], filters [13,14], splitters [15,16], polarizers [17], mirrors [18], and demultiplexers [19–25].

In addition to PhCs, several types of plasmonic demultiplexers for integrated technology have been proposed, such as demultiplexers based on metal–insulator–metal (MIM) waveguide [26]. Such systems have the potential to exceed the classical diffraction limit which makes them suitable for use in several nanoscale devices. Demultiplexers based on MIM systems can have a very important resolution of about a few nm. In this case, the systems contain an MIM waveguide with resonators or nanocavities [26–34].

In the case of demultiplexers based on photonic crystals, we can filter a frequency in the bandgap utilizing the resonant cavity which allows the localization of a particular resonant state to be filtered [2,23,35–41]. A cavity is one of the elements for which the PhC was developed and can filter very effectively a given wavelength out of different wavelengths. There are different resonance wavelengths for a particular cavity which allows an efficient trapping of those wavelengths. Another method consists of grafting resonators on the waveguides to have transmission zeros and resonances [10,42–48].

The coupling of photonic waveguides with one or more resonators has been studied and applied to obtain Fano- and EIT-type resonances [49–58]. This process can also be applied to the design of a wavelength photonic crystal demultiplexer [59–62]. The device utilizes electromagnetic induced transparency (EIT) and Fano-like resonances based on a photonic crystal waveguide managed to achieve high resolutions with relatively high transmittance and high quality factor.

To achieve a high resolution of wavelength demultiplexing, the separation of wavelengths must be significantly larger than the full width at half maximum (FWHM), which limits the performance of these functional devices [28,42,43]. This resolution of the demultiplexer is essential to increase the number of channels that can be multiplexed over a given band. In general, there is a trade-off between FWHM and transmittance [30].

In Chapter 9, we have shown the possibility of existence of Fano, electromagnetic induced transparency (EIT), and electromagnetic induced absorption (EIA) resonances in a simple photonic circuit made of a waveguide with two grafted stubs placed at the same site [52] or at two different sites [51]. Such resonances are obtained by detuning the lengths of two stubs of lengths d_1 and d_2 (i.e., $\delta = d_2 - d_1$) grafted at the same site or separated by a segment of length d_3. These resonances can be either squeezed between two transmission zeros induced by the two stubs, giving rise to EIT-like resonances (here the two stubs should be chosen slightly different), or falling near to two transmission zeros induced by identical stubs, giving rise to Fano-like resonances.

In this chapter we propose two Y-shaped acoustic demultiplexers with two different configurations: the first structure is based on the cross-shape structure [63] (Fig. 10.1), and the second is based on the U-shape structure (Fig. 10.9). The demultiplexers in these two devices are based on EIT and Fano resonances. The existence of transmission zeros around Fano and EIT resonances enables us to derive in closed form the expressions of the geometrical parameters for selective transfer of a single propagating mode through one output line, keeping the other line unaffected by the mode transmitted through the other output signal. We illustrate these analytical results by some numerical solutions for a simple structure made out of one-dimensional photonic waveguides. In addition, the exact results are supported by approximate results (Taylor expansion) around the resonances, showing that the transmission coefficients can be written following a Fano-like form where the position, width, and Fano parameters of the resonances can be deduced explicitly as functions of the detuning $\delta = d_2 - d_1$ between the stubs. Finally, an experimental demonstration of this filtering between output lines is given for the wave propagation in coaxial cables in the radio-frequency regime.

The demultiplexers proposed in this study have several advantages over those based on photonic crystals, such as (i) the simplicity of the device manufacturing where only two resonators are needed on each output line instead of a periodic structure with a given defect, (ii) the simplicity of the structure enables a full analytical calculation which allows deducing the exact expressions of the different lengths of the waveguides to achieve a perfect demultiplexing, and (iii) the possibility of increasing the quality factor of the filtered resonances to infinite values by detuning the lengths of the two stubs. The latter property is a feature of Fano and induced transparency resonances that does not exist in standard photonic crystals with defects, in which filtering is performed using finite width Breit–Wigner resonances [64]. It should be pointed out that the validity of our results is subject to the requirement that the cross-section of the slender tubes is negligible compared to their length and to the propagation wavelength.

10.2 Demultiplexer based on a cross structure

10.2.1 Transmission and reflection coefficients through a demultiplexer

Let us consider the structure depicted in Fig. 10.1(a). This structure consists of an input line and two output lines, all attached to point 1. The first output line contains two stubs of lengths d_1 and d_2 inserted at the same site 2 at a distance d_5 from entrance 1. Similarly, the second output line contains two stubs of lengths d_3 and d_4 inserted at the same site 3 at a distance d_6 from entrance 1 (Fig. 10.1(a)). Fig. 10.1(b) illustrates the experimental set-up showing the Y-shaped system with coaxial cables connected to the vector network analyzer. The boundary conditions at the end of the stubs are $H = 0$. The different lengths d_i of the cables are indicated in accordance with Fig. 10.1(a). When one output line is connected to the analyzer entrance, the second output has 50 Ω impedance matching in order to avoid any reflection of the wave at the end of this line.

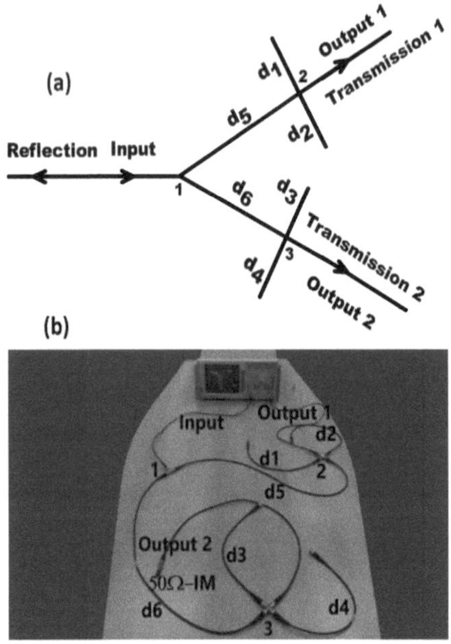

Figure 10.1 (a) Schematic representation of a Y-shaped demultiplexer with one input line and two output lines. Along the first output line, two stubs of lengths d_1 and d_2 are inserted at the same position on the waveguide at a distance d_5 from the input line. Along the second output line, two stubs of lengths d_3 and d_4 are inserted at the same position on the waveguide at a distance d_6 from the input line. (b) Experimental set-up showing the Y-shaped system with coaxial cables connected to the vector network analyzer. The boundary conditions at the end of the stubs are $H = 0$. The different lengths of the cables are indicated in accordance with (a). When one output line is connected to the analyzer entrance, the second output has 50 Ω impedance matching in order to avoid any reflection of the wave at the end of this line.

In order to calculate the transmission and reflection coefficients, we should first build the expression giving the inverse of the Green's function of the whole system (Fig. 10.1(a)) in the interface space $M = \{1, 2, 3\}$, this expression can be obtained from a linear superposition of the above inverse Green's functions of the constituents. Therefore, we recall hereafter the inverse of the Green's function of each piece, namely:

(i) The inverse of the surface Green's function of a semiinfinite waveguide constituting the input and the output lines. Its expression is given by [65]

$$g_i^{-1}(0,0) = j\omega/Z, \tag{10.1}$$

where Z is the characteristic impedance of the semiinfinite wire i.

(ii) The inverse of the surface Green's functions of the dangling resonators (stubs) grafted vertically at a given site $\{0\}$ when the boundary conditions at the ends of the stubs are $H = 0$ (perfect magnetic conductor) at their ends. Its expression is given by [65]

$$g_i^{-1}(0,0) = -\omega S_i / Z C_i, \tag{10.2}$$

where $C_i = \cos(kd_i)$, $S_i = \sin(kd_i)$, and $k = \omega\sqrt{\varepsilon}/c$.

(iii) The inverse of the Green's function of a wire of length d_i in its space of interface $M_i = \{0, d_i\}$ is given by [65]

$$g_i^{-1}(M,M) = \begin{pmatrix} -\frac{\omega C_i}{Z S_i} & \frac{\omega}{Z S_i} \\ \frac{\omega}{Z S_i} & -\frac{\omega C_i}{Z S_i} \end{pmatrix} \tag{10.3}$$

From Eqs. (10.1), (10.2), and (10.3), one can build easily the inverse of the Green's function in the space of interfaces $M = \{1, 2, 3\}$ of the structure depicted in Fig. 10.1(a), namely

$$g^{-1}(M,M) = \frac{-\omega}{Z} \begin{pmatrix} \frac{C_5}{S_5} + \frac{C_6}{S_6} - j & -\frac{1}{S_5} & -\frac{1}{S_6} \\ -\frac{1}{S_5} & \frac{C_5}{S_5} + \frac{S_1}{C_1} + \frac{S_2}{C_2} - j & 0 \\ -\frac{1}{S_6} & 0 & \frac{C_6}{S_6} + \frac{S_3}{C_3} + \frac{S_4}{C_4} - j \end{pmatrix}. \tag{10.4}$$

From the inverse of Eq. (10.4), one can deduce in closed from the expressions of the transmission and reflection coefficients. Indeed, the reflection amplitude wave in the input line of the demultiplexer (Fig. 10.1(a)) is given by [65] $r = -1 - 2F_s g(1,1)$, or equivalently

$$r = -\frac{\xi_1 + j\xi_2}{\chi_1 + j\chi_2} \tag{10.5}$$

where

$$\xi_1 = C_1C_2C_3C_4(S_5S_6 - C_5C_6) + C_1C_2S'C_6S_5 + C_3C_4SC_5S_6 + SS'S_5S_6, \tag{10.6}$$

$$\xi_2 = C_1C_2C_3C_4S_0 + C_1C_2S'C_5C_6 + C_3C_4SC_5C_6 - S_0SS', \tag{10.7}$$

$$\begin{aligned} \chi_1 = {} & 3C_1C_2C_3C_4(S_5S_6 - C_5C_6) + C_1C_2S'(S_0 + C_5S_6) \\ & + C_3C_4S(S_0 + C_6S_5) - SS'S_5S_6, \end{aligned} \tag{10.8}$$

$$\begin{aligned} \chi_2 = {} & C_1C_2C_3C_4(3C_6S_5 + 3C_5S_6) + (C_5C_6 - 2S_5S_6)(C_1C_2S' + C_3C_4S) \\ & - S_0SS', \end{aligned} \tag{10.9}$$

and $S = \sin(k(d_1 + d_2))$, $S' = \sin(k(d_3 + d_4))$, $S_0 = \sin(k(d_5 + d_6))$.

The expression of the reflection R in the input line is given by

$$R = |r|^2. \tag{10.10}$$

Also, the transmission amplitude waves in the output lines 1 and 2 (Fig. 10.1(a)) are given respectively by [65] $t_1 = (2j\omega/Z)g(1,2)$ and $t_2 = (2j\omega/Z)g(1,3)$, or equivalently

$$t_1 = \frac{2C_1C_2(-C_6C_3C_4 + S'S_6 + jC_3C_4S_6)}{\chi_1 + j\chi_2} \tag{10.11}$$

and

$$t_2 = \frac{2C_3C_4(-C_5C_1C_2 + SS_5 + jC_1C_2S_5)}{\chi_1 + j\chi_2}. \tag{10.12}$$

The transmission coefficients in the two output lines are given respectively by

$$T_1 = |t_1|^2 \tag{10.13}$$

and

$$T_2 = |t_2|^2. \tag{10.14}$$

The transmission and reflection coefficients satisfy the energy conservation: $T_1 + T_2 + R = 1$. The closed form expressions of the transmission and reflection coefficients (Eqs. (10.5)–(10.14)) can be used by any reader interested in this subject to reproduce the theoretical calculations without going into the details of the calculations.

10.2.2 *Analytical and numerical results*

Now we are able to specify the system parameters to achieve complete transmission in both lines with closed frequencies. Indeed, from Eqs. (10.5), (10.11), and (10.12), one can show easily that in order to realize $|T_1| = 1$, $T_2 = 0$, and $R = 0$, one should have $C_3C_4 = 0$ (i.e., $C_3 = 0$ or $C_4 = 0$), $S = 0$, and $C_6 = 0$. Similarly, in order to realize $|T_2| = 1$, $T_1 = 0$, and $R = 0$, one should have $C_1C_2 = 0$ (i.e., $C_1 = 0$ or $C_2 = 0$), $S' = 0$, and $C_5 = 0$. Now, in order to realize both conditions (i.e., $|T_1| = 1$ and $|T_2| = 1$) at two different but closed frequencies, one can show after some algebraic calculation that the six lengths d_1, d_2, d_3, d_4, d_5, and d_6, should satisfy the following conditions:

$$d_3 = \frac{d_1 + d_2}{2}, \tag{10.15}$$

$$d_4 = \frac{3d_2 - d_1}{2}, \tag{10.16}$$

$$d_5 = d_2, \tag{10.17}$$

$$d_6 = d_3. \tag{10.18}$$

As we have demonstrated in Chapter 9, in order to realize an EIT resonance, one should take the lengths of the two stubs along the waveguide slightly different, i.e., $\delta = d_2 - d_1 \neq 0$. Also, along this study we shall fix the lengths of the two stubs, i.e., $d_0 = d_2 + d_1$, which fixes the position of the EIT resonance along the output 1.

Therefore, the above conditions (Eqs. (10.15), (10.16), (10.17), and (10.18)) become respectively:

$$d_1 = \frac{d_0}{2} - \frac{\delta}{2}, \tag{10.19}$$

$$d_2 = d_5 = \frac{d_0}{2} + \frac{\delta}{2}, \tag{10.20}$$

$$d_3 = d_6 = \frac{d_0}{2}, \tag{10.21}$$

$$d_4 = \frac{d_0}{2} + \delta. \tag{10.22}$$

In order to illustrate the above analytical results, we present in Fig. 10.2 the variation of the transmission coefficients T_1, T_2 and the reflection coefficient R versus the dimensionless frequency $\Omega = \omega d_0\sqrt{\varepsilon}/c$ for different values of δ around $\delta = 0$. Fig. 10.2 clearly shows that when the transmission in the first output (blue line, dark gray in print version) is unity ($T_1 = 1$), the transmission in the second output T_2 (red line, light gray in print version) and the reflection R (black line) vanish (i.e., $T_2 = R = 0$). Similarly, when the transmission in the second output (red line, light gray in print version) is unity ($T_2 = 1$), the transmission in the first output T_1 (blue line, dark gray in print version) and the reflection R (black line) vanish (i.e., $T_1 = R = 0$). As mentioned in Chapter 9, when the total length $d_0 = d_1 + d_2$ of the two stubs is fixed, the EIT resonance in the first output falls at the same frequency ($\Omega = \pi$) for all the values of δ, its width decreases when δ decreases and vanishes for $\delta = 0$ (Fig. 10.2(c)). The hidden resonances in Fig. 10.2(c) are known as bound in continuum (BIC) states [66]. Such modes, characterized by an infinite lifetime, represent confined states (eigenmodes) in the vertical stubs and do not interact with the horizontal waveguides. Also, the shape and width of the EIT resonance slightly changes when δ becomes negative (i.e., when permuting the two stubs 1 and 2). However, the position and width of the resonance in the second output depend strongly on δ. Indeed, as the first EIT resonance exhibits two transmission zeros around $\Omega = \pi$, the position of the second EIT resonance falls below $\Omega = \pi$ for $\delta < 0$ at the left of $\Omega = \pi$ resonance, crosses the first resonance at $\delta = 0$, and reappears above $\Omega = \pi$ for $\delta > 0$ at the right of $\Omega = \pi$ resonance (Fig. 10.2). This is illustrated in the inset of Fig. 10.3 where the crossing of the resonances occurs at $\delta = 0$.

The two filtered resonances exhibit an EIT-like shape, i.e., a resonance squeezed between two transmission zeros induced by the stubs. Indeed, a Taylor expansion of the transmission amplitude (Eqs. (10.11) and (10.13)) around the resonance at $\Omega = \pi$ (i.e., $\Omega = \pi + \varepsilon$) enables us to write T_1 following the EIT-like resonance, namely

$$T_1 = A\frac{(\varepsilon + q_1\Gamma)^2(\varepsilon + q_2\Gamma)^2}{\varepsilon(\varepsilon + \beta) + \Gamma^2}, \tag{10.23}$$

where $A = \frac{(\frac{1}{4} - \frac{\Delta^2}{\pi^2})^2}{\frac{1}{4} + \frac{5\Delta^2}{8} + \frac{\Delta^3}{2\pi^2}}$, $\beta = \frac{(\frac{4\Delta^4}{\pi} - \Delta^5 + \frac{6\Delta^6}{\pi})}{\frac{1}{4} + \frac{5\Delta^2}{8} + \frac{\Delta^3}{2\pi^2}}$, and $\Delta = -\pi\delta/2d_0$; Γ is the full width at half maximum of the EIT resonance falling at $\varepsilon = 0$ (i.e., $\Omega = \pi$); its expression is

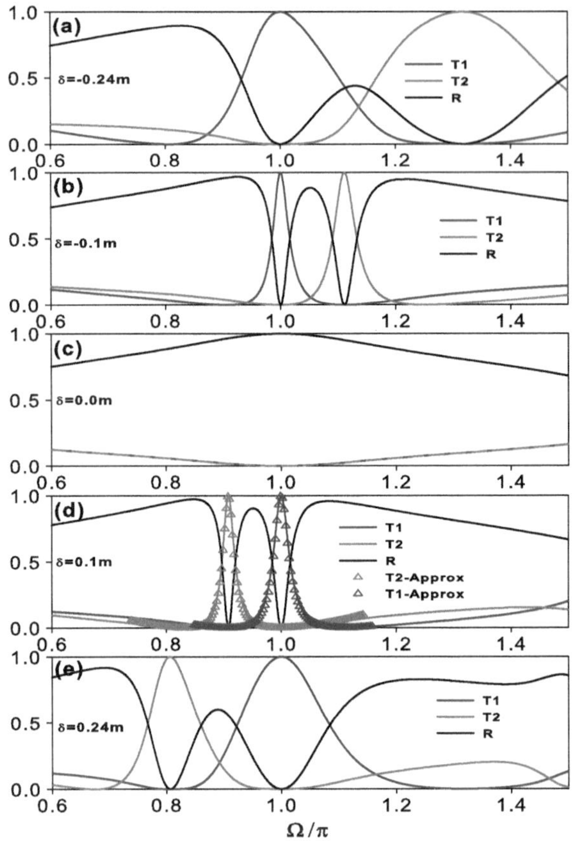

Figure 10.2 Variation of the intensity of the transmitted signal in output 1 (blue line, dark gray in print version), output 2 (red line, light gray in print version), and the reflected signal in the input of the demultiplexer (black line) versus the dimensionless frequency $\Omega = \omega d_0 \sqrt{\varepsilon}/c$ for different values of $\delta = d_2 - d_1$ and for $d_0 = d_1 + d_2$ fixed. Triangles in Fig. 10.2(d) correspond to Taylor expansions.

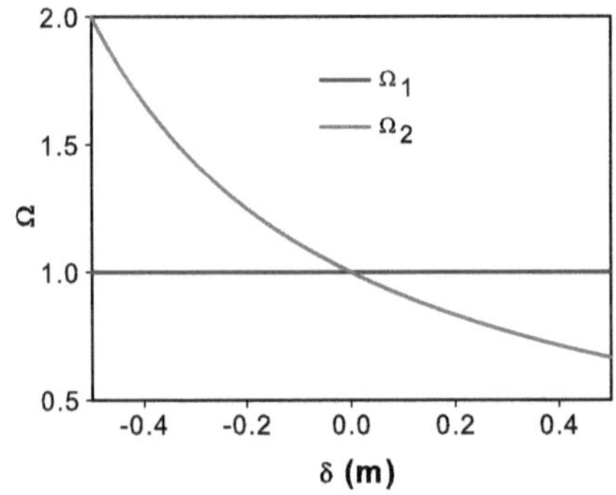

Figure 10.3 Variation of the positions of the two resonances in Fig. 10.2 as a function of δ.

given by

$$\Gamma = \frac{\Delta^2}{(\frac{1}{4} + \frac{5\Delta^2}{8} + \frac{\Delta^3}{2\pi^2})^{1/2}}, \tag{10.24}$$

$q_1 = \frac{\Delta}{\Gamma(\frac{1}{2}+\frac{\Delta}{\pi})}$, and $q_2 = -\frac{\Delta}{\Gamma(\frac{1}{2}-\frac{\Delta}{\pi})}$ are the Fano parameters; they give qualitatively the strength of the interference between the bound state and the propagating continuum states [67,68]. One can notice that when increasing δ from 0, Γ increases (i.e., the quality factor decreases) and q_1 and q_2 decrease (see below). The approximate expression for the second transmission coefficient T_2 is given by

$$T_2 = B\frac{(\varepsilon + q_1'\Gamma)^2(\varepsilon + q_2'\Gamma')^2}{\varepsilon(\varepsilon + \beta') + \Gamma'^2}, \tag{10.25}$$

where $B = \frac{(\frac{1}{4}-\frac{\Delta'^2}{\pi^2})^2}{\frac{1}{4}+\frac{5\Delta'^2}{8}+\frac{\Delta'^3}{2\pi^2}}$, $\beta' = \frac{(\frac{4\Delta'^4}{\pi}-\Delta'^5+\frac{6\Delta'^6}{\pi})}{\frac{1}{4}+\frac{5\Delta'^2}{8}+\frac{\Delta'^3}{2\pi^2}}$, and $\Delta' = -\pi\delta'/2d_0'$ with $d_0' = d_3 + d_4$. The width of the resonance Γ' is given by

$$\Gamma' = \frac{\Delta'^2}{(\frac{1}{4} + \frac{5\Delta'^2}{8} + \frac{\Delta'^3}{2\pi^2})^{1/2}}, \tag{10.26}$$

$q_1' = \frac{\Delta'}{\Gamma'(\frac{1}{2}+\frac{\Delta'}{\pi})}$ and $q_2' = -\frac{\Delta'}{\Gamma'(\frac{1}{2}-\frac{\Delta'}{\pi})}$ are the Fano parameters.

It is worth noticing from Eqs. (10.19)–(10.22) that $d_4 - d_3$ keeps the same value as $d_2 - d_1$. An example of the results of the approximate expression (Eq. (10.23)) is shown by the red triangles (light gray in print version) in Fig. 10.2(d) for $\delta = 0.1d_0$ and clearly shows that the resonance exhibits an EIT shape with $q_1 = -7.4$, $q_2 = 6.06$, and width $2\Gamma = 0.03\pi$. As a matter of comparison, in periodic photonic crystals, a perturbation (i.e., a defect) is often introduced into the system in order to create the resonance state in the gap [64]. However, the above calculation shows that, without introducing any perturbation in the structure, one can find a well-defined EIT resonance with a width 2Γ and coupling parameters q_1 and q_2 that can be adjusted by tailoring the lengths of the resonators (i.e., δ).

The full width at half maximum (or equivalently the quality factor) of the resonances depends strongly on the difference between the lengths of the stubs (i.e., $\delta = d_2 - d_1 = d_4 - d_3$). This is illustrated in Fig. 10.4, where we have plotted the quality factor of the resonance at $\Omega = \pi$ which is defined by $Q = \pi/2\Gamma$ where Γ is given by Eq. (10.24). One can notice that the quality factor of the resonance depends slightly on the sign of δ and as predicted diverges when δ vanishes. The above analytical calculations (Eqs. (10.23) and (10.25)) clearly show that despite the existence of a bifurcation of the input line on two output lines, the transmission through each line can be written following a Fano line shape as it is the case for a single line [52]. Also in order to understand better the scattering properties of the filtered resonances, we have given the analytical expressions of Fano parameters q and quality factors Q of the EIT resonances for each line.

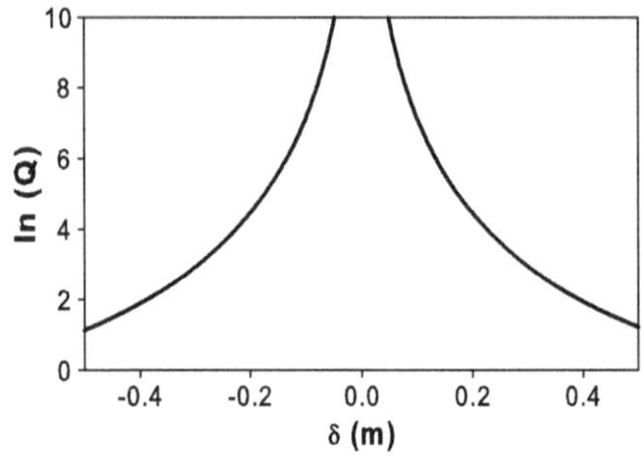

Figure 10.4 Variation of the logarithm of the quality factor Q versus δ of the transmitted signal in the first output.

Now, if one of the above conditions (Eqs. (10.15)–(10.18)) is not satisfied, then the transmission in one line may be affected by the geometry of the second line. An example is given in Fig. 10.5, where we have chosen the same parameters as in Fig. 10.2(a) except that $d_6 \neq d_3 = 0.5d_0$. Indeed, in order to clarify the effect of the detuning between d_6 and d_3 on the quality of the demultiplexing, we have taken d_6 and d_3 slightly different (see, for example, Fig. 10.5(b) for $d_3 = 0.5d_0$ and $d_6 = 0.55d_0$), then the results are less affected, and one obtains almost the same results as in the perfect case, i.e., $d_6 = d_3 = 0.5d_0$ (Fig. 10.5(a)). However, when d_6 is taken greater than $0.55d_0$ (see Fig. 10.5(c) for $d_6 = 0.6d_0$), then one obtains an incomplete transmission through the first line around $\Omega = \pi$ (blue curve, dark gray in print version). For $d_6 > 0.6d_0$, the transmission through the first outline becomes less than unity as it is the case in Fig. 10.5(d) for $d_6 = 0.75d_0$. One can notice that the transmission in the first output (blue line, dark gray in print version) does not reach unity at $\Omega = \pi$, and at the same time the reflection coefficient (black line) does not reach zero.

In order to analyze the spatial localization of the different modes that can be filtered or stopped by the demultiplexer, we have calculated the displacement field along the two output lines of the system [65]. Fig. 10.6 gives the square modulus of the electric field $|E|^2$ for the resonance $\Omega = \pi$ with $\delta = 0.1$ m (Fig. 10.2(d)). This mode corresponds to a filtered mode (full curve) in the first line and a stopped mode (dashed curve) in the second line (Fig. 10.2(d)). Fig. 10.6 clearly shows that the mode $\Omega = \pi$ is transferred along the output 1, whereas it is stopped along the output 2 (red curves, light gray in print version). The transfer of this mode along the output 1 is due to the excitation of both stubs of lengths d_1 and d_2 along this line as it is illustrated, whereas its stopping along the second line is due to the excitation of the stationary mode of only the stub of length d_4 (blue curves, dark gray in print version). Similar results are obtained for the second EIT resonance mode, but this time the transfer occurs along the second line through the excitation of its double stubs of lengths d_3 and d_4, whereas the wave is stopped along the first line as a consequence of the excitation of the mode of one of its stubs of length d_2. These results clearly show how the lengths of the finite guides constituting the demultiplexer should be chosen appropriately in order to transfer a wavelength in one line while keeping the other line unaffected.

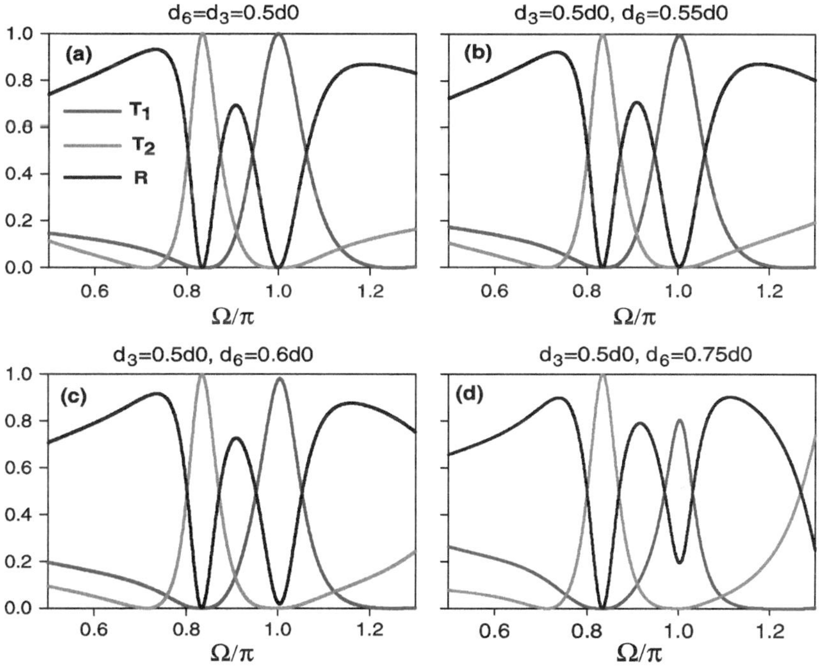

Figure 10.5 Same as in Fig. 10.2(d), except that $d_6 \neq d_3$.

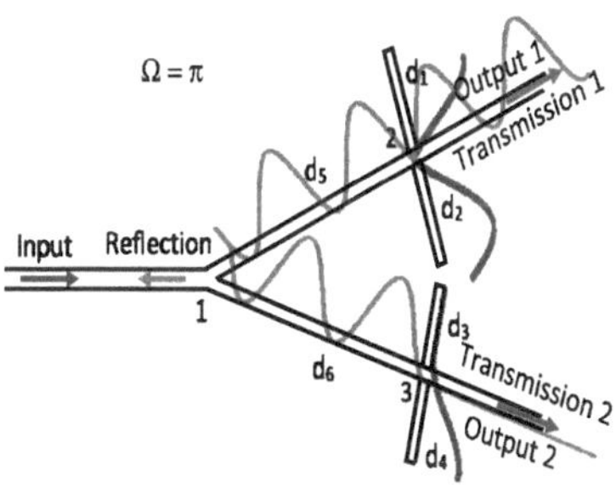

Figure 10.6 Square modulus of the electric field $|E|^2$ (in arbitrary units) versus the space position along the output 1 and the output 2 for $\Omega = \pi$ and $\delta = 0.1$ m. The electric field along the two branches and the stubs are indicated by red (light gray in print version) and blue (dark gray in print version) curves, respectively.

10.2.3 Experimental results

In order to give an experimental demonstration of the above analytical results, we performed an experiment in the radio-frequency domain using standard coaxial cables with the same characteristic impedances and permittivities [65]. The studied coaxial cable circuit consists of segments of RG-58/U coaxial cable of different lengths (Fig. 10.1(b)) with characteristic impedance $Z_c = 52\ \Omega$. The cables were filled with polyethylene ($\varepsilon' = 2.3$) corresponding to a nominal propagation speed of $0.66c$. The segments of coaxial cables were connected by standard BNC T connectors (Fig. 10.1(b)). The experimental set-up with a brief introduction to the experimental procedure are summarized in Chapter 9. The scattering matrix S_n of the 1D comb-like photonic crystal was measured in the frequency range 1–200 MHz by means of a broadband vector network analyzer Agilent PNA-X N5242A. The input source power is set to −20 dB. The number of acquisition points and the intermediate frequency bandwidth (IFBW) are set to 5000 points and 100 Hz, respectively, to yield a good compromise between the measurement accuracy and the acquisition time. The network analyzer was calibrated in advance using electronic calibration kit (Ecal). In particular, a Flush Thru standard is considered in transmission calibration to yield zero-length, and consequently no delay and no loss. The experimental results include both the magnitude and phase-shift of transmission ($S_{12} = S_{21}$) and reflection ($S_{11} = S_{22}$) coefficients. The attenuation inside the coaxial cables was simulated by introducing a complex dielectric constant ($\varepsilon' + j\varepsilon''$). The attenuation coefficient α'' can be expressed as $\alpha'' = \varepsilon''\omega/c\sqrt{\varepsilon'}$. The attenuation specification data supplied by the manufacturer of the coaxial cables in the frequency range of 10 to 200 MHz can be approximately fitted with the expression $\ln(\alpha'') = \gamma + \eta \ln(\omega)$, where γ and η are two constants. From this fitting procedure, a useful expression for ε'' as a function of frequency can be obtained under the form $\varepsilon'' = 0.017 f^{-0.5}$ where the frequency f is expressed in hertz. The boundary conditions at the upper ends of the stubs in Fig. 10.1(b) are those of a perfect magnetic conductor (i.e., vanishing magnetic field). It should be pointed out that the validity of our results is subject to the requirement that the cross-section of the waveguide is negligible compared to its length and to the propagation wavelength. The assumption of monomode propagation is then satisfied. The electric and magnetic fields inside the insulator between the inner and outer cylindrical conductors of the coaxial cable are polarized perpendicularly to the axis of the cable and propagate along this axis. Therefore, these fields can be treated as scalars.

Fig. 10.7 gives the theoretical (solid line) and experimental (open circles) transmission coefficients versus the frequency along ports 1 (blue curves, dark gray in print version) and 2 (red curves, light gray in print version) for $\delta = 0.2$ m. The lengths of the cables are chosen such that $d_1 = 0.40$ m, $d_2 = 0.60$ m (i.e., $d_0 \simeq 1$ m and $\delta \simeq 0.2$ m). The lengths of the other cables are chosen following the conditions in Eqs. (10.19)–(10.22), namely $d_5 = 0.60$ m, $d_3 = d_6 = 0.5$ m, and $d_4 = 0.70$ m. The experiments are in good agreement with the theoretical results. The filtered resonance in each line (around $\Omega_1 = 0.82\pi$ and $\Omega_2 = \pi$) does not reach unity because of the attenuation in the cables. However, we see clearly that when the transmission is maximal in one line, the corresponding transmission in the other line and the reflection in the input line vanish simultaneously.

From the expressions of r, t_1, and t_2 (Eqs. (10.5), (10.11), (10.12)), one can deduce also the phases φ_r, φ_{t1}, and φ_{t2} of the reflection and transmission coefficients or their first derivatives with respect to the pulsation ω, called delay times. These quantities

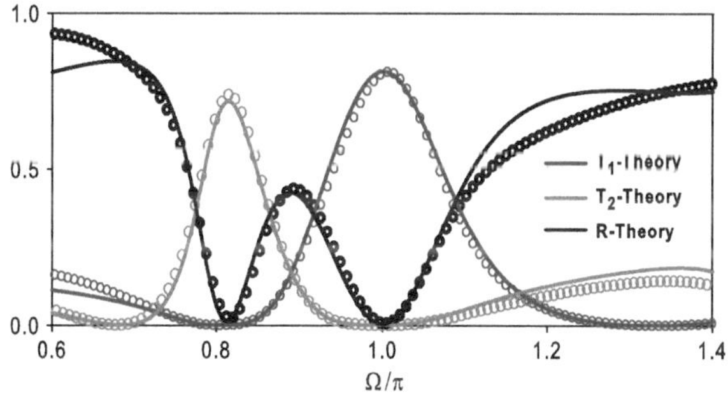

Figure 10.7 Theoretical (solid line) and experimental (open circles) transmission coefficients versus the frequency along ports 1 (blue curves, dark gray in print version) and 2 (red curves, light gray in print version). The reflection coefficient is given by black curves. The lengths of the cables are chosen such that $d_1 = 0.375$ m, $d_2 = d_5 = 0.62$ m, $d_3 = d_6 = 0.5$ m and $d_4 = 0.725$ m. These parameters are in accordance with Eqs. (10.20)–(10.23).

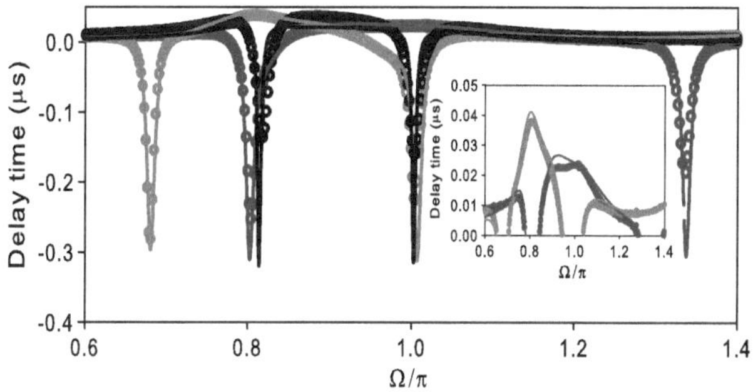

Figure 10.8 Theoretical (solid line) and experimental (open circles) delay time versus the frequency for the transmissions and reflection coefficients described in Fig. 10.7. The insert shows a zoom of the delay time around the transmission resonances in the two output lines.

are defined respectively by [69–71]

$$\tau_r = \frac{d\varphi_r}{d\omega}, \quad \tau_{t1} = \frac{d\varphi_{t1}}{d\omega}, \quad \tau_{t2} = \frac{d\varphi_{t2}}{d\omega}. \tag{10.27}$$

An example of the delay times associated to the transmission and reflection amplitudes are plotted in Fig. 10.8. As predicted, the transmission zeros give rise to a phase drop of π and therefore negative delay times [69–71]. The two resonances in each figure show a positive delay time (see the insert in Fig. 10.8) corresponding to the time of capture of these modes in the two stubs along each output line. Also, the transmission and reflection coefficients around $\Omega = 0.82\pi(T_1, R)$ and $\Omega = \pi(T_2, R)$ (Fig. 10.7) do not vanish at exactly the same frequency as it is displayed in Fig. 10.8 where the corresponding negative delay times around $\Omega = 0.82\pi$ (blue [dark gray in print version] and black curves) and $\Omega = \pi$ (red [light gray in print version] and black curves) fall close to each other, but not at the same reduced frequency. The small shift between the negative peaks is due to the experimental accuracy in determining the length of the cables. Indeed, the error in the coaxial cable length is about 0.5 cm, this error has a very small effect on the position of the transmission and reflection zeros as it can be observed in the delay time spectra (Fig. 10.8) around $\Omega = 0.82\pi$ and $\Omega = \pi$. However, the transmission and reflection experimental amplitudes (Fig. 10.7) still remain robust within this range of error to realize a good demultiplexing.

10.3 Demultiplexer based on a U-structure

As mentioned in Chapter 9, U-shaped photonic resonators along a waveguide were the subject of several studies. In particular, it was shown that such structure can present two types of resonances: EIT-type resonances when the two resonators have different lengths, and Fano resonances when the two resonators have the same lengths. In this section we consider a U-shaped demultiplexer composed of an input line and two output lines, all connected at the same point 1 (Fig. 10.9). On the first output line, we connect two lateral stubs of lengths d_1 and d_2 separated by a distance d_0. The stub of length d_1 is inserted at the site 2 at a distance d_5 from the input 1, and the stub of length d_2 is inserted at the site 3 at a distance $d_5 + d_0$ from input 1. Similarly, the second output line contains two lateral stubs of lengths d_3 and d_4 separated by a distance d_0'. The stub of length d_3 is inserted at the site 4 at the distance d_6 from the input 1, and the stub of length d_4 is inserted at the site 5 at a distance $d_6 + d_0'$ from the input 1 (Fig. 10.9).

The calculation of the transmission and reflection coefficients for the demultiplexer based on a U-shaped structure can be performed in the same way as in Section 10.2

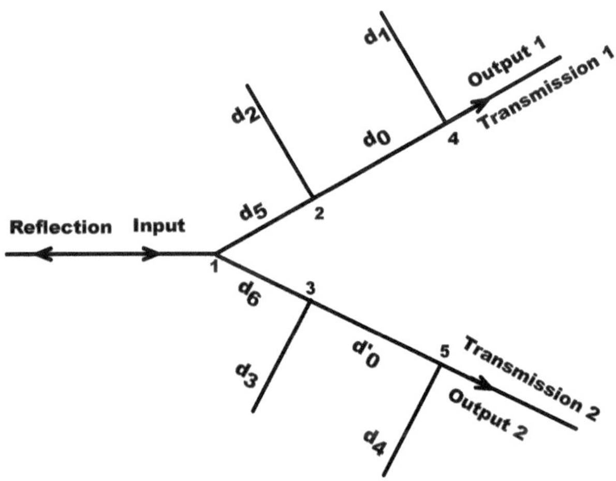

Figure 10.9 Schematic representation of a Y-shaped demultiplexer with one input line and two output lines containing a U-shaped structure. Two resonators of lengths d_1 and d_2, separated by a segment of length d_0, are inserted along the first waveguide at a distance d_5 from the input waveguide. The second waveguide is composed of two resonators of lengths d_3 and d_4, separated by a segment of length d_0' and inserted along the second waveguide at a distance d_6 from the input waveguide.

using the Green's function method [65]. The whole system is constituted from four finite segments of lengths d_0, d'_0, d_5, and d_6, four stubs of lengths d_1, d_2, d_3, and d_4 and three semiinfinite waveguides. In this structure the resonators are grafted at two different positions along each output waveguide. In this case, the interface space between all the connecting tubes is defined as $M = \{1, 2, 3, 4, 5\}$.

The expression giving the inverse of the Green's function of the whole system presented in Fig. 10.9 can be obtained from a linear superposition of the inverses of the Green's functions of the constituents mentioned in Section 10.2 in the interface space $M = \{1, 2, 3, 4, 5\}$, namely

$$g^{-1}(M,M) = \frac{-\omega}{Z}\begin{pmatrix} \frac{C_5}{S_5}+\frac{C_6}{S_6}-j & -\frac{1}{S_5} & 0 & -\frac{1}{S_6} & 0 \\ -\frac{1}{S_5} & \frac{C_5}{S_5}+\frac{C_0}{S_0}+\frac{S_2}{C_2} & -\frac{1}{S_0} & 0 & 0 \\ 0 & -\frac{1}{S_0} & \frac{C_0}{S_0}+\frac{S_1}{C_1}-j & 0 & 0 \\ -\frac{1}{S_6} & 0 & 0 & \frac{C_6}{S_6}+\frac{C_{0'}}{S_{0'}}+\frac{S_3}{C_3} & -\frac{1}{S_{0'}} \\ 0 & 0 & 0 & -\frac{1}{S_{0'}} & \frac{C_{0'}}{S_{0'}}+\frac{S_4}{C_4}-j \end{pmatrix}. \tag{10.28}$$

Similarly, the expressions of the transmission coefficients t_1 and t_2 in the two outputs, as well as the reflection coefficient r in the input, are given by $t_1 = (2j\omega/Z)g(1,4)$, $t_2 = (2j\omega/Z)g(1,5)$, and $r = -1 - (2j\omega/Z)g(1,1)$, respectively. By inverting analytically the matrix in Eq. (10.28) and truncating the elements $g(1,4)$, $g(1,5)$, and $g(1,1)$, one can obtain the expressions of t_1, t_2, and r. We can proceed numerically to invert the 5×5 matrix in Eq. (10.28) and get the elements $g(1,4)$, $g(1,5)$, and $g(1,1)$. However, the analytical expressions become very cumbersome, therefore we shall avoid giving these expressions and limit ourselves to only the conditions that should be satisfied by all the guides in order to realize perfect demultiplexing. In addition, we shall distinguish two different cases depending on whether the stubs along each output line are slightly different, which gives rise to EIT resonances, or identical, which gives rise to Fano resonances.

10.3.1 Demultiplexer based on EIT resonances

As already demonstrated in Chapter 9, in order to get EIT resonance with such structures, we should take the lengths of the two stubs along one guide slightly different (i.e., $\delta = d_2 - d_1 \neq 0$). In addition, along this study we will fix the total length of the two stubs along the first output line such that $d_0 = d_2 + d_1 = 1$ m, which fixes the position of the EIT resonance along this line. Now, in order to achieve total transmission for two neighboring frequencies along the two output lines, the lengths of the different guides should be taken as follows:

$$d_1 = \frac{d_0}{2} + \frac{\delta}{2}, \tag{10.29}$$

$$d_2 = d_5 = \frac{d_0}{2} - \frac{\delta}{2}, \tag{10.30}$$

$$d_3 = d_6 = \frac{d_0}{2}, \tag{10.31}$$

$$d_4 = \frac{d_0}{2} - \delta, \tag{10.32}$$

$$d_0' = d_3 + d_4 = d_0 - \delta. \tag{10.33}$$

Fig. 10.10 presents the variation of the transmission coefficients T_1 and T_2, and the reflection coefficient R, as a function of the reduced frequency Ω for different values of δ around $\delta = 0$ and for $d_0 = d_1 + d_2 = 1$ m. We can see clearly that for each δ, when the transmission along the first output line (blue curve, dark gray in print version) reaches unity ($T_1 = 1$), the transmission along the second output line T_2 (red curve, light gray in print version) and the reflection R (black curve) cancel out each other (i.e., $T_2 = R = 0$). Similarly, when the transmission along the second line (discontinuous curve) reaches unity ($T_2 = 1$), the transmission along the first line T_1 (continuous curve) and the reflection R (dashed curve) vanish (i.e., $T_1 = R = 0$). As mentioned above, the EIT resonance in the first output line falls at the same frequency

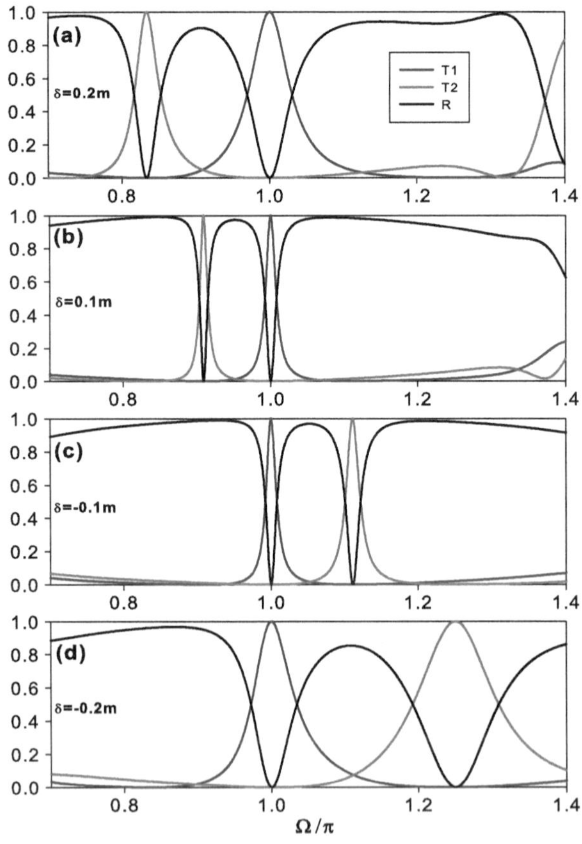

Figure 10.10 Variation of the transmission along the output 1 (blue curve, dark gray in print version), the output 2 (red curve, light gray in print version), and the reflection in the input (black curve) of the demultiplexer as a function of the reduced frequency Ω for different values of δ and for $d_0 = d_1 + d_2 = 1$ m.

for all δ values, its width decreases as δ decreases and disappears for $\delta = 0$, giving rise to bound in continuum states [66]. In addition, the shape and width of the EIT resonance change slightly when δ becomes negative (i.e., for a permutation of both stubs 1 and 2). The frequency position and width of the resonance along the second line strongly depend on δ.

Indeed, as the first EIT resonance has two transmission zeros around $\Omega = \pi$, the position of the second EIT resonance falls above $\Omega = \pi$ for $\delta < 0$, crosses the first resonance at $\delta = 0$, and reappears below $\Omega = \pi$ for $\delta > 0$. This behavior is similar to that obtained in Figs. 10.2 and 10.3 for the cross structure where the crossing between the two resonances takes place for $\delta = 0$. Also, the variation of the quality factor Q of the two EIT resonances in the two output lines as a function of δ follows the same behavior as in Fig. 10.4. In particular, Q increases rapidly when δ decreases and tends towards infinity when δ approaches zero.

10.3.2 Demultiplexer based on Fano resonances

As mentioned in Chapter 9, in order to achieve a Fano resonance, one should take both stubs with identical lengths, but slightly different from $\frac{d_0}{2}$ (i.e., $d_1 = d_2 \neq d_0/2$) along the first output line. Similarly, we should take $d_3 = d_4 \neq d_0'/2$ along the second output line. The explicit expressions of the eight different lengths d_1, d_2, d_3, d_4, d_5, d_6, d_0, and d_0' should satisfy the following equations in order to obtain a total transmission along one output line, keeping the other line unaffected:

$$d_1 = d_2 = \frac{d_0}{2} + \delta', \tag{10.34}$$

$$d_5 = d_2, \tag{10.35}$$

$$d_3 = d_4 = \frac{d_0}{2} + \frac{\delta'}{2}, \tag{10.36}$$

$$d_0' = d_0 + 3\delta', \tag{10.37}$$

$$d_6 = d_3, \tag{10.38}$$

where δ' represents the detuning between the lengths of the different guides constituting the demultiplexer. Fig. 10.11 gives the variation of transmission coefficients T_1 and T_2, and reflection coefficient R, as a function of the reduced frequency Ω for different values of δ' around $\delta' = 0$ and for $d_0 = 1$ m. From Fig. 10.11 one can see that both resonances are of Fano type, that is, resonances near a transmission zero. In addition, both resonances present different asymmetric line shapes (i.e., opposite Fano parameters [49,50]) in order to achieve a full transmission in one line and no signal in the other line. We can notice that for $\delta' = 0.15$ m, for example (Fig. 10.11(a)), the transmission along the first output (blue curve, dark gray in print version) is unity ($T_1 = 1$) at $\Omega = 0.87\pi$, the transmission along the second output T_2 (red curve, light gray in print version) and the reflection R (black curve) vanish (i.e., $T_2 = R = 0$). Similarly, when the transmission along the second line (red curve, light gray in print version) reaches unity ($T_2 = 1$) at $\Omega = 0.76\pi$, the transmission along the first line

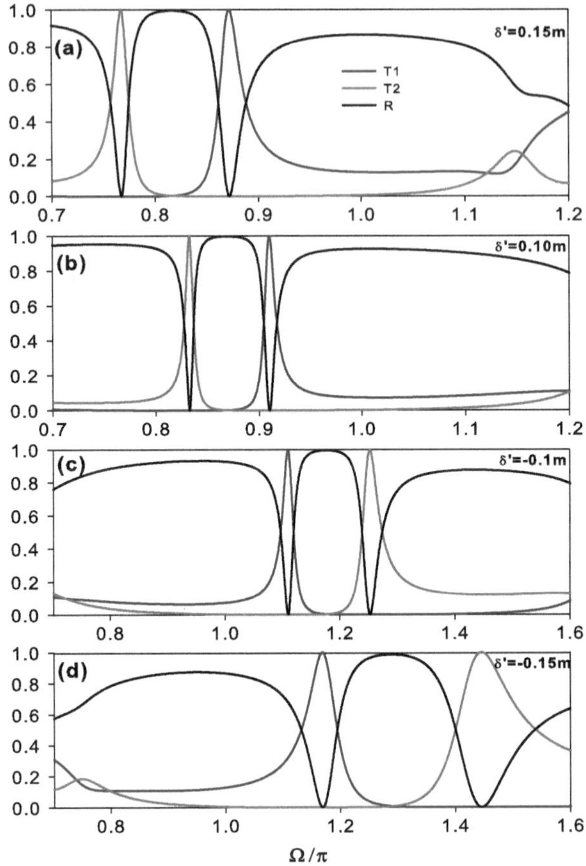

Figure 10.11 Variation of the transmission spectra along the output 1 (blue, dark gray in print version), the output 2 (red curve, light gray in print version), and the reflection in the input (black curve) of the demultiplexer as a function of the frequency f for different values of δ' and for $d_0 = 1$ m.

T_1 (blue curve, dark gray in print version) and the reflection R (black curve) vanish (i.e., $T_1 = R = 0$). Also, there exists a frequency between the two resonances (around $\Omega = 0.815\pi$) for which the reflection along the input line reaches almost unity ($R \simeq 1$), while the transmission along the first line T_1 (blue curve, dark gray in print version) and the transmission along the second line T_2 (red curve, light gray in print version) vanish (i.e., $T_1 \simeq T_2 \simeq 0$). This behavior does not exist in the case of EIT resonances (Fig. 10.10).

In addition, we can see that for $\delta' < 0$, the resonance in the second line falls below that in the first line (Fig. 10.11(a)), and when δ' increases, both resonances fall close to each other and their widths decrease (Fig. 10.11(b)). For $\delta' = 0$, both resonances fall at the same frequency (at $\Omega = \pi$), their widths vanish, giving rise to bound in continuum states [66]. For $\delta' > 0$, the resonance in the second line falls above that in the first line (Fig. 10.11(c)); its width increases when δ' increases (Fig. 10.11(d)).

These results are summarized in Fig. 10.12, where we have plotted the frequencies of both resonances as a function of δ'. We can see that contrary to Fig. 10.3, the frequencies of both Fano resonances depend on δ'. Also, the quality factors of the two resonances (not shown here) are almost identical and tend to infinity when δ' tends to zero. Finally, the amplitude of the filtered resonances in Figs. 10.10 and 10.11 can be affected considerably when loss is taken into consideration in particular for narrow resonances as in Fig. 10.7 (see also the next section).

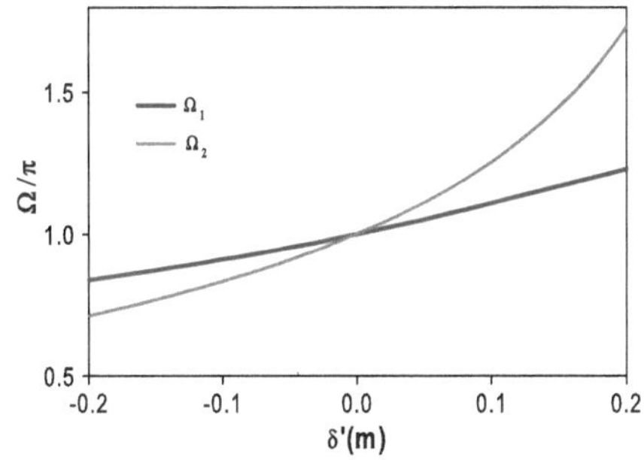

Figure 10.12 Variation of the frequencies of the two Fano resonances as a function of δ'.

10.3.3 Experimental results

As mentioned above, the behavior of the transmission spectra based on EIT resonances in the U-demultiplexer (Fig. 10.10) is quite similar to that corresponding to cross-demultiplexer (Fig. 10.2), and so are the experimental spectra (Fig. 10.7). Therefore, we shall present just the results for the U-demultiplexer based on Fano resonances. Fig. 10.13 gives the theoretical transmission and reflection coefficients (continuous lines) and experimental results (empty circles) as a function of the reduced frequency Ω along the output 1 (blue curves, dark gray in print version) and output 2 (red curves, light gray in print version). The reflection coefficient is given by the black curves. The lengths of the cables are chosen such that $d_0 = 1$ m, $d_1 = d_2 = 0.4$ m, $d_2 = d_5$, $d_3 = d_4 = 0.45$ m, $d_6 = d_3$, and $d_0' = 0.7$ m (i.e., $\delta' = -0.1$ m) which are very close to the conditions required in Eqs. (10.34)–(10.38). The experiments are in good agreement with the theoretical results. The filtered resonance in each line (around $\Omega = 1.08\pi$ and $\Omega = 1.24\pi$) does not reach unity because of the dissipation in the cables. However, we see clearly that when the transmission is maximal in one line, the corresponding transmission in the other line vanishes; however, the reflection in the input line presents a dip slightly different from zero. In order to achieve a good demultiplexing, we should take δ' far from zero in order to avoid the effect of the dissipation, but this will be at the expense of the quality factors of the resonances.

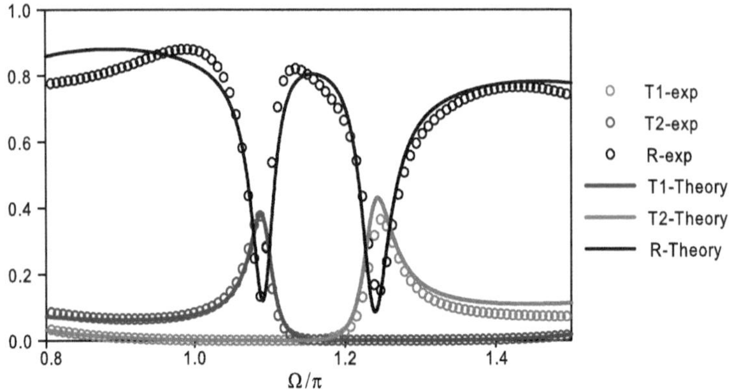

Figure 10.13 (a) Theoretical (solid line) and experimental (open circles) transmission coefficients versus the frequency along ports 1 (blue curves, dark gray in print version) and 2 (red curves, light gray in print version). The reflection coefficient is given by black curves. The lengths of the cables are chosen such that $d_1 = 0.375$ m, $d_2 = d_5 = 0.62$ m, $d_3 = d_6 = 0.5$ m, and $d_4 = 0.725$ m. These parameters are in accordance with Eqs. (10.34)–(10.38).

10.4 Conclusion

In this chapter we have studied a Y-shaped photonic demultiplexer based on two different configurations. The first demultiplexer is based on a cross structure; in this case we used the EIT-type resonances presented by such structures to realize a perfect demultiplexing. The demultiplexer consists on a Y-shaped waveguide with an input line and two output lines. Each line contains two grafted stubs at a given position far from the input line. The waveguides are made of standard coaxial cables and the experiments are performed in the radio-frequency domain. We derived closed-form expressions of the waveguide lengths to achieve selective transfer of a single propagating mode through one line, keeping the other line unaffected. The position and the width of the resonances depend on the different lengths of the waveguides which should be chosen appropriately. In addition, we have shown that the transmission can be written following the Fano-like form, which enables us to deduce the position, width (or the quality factor), and Fano parameters of the resonances as functions of δ. In the second configuration we have studied the U-shaped structure for the design of the demultiplexer. In this case, we have shown two possibilities of demultiplexing by using either EIT symmetrical resonances or Fano asymmetrical resonances. In both cases we have determined analytically the expressions of the lengths of the different guides to get the total transmission in one output line, keeping the other lines unperturbed. For both demultiplexers, we have shown that the frequencies of the filtered resonances, as well as their widths (i.e., the quality factors), can be tuned by appropriately choosing the lengths of the different waveguides constituting these systems. The confinement of the filtered and stopped resonances along each line is shown through an analysis of the electric field. The theoretical results were confirmed by experiments using coaxial cables, and a good agreement theory/experiment was achieved.

Besides the facility of fabrication design and the manipulation of the analytical calculation, the demultiplexers proposed in this study exhibit the possibility to tune the quality factors of the filtered resonances close to infinity by detuning the two stubs. This property is a feature of Fano and EIT resonances and does not exist in standard one-dimensional photonic crystals using Breit–Wigner resonances with finite width. Finally, let us mention that these results may have potential applications in the telecommunication domain around $\lambda = 1500$ nm as it is shown recently [72] in plasmonic structures based on metal-insulator-metal nano-waveguides. Also, the possibility of existence of EIT resonances has been shown also in acoustic waveguides [73–75] and pillared metasurfaces [76–79]. Therefore, the results presented in this chapter can be transposed straightforwardly to acoustic demultiplexers based on waveguide tubes and Helmholtz resonators [73–75]. These results have been presented both theoretically and experimentally in recent works [80,81].

Acknowledgments

The authors would like to thank Abdelkader Mouadili, Ali Soltani, and Kamal Haddadi for their fruitful contribution to the original papers from which this chapter is inspired.

References

[1] A.E. Akosman, M. Mutlu, H. Kurt, E. Ozbay, Dual-frequency division demultiplexer based on cascaded photonic crystal waveguides, Physica B 407 (2012) 4043.
[2] S. Bouamami, R. Naoum, Compact WDM demultiplexer for seven channels in photonic crystal, Optik 124 (2013) 16.
[3] E. Centeno, B. Guizal, D. Felbacq, Multiplexing and demultiplexing with photonic crystals, J. Opt. A, Pure Appl. Opt. 1 (1999) L10.
[4] C.A. Brackett, Dense wavelength division multiplexing networks: principles and applications, IEEE J. Sel. Areas Commun. 8 (1990) 948.
[5] C. Dragone, Efficient $N \times N$ star couplers using Fourier optics, J. Lightwave Technol. 7 (1989) 479.
[6] H. Takahashi, S. Suzuki, I. Nishi, Photonic crystals for micro lightwave circuits using wavelength-dependent angular beam steering, J. Lightwave Technol. 12 (1994) 989.
[7] K. Hill, Y. Fujii, D.C. Johnson, B.S. Kawasaki, Photosensitivity in optical fiber waveguides: application to reflection filter fabrication, Appl. Phys. Lett. 32 (1978) 647.
[8] R. Fan, R. Villeneuve, J.D. Joannopoulos, H.A. Haus, Channel drop tunneling through localized states, Phys. Rev. Lett. 80 (1998) 960.
[9] H.A. Haus, S. Fan, R. Villeneuve, J.D. Joannopoulos, Channel drop filters in photonic crystals, Opt. Express 3 (1998) 4.
[10] M.R. Rakhshani, M.A. Mansouri-Birjandi, Design and simulation of wavelength demultiplexer based on heterostructure photonic crystals ring resonators, Physica E 50 (2013) 97.
[11] A. Mansouri-Birjandi, M. Ghadrdan, All-optical ultra compact photonic crystal switch based on nonlinear micro ring resonators, Int. Res. J. Appl. Basic Sci. 4 (2013) 972.

[12] J.D. Joannopoulas, R.D. Mead, J.N. Winn, Photonic Crystals: Molding the Flow of Light, Princeton Univ. Press, Princeton, NJ, 1995.

[13] P. Jiang, C. Ding, X. Hu, Q. Gong, Tunable double-channel filter based on two-dimensional ferroelectric photonic crystals, Phys. Lett. A 363 (2007) 332.

[14] X. Hu, Z. Liu, Q. Gong, Tunable multichannel filter in photonic crystal heterostructure containing permeability-negative materials, Phys. Lett. A 372 (2008) 333.

[15] X. Chen, Z. Qiang, D. Zhao, Y. Wang, H. Li, Y. Qiu, W. Zhou, Polarization beam splitter based on photonic crystal self-collimation Mach–Zehnder interferometer, Opt. Commun. 284 (2011) 490.

[16] Z. Chen, W. Wang, L. Cui, L. Yu, G. Duan, Y. Zhao, J. Xiao, Spectral splitting based on electromagnetically induced transparency in plasmonic waveguide resonator system, Plasmonics 3 (2015) 10.

[17] N. Saidani, W. Belhadj, F. Abdelmalek, H. Bouchriha, Detailed investigation of self-imaging in multimode photonic crystal waveguides for applications in power and polarization beam splitters, Opt. Commun. 285 (2012) 3487.

[18] D. Yang, H. Tian, Y. Ji, High-bandwidth and low-loss photonic crystal power-splitter with parallel output based on the integration of Y-junction and waveguide bends, Opt. Commun. 285 (2012) 3752.

[19] M.R. Rakhshani, M.A. Mansouri-Birjandi, Heterostructure four channel wavelength demultiplexer using square photonic crystals ring resonators, J. Electromagn. Waves Appl. 26 (2012) 1700.

[20] T. Sreenivasulu, K. Bhowmick, S.A. Samad, T.I.R. Yadunath, T. Badrinarayana, G. Hegde, T. Srinivas, Photonic crystal ring resonator-based four-channel dense wavelength division multiplexing demultiplexer on silicon on insulator platform: design and analysis, Opt. Eng. 57 (2018) 4.

[21] V. Fallahi, M. Seifouri, S. Olyaee, H. Alipour-Banaei, Four-channel optical demultiplexer based on hexagonal photonic crystal ring resonators, Opt. Rev. 24 (2017) 605.

[22] D. Melati, A. Alippi, A. Annoni, N. Peserico, A. Melloni, Integrated all-optical MIMO demultiplexer for mode- and wavelength-division-multiplexed transmission, Opt. Lett. 42 (2017) 2.

[23] K. Venkatachalam, D.S. Kumar, S. Robinson, Investigation on 2D photonic crystal-based eight-channel wavelength-division demultiplexer, Photonic Netw. Commun. 34 (2017) 1.

[24] S. Naghizade, S.M. Sattari-Esfahlan, Tunable high performance 16-channel demultiplexer on 2D photonic crystal ring resonator operating at telecom wavelengths, J. Opt. Commun. (ISSN 0173-4911) (2017), https://doi.org/10.1515/joc-2017-0199, eISSN 2191-6322.

[25] F. Mehdizadeh, M. Soroosh, H. Alipour-Banaei, E. Farshidi, All optical 2-bit analog to digital converter using photonic crystal based cavities, Opt. Quantum Electron. 49 (2017) 38.

[26] A. Noual, A. Akjouj, Y. Pennec, J.N. Gillet, B. Djafari-Rouhani, Modeling of two-dimensional nanoscale Y-bent plasmonic waveguides with cavities for demultiplexing of the telecommunication wavelengths, New J. Phys. 11 (2009) 103020.

[27] J. Chen, Z. Li, J. Li, Q. Gong, Compact and high-resolution plasmonic wavelength demultiplexers based on Fano interference, Opt. Express 19 (2011) 9976.

[28] Z. Chen, R. Hu, L. Cui, L. Yu, L. Wang, J. Xiao, Plasmonic wavelength demultiplexers based on tunable Fano resonance in coupled-resonator systems, Opt. Commun. 320 (2014) 6.

[29] H. Liu, Y. Gao, B. Zhu, G. Ren, S. Jian, A T-shaped high resolution plasmonic demultiplexer based on perturbations of two nanoresonators, Opt. Commun. 334 (2015) 164.

[30] A. Azzazi, M.A. Swillam, Nanoscale highly selective plasmonic quad wavelength demultiplexer based on a metal–insulator–metal, Opt. Commun. 344 (2015) 106.

[31] H. Lu, X. Liu, Y. Gong, D. Mao, L. Wang, Enhancement of transmission efficiency of nanoplasmonic wavelength demultiplexer based on channel drop filters and reflection nanocavities, Opt. Express 19 (5) (2011) 1288.
[32] J. Tao, X.G. Huang, J.H. Zhu, A wavelength demultiplexing structure based on metal–dielectric–metal plasmonic nano-capillary resonators, Opt. Express 18 (2010) 11111.
[33] M.T.H. Azar, M. Zavvari, A. Arashmehr, Y. Zehforoosh, P. Mohammadi, Design of a high-performance metal–insulator–metal plasmonic demultiplexer, J. Nanophotonics 11 (2017) 2.
[34] Q. Ma, G. Liu, Y. Chen, Q. Zhao, S. Yang, J. Guo, W. Cai, Single reflection nanocavity enhanced transmission efficiency of a nanoplasmonic wavelength demultiplexer, J. Opt. Technol. 85 (2018) 6.
[35] A.E. Akosman, M. Mutlu, H. Kurt, E. Ozbay, Compact wavelength de-multiplexer design using slow light regime of photonic crystal waveguides, Opt. Express 19 (2011) 24129.
[36] X. Zhang, Q. Liao, T. Yu, N. Liu, Y. Huang, Novel ultracompact wavelength division demultiplexer based on photonic band gap, Opt. Commun. 285 (2012) 274.
[37] S. Rawal, R.K. Sinha, Design, analysis and optimization of silicon-on-insulator photonic crystal dual band wavelength demultiplexer, Opt. Commun. 282 (2009) 3889.
[38] A. Rostami, F. Nazari, H. Alipour Banaei, A. Bahrami, A novel proposal for DWDM demultiplexer design using modified-T photonic crystal structure, Photonics Nanostruct. Fundam. Appl. 8 (2010) 14.
[39] A. Rostami, H. Alipour Banaei, F. Nazari, A. Bahrami, An ultra compact photonic crystal wavelength division demultiplexer using resonance cavities in a modified Y-branch structure, Optik 122 (2011) 1481.
[40] K. Goodarzi, A. Mir, Design and analysis of an all-optical demultiplexer based on photonic crystals, Infrared Phys. Technol. 68 (2015) 193.
[41] N.D. Gupta, V. Janyani, Dense wavelength division demultiplexing using photonic crystal waveguides based on cavity resonance, Optik 125 (2014) 19.
[42] Michela Svaluto Moreolo, Flavia Silvestri, Mariateresa Armellino, Kurt Hingerl, Gabriella Cincotti, Optimization of a 2D photonic crystal add/drop multiplexer based on contra-directional coupling, Nanostruct. 4 (2006) 155.
[43] R. Selim, D. Pinto, S.S.A. Obayya, Novel fast photonic crystal multiplexer-demultiplexer switches, Opt. Quantum Electron. 42 (2011) 425.
[44] M. Djavid, F. Monifi, A. Ghaffari, M.S. Abrishamian, Heterostructure wavelength division demultiplexers using photonic crystal ring resonators, Opt. Commun. 281 (2008) 4028.
[45] B.E. Little, S.T. Chu, J.V. Hryniewicz, P.P. Absil, Filter synthesis for periodically coupled microring resonators, Opt. Lett. 25 (2000) 344.
[46] Liao Qing-Hua, Fan Hong-Ming, Chen Shu-Wen, Wang Tong-Biao, Yu Tian-Bao, Huang Yong-Zhen, The design of large separating angle ultracompact wavelength division demultiplexer based on photonic crystal ring resonators, Opt. Commun. 331 (2014) 160.
[47] Li Jiu-Sheng, Liu Han, Zhang Le, Compact four-channel terahertz demultiplexer based on directional coupling photonic crystal, Opt. Commun. 350 (2015) 248.
[48] Ying Tan, Hao Wu, Shipeng Wang, Chenlei Li, Daoxin Dai, Silicon-based hybrid demultiplexer for wavelength- and mode-division multiplexing, Opt. Lett. 43 (2018) 1962.
[49] U. Fano, Effects of configuration interaction on intensities and phase shifts, Phys. Rev. 124 (1961) 1866.
[50] For a review, see M. Fleischhauer, A. Imamoglu, J.P. Marangos, Electromagnetically induced transparency: optics in coherent media, Rev. Mod. Phys. 77 (2005) 633.
[51] A. Mouadili, E.H. El Boudouti, A. Soltani, A. Talbi, A. Akjouj, B. Djafari-Rouhani, Theoretical and experimental evidence of Fano-like resonances in simple monomode photonic circuits, J. Appl. Phys. 113 (2013) 164101.

[52] A. Mouadili, E.H. El Boudouti, A. Soltani, A. Talbi, B. Djafari-Rouhani, A. Akjouj, K. Haddadi, Electromagnetically induced absorption in detuned stub waveguides: a simple analytical and experimental model, J. Phys. Condens. Matter 26 (2014) 505901.
[53] J. Zhao, S. Han, H. Lin, H. Yang, Tailoring the multiple electrically resonant transparency through bi-layered metamaterial-induced coupling oscillators, J. Phys. D, Appl. Phys. 48 (2015) 455502.
[54] B. Wei, S. Jian, Analogue of electromagnetically-induced-transparency based on graphene nanotube waveguide, J. Phys. D, Appl. Phys. 50 (2017) 355101.
[55] B. Wang, Q. Zeng, S. Xiao, C. Xu, L. Xiong, H. Lv, J. Du, H. Yu, Low-power, ultrafast, and dynamic all-optical tunable plasmon induced transparency in two stub resonators side-coupled with a plasmonic waveguide system, J. Phys. D, Appl. Phys. 50 (2017) 455107.
[56] Y. Wang, Z. Li, F. Hu, Analog of electromagnetically induced transparency at terahertz frequency based on a bilayer-double-H-metamaterial, J. Phys. D, Appl. Phys. 51 (2018) 025103.
[57] Kun Zhang, Cheng Wang, Ling Qin, Ru-Wen Peng, Di-Hu Xu, Xiang Xiong, Mu Wang, Dual-mode electromagnetically induced transparency and slow light in a terahertz metamaterial, Opt. Lett. 39 (2014) 3539.
[58] Lei Zhu, Liang Dong, Jing Guo, Fan-Yi Meng, Xun Jun He, Chun Hui Zhao, Qun Wu, A low-loss electromagnetically induced transparency (EIT) metamaterial based on coupling between electric and toroidal dipoles, RSC Adv. 7 (2017) 55897.
[59] C. Jin, S. Han, X. Meng, B. Cheng, D. Zhang, Demultiplexer using directly resonant tunneling between point defects and waveguides in a photonic crystal, J. Appl. Phys. 91 (2002) 4771.
[60] K. Ogusu, K. Takayama, Transmission characteristics of photonic crystal waveguides with stubs and their application to optical filters, Opt. Lett. 32 (2007) 2185.
[61] Z. Qiang, W. Zhou, R.A. Soref, Optical add-drop filters based on photonic crystal ring resonators, Opt. Express 15 (2007) 1823.
[62] Z. Chen, M. Chen, J. Li, L. Wang, R. Jiao, G. Duan, L. Yu, J. Xiao, Symmetry breaking induced mode splitting based on a plasmonic waveguide system, J. Phys. D, Appl. Phys. 49 (2016) 145109.
[63] A. Mouadili, E.H. El Boudouti, A. Soltani, A. Talbi, K. Haddadi, A. Akjouj, B. Djafari-Rouhani, Photonic demultiplexer based on electromagnetically induced transparency resonances, J. Phys. D 52 (2019) 075101.
[64] E.H. El Boudouti, N. Fettouhi, A. Akjouj, B. Djafari-Rouhani, A. Mir, J.O. Vasseur, L. Dobrzynski, J. Zemmouri, Experimental and theoretical evidence for the existence of photonic bandgaps and selective transmissions in serial loop structures, J. Appl. Phys. 95 (2004) 1102.
[65] J.O. Vasseur, A. Akjouj, L. Dobrzynski, B. Djafari-Rouhani, E.H. El Boudouti, Photon, electron, magnon, phonon and plasmon mono-mode circuits, Surf. Sci. Rep. 54 (2004) 1.
[66] C.W. Hsu, B. Zhen, A.D. Stone, J.D. Joannopoulos, M. Soljačić, Bound states in the continuum, Nat. Rev. Mater. 1 (2016) 16048.
[67] C.H. Raymond Ooi, C.H. Kam, Controlling quantum resonances in photonic crystals and thin films with electromagnetically induced transparency, Phys. Rev. B 81 (2010) 195119.
[68] B.B. Li, Y.F. Xiao, C.L. Zou, X.F. Jiang, Y.C. Liu, F.W. Sun, Y. Li, Q. Gong, Experimental controlling of Fano resonance in indirectly coupled whispering-gallery microresonators, Appl. Phys. Lett. 100 (2012) 021108.
[69] M. Buttiker, R. Landauer, Traversal time for tunneling, Phys. Rev. Lett. 49 (1982) 1739.
[70] E.H. Hauge, J.A. Stovneng, Tunneling times: a critical review, Rev. Mod. Phys. 61 (1989) 917.

[71] M.L.H. Lahlaouti, A. Akjouj, B. Djafari-Rouhani, L. Dobrzynski, M. Hammouchi, E.H. El Boudouti, A. Nougaoui, B. Kharbouch, Theoretical analysis of the density of states and phase times: application to resonant electromagnetic modes in finite superlattices, Phys. Rev. B 63 (2001) 035312.

[72] M. Amrani, S. Khattou, A. Noual, E.H. El Boudouti, B. Djafari-Rouhani, Plasmonic demultiplexer based on induced transparency resonances: analytical and numerical study, Lect. Notes Electr. Eng. 681 (2021) 239.

[73] E.H. El Boudouti, T. Mrabti, H. Al-Wahsh, B. Djafari-Rouhani, A. Akjouj, L. Dobrzynski, Transmission gaps and Fano resonances in an acoustic waveguide: analytical model, J. Phys. Condens. Matter 20 (2008) 255212.

[74] A. Santillan, S.I. Bozhevolnyi, Acoustic transparency and slow sound using detuned acoustic resonators, Phys. Rev. B 84 (2011) 064304.

[75] A. Merkel, G. Theocharis, O. Richoux, V. Romero-Garcìa, V. Pagneux, Control of acoustic absorption in one-dimensional scattering by resonant scatterers, Appl. Phys. Lett. 107 (2015) 244102.

[76] Y. Jin, E.H. El Boudouti, Y. Pennec, B. Djafari-Rouhani, Tunable Fano resonances of Lamb modes in a pillared metasurface, J. Phys. D, Appl. Phys. 50 (2017) 425304.

[77] Y. Jin, Y. Pennec, B. Djafari-Rouhani, Acoustic analogue of electromagnetically induced transparency and Autler–Townes splitting in pillared metasurfaces, J. Phys. D, Appl. Phys. 51 (2018) 494004.

[78] A. Elayouch, M. Addouche, M. Farhat, M. Amin, H. Bağci, A. Khelif, Density-near-zero using the acoustically induced transparency of a Fano acoustic resonator, Europhys. Lett. 116 (2016) 46004.

[79] M. Amin, A. Elayouch, M. Farhat, M. Addouche, A. Khelif, H. Bağci, Acoustically induced transparency using Fano resonant periodic arrays, J. Appl. Phys. 118 (2015) 164901.

[80] A. Mouadili, E.H. El Boudouti, B. Djafari-Rouhani, Acoustic demultiplexer based on Fano and induced transparency resonances in slender tubes, Eur. Phys. J. Appl. Phys. 90 (2020) 10902.

[81] T. Gu, Y. Cheng, Z. Wen, E.H. El Boudouti, Y. Jin, Y. Li, B. Djafari-Rouhani, Induced transparency based subwavelength acoustic demultiplexers, Phys. Rev. B (2020) (submitted).

Photonic monomode circuits: comb structures

11

Abdellatif Akjouj[a], *El Houssaine El Boudouti*[b], *Leonard Dobrzyński*[a], *and Bahram Djafari-Rouhani*[a]
[a]Institute of Electronic, Microelectronics and Nanotechnology (IEMN), UMR CNRS 8520, Department of Physics, Faculty of Sciences and Technologies, Lille University, Villeneuve d'Ascq Cedex, France, [b]Laboratoire de Physique de la Matière et de Rayonnement, Département de Physique, Faculté des Sciences, Université Mohamed I, Oujda, Morocco

11.1 Introduction

Photon circuits are light conducting networks formed by the junction of several light waveguide channels. Such circuits made out of optical fibers are already widely used in telecommunications. Micrometric networks are also under investigation. Most advanced surface technologies enable producing such microdevices. In particular, there is considerable recent interest in photonic crystal waveguides. By a photonic crystal we mean a crystal build out of a periodic arrangement of inclusions of one material inside another. These waveguide structures can be formed, for example, in photonic crystals by the addition or the removal of a line of site impurities to the photonic crystal. Since the first proposals [1,2] of photonic crystals for inhibited spontaneous light emission and localization of photons [3,4], there was an explosion of investigations for such novel materials [5–8] and in particular for their utilization in the fabrication of microphotonic waveguides [9–12]. Other surface techniques enable also manufacturing microphotonic circuits. For example, microdevices are fabricated in AlGaAs–GaAs material grown by molecular beam epitaxy on GaAs substrates [13], combined with electron beam lithography and ion beam etching. One general feature of most of these photon circuits is that they are made out of monomode wires. By monomode wires we mean wires such that their lateral dimensions are small compared to the wavelength of the propagating waves. Only one type of mode can propagate in them and this mode behaves as that of a one-dimensional structure. In addition, the usual TE (transverse electric) and TM (transverse magnetic) well-known electromagnetic modes are degenerate in such wires and have the same frequency. This simplifies greatly the understanding of the propagation of electromagnetic excitations in monomode circuits, enables studying analytically most of their properties, and helps design microdevices which can then be realized with the help of current numerical simulations and surface technologies. This allows applications to selective frequency filters and efficient waveguides [13].

In 1987, E. Yablonovitch [1], inspired by experiments on Rydberg atoms and Penning-trapped electrons [14,15], suggested that a periodic dielectric structure possessing an electromagnetic bandgap could lead to inhibited spontaneous emission.

Photonics. https://doi.org/10.1016/B978-0-12-819388-4.00021-6

The spontaneous emission inhibition in such gaps could be utilized to enhance the performance of semiconductor lasers and other quantum devices. Moreover, John [2] highlighted the relationship between forbidden electromagnetic bandgaps and the localization of photons. In addition to these proposals, the problem of propagation of electromagnetic waves in composite media has received a great deal of attention. Of particular interest is the existence of photonic gaps in the electromagnetic band structure of artificial materials named photonic crystals.

At the outset, theoretical and experimental works focused on three-dimensional photonic crystals. By employing the full vector Maxwell's equations, the existence of a pseudogap in the photonic band structure of a face-centered-cubic lattice of dielectric spheres has been established [16,17]. It was proved that spheres arranged in a diamond structure [18], as well as non-spherical dielectric inclusions placed on an FCC lattice [19], possess absolute bandgaps. Complete bandgaps were further obtained in two-dimensional photonic crystals constituted of periodic arrays of dielectric rods embedded in a dielectric background. Square [20,21], triangular [22–25], and hexagonal [26,27] lattices were investigated. The propagation of electromagnetic waves in one-dimensional systems, such as superlattices [28–31], has also been studied extensively during the last two decades.

In all these composite systems, the contrast in dielectric properties between the constituent materials and the composition of the inhomogeneous material are emerging as critical parameters in determining the existence of gaps [32]. These numerous studies open a question regarding the occurrence of bandgaps for electromagnetic waves in homogeneous systems by tailoring their geometry.

It is worth mentioning that several stubbed circuit structures have been proposed in the last three decades to study different elementary excitations, such as photons, phonons, electrons, magnons, and plasmons [33]. These structures can be arranged through just a few stubs or periodic systems or even quasiperiodic systems. In the case of photons, phonons, magnons, and plasmons in two stubs placed at the same site or at two different sites along a waveguide, different phenomena have been demonstrated both theoretically and experimentally, such as Fano and electromagnetic induced transparency (EIT) resonances [34–49], Autler splitting resonance [50], perfect absorption [43], and demultiplexing [51–53]. In the case of periodic structures, photons, phonons, magnons and electrons have been intensively investigated to show Bragg and non-Bragg gaps, surface and cavity modes, resonant tunneling, and metamaterial properties [54–75]. In the case of quasiperiodic structures such as Fibonacci systems, electrons and photons have been studied to show the scaling law and the fragmentation of the bands for high generations [76–78].

In this chapter, we review first the appealing possibility of designing a comb-like structure of one-dimensional waveguides exhibiting stopbands [54]. This structure is composed of a backbone (or substrate) waveguide along which finite side branches are grafted periodically. This study is conducted within the framework of the interface response theory [79–82] which we recall briefly in Section 11.2. This theory allows the calculation of the Green's functions of a network structure in terms of the Green's functions of its elementary constituents. Three network structures are then considered, namely a single side branch on an infinite one-dimensional waveguide, an infinite periodic comb-like structure, and a finite comb with two semiinfinite leads. The first one is shown in Section 11.3.1 to give rise to well defined zeros of transmission due to resonances between the side branch and the backbone. These resonances are enlarged to absolute gaps in the limit of an infinite periodic comb. It is also shown in Section 11.3.2 that additional gaps form as a result of the periodicity. In Section 11.3.3, we calculate the transmission coefficient for electromagnetic waves of a finite comb. Despite its finite size, this device retains most of the features of the infinite periodic one. This work [54] demonstrates the possibility of designing simple homogeneous networks of one-dimensional waveguides with absolute bandgaps. In Section 11.4, we add to each side branch several structures in order to obtain star combs which may exhibit giant gaps and discuss the results obtained with another boundary condition at the ends of the dangling branches, namely, the vanishing of the electric field. In Section 11.5, we address defect modes in one-dimensional comb structures. Section 11.6 is devoted to surface modes. Section 11.7 concerns multiplexers able to transmit selectively, in one given direction, one precise wavelength from one wire to another. The effects of absorption and amplification in the above devices are discussed in Section 11.8.

11.2 Interface response theory

We will address the propagation of electromagnetic waves in composite systems composed of one-dimensional continuous segments (or branches) grafted on different substrates. This study is performed with the help of the interface response theory [79–83] which permits calculating the Green's function of any composite material. In what follows, we present the basic concepts and the fundamental equations of this theory.

Let us consider any composite material contained in its space of definition D and formed out of N different homogeneous pieces situated in their domains D_i $(1 \leq i \leq N)$. Each piece is bounded by an interface M_i, adjacent in general to J other pieces through subinterface domains M_{ij} $(1 \leq j \leq J)$. The ensemble of all these interface spaces M_i is called the interface space M of the composite material.

11.2.1 Discrete theory

We consider first the discrete theory designed for problems using matrix formulations for linear Hamiltonians. The starting point is an infinite homogeneous material i described by an infinite matrix $[(E + j\varepsilon)I - H_i]$, where E stands for the eigenenergy, I is the identity matrix, $j = \sqrt{-1}$, and ε is an infinitesimally small positive number. The inverse of this matrix is called the corresponding Green's function G_i and

$$[(E + j\varepsilon)I - H_i]G_i = I, \tag{11.1}$$

where I is the identity matrix. One cuts out of this medium a finite piece with free surfaces in its space D_i with the help of a cleavage operator V_i in the interface space M_i.

One defines A_{si} as the truncated part within D_i of $A_i = V_i G_i$. In the same manner, one constructs also G_{si} out of the truncated part of the G_i. One defines then block diagonal matrices G and A_s by juxtaposition of respectively all the G_{si} and A_{si} defined for N different homogeneous materials indexed by i. A composite material is then constructed by assembling such finite media with the help of a coupling operator V_I defined in the whole interface space M. One defines then in the whole space D of the composite the matrices

$$A = A_s + V_I G \tag{11.2}$$

and in the interface space M,

$$\Delta(M, M) = I(M, M) + A(M, M). \tag{11.3}$$

The elements of the Green's function $g(D, D)$ of any composite material can be obtained from [79,81]

$$g(D, D) = G(D, D) - G(D, M)[\Delta(M, M)]^{-1} A(M, D). \tag{11.4}$$

The new interface states can be calculated from [79,80]

$$\det[\Delta(M, M)] = 0, \tag{11.5}$$

showing that, if one is interested in calculating the interface states of a composite, one only needs to know $\Delta(M, M)$ in the interface space M.

The density of states n_i corresponding to H_i can then be obtained from the imaginary part of the trace of G_i, namely,

$$n_i(E) = -\frac{1}{\pi} \operatorname{Im} \operatorname{Tr} G_i(E). \tag{11.6}$$

Moreover, if $U(D)$ [84] represents an eigenvector of the reference system formed by all the infinite materials i, Eq. (11.4) enables one to calculate the eigenvectors $u(D)$ of the composite material

$$u(D) = U(D) - U(M)[\Delta(M, M)]^{-1} A(M, D). \tag{11.7}$$

In Eq. (11.7), $U(D)$, $U(M)$, and $u(D)$ are row-vectors. Eq. (11.7) enables also to calculate all the waves reflected and transmitted by the interfaces, as well as the reflection and the transmission coefficients of the composite system. In this case, $U(D)$ must be replaced by a bulk wave launched in one homogeneous piece of the composite material [84].

11.2.1.1 Continuous theory

We consider now the continuous theory designed for problems using differential formulations for linear Hamiltonians. The elements of the Green's function $g(D, D)$ of any composite material can now be obtained from [79,81,85]

$$g(D,D) = G(D,D) - G(D,M)[G(M,M)]^{-1}G(M,D) \\ + G(D,M)[G(M,M)]^{-1}g(M,M)[G(M,M)]^{-1}G(M,D), \tag{11.8}$$

where $G(D,D)$ is the block diagonal Green's function of the reference system and $g(M,M)$ are the interface elements of the Green's function of the composite system. The inverse $[g(M,M)]^{-1}$ of $g(M,M)$ is obtained for any points in the interface space $M = \left[\bigcup M_i\right]$ as a superposition of the different $[g_i(M_i,M_i)]^{-1}$ [79–82], being the inverse of the $[g_i(M_i,M_i)]$ for each constituent i of the composite system. The latter quantities are given by the equation

$$[g_i(M_i,M_i)]^{-1} = \Delta_i(M_i,M_i)[G_i(M_i,M_i)]^{-1}, \tag{11.9}$$

where

$$\Delta_i(M_i,M_i) = I(M_i,M_i) + A_i(M_i,M_i), \tag{11.10}$$

with I being the unit matrix,

$$A_i(x,x') = \int V_{0i}(x'')G_i(x'',x')dx'', \tag{11.11}$$

as well as $x', x'' \in D_i$, and $x \in M_i$.

In Eq. (11.11), the cleavage operator V_{0i} acts only in the surface domain M_i of D_i and cuts the finite or semiinfinite size block out of the infinite homogeneous medium [79–82]; A_i is called the surface response operator of block i.

The new interface states can be calculated from [79,80]

$$\det[g(M,M)]^{-1} = 0, \tag{11.12}$$

showing that, if one is interested in calculating the interface states of a composite, one only needs to know the inverse of the Green's function of each individual block in the space of their respective surfaces.

Moreover, if $U(D)$ [84] represents an eigenvector of the reference system, Eq. (11.8) enables one to calculate the eigenvectors $u(D)$ of the composite material

$$u(D) = U(D) - U(M)[G(M,M)]^{-1}G(M,D) \\ + U(M)[G(M,M)]^{-1}g(M,M)[G(M,M)]^{-1}G(M,D). \tag{11.13}$$

In Eq. (11.13), $U(D)$, $U(M)$, and $u(D)$ are row-vectors. Eq. (11.13) enables also to calculate all the waves reflected and transmitted by the interfaces, as well as the reflection and the transmission coefficients of the composite system. In this case, $U(D)$ must be replaced by a bulk wave launched in one homogeneous piece of the composite material [84].

11.2.2 Inverse surface Green's functions of the elementary constituents

We report here the expression of the Green's function of a homogeneous isotropic infinite one-dimensional dielectric medium. For the sake of simplicity, we restrict ourselves to nonmagnetic media. We give also the inverse of the surface Green's function for the semiinfinite medium with a free end and for a finite medium of length d.

11.2.2.1 Green's function of an infinite medium

We consider an infinite one-dimensional medium i associated to the Cartesian coordinate x. When the lateral dimensions of such a medium are small compared to the wavelength of the electromagnetic waves, their classical propagation equation becomes three times degenerate and reads

$$\left(\frac{\omega^2\varepsilon_i}{c^2}+\frac{\partial^2}{\partial x^2}\right)G_i(x-x')=\delta(x-x'), \tag{11.14}$$

where rather than considering the electric or magnetic field at point x we introduced the Green's function $G_i(x-x')$ between two points of this medium.

It has been established [86–89] that this Green's function is

$$G_i(x,x')=\frac{e^{j\alpha_i|x-x'|}}{2j\alpha_i}. \tag{11.15}$$

For electromagnetic waves with which we are dealing in this chapter, α_i is given as

$$\alpha_i=\frac{\omega}{c}\sqrt{\varepsilon_i(\omega)}, \tag{11.16}$$

for the transverse electric (TE) and transverse magnetic (TM) polarizations, where ω is the wave pulsation, c the speed of light in vacuum, $\varepsilon_i(\omega)$ the relative permittivity for homogeneous isotropic dielectric medium i, and $j=\sqrt{-1}$.

Let us emphasize that α_i has the value of the propagation vector in the infinite lossless line of type i and that

$$\alpha_i=\frac{2\pi}{\lambda_i}, \tag{11.17}$$

where λ_i is the wavelength in the line i.

11.2.2.2 Inverse surface Green's functions of the semiinfinite medium

One considers a semiinfinite monomode guide i with a "free surface" such that the magnetic field $H=0$ at its extremity $x=0$. In this case [80], the cleavage operator (Eq. (11.11)) is

$$V_i(x)=\delta(x)\frac{\partial}{\partial x}, \tag{11.18}$$

and then

$$[g_i(0,0)]^{-1} = j\alpha_i. \tag{11.19}$$

11.2.2.3 *Inverse surface Green's functions of the finite medium*

One considers a finite monomode guide of width d_i bounded by two free surfaces at $x = 0$ and $x = d_i$ such that the magnetic field H vanishes on these extremities. In this case [83],

$$[g_i(M_i, M_i)]^{-1} = \frac{\alpha_i}{\sin(\alpha_i d_i)} \begin{pmatrix} -\cos(\alpha_i d_i) & 1 \\ 1 & -\cos(\alpha_i d_i) \end{pmatrix}. \tag{11.20}$$

One can see that in the interface domain M_i corresponding to interfaces $x = 0$ and $x = d_i$, the surface Green's function is a 2×2 square matrix. In order to study elementary electromagnetic excitations, we will calculate the Green's function for different composite devices composed of finite segments grafted on a one-dimensional waveguide.

11.3 Comb structures

11.3.1 *Propagation of electromagnetic waves in an infinite line with one grafted finite line*

One considers a quasi-one-dimensional composite system formed out of a finite segment of length d_2 (with the $H = 0$ boundary condition) grafted on an infinite waveguide line (see Fig. 11.1(a)). In order to calculate the corresponding interface Green's function, we construct this system out of two semiinfinite lines constituted of the same dielectric material 1 and a segment of dielectric material 2 of finite length d_2. These three blocks are coupled at their ends (see Fig. 11.1(b)). For the two semiinfinite lines and for the finite segment, the interface domains correspond to site 0 and sites 0 and $1'$, respectively.

The inverse surface Green's functions $[g_i(M_i, M_i)]^{-1}$ for the two semiinfinite lines and for the finite segment are given by Eqs. (11.19) and (11.20) with $i = 1$ and $i = 2$, respectively ($d_2 = d$). Media 1 and 2 are one-dimensional and the parameters α_i ($i = 1, 2$) are given by Eqs. (11.16) and (11.17).

In this case, the interface domain of the composite system can be reduced to site 0 and the finite segment contribution to the surface Green's function of the composite system takes then the form

$$g_2(0,0) = \frac{1}{\alpha_2 \tan(\alpha_2 d)}. \tag{11.21}$$

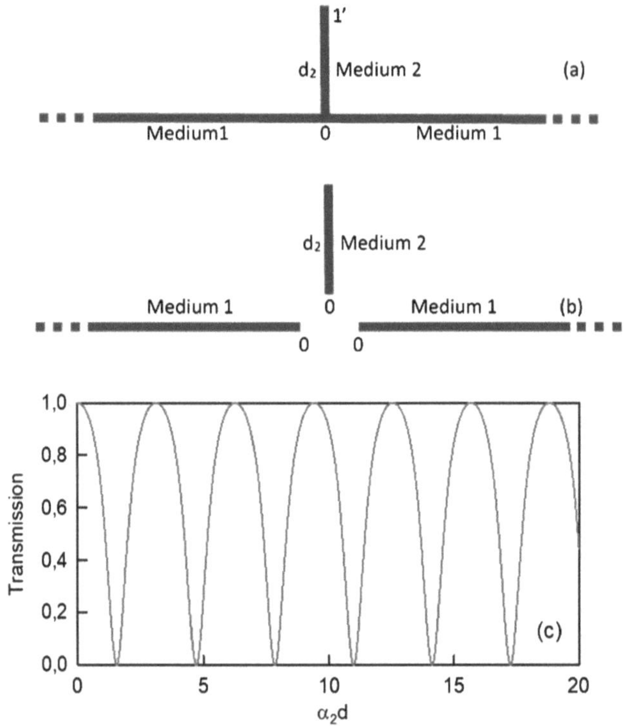

Figure 11.1 (a) Waveguide with a single grafted segment of length $d_2 = d$. (b) Elementary constituents of the waveguide with a single grafted segment of length $d_2 = d$. (c) Transmission coefficient versus the dimensionless quantity $\alpha_2 d$ for a waveguide with a single grafted segment of length d in the case of identical media 1 and 2 ($\alpha_2 = \omega\sqrt{\varepsilon_2}/c$ where ε_2 is the relative dielectric permittivity of medium 2).

Superposing these different linear contributions, one deduces [54] that the inverse interface Green's function of the composite system is

$$g^{-1}(0,0) = 2j\alpha_1 + \alpha_2 \tan(\alpha_2 d). \tag{11.22}$$

Eq. (11.13) allows us to calculate the transmission coefficient of this composite system. Consider, $U(x) = e^{j\alpha_1 x}$, a bulk propagating wave coming from $x = -\infty$. Using this incident wave in Eq. (11.13), one obtains the transmitted wave for x greater than 0 and the transmission coefficient as

$$T = \frac{1}{1 + \frac{1}{4}\sqrt{\frac{\varepsilon_2}{\varepsilon_1}}\tan^2(\alpha_2 d)}. \tag{11.23}$$

We observe that this coefficient equals to zero when

$$\alpha_2 = \left(m + \frac{1}{2}\right)\left(\frac{\pi}{d}\right), \tag{11.24}$$

where m is a positive integer. The variations of T versus the dimensionless quantity $\alpha_2 d$ are reported in Fig. 11.1(c) in the case of identical media 1 and 2 for the backbone and for the branch. Here T is equal to zero for $\alpha_2 d$ being the odd multiple of $\frac{\pi}{2}$ and reaches its maximum value of 1 for $\alpha_2 d$ being multiple of π. For this composite system, there exists an infinite set of forbidden frequencies ω_g such as

$$\omega_g = \frac{c}{\sqrt{\varepsilon_2(\omega)}}\left(m + \frac{1}{2}\right)\left(\frac{\pi}{d}\right), \tag{11.25}$$

corresponding to the eigenmodes of the grafted finite segment.

This grafted segment behaves as a resonator, and this simple composite system filters out the frequencies ω_g. This phenomenon is related to the resonances associated with the finite additional path offered to the wave propagation.

We can now consider more complex structures of the composite system containing a larger number of side branches.

11.3.2 One dimensional infinite backbone with periodic array of finite segments: infinite comb ($H = 0$ boundary condition)

We treat the case of a comb-like structure composed of finite segments (medium 2) of length d_2 grafted periodically with lattice spacing d_1 on an infinite substrate (medium 1) (see Fig. 11.2). Let us first, write the surface Green's function of this composite system. The infinite line can be modeled as an infinite number of finite segments of length d_1 in the direction x, each one being pasted to two neighbors. The interface domain is constituted of all the connection points between finite segments. In what follows, these connection points will be called sites and each site on the infinite chain will be defined by the integer n such as $-\infty < n < \infty$. On each site n, a finite segment of length d_2 is connected. The respective contributions of media 1 and 2 to the inverse surface Green's function of the composite system are given by Eqs. (11.20) and (11.21), respectively. The inverse surface Green's function of the composite system is then obtained as an infinite banded matrix $[g_\infty(M, M)]^{-1}$ defined in the interface domain constituted of all the sites n. The diagonal and off-diagonal elements of this matrix are respectively given by $[-2\alpha_1 \cot(\alpha_1 d_1) + \alpha_2 \tan(\alpha_2 d_2)]$ and $[\alpha_1 \sin(\alpha_1 d_1)]$.

Taking advantage of the translational periodicity of this system, this matrix can be Fourier transformed as

$$[g_\infty(k, M, M)]^{-1} = \frac{2\alpha_1}{\sin(\alpha_1 d_1)}\left[-\xi + \cos(kd_1)\right], \tag{11.26}$$

where k is the complex one-dimensional propagation vector in this infinite comb (note that k is real inside the passing bands of this comb and complex within the gaps) and

$$\xi = \cos(\alpha_1 d_1) - \frac{1}{2}\sqrt{\frac{\varepsilon_2}{\varepsilon_1}}\sin(\alpha_1 d_1)\tan(\alpha_2 d_2). \tag{11.27}$$

The dispersion relation of the infinite periodic comb-like waveguide is obtained from Eqs. (11.12) and (11.26) and is expressed in the simple form

$$\cos(kd_1) = \xi. \tag{11.28}$$

There exist forbidden frequencies at the zeros of transmission (see Eqs. (11.24) and (11.25)) of an infinite substrate with a single grafted segment. On the other hand, in the k-space, the corresponding Green's function is

$$g_\infty(k, M, M) = -\frac{\sin(\alpha_1 d_1)}{2\alpha_1[\xi - \cos(kd_1)]}. \tag{11.29}$$

After applying the inverse Fourier transformation, Eq. (11.29) gives [90]

$$g_\infty(n, n') = \frac{\sin(\alpha_1 d_1)}{\alpha_1} \frac{t^{|n-n'|+1}}{t^2 - 1}, \tag{11.30}$$

where the integers n and n' refer to the sites ($-\infty < n < \infty$) on the infinite line. The parameter t is defined as

$$t = e^{jkd_1}, \tag{11.31}$$

or equivalently

$$t = \begin{cases} \xi - \sqrt{\xi^2 - 1}, & \xi > 1, \\ \xi + \sqrt{\xi^2 - 1}, & \xi < -1, \\ \xi + j\sqrt{1 - \xi^2}, & -1 < \xi < 1, \end{cases} \tag{11.32}$$

with

$$t + \frac{1}{t} = 2\xi \tag{11.33}$$

and

$$|t| < 1. \tag{11.34}$$

Note that the signs in Eqs. (11.31) and (11.32) are determined by the facts that inside the passing bands the density of states is positive and outside the passing bands the Green's function has to decay when the distances increase.

We now focus on the dispersion relation of this composite system. Eq. (11.28) can be written explicitly as

$$\cos(\alpha_1 d_1) - \frac{1}{2}\sqrt{\frac{\varepsilon_2}{\varepsilon_1}} \sin(\alpha_1 d_1) \tan(\alpha_2 d_2) = \cos(kd_1). \tag{11.35}$$

Let us first consider the particular case where media 1 and 2 are identical (i.e., $\alpha_2 = \alpha_1$) with $d_1 = d_2$. In this very simple case, the resolution of Eq. (11.35) is strictly analytical. The limits of the lowest bands are given by the values of $\alpha_1 d_1$ such that

$$\frac{3\cos^2(\alpha_1 d_1) - 1}{2\cos(\alpha_1 d_1)} = \pm 1, \tag{11.36}$$

i.e., $\alpha_1 d_1 = 0$, 1.2309, 1.9106 and π.

We then deduce the gap widths

$$\Delta\omega = \frac{c}{\sqrt{\varepsilon_1(\omega)}} \frac{\Delta(\alpha_1 d_1)}{d_1}. \tag{11.37}$$

Fig. 11.2(b) represents the band structure of the infinite comb composite in terms of $\alpha_1 d_1$ versus kd_1 in the case of $\varepsilon_2 = \varepsilon_1$ and $d_2 = d_1$ for $-\pi < kd_1 < +\pi$. There exist two gaps of identical width between the first and the second band and between the third and the fourth band. These gaps appear around values of $\alpha_1 d_1$ corresponding to odd multiple of $\pi/2$ associated with the zeros of transmission of the grafted resonators. The second and the third bands meet at $kd_1 = \pm\pi$ and there is no gap between these two bands. Along the $\alpha_1 d_1$ axis, the band structure repeats periodically with 2π period.

Eq. (11.37) gives the width of the gaps in the electromagnetic band structure of this particular one-dimensional composite system. These gaps appear around the frequencies

$$\omega_g = \frac{c}{\sqrt{\varepsilon_1(\omega)}} \left(m + \frac{1}{2}\right)\left(\frac{\pi}{d_1}\right), \tag{11.38}$$

where m is a positive integer. Eq. (11.38) shows that according to the value of d_1, the first forbidden band ($m = 0$) exists in different frequency domains of the electromagnetic spectrum. More precisely, if one supposes a dielectric constant, $\varepsilon_1 = 9$, which corresponds to a material (alumina composite) often used in dielectric composite studies [91], the first forbidden band appears in the microwave domain for 0.05 mm $< d_1 <$ 25 mm, in the infrared domain for 0.05 μm, $< d_1 <$ 50 μm and in the range of visible electromagnetic radiations for 33.3 nm $< d_1 <$ 50 nm.

In order to study the influence of the geometry of the comb-like system on its electromagnetic band structure, we compute the band structure for $\varepsilon_2 = \varepsilon_1$ and $d_2 \neq d_1$. For instance, Fig. 11.2(c) shows the band structure for $\varepsilon_2 = \varepsilon_1$ and $d_2/d_1 = 0.3$. In the low frequency domain, one observes gaps between the first and the second band, the second and the third band, the third and the fourth band. Contrary to the second gap which is associated with a zero of transmission of a single resonator, the first and the third gaps appear at $\alpha_1 d_1$ values different from an odd multiple of $\pi/2$. Therefore, these gaps must result from the superlattice nature of our quasi-one-dimensional waveguide with periodic side branches.

The study of the electromagnetic band structure of an infinite two-dimensional waveguide was also performed [54]. This brush-like structure is composed of finite segments of identical length (medium 2) grafted on each site of a periodic square array of one-dimensional waveguides (medium 1). Wide absolute bandgaps extending throughout the two-dimensional Brillouin zone were obtained. In the particular case

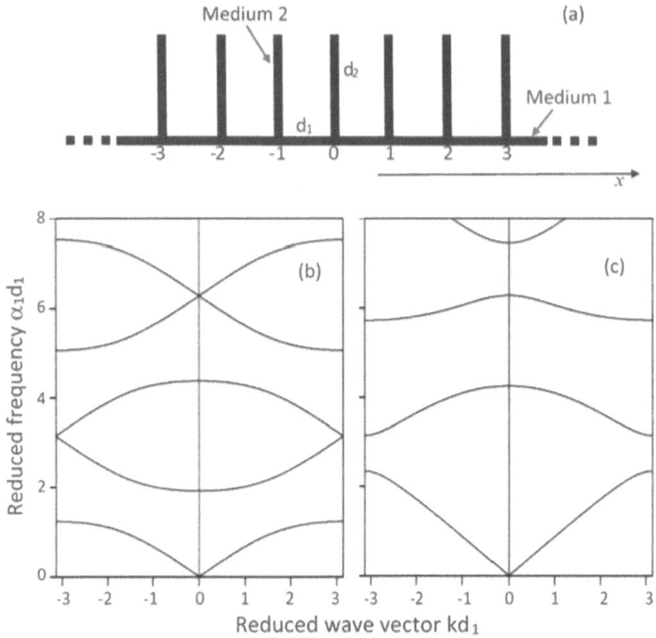

Figure 11.2 (a) Waveguide with periodic array of grafted segments of length d_2 distant one from each other by a length d_1. Each medium i ($i = 1$ for the backbone and $i = 2$ for the side branches) is characterized by the relative permittivity ε_i. (b) Electromagnetic band structure of the infinite periodic comb with ($\varepsilon_2/\varepsilon_1 = 1$ and $d_2/d_1 = 1$). The plot is given in terms of $\alpha_1 d_1$ ($\alpha_1 = \omega\sqrt{\varepsilon_1}/c$) versus the dimensionless quantity kd_1 ($-\pi < kd_1 < +\pi$) where k is the modulus of the propagation vector. One observes two absolute gaps of identical width between the first and the second band and between the third and the fourth band. (c) Same as in Fig. 11.2(b) but with $\varepsilon_2/\varepsilon_1 = 1$ and $d_2/d_1 = 0.3$. In this case one observes three absolute gaps between the first and the second band, the second and the third band, the third and the fourth band.

of two identical constituent materials with a network periodicity equal to the length of the grafted segments, these gaps were narrower than in the corresponding comb-like structure.

11.3.3 Transmission coefficient of the finite comb

Infinite electromagnetic combs are not physically realizable. We investigate in this section a finite comb with a behavior similar to the infinite one.

We consider in this section the quasi-one-dimensional electromagnetic comb represented in Fig. 11.3. This composite system is constructed out of a finite comb cut out of the infinite periodic system of Fig. 11.2 which is subsequently connected at its ends to two semiinfinite leading lines. The finite comb is therefore composed of N segments (medium 2) of length d_2 grafted periodically with a lattice spacing d_1 on a finite line (medium 1). For the sake of simplicity, the semiinfinite leads are assumed to be constituted of the same material as medium 1. We calculate analytically the transmission coefficient of a bulk electromagnetic wave coming from $x = -\infty$.

The system depicted in Fig. 11.3 is constructed from the infinite comb of Fig. 11.2. In a first step, one suppresses the segments linking sites 0 and 1, and sites N and $N+1$. For this new system composed of a finite comb and two semiinfinite combs, the inverse interface Green's function, $[g_t(M,M)]^{-1}$, is an infinite banded matrix defined in the interface domain of all the sites n, $-\infty < n < +\infty$. This matrix is similar to the one associated with the infinite comb. Only a few matrix elements differ, namely, those associated with the sites $n=0$, $n=1$, $n=N$ and $n=N+1$.

The cleavage operator $V_{cl} = [g_t(M,M)]^{-1} - [g_\infty(M,M)]^{-1}$ [79], is the following 4×4 square matrix defined in the interface domain constituted of sites 0, 1, N, and $N+1$

$$V_{cl} = \frac{\alpha_1}{\sin(\alpha_1 d_1)} \begin{pmatrix} \cos(\alpha_1 d_1) & -1 & 0 & 0 \\ -1 & \cos(\alpha_1 d_1) & 0 & 0 \\ 0 & 0 & \cos(\alpha_1 d_1) & -1 \\ 0 & 0 & -1 & \cos(\alpha_1 d_1) \end{pmatrix}. \tag{11.39}$$

In a second step, two semiinfinite leads constituted of the same material as medium 1 are connected to the extremities $n=1$ and $n=N$ of the finite comb. With the help of the interface response theory of discrete systems [81], one deduces that the perturbing operator $V_p(M,M)$ allowing the construction of the system of Fig. 11.3(a) from the infinite comb is then defined as the 4×4 square matrix

$$V_p = \frac{\alpha_1}{\sin(\alpha_1 d_1)} \begin{pmatrix} \cos(\alpha_1 d_1) & -1 & 0 & 0 \\ -1 & e^{j\alpha_1 d_1} & 0 & 0 \\ 0 & 0 & e^{j\alpha_1 d_1} & -1 \\ 0 & 0 & -1 & \cos(\alpha_1 d_1) \end{pmatrix}. \tag{11.40}$$

On the other hand, using Eq. (11.30), one can write the elements of the interface Green's function of the infinite comb for n and $n' = 0$, 1, N, and $N+1$ in the form of a 4×4 square matrix $g_\infty(M,M)$

$$g_\infty(M,M) = \frac{\sin(\alpha_1 d_1)}{\alpha_1} \frac{t}{t^2-1} \begin{pmatrix} 1 & t & t^N & t^{N+1} \\ t & 1 & t^{N-1} & t^N \\ t^N & t^{N-1} & 1 & t \\ t^{N+1} & t^N & t & 1 \end{pmatrix}. \tag{11.41}$$

Using Eqs. (11.40) and (11.41), one obtains the matrix operator

$$\Delta_s(M,M) = I(M,M) + V_p(M,M) g_\infty(M,M), \tag{11.42}$$

in the space M of sites 0, 1, N, and $N+1$. For the calculation of the transmission coefficient, we only need the matrix elements $\Delta_s(1,1)$, $\Delta_s(1,N)$, $\Delta_s(N,1)$ and $\Delta_s(N,N)$ which can be set in the form of a 2×2 matrix

$$\Delta_s(M,M) = \begin{pmatrix} 1 + A_s t & A_s t^N \\ A_s t^N & 1 + A_s t \end{pmatrix}, \tag{11.43}$$

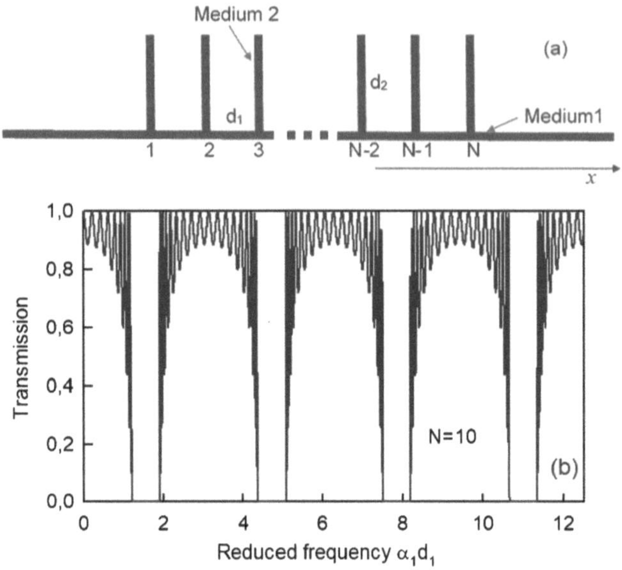

Figure 11.3 (a) Waveguide with a finite number N of grafted branches arranged periodically. (b) Transmission coefficient for the finite comb with $\varepsilon_2/\varepsilon_1 = 1$ and $d_2/d_1 = 1$ and $N = 10$ grafted branches.

with

$$A_s = -\frac{t - e^{j\alpha_1 d_1}}{t^2 - 1}. \tag{11.44}$$

We now calculate the transmission coefficient with a bulk electromagnetic wave coming from $x = -\infty$, $u(x) = e^{j\alpha_1 x}$. Substituting this incident wave in Eq. (11.13) and considering Eqs. (11.15) and (11.42), we obtain the transmitted wave as

$$u(x) = 2j\sin(\alpha_1 d_1)\frac{t^N}{t^2 - 1}\frac{e^{j\alpha_1[x-(N-1)d_1]}}{|\Delta_s(M, M)|}, \tag{11.45}$$

where $|\Delta_s(M, M)|$ is the determinant of $\Delta_s(M, M)$.

One deduces that the transmission coefficient is

$$T = \left|\frac{2\sin(\alpha_1 d_1)(t^2 - 1)t^N}{(1 - te^{j\alpha_1 d_1})^2 - t^{2N}(t - e^{j\alpha_1 d_1})^2}\right|^2. \tag{11.46}$$

One can easily check that for $N = 1$, which corresponds to the single grafted segment on an infinite line (see Fig. 11.1(a)), Eq. (11.46) leads to Eq. (11.23). In what follows, we study the variation of T versus $\alpha_1 d_1$ for different values of N in various finite combs.

Fig. 11.3(b) represents the variation of the transmission coefficient T versus $\alpha_1 d_1$ for identical media ($\varepsilon_2 = \varepsilon_1$), $d_2 = d_1$, and $N = 10$. Despite the finite number of grafted segments, T approaches zero in regions corresponding to the observed gaps in

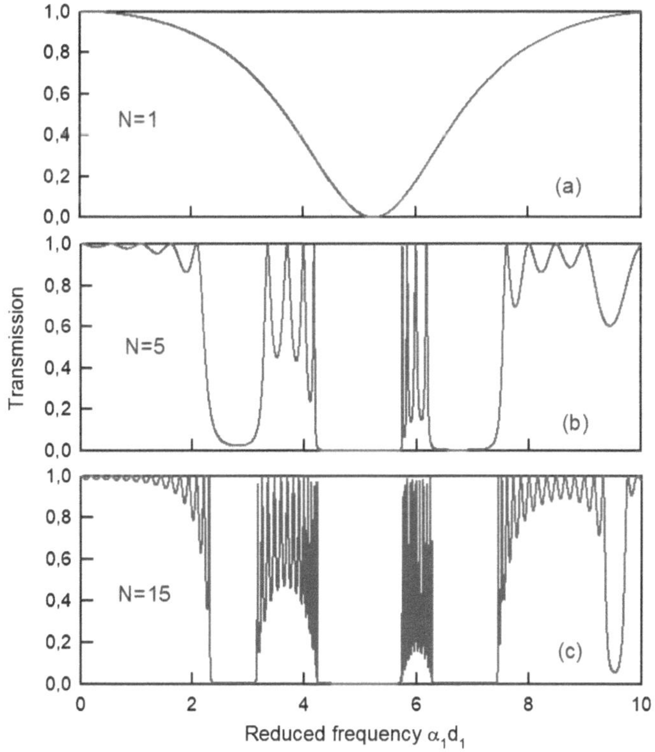

Figure 11.4 Transmission coefficient for the finite comb with $\varepsilon_2/\varepsilon_1 = 1$ and $d_2/d_1 = 0.3$: (a) $N = 1$ grafted branch, (b) $N = 5$ grafted branches, and (c) $N = 15$ grafted branches.

the electromagnetic band structure of Fig. 11.2(b). Next, we analyze the evolution of T as the number of grafted segments N increases in the case of $\varepsilon_2 = \varepsilon_1$ and $d_2/d_1 = 0.3$. The first zero of transmission is situated at $\alpha_1 d_1 = \pi/0.6$ as shown in Fig. 11.4(a). As the number of grafted segments increases to $N = 5$, a gap forms around the zero of transmission and the coefficient of transmission is also strongly reduced around $\alpha_1 d_1 = \pi, 2\pi, 3\pi, \ldots$ (see Fig. 11.4(b)). The depressions increase by increasing the number of the grafted segments. In Fig. 11.4(c), for $N = 15$, these gaps have appeared at π and 2π. At $\alpha_1 d_1 = 3\pi$, T has not yet converged to zero for $N = 15$. One would need to increase the number of grafted segments to open a nearly absolute gap in this vicinity. These results parallel the calculated electromagnetic band structure of the infinite superlattice shown in Fig. 11.2(c). Therefore, the convergence, concerning the limits of the bandgaps, can be achieved in general for a reasonably small number of side branches, usually N less than 10 to 20.

The evolution of T for a finite quasi-one-dimensional comb-like structure suggests the possibility of designing electromagnetic devices mimicking the behavior of the periodic infinite comb-like waveguide. These devices may serve as frequency filters with wide stopbands.

11.4 Star combs

Now we review the electromagnetic transmission through a one-dimensional waveguide along which stars of N' side branches are grafted at N equidistant sites, with N and N' being integers. Analytic expressions were reported [92] for the corresponding band structure and for the transmission coefficient for any value of N and N' with the $H = 0$ and $E = 0$ boundary conditions at the extremities of the resonators. The generic behaviors enabling us to obtain quasiquantized photonic band structures of these star waveguides was confirmed by simple experiments using coaxial cables.

11.4.1 Analytical calculations

We present first the analytic expression for an infinite ($N \to \infty$) star waveguide composed of N' side branches grafted at each site of the one-dimensional waveguide (Fig. 11.5). Each medium i ($i = 1$ for the backbone and $i = 2$ for the side branches) is characterized by the relative permittivity ε_i. The dispersion relations, for such infinite star wave guides, are easily obtained in the same manner as in Eq. (11.35) to be

$$\cos(kd_1) = \cos(\alpha_1 d_1) + \frac{N'}{2}\sqrt{\frac{\varepsilon_2}{\varepsilon_1}}\sin(\alpha_1 d_1)\cot(\alpha_2 d_2), \tag{11.47}$$

when the electric field E vanishes at the free ends of the side branches and d_1 and d_2 being respectively the lattice spacing and the length of the resonators, and

$$\cos(kd_1) = \cos(\alpha_1 d_1) - \frac{N'}{2}\sqrt{\frac{\varepsilon_2}{\varepsilon_1}}\sin(\alpha_1 d_1)\tan(\alpha_2 d_2), \tag{11.48}$$

when the magnetic field $H = 0$ at the free ends of the side branches.

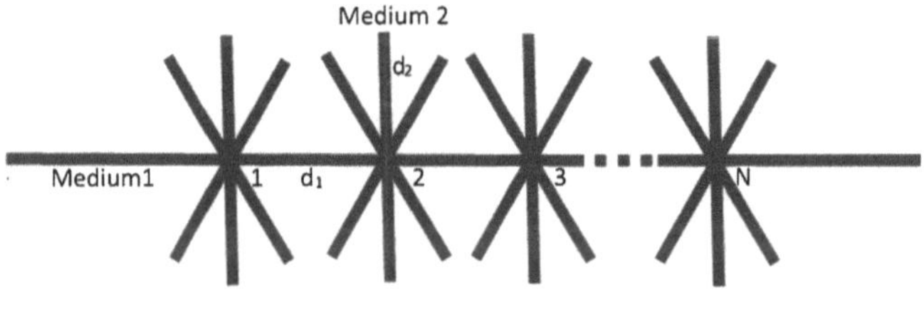

Figure 11.5 Waveguide with a periodic array of grafted stars distant from each other by a length d_1. Each star is composed of N' identical side branches of length d_2. Each medium i ($i = 1$ for the backbone and $i = 2$ for the side branches) is characterized by the relative permittivity ε_i.

The narrowing of photonic passbands when increasing the number N' of grafted branches in each star can readily be observed from Eqs. (11.47) and (11.48). Indeed, the passing bands are obtained when the right-hand sides of these equations fall in the range $[-1, +1]$. For high values of N', this condition can only be satisfied in narrow frequency domains with the mean frequencies of the bands given by $\sin(\alpha_1 d_1) = 0$ or $\cos(\alpha_2 d_2) = 0$ for Eq. (11.47) and $\sin(\alpha_1 d_1) = 0$ or $\sin(\alpha_2 d_2) = 0$ for Eq. (11.48).

For example, if we assume that both constituent materials in the star waveguide are identical by taking $\varepsilon_1 = \varepsilon_2 = \varepsilon$ with $d_1 = d_2 = d$, the frequencies of the bands are obtained from a simple Taylor expansion of Eq. (11.47) with respect to $1/N'$ to be

$$\omega \approx \frac{c}{d\sqrt{\varepsilon}}\left[\left(n+\frac{1}{2}\right)\pi + \frac{2(-1)^{n+1}}{N'+2}\cos(kd)\right], \quad n = 0, 1, 2, \ldots \tag{11.49}$$

This expression shows clearly that the almost flat photonic bands are lying around

$$\omega = (2n+1)\pi c/2d\sqrt{\varepsilon} \tag{11.50}$$

and that their widths are proportional to $4c/[(N'+2)d\sqrt{\varepsilon}]$.

The width of the gaps is of the order of $\pi c/d\sqrt{\varepsilon}$, except for the first one whose width is half of that of all other gaps. Let us stress out that this first forbidden band starts at $\omega = 0$.

The ratio between the gap and passing-band widths is equal to $(N'+2)\pi/4$ for a large number N' of grafted side branches at each site.

A similar analysis of Eq. (11.48) provides

$$\omega \approx \frac{c}{d\sqrt{\varepsilon}}\left(n\pi \pm \sqrt{\frac{2}{N'+1}}\left[1+(-1)^{n+1}\cos(kd)\right]^{\frac{1}{2}}\right), \quad n = 0, 1, 2, \ldots \tag{11.51}$$

In this case, the ratio between the gap and passband widths is equal to $(N'+1)^{\frac{1}{2}}\pi/4$ for large N'.

Let us also mention that tuning the boundary condition at the ends of the resonators enables to shift the periodic quasiquantized photonic bands to any other values such that for large values of N',

$$\omega \approx \omega_0 + \frac{n\pi c}{d\sqrt{\varepsilon}}, \tag{11.52}$$

where ω_0 can be adjusted to any value such that $0 \le \omega_0 \le \frac{\pi c}{2d\sqrt{\varepsilon}}$.

Coming back to the quasi-one-dimensional star waveguide, it was shown [92] that the transmission factor T through such a composite formed out of N nodes on which N' identical side branches are grafted has the same expression as that given above by Eq. (11.46), with $t = e^{jkd_1}$ and k being defined by either Eq. (11.47) or Eq. (11.48).

11.4.2 Flat bands and bound in continuum states

Another interesting point to notice in the dispersion relation (Eq. (11.48)) is the possibility to obtain flat bands when the lengths of the segments and stubs with $H = 0$ boundary conditions on their ends are chosen commensurate. Indeed, when the denominator of Eq. (11.48) vanishes (i.e., $\cos(\alpha_2 d_2) = 0$) then the Bloch wavevector diverges giving rise to the so-called hybrid or local resonance bandgaps. Now, if the numerator and denominator of Eq. (11.48) vanish simultaneously ($\cos(\alpha_2 d_2) = 0$ and

$\sin(\alpha_1 d_1) = 0$), then one obtains flat bands. If the waveguides are made of the same material (i.e., $\varepsilon_1 = \varepsilon_2$), the two latter equations are satisfied if d_1 and d_2 are chosen commensurate such that

$$\frac{d_2}{d_1} = \frac{2m_2 + 1}{2m_1}, \tag{11.53}$$

where m_1 and m_2 are integers.

Then the reduced frequencies of the flat bands are given by

$$\alpha_1 d_1 = m_1 \pi. \tag{11.54}$$

The modes inside such bands are confined in the stubs and do not interact with the horizontal waveguide giving rise to the so-called bound in continuum (BIC) states [93,94]. In a finite system, these modes cannot be detected in the transmission spectra (see below) as their widths reduce to zero.

Similarly when the boundary conditions on the stub ends are $E = 0$, then the flat bands are given by vanishing simultaneously the numerator and denominator of Eq. (11.47) (i.e., $\sin(\alpha_2 d_2) = 0$ and $\sin(\alpha_1 d_1) = 0$), then one obtains the following relations between d_1 and d_2 if the cables are made of the materials (i.e., $\varepsilon_1 = \varepsilon_2$), namely

$$\frac{d_2}{d_1} = \frac{m_2}{m_1}. \tag{11.55}$$

Then the reduced frequencies of the flat bands are given by

$$\alpha_1 d_1 = m_1 \pi. \tag{11.56}$$

11.4.3 Path states

Following a similar approach to the one developed for loop chains in Section 5.10 of Chapter 5 of this Photonics book, path states start to be investigated for the finite comb system presented on Fig. 11.6. For this system ($\alpha_1 = \alpha_2 = \alpha$, $d_1 = 2d$, $d_2 = d$, and $N' = 2$) the path states are given by $SC^{2N+1} = 0$, where $S = \sin(\alpha d)$ and $C = \cos(\alpha d)$. The $C = 0$ states are $2N + 1$ times degenerate. They form flat bands. Such flat bands exist in the dispersion curves, but are not visible by the experimental and theoretical transmission coefficients. It is, however, possible to activate each individual state of such flat bands with leads not connected to the robust zeros of their states. This is a suggestion for novel experimental investigations.

11.4.4 Experimental and numerical results

We recall numerical results for the transmission spectrum T through a star comb and compare the results with the photonic band structure of this system in the limit of $N \to \infty$, when it becomes a periodic system. The calculated transmission coefficients

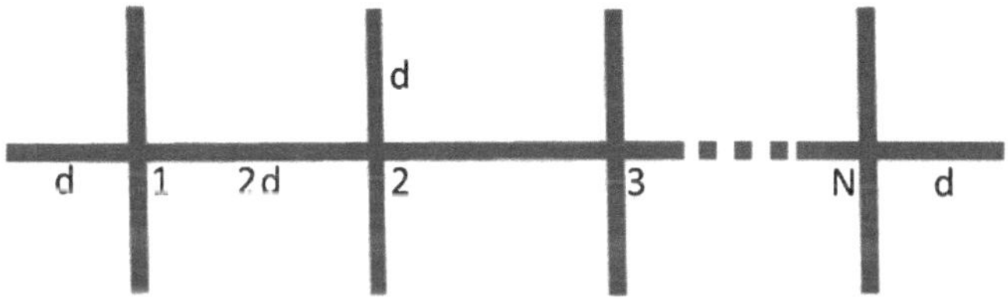

Figure 11.6 A finite structure made out of $2N$ stubs of length d grafted periodically to a finite backbone guide. The boundary condition on the stub ends is $H = 0$.

are also compared with the experimental results of the transmission intensity for various values of N' and d_i and the two different boundary conditions at the extremities of the side branches.

In Fig. 11.7(a), we present the five lowest dispersion curves of the photonic band structure for the star waveguide of Fig. 11.5, when the electric field $E = 0$ at the free ends of the side branches. We have chosen $\varepsilon_2 = \varepsilon_1$, $d_2 = d_1$, and $N' = 10$. The most interesting feature of the band structure in this system is the quasiquantized passbands. Two flat bands corresponding to bound in continuum states (as discussed in Section 11.4.2, Eqs. (11.55) and (11.56)) appear at the reduced frequencies π and 2π. They cannot be observed in the transmission spectrum shown in panel (c). Perhaps the other dominant feature is the gap below the lowest frequency band which exists for any values of $\varepsilon_2/\varepsilon_1$, d_2/d_1, and N'. The width of this gap is found to increase monotonically with increasing value of N'. In Fig. 11.7(b), we plotted the frequency dependence of the transmission for 10 sites (i.e., $N = 10$).

We now look at the boundary condition when the magnetic field H vanishes at the free ends of the side branches. In Fig. 11.7(c), we present the photonic band structure for $\varepsilon_2 = \varepsilon_1$, $d_2/d_1 = 0.4$, and $N' = 10$. From this result we see that the photonic band structure possesses no gap below the lowest band and the three lowest frequencies of the quasiquantized photonic bands appear at $\omega d_1/c = \pi$, 4π, and 6π. In Fig. 11.7(d), we plot the frequency dependence of the transmission coefficient of 10 sites (i.e., $N = 10$). Again, a comparison of the frequency dependence of the transmission coefficient with the corresponding band structure shows that it predicts the positions of the allowed bands, including the nearly dispersionless bands, and the bandgaps, but lacks the details present in the dispersion curves. The detailed form of the band structures displayed in Figs. 11.7(a) and 11.7(c) are in complete agreement with the predictions of the qualitative analysis of Eqs. (11.49) and (11.51).

In Fig. 11.8 we emphasized the shrinking of the photonic passbands when increasing the number N' of the side branches. These bands are obtained by cutting the transmission coefficient at -30 dB.

In the experimental setup, the one-dimensional waveguide was the standard 50 Ω coaxial cable and the transmission measurements were realized by using a tracking generator coupled to a spectrum analyzer. Fig. 11.9 presents the variation of the transmission coefficient versus the frequency for $d_1 = d_2 = 1$ m, $\varepsilon_1 = \varepsilon_2 = 2.3$, and $N = 10$, considering the $E = 0$ boundary condition. Figs. 11.9(a) and 11.9(b) present theoretical results for $N' = 1$ and 8, respectively. The corresponding experimental re-

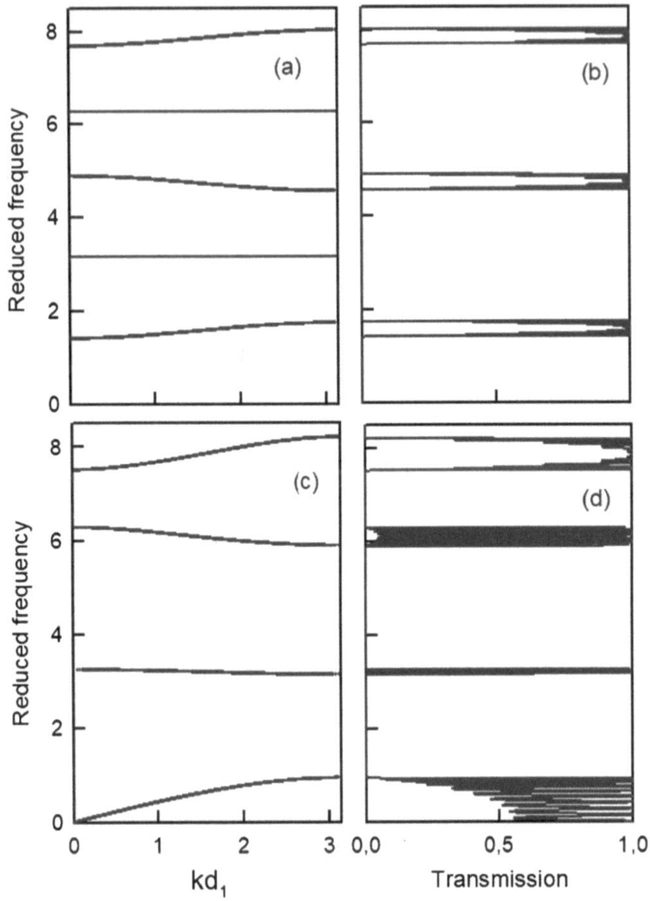

Figure 11.7 (a) The photonic band structure of the infinite one-dimensional system depicted in Fig. 11.5, for the $E = 0$ boundary condition, for $\varepsilon_2/\varepsilon_1 = d_2/d_1 = 1$ and $N' = 10$. Two flat bands corresponding to bound in continuum states (as discussed in Section 11.4.2) appear at the reduced frequencies π and 2π. They cannot be observed in the transmission spectrum shown in panel (c). (b) The frequency dependence of the transmission T of a periodic system of 10 sites, for $\varepsilon_2/\varepsilon_1 = d_2/d_1 = 1$ and $N' = 10$. (c)–(d) Same as in (a)–(b) but for $H = 0$ boundary condition and $d_2/d_1 = 0.4$ and $\varepsilon_2/\varepsilon_1 = 1$.

sults are presented, respectively, in Figs. 11.9(c) and 11.9(d). In this case, we note the existence of the gap below the lowest frequency band. One also observes that the width of the photonic passbands decreases for increasing N'. The width of these bands at -30 dB is equal to 55 MHz for $N' = 1$, and 12 MHz for $N' = 8$. Let us notice that these photonic bands become quasiquantized for $N' \geq 10$. The attenuation observed in experimental results increases almost linearly as a function of frequency. This attenuation factor is of the order of 50 dB/100 m in the frequency range from 10 to 500 MHz. We can also note that the photonic band structure is very sensitive to the ratio d_2/d_1; for $d_2 \neq d_1$, we observe experimentally and theoretically the presence of passbands inside the gaps with larger widths at high frequencies. Concerning the width

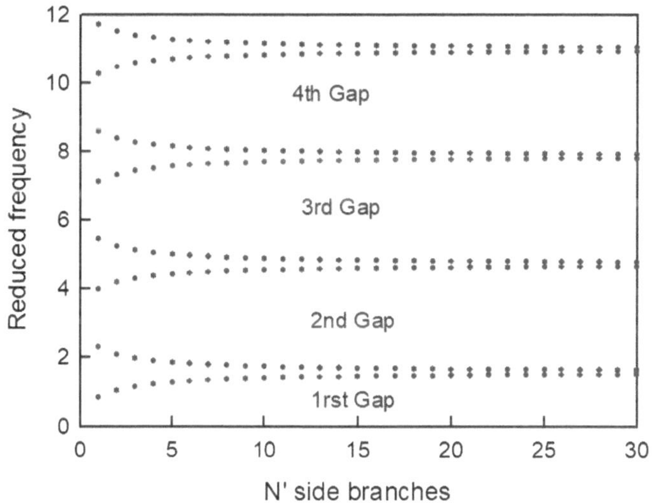

Figure 11.8 Variation of the width of the lowest photonic bandgaps versus N' for $E = 0$ boundary condition with $d_2 = d_1$, $\varepsilon_2 = \varepsilon_1$, and $N = 10$.

of the gap below the first frequency band, it increases or decreases with the variation of the ratio d_2/d_1. The behavior of the transmission intensity is different for the other boundary condition, i.e., the vanishing of the magnetic field H at the free ends of the side branches. In this case, we note the absence of the gap below the lowest frequency band and the decrease of the passbands with N' is less spectacular. The appearance of narrow passbands under the $H = 0$ boundary condition is better realized when $d_2 \neq d_1$ (see Figs. 11.7(c) and 11.7(d)).

In Fig. 11.10, we plotted the transmission coefficient versus the frequency for $d_1 = 1$ m, $d_2 = 0.03$ m, $\varepsilon_1 = \varepsilon_2 = 2.3$, $N' = 1$, and $N = 10$. Figs. 11.10(a) and 11.10(b) present the theoretical and experimental results for the $E = 0$ boundary condition. In this case, one observes a large gap (90 MHz) below the lowest frequency band and increasing width of the passbands when going to higher bands. Figs. 11.10(c) and 11.10(d) present the theoretical and experimental results for the $H = 0$ boundary condition. Now, at the scale of the figure, the transmission is almost unity except at frequencies of a few narrow gaps. Let us notice the attenuation of the transmission intensity and the widening of the passbands in experimental results. The origin of these phenomena is the attenuation in the coaxial cables which increase for high frequencies as will be reviewed in Section 11.8.

11.5 Defect modes in comb structures

We review now the existence of defect modes in the photonic band structure of a comb-like waveguide made of dangling side branches grafted periodically along an infinite monomode waveguide. The defect modes result from the presence of defective

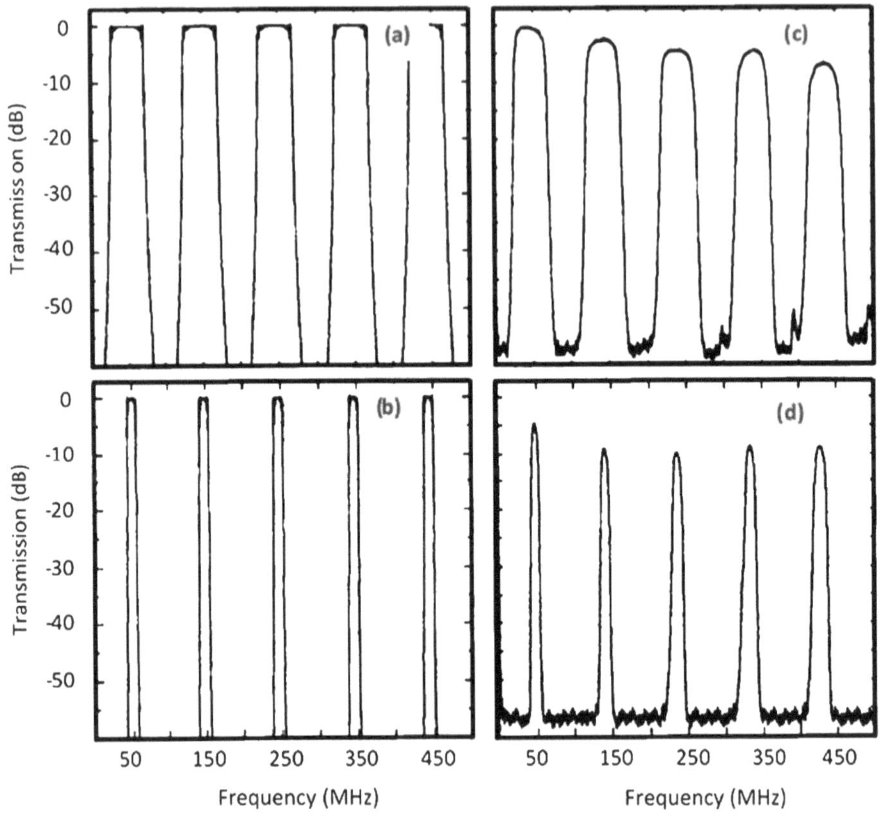

Figure 11.9 Transmission coefficient versus the frequency for $E = 0$ boundary condition with $N = 10$, $d_1 = d_2 = 1$ m, and $\varepsilon_1 = \varepsilon_2 = 2.3$. Figures (a) and (b) (resp. (c) and (d)) present theoretical (resp. experimental) results for $N' = 1$ and 8.

branches in the comb and may occur in the stopbands. The localized states appear as very narrow peaks in the transmission spectrum of finite comb-like waveguides composed of a finite number of grafted branches. The behavior of the localized states is analyzed as a function of the length, position and number of the defective branches. These theoretical results were confirmed by experiments using coaxial cables in the frequency range of a few hundreds of MHz.

Some studies have addressed the problem of the emergence of localized states in the photonic bandgaps by introducing defects in the periodic dielectric structure. In one-dimensional photonic crystals, i.e., classical Bragg mirrors [95,96], defects may be layers of different nature than the other slabs. In two- and three-dimensional photonic crystals, made of a periodic distribution of dielectric cylindrical or spherical inclusions embedded in a dielectric matrix, defects modes were observed by removing or adding some inclusions [97–99] or by changing the characteristics (material or diameter) of several inclusions [100]. These properties were also investigated in quasi-one-dimensional photonic crystals [11]. Photonic crystals possessing localized

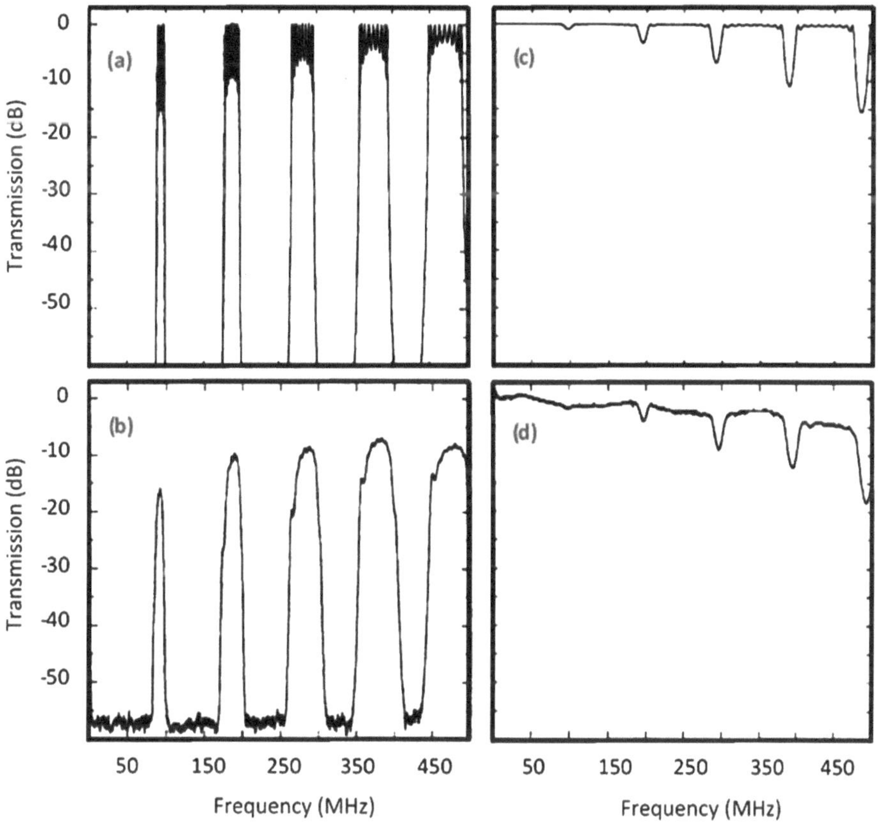

Figure 11.10 Transmission coefficient versus the frequency for $\varepsilon_2/\varepsilon_1 = 1$, $d_2/d_1 = 0.03$, $N' = 1$, and $N = 10$. Figures (a) and (b) (resp. (c) and (d)) present theoretical and experimental results for $E = 0$ (resp. $H = 0$) boundary condition.

modes in the forbidden bands should have several applications to optical devices such as selective frequency filters or very efficient waveguides.

In this Section 11.5, we focus on the existence of localized modes inside gaps when defective branches of different length and constituted of a different material are inserted in the star waveguide described in the above section [101–103].

11.5.1 Model

The localized states in the band structure are obtained, for example, by removing, on one site of the photonic crystal depicted in Fig. 11.5, N_2' resonators and replacing them by N_3' branches of length d_3 ($\neq d_2$), constituted of a material of relative dielectric permittivity ε_3 ($\neq \varepsilon_2$). Using the interface response theory [79,81], one can calculate analytically the frequencies of the localized states in the case of an infinite comb, i.e., with an infinite number of nodes ($N \to \infty$). They are given by the relation

$$1+\chi\frac{t}{t^2-1}=0. \tag{11.57}$$

In the latter equation, t is given by Eq. (11.31) together with (11.47) or (11.48). The parameter χ characterizes the perturbation introduced in the comb by the defective resonators and is defined as

$$\chi=\frac{\sin(\alpha_1 d_1)}{\alpha_1}\left(N_2'\alpha_2\cot(\alpha_2 d_2)-N_3'\alpha_3\cot(\alpha_3 d_3)\right), \tag{11.58}$$

when the electric field $E=0$ vanishes at the free ends of all the resonators, and as

$$\chi=-\frac{\sin(\alpha_1 d_1)}{\alpha_1}\left(N_2'\alpha_2\tan(\alpha_2 d_2)-N_3'\alpha_3\tan(\alpha_3 d_3)\right), \tag{11.59}$$

when the boundary condition $H=0$ has been taken into account.

The transmission coefficient through a defective star waveguide with a finite number N of nodes can be obtained as

$$T=4\sin^2(\alpha_1 d_1)\left|\frac{(t^2-1)^2 t^N}{(A^2-B^2t^{2N})(t^2-1)+\chi\psi}\right|^2, \tag{11.60}$$

where

$$\psi=t\left(A^2+B^2t^{2N}\right)-AB\left[t^{2(N-n+1)}+t^{2n}\right], \tag{11.61}$$

$$A=1-te^{j\alpha_1 d_1}, \tag{11.62}$$

and

$$B=t-e^{j\alpha_1 d_1}. \tag{11.63}$$

In Eq. (11.60), the integer n ($1\le n\le N$) stands for the location of the defects in the finite comb. One can easily check that for $\chi=0$ (i.e., $N_2'=N_3'$, $d_2=d_3$, and $\varepsilon_2=\varepsilon_3$), which corresponds to the comb without defect, Eq. (11.60) leads to Eq. (11.46). One can also readily verify that the same transmission coefficient is obtained for two symmetrical locations of the defect with respect to the middle of the comb.

11.5.2 Applications and discussion

We shall now numerically illustrate the above analytic results for comb-like waveguides composed of identical constituents, i.e., $\varepsilon_1=\varepsilon_2$, with the same characteristics lengths $d_1=d_2$ and the boundary condition $E=0$.

In Fig. 11.11, we present the band structures and transmission coefficients of the waveguide with and without defects. The top and bottom panels, respectively, describe the situations where the number of grafted side branches at every node is $N' = 1$ and $N' = 4$. Fig. 11.11(a) shows the six lowest dispersion curves of the photonic band structure for $N \to \infty$ and $N' = 1$. The plots are given in terms of the reduced frequency $\Omega = \omega d_1 \sqrt{\varepsilon_1}/c$ versus the reduced wave vector kd_1. In Fig. 11.11(b), the variations of the transmission coefficient T versus the reduced frequency for a finite number of nodes, namely $N = 5$, are plotted. Despite the finite number of resonators, one can notice that T approaches zero in regions corresponding to the observed gaps in the electromagnetic band structure of Fig. 11.11(a). The flat bands in Fig. 11.11(a) do not contribute to the transmission since the corresponding eigen-functions vanish at every node of the comb structure. These flat bands satisfy Eqs. (11.55) and (11.56). It was shown [101,102] that increasing the number N of nodes to 10 or more does not significantly modify the edges of the forbidden bands. In Fig. 11.11(c), the resonator located at the middle node, namely $N = 3$, has been replaced by a defective branch of length $d_3 = 0.7d_1$ and of relative dielectric constant $\varepsilon_3 = \varepsilon_1$. In the frequency range displayed in Fig. 11.11(c), there are five peaks associated with the localized states; those falling in the middle of a gap being much narrower than those situated in the vicinity of a band edge. The transmission inside a passband is also significantly affected by the presence of the defect as can be observed in the second passband of Fig. 11.11(c), where T is totally depressed. Fig. 11.11(d), where the defect is located on the second or on the fourth node, highlights the influence of the location of the defect site on the transmission coefficient and in particular on the intensities of the peaks associated with the localized modes.

The bottom panel of Fig. 11.11 contains similar results to those of the top panel when several ($N' = 4$) side branches are grafted at every node. This panel illustrates the narrowing (widening) of the passbands (forbidden bands) when increasing N'. Fig. 11.11(g) (Fig. 11.11(h)) shows the variation of the transmission coefficient when the four resonators located on the middle (second or fourth) node have been replaced by one defective branch ($d_3 = 0.7d_1$ and $\varepsilon_3 = \varepsilon_1$). In this case, some of the peaks associated with the defect modes are more clearly detached from the band edge than in Fig. 11.11(c), especially those around the second passband. One also observes that these peaks are much narrower in Fig. 11.11(g) than in Fig. 11.11(c).

The influence of the number of grafted defects on the localized modes was also studied. The top panel of Fig. 11.12 shows the variations of the transmission coefficient through a nondefective comb whose parameters are $N = 5$ and $N' = 1$. The three other panels present the transmission when the resonator grafted on the middle node has been replaced by $N_3' = 1$, 2, and 4 defective branches of length $d_3 = 0.7d_1$. The transmission factor in the passbands is more and more depressed as N_3' increases. The localized states are getting closer to the middle of the gaps when increasing N_3'. These states become more and more confined, i.e., the quality factor of the peaks defined as the ratio between the central frequency and the full width at half maximum increases.

Fig. 11.12(e) focuses on the peak associated with the localized state appearing above $\Omega \approx 9$ in Figs. 11.12(b), 11.12(c), and 11.12(d). This figure shows the shifting

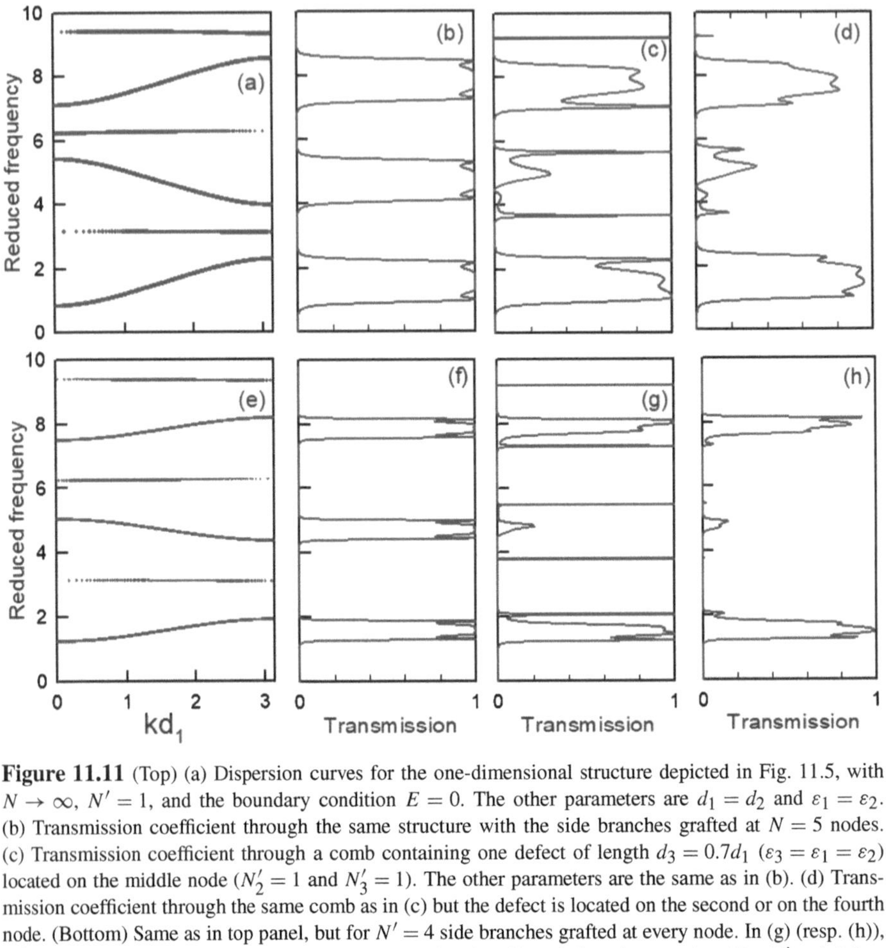

Figure 11.11 (Top) (a) Dispersion curves for the one-dimensional structure depicted in Fig. 11.5, with $N \to \infty$, $N' = 1$, and the boundary condition $E = 0$. The other parameters are $d_1 = d_2$ and $\varepsilon_1 = \varepsilon_2$. (b) Transmission coefficient through the same structure with the side branches grafted at $N = 5$ nodes. (c) Transmission coefficient through a comb containing one defect of length $d_3 = 0.7d_1$ ($\varepsilon_3 = \varepsilon_1 = \varepsilon_2$) located on the middle node ($N_2' = 1$ and $N_3' = 1$). The other parameters are the same as in (b). (d) Transmission coefficient through the same comb as in (c) but the defect is located on the second or on the fourth node. (Bottom) Same as in top panel, but for $N' = 4$ side branches grafted at every node. In (g) (resp. (h)), $N_2' = 4$ resonators grafted at the middle (resp. second or fourth) node have been replaced by $N_3' = 1$ defective branch ($d_3 = 0.7d_1$, $\varepsilon_3 = \varepsilon_1$).

in frequency of this peak as well as the increase of its quality factor with increasing N_3'. From Figs. 11.11 and 11.12, one can conclude that an increase of N' or N_3' in the comb gives rise to narrower peaks.

Fig. 11.13(a) presents the reduced frequencies of the localized modes as a function of the ratio d_3/d_1 for a fixed value of ε_3 when one resonator in an infinite comb has been replaced by one defective branch. The hatched areas correspond to the passbands of the perfect comb with $N' = 1$. One observes that the localized modes emerge from the passbands, decrease in frequency by increasing d_3, and finally merge into a lower passband. At each frequency, there is a periodical repetition of the localized state as a function of d_3. This figure clearly shows that the frequencies of the localized states inside the forbidden bands strongly depend on the physical characteristics of the

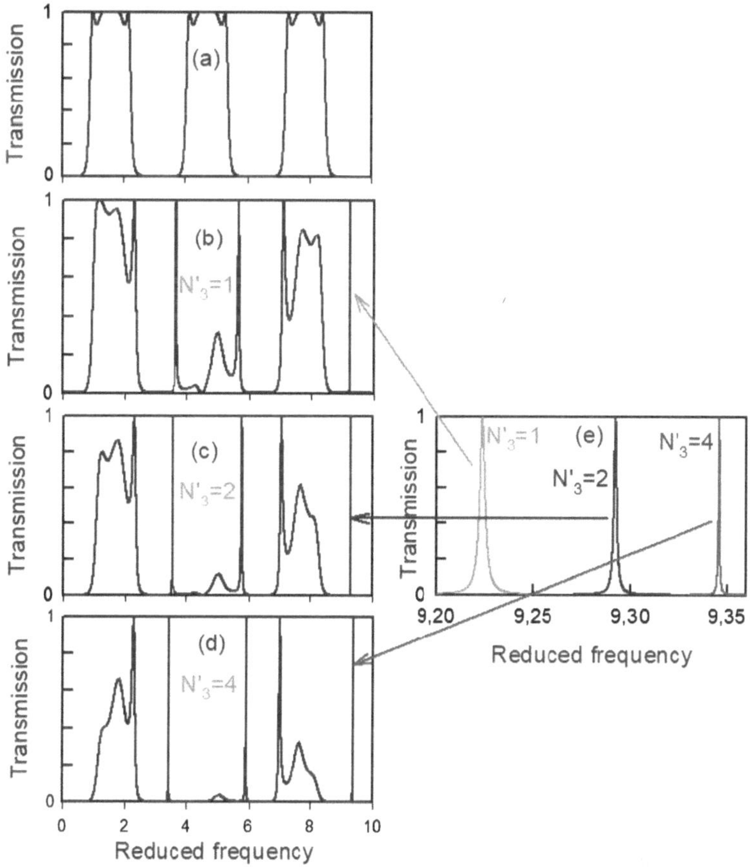

Figure 11.12 Transmission coefficients versus the reduced frequency through four combs without defect (a) and with defects ((b), (c), and (d)). The results are illustrated for $N = 5$, $N' = 1$, $\varepsilon_1 = \varepsilon_2$, $d_1 = d_2$ with the boundary condition $E = 0$. In (b), (c), and (d), the resonator grafted at the middle node has been replaced by $N_3' = 1, 2$, and 4 defective branches ($d_3 = 0.7d_1$ and $\varepsilon_1 = \varepsilon_3$). (e) The shape of the peak associated with the localized state appearing in Figs. 11.12(b), 11.12(c), and 11.12(d) around $\Omega \approx 9$ is illustrated at a larger scale to show the increase in the quality factor with increasing N_3'.

defective side branch. For example, one observes in Fig. 11.13(a) that the localized states occur in the middle of the second forbidden band for $d_3/d_1 \approx 0.5, 1, 1.5, \ldots$ Fig. 11.13(b) presents similar results for an infinite comb with $N' = 4$ and the same other parameters. The defect modes were obtained by replacing four resonators in one node by a defective branch of length d_3 or made of a material of relative dielectric constant ε_3.

The results reviewed here were obtained with the boundary condition $E = 0$ at the free ends of the resonators. Qualitatively similar results are also obtained by using the boundary condition $H = 0$ at the free extremities of the side branches, although the gaps are narrower in this case for $d_1 = d_2$ and $\varepsilon_1 = \varepsilon_2$ [54,92].

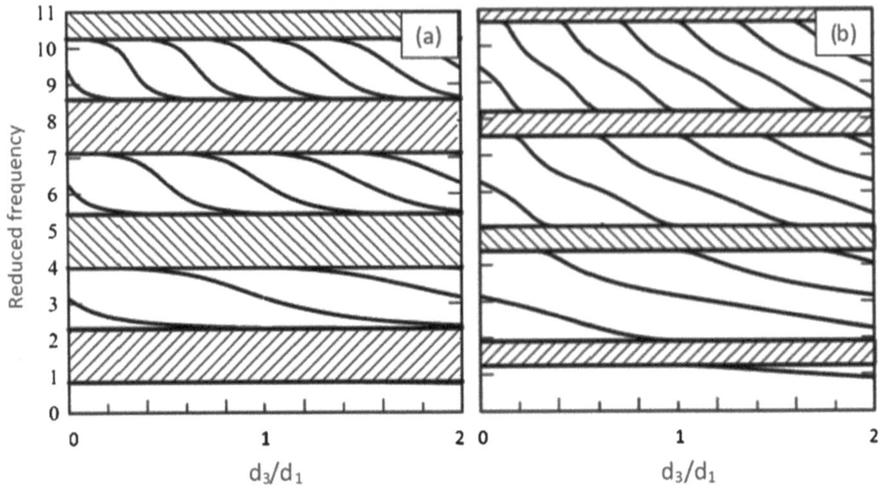

Figure 11.13 Reduced frequencies of the localized modes associated with $N_3' = 1$ defective side branch inserted in an infinite comb ($N \to \infty$). The characteristics of the combs are $N' = 1$, $\varepsilon_1 = \varepsilon_2$, $d_1 = d_2$, and the boundary condition is $E = 0$. (a) The length d_3 of the defective branch varies and its relative dielectric permittivity ε_3 is fixed and equal to ε_1. (b) Same as in (a) but for $N' = 4$ side branches grafted at every node; $N_2' = 4$ resonators grafted at one node of the infinite comb have been replaced by $N_3' = 1$ defective branch (d_3, ε_3).

To check the validity of the above theoretical predictions in a frequency range up to 500 MHz, experiments were performed, where both the one-dimensional waveguide and the side branches are constituted by standard 50 Ω coaxial cables. The transmission measurements have been realized by using a tracking generator coupled to a spectrum analyzer. The experimental measurements were compared with the theoretical predictions for combs without and with defects. As predicted by the theoretical results, an increase of N' or N_3' in the comb gives rise to a narrowing of the peaks associated with the localized states. For comb-like waveguides with N' and $N_3' \geq 4$, the resolution of the experimental setup does not allow detecting these very sharp peaks. Then the experimental investigations were limited to the propagation of electromagnetic waves in defective combs with N' and $N_3' \leq 2$.

The upper panel of Fig. 11.14 shows the theoretical and experimental variations of the transmission coefficient through a perfect comb whose parameters are $N = 5$, $N' = 1$, $\varepsilon_1 = \varepsilon_2 = 2.3$, and $d_1 = d_2 = 1$ m; the boundary condition being $E = 0$ at the free ends of the grafted resonators. The width and the frequency domains of the pass- and stopbands are quite similar in both spectra. The attenuation of the amplitude in the experimental signal with growing frequency, observed in Fig. 11.14(b), is well-known in this kind of coaxial cables. The variations of T when the coaxial cable of length $d_2 = 1$ m grafted on the middle node is replaced by another cable of length $d_3 = 1.175$ m are presented in the middle panel. The localized modes appear in the experimental spectrum as peaks whose frequencies agree quite well with the theoretical prediction. The widening of these peaks with increasing frequency is also observed in both spectra. Because of the resolution of the experimental setup, the amplitude of

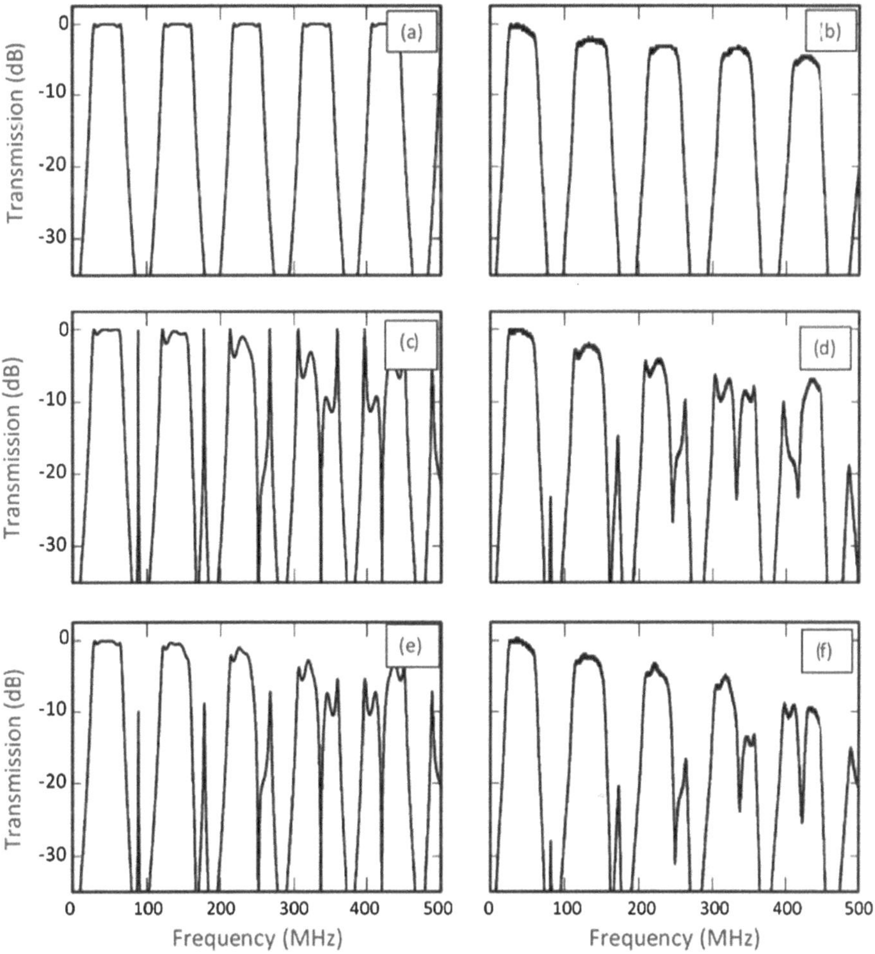

Figure 11.14 Comparison between the theoretical ((a), (c), and (e)) and experimental ((b), (d), and (f)) transmission coefficients through (a) and (b): a perfect comb of characteristics $N = 5$, $N' = 1$, $\varepsilon_1 = \varepsilon_2 = 2.3$, $d_2 = d_1 = 1$ m with the boundary condition $E = 0$ ($N_2' = N_3' = 1$). (c)–(d) The same comb containing one defect of length $d_3 = 1.175$ m ($\varepsilon_3 = \varepsilon_1 = 2.3$) located on the middle node. (e)–(f) The same comb but with the defect located on the fourth node.

the peaks appears weaker in the experimental spectrum than in the theoretical one. The presence of the defective branch also affects the transmission in the third, fourth, and fifth passbands. Finally, the bottom panel shows the behavior of T when the same defect branch is located on the fourth node. This behavior is similar to the previous one except for the strongest attenuation of the transmission in the fourth and fifth passbands, as well as the weaker amplitude of the peaks associated with the localized states.

Other experiments have been performed on the above-described comb structure with one *defect* coaxial cable grafted on the middle node. Transmission has been mea-

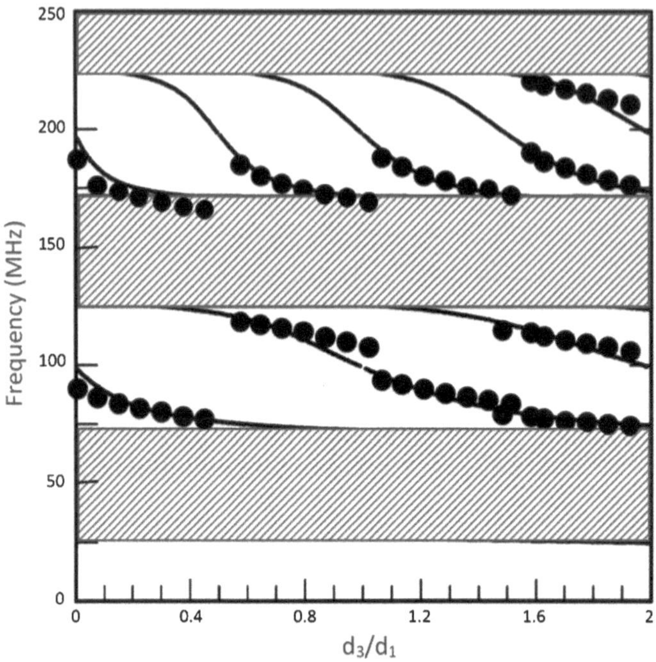

Figure 11.15 Comparison between the theoretical and the experimental frequencies of the localized states associated with the presence of a defect of length d_3 in a comb structure ($N_2' = N_3' = 1$): *(thin line)* calculated frequencies in an infinite comb ($N \to \infty$), (*filled dots*) measured frequencies in a finite comb with $N = 5$. The other parameters are $N' = 1$, $d_1 = d_2 = 1$ m, $\varepsilon_1 = \varepsilon_2 = \varepsilon_3 = 2.3$, and the boundary condition is $E = 0$. The shaded areas correspond to the passbands of the infinite comb structure.

sured for various values of d_3. The frequencies of the localized states associated with the defective resonator were measured using the spectrum analyzer. For each value of d_3, the experimental frequencies measured on a finite comb were compared with the theoretical frequencies calculated for a defect in infinite comb structure. The results are displayed in Fig. 11.15. The theoretical and experimental values are in reasonable agreement. The shift in frequency observed for a few points may be attributed to the finite number of resonators taken into account in the experimental setup. Similar results were obtained with star waveguides possessing two grafted resonators at each node. The forbidden bands are slightly wider in this case.

The presence of defective branches in the comb gives rise to localized states inside the forbidden bands. These defect modes appear as very narrow peaks of strong amplitude in the transmission spectrum. The localized states are very sensitive to the length of the defects, to their location in the comb, as well as to their number. Experimental measurements using coaxial cables have confirmed the theoretical predictions for frequencies up to 500 MHz. Recent works show clearly that the manufacturing of such comb structures at a micrometric scale becomes realizable with the new technological developments using high-resolution electron beam lithography [11,104].

11.6 Surface modes in comb structures

In this section, we review the existence and behavior of surface localized modes in a one-dimensional comb-like photonic bandgap structure [105]. The structure is composed as above of a waveguide, along which dangling side branches are grafted periodically. A theoretical rule was demonstrated about the existence of surface states in such one-dimensional periodic structures, namely, when one considers together two semiinfinite structures obtained by cutting an infinite one along a plane perpendicular to the axis of the structure, one always has as many localized surface states as mini-gaps. These surface modes were observed by measuring the transmission through a guide along which a finite comb is attached vertically. The surface modes of the comb were obtained as maxima or minima of the transmission spectrum. The experiments were performed by using coaxial cables in the frequency range of a few hundreds MHz.

11.6.1 Analytical calculations

The comb-like structure was composed of an infinite one-dimensional waveguide or backbone (medium 1), along which N' dangling side branches of length d_2 (medium 2) are grafted periodically with spacing period d_1 at N sites, N and N' being integers (see Fig. 11.16(a)). The semiinfinite comb structure with a different segment between the sites $n = 0$ and $n = 1$ at the surface (Fig. 11.16(b)) is obtained by removing first the segment laying between the sites $n = -1$ and $n = 0$ and the N' side branches on the site $n = 0$ of the infinite comb (Fig. 11.16(a)). Then, one replaces the segment lying between the sites $n = 0$ and $n = 1$ by a different segment characterized by the dielectric constant ε_s and the length d_s. The Green's function of the final system has its inverse $d^{-1}(M_s, M_s)$ in the semiinfinite interface space M_s ($n \geq 0$) represented by a semiinfinite tridiagonal matrix. By standard diagonalization of such a semiinfinite tridiagonal matrix, one obtains for $n, n' \geq 1$,

$$d(n,n') = \frac{\sin(\alpha_1 d_1)}{\alpha_1} \frac{t^{|n-n'|+1}}{t^2-1} - \frac{\sin(\alpha_1 d_1)}{\alpha_1} \frac{t^{n+n'-1}}{t^2-1} \frac{Y}{\Delta}, \tag{11.64}$$

where

$$Y = \cos(\alpha_1 d_1) + \frac{\alpha_s}{\alpha_1} \frac{\sin(\alpha_1 d_1)\sin(\alpha_s d_s)}{\cos(\alpha_s d_s)} - t, \tag{11.65}$$

$$\Delta = \cos(\alpha_1 d_1) + \frac{\alpha_s}{\alpha_1} \frac{\sin(\alpha_1 d_1)\sin(\alpha_s d_s)}{\cos(\alpha_s d_s)} - \frac{1}{t}, \tag{11.66}$$

and t is given by Eqs. (11.31) and (11.47) for the $E = 0$ boundary condition at the ends of the resonators.

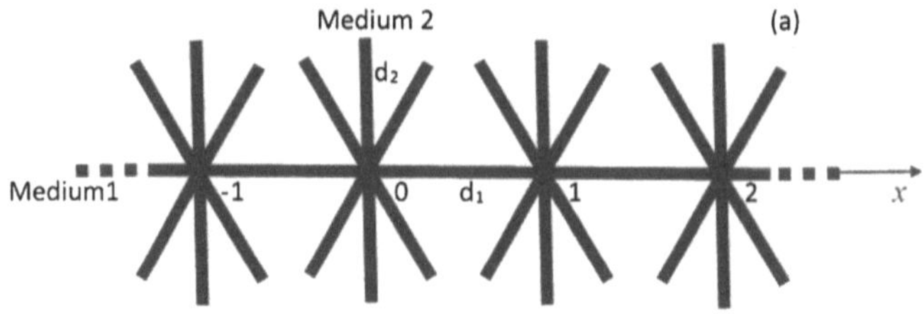

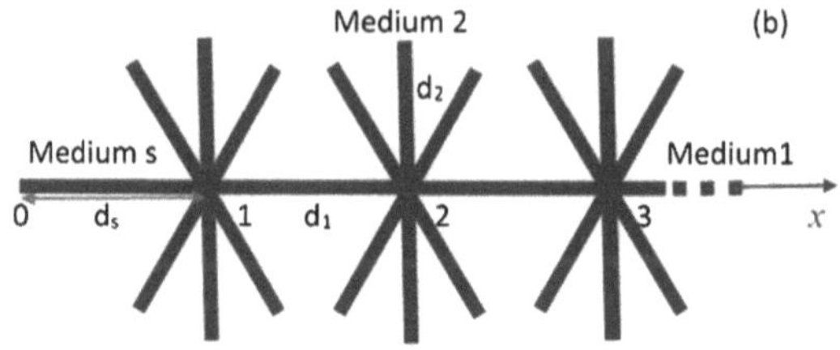

Figure 11.16 (a) Schematic of the one-dimensional waveguide with a periodic array of N' grafted segments (dangling side branches) of length d_2, distant from each other by a length d_1. (b) Semiinfinite comb structure terminated with a segment of length d_s ($d_s \leq d_1$).

The interface elements of the surface segment can be worked out in the following form:

$$d(0,0) = \frac{1}{\Delta}\left\{\frac{\sin(\alpha_1 d_1)}{\alpha_1} - \frac{\sin(\alpha_s d_s)}{\alpha_s}\left(\cos(\alpha_1 d_1) - \frac{1}{t}\right)\right\}, \tag{11.67}$$

$$d(0,1) = d(1,0) = \frac{\sin(\alpha_1 d_1)}{\alpha_1 \Delta \cos(\alpha_s d_s)}, \tag{11.68}$$

and

$$d(1,1) = \frac{\sin(\alpha_1 d_1)}{\alpha_1 \Delta}. \tag{11.69}$$

From the knowledge of the interface matrix elements, one can obtain the elements of the Green's function between any two points of the whole system (Eq. (11.8)). When the denominator of the Green's function d vanishes for the frequency laying inside the gaps of the infinite comb, one obtains localized states within the surface segment which decay exponentially inside the bulk of the comb. The explicit expression giving the frequency ω of the localized states is

$$\begin{aligned} N'\alpha_2 \sin(\alpha_2 d_2)\cos(\alpha_1 d_1) + \alpha_1 \sin(\alpha_1 d_1)\cos(\alpha_2 d_2) \\ + N'\frac{\alpha_2\alpha_3 \sin(\alpha_1 d_1)}{\alpha_1}\frac{\sin(\alpha_2 d_2)\sin(\alpha_s d_s)}{\cos(\alpha_s d_s)} \\ + \frac{\alpha_s^2 \sin^2(\alpha_s d_s)\sin(\alpha_1 d_1)\cos(\alpha_2 d_2)}{\alpha_1 \cos^2(\alpha_s d_s)} = 0, \end{aligned} \tag{11.70}$$

together with the condition

$$\left|\cos(\alpha_1 d_1) + \frac{\alpha_s}{\alpha_1}\frac{\sin(\alpha_1 d_1)\sin(\alpha_s d_s)}{\cos(\alpha_s d_s)}\right| > 1. \tag{11.71}$$

Condition (11.71) ensures that the wave is decaying when penetrating into the comb structure far from the surface.

From the above general equations (Eqs. (11.70) and (11.71)), one can deduce the case when the surface segment is of the same nature as the interior segments ($\alpha_s = \alpha_1$) but with variable length $0 \le d_s \le d_1$. In the particular cases when $d_s = 0$ and $d_s = d_1$, the expression giving the surface states becomes

$$N'\alpha_2 \cos(\alpha_1 d_1)\sin(\alpha_2 d_2) + \alpha_1 \cos(\alpha_2 d_2)\sin(\alpha_1 d_1) = 0. \tag{11.72}$$

However, the condition (Eq. (11.71)) ensuring the vanishing of these surface states is never satisfied in the case $d_s = 0$ (i.e., $\sin(\alpha_s d_s) = 0$) as $|\cos(\alpha_1 d_1)|$ cannot exceed unity. On the contrary, this condition is usually fulfilled in the case $d_s = d_1$ as $|\cos(\alpha_1 d_1)| < 1$.

A particularly interesting result can be obtained from the calculation of the variation of the total density of states $\Delta n(\omega^2)$ between two semiinfinite complementary combs with surface segments of lengths d_s and $d_1 - d_s$, obtained by the cleavage of the infinite one within one segment. Indeed, it is possible to show by using standard transformation of the trace of the response functions that $\Delta n(\omega^2)$ can be obtained from the knowledge of the elements $d_1(n, x_3 = d_s; n, x_3 = d_s)$ and $d_2(n, x_3 = d_1 - d_s; n, x_3 = d_1 - d_s)$ of the surface response function of the two complementary semiinfinite comb structures, namely,

$$\Delta n(\omega^2) = \frac{1}{\pi}\arg\left\{\det\left[d_1(n, x_3 = d_s; n, x_3 = d_s) + d_2(n, x_3 = d_1 - d_s; n, x_3 = d_1 - d_s)\right]\right\}. \tag{11.73}$$

Using the expression given above for these elements of the response function (Eq. (11.73)), one finds that $\Delta n(\omega^2)$ is zero inside the bulk bands of the comb. Besides, we show also that at all edges of these bulk bands, $\Delta n(\omega^2)$ displays δ functions of weight $-1/2$. This condition, together with the necessary conservation of the total number of states, enables us to conclude that when one considers together the two semiinfinite combs obtained by the cleavage of an infinite one, one always has as many localized surface states as there are minigaps.

It is worth to notice that the above results are given for the case when $H = 0$ at the extremity of the side branches and at the end of the surface segment. The other boundary condition $E = 0$ at the extremity of the side branches and/or at the end of the surface segment could be also considered by permuting $\cos(\alpha_2 d_2)$ and $-\sin(\alpha_2 d_2)$ and/or $\cos(\alpha_s d_s)$ and $-\sin(\alpha_s d_s)$ in the above Eqs. (11.64)–(11.66) and (11.70)–(11.71).

In relation with the experiments using coaxial cables, all the materials involved in the structure were assumed to be identical (which means $\varepsilon_1 = \varepsilon_2 = \varepsilon_s = 2.3$), the number of dangling side branches at every site was $N' = 1$, and the boundary condition $H = 0$ was applied to the end of the resonators. The length d_1 was chosen to be 1 m, whereas d_2 and d_s were taken to be variable lengths. Finally, a small imaginary part was added to the dielectric constant of the materials in order to take into account the energy dissipation during the propagation along the coaxial cables.

11.6.2 Bulk and surface modes

In Fig. 11.17, are presented, for $d_2 = 0.5$ m, the dispersion curves (Fig. 11.17(a)) (frequency ω versus the wave vector k) for an infinite comb, as well as the transmission coefficient (Fig. 11.17(b)) through a finite comb containing four sites ($N = 4$). The theoretical transmission coefficient can be directly compared with the experimental measurements using a tracking generator coupled to a spectrum analyzer in the frequency range of 10 to 500 MHz, all the constituents in the waveguide and in the resonators being standard 50 Ω coaxial cables assembled with metallic T-shaped

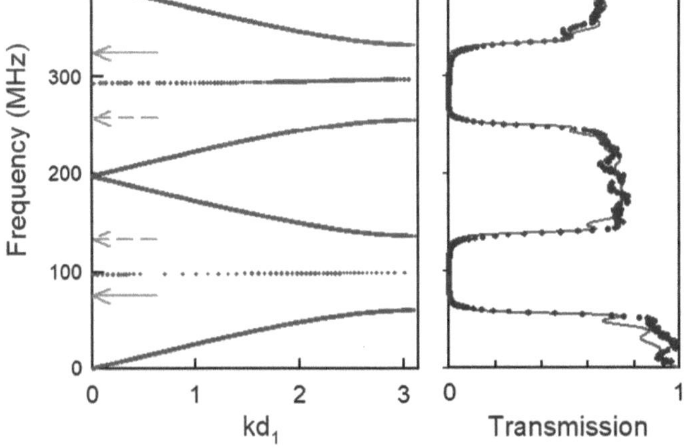

Figure 11.17 (a) Dispersion curves (frequency versus dimensionless wave vector) for the one-dimensional structure depicted in Fig. 11.16(a), with $d_1 = 1$ m, $d_2 = 0.5$ m, $N' = 1$, and the boundary condition $H = 0$. The horizontal solid lines indicate the frequencies of the flat bands. The arrows give the frequencies of the surface modes inside the minigaps of the comb when the length of the surface segment is $d_s = 0.7$ m (full arrows) or $d_s = 0.3$ m (dashed arrows). (b) Transmission coefficient T versus frequency through the same structure with grafted resonators at $N = 4$ sites. The solid and dotted curves are theoretical and experimental results, respectively.

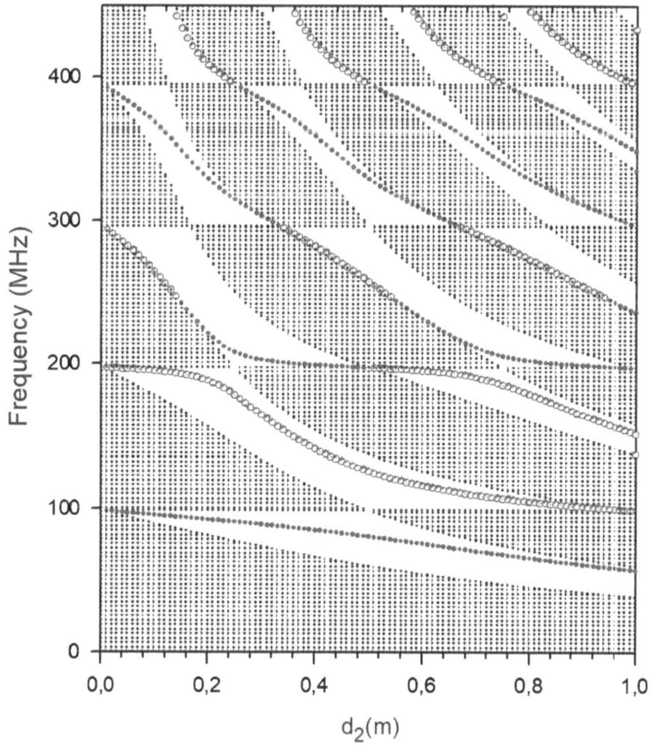

Figure 11.18 Projected band structure and surface modes of the comb-like structure as a function of the length d_2 of the resonators. The shaded areas represent the bulk bands. The filled and empty circles, respectively, indicate the surface modes for two complementary semiinfinite combs terminated by a surface segment of length $d_s = 0.7$ m and $d_s = 0.3$ m.

connectors. The band structure and the transmission coefficients exhibit passbands separated by stopbands. The latter originate both from the periodicity of the structure, related to the length d_1, and from the resonators, related to the length d_2. In particular, the resonators give rise to zeros of transmission through the waveguide at frequencies such that $\omega d_2 \sqrt{\varepsilon}/c = m\pi/2$, m being an integer, when the boundary condition $H = 0$ is applied at the end of the resonators. The transmission coefficient exhibits small oscillations in the passbands and vanishes inside the gaps. In the example shown in Fig. 11.17 for $d_2 = 0.5$ m, it is worth to mention that there are also flat bands (indicated by horizontal lines) in the dispersion curves which occur at the same frequencies as the above zeros of transmission (this coincidence results from the choice of $d_2 = d_1/2$). These flat bands satisfy Eqs. (11.53) and (11.54). The small features appearing in the transmission spectrum inside the gaps are manifestation of these flat bands. They occur because the actual length of the resonators is slightly different from 0.5 m and the corresponding bands have a small width instead of being totally flat (see also Fig. 11.18).

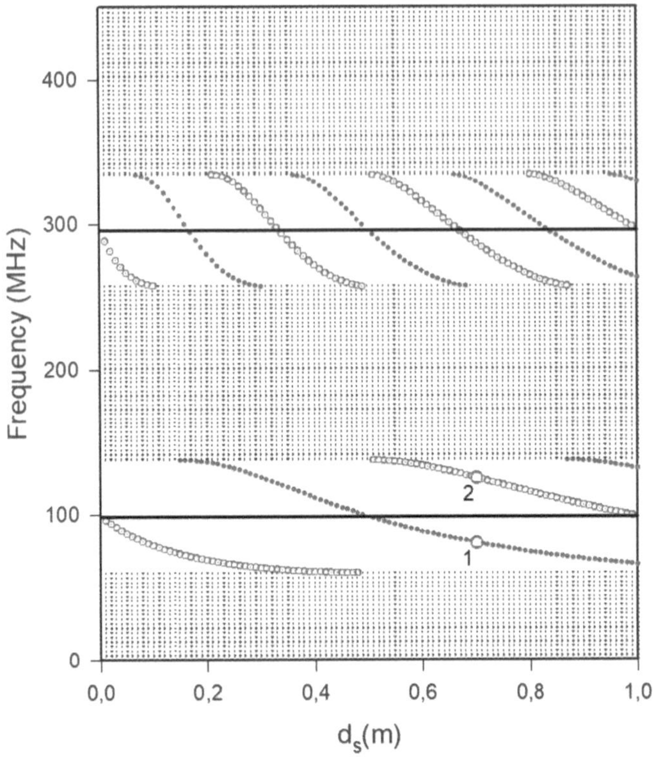

Figure 11.19 Frequencies of the surface modes of the semi-infinite comb versus the length d_s of the surface segment, for $d_2 = 0.5$ m. The filled and empty circles, respectively, give the dispersion curves for the boundary condition $H = 0$ and $E = 0$ at the surface. The shaded areas show the first three bulk bands of the comb. The horizontal solid lines in the middle of the gaps indicate the frequencies of the flat bands. The local densities of states associated with the modes 1 and 2 are presented in Fig. 11.20.

In Fig. 11.17(a) the frequencies of the surface modes inside the gaps are also presented (by arrows), when the length of the segment at the surface is $d_s = 0.7$ m (full arrows) or $d_s = 0.3$ m (dashed arrows), i.e., for two complementary systems obtained from cutting an infinite comb into two parts. One can observe the existence of one surface state in each gap, this mode being associated with the surface of one of the two semiinfinite combs, in accordance with the analytical calculations of Section 11.6.1.

A more detailed investigation of the surface modes dispersion curves is summarized in Figs. 11.18 and 11.19. Fig. 11.18 displays the so-called projected band structure of the bulk bands and surface modes as a function of the length d_2 when the magnetic field H vanishes at the free ends of the resonators and at the free surface of the comb. The shaded areas represent bulk bands, where waves are allowed to propagate in the comb; these areas are separated by minigaps. Inside these gaps, we represented surface modes corresponding to two complementary semiinfinite comb structures obtained by cleaving the infinite comb within one segment, such that the length of the remaining surface segment is, respectively, $d_s = 0.3$ m (empty circles)

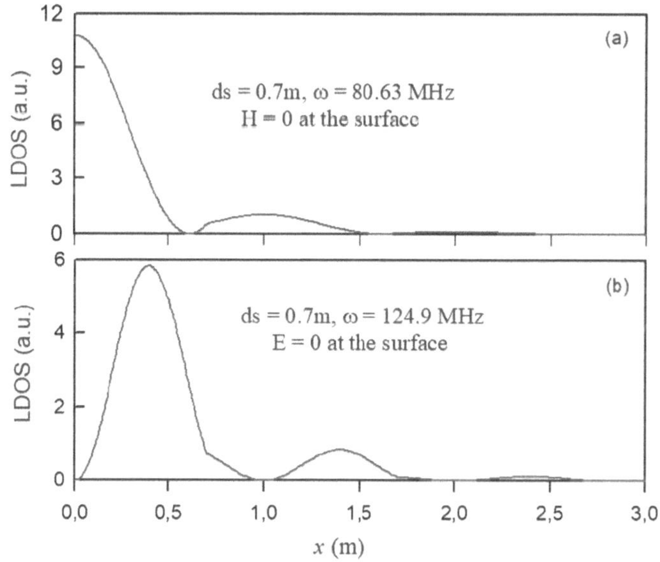

Figure 11.20 Spatial representation of the local density of states (LDOS) (in arbitrary units) for the modes labeled 1 and 2 in Fig. 11.19.

and $d_s = 0.7$ m (filled circles) in each semiinfinite part. As mentioned above, one obtains as many surface states as there are gaps. Moreover, there is one surface state in each gap associated with either of the complementary semiinfinite comb structures. Let us notice that the flat bands indicated in Fig. 11.17(a) occur at frequencies where some of the bulk bands close, i.e., their widths reduce to zero.

In Fig. 11.19, we presented, for $d_2 = 0.5$ m, the dispersion of the surface modes as a function of the length d_s of the surface segment. The results are presented for both $H = 0$ (filled circles) and $E = 0$ (empty circles) boundary conditions at the free surfaces. When d_s increases, the frequencies of the existing surface states decrease until the corresponding branches merge into the bulk bands and become resonant states. At the same time, new localized branches are extracted from the bulk bands. One observes, in particular, that for all combinations of two complementary structures such that $d_s^{(1)} + d_s^{(2)} = d_1$, one always has a surface state in each gap. The frequencies of the flat bands are indicated here by horizontal lines inside the mini-gaps. To illustrate the spatial localization of the surface modes, are sketched in Fig. 11.20 the local density of states as a function of the space position x for the surface modes labeled 1 and 2 in Fig. 11.19. The local density of states reflects the square modulus of the electric field and clearly shows that the surface modes are indeed localized in the surface segment of the comb structure and decay exponentially over a few segments away from the surface. Finally, let us mention that the bulk bands and the surface modes will exhibit different, although qualitatively similar, behavior if the boundary condition at the end of the resonators is $E = 0$ instead of being $H = 0$.

11.6.3 Experimental observation of the surface modes

In order to obtain an experimental evidence of the above theoretical predictions, it is necessary to imagine an experiment capable of detecting the surface modes. A measure of the transmission coefficient T in the geometry depicted in Fig. 11.21(a) in which a finite comb is attached vertically to an infinite guide, was proposed [105]. Indeed, the maxima and minima of T will contain information about the surface modes of the isolated comb, i.e., the comb before being attached to the guide. This result can be explained in the following way. Let ω_i be the eigenfrequencies of the isolated finite comb when the boundary condition $H = 0$ is applied to its bottom end (for a given choice of the boundary condition at the upper end). Similarly, let ω_i' be the eigenfrequencies of the isolated comb when the boundary condition $E = 0$ is applied to its bottom end. One can demonstrate that in the geometry of Fig. 11.21(a), the transmission coefficient T through the guide will reach unity at the frequencies ω_i, whereas it falls to zero at the frequencies ω_i'. Now the surface modes of the comb are necessarily among the eigenfrequencies and therefore will belong to the set ω_i or ω_i', depending on whether the boundary condition at the bottom end of the isolated comb is $H = 0$ or $E = 0$. Therefore, one can expect the following behavior of the transmission coefficient T as a function of the frequency: T will oscillate between 0 and 1, and its maxima and minima will respectively correspond to the eigenfrequencies of the isolated comb with an appropriate boundary condition at the bottom end. This means that T will display several oscillations in any miniband of the comb structure (depending on the number of sites contained in the comb), whereas inside the gaps a maximum or a minimum of T will be related to a surface localized mode of the isolated comb (with the appropriate boundary condition). It is worth noticing that such states may exceptionally not appear in the transmission coefficient if they correspond to a bound in continuum states.

The above consideration can be observed in Fig. 11.21(b) which displays both the experimental and theoretical transmission coefficient, for various lengths d_s of the bottom (surface) segment in the comb, and with $d_2 = 0.5$ m. Let us notice that, in this example, the upper end of the comb does not support any surface state. One can notice several oscillations of the transmission coefficient inside the minibands of the comb structure (see Fig. 11.17 for comparison). On the other hand, inside the minigap shown in Fig. 11.21(b), the transmission coefficient contains 0, 1, or 2 maxima and 0 or 1 minima, depending on the value of d_s. The maxima (indicated by arrows) and the minima are related to the surface states of the isolated comb when the boundary condition $H = 0$ or $E = 0$ is applied to its bottom end, respectively. These results are totally in agreement with the dispersion curves of the surface modes given in Fig. 11.19. From the maxima and minima of the transmission coefficient, the eigenfrequencies of the isolated finite comb are obtained and presented in Figs. 11.21(c) and 11.21(d), with the boundary condition $H = 0$ and $E = 0$, respectively, as a function of the length d_s. One can notice that the bulk modes are weakly dependent upon d_s due to the finite length of the whole structure, while the surface modes display a strong variation with this length. The dispersion curves of the surface modes are quite comparable with those given in Fig. 11.19.

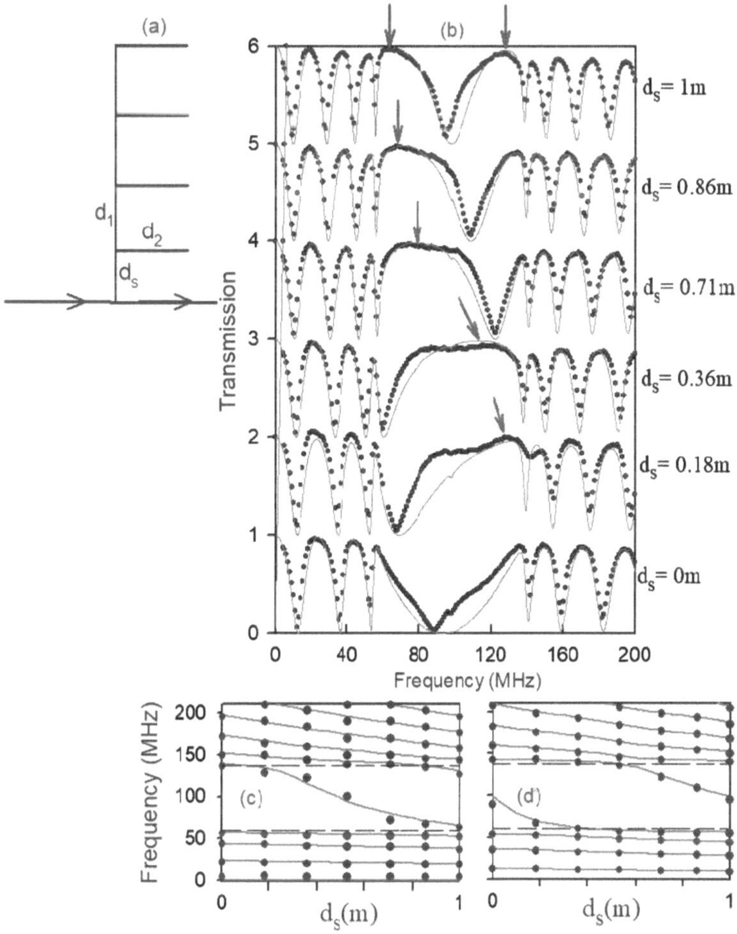

Figure 11.21 (a) Schematic of the vertical comb with a surface segment of length d_s in contact with the infinite guide. (b) Transmission spectra for various lengths of the surface segment such that $0 \leq d_s \leq$ 1 m. Solid (dotted) curves are theoretical (experimental) results. The arrows indicate the frequencies of the surface modes (for the boundary condition $H = 0$ at the surface) obtained from the dispersion curves in Fig. 11.19. (c) Eigenfrequencies of the isolated comb, with the boundary condition $H = 0$ at the surface, versus the length of the surface segment. The dots are taken from the maxima of the transmission spectrum. (d) Same as (c) with the boundary condition $E = 0$ at the surface. The dots are taken from the minima of the transmission coefficient.

11.7 Optical add–drop multiplexers

The directional transfer of photons [106–108] between two wires was intensively investigated in the last years. Applications of such transfer processes are in particular important for wavelength multiplexing in communication devices. In order to allow a directional transfer of one photon with a given energy between two monomode wires, one has to build an appropriate coupling structure, to permit all incoming particles of different energies to continue traveling without perturbation along the input wire and transferring successfully one into the output wire in a given direction. The criteria for such transfers were given on the basis of symmetry arguments and illustrated by applications to structures designed in photonic materials by rod removal and perturbation [107]. Other structures [108] made out of resonating finite wires were proposed. All the above structures require very precise definition of their parameters. A simpler structure is presented here [109] which gives closed-form expressions for all the parameters that are necessary for the device fabrication. Another advantage of this structure is that it requires less tuning accuracy as the former. We present it in what follows.

11.7.1 Theory and examples

Let us consider the system presented in Fig. 11.22. The two continua are the two infinite lines passing by, respectively, points $(1, 2)$ and $(3, 4)$. The distance d_0 between points 1 and 2 is the same as between points 3 and 4. Four identical monomode structures are branched between points $(1, 5)$, $(5, 4)$, $(2, 6)$, and $(6, 3)$. These structures have one finite wire of length d_2 branched in the middle of the lines of length $2d_1$ situated between the above-given points. The electric field E associated with the particles traveling in this structure is supposed to vanish at the free end of the dangling wires. Such structures enable transmission gaps to be opened in the desired frequency range in the same manner as in similar stub structures (see Sections 11.3 and 11.4). Between points 5 and 6 is fixed a waveguide of length $3d_1$ with two finite wires of length d_2 located at distances d_1 and $2d_1$ from point 5. They act as a resonant cavity with a localized mode in the above-mentioned gap.

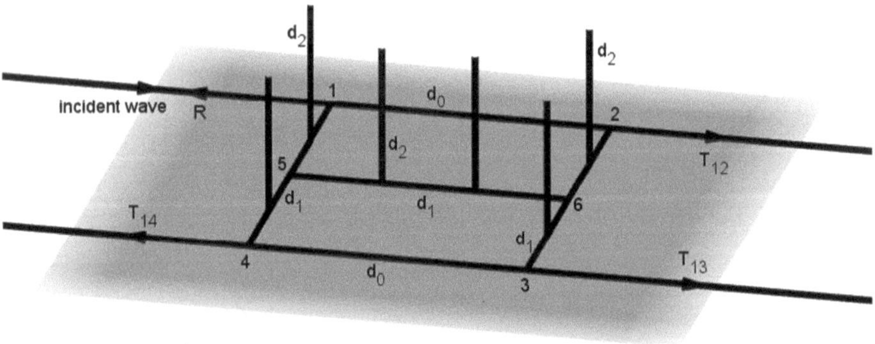

Figure 11.22 Schematic illustration of the one dimensional photonic add–drop multiplexer design.

The reflection R and transmission coefficients T_{1n}, $n = 2, 3, 4$, as a function of the wavelength λ associated to the particles traveling in the monomode wires were found to be related to the elements of the Green's function g of this system by the following expressions

$$R = |1 - 2j\alpha g(1, 1)|^2 \tag{11.74}$$

and

$$T_{1n} = |2j\alpha g(1,n)|^2, \quad n = 2, 3, 4. \tag{11.75}$$

These Green's function elements can be obtained in the following way. Onc calculates first the $[g(0,0)]^{-1}$ associated to one end of a finite line of length d, using Eq. (11.20). By linear superposition of the corresponding elements associated to those of the finite lines of length d_1 and d_2, one deduces the 3×3 matrix $[g(M, M)]^{-1}$ associated to the finite structure with free ends situated in particular between points 1 and 5. The corresponding 2×2 matrix $[g(M, M)]^{-1}$ associated only to the two ends of this element was finally found to be

$$[g_i(M, M)]^{-1} = \begin{pmatrix} A_5 & B_5 \\ B_5 & A_5 \end{pmatrix}, \tag{11.76}$$

where

$$A_5 = A_1 - \frac{B_1^2}{2A_1 + A_2}, \tag{11.77}$$

$$B_5 = -\frac{B_1^2}{2A_1 + A_2}, \tag{11.78}$$

with

$$A_i = -\left[\tan\left(\frac{2\pi d_i}{\lambda}\right)\right]^{-1}, \quad i = 1 \text{ and } 2, \tag{11.79}$$

and

$$B_1 = \left[\sin\left(\frac{2\pi d_1}{\lambda}\right)\right]^{-1}. \tag{11.80}$$

A similar procedure enables to obtain the 2×2 matrix $[g(0,0)]^{-1}$ associated with the two free ends of the element with two grafted resonators (situated between points 5 and 6)

$$[g_i(M, M)]^{-1} = \begin{pmatrix} A_6 & B_6 \\ B_6 & A_6 \end{pmatrix}, \tag{11.81}$$

where

$$A_6 = A_1 - \frac{1}{2}\left(\frac{B_1^2}{2A_1 + A_2 + B_1} + \frac{B_1^2}{2A_1 + A_2 - B_1}\right) \tag{11.82}$$

and

$$B_6 = \frac{1}{2}\left(\frac{B_1^2}{2A_1 + A_2 - B_1} - \frac{B_1^2}{2A_1 + A_2 + B_1}\right). \tag{11.83}$$

With these results and that given by Eq. (11.19) in hand, one constructs then a 6×6 matrix $[g(M, M)]^{-1}$ associated with the multiplexer of Fig. 11.22. Using its symmetries, one inverts easily this matrix and obtains the needed Green's function elements. The reflection and transmission coefficients as a function of the wavelength λ associated to the particles traveling in the monomode wires are then found to be given by the following expressions:

$$R = |z_1 + z_2 + z_3 + z_4 - 1|^2, \tag{11.84}$$

$$T_{12} = |z_1 + z_2 - z_3 - z_4|^2, \tag{11.85}$$

$$T_{13} = |z_1 - z_2 + z_3 - z_4|^2, \tag{11.86}$$

$$T_{14} = |z_1 - z_2 - z_3 + z_4|^2, \tag{11.87}$$

where

$$z_n = \frac{j}{2}(y_n + j)^{-1}, \quad n = 1, 2, 3, 4, \tag{11.88}$$

$$y_1 = y_2 - \frac{2B_1^4}{(2A_1 + A_2)^2\left(3A_1 - \frac{2B_1^2}{2A_1+A_2} - \frac{B_1^2}{2A_1+A_2+B_1}\right)}, \tag{11.89}$$

$$y_2 = \tan\left(\frac{\pi d_0}{\lambda}\right) + A_1 - \frac{B_1^2}{2A_1 + A_2}, \tag{11.90}$$

$$y_3 = -\left[\tan\left(\frac{\pi d_0}{\lambda}\right)\right]^{-1} + A_1 - \frac{B_1^2}{2A_1 + A_2}, \tag{11.91}$$

and

$$y_4 = y_3 - \frac{2B_1^4}{(2A_1 + A_2)^2\left(3A_1 - \frac{2B_1^2}{2A_1+A_2} - \frac{B_1^2}{2A_1+A_2-B_1}\right)}. \tag{11.92}$$

It was found that, in order to obtain a complete transfer of a particle with a propagating vector $k_0 = 2\pi/\lambda_0$ from site 1 to site 3 (namely, $T_{13} = 1$ and $R = T_{12} = T_{14} = 0$), the structure given by Fig. 11.22 must be such that for $\lambda = \lambda_0$,

$$y_1 = y_3 = -\frac{1}{y_2} = -\frac{1}{y_4}. \tag{11.93}$$

With the above conditions, one provides the following expressions for the characteristic lengths:

$$d_0 = (1 + 4n_0)\frac{\lambda_0}{4}, \tag{11.94}$$

$$d_1 = (1 + 4n_1 \mp \delta)\frac{\lambda_0}{4}, \tag{11.95}$$

and

$$d_2 = (2n_2 \pm \delta)\frac{\lambda_0}{4}, \tag{11.96}$$

where

$$\delta = \left[\frac{3(1 + 4n_1 + 2n_2)}{\pi Q}\right]^{1/2} \tag{11.97}$$

is supposed to be small compared to 1, $n_i = 0, 1, 2, 3, \ldots$, and Q is the quality factor associated with the line-width of the transferred signal defined by

$$Q = \frac{\lambda_0}{2(\lambda' - \lambda_0)}, \tag{11.98}$$

where λ' is the wavelength at which $T_{13}(\lambda') = 0.5$.

In the same manner, one may define a quality factor Q' for the envelope of the direct transmission T_{12} (not to be confused with the dip in T_{12} due to the transferred particle)

$$Q' = \frac{\lambda_0}{2(\lambda'' - \lambda_0)}, \tag{11.99}$$

where λ'' is the wavelength at which $T_{12}(\lambda'') = 0.5$, such that $|\lambda'' - \lambda_0| \gg |\lambda' - \lambda_0|$. With a good approximation,

$$Q' = \frac{\pi}{2}(1 + 2n_0 + 2n_1 + n_2). \tag{11.100}$$

For multiplexing applications, one needs small values of Q', in order not to perturb the particles with wavelengths close to λ_0. So, as described above, this system enables completely and in closed-form determining all the wire lengths for a complete channel transfer, once the wavelength λ_0 and the order of magnitude of the quality factor Q are chosen. Relations (11.94)–(11.96) give us the values of d_0, d_1, and d_2. Note that several solutions are possible due to the different values of the integers n_0, n_1, and n_2. Note also that the $-\delta$ value in Eq. (11.95) is associated with the $+\delta$ in Eq. (11.96), and vice versa. Let us stress that in order to have a multiplexer with a large unperturbed region for the directly transmitted particle, one has to choose the lengths d_0, d_1, and d_2 as small as possible. The smallest possible solution for a given λ_0 depends also on the width h of the monomode wires. It is possible to estimate that h has to be smaller than 0.5 μm for a vacuum wavelength $\lambda_0' = 1.5$ μm, and a relative dielectric constant ε such that $\sqrt{\varepsilon} = 2.8$, in order for the wires to be mono-mode. Current technologies enable to prepare by electron lithography wires such that 0.2 μm $< h <$ 0.5 μm. With a width of 0.2 μm, one may choose $n_0 = n_1 = n_2 = 1$ in Eqs. (11.94)–(11.96) and obtain a quality factor Q of the order of 1000 (Eq. (11.97)) with $d_0 = 0.67$ μm, $d_1 = 0.68$ μm, and $d_2 = 0.26$ μm. We give these values with an accuracy of the order of 0.01 μm, which is currently available.

Let us note that once these lengths are fixed, the wavelength and the quality factor of the transferred particle can be obtained from Eqs. (11.95) and (11.96) and (11.97) to be

$$\lambda_0 = \frac{4(d_1 + d_2)}{1 + 4n_1 + 2n_2} \tag{11.101}$$

and

$$Q = \frac{3}{\pi} \frac{(1 + 4n_1 + 2n_2)(d_1 + d_2)^2}{[2n_2 d_1 - (1 + 4n_1)d_2]^2}. \tag{11.102}$$

As can be seen on Fig. 11.23 obtained by an exact calculation of T_{12}, T_{13}, and T_{14} from Eqs. (11.79)–(11.80) and (11.84)–(11.92) as a function of the vacuum wavelength $\lambda' = \lambda\sqrt{\varepsilon}$, the transferred particle wavelength $\lambda_0\sqrt{\varepsilon}$ and the quality factor Q are well estimated by Eqs. (11.101) and (11.102), as well as the quality factor Q' (Eq. (11.100)), giving the unperturbed wavelength domain in T_{12}. Figs. 11.23(c) and 11.23(d) show mainly how an increase of the device size modifies the region of unperturbed transmission T_{12}. A comparison between Figs. 11.23(e) and 11.23(f) as well as Figs. 11.23(c) and 11.23(d) shows the effect of an increase in the width of the transferred frequency peak. If one desires to shift λ_0, it is possible to adapt a heating device under the structure. Indeed, when heating, all the distances will increase in the same proportion, and λ_0 will increase proportionally to $(d_1 + d_2)$ as shown in Eq. (11.101). Another way to tune this multiplexer is to change the relative dielectric constant ε, as Eqs. (11.94)–(11.96) giving the distances are proportional to $\lambda_0'/\sqrt{\varepsilon}$. This can be done by isolation techniques or by the application of an electric field.

11.7.2 Experimental procedure and realization

Using the same experimental setup as in the experiments described above, the transmitted signals at points 2, 3, and 4 of a similar structure made of coaxial cables were measured [108]. This structure was different from the one described above only by

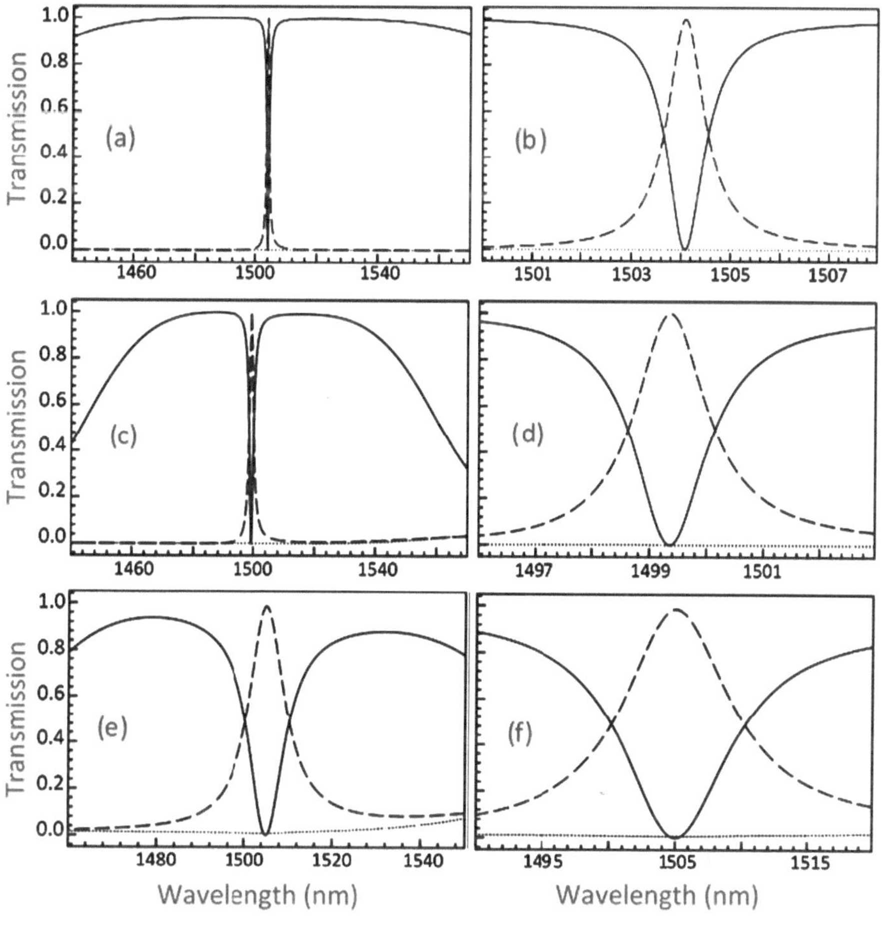

Figure 11.23 (a)–(b) Variation of the intensity of the transmitted signal T_{12} from site 1 to site 2 (solid line), and of the forward signal T_{13} (long dashed line), in the structure shown in Fig. 11.22 versus the light wavelength in vacuum. The dots represent the signal intensity in the backward direction T_{14}. These theoretical results were obtained for $d_0 = 0.67$ μm, $d_1 = 0.68$ μm, and $d_2 = 0.26$ μm, $\sqrt{\varepsilon} = 2.8$, and Q of the order of 1000. Only the scale on the horizontal axis varies from (a) to (b). (c)–(d) Same as (a)–(b), but for $d_0 = 1.21$ μm, $d_1 = 1.22$ μm, and $d_2 = 0.52$ μm. (e)–(f) Same as (c)–(d), but with Q of the order of 100, when $d_0 = 1.21$ μm, $d_1 = 1.25$ μm, and $d_2 = 0.49$ μm.

the line branched between points 5 and 6; only one resonator of length d_4 was put in the middle of a wire of length $2d_3$. Fig. 11.24(a) shows the theoretical variations of the transmission coefficients, while Fig. 11.24(b) presents the experimental measurements. Although the radiative losses in the coaxial cables are negligible, the attenuation of the electromagnetic waves in the cables is important. Attenuation in the coaxial cables was simulated in the computations by a complex relative dielectric permittivity, $\varepsilon = \varepsilon' - j\varepsilon''$. The imaginary part of ε was obtained from the attenuation specification data supplied by the manufacturer of the coaxial cables as a function

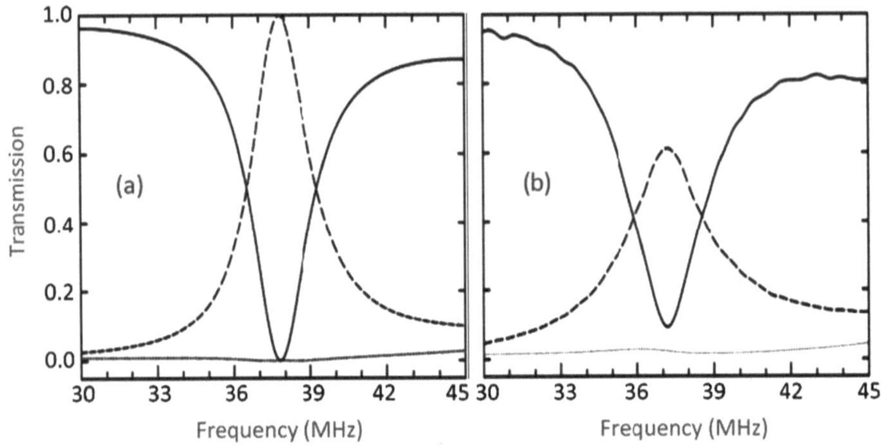

Figure 11.24 Variation of the intensity of the transmitted signal from site 1 to site 2 (solid line), and of the forward signal (long dashed line), in the structure described in the text versus the frequency. The dots represent the signal intensity in the backward direction. Comparison between theoretical (a) and experimental (b) results for $d_0 = 1.3$ m, $d_1 = 1$ m, $d_2 = 0.4$ m, $d_3 = 1.10$ m, and $d_4 = 0.15$ m with the boundary condition $E = 0$.

of frequency f, i.e., $\varepsilon'' = 0.017 f^{-0.5}$ with f in units of Hz. Due to this absorption phenomenon, the amplitude of a narrow peak of selective transmission would become very low. Therefore, the conditions of a relatively large resonant peak was chosen. One observes that taking into account the attenuation in the coaxial cables, measurements validate fairly well the theoretical prediction.

Designing a resonant tunneling device working at optical frequencies necessitates the manufacturing of propagation channels with lengths of the order of the micrometer. The high selectivity of the device requires a precision on the lengths of the order of the nanometer. Recent technological advances in the microfabrication technology allow manufacturing of such channels.

In the experimental realization of the structure studied in Section 11.7.1, the different lengths d_0, d_1, and d_2 can be made out of the same mask. So one expects that the relative imprecision between the two d_0, the eleven d_1 and the six d_2 of this structure can be neglected in front of the imprecision of the length d_0, d_1, and d_2 of the corresponding masks. So, once obtained the values of d_0, d_1, and d_2 from Eqs. (11.94)–(11.96), one varied them by a random amount and computed the transmission factors from the exact Eqs. (11.84)–(11.92). This procedure shows that the structure studied here operates with a precision of the order of 0.01 μm on the different lengths d_0, d_1, and d_2, for quality factors Q of a few thousands. Let us also remember that if the errors on d_1 and d_2 have the same sign, the shift in λ_0 may be of the order of 0.02 μm (see Eq. (11.101)). But as explained above λ_0 may be tuned back to its required value with the help of a heating device or by adjustment of the relative dielectric constant with an electric field.

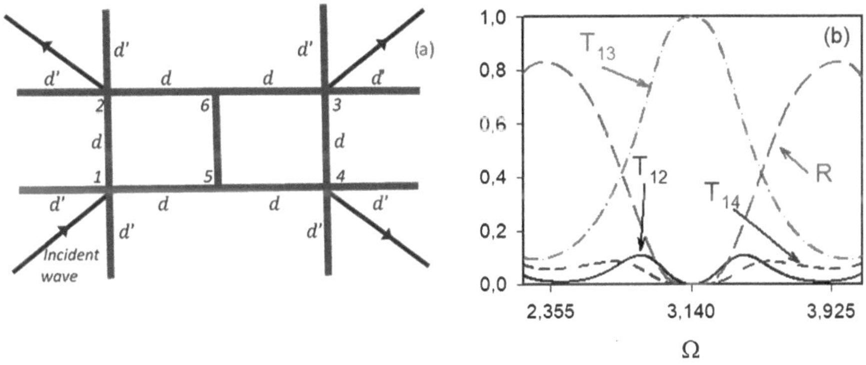

Figure 11.25 (a) Path multiplexer for $N = 2$. (b) Variation of the transmission and reflection coefficients in different input–output of the multiplexer, T_{12} (black curve), T_{13} (red dashed-dotted curve, light gray in print version), T_{14} (blue dashed curve, dark gray in print version) and reflection (green dashed curve, mid gray in print version), for $d = d_0/2$ and $d' = d_0$.

11.7.3 Four port path systems

11.7.3.1 Finite system

Consider two closed loops of length $4d$ sharing one common part of length d, with $N(= 1, 2, 3, \ldots)$ stubs of length d' grafted to ports 1, 2, 3, and 4 of this system, see Fig. 11.25.

11.7.3.2 Reference system

In order to build the system described by Fig. 11.25(a), it is possible to use as reference system $4N$ open loops of length d' whose states are given by $(\alpha S')^{4N} = 0$, $N = 1, 2, 3, \ldots, 7$ open loop of length d whose states are given $[\alpha S]^7$ and an interface space $M \equiv (1, 2, 3, 4, 5, 6)$.

Define λ as the wavelength, $\alpha = 2\pi/\lambda$, $S = \sin(\alpha d)$, $C = \cos(\alpha d)$, $S = \sin(\alpha d')$, and $C' = \cos(\alpha d')$.

The reference system states are given by

$$(\alpha S')^{4N}(\alpha S)^7 = 0, \tag{11.103}$$

and the determinant of the inverse of the corresponding $g_R(M, M)$ is

$$[g_R(M, M)]^{-1} = \left(\frac{\alpha S'}{C'}\right)^{4N} (-\alpha^2)^7. \tag{11.104}$$

11.7.3.3 Path states by inspection

As for the preceding examples given in Chapters 1–6 of this Photonics book, it is possible to find the path state eigenwavelengths to be given here by

$$S^2 (C')^{4(N-1)} = 0. \tag{11.105}$$

The eigenfunctions are made out of the usual sine functions

$$E(x) = \sin\left[2\pi p \frac{x}{d}\right], \quad p = 1, 2, 3, \ldots, \tag{11.106}$$

having robust zeros on the six space points of the interface space and

$$E(x) = \sin\left[2\pi p \frac{x'}{d'}\right], \quad p = 1, 2, 3, \ldots, \tag{11.107}$$

having robust zeros on the four space port points of the interface space.

11.7.3.4 All states from the interface response theory

In order to obtain the final system depicted by Fig. (11.25), one proceeds as usual within the interface response theory. The inverse of the interface space response function is here

$$[g(M,M)]^{-1} = \frac{\alpha}{SC'} \begin{pmatrix} A_0 & C' & 0 & 0 & C' & 0 \\ C' & A_0 & 0 & 0 & 0 & C' \\ 0 & 0 & A_0 & C' & 0 & C' \\ 0 & 0 & C' & A_0 & C' & 0 \\ C' & 0 & 0 & C' & -3CC' & C' \\ 0 & C' & C' & 0 & C' & -3CC' \end{pmatrix}, \tag{11.108}$$

where

$$A_0 = -2CC' + NSS'. \tag{11.109}$$

The determinant of the above matrix is

$$|[g(M,M)]^{-1}| = \frac{16\alpha^6}{S^6 C'^4}(A_0^2 - C'^2)[A_0(1-3C) - C'(3C+1)] \\ [-A_0(1+3C) + C'(3C-1)]. \tag{11.110}$$

With the above results and the conservation of the number of states between the reference and the final system, one obtains that the final system states are given by

$$\frac{S}{\alpha}C'^{4(N-1)}\left(A_0^2 - C'^2\right)\left[A_0(1-3C) - C'(3C+1)\right]\left[-A_0(1+3C) + C'(3C-1)\right] \\ = \frac{S^2}{\alpha}C'^{4(N-1)}\left(A_0^2 - C'^2\right)\left[C'C(\alpha/2)(1-6C) + NS'S(\alpha/2)(1+3C)\right] \\ \times\left[-C'S(\alpha/2)(1+6C) + NS'C(\alpha/2)(1-3C)\right] = 0. \tag{11.111}$$

In this result $C(\alpha/2) = \cos(\alpha d/2)$ and $S(\alpha/2) = \sin(\alpha d/2)$.

11.7.3.5 State number conservation

The $(S^2(C')^{4(N-1)} = 0)$ states were obtained above from the confined state theorem. The two universal $(1/\lambda = 0)$ states are obtained, one from $S^2/\alpha = 0$ and the other from the $\alpha = 0$ limit of the last factor of the above equation.

11.7.4 The system with the four semiinfinite leads added to interface points $M(1, 2, 3, 4)$

In the preceding subsection of this chapter a similar model is presented with stubs grafted to other sites.

In order to obtain the reflection and transmission with the present model, one has here the same general equations (80)–(83) as above, function of different values of the z_i, which are here

$$z_1 = \frac{SC'(1-3C)}{4\alpha\Delta_1}, \tag{11.112}$$

$$z_2 = \frac{SC'}{4\alpha(A+C')}, \tag{11.113}$$

$$z_3 = \frac{SC'}{4\alpha(A-C')}, \tag{11.114}$$

$$z_4 = \frac{-SC'(1+3C)}{4\alpha\Delta_2}, \tag{11.115}$$

where

$$A = A_0 + jSC', \tag{11.116}$$

$$\Delta_1 = 2(A+C')(1-3C) - 4C', \tag{11.117}$$

and

$$\Delta_2 = -2(A-C')(1+3C) - 4C'. \tag{11.118}$$

In order to illustrate the above analytical results, we present in Fig. 11.25(b) the variation of the transmission coefficients T_{12}, T_{13}, T_{14}, and the reflection coefficient R versus the dimensionless frequency $\Omega = \omega d_0\sqrt{\varepsilon}/c$ for $d = d_0/2$ and $d' = d_0$. Fig. 11.25(b) clearly shows that when the transmission in the output $(1, 3)$ (dashed red curve, light gray in print version) is unity (i.e., $T_{13} = 1$), the transmission in the two outputs $(1, 2)$ and $(1, 4)$ (black and dashed blue [dark gray in print version] curves), and reflection R (dashed green curve, mid gray in print version) vanish at the same frequency $\Omega = \pi$ (i.e., $T_{12} = T_{14} = R = 0$). From these results, one concludes that there is a total transfer of a mode from the output $(1, 2)$ to the output $(1, 3)$ without any disturbance in the output $(1, 4)$ and in the reflection R.

11.8 Effects of absorption

In this last Section 11.8, we review [110] the effect of absorption on the amplitude and the phase of transmission and reflection coefficients in the star comb structure with a defective branch inserted inside as described in Section 11.5. More specifically, we will focus on the amplitude and phase of transmission and reflection of the defect mode induced by the defect branch when loss is taken into account. The two latter quantities are introduced into the system by adding a negative imaginary part ε'' to the dielectric constant ε' in the defect branch. For the defect branch with loss ($\varepsilon'' < 0$), we will see that, as $|\varepsilon''|$ increases, the absorption goes through a maximum and then falls to zero. However, the transmission (reflection) decreases (increases) monotonically. As mentioned in [111], this property may be used as an on/off switching device, depending on the strength of the loss. Also, by increasing $|\varepsilon''|$, the phase time (i.e., the time spent by the photon in the defect branch) decreases, giving rise to a delocalization of the defect mode. An interesting result is obtained around the transmission zeros, induced by the comb structure, where negative phase times and, therefore, negative group velocities are obtained. On the contrary, when the defect branch is an amplifying medium with a gain ($\varepsilon'' > 0$), we show that, as ε'' increases (and within the validity of the model), transmission and reflection both increase together. The phase time increases as well, giving rise to a strong localization of the defect mode.

A frequency selective multiplexer was proposed [110] using three comb structures attached to a lead, in the shape of a tree with different defective branches to control the filtering of desirable frequencies within each comb. In the presence of absorption and amplification, this structure can be used as an on/off switching device to select in which comb the transmission would be allowed or prohibited.

This section is organized as follows. First, we recall the main analytical expressions for the transmission and reflection amplitudes of the periodic comb structure with a defect branch. The next two subsections are devoted to a numerical discussion of these expressions, when absorption and amplification are introduced into this system with an application to a multiplexer constructed from these structures.

11.8.1 Theoretical model

The details of the analytical calculations of the transmission (t_r) and reflection (r) coefficients through finite size star waveguides, depicted in Fig. 11.26(a) with or without a defect branch, were given above (Section 11.5). A defect branch may be inserted in the periodic structure of Fig. 11.26(a) by removing, on the site n ($1 \leq n \leq N$), N_2' resonators and replacing them by N_3' branches of length d_3 ($\neq d_2$), composed of a material with relative dielectric permittivity ε_3 ($\neq \varepsilon_2$). Within the framework of interface response theory [81], the expressions of t_r and r in the structure with such a defect and for the $E = 0$ boundary condition at the ends of the resonators were found to be

$$t_r = \frac{2j\sin(\alpha_1 d_1)(t^2 - 1)^2 t^N}{(A^2 - B^2 t^{2N})(t^2 - 1) + \chi\psi} \tag{11.119}$$

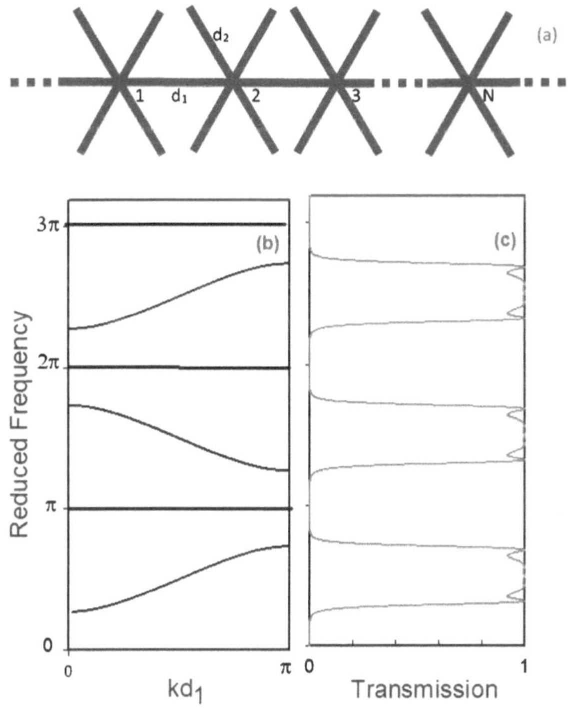

Figure 11.26 (a) Periodic waveguide with N stars of N' ($N' = 1$ here) grafted branches. (b) Dispersion curves (reduced frequency $\Omega = \omega d_1\sqrt{\varepsilon'}/c$ versus reduced wavevector kd_1) for the structure depicted in (a) with $N \to \infty$, $N' = 1$ and boundary condition $E = 0$ at the ends of the resonators. Other parameters are $d_1 = d_2$ and $\varepsilon_1 = \varepsilon_2 = \varepsilon'$. (c) Transmission coefficient through the same structure with side branches grafted at $N = 5$ nodes.

and

$$r = -1 + \left\{ \frac{2j\sin(\alpha_1 d_1)}{(A^2 - B^2 t^{2N})(t^2 - 1) + \chi\psi} \right\} \times \left\{ (At - Bt^{2N})(t^2 - 1) - \chi[A(t^{2n} - t^2) + Bt^{2N+1}(t^{2(1-n)} - 1)] \right\}, \tag{11.120}$$

where ψ, A, and B, are given by Eqs. (11.61)–(11.63), χ by Eq. (11.58), and

$$t = e^{jkd_1}, \tag{11.121}$$

with k being is the wave number of the infinite structure given by Eq. (11.47).

These equations remain valid in the case of the $H = 0$ boundary condition by using Eq. (11.48) rather than Eq. (11.47) and Eq. (11.59) rather than Eq. (11.58). Also, the expressions of t_r and r for the periodic structure without a defect can be obtained by assuming $d_3 = d_2$ and $N_2' = N_3'$ in Eq. (11.58) (i.e., $\chi = 0$ in Eqs. (11.119) and (11.120)).

To model the microscopic processes of absorption and amplification, continuum electromagnetic theory adds an imaginary part ε'' to the dielectric permittivity ε'. The sign of ε'' determines whether the system is locally absorbing ($\varepsilon'' < 0$) or amplifying ($\varepsilon'' > 0$). Investigations of different situations, where ε'' has been added to the dielectric constant in different parts of the structure, showed that, for reasonable values of ε'', the essential effect appears on the defect modes induced by the defect branch. Therefore, in what follows an imaginary part ε'' is introduced only in the defect branch ($i = 3$). In addition, all the wires constituting the structure were made of the same material (i.e., $\varepsilon_1 = \varepsilon_2 = \mathrm{Re}(\varepsilon_3) = \varepsilon'$).

11.8.2 Numerical results and discussion

We illustrate numerically the above analytical results for a comb-like structure with the same characteristic lengths $d_1 = d_2$, and when the boundary condition $E = 0$ is applied at the end of the resonators. For the sake of simplicity, the number of resonators on each node was fixed at one (i.e., $N' = N'_2 = N'_3 = 1$). Before addressing the effects of absorption and amplification on the transmission and filtering in these structures, let us recall briefly the bandgap structure of the perfect periodic system without defect and in the absence of loss and gain.

In Fig. 11.26(b), we presented the dispersion curves (the reduced frequency $\Omega = \omega d_1\sqrt{\varepsilon'}/c$ versus the reduced wave vector kd_1) for an infinite comb, as well as the transmission rate T ($T = |t_r|^2$) (Fig. 11.26(c)) through a finite comb containing $N = 5$ sites. The band structure, as well as the transmission rate, exhibits passbands separated by stopbands. The latter originate both from the periodicity of the structure, related to the length d_1, and from the resonant behavior of the side branches, related to the length d_2. In particular, the resonators give rise to zeros of transmission through the waveguide at the frequencies, satisfying $\sin(\omega d_2\sqrt{\varepsilon'}/c) = 0$ (i.e., $\omega d_2\sqrt{\varepsilon'}/c = 0, \pi, 2\pi, \ldots$). Despite the finite number of resonators, we see that T exhibits small oscillations in the passbands and vanishes inside the gaps predicted in the band structure of Fig. 11.26(b). It is also worth mentioning that, due to the choice of the boundary condition $E = 0$ at the end of resonators, the band structure in Figs. 11.26(b) and 11.26(c) displays a forbidden band that starts at zero frequency. Finally, the horizontal lines in Fig. 11.26(b) correspond to flat bands that occur at the frequencies of zero transmission ($\omega d_2\sqrt{\varepsilon'}/c = 0, \pi, 2\pi, \ldots$). Indeed, the eigenfunctions associated with these frequencies vanish at every node of the comb structure (i.e., they correspond to the stationary modes of every segment) and, therefore, do not contribute to the transmission. The occurrence of flat bands is related to particular choices of the ratio d_1/d_2, this ratio being chosen equal to one in the example of Fig. 11.26. If d_2 is chosen to be slightly different from d_1, each flat band in Fig. 11.26 transforms into a very narrow band in the vicinity of a zero of transmission [105].

It is well known [98,101,112,113] that the introduction of defects into photonic crystals gives rise to defect modes in the gaps of the structure. Fig. 11.27(a) shows the variation of the transmission rate T (full curves) and the reflection rate R ($R = |r|^2$) (dashed curves) versus the reduced frequency for the structure depicted in Fig. 11.26(a), where the resonator located at the middle node (namely, $n = 3$) has been

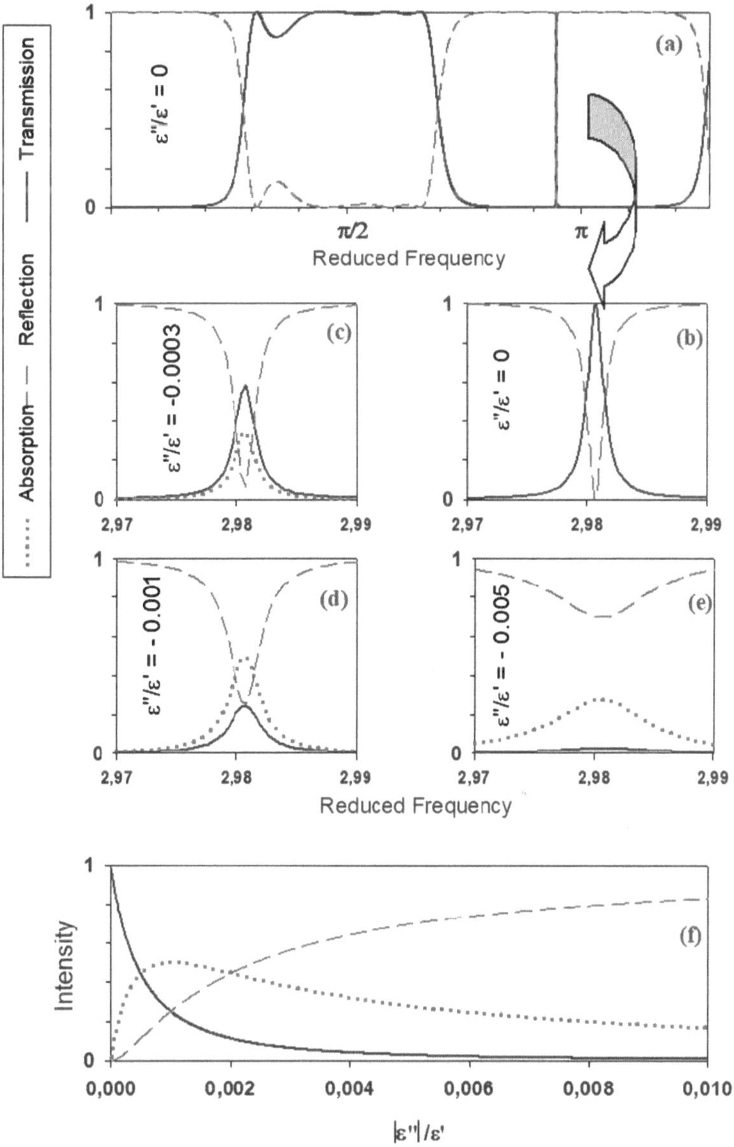

Figure 11.27 (a)–(e) Transmission (T) (solid curve), reflection (R) (dashed curve), and absorption (α) (dotted curve) for a finite comb containing one defect of length $d_3 = 1.1d_2$ located on middle node $n = 3$. Permittivities in waveguide and resonators are kept real as in Fig. 11.26(c), $\varepsilon_1 = \varepsilon_2 = \varepsilon'$, while permittivity in defect branch is supposed to be complex $\varepsilon_3 = \varepsilon' - j\varepsilon''$, where $\varepsilon'' < 0$ represents loss. (f) Variation of T, R, and α as function of $|\varepsilon''|/\varepsilon'$ at frequency of the defect mode.

replaced by a defective branch of length $d_3 = 1.1d_2$. The narrow peak in the middle of the first gap (at $\Omega \simeq 3$) corresponds to a defect mode introduced by the defect branch.

The frequency of the defect mode inside the gap depends mainly on the length of the resonator defect, whereas the width at half maximum in the transmission spectrum is also dependent upon the number of nodes in the structure [101]. Since the effect of absorption has not been taken into account in Fig. 11.27(a), we have $R + T = 1$ in the whole range of frequency. Around the frequency of the defect mode (see the enlarged Fig. 11.27(b)), the transmission T reaches the value of one, whereas R falls down to zero. This result is related to the symmetry of this structure, in which the defect mode is located at the middle of the comb. Indeed, it was shown [101] that the maximum of T remains below 1 when the defect is located in an asymmetric position.

11.8.2.1 Effect of absorption

In the presence of dissipation in the defect branch ($\varepsilon'' < 0$), the transmission and reflection spectra are dramatically affected around the frequency of the localized mode, even for small values of the absorption coefficient (see Figs. 11.27(c)–(e)). At the same time, the effect of absorption is much less significant in the frequency range of the passbands. In Figs. 11.27(c) to 11.27(e), we present, for increasing values of the absorption coefficient (ε''), the transmission (solid curve) and reflection (dashed curve) spectra, as well as the dissipation rate (dotted curve) ($\alpha = 1 - R - T$) around the frequency of the defect mode at $\Omega = 2.9805$. We notice that T decreases monotonically, while R increases. However, the absorption α reaches a maximum of 0.5 for a threshold value of $|\varepsilon''_{th}|/\varepsilon' \simeq 10^{-3}$, and then decreases for increasing values of $|\varepsilon''|$. These results are summarized in Fig. 11.27(f), where the values of T, R, and α, taken at the maximum of transmission, are presented versus $|\varepsilon''|$.

Defect modes in two- and three-dimensional photonic crystals [113] have shown similar dependencies for the absorption and transmission. However, the reflection, instead of increasing monotonically, exhibits a minimum at the same value of ε'', where the absorption has a maximum. This effect has been proposed for use in the fabrication of antireflection coatings [113]. In the case of the one dimensional photonic crystal, reviewed here, there is a competition between absorption and reflection: for $|\varepsilon''| < |\varepsilon''_{th}|$, both the absorption and reflection increase together with a maximum of absorption at $\varepsilon'' = \varepsilon''_{th}$ (Fig. 11.27(f)). Then, for larger values of $|\varepsilon''|$ ($|\varepsilon''| > |\varepsilon''_{th}|$), reflection becomes the dominant effect and absorption decreases to zero. This effect could be used in the fabrication of an on/off switch in an add/drop filter by tuning the optical loss in a resonator [111]. We now discuss the above effects as a function of the number N of nodes in the comb structure. First, we assume that there is no dissipation. Fig. 11.28(a) displays the transmission spectra T, for different values of N, around the frequency of the defect mode. The peak frequency and the width of the spectrum are significantly dependent upon N, especially for small values of N. By increasing N, the peak frequency goes to the frequency of the defect mode in an infinite structure, namely $\Omega = 2.978$. In addition, the localization of the mode increases, or the width at half maximum Δ of the peak decreases, giving rise to a high value of the quality factor $Q = \Omega/\Delta$. The effect of dissipation in the defect side branch is illustrated in Figs. 11.28(b)–(d). In Fig. 11.27(f), for different choices of the number N, the values of T, R, and α (taken at the maximum of transmission) as functions of $|\varepsilon''|$ are

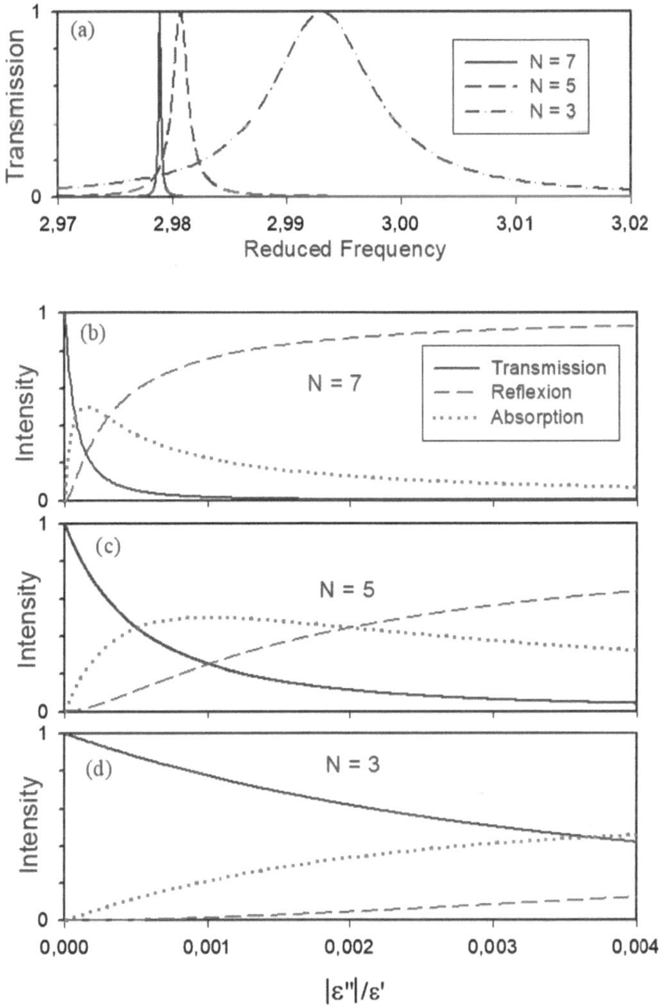

Figure 11.28 (a) Transmission as function of reduced frequency around defect mode for different values of number N of nodes in the structure. Notice that increasing N produces a decrease in width of peak, whereas peak position tends to frequency of defect branch inserted in infinite comb. (b)–(d) T, R, and α as functions of $|\varepsilon''|/\varepsilon'$ for different values of N. Notice that maximum of α occurs at smaller values of $|\varepsilon''|/\varepsilon'$ as far as N increases.

reported. By increasing the number of nodes in the comb structure, the maximum of the dissipation rate (which is equal to 0.5) occurs for smaller values of dissipation $|\varepsilon''|$. At the same time, the transmission rapidly drops to zero and the reflection becomes unity. One can conclude that, for an on/off switching device, a large size comb structure would be more appropriate because a slight modification of ε'' from zero will drop the transmission from 1 to 0 and enhance the reflection from 0 to 1 without introducing a significant dissipation.

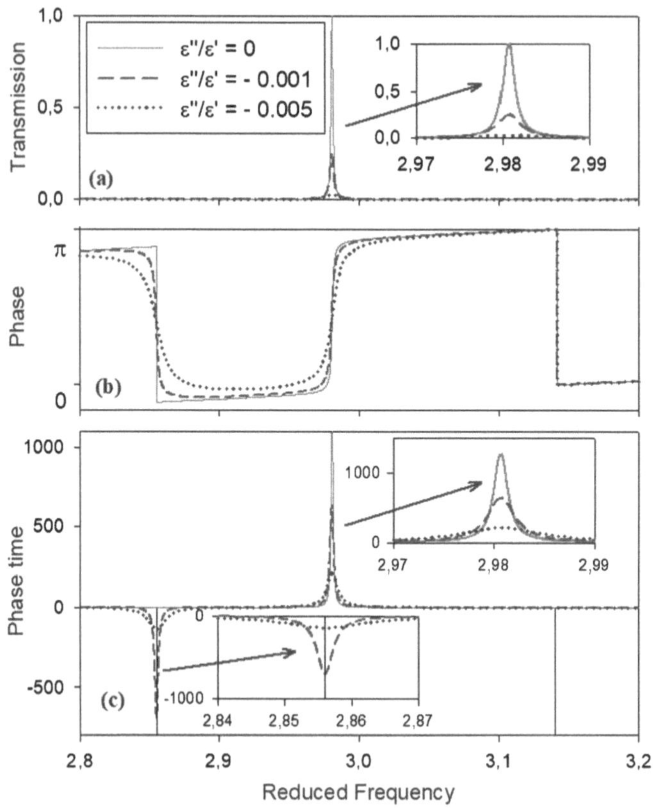

Figure 11.29 Variation of transmission (a), phase (b), and phase time (c) as functions of reduced frequency. Besides phase jumps of π around defect mode at $\Omega = 2.98$, phase exhibits jump of $-\pi$ at $\Omega = 2.856$ and at $\Omega = \pi$ associated with transmission zeros induced, by the defect branch and the other resonators, respectively.

Another interesting quantity, which depends on the absorption, is the phase of the transmission coefficient [114] or, equivalently, the phase time defined as the derivative of the phase with respect to the frequency. The latter quantity is considered as the time spent by the photon in the defect region (i.e., time of photon capture). Fig. 11.29 displays the transmission (Fig. 11.29(a)), the phase (Fig. 11.29(b)), and the phase time (Fig. 11.29(c)) as functions of the frequency around the defect mode for a few values of $|\varepsilon''|$. For $\varepsilon'' = 0$ (no dissipation), the defect mode gives rise to a positive phase jump of π around the defect mode at $\Omega \simeq 2.98$ and a delta peak in the phase time. Besides the phase jump at the defect mode, there are two negative π phase jumps at $\Omega \simeq 2.856$ and $\Omega = \pi$. They correspond to zeros of transmission induced respectively by the defect branch and by the other resonators of the comb structure (see the discussion in connection with Fig. 11.26(b)). Therefore, the phase time displays two negative delta peaks, whose frequencies are indicated by vertical lines in Fig. 11.29(c). In the presence of dissipation, the phase and the corresponding phase time are significantly

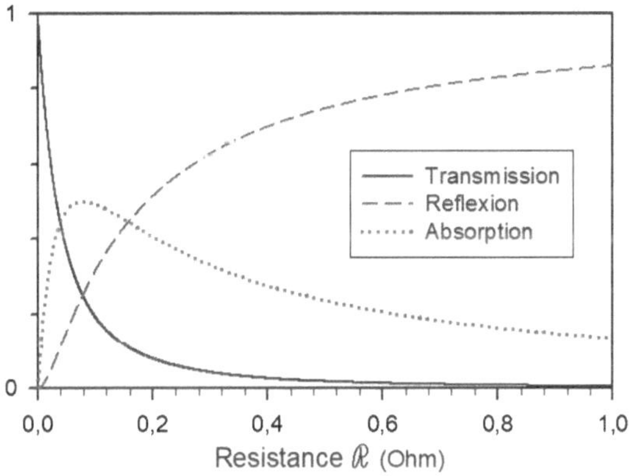

Figure 11.30 T, R, and α as functions of a resistance $\Re$ connected at the extremity of defect coaxial cable of length $d_3 = 1.1$ m inserted in the middle of a comb structure made of $N = 5$ coaxial cables of lengths $d_1 = d_2 = 1$ m. Dielectric constant of cables is considered to be $\varepsilon' = 2.3$.

affected around the frequency of the defect mode and around the zero of transmission associated with the defect side branch, while they remain essentially unchanged at the zero of transmission associated with the nondefective resonators at $\Omega = \pi$. Indeed, in this calculation, the dissipation phenomenon is only introduced in the defect branch. Therefore, two of the peaks in the phase time (one positive and one negative) transform into Lorentzian-shape resonances, whose widths increase with $|\varepsilon''|$. From the above discussion, we conclude that (i) the absorption diminishes the time spent by the photon inside the defect branch, (ii) the dissipation enables the observation of the negative phase time around a transmission zero by widening it. Let us mention that a negative phase time corresponds to a negative group velocity v_g, since v_g is related to the phase time τ by $v_g = L/\tau$, where L is the size of the structure.

To end this section, we note that one way to introduce intrinsic loss in the defect resonator is to change the boundary condition at the end of the defect branch. For instance, if the comb is composed of coaxial cables, as in [92,101], one can connect the end of the defect resonator to a resistor of given impedance $\Re$. It can be shown [115] that this is equivalent to connecting the end of the defect resonator to a semiinfinite coaxial cable with characteristic impedance $\Re$ (the characteristic impedance of all the other coaxial cables is assumed to be 50 Ω). Note the qualitative similarity of the results presented in Fig. 11.30 with those of Figs. 11.27(f) and 11.28, where the dissipation is an intrinsic property of the resonator constituting the defect branch.

11.8.2.2 Application of absorption to demultiplexer with on/off switching device

A frequency selective demultiplexer, using the above photonic bandgap structures, was proposed [110]. The demultiplexer consists of an input waveguide with three output channels, each channel being formed by a comb structure with a defect branch (see the inset of Fig. 11.31(b)). We assume that the comb structures in the different channels are identical and characterized by the same parameters as those studied before

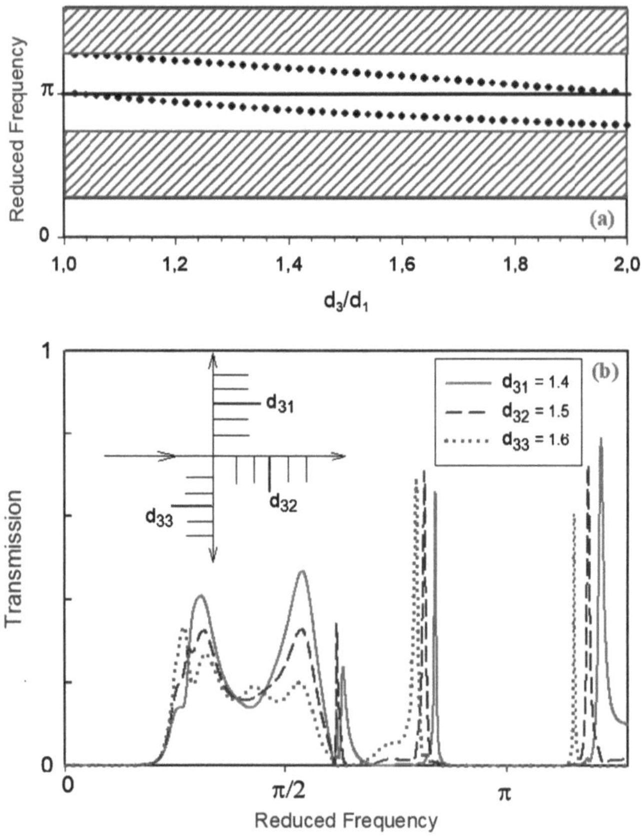

Figure 11.31 (a) Deduced frequencies Ω of localized modes versus reduced length d_3/d_1 of the defect branch inserted in an infinite comb. Hatched areas represent bulk bands separated by gap. Horizontal solid line at $\Omega = \pi$ in the middle of large gap corresponds to a flat band. Filled circles are defect modes induced by the defect branch. (b) Transmission through the demultiplexer constituted by one input waveguide and three output channels. Each channel is formed by a finite comb with its own defect as indicated in insert. Reduced lengths of three defect branches are indicated as d_{3i} in units of d_1 in channel $i = 1, 2, 3$.

(Fig. 11.26(c)) except for the length of the defect resonator. Therefore, the three structures will have the same stopbands (Fig. 11.31(b)) but different defect modes, which can produce selective transmission at three different frequencies.

First, we show in Fig. 11.31(a) the variation of the frequencies of the localized modes associated with a defect resonator in an infinite comb structure versus the reduced length d_3/d_1 of this resonator. In the bandgap extending from $\Omega = 2.301$ to $\Omega = 3.983$, there are two localized branches that decrease monotonically by increasing the length of the defect resonator. Now, we consider the demultiplexing structure, depicted in the inset of Fig. 11.31(b), in which each of the three comb structures contains a defect branch of length $d_3/d_1 = 1.4$, 1.5, and 1.6, respectively. The transmission spectra through the three channels (Fig. 11.31(b)) shows that, in the frequency range of the bandgap, each channel selectively transmits two frequencies corresponding to

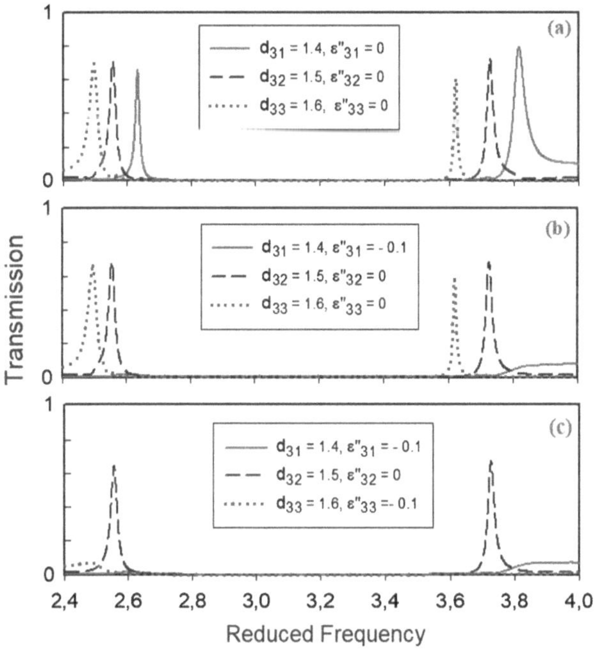

Figure 11.32 (a) Transmission through the three channels of the demultiplexer depicted in Fig. 11.31(b). Media are without gain or loss. (b) Defect branch in the first channel is lossy. (c) Defect branches in the first and the third channels are lossy. These examples show how one can control transmission and filtering in these photonic circuits. Reduced dielectric constants of the three defect branches are indicated as ε''_{3i} in unit of ε' for channel $i = 1, 2, 3$.

the localized modes of that channel without perturbing the other channels. Similar results were found using microwave photonic bandgap structures consisting of 2D square lattices with square cells [116].

By using the properties of absorption and amplification, in the different defect branches, it should be possible to reduce or to enhance the transmission through one channel without perturbing the others. A few examples of such possibilities are sketched in Fig. 11.32. Fig. 11.32(a) reproduces the result of Fig. 11.31(b), where the side branches are without loss or gain. Introducing a loss in the first channel with $d_{31} = 1.4$ (Fig. 11.32(b)) eliminates the transmission through this channel. Similarly, making the defect branch of length $d_{32} = 1.6$ dissipative (Fig. 11.32(c)), eliminates the transmission through the third channel.

11.9 Summary and perspectives

The results presented in this chapter look like a starting point for many applications in comb-like monomode circuits. Its backbone is a simple analytical theory for the propagation of coherent electromagnetic waves in monomode circuits. This simple approach starts to be complemented by more realistic numerical simulations and by experiments for some long-wavelength waves. A lot remains to be done for microwaves, especially those used in the telecommunication domain.

We have presented the propagation of electromagnetic waves in simple networked waveguides with a comb-like structure in Section 11.3. There exist zeros in the transmission spectrum of a quasi-one-dimensional waveguide with a single grafted side branch. These zeros of transmission give rise to absolute bandgaps in the electromagnetic band structure of an infinite periodic comb. Additional gaps form due to the periodic nature of the structure. Moreover, these features are still present in the transmission spectrum of a simple device constructed from a finite comb. In these systems, the gap width is controlled by geometrical parameters including the length of the side branches, the periodicity of the comb as well as the contrast in dielectric properties of side branch material and the backbone. Nevertheless, the electromagnetic band structure exhibits relatively wide gaps for homogeneous systems where the branches and the substrate are constituted of the same material. We have also shown that devices composed of finite numbers of grafted side branches exhibit a behavior similar to that of an infinite periodic comb.

We reviewed in Section 11.4 the propagation of electromagnetic waves in simple networked waveguides with a comb-like structure. Quasiquantized passing bands in the transmission spectrum of a quasi-one-dimensional waveguide with N' grafted side branches were investigated. These generic and universal behaviors of the star waveguides valid in different frequency domains (microwave, infrared, and visible) of electromagnetic spectrum, as different manufacturing techniques permit the fabrication of extremely thin wires [117–119], was confirmed by simple experiments using coaxial cables for frequencies up to a few hundreds of MHz. In these systems, the gap width is controlled by geometrical parameters, including the length and the number N' of the side branches, the periodicity of the comb, as well as the contrast in dielectric properties of side branch material and the backbone.

The generalization to two- and three-dimensional photonic networks constructed in the same manner as the above quasi-one-dimensional one is also straightforward. Indeed, one can imagine a two- or three-dimensional network of connected waveguides, with one or several grafted branches at every site. For example, in the case of a network having a square lattice symmetry, the dispersion relation (11.47) will be replaced by

$$\cos(k_x d_1) + \cos(k_y d_1) = 2\cos(\alpha_1 d_1) + \frac{N'}{2}\sqrt{\frac{\varepsilon_2}{\varepsilon_1}}\sin(\alpha_1 d_1)\cot(\alpha_2 d_2), \quad (11.122)$$

where k_x and k_y are the x and y components of the wave vector. Similarly to Eq. (11.47), this dispersion relation easily reveals that bandgaps do exist as a consequence of the dangling side branches, while the band structure does not contain any gap in the absence of the side branches.

In Section 11.5, we have reported that the presence of defective branches in the comb gives rise to localized states inside the forbidden bands. These defect modes appear as very narrow peaks of strong amplitude in the transmission spectrum. The localized states are very sensitive to the length of the defects, to their location in the comb, as well as to their number. Experimental measurements using coaxial cables have confirmed the theoretical predictions for frequencies up to 500 MHz. Recent works show clearly that the manufacturing of such comb structures at a micrometric scale becomes realizable with the new technological developments using high-resolution electron beam lithography [11,104].

We have reviewed in Section 11.6, both theoretically and experimentally, the existence of surface electromagnetic states associated with the termination of a comb-like structure composed of resonators grafted on a one-dimensional waveguide. In particular, it was demonstrated that when one considers the two semiinfinite comb structures obtained by the cleavage of an infinite one along a plane perpendicular to the waveguide, there exists as many localized surface states as there are minigaps. These surface modes may be detected experimentally by measuring the transmission coefficient through a guide along which the comb is attached vertically.

In Section 11.7, we have investigated the tunneling between two monomodes continua coupled by a monomode structure. We have conceived a model system allowing a complete analytical and experimental study of the conditions for a selective transfer of propagating state from one continuum to the other leaving all the other neighbor states unaffected. Our theoretical model enables us to determine completely the geometric parameters of the monomode structure for a complete channel transfer at a specific frequency, once this frequency has been chosen. The frequency domain where the channel drop tunneling occurs only depends on the characteristic lengths of the constituents of the model system. Our theoretical model is then universal and valid in all the frequency domains of the electromagnetic spectrum.

In Section 11.8, we have reviewed the effect of absorption on transmission and filtering in photonic bandgap materials made of dangling side branches grafted periodically along a monomode waveguide with an embedded defect branch. The defect modes induced by the defect branch are found to be very sensitive to small loss introduced in the system as an imaginary part ε'' added to the dielectric constant ε'. In the presence of loss, the transmission (reflection) decreases (increases), while the absorption passes through a maximum for certain value of ε'', labeled ε''_{th}. The value of ε''_{th} depends on the size of the structure; the larger the size, the lower the values of $|\varepsilon''_{th}|$. The property of absorption has been proposed as a means of realizing an on/off switching device in these structures in accordance with [111]. Finally, we proposed a frequency selective demultiplexer, built of the above photonic bandgap structures, and suggested the absorption property for realizing an on/off switching of selected frequencies. Moreover, the above formalism developed for electromagnetic waves can be easily transposed to electrons and magnons, as well as to acoustic waves.

Finally, let us mention that different simulation tools, based on the use of lines theory and electromagnetic wave equations, were used in Chapter 22 to find some of the properties demonstrated in Chapters 8, 9, and 10 using the formalism of Green's function. The advanced design system (ADS) tool is a simulator like spice, cadence. It focuses on the RF and microwave design, so most of its devices on the library are microwave devices and the models used in RF domain are based on lines theory and lumped elements. Also, finite element method (FEM) based on the software Comsol Multiphysics can be used to solve the transmission line and electromagnetic wave equations for various photonic configurations. The ADS tool enables calculating the complete scattering matrix, while Comsol enables calculating the band diagram, scattering matrix, and the electrical field distribution.

Acknowledgments

The authors would like to thank Jérôme Vasseur, Jaouad Zemmouri, and Pierre Deymier for their fruitful contributions to the original papers from which this chapter is inspired. We would like to thank also Abdelkader Mouadili who has plotted several figures in this chapter.

References

[1] E. Yablonovitch, Inhibited spontaneous emission in solid-state physics and electronics, Phys. Rev. Lett. 58 (1987) 2059.
[2] S. John, Strong localization of photons in certain disordered dielectric superlattices, Phys. Rev. Lett. 58 (1987) 2486.
[3] J. Martorell, N.M. Lawandy, Observation of inhibited spontaneous emission in a periodic dielectric structure, Phys. Rev. Lett. 65 (1990) 1877.
[4] E. Yablonovitch, Photonic band-gap structures, J. Opt. Soc. Am. B 10 (1993) 283.
[5] J.D. Joannopoulos, R.D. Meade, J.N. Winn, Photonic Crystals, Princeton University Press, Princeton, 1995.
[6] C.M. Soukoulis (Ed.), Photonic Band Gaps and Localization, Plenum, New York, 1992.
[7] C.M. Soukoulis (Ed.), Photonic Band Gap Materials, Kluwer, Dordrecht, 1996.
[8] J.M. Lourtioz, V. Berger, J-M. Gérard, D. Maystre, A. Tchelnokov, Les cristaux photoniques (ou la lumière en cage), Hermes Science Publications, Paris, 2003.
[9] A.R. McGurn, Photonic crystal circuits: a theory for two- and three-dimensional networks, Phys. Rev. B 61 (2000) 13235.
[10] J.D. Joannopoulos, P.R. Villeneuve, S. Fan, Photonic crystals, Solid State Commun. 102 (1997) 165.
[11] S. Foresi, P.R. Villeneuve, J. Ferrera, E.R. Thoen, G. Steinmeyer, S. Fan, J.D. Joannopoulos, L.C. Kimerling, H.I. Smith, E.P. Ippen, Photonic crystals: putting a new twist on light, Nature (London) 390 (1997) 143.
[12] Z.Q. Zhang, C.C. Wong, K.K. Fung, Y.L. Ho, W.L. Chan, S.C. Kan, T.L. Chan, N. Cheung, Observation of localized electromagnetic waves in three-dimensional networks of waveguides, Phys. Rev. Lett. 81 (1998) 5540.
[13] J.V. Hryniewicz, P.P. Absil, B.E. Little, R.A. Wilson, P.T. Ho, Higher order filter response in coupled microring resonators, IEEE Photonics Technol. Lett. 12 (2000) 320.
[14] D. Kleppner, Inhibited spontaneous emission, Phys. Rev. Lett. 47 (1981) 233.
[15] R.G. Hulet, E.S. Hilfer, D. Kleppner, Inhibited spontaneous emission by a Rydberg atom, Phys. Rev. Lett. 55 (1985) 2137.
[16] K.M. Leung, Y.F. Liu, Full vector wave calculation of photonic band structures in face-centered-cubic dielectric media, Phys. Rev. Lett. 65 (1990) 2646.
[17] Z. Zhang, S. Satpathy, Electromagnetic wave propagation in periodic structures: Bloch wave solution of Maxwell's equations, Phys. Rev. Lett. 65 (1990) 2650.

[18] K.M. Ho, C.T. Chan, C.M. Soukoulis, Existence of a photonic gap in periodic dielectric structures, Phys. Rev. Lett. 65 (1990) 3152.
[19] E. Yablonovitch, T.J. Gmitter, K.M. Leung, Photonic band structure: the face-centered-cubic case employing nonspherical atoms, Phys. Rev. Lett. 67 (1991) 2295.
[20] M. Plihal, A. Shambrook, A.A. Maradudin, P. Sheng, Two-dimensional photonic band structures, Opt. Commun. 80 (1991) 199.
[21] P.R. Villeneuve, M. Piché, Photonic band gaps in two-dimensional square lattices: square and circular rods, Phys. Rev. B 46 (1992) 4973.
[22] M. Plihal, A.A. Maradudin, Photonic band structure of two-dimensional systems: the triangular lattice, Phys. Rev. B 44 (1991) 8565.
[23] P.R. Villeneuve, M. Piché, Photonic band gaps in two-dimensional square and hexagonal lattices, Phys. Rev. B 46 (1992) 4969.
[24] R.D. Meade, K.D. Brommer, A.D. Rappe, J.D. Joannopoulos, Existence of a photonic band gap in two dimensions, Appl. Phys. Lett. 61 (1992) 495.
[25] For two- and three-dimensional photonic crystals, see also the special issue of J. Mod. Opt. 41 (2) (1994).
[26] D. Cassagne, C. Jouanin, D. Bertho, Photonic band gaps in a two-dimensional graphite structure, Phys. Rev. B 52 (1995) R2217.
[27] D. Cassagne, C. Jouanin, D. Bertho, Phys. Rev. B 53 (1996) 7134.
[28] A. Yariv, P. Yeh, Optical Waves in Crystals, Wiley, New York, 1984, p. 155.
[29] P. Yeh, Optical Waves in Layered Media, Wiley, New York, 1988.
[30] D.R. Smith, R. Dalichaouch, N. Kroll, S. Schultz, S.L. McCall, P.M. Platzmann, Photonic band structure and defects in one and two dimensions, J. Opt. Soc. Am. B 10 (1993) 314.
[31] M.L. Bah, A. Akjouj, E.H. El Boudouti, B. Djafari-Rouhani, L. Dobrzynski, Surface and interface optical waves in superlattices: transverse electric localized and resonant modes, J. Phys. Condens. Matter 8 (1996) 4171.
[32] E.N. Economou, M.M. Sigalas, Classical wave propagation in periodic structures: cermet versus network topology, Phys. Rev. B 48 (1993) 13434.
[33] J.O. Vasseur, A. Akjouj, L. Dobrzynski, B. Djafari-Rouhani, E.H. El Boudouti, Photon, electron, magnon, phonon and plasmon mono-mode circuits, Surf. Sci. Rep. 54 (2004) 1.
[34] U. Fano, Effects of configuration interaction on intensities and phase shifts, Phys. Rev. 124 (1961) 1866.
[35] For a review, see A.E. Miroshnichenko, S. Flach, Y.S. Kivshar, Fano resonances in nanoscale structures, Rev. Mod. Phys. 82 (2010) 2257.
[36] M. Fleischhauer, A. Imamoglu, J.P. Marangos, Electromagnetically induced transparency: optics in coherent media, Rev. Mod. Phys. 77 (2005) 633.
[37] S.E. Harris, Electromagnetically induced transparency, Phys. Today 50 (1997) 36.
[38] E.H. El Boudouti, T. Mrabti, H. Al-Wahsh, B. Djafari-Rouhani, A. Akjouj, L. Dobrzynski, Transmission gaps and Fano resonances in an acoustic waveguide: analytical model, J. Phys. Condens. Matter 20 (2008) 255212.
[39] A. Mouadili, E. El Boudouti, A. Soltani, A. Talbi, A. Akjouj, B. Djafari-Rouhani, Theoretical and experimental evidence of Fano-like resonances in simple monomode photonic circuits, J. Appl. Phys. 113 (2013) 164101.
[40] A. Mouadili, E. El Boudouti, A. Soltani, A. Talbi, B. Djafari-Rouhani, A. Akjouj, K. Haddadi, Electromagnetically induced absorption in detuned stub waveguides: a simple analytical and experimental model, J. Phys. Condens. Matter 26 (2014) 505901.
[41] Arturo Santillan, S.I. Bozhevolnyi, Acoustic transparency and slow sound using detuned acoustic resonators, Phys. Rev. B 84 (2011) 064304.

[42] W. Tan, C.Z. Yang, H.S. Liu, Z.G. Wang, H.Q. Lin, H. Chen, Manipulating classical waves with an analogue of quantum interference in a V-type atom, Eur. Phys. J. 97 (2012) 24003.

[43] A. Merkel, G. Theocharis, O. Richoux, V. Romero-Garcia, V. Pagneux, Control of acoustic absorption in one-dimensional scattering by resonant scatterers, Appl. Phys. Lett. 107 (2015) 244102.

[44] H. Al-Wahsh, E.H. El Boudouti, B. Djafari-Rouhani, A. Akjouj, T. Mrabti, L. Dobrzynski, Evidence of Fano-like resonances in mono-mode magnetic circuits, Phys. Rev. B 78 (2008) 075401.

[45] X. Piao, S. Yu, S. Koo, K. Lee, N. Park, Fano-type spectral asymmetry and its control for plasmonic metal–insulator–metal stub structures, Opt. Express 19 (2011) 10907.

[46] X. Piao, S. Yu, N. Park, Control of Fano asymmetry in plasmon induced transparency and its application to plasmonic waveguide modulator, Opt. Express 20 (2012) 18994.

[47] G. Wang, H. Lu, X. Liu, Dispersionless slow light in MIM waveguide based on a plasmonic analogue of electromagnetically induced transparency, Opt. Express 20 (2012) 20902.

[48] Qiong-gan Zhu, Zhi-guo Wang, The Green's function method for metal–dielectric–metal SPP waveguide network, Europhys. Lett. 103 (2013) 17004.

[49] Qiong-gan Zhu, Zhi-guo Wang, Analytical method for metal–insulator–metal surface plasmon polaritons waveguide networks, Opt. Express 27 (2019) 303.

[50] Yi Cheng, Yabin Jin, Yukun Zhou, Tong Hao, Yong Li, Distinction of acoustically induced transparency and Autler–Townes splitting by Helmholtz resonators, Phys. Rev. Appl. 12 (2019) 044025.

[51] A. Mouadili, E.H. El Boudouti, A. Soltani, A. Talbi, K. Haddadi, A. Akjouj, B. Djafari-Rouhani, Photonic demultiplexer based on electromagnetically induced transparency resonances, J. Phys. D 52 (2019) 075101.

[52] A. Mouadili, E.H. El Boudouti, A. Akjouj, H. Al-Wahsh, B. Djafari-Rouhani, L. Dobrzynski, Y-shaped magnonic demultiplexer using induced transparency resonances, AIP Adv. 9 (2019) 035011.

[53] A. Mouadili, E.H. El Boudouti, B. Djafari-Rouhani, Acoustic demultiplexer based on Fano and induced transparency resonances in slender tubes, Eur. Phys. J. Appl. Phys. 90 (2020) 10902.

[54] J.O. Vasseur, P.A. Deymier, L. Dobrzynski, B. Djafari-Rouhani, A. Akjouj, Absolute band gaps and electromagnetic transmission in quasi-one-dimensional comb structures, Phys. Rev. B 55 (1997) 10434.

[55] H. Al-Wahsh, A. Akjouj, B. Djafari-Rouhani, L. Dobrzynski, Magnonic circuits and crystals, Surf. Sci. Rep. 66 (2011) 29.

[56] S. Khattou, M. Amrani, A. Mouadili, E.H. El Boudouti, A. Talbi, A. Akjouj, B. Djafari-Rouhani, Comparison of density of states and scattering parameters in coaxial photonic crystals: theory and experiment, Phys. Rev. B 102 (2020) 165310.

[57] A. Mouadili, A. Akjouj, E.H. El Boudouti, L. Dobrzynski, Magnetic demultiplexer circuit with four channels, J. Exp. Theor. Phys. 130 (2020) 859.

[58] G.H. Cocoletzi, L. Dobrzynski, B. Djafari-Rouhani, H. Al-Wahsh, D. Bria, Electromagnetic wave propagation in quasi-one-dimensional comb-like structures made up of dissipative negative-phase-velocity materials, J. Phys. Condens. Matter 18 (2006) 3683.

[59] M.S. Kushwaha, A. Akjouj, B. Djafari-Rouhani, L. Dobrzynski, J.O. Vasseur, Acoustic spectral gaps and discrete transmission in slender tubes, Solid State Commun. 106 (1998) 659.

[60] A. Akjouj, L. Dobrzynski, B. Djafari-Rouhani, J.O. Vasseur, M.S. Kushwaha, Giant electronic stop bands in one-dimensional comb-like structures, Europhys. Lett. 41 (1998) 321.

[61] H. Al-Wahsh, A. Akjouj, B. Djafari-Rouhani, J.O. Vasseur, L. Dobrzynski, P.A. Deymier, Large magnonic band gaps and defect modes in one-dimensional comb-like structures, Phys. Rev. B 59 (1999) 8709.

[62] J.O. Vasseur, A. Akjouj, B. Djafari-Rouhani, L. Dobrzynski, Resonant tunneling between two monomode electronic waveguides, Surf. Sci. 454 (2000) 428.

[63] R. Kucharczyk, L. Dobrzynski, B. Djafari-Rouhani, A. Akjouj, A nanometric electron multiplexer, Physica E 24 (2004) 355.

[64] L. Dobrzynski, B. Djafari-Rouhani, A. Akjouj, J.O. Vasseur, M.L.H. Lahlaouti, Electron channel drop tunnelling, J. Phys. Condens. Matter 11 (1999) L247.

[65] L. Dobrzynski, B. Djafari-Rouhani, A. Akjouj, J.O. Vasseur, J. Zemmouri, Resonant tunnelling of acoustic waves between two slender tubes, Europhys. Lett. 46 (1999) 467.

[66] W. Porod, Z. Shao, C.S. Lent, Transmission resonances and zeros in quantum waveguides with resonantly coupled cavities, Appl. Phys. Lett. 61 (1992) 1350.

[67] Z. Shao, W. Porod, C.S. Lent, Transmission resonances and zeros in quantum waveguide systems with attached resonators, Phys. Rev. B 49 (1994) 7453.

[68] J.R. Shi, B.Y. Gu, Quantum waveguide transport with side-branch structures: a recursive algorithm, Phys. Rev. B 55 (1997) 4703.

[69] J.B. Xia, Quantum waveguide theory for mesoscopic structures, Phys. Rev. B 45 (1992) 3593.

[70] P.S. Deo, A.M. Jayannavar, Quantum waveguide transport in serial stub and loop structures, Phys. Rev. B 50 (1994) 11629.

[71] J. Fey, W.M. Robertson, Compact acoustic bandgap material based on a subwavelength collection of detuned Helmholtz resonators, J. Appl. Phys. 109 (2011) 114903.

[72] Zhi Guo Wang, Sam Hyeon Lee, Chul Koo Kim, Choon Mahn Park, Kyun Nahm, S.A. Nikitov, Acoustic wave propagation in one-dimensional phononic crystals containing Helmholtz resonators, J. Appl. Phys. 103 (2008) 064907.

[73] Zhi Guo Wang, Sam Hyeon Lee, Chul Koo Kim, Choon Mahn Park, Kyun Nahm, S.A. Nikitov, Effective medium theory of the one-dimensional resonance phononic crystal, J. Phys. Condens. Matter 20 (2008) 055209.

[74] N. Fang, D.J. Xi, J. Xu, M. Ambati, W. Srituravanich, C. Sun, X. Zhang, Ultrasonic metamaterials with negative modulus, Nat. Mater. 5 (2006) 452.

[75] G.J. Jin, Z.D. Wang, A. Hu, S.S. Jiang, Quantum waveguide theory of serial stub structures, J. Appl. Phys. 85 (1999) 1597.

[76] A. Nomata, S. Horie, Self-similarity appearance conditions for electronic transmission probability and Landauer resistance in a Fibonacci array of T stubs, Phys. Rev. B 76 (2007) 235113.

[77] S. Chattopadhyay, A. Chakrabarti, Electronic transmission in quasiperiodic serial stub structures, J. Phys. Condens. Matter 16 (2004) 313.

[78] H. Aynaou, A. Mouadili, N. Ouchani, E.H. El Boudouti, A. Akjouj, B. Djafari-Rouhani, Scaling law, confined and surface modes in photonic Fibonacci stub structures: theory and experiment, Appl. Phys. 10 (2020) 7767.

[79] L. Dobrzynski, Interface response theory of continuous composite materials, Surf. Sci. 180 (1987) 489.

[80] L. Dobrzynski, Interface response theory of electromagnetism in composite dielectric materials, Surf. Sci. 180 (1987) 505.

[81] L. Dobrzynski, Interface response theory of discrete composite systems, Surf. Sci. Rep. 6 (1986) 119.

[82] L. Dobrzynski, Interface response theory of continuous composite systems, Surf. Sci. Rep. 11 (1990) 139.
[83] L. Dobrzynski, Interface response theory of electromagnetism in dielectric superlattices, Phys. Rev. B 37 (1988) 8027.
[84] L. Dobrzynski, H. Puszkarski, Eigenvectors of composite systems. I. General theory, J. Phys. Condens. Matter 1 (1989) 1239.
[85] F. Garcia-Moliner, J. Rubio, A new method in the quantum theory of surface states, J. Phys. C 2 (1969) 1789.
[86] M.G. Cottam, A.A. Maradudin, in: Surface Excitations, in: Modern Problems in Condensed Matter Sciences, vol. 9, North-Holland, Amsterdam, 1984.
[87] A.A. Maradudin, D.L. Mills, Scattering and absorption of electromagnetic radiation by a semi-infinite medium in the presence of surface roughness, Phys. Rev. B 11 (1975) 1392.
[88] A. Akjouj, B. Sylla, L. Dobrzynski, Introduction a la théorie des systemes composites: exemples simples de matériaux lamellaires, Ann. Phys. 18 (1993) 363.
[89] M.L. Bah, A. Akjouj, L. Dobrzynski, Response functions in layered dielectric media, Surf. Sci. Rep. 16 (1992) 95.
[90] E.N. Economou, Green's functions in Quantum Physics, Springer-Verlag, 1983, p. 79.
[91] S.L. McCall, P.M. Platzmann, R. Dalichaouch, D. Smith, S. Schultz, Microwave propagation in two-dimensional dielectric lattices, Phys. Rev. Lett. 67 (1991) 2017.
[92] L. Dobrzynski, A. Akjouj, B. Djafari-Rouhani, J.O. Vasseur, J. Zemmouri, Giant gaps in photonic band structures, Phys. Rev. B 57 (1998) R9388.
[93] J. Von Neuman, E. Wigner, Über merkwürdige diskrete Eigenwerte, Phys. Z. 30 (1929) 465.
[94] C.W. Hsu, B. Zhen, A.D. Stone, J.D. Joannopoulos, M. Soljacic, Bound states in the continuum, Nat. Rev. Mater. 1 (2016) 16048.
[95] R.P. Stanley, R. Houdré, U. Oesterle, M. Ilegems, C. Weisbuch, Impurity modes in one-dimensional periodic systems: the transition from photonic band gaps to microcavities, Phys. Rev. A 48 (1993) 2246.
[96] R. Wang, J. Dong, D.Y. Xing, Defect studies in a one-dimensional photonic band gap structure, Phys. Status Solidi B 200 (1997) 529.
[97] M. Sigalas, C.M. Soukoulis, E.N. Economou, C.T. Chan, K.M. Ho, Photonic band gaps and defects in two dimensions: studies of the transmission coefficient, Phys. Rev. B 48 (1993) 14121.
[98] E. Yablonovitch, T.J. Gmitter, R.D. Meade, A.M. Rappe, K.D. Brommer, J.D. Joannopoulos, Donor and acceptor modes in photonic band structure, Phys. Rev. Lett. 67 (1991) 3380.
[99] A. Mekis, J.C. Chen, I. Kurland, S. Fan, P.R. Villeneuve, J.D. Joannopoulos, High transmission through sharp bends in photonic crystal waveguides, Phys. Rev. Lett. 77 (1996) 3787.
[100] K. Sakoda, H. Shiroma, Numerical method for localized defect modes in photonic lattices, Phys. Rev. B 56 (1997) 4830.
[101] J.O. Vasseur, B. Djafari-Rouhani, L. Dobrzynski, A. Akjouj, J. Zemmouri, Defect modes in one-dimensional comblike photonic waveguides, Phys. Rev. B 59 (1999) 13446.
[102] J.O. Vasseur, M. Pecquery, B. Djafari-Rouhani, L. Dobrzynski, A. Akjouj, J. Zemmouri, N. Fettouhi, E.H. El Boudouti, in: C.M. Soukoulis (Ed.), Photonic Crystals and Light Localization in the 21st Century, Kluwer Academic Publishers, 2001, p. 181.
[103] B. Djafari-Rouhani, J.O. Vasseur, A. Akjouj, L. Dobrzynski, M.S. Kushwaha, P.A. Deymier, J. Zemmouri, Giant stop bands and defect mode in one-dimensional waveguide, Prog. Surf. Sci. 59 (1998) 255.

[104] G. Feiertag, W. Ehrfeld, H. Freimuth, H. Kolle, H. Lehr, M. Schmidt, M.M. Sigalas, C.M. Soukoulis, G. Kiriakidis, T. Pedersen, J. Kuhl, W. Koenig, Synthesis and characterization of highly-conducting nitrogen-doped ultrananocrystalline diamond films, Appl. Phys. Lett. 71 (1997) 1441.

[105] B. Djafari-Rouhani, E.H. El Boudouti, A. Akjouj, L. Dobrzynski, J.O. Vasseur, A. Mir, N. Fettouhi, J. Zemmouri, Surface states in one-dimensional photonic band gap structures, Vacuum 63 (2001) 177.

[106] S.S. Orlov, A. Yariv, S. Van Essen, Coupled-mode analysis of fiber-optic add–drop filters for dense wavelength-division multiplexing, Opt. Lett. 22 (1997) 688.

[107] S. Fan, P.R. Villeneuve, J.D. Joannopoulos, A. Haus, Channel drop tunneling through localized states, Phys. Rev. Lett. 80 (1998) 960.

[108] L. Dobrzynski, B. Djafari-Rouhani, A. Akjouj, J.O. Vasseur, J. Zemmouri, Resonant tunneling between two continua, Phys. Rev. B 60 (1999) 10628.

[109] L. Dobrzynski, A. Akjouj, B. Djafari-Rouhani, J.O. Vasseur, M. Bouazaoui, J.P. Vilcot, A. Beaurain, S. McMurtry, Directional photon transfer between two wires, Phys. Rev. E 67 (2003) 057603.

[110] B. Djafari-Rouhani, E.H. El Boudouti, A. Akjouj, J.O. Vasseur, L. Dobrzynski, Transmission and filtering in photonic circuits: effects of absorption and amplification, Prog. Surf. Sci. 74 (2003) 389.

[111] S. Fan, P.R. Villeneuve, J.D. Joannopoulos, H.A. Haus, Loss-induced on/off switching in a channel add/drop filter, Phys. Rev. B 64 (2001) 245302.

[112] E.O. Ozbay, G. Tuttle, M. Sigalas, C.M. Soukoulis, K.M. Ho, Defect structures in a layer-by-layer photonic band-gap crystal, Phys. Rev. B 51 (1995) 13961.

[113] M. Sigalas, K.M. Ho, R. Biswas, C.M. Soukoulis, Theoretical investigation of defects in photonic crystals in the presence of dielectric losses, Phys. Rev. B 57 (1998) 3815.

[114] J. Heinrichs, Light amplification and absorption in a random medium, Phys. Rev. B 56 (1997) 8674.

[115] The demonstration can be found for instance in the text book: Fred Gardiol, Traité d'électricité, Electromagnétisme, vol. III, Polytechniques et Universitaires Romandes, 2002.

[116] S.S. Oh, C.-S. Kee, J.-E. Kim, H.Y. Park, T. Kim, I. Park, H. Lim, Duplexer using microwave photonic band gap structure, Appl. Phys. Lett. 76 (2000) 2301.

[117] A. Blondel, J.P. Meier, B. Boudin, J.P. Ansermet, Giant magnetoresistance of nanowires of multilayers, Appl. Phys. Lett. 65 (1994) 3019.

[118] J. de la Figuera, M.A. Huerta-Garnica, J.E. Prioto, C. Ocal, R. Miranda, Fabrication of magnetic quantum wires by step-flow growth of cobalt on copper surfaces, Appl. Phys. Lett. 66 (1995) 1006.

[119] W.Z. Li, S.S. Xie, L.X. Qian, B.H. Chang, B.S. Zou, W.Y. Zhou, R.A. Zhao, J. Wang, Large-scale synthesis of aligned carbon nanotubes, Science 274 (1996) 1701.

Serial loop structures: photonic bandgaps, confined, cavity, and surface modes

12

El Houssaine El Boudouti[a], Bahram Djafari-Rouhani[b], Abdellatif Akjouj[b], and Leonard Dobrzyński[b]
[a]Laboratoire de Physique de la Matière et de Rayonnement, Département de Physique, Faculté des Sciences, Université Mohammed I, Oujda, Morocco, [b]Institute of Electronic, Microelectronics and Nanotechnology (IEMN), UMR CNRS 8520, Department of Physics, Faculty of Sciences and Technologies, Lille University, Villeneuve d'Ascq Cedex, France

12.1 Introduction

The problem of propagation of electromagnetic waves in artificial periodic dielectric materials received a great deal of attention in the last three decades [1,2]. Of particular interest is the existence of photonic bandgaps (PBG) in the band structures of such materials called photonic crystals (PCs). The PBGs are analogs, for electrons, to the electronic bandgaps in the band structure of semiconductor crystals. These structures present unusual properties which can be exploited in the control and guidance of the propagation of light [3,4]. Of particular interest is the existence of PBG in the band structure of one-dimensional (1D) structures with a variety of geometries [5–9].

In Chapter 11, we have given a comprehensive study about the propagation of electromagnetic waves in quasi-1D systems [7,8]. These structures, called *star waveguides*, are composed of an infinite backbone along which stars of N' finite branches are grafted at N equidistant sites, N and N' being integers. The stopbands originate both from the periodicity of the system determined by the distance between two neighboring sites and from the eigenmodes of the dangling side branches which play the role of resonators. The gap widths depend also on the boundary conditions at the free ends of the side branches, namely the vanishing of either the electric field ($E = 0$) or the magnetic field ($H = 0$). In such systems the propagation is monomode, provided the two characteristic lengths (the periodicity and the resonator length) are much greater than the backbone and the side branch diameters [7,8]. It is worthwhile to point out that this star waveguide structure exhibits relatively broad forbidden bands even if the backbone and the resonators are made of the same material [7]; in other words, the existence of the gaps does not require a contrast between the parameters of the two constituents. This property offers the possibility of engineering bandgaps in homogeneous materials. Moreover, localized modes may appear in the forbidden bands when a defect is introduced in the star waveguide periodic structure. Applications as selective frequency filters and efficient waveguides [9] can be expected.

Photonics. https://doi.org/10.1016/B978-0-12-819388-4.00022-8

Besides comb structures, the *serial loop structure* (SLS) made of a periodic repetition of segments and loops has been also extensively studied in the last years [10]. Zhang et al. [11] studied theoretically and experimentally 3D PBG systems constructed by segments and loops made of coaxial cables of 1D waveguides arranged in a diamond structure. The transmission measurement shows a wide gap, but the presence of the free ends of the structure induces a surface state, which makes the gap not truly forbidden for all states. In addition, besides the Anderson localized states observed in random structures, defect modes are introduced in the gap by changing the length of one loop in an ordered network. However, the peak associated with the defect mode is not well-defined because of the dissipation inside the cables. Schneider et al. [12] and Haché et al. [13] have treated photonic circuits constituted by an alternative repetition of two different coaxial cables, indicating that the behaviors observed in these 1D crystals are qualitatively comparable to those of photonic systems of higher dimension as well.

In this chapter, we show that a one-dimensional (1D) structure made of segments and loops may exhibit large gaps in which the existence of defect modes can be clearly observed [14]. This SLS may be modeled as an infinite number of unit cells pasted together; each cell is made of a finite segment of length d_1 connected to a loop of length $d_2 + d_3$ (each loop is composed of two wires of different lengths $d_2 \neq d_3$, see Fig. 12.1(a)). The propagation of electromagnetic waves through SLS is also assumed to be monomode. The simple analytical expressions obtained enable us to explain clearly the origin of the bandgaps in terms of the different lengths of segments and loops. Indeed, among the SLS, one can distinguish two different structures: (i) the symmetric SLS (i.e., $d_2 = d_3$) in which the bandgaps result only from the periodicity $d = d_1 + d_2$ of the SLS and (ii) the asymmetric SLS (i.e., $d_2 \neq d_3$) in which the bandgaps result from both the periodicity and the transmission zeros in these systems. As we shall discuss in Section 12.2, these zeros of transmission result, on the one hand, from a destructive interference between the waves traveling on the two arms of the loop which have different lengths, similarly to a Mach–Zehnder interferometer and, on the other hand, from the internal resonant modes of the loop corresponding to waves traveling around the loop. In particular, the phase time measurements show different behaviors in these two cases in accordance with the experiments on phase coherent transport through a quantum dot [15,16]. The phase time measurements also enable us to determine the group velocity, which can exceed the speed of light in vacuum or even become negative. The existence of such supraluminal or negative delay phenomena has attracted much attention in different studies [17–19], with the challenge to attain a pulse advancement comparable to the pulse width with a low level of pulse distortion. The SLS may also show additional features in comparison with star waveguides such as the existence of larger gaps and the absence of the boundary condition at the end of the side branches which could be of potential interest in optical waveguide structures. These features are essentially due to the loop structure which is quite different from the case of a simple resonator [7,8]. The introduction of defects inside the SLS structure may also give rise to well-defined defect modes inside the bandgaps which can be observable in the transmission spectra.

In addition to bandgaps and defect modes, it is interesting to investigate also the eigenmodes (i.e., confined modes) in a finite SLS with different boundary conditions on its both ends, as well as the surface modes induced by its extremities. There exist several papers in the literature dealing with confined and surface modes in photonic and phononic crystals. The study of electromagnetic wave propagation in finite size one-dimensional (1D) multilayer PC was largely developed theoretically and experimentally [20–32]. In general, the finite size structure is deposited on an homogeneous substrate with or without a buffer layer or encapsulated with a cap layer [24]. The results obtained show that, besides the standing waves in the finite size structure, there exist additional modes inside the bandgaps induced by the surface termination of the periodic structure.

Several years ago, El Boudouti et al. [33] have demonstrated that in the case of transverse elastic waves, the creation of two semiinfinite PCs from the cleavage of an infinite N-layer PC gives rise to one surface mode per gap for any value of the wave vector parallel to the interfaces. This mode belongs to one or the other of the two complementary PCs. These results were confirmed experimentally in a two-layer elastic PC [34,35]. These demonstrations have been extended to transverse elastic waves [36], as well as sagittal elastic waves [37] in solid–solid and solid–fluid finite-size PCs, respectively. Similar conclusions have been derived for phonons in a linear multiatomic chain, i.e., when the unit cell is constituted by several atoms differing either by their mass or the spring constant that link them together [38]. Actually, this is straightforward in the simple cases of a mono- and biatomic linear chain. Some experiments [39,40] on linear chains composed of welded steel spheres of the same or different diameters have been explained in this framework. Also, it was demonstrated theoretically and experimentally that the same conclusions hold in 1D systems made of coaxial cables with different geometries such as periodic comb-like [41], as well as periodic and quasiperiodic serial loop structures [42,43].

The analogy between surface elastic waves in multilayered structures and surface electromagnetic waves in coaxial cables is straightforward in two limiting cases, namely when the boundary conditions at the ends of the coaxial cables are either $E = 0$ or $H = 0$. Let us recall that these two conditions mean, respectively, that the two constituting conductors of the coaxial cables are, or are not, short-circuited. However, this analogy is not fulfilled in the case of multilayered optical structures where the boundary conditions at the surface are neither $E = 0$ nor $H = 0$. Indeed, when dealing with layered media, the PC is in contact with a dielectric homogeneous medium (like vacuum, for example) and therefore, the continuity of the transverse component of H and the normal component of the displacement field D should be satisfied at the surface. For these reasons, coaxial cables are good candidates for highlighting general rules about confined and surface electromagnetic modes in finite size 1D structures in the above mentioned cases (i.e., $E = 0$ or $H = 0$). Of course, these rules do not apply for optical multilayered media [26]. Also, it was shown that coaxial cables present an easily realizable experimental approach to the study of wave interference phenomena like the bandgap structures with or without defect modes [12,14], superluminal and subluminal effects [13,14,44,45], and standing waves [41,42].

In Section 12.3 of this chapter, we consider a finite-size SLS photonic crystal made of N unit cells (Figs. 12.13(a)–(b)). The left and right surfaces of one unit cell are in general different (Fig. 12.13(b)), we call them *complementary surfaces*; and the boundary conditions at both ends of the finite structure are of type $H = 0$. Our goal is to show analytically and experimentally the existence of $N - 1$ modes in each allowed band and one additional mode per gap induced by one of the two complementary surfaces surrounding the structure [43]. We demonstrate that these modes are those of a unique single cell (i.e., $N = 1$). In the particular case where the cells are symmetrical (i.e., with the same surface termination), we show that the surface modes fall at the band edges. For such structures, N confined loop modes exist also when their wavelength is commensurable with the two loop lengths d_2 and d_3 shown in Fig. 12.1(a). These modes form a flat band. Such confined modes appear in Section 5.10 of Chapter 5 of this Photonics book. When the above condition is not fulfilled, these N modes become usually one almost-flat band. There exist also two degenerate modes of zero frequency: ground state and invariant by translation state.

Contrary to the usual results where the defect modes induced by the two surfaces surrounding the finite structure depend strongly on its size [24,32], we show that the modes induced by the two complementary surfaces of the finite structure (Fig. 12.12(b)) are independent of N and coincide exactly with the surface modes associated to two complementary semiinfinite PCs (Fig. 12.12(c)) obtained from the cleavage of an infinite PC (Fig. 12.12(d)) [33,41,42]. So, even the standing waves of only one cell ($N = 1$, Fig. 12.12(a)) give the surface modes of two complementary PCs (Fig. 12.12(c)). It is worth noticing that different results (eigenmodes and spatial confinement) are obtained if the boundary conditions on both ends of the finite size structure (Fig. 12.12(b)) are of type $E = 0$ instead of $H = 0$ [41]. However, the same rules apply to confined and surface modes in these structures. For the sake of briefness, we shall avoid giving the results concerning the case $E = 0$. Let us mention that similar results to those presented here are found theoretically by Ren at al. [46,47] for the complete confinement in the whole structure space of electronic states in 1D crystal of finite size.

Among different theoretical models, the transfer-matrix [48] and the Green's function methods [49] are quite suitable to deduce the eigenmodes, eigenvectors, as well as the transmission and reflection coefficients in composite media. In addition to these quantities, the Green's function approach also enables deducing easily the local and total densities of states [24,26,32,33,42]. The experiments are realized on a PC based on coaxial BNC connectors with different impedances called coaxial PC. The propagation in these structures is monomode [49], and one can obtain very accurate experimental results that may be fitted with a simple 1D theoretical model. The interference of the multiple reflected waves leads to the same phenomena in the radio-frequency range as for light propagation through a photonic crystal [50–52].

The theoretical model developed here falls within the framework of the interface response theory of continuous media [53]. In Section 12.2 we give the analytical expressions and experimental demonstrations for the band structure of an infinite SLS and for the transmission coefficient through a finite SLS with and without a defect. In Section 12.3 we give a short summary of the theoretical model and the main analytical, numerical, and experimental results about confined and surface modes in finite size coaxial structures with symmetric and asymmetric cells. The experimental demonstration of our theoretical predictions is performed by measuring the transmission spectra through finite SLS constituted by coaxial cables. The typical length of the cables is of the order of a meter and therefore the frequencies fall in the range of 10 to 400 MHz. Finally, a summary and some conclusions are drawn in Section 12.4.

12.2 Bandgaps and selective transmission

12.2.1 Theoretical model: Green's function approach

Our theoretical analysis is performed with the help of the interface response theory of continuous media which allows calculating the Green's function of any composite material. In what follows, we present the basic concept and the fundamental equations of this theory [53].

Let us consider any composite material contained in its space of definition D and formed out of N different homogeneous pieces located in their domains D_i. Each piece is bounded by an interface M_i, adjacent in general to j $(1 \leq j \leq J)$ other pieces through subinterface domains M_{ij}. The ensemble of all these interface spaces M_i will be called the interface space M of the composite material.

The elements of the Green's function $g(D, D)$ of any composite material can be obtained from [53]

$$g(D, D) = G(D, D) - G(D, M)G^{-1}(M, M)G(M, D) \\ + G(D, M)G^{-1}(M, M)g(M, M)G^{-1}(M, M)G(M, D), \qquad (12.1)$$

where $G(D, D)$ is the reference Green's function formed out of truncated pieces in D_i of the bulk Green's functions of the infinite continuous media and $g(M, M)$, the interface element of the Green's function of the composite system.

The knowledge of the inverse of $g(M, M)$ is sufficient to calculate the interface states of a composite system through the relation [53]

$$\det[g^{-1}(M, M)] = 0. \qquad (12.2)$$

Moreover, if $U(D)$ represents an eigenvector of the reference system, Eq. (12.1) enables the calculation of the eigenvectors $u(D)$ of the composite material as

$$u(D) = U(D) - U(M)G^{-1}(M, M)G(M, D) \\ + U(M)G^{-1}(M, M)g(M, M)G^{-1}(M, M)G(M, D). \qquad (12.3)$$

In Eq. (12.3), $U(D)$, $U(M)$, and $u(D)$ are row vectors. Eq. (12.3) provides a description of all the waves reflected and transmitted by the interfaces, as well as the reflection and transmission coefficients of the composite system. In this case, $U(D)$ is a bulk wave launched in one homogeneous piece of the composite material [54].

12.2.2 Inverse surface Green's functions of the elementary constituents

We consider an infinite homogeneous isotropic dielectric wire i characterized by its characteristic impedance Z_i. The Fourier transformed Green's function between two points x and x' of this wire is [49]

$$G_i(x, x') = -\frac{e^{-\alpha_i |x - x'|}}{2F_i}, \tag{12.4}$$

where

$$\alpha_i = j\frac{\omega}{c}\sqrt{\varepsilon_i(\omega)} \tag{12.5}$$

and

$$F_i = -\frac{j\omega}{Z_i}, \tag{12.6}$$

with ω being the angular frequency of the wave, c the speed of light in vacuum, $\varepsilon_i(\omega)$ the relative permittivity, and $j = \sqrt{-1}$.

Before addressing the problem of SLS, it is helpful to know the surface elements of the Green's function of a finite wire of length d_i and of a semiinfinite wire. The finite wire is bounded by two free surfaces located at $x = -d_i/2$ and $x = +d_i/2$. These surface elements can be written in the form of a 2×2 matrix $g_i(M, M)$, within the interface space $M_i \equiv \{-d_i/2, +d_i/2\}$. The inverse of this matrix takes the following form [49]:

$$[g_i(M, M)]^{-1} = \begin{pmatrix} -\frac{F_i C_i}{S_i} & \frac{F_i}{S_i} \\ \frac{F_i}{S_i} & -\frac{F_i C_i}{S_i} \end{pmatrix} \tag{12.7}$$

where $C_i = \cosh(\alpha_i d_i)$ and $S_i = \sinh(\alpha_i d_i)$. The inverse of the surface element of a semiinfinite wire is given by [49]

$$[g_i(0, 0)]^{-1} = -F_i. \tag{12.8}$$

12.2.3 Dispersion relations and transmission coefficients

The one-dimensional serial loop waveguide can be considered as an infinite number of unit cells pasted together (see Fig. 12.1(a)). Each cell is composed of a finite wire (*medium 1*) of length d_1 in the direction x, connected to a loop "ring" (*medium 2*) of length $d_2 + d_3$ (each loop is constructed of two wires of lengths d_2 and d_3, respectively). The interface domain is made of all the connection points between finite segments and loops. A space position along the x axis in medium i belonging to the unit cell n is indicated by (n, i, x), where n the "cell number" is an integer such that $-\infty \leq n \leq +\infty$, i the medium index ($i = 1, 2, 3$) and $-\frac{d_i}{2} \leq x \leq +\frac{d_i}{2}$. Due to the 1D

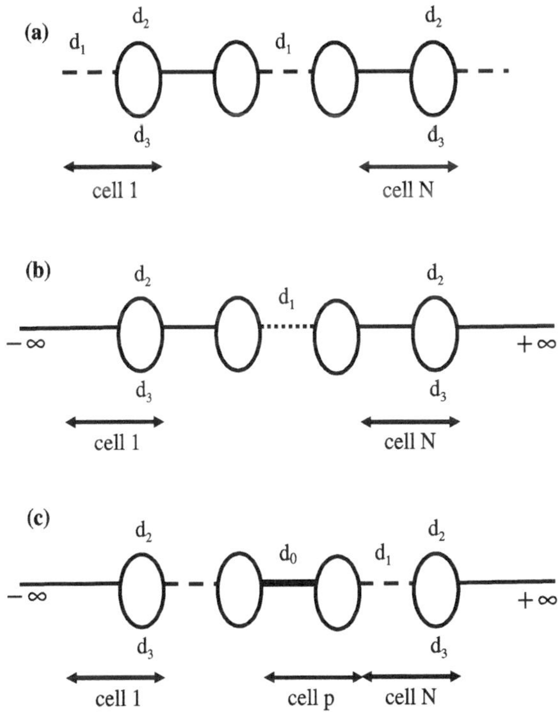

Figure 12.1 (a) Schematic of the one-dimensional serial loop structures studied in this chapter. The media are labeled by an index i, with $i = 1$ for the finite branch and 2 and 3 for the arms of the loop. Each loop has a circumference length $L = d_2 + d_3$ and is distant by d_1 from the neighboring loops. Each cell is composed of a finite branch and the loop connected to its right. (b) Waveguide with finite N loops separated by a length d_1 and connected at its extremities to two semiinfinite leads. (c) Same as in (b) except that a defect branch of length d_0 (heavy line) is introduced in the cell p.

spatial periodicity of the system in the direction x, one can define a Bloch wave vector k along the axis of the waveguide.

Within the total interface space of the infinite SLS, the inverse of the matrix giving all the interface elements of the Green's function g is an infinite tridiagonal matrix formed by linear superposition of the elements $[g_i(M,M)]^{-1}$ (Eq. (12.7)). The explicit expression of the Green's function elements in the interface space is given as [55]:

$$g(n,1,+d_1/2;n',1,+d_1/2) = g(n,1,-d_1/2;n',1,-d_1/2) = Y_1 \frac{t^{|n-n'|+1}}{t^2-1}, \tag{12.9}$$

$$g(n,1,+d_1/2;n',1,-d_1/2) = \frac{S_2 S_3}{F_2 S_3 + F_3 S_2} \frac{t^{|n-n'|+1}}{t^2-1} + \frac{S_1}{F_1} \frac{t^{|n-n'+1|+1}}{t^2-1}, \tag{12.10}$$

$$g(n,1,-d_1/2;n',1,+d_1/2)=\frac{S_2S_3}{F_2S_3+F_3S_2}\frac{t^{|n-n'|+1}}{t^2-1}+\frac{S_1}{F_1}\frac{t^{|n-n'-1|+1}}{t^2-1}, \tag{12.11}$$

where the integers n and n' refer to the cell numbers ($-\infty < n, n' < +\infty$) and

$$Y_1=\frac{C_1S_2S_3+\frac{S_1}{F_1}(F_2C_2S_3+F_3C_3S_2)}{F_2S_3+F_3S_2}. \tag{12.12}$$

The parameter t is defined by

$$t=e^{jkd}, \tag{12.13}$$

and d is the period of the structure.

The dispersion relation of the infinite serial loop waveguide is obtained from Eq. (12.2). The stopbands are determined by the condition $|\xi| > 1$, whereas the allowed bands are given by the condition $|\xi| < 1$, where

$$\xi=\cos(kd)=\frac{1}{F_2S_3+F_3S_2}\left\{S_1S_2S_3\left(\frac{F_1^2+F_2^2+F_3^2}{2F_1}\right)+F_2C_1C_2S_3+\right.$$
$$\left.+F_3C_1C_3S_2+\frac{F_2F_3}{F_1}S_1(C_2C_3-1)\right\}. \tag{12.14}$$

The bandgaps of the structure may be also obtained from the transmission coefficient through a finite-size serial loop structure. The finite SLS containing N equidistant loops is cut out of the infinite periodic system illustrated in Fig. 12.1(a), and this piece is subsequently connected at its extremities $((1,1,+d_1/2),(N+1,1,-d_1/2))$ to two semiinfinite leading lines (Fig. 12.1(b)). An incident bulk electromagnetic wave $e^{-\alpha_s x}$ of unit amplitude launched from $x=-\infty$ is scattered by the interfaces between the dissimilar wires constituting the system. The transmission function is then obtained with the help of Eqs. (12.3), (12.4), and (12.7) as [10]

$$t_r=\frac{2F_sB(N)}{[A(N)-F_s]^2-B^2(N)}, \tag{12.15}$$

where $A(N)$ and $B(N)$ are given by

$$A(N)=\frac{Y_1Y_3-t^{2N-2}Y_2Y_4}{Y_1^2-t^{2N-2}Y_2^2} \tag{12.16}$$

and

$$B(N)=\frac{Y_1Y_4-Y_2Y_3}{Y_1^2-t^{2N-2}Y_2^2}t^{N-1}, \tag{12.17}$$

where

$$Y_2 = \frac{S_2 S_3}{F_2 S_3 + F_3 S_2} t + \frac{S_1}{F_1}, \tag{12.18}$$

$$Y_3 = -\frac{1}{t} + \frac{F_2 C_1 C_2 S_3 + F_3 C_1 C_3 S_2 + F_1 S_1 S_2 S_3}{F_2 S_3 + F_3 S_2}, \tag{12.19}$$

and

$$Y_4 = C_1 - \frac{F_2 C_2 S_3 + F_3 C_3 S_2}{F_2 S_3 + F_3 S_2} t. \tag{12.20}$$

The transmission function can be written in an explicit complex form as

$$t_r = a + jb = \sqrt{T} e^{j\varphi} \tag{12.21}$$

where T is the transmission coefficient, $\varphi = \arctan(b/a) \pm l\pi$ is the total phase associated with the transmission field, and l is an integer. The first derivative of φ with respect to the frequency is indicative of the delay time taken by the wave to go through the structure. This quantity, called phase time, is defined by [10,56–60]

$$\tau = \frac{d\varphi}{d\omega}. \tag{12.22}$$

On the other hand, a defect may be created in the SLS by replacing a segment d_1 in one cell ($n = p$) by a segment of different length $d_0 \neq d_1$ (Fig. 12.1(c)). Within the Green's function formalism, one can calculate analytically the corresponding localized state frequencies. These frequencies satisfy the equation

$$\left\{1 + \frac{t}{t^2 - 1}(Y_1 - Y_5)[-F_0(C_0 + 1)/S_0 + F_1(C_1 + 1)/S_1]\right\} \times \left\{1 + \frac{t}{t^2 - 1}(Y_1 + Y_5)[-F_0(C_0 - 1)/S_0 + F_1(C_1 - 1)/S_1]\right\} = 0, \tag{12.23}$$

where

$$Y_5 = \frac{S_2 S_3}{F_2 S_3 + F_3 S_2} + \frac{S_1}{F_1} t. \tag{12.24}$$

The transmission function through an SLS composed of N cells and containing a defect located in the cell p ($1 < p \leq N$) can be obtained as

$$t_r' = \frac{2F_s(F_0/S_0)B(N)B(p)}{Y(N)Y(p) - (F_0/S_0)^2(A(N) - F_s)(A(p) - F_s)}, \tag{12.25}$$

where $Y(q)$ is given by

$$Y(q) = B^2(q) - [A(q) - F_1]\left[A(q) - \frac{F_0 C_0}{S_0}\right], \tag{12.26}$$

with $q \equiv N$ or p. The other parameters in Eq. (12.26) are given earlier.

From the previous general equations, one can deduce two particular cases corresponding to specific lengths of the loop:

(i) The *tangent* SLS which corresponds to $d_2 \neq 0$ and $d_3 = 0$. In this case, ξ (Eq. (12.14)) becomes

$$\xi = C_1 + \frac{F_2}{F_1}\frac{S_1 S_2'}{C_2'}, \tag{12.27}$$

where $S_2' = \sinh(\alpha_2 d_2/2)$ and $C_2' = \cosh(\alpha_2 d_2/2)$. This relation is equivalent to the dispersion relation of a *star waveguide* along which two resonators of length $d_2/2$ are grafted periodically. This system was largely investigated both theoretically and experimentally (see Chapter 11) [7,8]. The forbidden bands in the dispersion curves result from the periodicity of the structure and the zeros of transmission due to the resonators.

(ii) A symmetric SLS characterized by identical arms of the loop (i.e., $d_2 = d_3$). In this case, the dispersion relation becomes

$$\xi = C_1 C_2 + \frac{1}{2}\left(\frac{F_1}{2F_2} + \frac{2F_2}{F_1}\right) S_1 S_2. \tag{12.28}$$

This relation is analogous to the dispersion relation of a PC constituted out of two different layers [61,62]. The two layers should be of lengths d_1 and d_2 and are characterized by their impedances F_1 and $2F_2$, respectively. The bandgaps result only from the periodicity of the system. Several theoretical studies have dealt in the past with electromagnetic waves in superlattices with defects [63–66]. The experimental evidence of the existence of defect modes in 1D photonic crystals constructed by connecting segments of coaxial cables of different characteristic impedances was presented by Schneider et al. [12]. However, in all these studies the contrast between the dielectric constants in the two constituting layers is a critical parameter for the stopband to exist. In this chapter, we show that symmetric SLS may exhibit large gaps and defect modes even if the loops and the segments are made of the same materials which could be of potential interest in optical waveguide structures.

In the next section, we shall show the origin of the gaps in SLSs and the evolution of the bandgap structure when the lengths of the wires constituting the SLS vary between the two limiting cases (i) and (ii) just mentioned.

12.2.4 Numerical and experimental results

12.2.4.1 A single loop

Before illustrating these analytical results by numerical calculations confirmed experimentally with coaxial cables, let us consider the elementary case of a waveguide consisting of a single loop. From Eq. (12.15), one can derive the analytical expression of the transmission coefficient of a single homogeneous loop by setting $N = 1$. For the sake of simplicity and comparison with coaxial cable experiments, we have limited ourselves to the case where the media constituting the SLS are made of identical materials (i.e., $\varepsilon = \varepsilon_1 = \varepsilon_2 = \varepsilon_3 = \varepsilon_4$). The transmission and reflection amplitudes can be written respectively as

$$t = \frac{4\sin[\alpha'(d_3+d_2)/2]\cos[\alpha'(d_3-d_2)/2]}{4\sin^2[\alpha'(d_3+d_2)/2]+\sin(\alpha' d_3)\sin(\alpha' d_2)+j4\sin[\alpha'(d_3+d_2)/2]\cos[\alpha'(d_3-d_2)/2]}, \tag{12.29}$$

$$r = -\frac{4\sin^2[\alpha'(d_3+d_2)/2]+\sin(\alpha' d_3)\sin(\alpha' d_2)}{4\sin^2[\alpha'(d_3+d_2)/2]+\sin(\alpha' d_3)\sin(\alpha' d_2)+j4\sin[\alpha'(d_3+d_2)/2]\cos[\alpha'(d_3-d_2)/2]}, \tag{12.30}$$

where $\alpha' = \frac{\omega\sqrt{\varepsilon}}{c}$. Then, the transmission and reflection coefficients are given respectively by

$$T = |t|^2, \tag{12.31}$$

$$R = |r|^2. \tag{12.32}$$

One can see that in absence of dissipation, we can check easily that $R + T = 1$. However, in the presence of loss, then the absorption coefficient can be deduced from the equation $A = 1 - R - T$.

The transmission (Eq. (12.29)) coefficient is equal to zero when $\sin(\alpha' L/2) \times \cos(\alpha' \Delta L/2) = 0$ where $\Delta L = d_3 - d_2$ is the difference between the lengths of the arms of the loop and $L = d_2 + d_3$ is the total length of the loop. The zero transmission frequencies are thus given by

$$f_m = \frac{c}{\sqrt{\varepsilon}}(m + \frac{1}{2})\frac{1}{\Delta L} \tag{12.33}$$

and

$$f_{m'} = \frac{c}{\sqrt{\varepsilon}}\frac{m'}{L}, \tag{12.34}$$

where m and m' are integers. It is worth noticing that for frequencies given by (Eq. (12.33)), the waves traveling on both paths of the loop are out of phase and can interfere destructively, similarly to a Mach–Zehnder interferometer. On the other hand, the frequencies given by (Eq. (12.34)) correspond to the eigenmodes of a loop alone. In the particular case of a symmetric loop ($d_2 = d_3$, $\Delta L = 0$), the transmission coefficient becomes

$$T = \frac{16}{25 - 9\cos^2(\alpha' d_2)}. \tag{12.35}$$

In contrast to the case of an asymmetric single loop, the transmission through a symmetric loop never becomes zero.

A great deal of theoretical work has been devoted recently [10,56,67–69] to the understanding of the relation between the phase of the transmission and the density of states depending on whether the system exhibits or not zeros of transmission. Some of these researches tried to explain by continuum [67,68] and lattice [69] scattering models the phase jumps observed in few experiments realized on mesoscopic systems [15,16]. Here, we shall show theoretically and verify experimentally, through a simple model the existence of phase jump in the transmission of a single loop as far as the constituting wires of the loop have different lengths. In the particular case when the wires become identical, the phase increases monotonically as a function of the frequency. These results are summarized in Fig. 12.2 where we have plotted the amplitude (Fig. 12.2(a),(d)), the phase (Fig. 12.2(b),(e)), and the phase time (Fig. 12.2(c),(f)) for symmetric (right panel) and asymmetric (left panel) loops. The lengths of the wires of the asymmetric (symmetric) loop are $d_2 = 0.355$ m and $d_3 = 0.655$ m ($d_2 = d_3 = 0.5$ m), whereas the dielectric permittivity is taken to be $\epsilon = 2.3$. The solid curves represent the theoretical results whereas the open circles correspond to the experimental ones. The experiments were performed using standard 50 Ω coaxial cables assembled together with metallic T-shaped connectors. The cross-section of the cables is negligible compared to their length and to the propagation wavelength, so that the assumption of monomode propagation is satisfied. The transmission measurements have been realized by using a tracking generator coupled to a spectrum analyzer in the frequency range of 10–400 MHz. The attenuation inside the coaxial cables was simulated by introducing a complex relative dielectric permittivity ($\varepsilon = \varepsilon' - j\varepsilon''$). The imaginary part ε'' was obtained from the attenuation specification data supplied by the manufacturer of the coaxial cables as a function of frequency, i.e., $\varepsilon'' = 0.017 f^{-0.5}$ with f in hertz. Two successive minima are seen in the transmission through an asymmetric loop (Fig. 12.2(a)). The first, obtained for the frequency 193.5 MHz, corresponds to $m' = 1$ in Eq. (12.34). The second minimum at 340.3 MHz is associated with $m = 0$ in Eq. (12.33). The transmission zeros give rise to abrupt phase change by π in the phase of the transmission coefficient (Fig. 12.2(b)) or to delta peaks in the transmission phase time (Fig. 12.2(c)). It is worth noticing that the phase jump is slightly enlarged and becomes less than π because of the dissipation in the cables. These results agree with experiments of Yacoby et al. [15] and Schuster et al. [16] in which the conductance through a quantum dot shows a phase change by π between each pair of adjacent in-phase resonances. In contrast to Fig. 12.2(a), Fig. 12.2(d) clearly shows that the transmission through a symmetric loop does not reach zero and therefore the corresponding phase (Fig. 12.2(e)) increases monotonically. These results clearly show, in accordance with the theoretical works [56,67–69],

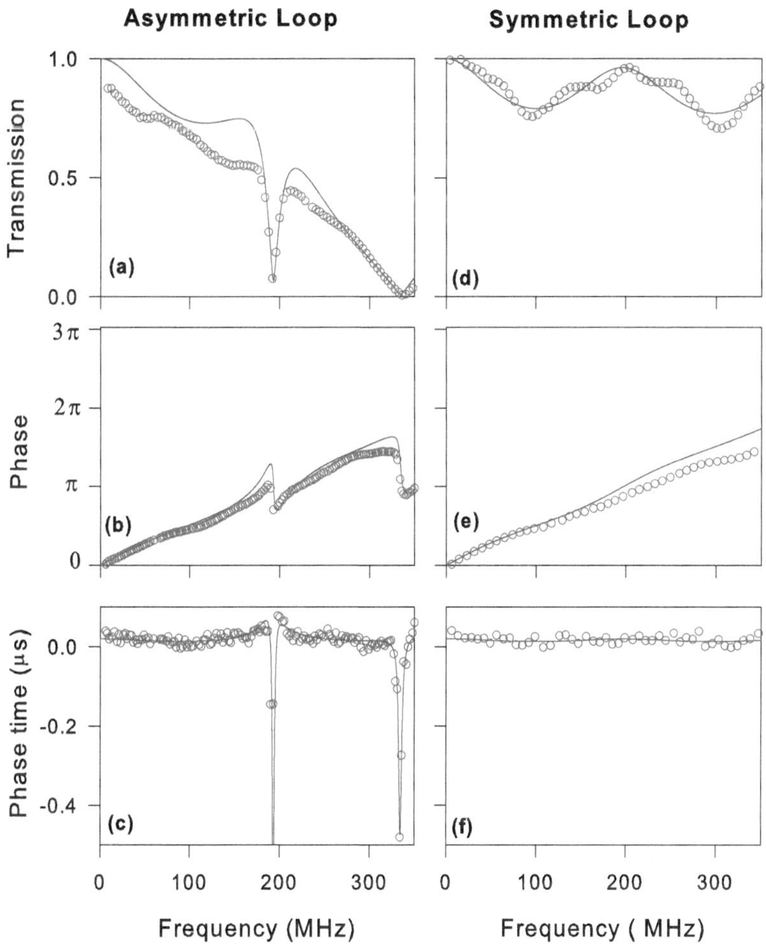

Figure 12.2 Transmission coefficient (a), (d), phase (b), (e), and phase time (c), (f) as a function of the frequency for an asymmetric single loop ($d_2 = 0.36$ m and $d_3 = 0.65$ m) and a symmetric single loop ($d_2 = d_3 = 0.5$ m). The blue lines (dark gray in print version) are the theoretical results, whereas the open circles represent experimental data.

that the phase time measurements in a symmetric loop reflect the DOS in this system, whereas in an asymmetric loop these two quantities are different because of the transmission zeroes.

An interesting result that can be deduced from the phase time τ (Figs. 12.2(c) and 12.2(f)) is the group velocity [66] $v_g = \ell/\tau$, where ℓ is the total length of the finite structure. In contrast to a symmetric loop, an asymmetric loop may give rise to negative group velocity around the transmission zeroes because of the negative peaks in the phase time spectrum. In the example of Fig. 12.2(c), the value of v_g becomes as small as $-0.05c$ around the peak situated at 340 MHz.

The idea of negative group velocity amounts to several decades with the challenge of realizing pulse advancement comparable to pulse width, with a low level of pulse distortion. Indeed, it is well known that inside an absorption line the refractive index, or equivalently the phase of transmission (Fig. 12.2(b)), may take a steep drop [70], resulting in an anomalous dispersion, and consequently a light pulse propagation at a group velocity faster than c or even negative [71,72] in absorbing, homogeneous dielectric materials. Also, negative group velocity has been found both theoretically and experimentally in media made of atomic vapor cells in the presence of gain [17,18,73]. In all these works, it was clearly pointed out that such superluminal behavior is not at odds with either causality or Einstein's theory of special relativity, but it exclusively results from interference between the different frequency components of the pulse in an anomalous dispersion region [19]. In addition, as argued by several authors [18,74–76], it is not the group velocity, but rather the front velocity that must not exceed c. In circuits, a discussion of negative group velocity in a simple bandpass filter [77] clearly demonstrated negative group delays. However, for a filter, the concept of group velocity is not defined. Recently, Munday and Robertson [78] showed experimentally negative group velocity in 1D photonic circuit made of alternating two impedance-different coaxial cables. However, from the theoretical point of view [56,79], it has been shown that for lossless, 1D filter, negative phase time [56], or equivalently negative group velocity [79], are not possible. These results let the authors suppose that the dissipation inside the coaxial cables may be at the origin of the observed negative group velocity. However, other experiments on coaxial cables made of alternating different cables [13] or SLS [14] do not show such negative group velocity. On the other hand, the results presented in Fig. 12.2(c) clearly show both theoretically and experimentally that a single asymmetric loop made of coaxial cables exhibits negative phase time τ around the transmission zeros of the structure. Therefore, we believe that this structure could be a good candidate for obtaining negative group velocity. Indeed, similar structures similar to that presented here have been studied both in photonic [80] and phononic [81] circuits as a Mach–Zehnder interferometer to observe superluminal and negative group velocities.

The loop structure can exhibit an interesting mode called bound in continuum (BIC) state [82,83] when the latter is inserted between two semiinfinite wires. These states, also called trapped modes, are defined as bound states confined in some parts of the system even though they coexist with a continuous spectral range of radiating waves that can transport energy away. This is a counterintuitive idea which was originally proposed several decades before in quantum mechanics [82], but found considerable interest in the recent literature [83]. They have been shown to exhibit potential applications as they are characterized by a high quality factor when they are revealed by a small amount of perturbation in the structure. In the case of the loop, this typical mode is one of its eigenmodes (see below) which do not interact with the continuum modes of the wires at its both ends and remains a stationary mode in its circumference. In order to show the possibility of existence of such modes, we start with the dispersion relation of the open system. This quantity can be obtained from vanishing of the denominator of the Green's function, or equivalently the transmission coefficient

(Eq. (12.29)), namely

$$\sin^2[\alpha'(d_3 + d_2)/2] + \sin(\alpha' d_3)\sin(\alpha' d_2) + j4\sin[\alpha'(d_3 + d_2)/2]\cos[\alpha'(d_3 - d_2)/2] = 0. \tag{12.36}$$

In general, it is not obvious that both the real and imaginary parts in Eq. (12.36) can vanish together, however, if both arms of the loop are commensurate (i.e., $md_2 = nd_3$, m and n are integers different from zero), then Eq. (12.36) can be factorized by a real quantity $\sin(\alpha' L/(m+n))$ or $\sin(\alpha' L/2(m+n))$ depending on whether $n+m$ is even or odd, respectively. Therefore, the BIC modes are obtained from vanishing of the two latter quantities. For example, in the particular case where $d_2 = d_3$ (i.e., $m = n = 1$), then Eq. (12.36) becomes

$$\sin(\alpha' L/2)\left[5\sin(\alpha' L/2) + j4\cos(\alpha' \Delta L/2)\right] = 0, \tag{12.37}$$

therefore, the BIC modes are given by

$$\sin(\alpha' L/2) = 0. \tag{12.38}$$

The imaginary part in Eq. (12.37) is due to the interaction of the loop structure with the semiinfinite wires at its both extremities. One can see that the eigenmodes of the loop alone are given by $\sin(\alpha' L/2)^2 = 0$ which corresponds to two degenerate modes (clockwise and counterclockwise) of the loop. Therefore, Eq. (12.37) shows that one of the two degenerate modes becomes a BIC state, i.e., a stationary state in the loop without interaction with the continuum modes of the surrounding wires, whereas the other mode interacts strongly giving rise to large Fabry–Perot resonances. In order to get a signature about the existence of such BIC modes, we have to detune slightly the lengths of the two arms by a small amount in order to render the BIC mode, a quasi-BIC mode. These results are shown in Fig. 12.3 where we have plotted the transmission, reflection, absorption, and DOS versus the dimensionless frequency Ω/π (where $\Omega = \alpha' L/2$) for $d_2 = d_3$ (Fig. 12.3(a), lower panel), $d_2 = 1.08d_3$ (Fig. 12.3(b), middle panel), and $d_2 = 1.14d_3$ (Fig. 12.3(c), upper panel). We can notice that around $\Omega = \pi, 2\pi, \ldots$ the dips in the transmission spectra or the peaks in the reflection and absorption spectra disappear when d_2 tends towards d_3, giving rise to BIC modes. This result is also clearly visible in the DOS spectra where we can notice for $d_2 = d_3$ (Fig. 12.3(a)) a very wide resonance and a hidden BIC mode, while for $d_2 \neq d_3$, the BIC mode is transformed into a resonance (pseudo-BIC), and at the same time the other resonance remains of the Fabry–Perot type.

In the next section, we shall discuss the bandgap structures, as well as the amplitudes and phase times of the transmission, when several SLS are connected together periodically by finite segments with or without a defect [14,84].

12.2.4.2 Band structure of SLS

Now we turn to the analytical and experimental results for the transmission of electromagnetic waves through SLS made of several loops. When the media constituting

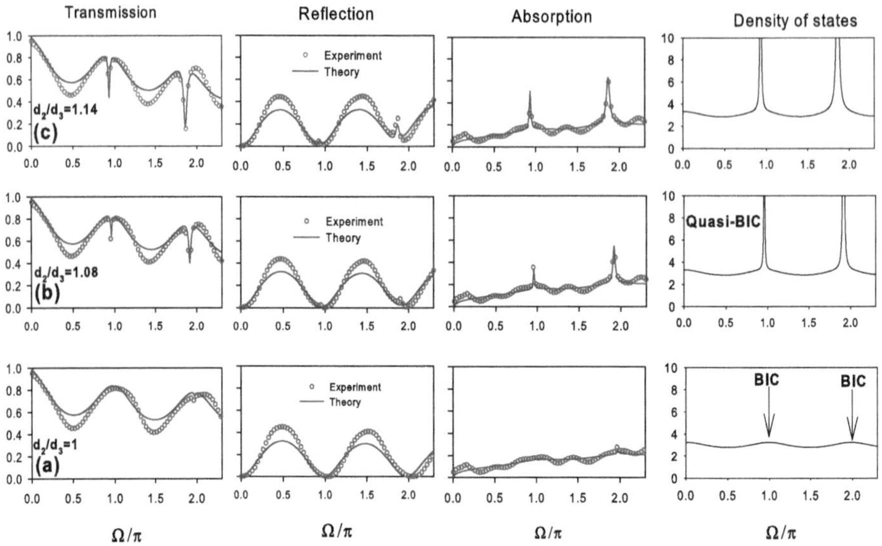

Figure 12.3 Transmission, reflection, absorption, and DOS versus the dimensionless frequency Ω/π (where $\Omega = \alpha' L/2$) for a loop with the lengths of its arms such that $d_2 = d_3$ (a) (lower panel), $d_2 = 1.08d_3$ (b) (middle panel), and $d_2 = 1.14d_3$ (c) (upper panel). The blue lines (dark gray in print version) are the theoretical results, whereas the open circles represent experimental data.

the SLS are made of identical materials, i.e., $\varepsilon_1 = \varepsilon_2 = \varepsilon_3 = \varepsilon_4 = \varepsilon$, the dispersion relation (Eq. (12.14)) can be rewritten in the following simplified form:

$$\xi = \frac{1}{2\sin(\frac{\alpha' L}{2})\cos(\frac{\alpha' \Delta L}{2})} \times \left\{ \sin(\alpha' d_1)\left[\frac{5}{4}\cos(\alpha' L) - \frac{1}{4}\cos(\alpha' \Delta L) - 1\right] + \cos(\alpha' d_1)\sin(\alpha' L) \right\}, \tag{12.39}$$

where $\Delta L = d_2 - d_3$, $L = d_2 + d_3$, and $\alpha' = \frac{\omega\sqrt{\varepsilon}}{c}$.

Fig. 12.4 displays the projected band structure, i.e., the frequency (in MHz) versus ΔL of an infinite SLS for given values of L and d_1, namely $L = 1.06$ m and $d_1 = 0.5$ m. For each value of ΔL, the shaded areas correspond to frequencies for which $|\xi| < 1$ and represent bulk bands. These areas are separated by minigaps. Inside these gaps, the dashed lines show the frequencies for which the denominator of ξ (Eq. (12.39)) vanishes. The dashed horizontal and curved lines correspond to the vanishing of $\sin(\alpha' L/2)$ and $\cos(\alpha' \Delta L/2)$, respectively [85]. The vanishing of $\cos(\alpha' \Delta L/2)$ defines the frequencies at which the transmission through a single loop is equal to zero. The existence and width of the minigaps are influenced by both the periodicity of the structure and the zeros of transmission. Let us notice that some of the minigaps having a lozenge pattern display a large variation with ΔL. In the latter minigaps, one can observe the existence of very narrow (almost flat) minibands. These

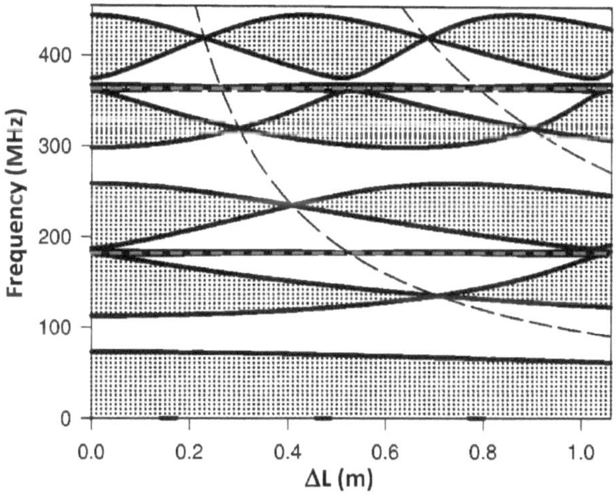

Figure 12.4 Projected band structure of the SLS as function of $\Delta L = d_2 - d_3$ for $d_1 = 0.5$ m and $L = d_2 + d_3 = 1.06$ m. The shaded areas represent the bulk bands. The dashed curves indicate the frequencies for which the denominator of ξ (Eq. (12.39)) vanishes.

minibands become totally flat (and coincide with the horizontal dashed lines) if L is taken to be equal to 1 m instead of 1.06 m. Indeed, for $L = 1$ m, we have $L = 2d_1$, and one can easily check that both the numerator and denominator of Eq. (12.39) vanish simultaneously at the frequencies represented by the dashed lines. The physical meaning of such flat bands is that the infinite structure possesses loop confined states (i.e., BIC modes) at these frequencies, while the transmission through the finite structure remains equal to zero, when the corresponding wavelength and d_1 are incommensurate. One also observes in Fig. 12.4 that, at certain values of ΔL (for instance, $\Delta L \simeq 0.25$ m), one can obtain a series of narrow minibands separated by large gaps. This is due to the fact that the points where the minibands close, align more or less vertically in such a way that a few successive bands may become very narrow.

In what follows, we shall detail the results concerning the case of symmetric SLS ($d_2 = d_3$) and then give some results about the case of an asymmetric SLS ($d_2 \neq d_3$). We also show at the end of this section that the width of the band gaps may be enlarged by coupling several SLS of different physical characteristics.

12.2.4.3 Transmission through symmetric SLS without a defect

First, we analyze the properties of the band structure and the transmission of a perfect SLS (without defect). Fig. 12.5(a) shows the first five dispersion curves (full lines) in the band structure of an infinite serial loop waveguide ($N \to \infty$) with the same characteristic lengths ($d_1 = d_2 = d_3 = 0.5$ m). One can observe a gap after the first passband and another gap between the third and the fourth bands. The gaps between the second and the third bands and between the fourth and the fifth bands (situated

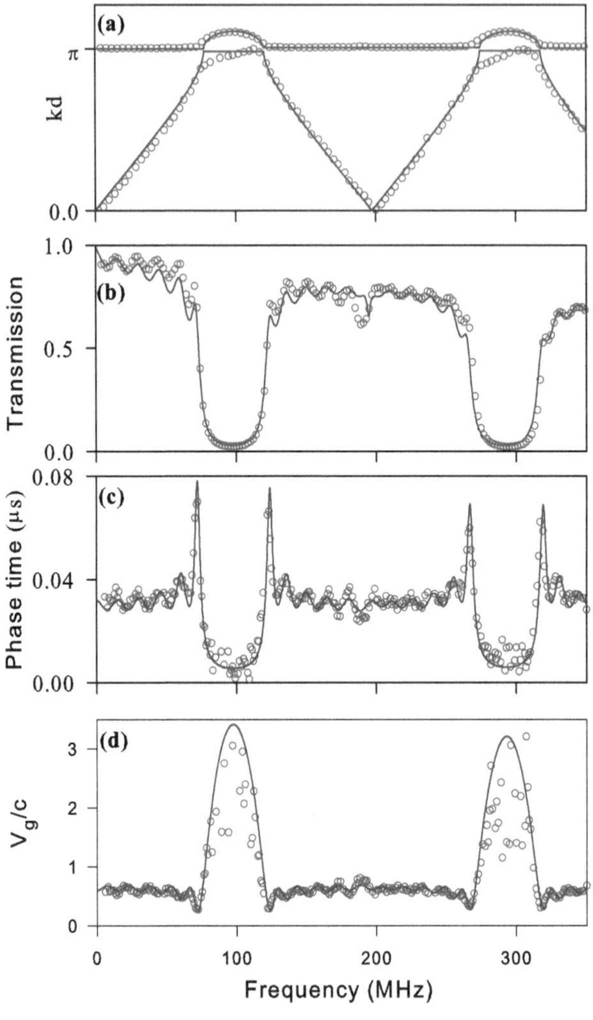

Figure 12.5 (a) Theoretical band structure of the infinite symmetric SLS with $d_1 = d_2 = d_3 = 0.5$ m and $\varepsilon = 2.3$. The experimental curve (open circles) is obtained from the amplitude (b) and the phase (c) of the transmission through a finite-size SLS. (b) Theoretical (solid line) and experimental (open circles) variations of the transmission coefficient through a finite SLS without defect and containing $N = 6$ loops with the same characteristic lengths as in (a). (c) Theoretical (solid line) and experimental (dots) variations of the delay time in the same conditions as in (b). Theoretical (solid line) and experimental (open circles) variations of the group velocity in the same conditions as in (b) and (c).

at $kd = 0$) close. The real and imaginary parts of the reduced wave vector kd are in very good agreement with the experimental results (dotted curves) obtained from the amplitude (Fig. 12.5(b)) and the phase (Fig. 12.5(c)) of the transmission through a finite-size SLS. Indeed, writing the transmission coefficient $t_r = |t|e^{j\Phi}$ in the form $t_r = e^{jk\ell}$, where ℓ is the total length of the finite structure, one obtains the effective wave number

$$k = \frac{\Phi}{\ell} - j\frac{\ln|t|}{\ell}. \tag{12.40}$$

One can notice that despite the finite number of serial loops ($N = 6$) used in the finite structure, the amplitude and the phase describe very well the bandgap structure of the infinite system $N \to \infty$ (Fig. 12.5(a)). The attenuation effect in coaxial cables induces a transmission depletion particularly at high frequencies. As mentioned earlier, the phase time in symmetric SLS is equivalent to the density of states (DOS) [56,67,68], therefore Fig. 12.5(c) is also an illustration of the DOS in these 1D photonic crystals. A large enhancement of the DOS is observed in the vicinity of the band edges. The number of oscillations in each band (that is related to the number of loops in the finite structure) gives the number of states. The behavior of the group velocity $v_g = L/\tau$ [66] is given in Fig. 12.5(d). For 1D PBG, v_g is equivalent to the inverse of DOS. It is well known that in infinite 1D systems, the density of modes approaches infinity at the band edge and the group velocity becomes very small [10]. In a finite system, however, the electromagnetic mode density is an oscillating function rather than a monotonous function (Fig. 12.5(c)). The enhancement of the phase time (or the DOS) at the band edges induces a small group velocity (Fig. 12.5(d)) which has been shown to be of potential interest in application to band-edge lasing and optical delay lines [86]. In addition, one can notice inside the passbands, the group velocity is equal to $0.66c$. However, inside the gaps, anomalous dispersion occurs and superluminal velocities are expected such that $3c < v_g < 3.5c$. These results are in accordance with those of Haché and Poirier [13]. The system studied here presents an advantage: all the segments in the periodic structure are made of the same material instead of being constructed by two different materials. Finally, let us mention a comprehensive analysis of the comparison between density of states and the scattering parameters in coaxial photonic crystals, is provided by Khattou et al. [10].

12.2.4.4 Transmission through symmetric SLS with a defect

Now, we study the transmission spectrum and the phase time through the structure when a segment of length $d_1 = 0.5$ m is replaced by a defect branch of length $d_0 = 1$ m in the middle of the SLS waveguide. The results are presented in Fig. 12.6 for three different values of N. First, one can notice that by increasing N, the transmission decreases more rapidly to zero at the band edges (Fig. 12.6(c)). The effect of the defect branch is to induce defect (or localized) modes in the forbidden bands. The frequencies of these modes are almost independent of N, whereas their intensities decrease progressively with increasing N (Fig. 12.6). This decrease of the intensities may appear to be in contradiction with the fact that, by increasing N, the localization degree of these modes around the defect increases and therefore the peaks associated with the defect modes become narrower. However, this behavior can be explained by the dissipation inside the coaxial cables that lead to a widening of the peaks due to the absorption phenomenon, but also to a decrease in the transmitted intensity due to an enhancement of the reflected intensity.

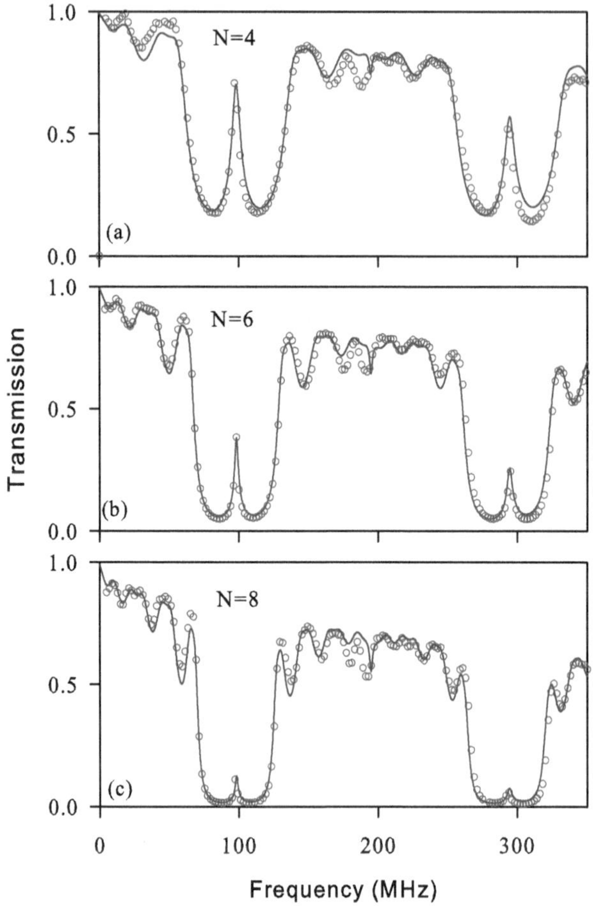

Figure 12.6 Theoretical (solid line) and experimental (open circles) variations of the transmission coefficient through a finite SLS of N loops containing a defect of length $d_0 = 1$ m located in the middle of the SLS (cell p) ($d_1 = d_2 = d_3 = 0.5$ m and $\varepsilon = 2.3$). (a) $N = 4$ and $p = 3$, (b) $N = 6$ and $p = 4$, (c) $N = 8$ and $p = 5$.

The evolution of the phase time versus the frequency is depicted in Fig. 12.7 for the same values of N as in Fig. 12.6. In contrast to the amplitude of the localized states in the transmission spectra of Fig. 12.6, the phase time associated with the defect modes increases with increasing N. Indeed, the intensity of the peaks in the phase time is related to the lifetime of the resonances and it reflects the time spent by the photon inside the defect (cavity) before its transmission [87]. Therefore, when N increases, the defect modes become more localized and the trapping time of photons increases. These results indicate that the characterization of localized modes in SLSs, in which the attenuation is not negligible, is easier with the phase time analysis than with the amplitude measurements. To give a better insight into the localization properties of photons inside the defect segment, Fig. 12.8 displays the phase time of the defect

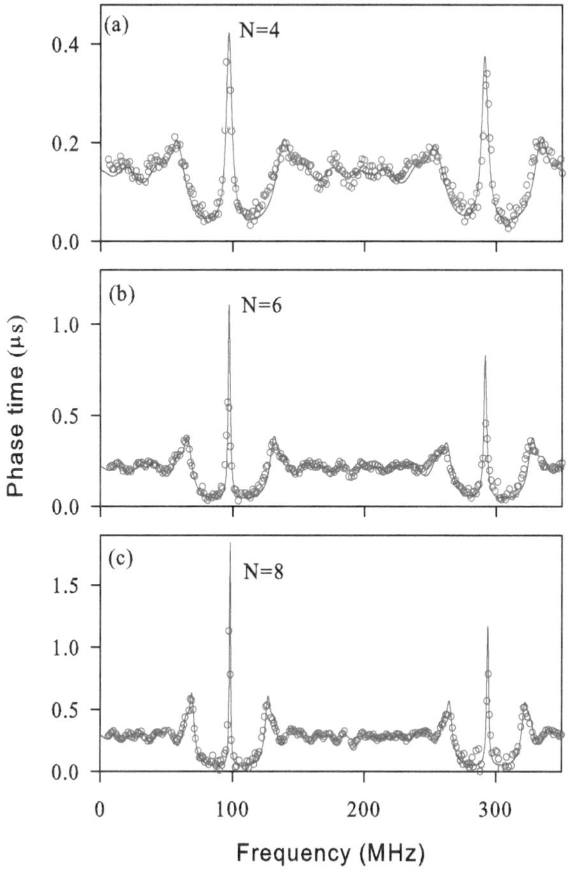

Figure 12.7 Theoretical (solid line) and experimental (open circles) variations of the delay time for a finite SLS of N loops containing a defect of length $d_0 = 1$ m located in the middle of the SLS (cell p) ($d_1 = d_2 = d_3 = 0.5$ m and $\varepsilon = 2.3$). (a) $N = 4$ and $p = 3$, (b) $N = 6$ and $p = 4$, (c) $N = 8$ and $p = 5$.

mode lying in the first gap of Fig. 12.7 for different values of N. One can notice that the lifetime of the resonances associated with defect modes increases from 0.42 to 1.7 μs when N increases from $N = 4$ to $N = 8$ (Fig. 12.8(a)). The integration of the phase time (Fig. 12.8(b)) shows a monotonic rise by almost π (one state) near the maximum of the resonance.

Finally, an analysis of the local DOS (Fig. 12.8(e), (f)) as a function of the space position x clearly shows that the defect mode is localized in the defect segment and rapidly decays in the vicinity of the defect. However, the localization is more pronounced when the SLS is larger. Let us mention that the local DOS represents the square modulus of the electric field in the structure. These results indicate that most of the time spent by the photon during the transmission process was inside the defect segment. Indeed, once a photon enters the defect region, it encounters two Bragg reflectors (the periodic parts) around it. Since the frequency of the photon is in the gap

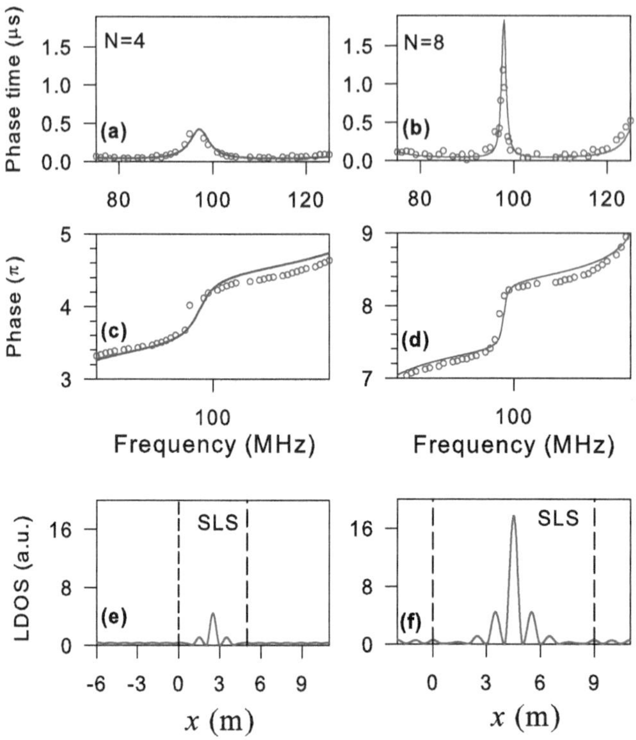

Figure 12.8 Theoretical (solid line) and experimental (open circles) variations of the phase time (a), (b) and the phase (c), (d) of the localized mode in the first gap of Fig. 12.7(a), (c). Variation of the local density of states (DOS) as function of the space position for the localized modes around 98 MHz in (a) and (b).

of the Bragg reflectors, the photon will be strongly reflected back to the defect region. The defect region acts like a Fabry–Perot cavity and the photon cannot escape from this region. The long time spent in the defect region results in very high energy density around the defect [66].

When the defect is inserted far from the middle of the SLS, the amplitude of the localized modes decreases progressively. This effect is depicted in Fig. 12.9 for $N = 6$. Indeed, the amplitude of the defect mode decreases when the defect lies between the second and the third loops (see Fig. 12.9(b)), and practically vanishes when the defect branch is inserted between the first and the second loops (see Fig. 12.9(c)). One can notice the same behavior when the defect is located between the fourth and the fifth loops and between the fifth and the sixth loops, respectively, because of the perfect symmetry of the SLS. These behaviors may be explained qualitatively as follows: when the defect is inserted in the middle of the SLS, the SLS behaves as two identical waveguides with $N/2$ loops connected with a defect. Because of the symmetry of the system, constructive interferences occur that lead to the enhancement of the amplitude of the transmission. However, when the defect branch lies far from the middle of the SLS, the SLS behaves as two linked waveguides with different number of loops (2, 4

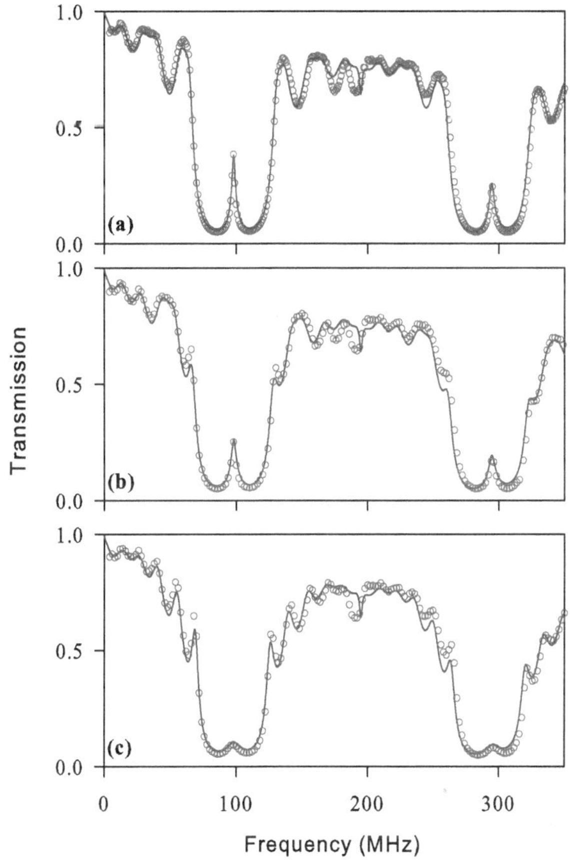

Figure 12.9 Theoretical (solid line) and experimental (open circles) variations of the transmission coefficient through a finite SLS of 6 loops containing a defect of length $d_0 = 1$ m located in the cell p ($d_1 = d_2 = d_3 = 0.5$ m and $\varepsilon = 2.3$): (a) $p = 4$, (b) $p = 3$, and (c) $p = 2$.

or 1, 5). Each of the linked waveguides contributes in its own way to the transmission of the SLS. Thus destructive effects are responsible for the decrease in amplitude when the defect is moved away from the middle of the SLS. This qualitative interpretation agrees with the evolution of the phase time corresponding to defect modes which decreases also when the defect is inserted far from the middle of the SLS (Fig. 12.10).

On the other hand, the transmission inside the passbands is affected by the presence of the defect. Indeed, the amplitude of the oscillations in the passbands is greater in the perturbed SLS (Figs. 12.6 and 12.9) than in the perfect one (Fig. 12.5) because the defect mode is detached from the bulk band (Fig. 12.11(a)). This behavior is particularly important when the defect is located far from the middle of the SLS (Figs. 12.9(b)–(c)). In these cases, the amplitude of the oscillations in the pass bands exhibits important variations due to destructive effects.

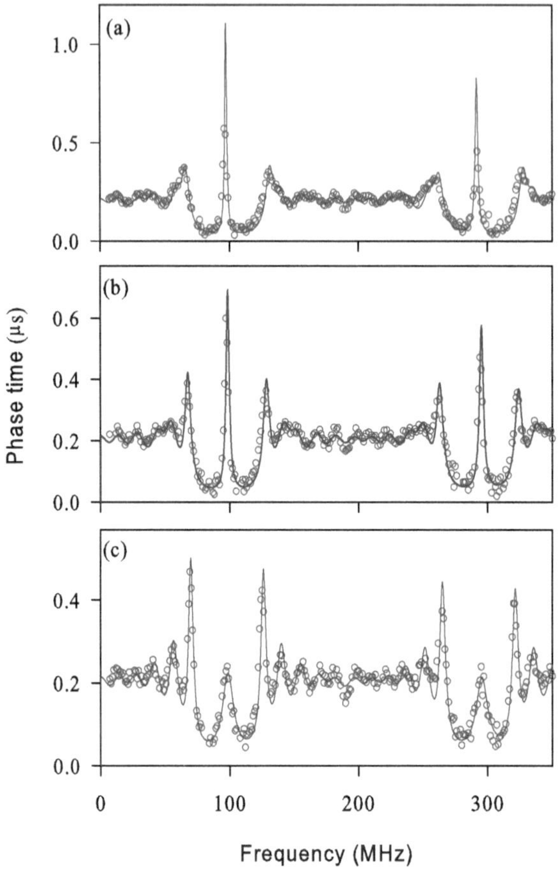

Figure 12.10 Theoretical (solid line) and experimental (open circles) variations of the delay time for a finite SLS of 6 loops containing a defect of length $d_0 = 1$ m located in the cell p ($d_1 = d_2 = d_3 = 0.5$ m and $\varepsilon = 2.3$): (a) $p = 4$, (b) $p = 3$, and (c) $p = 2$.

A comparison of Figs. 12.6 and 12.9 shows that the frequency of the defect mode does not depend on the defect location in the SLS; however, the frequencies are very sensitive to the defect length d_0 as it is shown in Fig. 12.11(a). The black areas correspond to the bulk bands. The solid lines represent the computed frequencies in an infinite SLS with a defect segment, whereas the open circles correspond to the experimental measurements. The localized modes emerge from the bulk band, decrease in frequency when d_0 increases, and finally merge into a lower bulk band. At each frequency, there is a periodic repetition of the localized states as function of d_0. One notes the overall good agreement between the experimental measurements and the theoretical results.

We have also studied the evolution of the power transmission at the frequency of the defect mode falling in the first gap as a function of the defect length d_0. Fig. 12.11(b) illustrates this evolution in the case of a defect inserted in the middle of an SLS con-

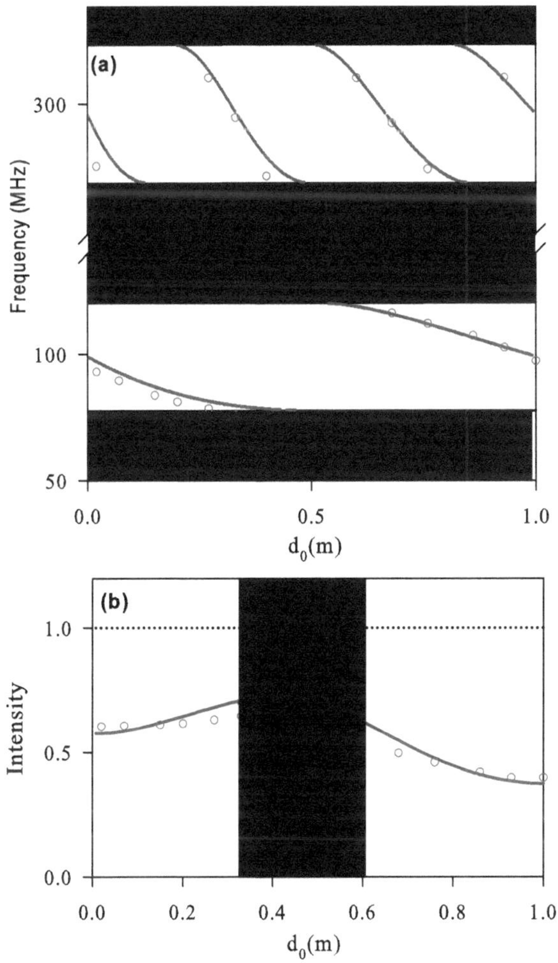

Figure 12.11 (a) Comparison between the theoretical and experimental frequencies of the localized states associated with the presence of a defect branch of length d_0, inserted in the cell p of an SLS: (solid lines) calculated frequencies in an infinite SLS ($N \to \infty$); (dots) measured frequencies in a finite SLS with $N = 6$ and $p = 4$. The other parameters are $d_1 = d_2 = d_3 = 0.5$ m and $\varepsilon = 2.3$. The black areas correspond to the bulk bands. (b) Variation of the intensity of the transmitted localized mode located in the first gap as function of the defect length d_0. The defect is inserted in the middle of an SLS with $N = 6$ and $p = 4$: (solid lines) calculated frequencies when the dissipation is taken into account; (open circles) measured frequencies; (dotted lines) calculated frequencies when the attenuation effect is neglected

structed of $N = 6$ loops. Let us recall that the power transmission is stronger when the defect is inserted closer to the middle of the waveguide (see Figs. 12.9 and 12.10). The black area in Fig. 12.11(b) corresponds to the situation in which the defect mode merges into a bulk band (Fig. 12.11(a)). The horizontal dashed line in Fig. 12.11(b) represents the computed intensity when the dissipation inside the cables is neglected. One can notice that the transmission is unity for all frequencies, this property is commonly verified in any symmetric composite system, whereas it may only happen under special conditions if the composite media is not symmetric. The full line corresponds to the computed intensity when the dissipation is taken into account. The transmission, which is now smaller than unity, increases progressively with the defect length until the localized mode merges with the first bulk band. With increasing d_0, a defect mode emerges from the second bulk band, its intensity decreases and reaches a stationary value. In Fig. 12.11(b), the open circles represent the experimental measurements whose evolution is in good agreement with the computed results.

12.2.4.5 *Transmission through asymmetric SLS*

An example of the dispersion curve in an infinite asymmetric SLS is depicted in Fig. 12.12(a) (full lines) for $\Delta L = d_2 - d_3 = 0.29$ m and $L = 1$ m. These curves are in good agreement with experimental measurements (open circles) obtained from the amplitude (Fig. 12.12(b)) and the phase (not shown) of the transmission through a finite size SLS made of $N = 4$ loops. In comparison with the symmetric SLS (Fig. 12.5(a)), asymmetric SLS may exhibit larger gaps (Figs. 12.12(a) and 12.12(b)). The width of the bandgaps discussed earlier may be enlarged by associating in tandem two or several successive SLSs that differ by their physical characteristics. An example of this association is shown in Fig. 12.12(d) when putting in tandem two SLS whose individual spectra are shown in Figs. 12.12(b) and 12.12(c). The parameters of these two SLS are respectively: (i) an SLS with $\Delta L = 0.29$ m and $L = 1$ m (i.e., $d_2 = 0.65$ m and $d_3 = 0.36$ m) (Fig. 12.12(b)) and (ii) an SLS with $\Delta L = L = 0.5$ m (i.e., $d_2 = 0.5$ m and $d_3 = 0$) (Fig. 12.12(c)). In these two cases $d_1 = 0.5$ m. Fig. 12.12(c) exhibits a large gap at frequencies lower than in Fig. 12.12(b), that is, between 100 and 300 MHz. The small feature appearing in the transmission spectrum around 180 MHz inside the large gap is associated with a flat band. It occurs because the actual length ΔL of the loops is slightly different from 0.5 m and the corresponding bands have a small width instead of being totally flat (see also Fig. 12.4 and the corresponding discussion). Now, by associating in tandem the serial loop structures (Figs. 12.12(b) and 12.12(c)), one obtains (Fig. 12.12(d)) an ultrawide gap in which the transmission vanishes over a large range of frequencies from 80 to 420 MHz. In this structure, the huge gap results from the superposition of the forbidden bands of the individual SLS (Figs. 12.12(b) and 12.12(c)). Theoretical and experimental results are in agreement within the limits of experimental precision.

12.3 Confined and surface modes in SLS

In this section, we demonstrate analytically and experimentally the existence and behavior of two types of modes in finite size one-dimensional coaxial PCs made of N cells with vanishing magnetic field on both sides. We highlight the existence of $N - 1$ confined modes in each band and one mode per gap associated to either one or the other of the two surfaces at both sides of the structure. We show that the latter modes are independent of N.

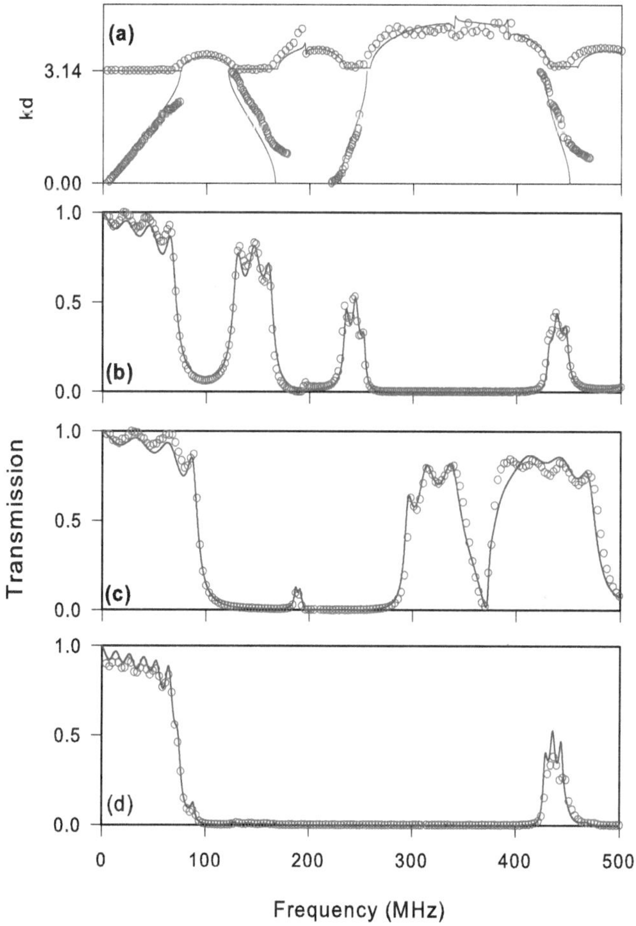

Figure 12.12 (a) Theoretical band structure of the infinite asymmetric SLS with $d_1 = 0.5$ m, $d_2 = 0.65$ m, and $d_3 = 0.36$ m (i.e., $\Delta_L = 0.29$ m). The experimental curve (open circles) is obtained from the amplitude (b) and the phase (not shown) of the transmission through a finite-sized SLS. (b) Theoretical (solid line) and experimental (open circles) variations of the transmission coefficient through a finite SLS containing $N = 4$ loops with the same characteristic lengths as in (a). (c) Same as in (b), but for $L = 0.5$ m and $\Delta_L = 0.5$ m (i.e., $d_2 = 0.5$ m and $d_3 \simeq 0$). (d) Same as in (b) and (c), but for a tandem built of the (b) and (c) structures.

12.3.1 Dispersion relations of infinite, finite, and semiinfinite SLS

Consider an infinite PC made of a periodic repetition of a given 1D cell (Fig. 12.13(d)). In this work, the cell is an SLS system composed of segments and loops (see below). Using the Green's function formalism, each cell is characterized by a 2×2 matrix constituted by the Green's function elements on the surfaces bounding the cell (called 0 and 1 in Fig. 12.13(a)). The boundary conditions on both sides of the cell are $H = 0$

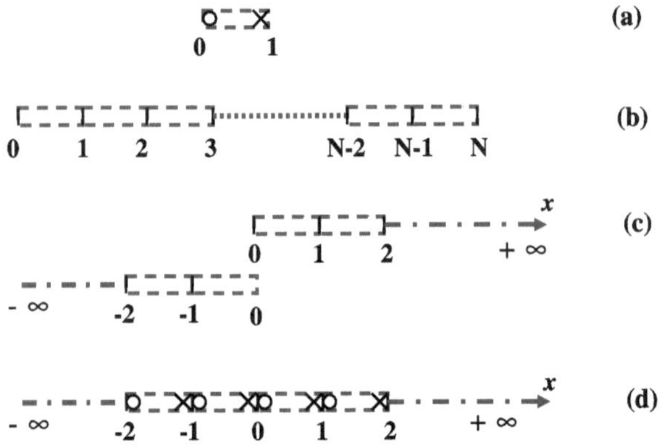

Figure 12.13 Schematic representation of (a) a finite cell bounded by the space of interfaces $M = \{0, 1\}$, the circle and the cross indicate the left and the right surfaces of the cell, respectively. (b) A finite PC constituted of N cells. (c) Two semiinfinite PCs obtained from the cleavage of an infinite PC (d) between two cells. Notice the similarities between the surfaces ending the two complementary PCs in (c) and those corresponding to a finite PC in (a)–(b).

(vanishing magnetic field). The inverse of the 2×2 matrix can be written explicitly as

$$[g(M, M)]^{-1} = \begin{pmatrix} a & b \\ b & c \end{pmatrix}, \tag{12.41}$$

where the space of interface $M = \{0, 1\}$. The four matrix elements are real quantities functions of the different parameters of the constituents of the cell (see below). It is worth to notice that in general $a \neq c$, however, if the cell is symmetrical (i.e., with the same surface terminations) then $a = c$. The eigenmodes of the elementary cell are given by Eq. (12.2), namely

$$ac - b^2 = 0. \tag{12.42}$$

The Green's function of the infinite PC made of a periodic repetition of a given cell (Fig. 12.13(d)) is obtained by a linear juxtaposition of the 2×2 matrices (Eq. (12.41)) in the interface domain of all the sites n. We obtain a tridiagonal matrix where the diagonal and off-diagonal elements of this matrix are given respectively by $a + c$ and b.

Taking advantage of the translational periodicity of this system along the x axis, this matrix can be Fourier transformed as [33]

$$[g(k, M, M)]^{-1} = 2b[\cos(kD) - \xi] \tag{12.43}$$

where k is the modulus of the one-dimensional reciprocal vector (Bloch wavevector), D is the period of the PC, and $\xi = -(a + c)/2b$.

The dispersion relation of the infinite periodic PC (Fig. 12.13(d)) is given by Eqs. (12.2) and (12.43), namely

$$\cos(kD) = -(a + c)/2b. \tag{12.44}$$

On the other hand, in the k space, the surface Green's function is

$$[g(k, M, M)] = \frac{1}{2b[\cos(kD) - \xi]}. \tag{12.45}$$

After inverse Fourier transformation, Eq. (12.45) gives

$$g(n, n') = \frac{1}{b}\frac{t^{|n-n'|+1}}{t^2 - 1}, \tag{12.46}$$

where n and n' denote the positions of the different interfaces between the cells and $t = e^{ikD}$.

Consider now a finite PC bounded by the two surfaces $n = 0$ and $n = N$ (Fig. 12.13(b)) with vanishing magnetic field on both ends. Following the same calculation procedure as in our previous works on acoustic waves in finite PCs [56,88], the 2×2 inverse matrix in the space of interface $M' = \{0, N\}$ of the finite PC can be written as

$$[g(M', M')]^{-1} = \begin{pmatrix} a' & b' \\ b' & c' \end{pmatrix}, \tag{12.47}$$

where

$$a' = a + bt - t^{2N}(a + \frac{b}{t}), \tag{12.48}$$

$$b' = -bt^N(t - \frac{1}{t}), \tag{12.49}$$

and

$$c' = -a - \frac{b}{t} + t^{2N}(a + bt). \tag{12.50}$$

The eigenmodes of the finite PC are given by Eqs. (12.2) and (12.47), namely $a'c' - b'^2 = 0$, or equivalently

$$(t + \frac{b}{a})(t + \frac{a}{b})(1 - t^{2N}) = 0. \tag{12.51}$$

Now, if the finite composite system is sandwiched (grafted) horizontally (vertically) between two homogeneous waveguides characterized by their impedance Z_s (see the insets of Figs. 12.14(c) and 12.14(e)), then an incident plane wave launched from the left waveguide gives rise to the transmission functions in the right waveguide as [42]

$$tr_h = \frac{2jb'/Z_s}{a'c' - b'^2 - (1/Z_s)^2 - j(a' + c')/Z_s} \tag{12.52}$$

and

$$tr_v = \frac{-2jc'/Z_s}{a'c' - b'^2 - 2jc'/Z_s}, \tag{12.53}$$

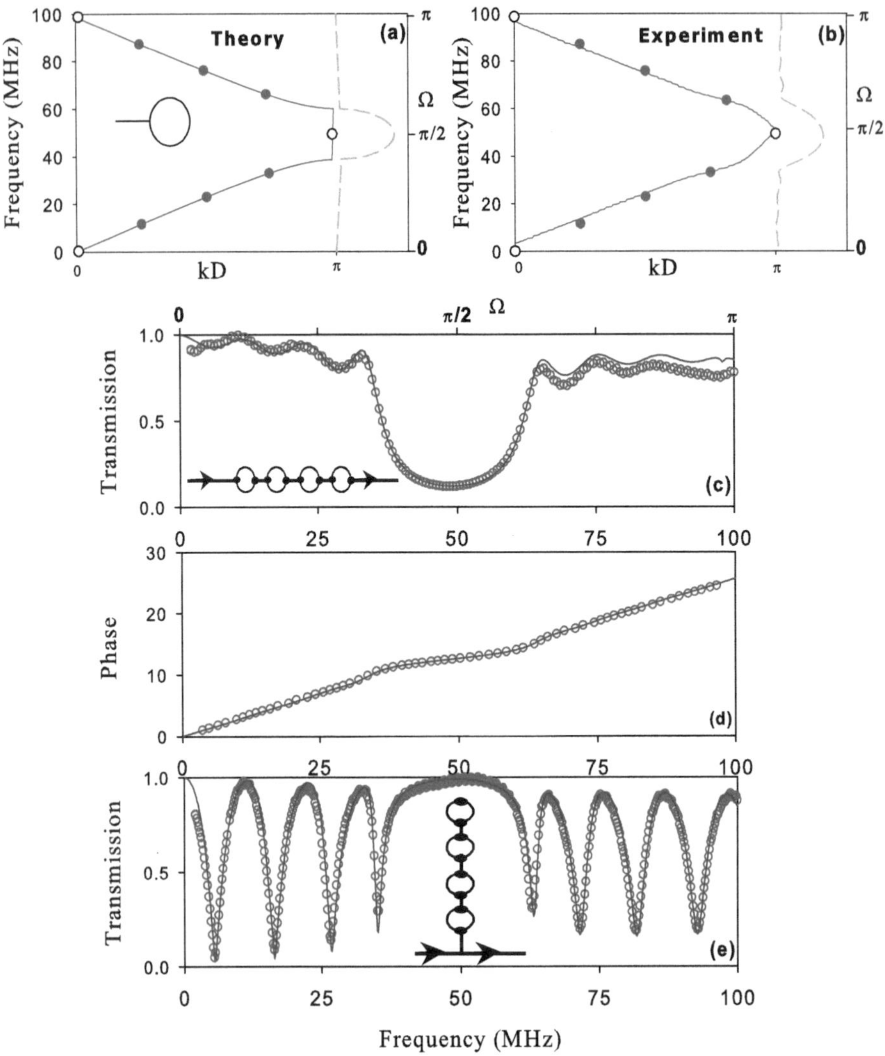

Figure 12.14 (a) Theoretical band structure of an infinite PC. Each cell is made of a standard coaxial cable of length $d_1 = 1$ m and impedance $Z = 50\ \Omega$ connected to a symmetric loop made of two identical coaxial cables of length $d_2 = 1$ m and impedance $Z = 50\ \Omega$ (see the inset). Solid and open circles correspond to the eigenmodes of a finite structure made of $N = 4$ cells. (b) Same as (a) but for the experimental results. (c) and (d) represent, respectively, theoretical (solid line) and experimental (open circles) variations of the transmission amplitude and phase through a finite PC containing $N = 4$ cells (see the inset of (c)) with the same characteristic lengths as in (a). (e) Same as (c) but for the finite structure grafted vertically along the guide (see the inset).

respectively, where h and v stand for horizontal and vertical insertion of the finite PC, respectively; while a', b', and c' are given by Eqs. (12.48), (12.49), and (12.50), respectively. The transmission function can be written in an explicit complex form as

$tr = \alpha + j\beta = |tr|e^{j\varphi}$, where $|tr|$ is the transmission coefficient, $\varphi = \arctan(\beta/\alpha) \pm m\pi$ is the phase associated with the transmission field and m is an integer.

12.3.2 General rules about confined and surface modes

For a finite SLS, it is expected to have N eigenmodes in each band. However, Eq. (12.51) shows that there are two types of eigenmodes in the finite structure:

1. If the wave vector k is real, which corresponds to an allowed band, then the eigenmodes of the finite PC are given by the third term in Eq. (12.51), namely

$$\sin(NkD) = 0, \tag{12.54}$$

which gives

$$kD = \frac{m\pi}{N}, \quad m = 1, 2, \ldots, N-1, \tag{12.55}$$

whereas the first and second terms in Eq. (12.51) cannot vanish in the bulk bands.

2. If the wave vector k is imaginary (modulo π), which corresponds to a forbidden band, then the eigenmodes correspond to surface modes and are given by the two first terms of Eq. (12.51), namely

$$t = -\frac{a}{b} \tag{12.56}$$

and

$$t = -\frac{b}{a}, \tag{12.57}$$

whereas the third term in Eq. (12.51) cannot vanish inside the gap since t should satisfy the condition

$$|t| < 1 \tag{12.58}$$

to ensure the decaying of surface modes from the surface.

In addition, we remark that if $N \to \infty$, the term t^{2N} vanishes and therefore the two expressions (Eqs. (12.56) and (12.57)) give the surface modes for two semiinfinite PCs with complementary surfaces (Fig. 12.13(c)).

Eqs. (12.56) and (12.57) can be written in a unique explicit form by replacing them in Eq. (12.44) and factorizing by the factor $\frac{1}{b}$; one obtains

$$ac - b^2 = 0. \tag{12.59}$$

Therefore, the surface modes of one semiinfinite PC are given by Eq. (12.59) together with the condition $|\frac{a}{b}| < 1$ (Eqs. (12.56) and (12.59)), whereas the surface modes of the complementary PC are given by Eq. (12.59) but with the condition $|\frac{b}{a}| < 1$ (Eqs. (12.57) and (12.59)). This result shows that if a surface mode appears on the surface of one PC, it does not appear on the other surface of the complementary PC. Moreover, Eq. (12.59) shows that the expression giving the surface modes for two complementary PCs is exactly the same expression as that giving the eigenmodes of one cell (Eq. (12.42)).

In addition to these results, let us recall briefly another result concerning the existence and behavior of surface waves in quasi-one-dimensional PC [42], namely the creation from the infinite PC of a surface with vanishing magnetic field gives rise to δ peaks of weight $-1/4$ in the density of states, at the edges of the PC bulk bands. Now, when one considers together the two complementary PCs: (i) the variation of density of states presents a loss of one half-mode at the limit of each band, i.e., one mode per band and (ii) the variation of density of states vanishes inside the bulk bands of the two complementary PCs. These two results associated with the necessary conservation of the total number of modes show that there must exist one surface mode of weight unity per gap to compensate the loss of one mode per band. These modes are associated with either one or the other surface of the two PCs.

From all the above results, one can deduce that a finite PC constituted of N cells gives rise to $N-1$ confined modes inside the bulk bands of the PC and one surface mode in each gap of the PC that may be attributed to one or the other of the two surfaces at both ends of the finite PC. The surface modes are independent of N and coincide with those of two semiinfinite PCs obtained by the cleavage of an infinite PC at the interface between two cells.

In the particular case of a symmetric cell, the surface modes inside the gaps move to the band gap edges. Indeed, in symmetric cells, $a = c$, and therefore Eq. (12.59) reduces to $a = \pm b$. This expression corresponds to a bandgap edge as $\cos(kd) = \pm 1$ (Eq. (12.44)). In what follows, we shall give a numerical and experimental confirmation of these results in the case of electromagnetic wave propagation in coaxial cables treated as quasi-one-dimensional waveguides.

12.3.3 Numerical and experimental results

12.3.3.1 Case of a finite periodic structure made of asymmetric cells

In what follows, we consider a finite periodic photonic crystal where each cell is made of a standard coaxial cable of length $d_A = 1$ m and impedance $Z = 50\ \Omega$, and a symmetric loop made of two identical coaxial cables of length $d_B = 1$ m and impedance $Z = 50\ \Omega$ (see the inset of Fig. 12.14(a)). We can show easily (see [14]) that the loop is equivalent to a coaxial cable of length d_B and half the impedance of the constituting cables (i.e., $Z = 25\ \Omega$). Therefore, we shall call $Z_A = 50\ \Omega$ and $Z_B = 25\ \Omega$ the impedances of the segment and the loop, respectively. Each cell becomes equivalent to a $50\ \Omega/25\ \Omega$ bisegment. One can notice that the two surfaces surrounding the cell are different and therefore the cell is asymmetric.

When applied to a PC made of a periodic repetition of a segment and a loop, the general dispersion relation (Eq. (12.44)) gives the well-known relation

$$\cos(kD) = C_A C_B - 0.5(Z_A/Z_B + Z_B/Z_A) S_A S_B, \quad (12.60)$$

where $C_{A,B} = \cos(\omega d_{A,B}\sqrt{\varepsilon}/c)$, $S_{A,B} = \sin(\omega d_{A,B}\sqrt{\varepsilon}/c)$, and $D = d_A + d_B$ is the period of the PC.

In the particular case where $d = d_A = d_B = 1$ m and $Z_A/Z_B = 2$, the above equation becomes simply

$$\cos(kD) = 1 - \frac{9}{4}\sin^2(\Omega), \tag{12.61}$$

where $\Omega = \omega d\sqrt{\varepsilon}/c$ is the reduced frequency.

The limits of the bandgaps are given by the successive sequences $kD = 0, \pi, \pi, 0, 0, \pi, \pi, 0, \ldots$, and therefore $\Omega \simeq 0, 0.39\pi, 0.61\pi, \pi, \pi, 1.39\pi, 1.61\pi, 2\pi, \ldots$

These results show that the bandgap structure is periodic every $\Omega = \pi$. The width of the successive bands is about 0.4π, the width of the gaps at the edge of the Brillouin zone is about 0.2π, whereas the width of the gaps at the center of the Brillouin zone vanishes. These results are confirmed theoretically (Fig. 12.14(a)) and experimentally (Fig. 12.14(b)) where we have plotted the dispersion curves (frequency versus kD) for the periodic structure depicted above. The experimental curves are obtained from the measurement of the amplitude $|t|$ (Fig. 12.14(c)) and the phase ϕ (Fig. 12.14(d)) of the transmission through a finite size structure inserted horizontally between two waveguides [14] (see the inset of Fig. 12.14(c)) as it is explained in Section 12.2. One can notice that despite the small number of cells ($N = 4$) used in the periodic structure, the amplitude and phase describe very well the band structure of the infinite system ($N \to \infty$) (Fig. 12.14(a)). As mentioned above, because of the periodicity of the bandgap structure, we limited ourselves to the reduced frequency region $0 \le \Omega \le \pi$.

Inside the first gap, $kD = \pi \pm j\kappa$ and the dispersion relation (Eq. (12.61)) becomes

$$\cosh(\kappa/2) = \frac{3}{2\sqrt{2}}\sin(\Omega). \tag{12.62}$$

Eq. (12.62) gives the imaginary part κ of the reduced wave vector kD inside the gaps which is responsible for the attenuation of the modes that may lie inside these gaps when a defect is inserted in the structure such as the surface [41,42]. From the above results, one can deduce that the center of the first gap is given by $\Omega = \pi/2$ and the value of κ at this frequency is $\kappa \simeq 0.69$ (Eq. (12.62)) as it is illustrated by the dashed curves in Figs. 12.14(a)–(b).

The theoretical transmission curves (amplitude and phase) are obtained from Eq. (12.52), whereas the transmission measurements have been realized by using the tracking generator coupled to a spectrum analyzer in the frequency range of 10 to 100 MHz. The experimental results are very well fitted by the 1D model using the Green's function method. One can notice in Fig. 12.14(c) that the attenuation inside the cables induces transmission depletion especially at high frequencies.

As concerns the eigenmodes of the finite PC, one can distinguish, as described in the section above, the bulk modes (Eq. (12.55)) lying inside the allowed bands and the surface modes (Eq. (12.59)) lying inside the forbidden bands. The expression giving the surface modes (Eq. (12.59)) can be written as [33]

$$Z_A C_A S_B + Z_B C_B S_A = 0, \tag{12.63}$$

together with the condition (Eq. (12.58))

$$|C_A C_B - \frac{Z_B}{Z_A} S_A S_B| < 1, \quad (12.64)$$

when the structure is terminated by a segment, and

$$|C_A C_B - \frac{Z_A}{Z_B} S_A S_B| < 1, \quad (12.65)$$

when the structure is terminated by a loop.

In the particular case considered here, $C_A = C_B = \cos(\Omega)$ and $S_A = S_B = \sin(\Omega)$. Therefore, Eq. (12.63) becomes simply

$$\sin(2\Omega) = 0, \text{ i.e., } \Omega = n\pi/2, \quad (12.66)$$

where n is an integer. If n is even (i.e., $\Omega = 0, \pi, 2\pi, \ldots$), then neither Eq. (12.64) nor Eq. (12.65) is fulfilled since the left-hand term in these equations is unity. As mentioned above, this situation corresponds to the center of the Brillouin zone (i.e., $kD = 0$, see Eq. (12.61)) where the bandgaps close. However, if n is odd (i.e., $\Omega = \pi/2, 3\pi/2, \ldots$), then only Eq. (12.64) is fulfilled since $Z_B < Z_A$, which means that all the surface modes appear on the surface of the structure terminated by a segment and no surface modes appear when the structure terminates with a loop. In Fig. 12.14(a), we have plotted by open circles the surface mode lying in the first gap at $\Omega = \pi/2$ (i.e., $f = 49.34$ MHz) as well as the frequencies lying at the bandgap edges $\Omega = 0$ and $\Omega = \pi$ (i.e., $f = 0$ and $f = 98.86$ MHz). These modes, given by Eq. (12.66), are independent of N. However, as predicted, there exist $N - 1 = 3$ modes in each band given by Eq. (12.55).

In order to give an experimental evidence of the eigenmodes of the finite PC, we measure the transmission coefficient through the structure grafted vertically along the guide (see the inset of Fig. 12.14(e)). Indeed, from Eqs. (12.51) and (12.53), one can notice that the maxima of the transmission (i.e., $tr_v = 1$) occur at the frequencies of the finite PC with vanishing magnetic field on both ends. An example of the transmission spectrum of a finite structure made of four loops is sketched in Fig. 12.14(e). The frequencies of the corresponding maxima are reported in Fig. 12.14(b). One can notice that the positions of the eigenmodes of the finite PC coincide exactly with those obtained from the theory (Fig. 12.14(a)).

Fig. 12.15 shows the variation of the eigenmodes of a finite PC as a function of the number of cells N. For $N = 1$ (one cell), the eigenmodes are given by Eq. (12.66), and we can distinguish the modes lying at the closing of the bandgaps (i.e., $\Omega = 0$ and $\Omega = \pi$) and the surface mode lying at the center of the bandgap (i.e., $\Omega = \pi/2$). When N increases, the above modes remain constant, whereas there exist $N - 1$ modes in each band for every value of N in accordance with the analytical results discussed earlier.

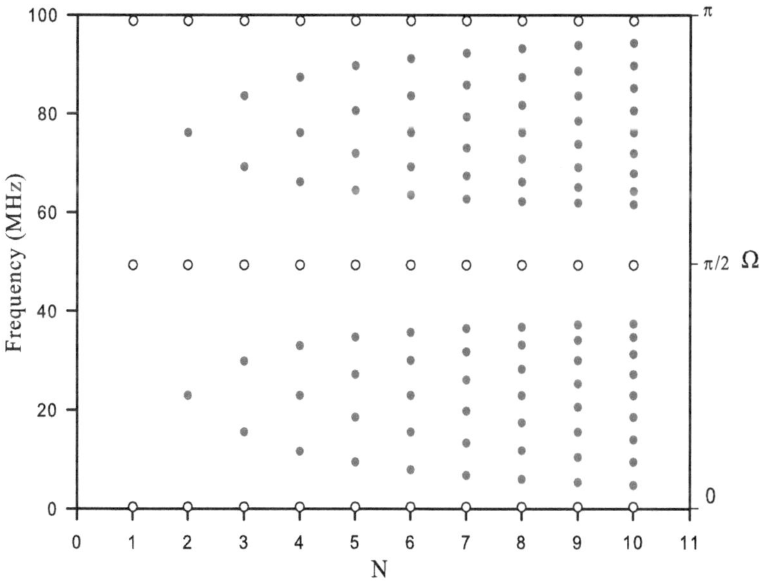

Figure 12.15 Variation of the eigenmodes of the finite PC as a function of the number of cells N. Open and solid circles have the same meaning as in Fig. 12.14.

The Green's function approach enables also deducing the local and total densities of states (LDOS). The details of these calculations are given in [33,42,56,88]. The LDOS reflect the behavior of the square modulus of the electric field inside the structure. An analysis of the LDOS as a function of the space position (Fig. 12.16) shows, as expected, that the surface mode lying in the first gap at $\Omega = \pi/2$ exhibits a strong localization at the surface terminated with a segment, with almost the same localization length regardless the length of the structure. Indeed, the localization length is defined as $\ell = D/|\kappa|$ where κ is the imaginary part of the reduced wave vector kD (Eq. (12.25)). As mentioned above, $\kappa \simeq 0.69$ for the surface mode lying at $\Omega = \pi/2$ and therefore $\ell \simeq 3$ m, which is in accordance with the results of Fig. 12.16.

12.3.3.2 Case of a finite structure with symmetric cells

In what follows, we consider a finite periodic structure made of symmetric cells. Each cell is composed of a symmetric loop of type B inserted between two identical segments of type A. Therefore, each cell becomes equivalent to a 50 Ω/25 Ω/50 Ω trisegment (see the inset of Fig. 12.17(a)). One can notice that the two surfaces surrounding the cell are equivalent, and therefore the cell is symmetrical. Also, when connected together, the cells give rise to finite PC made of loops separated by segments of length 2 m and the whole structure terminates by segments of length 1 m on both sides.

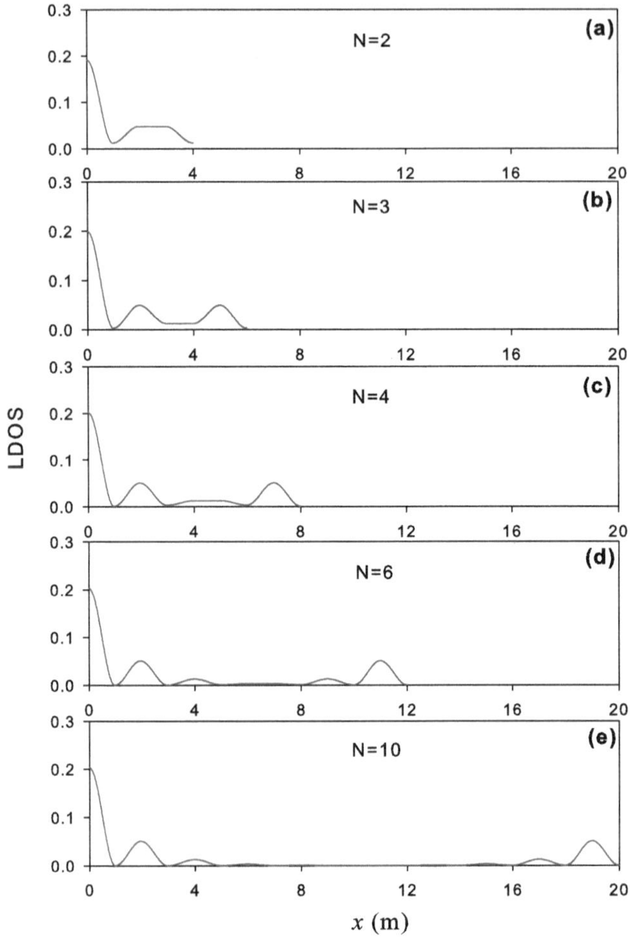

Figure 12.16 The local density of states (LDOS) (in arbitrary units) as a function of the space position x for the mode lying at the central gap frequency $\Omega = \pi/2$ (Fig. 12.3) for $N = 2, 3, 4, 6$, and 10 in (a) through (e), respectively. The finite PC is terminated by a segment and a loop at the left and the right of the structure, respectively.

In this case, the dispersion relation (Eq. (12.44)) becomes [33]

$$\cos(kD) = C_A^2 C_B - C_B^2 C_A - C_A S_A S_B (Z_A/Z_B + Z_B/Z_A), \tag{12.67}$$

where $D = 2d_A + d_B$. In the particular case where $d = d_A = d_B = 1$ m and $Z_A/Z_B = 2$, the above equation becomes simply

$$\cos(kD) = \frac{\cos(\Omega)}{2}\,(9\cos^2(\Omega) - 7). \tag{12.68}$$

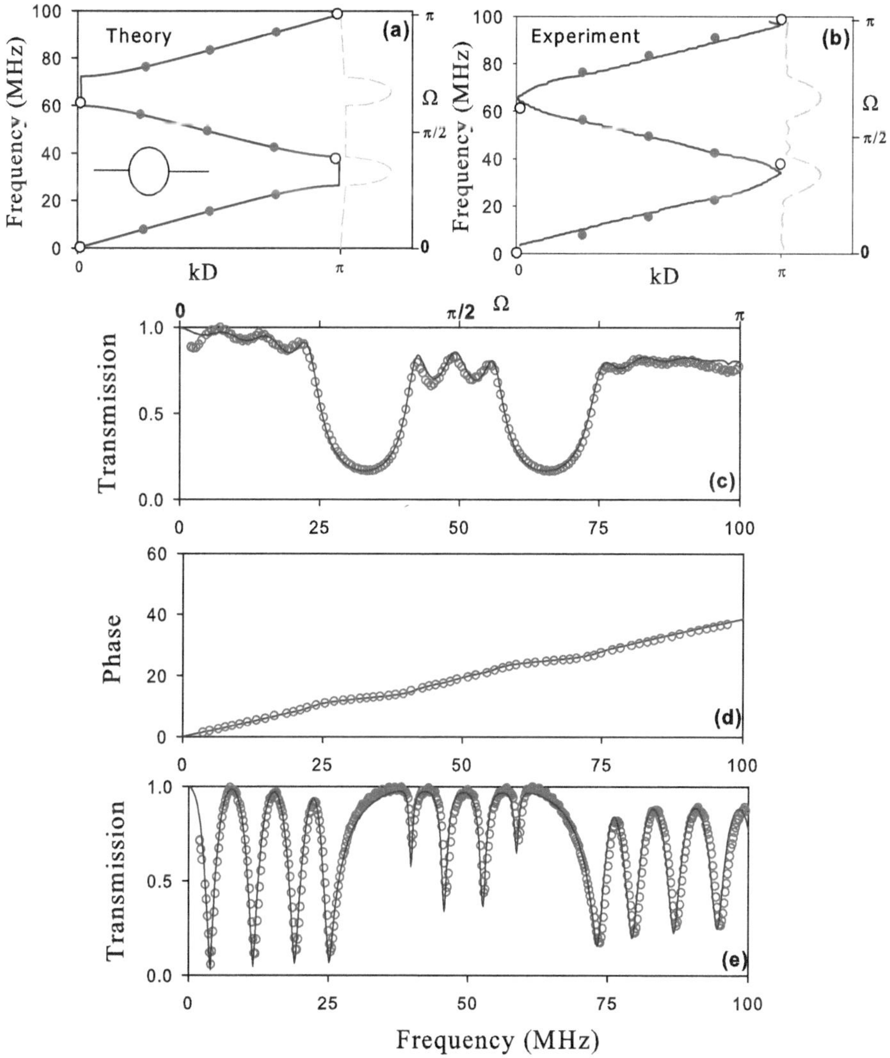

Figure 12.17 Same as in Fig. 12.14 but here each cell is made of a loop sandwiched between two identical segments (symmetric cell, see the inset of (a)).

The bandgap edges are given by $\cos(kD) = \pm 1$, namely $\cos(\Omega) = \pm 1, \pm 1/3$, and $\pm 2/3$. Therefore,

$$\Omega = 0, 0.27\pi, 0.39\pi, 0.61\pi, 0.73\pi, \pi, \ldots \tag{12.69}$$

Inside the two first gaps, $kD = \pi \pm j\kappa$ and $kD = \pm j\kappa$, respectively, and the dispersion relation (Eq. (12.68)) becomes

$$\cosh(\kappa) = \mp\frac{\cos(\Omega)}{2}\,(9\cos^2(\Omega) - 7). \tag{12.70}$$

Thus, one can deduce the reduced frequencies at the center of the first two gaps, namely

$$\cos(\Omega) = \frac{7\sqrt{7}}{9\sqrt{3}}, \text{ i.e., } \Omega \simeq 0.33\pi \text{ and } \Omega \simeq 0.67\pi, \tag{12.71}$$

as well as the corresponding values of κ ($\kappa \simeq 0.59$, see the dashed curves in Fig. 12.17(a)).

The dispersion relation giving the surface modes (Eq. (12.59)) can be written as follows [33]:

$$2C_A S_A C_B + S_B(C_A^2 \frac{Z_A}{Z_B} + S_A^2 \frac{Z_B}{Z_A}) = 0. \tag{12.72}$$

In the particular case considered here, $C_A = C_B = \cos(\Omega)$, $S_A = S_B = \sin(\Omega)$ and $Z_A/Z_B = 2$. Then, Eq. (12.72) becomes

$$\sin(\Omega)(9\cos^2(\Omega) - 1) = 0, \tag{12.73}$$

which leads to

$$\sin(\Omega) = 0, \text{ i.e., } \Omega = 0, \pi, 2\pi, \tag{12.74}$$

or

$$\cos(\Omega) = \pm 1/3, \text{ i.e., } \Omega = 0.39\pi, 0.61\pi, 1.39\pi, 1.61\pi, \ldots \tag{12.75}$$

However, the two latter equations give $\cos(kD) = \pm 1$ (Eq. (12.68)). Consequently, as mentioned above, the finite periodic PC with symmetric cells does not exhibit surface modes inside the bandgaps, but leads only to a constant frequency band edge modes. These results are similar to those found by Ren [46,47] for a complete confinement of electronic states in finite one-dimensional systems. Of course, in addition to the band edge modes, one can expect $N - 1$ modes in each band given by Eq. (12.55). However, for loop structures, the constant frequency band edge modes are a flat band of $N + 1$ modes. One of these modes is localized in the whole structure. Each of the other N modes is confined within one of the N closed loops and two $\Omega = 0$ modes.

Fig. 12.17(a) summarizes the numerical results corresponding to the analytical results detailed above, whereas Fig. 12.17(b), together with Figs. 12.17(c)–(e), summarizes the experimental results. One can notice a good agreement between theory and experiment concerning the bandgap structure, as well as the eigenmodes, of the finite PC. Among the different eigenmodes, one can distinguish the band-edge modes (N flat bands) plotted by open circles and the bulk band modes ($N - 1 = 3$) lying inside each allowed band. Fig. 12.18 gives the variation of the eigenmodes of the finite PC as a function of the number of cells N. One can notice the existence of band edge modes

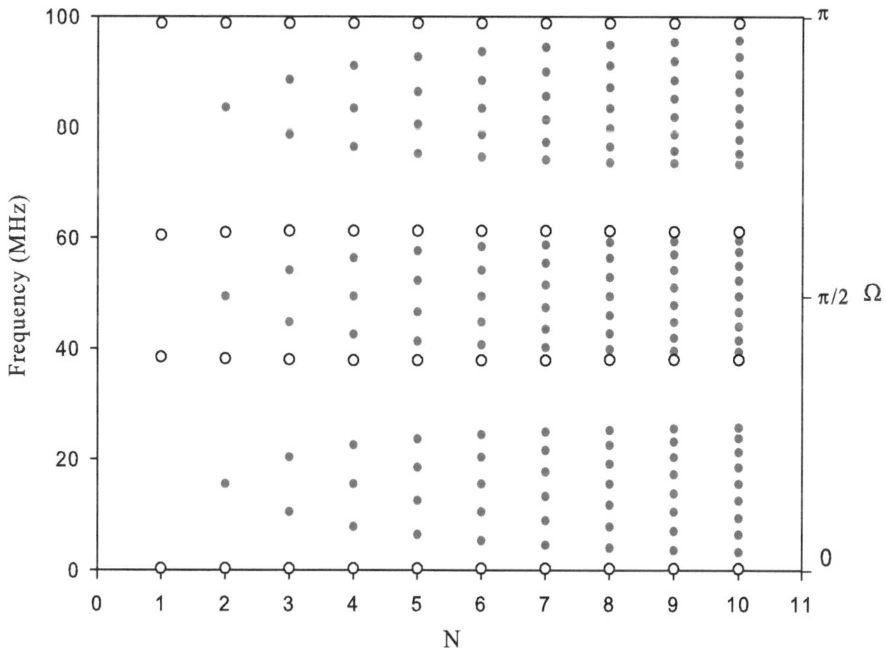

Figure 12.18 Variation of the eigenmodes of the finite PC as a function of the number of cells N for the structure described in Fig. 12.5.

(N flat bands) lying at a constant frequency independent of N (Eqs. (12.74)–(12.75)) and $N-1$ modes in each band and for each value of N. For $\Omega = 0$, there are also two modes, the ground state and that invariant by translation.

12.4 Summary and conclusions

In this chapter, we have considered bandgap structures, defect modes, as well as confined and surface modes in 1D serial loop structures. We have shown that such structures may exhibit very large photonic bandgaps. Compared to other 1D systems such as star waveguides, the observed gaps in SLS are significantly larger. The theoretical model assumed the cross-sections of the waveguide and of the loops small compared to their linear dimensions. This ensures monomode propagation of the electromagnetic waves. These serial loop structures may present large stopbands and are good candidates for photonic bandgap materials. Although they are limited to one dimension and are void of polarization effects, their properties are also described by Maxwell's equations, and some linear [12,89] and nonlinear [90] properties similar to those encountered in optical PBG materials have been reported. The measurements of the amplitude and the phase of the transmission coefficients through the finite-sized SLS enable us to deduce several properties on the wave propagation through such structures, namely dispersion curves, phase times, and therefore DOSs as well as group velocities.

In addition, we have shown that the presence of defect branches in the SLS gives rise to localized states inside the gaps. These defect modes appear as very narrow peaks of strong amplitude in the transmission and the phase time spectra. The localized states are very sensitive to the length of the defects, and to their position in the SLS. The phase time measurements in the SLS are in general different from the DOS, except for the symmetrical SLS in which these two quantities are equivalent. The experimental results are very well fitted by the 1D theoretical model using the Green's function method. Let us also stress that, in star waveguides, an important difficulty lies in the technical realization of the boundary condition at the free ends of the resonators, whereas this problem is avoided in SLS.

Also, we have presented a theoretical and experimental evidence of the existence of two types of modes in finite size 1D coaxial photonic crystal made of a periodic repetition of N cells. In particular, we have shown the existence of $N-1$ modes that match the bulk bands and one additional mode per gap if the cell is asymmetric. However, if the cell is symmetric, the additional mode falls at a constant frequency at the bulk band edges (within a flat band). These modes frequencies are independent of N. These results generalize our previous findings [33,41,42] on the existence of surface modes associated to two semiinfinite complementary PCs obtained from the cleavage of an infinite PC between two cells. The theoretical results are confirmed experimentally by using asymmetric cells made of a simple coaxial cable connected to a loop (constituted of two identical coaxial cables) and symmetric cells made of a loop inserted between two identical coaxial cables. The bandgap structure and the different eigenmodes of these photonic systems are obtained respectively from the measurement of the transmission coefficient through a finite size structure inserted horizontally between two waveguide cables and a finite size structure grafted vertically along a guide. The experimental results are in good agreement with theoretical calculations based on the formalism of the Green's function.

In this study, the lengths of the finite wires are of the order of the meter and the gaps occur in the hyperfrequency range. However, our theoretical model is in principle universal and is thus also valid in other frequency domains of the electromagnetic spectrum. For example, designing SLS working at optical frequencies requires characteristic lengths of the order of a micrometer. The manufacturing of such waveguides could be very useful once the problem of radiative and scattering losses is overcome by the amplification technique in making, for instance, filtering or multiplexing devices. It would even be more interesting for integrated structures working at optical frequencies. It was shown that the manufacturing of such serial loop structures at a submicrometric scale becomes realizable with the recent technological developments using high resolution electron beam lithography [91]. Nevertheless, the lateral size of such wires is in general of the same order of magnitude as the optical wavelengths, and a more thorough analysis of the scattering processes at the junctions would become desirable.

We report also in this chapter, for such structures, the existence of one confined mode in each loop, when its wavelength is commensurable with the two loop lengths

d_2 and d_3 shown in Fig. 12.1(a). These modes form a flat band with $N + 1$ modes. Another example of such confined modes appears in Section 5.10 of Chapter 5 of this Photonics book. When the above condition is not fulfilled, these N modes become usually one almost flat band.

Acknowledgments

The authors would like to thank Abdellah Mir, Noureddine Fettouhi, Youssef El Hassouani, Hassan Aynaou, Jérôme Vasseur, and Jaouad Zemmouri for their fruitful contribution in the original papers from which this chapter is inspired.

References

[1] E. Yablonovitch, Inhibited spontaneous emission in solid-state physics and electronics, Phys. Rev. Lett. 58 (1987) 2059.
[2] S. John, Strong localization of photons in certain disordered dielectric superlattices, Phys. Rev. Lett. 58 (1987) 2486.
[3] J. Martorell, N.M. Lawandy, Observation of inhibited spontaneous emission in a periodic dielectric structure, Phys. Rev. Lett. 65 (1990) 1877.
[4] E. Yablonovitch, Photonic band-gap structures, J. Opt. Soc. Am. B 10 (1993) 283.
[5] See for example: J.D. Joannopoulos, R.D. Meade, J.N. Winn, Photonic Crystals, Princeton University Press, Princeton, 1995;
C.M. Soukoulis (Ed.), Photonic Band Gaps and Localization, Plenum, New York, 1993;
C.M. Soukoulis (Ed.), Photonic Band Gap Materials, Kluwer, Dordrecht, 1996.
[6] J.D. Joannopoulos, P.R. Villeneuve, S. Fan, Photonic crystals, Solid State Commun. 102 (1997) 165, and references therein.
[7] L. Dobrzynski, A. Akjouj, B. Djafari-Rouhani, J.O. Vasseur, J. Zemmouri, Giant gaps in photonic band structures, Phys. Rev. B 57 (1998) R9388.
[8] J.O. Vasseur, B. Djafari-Rouhani, L. Dobrzynski, A. Akjouj, J. Zemmouri, Defect modes in one-dimensional comb-like photonic waveguides, Phys. Rev. B 49 (1999) 13446.
[9] S. Foresi, P.R. Villeneuve, J. Ferrera, E.R. Thoen, G. Steinmeyer, S. Fan, J.D. Joannopoulos, L.C. Kimerling, H.I. Smith, E.P. Ippen, Photonic-bandgap microcavities in optical waveguides, Nature (London) 390 (1997) 143.
[10] S. Khattou, M. Amrani, A. Mouadili, E.H. El Boudouti, A. Talbi, A. Akjouj, B. Djafari-Rouhani, Comparison of density of states and scattering parameters in coaxial photonic crystals: theory and experiment, Phys. Rev. B 82 (2020) 165310.
[11] Z.Q. Zhang, C.C. Wong, K.K. Fung, Y.L. Ho, W.L. Chan, S.C. Kan, T.L. Chan, N. Cheung, Observation of localized electromagnetic waves in three-dimensional networks of waveguides, Phys. Rev. Lett. 81 (1998) 5540.
[12] G.J. Schneider, S. Hanna, J.L. Davis, G.H. Watson, Defect modes in coaxial photonic crystals, J. Appl. Phys. 90 (2001) 2642.
[13] A. Haché, L. Poirier, Anomalous dispersion and superluminal group velocity in a coaxial photonic crystal: theory and experiment, Phys. Rev. E 65 (2002) 036608.
[14] E.H. El Boudouti, N. Fettouhi, A. Akjouj, B. Djafari-Rouhani, A. Mir, J.O. Vasseur, L. Dobrzynski, J. Zemmouri, Experimental and theoretical evidence for the existence of

photonic bandgaps and selective transmissions in serial loop structures, J. Appl. Phys. 95 (2004) 1102.

[15] A. Yacoby, M. Heiblum, D. Mahalu, M. Shtrikman, Coherence and phase sensitive measurements in a quantum dot, Phys. Rev. Lett. 74 (1995) 4047.

[16] R. Schuster, E. Buks, M. Heiblum, D. Mahalu, V. Umansky, H. Shtrikman, Phase measurement in a quantum dot via a double-slit interference experiment, Nature (London) 385 (1997) 417.

[17] L.J. Wang, A. Kusmich, A. Dogarin, Gain-assisted superluminal light propagation, Nature (London) 406 (2000) 277.

[18] A. Dogarin, A. Kusmich, L.J. Wang, Transparent anomalous dispersion and superluminal light-pulse propagation at a negative group velocity, Phys. Rev. A 63 (2001) 053806.

[19] B. Macke, B. Segard, Propagation of light-pulses at a negative group-velocity, Eur. Phys. J. D 23 (2003) 125.

[20] P. Yeh, Optical Waves in Layered Media, John Wiley Sons, New York, 1988.

[21] A.A. Bulgakov, V.R. Koturn, Surface optical oscillations in a limited stratified-periodic medium, Opt. Spectrosc. 56 (1984) 269.

[22] W.M. Robertson, M.S. May, Surface electromagnetic wave excitation on one-dimensional photonic band-gap arrays, Appl. Phys. Lett. 74 (1999) 1800.

[23] F. Ramos-Mendieta, P. Halevi, Electromagnetic surface modes of a dielectric superlattice: the supercell method, J. Opt. Soc. Am. B 14 (1997) 370.

[24] M.L.H. Lahlaouti, A. Akjouj, B. Djafari-Rouhani, L. Dobrzynski, M. Hammouchi, E.H. El Boudouti, A. Nougaoui, B. Kharbouch, Theoretical analysis of the density of states and phase times: application to resonant electromagnetic modes in finite superlattices, Phys. Rev. B 63 (2001) 035312.

[25] F. Villa, J.A. Gaspar-Armenta, F. Ramos-Mendieta, One-dimensional photonic crystals: equivalent systems to single layers with a classical oscillator like dielectric function, Opt. Commun. 216 (2003) 361.

[26] Y. El Hassouani, E.H. El Boudouti, H. Aynaou, B. Djafari-Rouhani, V.R. Velasco, Comment on "Sensitivity of surface states to the stack sequence of one-dimensional photonic crystals", J. Opt. A, Pure Appl. Opt. 9 (2007) 308.

[27] D.N. Chigrin, A.V. Lavrinenko, D.A. Yarotsky, V. Gaponenko, All-dielectric one-dimensional periodic structures for total omnidirectional reflection and partial spontaneous emission control, J. Lightwave Technol. 17 (1999) 2018.

[28] T. Okamoto, T. Kamiyama, I. Yamaguchi, All-optical spatial light modulator with surface plasmon resonance, Opt. Lett. 18 (1993) 1570.

[29] W.M. Robertson, Experimental measurement of the effect of termination on surface electromagnetic waves in one-dimensional photonic bandgap arrays, J. Lightwave Technol. 17 (1999) 2013.

[30] M. Shinn, W.M. Robertson, Surface plasmon-like sensor based on surface electromagnetic waves in a photonic band-gap material, Sens. Actuators B 105 (2005) 360.

[31] J.A. Gaspar-Armenta, F. Villa, Band-structure properties of one-dimensional photonic crystals under the formalism of equivalent systems, J. Opt. Soc. Am. B 21 (2004) 405.

[32] M.L.H. Lahlaouti, A. Akjouj, B. Djafari-Rouhani, L. Dobrzynski, Resonant and localized electromagnetic modes in finite superlattices, Phys. Rev. B 61 (2000) 2059.

[33] E.H. El Boudouti, B. Djafari-Rouhani, E.M. Khourdifi, L. Dobrzynski, Surface and interface elastic waves in superlattices: transverse localized and resonant modes, Phys. Rev. B 48 (1993) 10987;
E.H. El Boudouti, B. Djafari-Rouhani, A. Akjouj, L. Dobrzynski, Theory of surface and interface transverse elastic waves in N-layer superlattices, Phys. Rev. B 54 (1996) 14728.

[34] W. Chen, Y. Lu, H.J. Maris, G. Xiao, Picosecond ultrasonic study of localized phonon surface modes in Al/Ag superlattices, Phys. Rev. B 50 (1994) 14506.
[35] N.-W. Pu, J. Bokor, Study of surface and bulk acoustic phonon excitations in superlattices using picosecond ultrasonics, Phys. Rev. Lett. 91 (2003) 076101;
N.-W. Pu, Ultrafast excitation and detection of acoustic phonon modes in superlattices, Phys. Rev. B 72 (2005) 115428.
[36] S.-Y. Ren, Y.-C. Chang, Theory of confinement effects in finite one-dimensional phononic crystals, Phys. Rev. B 75 (2007) 212301.
[37] Y. El Hassouani, E.H. El Boudouti, B. Djafari-Rouhani, H. Aynaou, Sagittal acoustic waves in finite solid-fluid superlattices: band-gap structure, surface and confined modes, and omnidirectional reflection and selective transmission, Phys. Rev. B 78 (2008) 174306.
[38] L. Dobrzynski, E.H. El Boudouti, A. Akjouj, Y. Pennec, H. Al-Wahsh, G. Lévêque, B. Djafari-Rouhani, Phononics, Interface Transmission Tutorial Book Series, Elsevier, 2017.
[39] A.-C. Hladky, A. Devos, M. de Billy, Quantitative analysis of the vibration modes in a finite set of coupled spheres, J. Acoust. Soc. Am. 116 (2004) 117.
[40] A.-C. Hladky, G. Allan, M. de Billy, Localized modes in a one-dimensional diatomic chain of coupled spheres, J. Appl. Phys. 98 (2005) 054909.
[41] B. Djafari-Rouhani, E.H. El Boudouti, A. Akjouj, L. Dobrzynski, J.O. Vasseur, A. Mir, N. Fettouhi, J. Zemmouri, Surface states in one-dimensional photonic band gap structures, Vacuum 63 (2001) 177.
[42] Y. El Hassouani, H. Aynaou, E.H. El Boudouti, B. Djafari-Rouhani, A. Akjouj, V.R. Velasco, Surface electromagnetic waves in Fibonacci superlattices: theoretical and experimental results, Phys. Rev. B 74 (2006) 035314.
[43] E.H. El Boudouti, Y. El Hassouani, H. Aynaou, B. Djafari-Rouhani, A. Akjouj, V.R. Velasco, Two types of modes in finite size one-dimensional coaxial photonic crystals: general rules and experimental evidence, Phys. Rev. E 76 (2007) 026607.
[44] A. Haché, L. Poirier, Long-range superluminal pulse propagation in a coaxial photonic crystal, Appl. Phys. Lett. 80 (2002) 518.
[45] H. Aynaou, E.H. El Boudouti, Y. El Hassouani, A. Akjouj, B. Djafari-Rouhani, J. Vasseur, A. Benomar, V.R. Velasco, Propagation and localization of electromagnetic waves in quasiperiodic serial loop structures, Phys. Rev. E 72 (2005) 056601.
[46] S.Y. Ren, Complete quantum confinement of one-dimensional Bloch waves, Phys. Rev. B 64 (2001) 035322.
[47] S.Y. Ren, Two types of electronic states in one-dimensional crystals of finite length, Ann. Phys. 301 (2002) 22.
[48] E.L. Albuquerque, M.G. Cottam, Theory of elementary excitations in quasiperiodic structures, Phys. Rep. 376 (2003) 225, and references therein.
[49] J. Vasseur, A. Akjouj, B. Djafari-Rouhani, L. Dobrzynski, E.H. El Boudouti, Photon, electron, magnon, phonon and plasmon mono-mode circuits, Surf. Sci. Rep. 51 (2004) 1.
[50] M. del Mar Sanchez-Lopez, J.A. Davis, K. Crabtree, Coaxial cable analogs of multilayer dielectric optical coatings, Am. J. Phys. 71 (2003) 1314.
[51] A. Haché, A. Slimani, A model coaxial photonic crystal for studying band structures, dispersion, field localization, and superluminal effects, Am. J. Phys. 72 (2004) 916.
[52] J.A. Davis, D.A. Miller, M. del Mar Sanchez-Lopez, J. Cos, Multiple-beam interference with coaxial cable analogs of optical arrays, Am. J. Phys. 74 (2006) 1066.
[53] L. Dobrzynski, Interface response theory of continuous composite systems, Surf. Sci. Rep. 11 (1990) 139.
[54] L. Dobrzynski, J. Mendialdua, A. Rodriguez, S. Bolibo, M. More, Interface response theory of composite elastic media, J. Phys. France 50 (1989) 2563.

[55] A. Mir, Ph.D. thesis, University of Meknès, Morocco, 2002.

[56] M.L. Lahlaouti, A. Akjouj, B. Djafari-Rouhani, L. Dobrzynski, M. Hammouchi, E.H. El Boudouti, A. Nougaoui, Electromagnetic waves in finite superlattices with buffer and cap layers, J. Opt. Soc. Am. A 16 (1999) 7.

[57] M. Büttiker, R. Landauer, Traversal time for tunneling, Phys. Rev. Lett. 49 (1982) 1739.

[58] M. Büttiker, Larmor precession and the traversal time for tunneling, Phys. Rev. B 27 (1983) 6178.

[59] E.H. Hauge, J.A. Stovneng, Tunneling times: a critical review, Rev. Mod. Phys. 61 (1989) 917.

[60] V. Laude, P. Tournois, Superluminal asymptotic tunneling times through one-dimensional photonic bandgaps in quarter-wave-stack dielectric mirrors, J. Opt. Soc. Am. B 16 (1999) 194.

[61] M.L. Bah, A. Akjouj, E.H. El Boudouti, B. Djafari-Rouhani, L. Dobrzynski, Surface and interface optical waves in superlattices: transverse electric localized and resonant modes, J. Phys. Condens. Matter 8 (1996) 4171.

[62] Xin Wang, Xinhua Hu, Yizhou Li, Wulin Jia, Chun Xu, Xiaohan Liu, Jian Zia, Enlargement of omnidirectional total reflection frequency range in onedimensional photonic crystals by using photonic heterostructures, Appl. Phys. Lett. 80 (2002) 4291.

[63] R.P. Stanley, R. Houdré, U. Oesterle, M. Ilegems, C. Weisbuch, Impurity modes in one-dimensional periodic systems: the transition from photonic band gaps to microcavities, Phys. Rev. A 48 (1993) 2246.

[64] R. Wang, J. Dong, D.Y. Xing, Defect studies in a one dimensional photonic band gap structure, Phys. Status Solidi B 200 (1997) 529.

[65] M. Sigalas, C.M. Soukoulis, E.N. Economou, C.T. Chan, K.M. Ho, Photonic band gaps and defects in two dimensions: studies of the transmission coefficient, Phys. Rev. B 48 (1993) 14121.

[66] S. Zhu, N. Liu, H. Zheng, H. Chen, Time delay of light propagation through defect modes of one-dimensional photonic band-gap structures, Opt. Commun. B 174 (2000) 139.

[67] H.-W. Lee, Generic transmission zeros and in-phase resonances in time-reversal symmetric single channel transport, Phys. Rev. Lett. 82 (1999) 2358.

[68] T. Taniguchi, M. Buttiker, Friedel phases and phases of transmission amplitudes in quantum scattering systems, Phys. Rev. B 60 (1999) 13814, and references therein.

[69] A. Leavy Yeyati, M. Buttiker, Scattering phases in quantum dots: an analysis based on lattice models, Phys. Rev. B 62 (2000) 7307.

[70] L. Brillouin, Wave Propagation and Group Velocity, Academic, New York, 1960.

[71] S.G.B. Garret, D.E. McCumber, Propagation of a Gaussian light pulse through an anomalous dispersion medium, Phys. Rev. A 1 (1970) 305.

[72] S. Chu, S. Wong, Linear pulse propagation in an absorbing medium, Phys. Rev. Lett. 48 (1982) 738.

[73] A.M. Steinberg, R. Chiao, Dispersionless, highly superluminal propagation in a medium with a gain doublet, Phys. Rev. A 49 (1994) 2071.

[74] P.C. Peters, Does a group velocity larger than c violate relativity?, Am. J. Phys. 56 (1988) 129.

[75] R. Landauer, T. Martin, Barrier interaction time in tunneling, Rev. Mod. Phys. 66 (1994) 217.

[76] G. Diener, Superluminal group velocities and information transfer, Phys. Lett. A 223 (1996) 327.

[77] M.W. Mitchell, R.Y. Chiao, Causality and negative group delays in a simple bandpass amplifier, Am. J. Phys. 66 (1998) 14.

[78] J.N. Munday, W.M. Robertson, Negative group velocity pulse tunneling through a coaxial photonic crystal, Appl. Phys. Lett. 81 (2002) 2127.
[79] G.D. D'Aguanno, M. Centini, M. Scolara, C. Sibilia, M. Bloemer, C.M. Bowden, J.M. Haus, M. Bertolotti, Group velocity, energy velocity, and superluminal propagation in finite photonic band-gap structures, Phys. Rev. E 63 (2001) 036610.
[80] M.M. Sánchez-López, A. Sánchez-Meroño, J. Arias, J.A. Davis, I. Moreno, Observation of superluminal and negative group velocities in a Mach–Zehnder interferometer, Appl. Phys. Lett. 93 (2008) 074102.
[81] W.M. Robertson, J. Pappafotis, P. Flannigan, Sound beyond the speed of light: measurement of negative group velocity in an acoustic loop filter, Appl. Phys. Lett. 90 (2007) 014102.
[82] J. Von Neuman, E. Wigner, On remarkable discrete eigenvalues, Phys. Z. 30 (1929) 465.
[83] C.W. Hsu, B. Zhen, A.D. Stone, J.D. Joannopoulos, M. Soljacic, Bound states in the continuum, Nat. Rev. Mater. 1 (2016) 16048.
[84] A. Mir, A. Akjouj, J.O. Vasseur, B. Djafari-Rouhani, N. Fettouhi, E.H. El Boudouti, L. Dobrzynski, J. Zemmouri, Observation of large photonic band gaps and defect modes in one-dimensional networked waveguides, J. Phys. Condens. Matter 15 (2003) 1593.
[85] J-B. Xia, Quantum waveguide theory for mesoscopic structures, Phys. Rev. B 45 (1992) 3593.
[86] M.D. Tocci, M. Scalora, M.J. Bloemer, J.P. Dowling, C.M. Bowden, Measurement of spontaneous-emission enhancement near the one-dimensional photonic band edge of semiconductor heterostructures, Phys. Rev. A 53 (1996) 2799.
[87] M. Beyindir, E. Ozbay, Heavy photons at coupled-cavity waveguide band edges in a three-dimensional photonic crystal, Phys. Rev. B 62 (2000) R2247.
[88] E.H. El Boudouti, B. Djafari-Rouhani, Acoustic waves in finite superlattices, Phys. Rev. B 49 (1994) 4586;
M. Hammouchi, E.H. El Boudouti, A. Nougaoui, B. Djafari-Rouhani, M.L.H. Lahlaouti, A. Akjouj, L. Dobrzynski, Acoustic waves in finite superlattices: influence of buffer layers, Phys. Rev. B 59 (1999) 1999.
[89] R.D. Pradhan, G.H. Watson, Impurity effects in coaxial-connector photonic crystals: a quasi-one-dimensional periodic system, Phys. Rev. B 60 (1999) 2410.
[90] L. Poirier, A. Haché, Nonlinear coaxial photonic crystal, Appl. Phys. Lett. 78 (2001) 2626.
[91] J.V. Hryniewicz, P.P. Absil, B.E. Little, R.A. Wilson, P.T. Ho, Higher order filter response in coupled microring resonators, IEEE Photonics Technol. Lett. 12 (2000) 320.

Fibonacci loop structures: bandgaps, power law, scaling law, confined and surface modes

13

El Houssaine El Boudouti[a], *Bahram Djafari-Rouhani*[b], *Abdellatif Akjouj*[b], *and Leonard Dobrzyński*[b]
[a]Laboratoire de Physique de la Matière et de Rayonnement, Département de Physique, Faculté des Sciences, Université Mohammed I, Oujda, Morocco, [b]Institute of Electronic, Microelectronics and Nanotechnology (IEMN), UMR CNRS 8520, Department of Physics, Faculty of Sciences and Technologies, Lille University, Villeneuve d'Ascq Cedex, France

13.1 Introduction

Photonic crystals (PCs) have been a subject of great interest during the last three decades because of their interesting properties in the development of new optical circuits [1,2]. These systems, constituted by periodic arrangements (cells) of dielectric materials according to one (1D), two (2D), and three (3D) dimensions, present characteristic frequency domains in the dispersion curves where light can propagate (bulk bands) and frequency domains where light cannot propagate (gaps). These structures present unusual properties which can be exploited in the control and the guidance of light [3,4]. In their 1D version, PCs are well known as optical multilayers of alternating dielectric materials [5]. Such structures exhibit multiple reflections and destructive interference, giving rise to forbidden bands where light cannot propagate. This effect is used in designing optical devices such as omnidirectional optical mirrors [6–9], Fabry–Perot filters [5], and thermophotovoltaics [10]. Quasi-one-dimensional photonic crystals based on coaxial BNC connectors have been used to reproduce several effects related to periodic dielectric media in the radio-frequency region such as bandgap structures with or without cavity modes [11–13], superluminal effects [13,14], field localization [12], Fabry–Perot resonators [15], and nonlinear effects [16]. For a simple physical discussion of periodic and quasiperiodic phenomena, the coaxial PCs are more attractive than their optical counterparts since they require relatively simple equipment. In addition, the propagation in these structures is monomode [17], and one can obtain very accurate experimental results that may be fitted with a simple 1D theoretical model.

In real materials, the photonic crystals are often of finite size with free surfaces. Consequently, the surface modes can exist inside the gaps [18–22] as defect modes. The existence of surface modes can directly affect the performance and the efficiency of PC to prohibit the propagation of electromagnetic waves in certain frequency regions [18,20]. Therefore, it is interesting to study surface modes in PC. The study of surface electromagnetic waves in 1D multilayer PC (or superlattice) was largely

Photonics. https://doi.org/10.1016/B978-0-12-819388-4.00023-X

developed both theoretically and experimentally [21–31]. In particular, it was shown that surface waves in PC can present an alternative to surface plasmons which are typical with metal films. From the technological point of view, surface electromagnetic waves can present some applications in the design of optical devices such as sensors and modulators [27–31].

Photonic quasicrystals are intermediate between periodic and disordered systems [32–34]. The quasiperiodic structures are generally built from two blocks A and B. Quasiperiodic structures could be designed in many different sequences such as Fibonacci [32,35], Thue–Morse [36,37], Rudin–Shapiro [38], and double period [39,40]. The most well known of these systems is the Fibonacci structure which is constituted following the Fibonacci rule $S_{k+1} = S_k S_{k-1}$ with $S_1 = A$, $S_2 = AB$, and k being the generation number. This leads to the Fibonacci sequences (FS): $S_3 = ABA$, $S_4 = ABAAB$, $S_5 = ABAABABA, \ldots$ Since the introduction of the Fibonacci dielectric multilayers by Kohmoto et al. [41], the propagation of electromagnetic waves in quasiperiodic systems has been the subject of several theoretical and experimental investigations [42–54]. These structures may present several applications [34,55] in, e.g., optical microcavities [53], omnidirectional reflection [54], multiwavelength narrow-band optical filters, and wavelength division multiplexing systems [48]. Also, it was demonstrated [47] that these systems can provide an interesting alternative to regular photonic crystals for the realization of photonic devices, such as optical filters with a self-similar spectrum and a high wavelength selectivity in the band edge region. Because of their noninteracting nature, electromagnetic waves provide an excellent tool for probing the localization phenomena in comparison with other excitations, such as electronic waves.

As mentioned above, the study of the propagation and the localization of light in 1D photonic quasicrystals was largely developed during the last decade [42–54]. Most of these results concern the bulk properties of these structures, however, the study of surface modes has attracted less interest [32,56], in particular from the experimental point of view. In addition, the experimental analysis of the scaling law and the self-similarity of the transmission amplitude and the phase time is lacking in such systems. Let us recall that the study of surface modes in periodic 1D photonic structures composed of waveguides, along which serial loops are placed [57] or dangling side branches are grafted [58], is given in Chapters 12 and 11 respectively of the Photonics book.

In this chapter, we consider two types of quasiperiodic structures: either a given Fibonacci sequence (FS) or a periodic repetition of a given sequence called Fibonacci superlattice.

In the case of a Fibonacci superlattice [59], we focus onto the study of the existence and the localization of surface electromagnetic waves in quasiperiodic structures formed by two blocks A (segment) and B (loop) (Fig. 13.1) arranged according to the Fibonacci sequence, where block A is formed by a coaxial cable of length $d_A = 1$ m and impedance $Z_A = 50\ \Omega$ and block B is a symmetric loop of impedance $Z_B = 25\ \Omega$ and length $d_B = 1$ m. In particular, we study the surface waves in periodic superlattices where each cell is formed by loops and segments following the Fibonacci sequence. Our objective is threefold: (i) we generalize to the case of quasiperiodic structures a rule about the existence of surface modes, namely, when one considers two semiinfinite superlattices obtained by the cleavage of an infinite superlattice, there is exactly one surface mode in each gap. This mode is localized on the surface of one of the two semiinfinite superlattices. We show that these surface modes correspond to the modes of a single cell when their frequencies fall inside the gaps of the infinite superlattice; (ii) we show that in addition to the existence of surface modes related to the periodicity as in usual superlattices, there exist new surface modes related to the quasiperiodicity of the loops and segments inside each cell; (iii) we propose an original experimental method to observe these surface modes. However, before dealing with surface waves, it is helpful to discuss the bandgap structures of infinite Fibonacci superlattices and some related properties such as the power-law behavior of the measure of the energy spectrum.

In the case of a Fibonacci sequence [60], our aim is to reproduce Fibonacci structures similar to those used in the pioneering work of Merlin et al. [61] on layered media where each block is composed of two different layers (bilayer). However, because of the length of these structures and the attenuation in the cables, we were not able to check experimentally some peculiar features of FS such as the scaling behavior of the amplitude and the phase of the transmission coefficient. As mentioned above, in this chapter, we consider a simpler Fibonacci structure where blocks A and B are formed by just a segment and loop, respectively. This structure is equivalent to that used in our theoretical work on acoustic waves in Fibonacci phononic circuits [62] and the works of Kohmoto et al. [41,42] on layered optical media. Our aim in the second part of this chapter is twofold: (i) we give a theoretical and experimental evidence of the scaling behavior of the amplitude and the phase of the transmission coefficient through a finite FS inserted between two semiinfinite waveguides; (ii) we study a vertical structure formed by a given FS grafted along a guide. This structure enables us to deduce all the eigenmodes of the FS (assuming the vanishing of the magnetic field at the boundaries of the FS). Different effects related to these quasiperiodic media are observed, including localization and scaling properties.

This chapter is organized as follows: In Section 13.2, we give a short summary of the theoretical formalism and the main analytical results used in our numerical calculations. Section 13.3 gives the main numerical and experimental results about bulk and surface modes in Fibonacci superlattices. Section 13.4 is devoted to the scaling law and the self-similarity of the transmission spectra in Fibonacci sequences. The conclusions are summarized in Section 13.5.

13.2 Method of theoretical and numerical calculation

The Fibonacci superlattice is a simple periodic 1D structure where each unit cell is constituted by a quasiperiodic structure built up of two blocks A and B. For the structure considered here, block A is formed by a simple coaxial cable of length $d_A = 1$ m characterized by its characteristic impedance $Z_A = 50\ \Omega$ and its permittivity $\varepsilon = 2.3$, while block B is a symmetrical loop formed by two identical standard coaxial cables

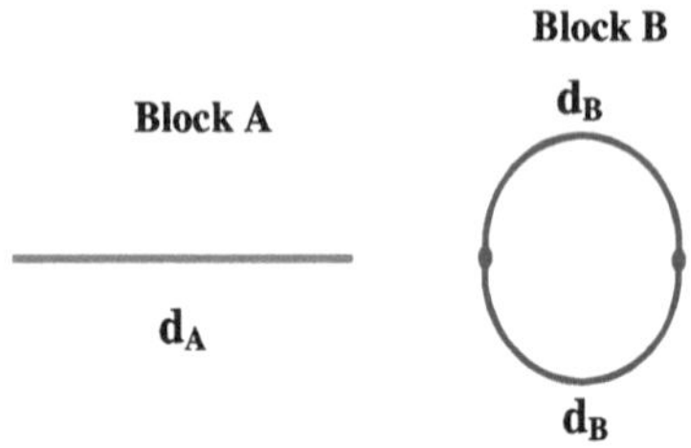

Figure 13.1 Schematic representation of the blocks A and B constituting the Fibonacci structure. Block A is a simple coaxial cable of length $d_A = d = 1$ m and impedance $Z = 50\ \Omega$. Block B is a symmetrical loop made up of two identical coaxial cables of the same length $d_B = d = 1$ m and impedance $Z = 50\ \Omega$.

of the same length (i.e., $d_B = d = 1$ m, Fig. 13.1). It is easy to show that each block B (loop) is equivalent to a coaxial cable of length $d = 1$ m and characterized by a half impedance $Z_B = Z_A/2 = 25\ \Omega$. This contrast between the characteristic impedances of the two blocks A and B gives rise to the opening of large gaps in the band structure of these serial loop structures [63]. Blocks A and B inside each cell of the superlattice follow the Fibonacci sequence:

$$S_{k+1} = S_k S_{k-1} \quad \text{with} \quad S_1 = [A],\ S_2 = [AB]. \tag{13.1}$$

Consequently, the Fibonacci generations are: $S_1 = [A]$, $S_2 = [AB]$, $S_3 = [ABA]$, $S_4 = [ABAAB]$, $S_5 = [ABAABABA]$, etc. The theoretical formalism used is based on the interface response theory [64]. In [65], El Boudouti et al. presented the analytical expressions of the dispersion relations of surface and interface transverse elastic waves in N-layer periodic superlattices, i.e., superlattices in which each cell is composed of N different layers. The expressions of the displacement fields and the local and total densities of states were also given. These expressions also remain valid to study the wave propagation in coaxial cables which are monomode circuits [17]; it is sufficient to replace the acoustic impedances by the electromagnetic impedances and the elastic constants by the dielectric permittivities. The calculation of the Green functions of an N-layer superlattice (Fig. 13.2(a)) can be carried out by using the 2×2 transfer matrix:

$$T_{1,\ldots,N} = \Delta_N \Delta_{N-1} \cdots \Delta_1, \tag{13.2}$$

where

$$\Delta_i = \begin{pmatrix} C_i & -jZ_i S_i \\ \frac{jS_i}{Z_i} & C_i \end{pmatrix}, \tag{13.3}$$

$S_i = \sinh(\alpha_i d_i)$ and $C_i = \cosh(\alpha_i d_i)$. $\alpha_i = \frac{-j\omega\sqrt{\varepsilon_i}}{c}$, $j = \sqrt{-1}$, ω is the angular frequency, ε_i the dielectric permittivity of the medium i, and c the speed of light. In our case, blocks A and B are characterized by $\varepsilon_A = \varepsilon_B = 2.3$ and $Z_A = 2Z_B = 50\ \Omega$ and the N layers (blocks) inside each cell respect the Fibonacci sequence (Fig. 13.2(a)). Consequently, the transfer matrices become (Eq. (13.2)): $T^{(1)} = \Delta_A$ for S_1, $T^{(2)} =$

$\Delta_B\Delta_A$ for S_2, $T^{(3)} = \Delta_A\Delta_B\Delta_A$ for S_3, $T^{(4)} = \Delta_B\Delta_A\Delta_A\Delta_B\Delta_A$ for S_4, etc. Then, a recurrence relation is obtained, $T^{(k+1)} = T^{(k)}T^{(k-1)}$, from which all the matrices $T^{(k)}$ can be obtained, where k is the number of the Fibonacci generation. The number of the blocks in each generation k is $F_k = F_{k-1} + F_{k-2}$ with $F_1 = 1$ and $F_2 = 2$. Let us notice also the useful property of the transfer matrix valid for any generation number k:

$$\det T^{(k)} = T_{11}T_{22} - T_{12}T_{21} = 1, \tag{13.4}$$

where T_{11}, T_{12}, T_{21}, and T_{22} are the elements of the transfer matrix $T^{(k)}$ (Eq. (13.2)).

13.2.1 Dispersion relations and Green's functions

The dispersion relation of an infinite superlattice (i.e., ω as a function of the wave vector Q) is given by [65]:

$$\cos(QD) = \frac{1}{2}(T_{11} + T_{22}) = \eta, \tag{13.5}$$

where $D = F_k$ (in meters) is the period of the superlattice. Each period is a Fibonacci sequence of order k. The calculation of the elements of the Green function $g(D, D)$ in the whole space of the superlattice requires first the knowledge of the elements $g(M_m, M_m)$ in the restricted space of the interfaces M_m (the index m indicates the block i belonging to the cell n, $m \equiv (n, i)$, and M_m represents the space of the two interfaces of the block i, $M_m = (n, i, \pm\frac{d_i}{2})$, $i = 1, 2, \ldots, N$). These elements are given by [65]:

$$g\left(n, i, -\frac{d_i}{2}; n', j, -\frac{d_j}{2}\right) = \begin{cases} (T_{12})_{i,\ldots,N,1,\ldots,i-1}\dfrac{t^{|n-n'|+1}}{t^2-1}, & i = j, \\ (T_{12})_{j,\ldots,N,1,\ldots,i-1}\dfrac{t^{|n-n'|+1}}{t^2-1} + (T_{12})_{i,\ldots,j-1}\dfrac{t^{|n-n'-1|+1}}{t^2-1}, & i < j \leqslant N, \\ (T_{12})_{i,\ldots,N,1,\ldots,j-1}\dfrac{t^{|n-n'|+1}}{t^2-1} + (T_{12})_{j,\ldots,i-1}\dfrac{t^{|n-n'+1|+1}}{t^2-1}, & j < i \leqslant N. \end{cases} \tag{13.6}$$

Here the different terms (T_{12}) are obtained from Eq. (13.2) with due account of the indices giving the order of the blocks beginning from the first and ending by the last and t is defined by:

$$t = \begin{cases} \eta + (\eta^2 - 1)^{1/2}, & \eta < -1, \\ \eta \pm i(1 - \eta^2)^{1/2}, & -1 < \eta < +1, \\ \eta - (\eta^2 - 1)^{1/2}, & \eta > 1. \end{cases} \tag{13.7}$$

Inside the bulk bands ($-1 < \eta < +1$), the sign $\pm$ in the expression of t has to be chosen such that $|t(\omega^2 + i\varepsilon)|_{\varepsilon \to 0}$ will be slightly smaller than one. The expression of

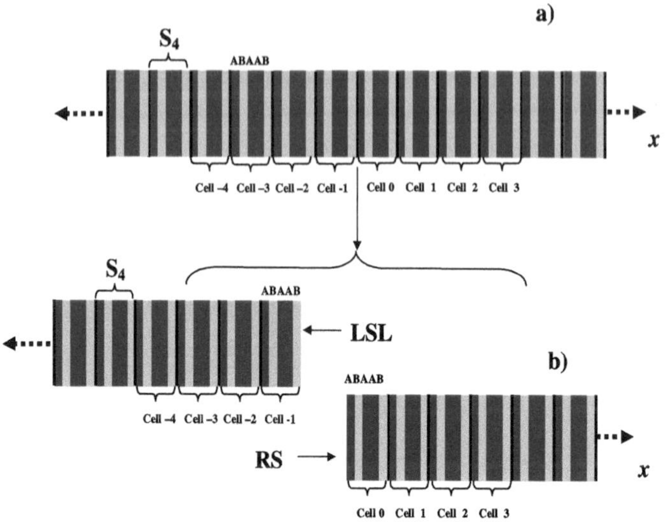

Figure 13.2 (a) Schematic representation of an infinite periodic superlattice where each cell is composed of a Fibonacci generation (for example, S_4). (b) Schematic representation of two semiinfinite superlattices obtained by the cleavage of the infinite superlattice. We call RSL and LSL the right- and left-hand side superlattices.

the Green function between any two points of the infinite superlattice is given by [65]:

$$\begin{aligned} g(n,i,x;n',i',x') &= \delta_{nn'}\delta_{ii'}U_i(x,x') \\ &\quad + \frac{1}{S_i S_i'}\left\{ \sinh(\alpha_i(\frac{d_i}{2}-x)); \sinh(\alpha_i(\frac{d_i}{2}+x))\right\} \\ &\quad \times g(M_m, M_{m'}) \left\{ \begin{array}{c} \sinh(\alpha_{i'}(\frac{d_{i'}}{2}-x')) \\ \sinh(\alpha_{i'}(\frac{d_{i'}}{2}+x')) \end{array} \right\}, \end{aligned} \tag{13.8}$$

where

$$\begin{aligned} U_i(x,x') &= -\frac{1}{2F_i}\exp[-\alpha_i|x-x'|] \\ &\quad + \frac{1}{2F_i S_i}\left\{ \sinh(\alpha_i(\frac{d_i}{2}-x'))\exp(-\alpha_i(\frac{d_i}{2}+x)) \right. \\ &\quad \left. + \sinh(\alpha_i(\frac{d_i}{2}+x'))\exp(-\alpha_i(\frac{d_i}{2}-x)) \right\}. \end{aligned} \tag{13.9}$$

In Eq. (13.8), the last three terms are the product of a 1×2 matrix by the 2×2 $g(M_m, M_{m'})$ matrix and by a 2×1 matrix; $g(M_m, M_{m'})$ is the 2×2 matrix formed out of the elements given by Eq. (13.6), for $m \equiv (n, i, \pm\frac{d_i}{2})$ and $m' \equiv (n', i', \pm\frac{d_i'}{2})$. Now, by cutting an infinite superlattice between two cells, one obtains two semiinfinite superlattices. We shall call them left superlattice (LSL) and right superlattice (RSL) (Fig. 13.2(b)). The dispersion relation of a semiinfinite superlattice with $H = 0$ surface boundary condition (Fig. 13.2(b)) is given by

$$T_{21} = 0, \tag{13.10}$$

where T_{21} is given by Eq. (13.2), together with the condition

$$|T_{22}| > 1 \ (|T_{22}| < 1) \quad \text{for the } RSL \ (LSL). \tag{13.11}$$

This last condition ensures that the wave is decaying when penetrating into the superlattice far from the surface. The Green function in the space of the interfaces of the semiinfinite superlattice is given by [65]

$$d(n, i, -\frac{d_i}{2}; n, j, -\frac{d_j}{2}) = g(n, i, -\frac{d_i}{2}; n, j, -\frac{d_j}{2}) + \frac{t^{n+n'+1}}{t^2-1}\frac{Y_i Y_j}{T_{21}}, \tag{13.12}$$

where $Y_1 = T_{22}$ for $i = 1$, and $Y_i = -t(T_{11})_{1,\ldots,i-1} + (T_{22})_{i,\ldots,N}$ for $i = 2, \ldots, N$. The Green's function in the whole space of the semiinfinite superlattice is given by Eq. (13.8) by replacing $g(M_n, M_{n'})$ (Eq. (13.6)) with $d(M_n, M_{n'})$ (Eq. (13.12)).

13.2.2 Density of states and displacement fields

The local density of states at a point x of the structure is given by

$$n(\omega; n, i, x) = -\frac{1}{\pi}\text{Im}\, d^+(\omega; n, i, x; n, i, x), \tag{13.13}$$

where $d^+(\omega) = \lim_{\varepsilon\to 0} d(\omega + i\varepsilon)$. The total density of states is obtained by integrating over x and summing over n and i the local density of states. We are particularly interested in the total density of states of the semiinfinite superlattice (Fig. 13.2(b)) from which the contribution of the infinite superlattice is subtracted (Fig. 13.2(a)).

This variation $\Delta_n(\omega)$ can be written as the sum of the variations of the densities of states $\Delta_i n(\omega)$ in any block i of the superlattice,

$$\Delta n(\omega) = \sum_{i=1}^{N} \Delta_i n(\omega), \tag{13.14}$$

where

$$\Delta_i n(\omega) = -\frac{Z_i}{\pi}\text{Im}\sum_{n=0}^{+\infty}\int_{-d_i/2}^{d_i/2}[d(n, i, x; n, i, x) - g(n, i, x; n, i, x)]dx, \tag{13.15}$$

d and g are the Green functions of the semiinfinite and the infinite superlattice, respectively. By replacing d and g with their respective values, one obtains:

$$\begin{aligned}\Delta_i(\omega) = \frac{Z_i\omega}{\pi}\text{Im}\,&\frac{t}{(t^2-1)^2}(\frac{1}{\alpha_i S_i^2 T_{21}})[(Y_i^2 + Y_{i+1}^2)(C_i S_i - \alpha_i d_i)\\ &+ 2Y_i Y_{i+1}(\alpha_i d_i C_i - S_i)], \qquad i = 1, 2, \ldots, N-1,\end{aligned} \tag{13.16}$$

$$\Delta_N(\omega) = \frac{Z_N\omega}{\pi}\text{Im}\,\frac{t}{(t^2-1)^2}(\frac{1}{\alpha_N S_N^2 T_{21}})[(Y_N^2+(tY_1)^2)(C_N S_N - \alpha_N d_N)$$
$$+2Y_N Y_1 t(\alpha_N d_N C_N - S_N)]. \tag{13.17}$$

The electric field associated to the surface modes of the semiinfinite N-layer superlattice (Eq. (13.10)) is given by [65]:

$$E_i(n,x) \propto \frac{1}{S_i}[Y_i \sinh\alpha_i(\frac{d_i}{2}-x) + Y_{i+1}\sinh\alpha_i(\frac{d_i}{2}+x)]e^{inqD}, \tag{13.18}$$
$$i = 1, 2, \ldots, N-1,$$

$$E_N(n,x) \propto \frac{1}{S_N}[Y_N \sinh\alpha_N(\frac{d_N}{2}-x) + tY_1\sinh\alpha_N(\frac{d_N}{2}+x)]e^{inqD}. \tag{13.19}$$

13.3 Fibonacci superlattice

In this section we study the bulk bands and surface modes in a Fibonacci superlattice composed of a periodic repetition of cells, where each cell is constituted of segments and loops arranged according to a Fibonacci sequence. In these structures, we generalize a theoretical rule on the surface modes, namely when one considers two semiinfinite superlattices obtained by the cleavage of an infinite superlattice, there exists exactly one surface mode in each gap. This mode is localized on the surface either of one or the other semiinfinite superlattice. We discuss the existence of various types of surface modes and their spatial localization. The experimental observation of these modes is carried out in an original way, namely by measuring the transmission through a guide along which a finite superlattice (i.e., constituted of a finite number of quasiperiodic cells) is grafted vertically. The surface modes appear as maxima of the transmission spectrum. These experiments are in good agreement with the theoretical model based on the formalism of the Green's function

13.3.1 Bulk modes

Fig. 13.3(a) shows the allowed and prohibited frequencies (distribution of the bandwidths) of a Fibonacci superlattice versus the generation number k going until the eighth generation (34 blocks). Higher generations are shown in Fig. 13.3(b) for frequencies around the center of the spectrum. In Fig. 13.3, the limits of the bulk bands are given by $QD = 0$ and $QD = \pi$ in Eq. (13.5) in such a way that one has an alternation from one band to another according to the sequence $0, \pi, \pi, 0, 0, \pi, \pi, 0, \ldots$ The number of the allowed bands is equal to the number of blocks F_k in each generation for $f < 100$ MHz. Actually, it is sufficient to discuss the band structure of the Fibonacci structure in the frequency range $f < 100$ MHz. Indeed, with our choice of the lengths of the cables ($d_A = d_B = d = 1$ m), the whole spectrum of the Fibonacci structure is repeated periodically every $\frac{\omega d\sqrt{\varepsilon}}{c} = \pi$ (i.e., $f \simeq 98.68$ MHz). One can notice that, as expected, the allowed bands become narrower as k increases, showing the localized character of the modes. Indeed, we have calculated the total width Δ_k of the allowed bands and shown that it is connected to the number of blocks F_k for each generation k according to the power law (see Fig. 13.3(c))

$$\Delta_k = F_k^{-\delta} \tag{13.20}$$

with $\delta \simeq 0.137$. For higher values of k, Δ_k (the Lebesgue measure of the energy bands) tends towards zero, consequently all the states become critical. This result, specific to quasiperiodic structures, was found in other types of excitations such as phonons or magnons [32]. One can also notice in Fig. 13.3(a), the existence of stable gaps around $f = 22, 35, 65$ and 78 MHz (labeled S_g), which appear for all the generation numbers and transient gaps (labeled T_g) around the central frequency $f_c = \frac{\omega_c d\sqrt{\varepsilon}}{c} = \frac{\pi}{2}$ (i.e., $f_c = 49.34$ MHz) for S_2, S_5, S_8, etc. [66], which appear every three generations. This is illustrated in Fig. 13.3(b) where a large zoom of the bandwidths around f_c up to the 16th generation is sketched. In Figs. 13.3(a) and 13.3(b), one can notice that the width of the stable gaps remains almost constant for all generation numbers, while the width of the transient gaps decreases as the number of generation increases. However, the widths of all these gaps increase, leading to a decreasing of the bulk bands widths following the power law (Eq. (13.20)).

In order to show experimentally the bulk band fragmentation versus the generation numbers, we measured the transmission spectrum through finite periodic structures placed between two semiinfinite waveguides [17] for different generations. Fig. 13.4 represents the theoretical (solid line) and experimental (open circles) transmission spectra for a periodic Fibonacci structure made up of four cells. Each cell corresponds to sequences S_2, S_3, S_4, and S_5. The horizontal solid lines correspond to the widths of the bulk bands of an infinite periodic Fibonacci structure. The transmission measurements have been realized by using the tracking generator coupled to a spectrum analyzer in the frequency range up to 100 MHz. The attenuation inside the coaxial cables was simulated by introducing a complex relative dielectric permittivity ($\varepsilon = \varepsilon' - j\varepsilon''$). The attenuation coefficient α'' can be expressed as $\alpha'' = \varepsilon''\omega/c$. On the other hand, the attenuation specification data supplied by the manufacturer of the coaxial cables in the frequency range up to 100 MHz can be approximately fitted with the expression $\ln(\alpha'') = a + b\ln(\omega)$, where a and b are two constants. From this fitting procedure, a useful expression for ε'' as a function of frequency can be obtained under the form $\varepsilon'' = 0.017 f^{-0.5}$ where the frequency f is expressed in hertz. The experimental results are very well fitted by the 1D model using the Green's function method. The results in Fig. 13.4 clearly show the fragmentation of the spectrum (subdivision of the bands) when one goes from a Fibonacci generation to a higher generation. Despite of the low number of cells (four) used in the finite structure, the transmission spectra describe very well the band structure of the infinite system. The attenuation effect in coaxial cables induces a transmission depletion, particularly at high frequencies. Also, this effect limits the study of high generation structures where the small gaps are not

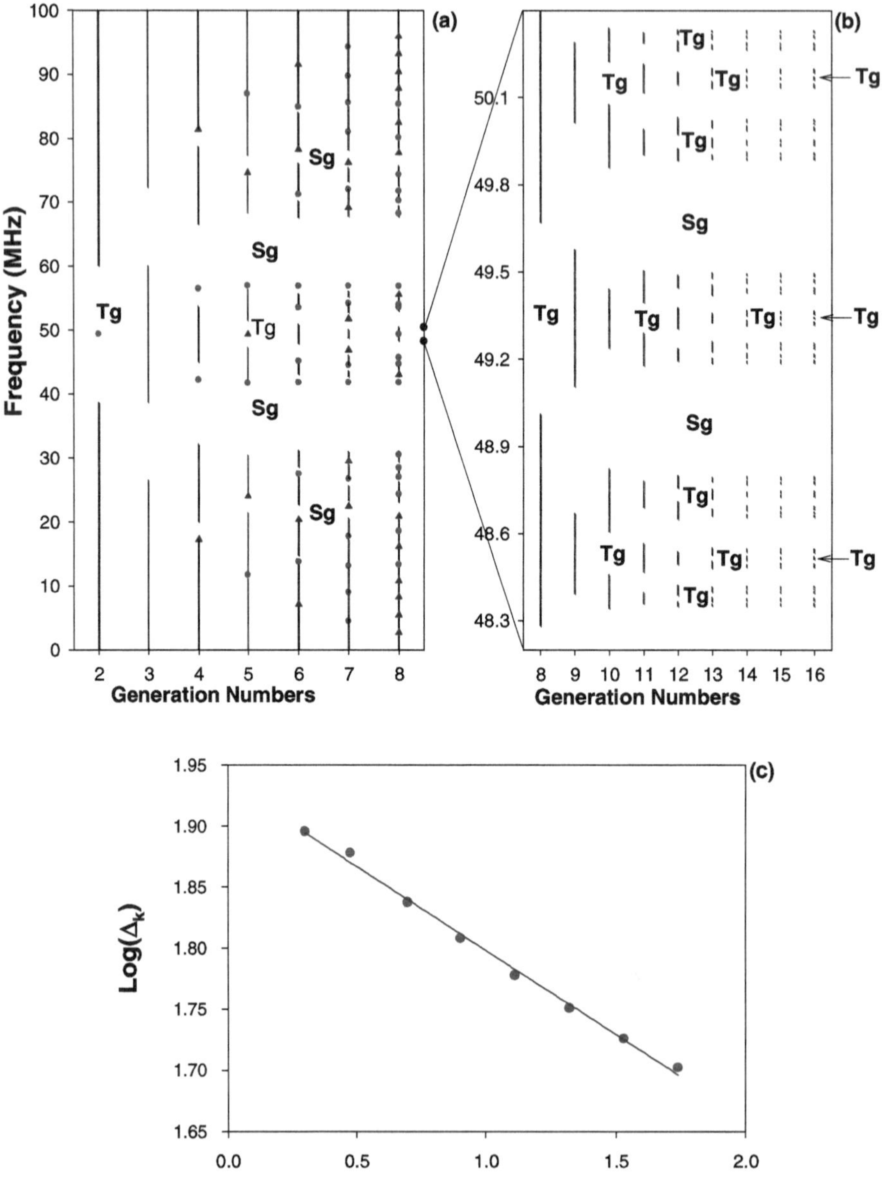

Figure 13.3 (a) Distribution of the bandwidths (vertical lines) as a function of the generation number; T_g and S_g indicate the positions of transient and stable gaps, respectively. Open circles and full triangles correspond to the surface modes of two complementary superlattices RSL and LSL (Fig. 13.2(b)), respectively (see discussion in the text). (b) Same as in (a) but for high generations in the frequency range [48.2, 50.4] MHz around the central frequency $f_c = 49.34$ MHZ in order to show the positions of the transient gaps (T_g) around f_c which appear every three generations. (c) Logarithmic presentation of the total width $\log(\Delta_k)$ of the allowed bands versus the number of blocks $\log(F_k)$ for each generation k.

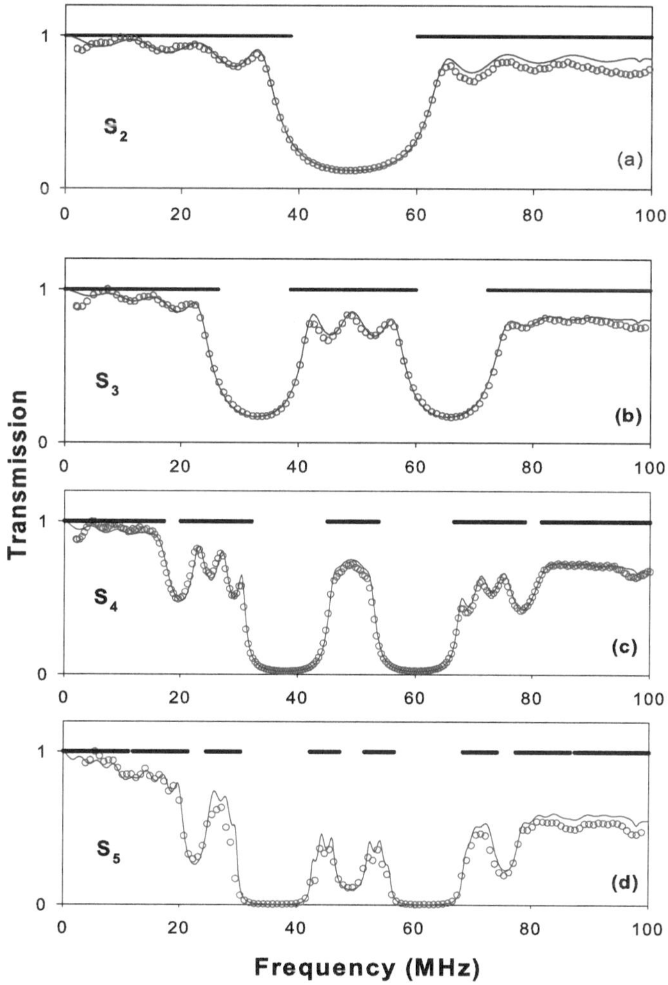

Figure 13.4 Transmission spectrum as a function of the frequency for a finite periodic structure made up of four cells. Each cell is a Fibonacci generation S_k $(k = 2, 3, 4, 5)$. Solid lines and open circles represent theoretical and experimental results, respectively. The horizontal segments correspond to the bandwidths of the infinite superlattices (Fig. 13.3(a)).

well defined in the transmission spectra (see Fig. 13.4(d)). For this reason, it is difficult to give an experimental verification of the power law (Eq. (13.20)) property of the Fibonacci structures studied here.

13.3.2 Surface modes: numerical results

Some years ago, El Boudouti et al. [65] have shown theoretically in the case of the transverse elastic waves that the creation of two semiinfinite superlattices from an infinite N-layer superlattice enables the creation of one surface mode per gap for any value of the wave vector parallel to the interfaces. This result was confirmed experimentally in a two-layer metallic superlattice Ag/Al [67]. These results have been confirmed also in a comb-like photonic structure made out by resonators grafted periodically along a guide [58] and serial loop structures [57]. In this section, we present an extension of this work to the case of a semiinfinite Fibonacci superlattice. In particular, as in our previous works, we consider two complementary semiinfinite superlattices obtained by the cleavage of an infinite superlattice (Fig. 13.2(b)). One can notice that when one goes from one generation to the next, the surface terminations of these two superlattices do not behave in the same way. For example, the cleavage of an infinite superlattice corresponding to the fourth Fibonacci generation $\ldots|ABAAB|ABAAB|ABAAB|ABAAB|\ldots$ gives rise to what we have called LSL and RSL semiinfinite superlattices:

$$\ldots|ABAAB|ABAAB| \quad \text{and} \quad |ABAAB|ABAAB|\ldots \tag{13.21}$$

In the same way, one obtains for the fifth Fibonacci generation

$$\ldots|ABAABABA|ABAABABA| \quad \text{and} \quad |ABAABABA|ABAABABA|\ldots \tag{13.22}$$

These results show that the terminations of the RSL remain unchanged, whereas the terminations of the LSL change from one generation to the next. However, the LSL terminations remain similar for even generations (with block B at the surface) and odd generations (with block A at the surface).

In Fig. 13.3(a), we represented the surface modes for two complementary semiinfinite superlattices for various generations until the eighth generation. The open circles give the surface modes of the RSL and the full triangles those of the complementary LSL. These modes are obtained from the dispersion relation (Eqs. (13.10) and (13.11)). One can notice the existence of one surface mode per gap, this mode is associated with the surface of one of the two semiinfinite periodic structures. To explain the origin of this result, we illustrated in Fig. 13.5, the variation of the density of states between the semiinfinite superlattice and the same volume of an infinite superlattice (Eqs. (13.16) and (13.17)) for both superlattices RSL (Fig. 13.5(a)) and LSL (Fig. 13.5(b)) corresponding to the fourth generation. The localized surface modes appear as delta peaks of weight unity. These peaks are enlarged artificially by adding a small imaginary part to the frequency ω. At the bulk band edges appear delta peaks of weight $-\frac{1}{4}$, these peaks do not display exactly the same weight because of the divergences which exist at the bulk band edges in the density of states of 1D systems. Now, when one considers together the two complementary superlattices (Fig. 13.5(c)): (i) the variation of density of states presents a loss of one half-mode at the limit of each band, i.e., one mode by band; (ii) the variation of density of states vanishes inside the bulk bands of the two complementary structures. These two results associated with the necessary conservation of the total number of modes show that there must exist a surface mode of weight unity per gap (Fig. 13.5(c)) to compensate the loss of one mode per band. It is worth noticing that the analytical expression giving the frequencies of

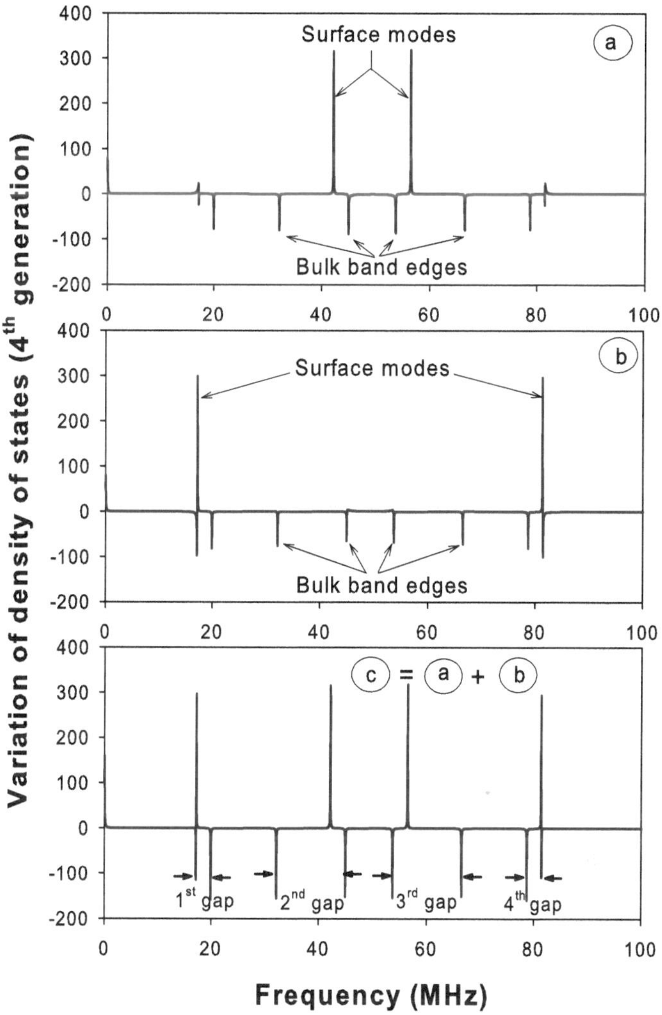

Figure 13.5 (a) Variation of the total density of states (in arbitrary units) between the semiinfinite superlattice RSL (Fig. 13.2(b)) and the same volume of the infinite superlattice (Fig. 13.2(a)) as a function of the frequency for the fourth generation. Positive δ peaks indicate the positions of surface modes, while the negative $-\frac{1}{4}\delta$ peaks correspond to the limits of the bulk bands. (b) Same as (a) for the LSL superlattice. (c) Same as (a), but for the two complementary superlattices.

the surface modes for two complementary semiinfinite superlattices (Eq. (13.10)) is exactly the same as that giving the standing waves of one unit cell with $H = 0$ surface boundary condition [65] (see also Section 13.4). In Fig. 13.3(a), one can distinguish two types of surface modes:

(i) The modes falling inside the stable gaps. These modes appear almost at the same frequency for all the generations of the RSL (open circles) and for one generation over two of the LSL (full triangles). This result is understandable since, as mentioned above, the terminations of the RSL remain unchanged as a function of the generation number, while the terminations of the LSL changes in an alternating way for even and odd generations. This result is well illustrated in Fig. 13.6 where we have magnified Fig. 13.3 in the frequency intervals $[10, 30]$ MHz (Fig. 13.6(a)) and $[40, 60]$ MHz (Fig. 13.6(b)), and going until the 11th generation. Let us notice that the frequency region $[70, 90]$ MHz is equivalent to $[10, 30]$ MHz since the frequency spectrum of Fig. 13.3(a) is symmetrical around the central frequency f_c. Let us mention that similar results were found for magnetostatic modes in Fibonacci structures made of magnetic and nonmagnetic layers [68].

(ii) The modes falling inside the transient gaps around f_c (Fig. 13.3(b)). These modes appear at the same frequency each six generations according to the splitting pattern presented in Fig. 13.7 for the RSL. It is seen that one "central" frequency is formed after six generations, out of which two surface modes are formed in the transient gaps (Fig. 13.3(b)) after two generations. Then the splitting pattern repeats. The same diagram is obtained for the LSL where a surface mode appears between two surface modes of the same frequency of the RSL.

An analysis of the electric field or equivalently the local density of states (Fig. 13.8), enables us to study the spatial localization of the two types of surface modes discussed above. The localization length is defined as $\ell = \frac{1}{|\chi|}$ where χ is the imaginary part of the Bloch wavevector Q obtained from Eq. (13.5). The surface modes falling inside the stable gaps present almost the same localization length (~ 8 m), as it is illustrated in Figs. 13.8(a) and 13.8(b) for the modes labeled 1 and 2 in Fig. 13.6(b) for the generations S_6 and S_8, respectively. On the other hand, the modes inside the transient gaps present a localization length proportional to the period D. Indeed, one can show analytically that $\ell = D/\log 2$ for the modes falling at the central frequency $f_c = 98.34$ MHz. The same expression was found also by Zijlstra et al. [69] for surface states in Fibonacci superlattices using a tight binding model. We check numerically that this value remains almost the same for the surface modes in the transient gaps lying at the vicinity of f_c (Fig. 13.7). This result shows that, contrary to the modes in the stable gaps, the localization length of the modes in the transient gaps increases with the generation number as it is illustrated in Figs. 13.8(c) and 13.8(d) for the modes labeled 3 and 4 in Fig. 13.7 for S_8 and S_{14}, respectively. The spatial localization of the two types of surface modes discussed above present different behaviors as a function of the space position x. The surface modes in the stable gaps decrease from the surface while penetrating into the bulk of the superlattice (Figs. 13.8(a) and 13.8(b)) as in ordinary periodic superlattices. However, the modes in the transient gaps present an important localization at the middle of each cell and a self-similar behavior of order six (see Figs. 13.8(c) and 13.8(d) for S_8 and S_{14}, respectively). These results show that the electric field around the central frequency is neither extended nor localized. These spectra, which are due to the quasiperiodicity of the blocks A and B inside each unit cell, can be characterized by the multifractal analysis. This concept has been successfully used to understand the critical wave functions in Fibonacci tight binding model [70] and k-component Fibonacci quantum

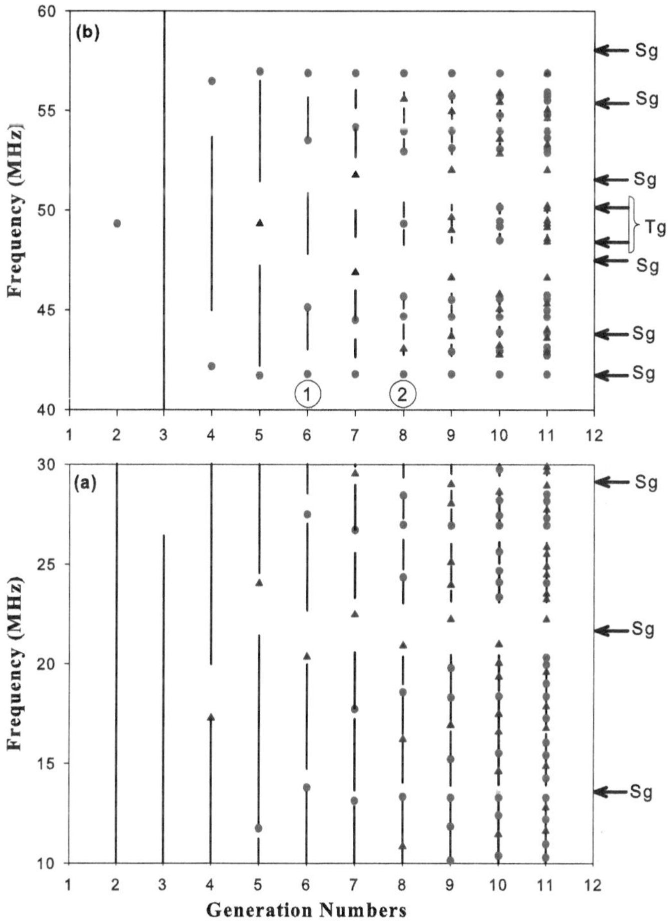

Figure 13.6 Same as in Fig. 13.3, but enlarged in the range of frequencies [10, 30] MHz (a) and [40, 60] MHz (b) in order to illustrate clearly the positions of the different surface modes in the stable and transient gaps. The positions of these gaps are indicated by arrows on the right side of these curves.

waveguide [71]. However, as far as we know, this concept has not been applied for localized surface modes lying in the transient gaps.

The multifractal analysis enables us to describe the statistical properties of measure. If one covers the support of the measure by a partition of boxes of size ε and one denotes by $p_i(\varepsilon)$ the probability (integrated measure) in the ith box, an exponent (singularity strength) α_i can be defined as

$$p_i(\varepsilon) \simeq \varepsilon^{\alpha_i}. \tag{13.23}$$

If we count the number of boxes $N(\alpha)d\alpha$ where the probability p_i has singularity strength between α and $\alpha + d\alpha$, then $f(\alpha)$ can be defined as the fractal dimension of

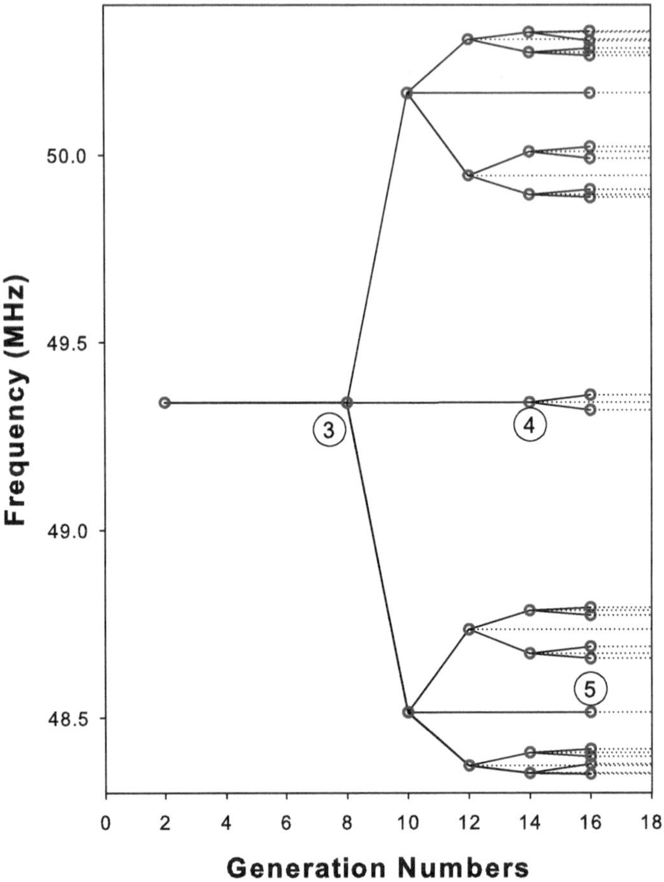

Figure 13.7 Splitting pattern of surface modes inside the transient gaps of Fig. 13.5 for the RSL superlattice around the central frequency $f_c = 49.34$ MHz (i.e., $\frac{\omega_c d\sqrt{\varepsilon}}{c} = \frac{\pi}{2}$).

the whole set of boxes of singularity strength α as

$$N(\alpha)d\alpha \sim \varepsilon^{-f(\alpha)}d\alpha. \tag{13.24}$$

Several models based on the calculation of the generalized fractal dimension D_q were elaborated [72–74] to calculate α and $f(\alpha)$ using a Legendre transform. However, the application of these methods presents several difficulties using the Legendre transform itself, in particular when $f(\alpha)$ exhibits discontinuities [75]. To solve this problem, a new precise mathematical method was proposed by Chabra et al. [76]. It allows the calculation of $f(\alpha)$ with an excellent numerical precision. This method was applied successfully for several elementary excitations in 1D systems [32,71], and it is the method used here. In this method, the first step in multifractal characterization is the probability measure p_i. In our case, we choose the probability p_i as the weight

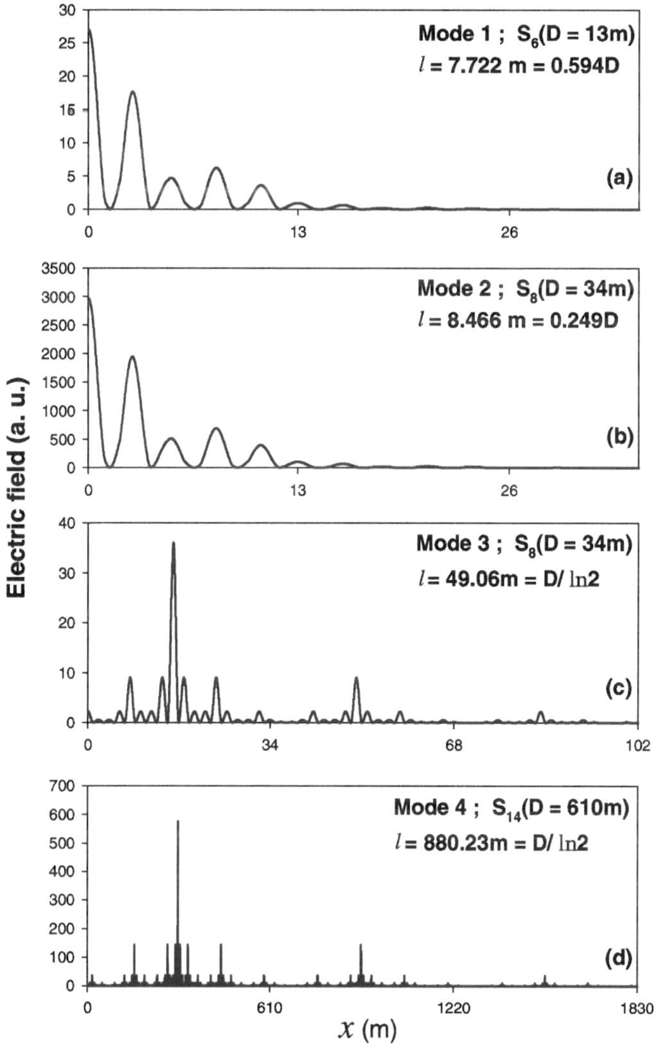

Figure 13.8 (a)–(b) Modulus of the electric field as a function of the space position for the modes labeled 1 and 2 in Fig. 13.6(b), respectively. (c)–(d) Same as in (a)–(b) for the modes labeled 3 and 4 in Fig. 13.7, respectively.

of the electric field in the density spectrum, i.e.,

$$p_i = \frac{|E_i|^2}{\sum_{j=1}^{N} |E_j|^2}, \tag{13.25}$$

where $N = F_k$ is the total number of blocks considered in one cell of the generation k and $|E_i|$ is the mean electric field in the block i. Then, we construct a partition function

$$Z(q) = \sum_{i=1}^{N} p_i^q, \tag{13.26}$$

where the parameter q provides a microscope for exploring the singular measure in different regions. The mulifractal spectrum $f(\alpha)$ is obtained by varying the parameter q in Eq. (13.26) and calculating [71]

$$\alpha = -\frac{1}{\log N}\frac{Z'(q)}{Z(q)}, \tag{13.27}$$

$$f(\alpha) = \frac{1}{\log N}(\log Z(q) - q\frac{Z'(q)}{Z(q)}), \tag{13.28}$$

with

$$Z'(q) = \frac{dZ}{dq} = \sum_{i=1}^{N} p_i^q \log p_i.$$

The $f(\alpha)$ singularity spectrum gives a precise description of the distribution of the electric field in the structure. Indeed, it is well established [70] that $f(\alpha)$ is reduced to a single point $f(\alpha = 1) = 1$ for extended modes and $f(\alpha = 0) = 0$ and $f(\alpha = \infty) = 1$ for localized modes. However, for self-similar modes, $f(\alpha)$ is expected to have a smooth function defined in a finite region $[\alpha_{\min}, \alpha_{\max}]$.

The numerical results of $f(\alpha)$ for the electric field associated to the modes labeled 4 and 5 in Fig. 13.7 are sketched in Fig. 13.9(a). These results clearly show that the modes lying in the transient gaps are neither extended nor localized, but possess multifractal properties which are characteristic of self-similar modes. In Fig. 13.9(a), $\alpha_{\min}$ and $\alpha_{\max}$ characterize the scaling property of the most concentrated and most rarefied regions of the density measure, respectively [32]. The quantity $\Delta_\alpha = \alpha_{\max} - \alpha_{\min}$ can be used as a parameter reflecting the randomness of the intensity measure. Contrary to localized modes lying in transient gaps, the electric field associated to surface modes lying in the stable gaps (Figs. 13.8(a)–(b)) exhibit the property of localized waves, namely $\alpha_{\min} \simeq 0$ and $\alpha_{\max}$ increases rapidly with the increasing generation number. This result is shown in Fig. 13.9(b) for the surface mode 1 at $\Omega = 41.788$ in Fig. 13.6(a).

13.3.3 Surface modes: experimental results

In order to give an experimental evidence of the existence of surface modes in Fibonacci superlattices, we propose an original method that consists to attach vertically a finite periodic Fibonacci structure to a horizontal waveguide (Fig. 13.10). Indeed, the transmission coefficient through such a structure may give us information on all the eigenmodes of the finite structure with $H = 0$ surface boundary condition. Let $g^{-1}(M, M)$ be the inverse of the Green's function of the finite structure in the space

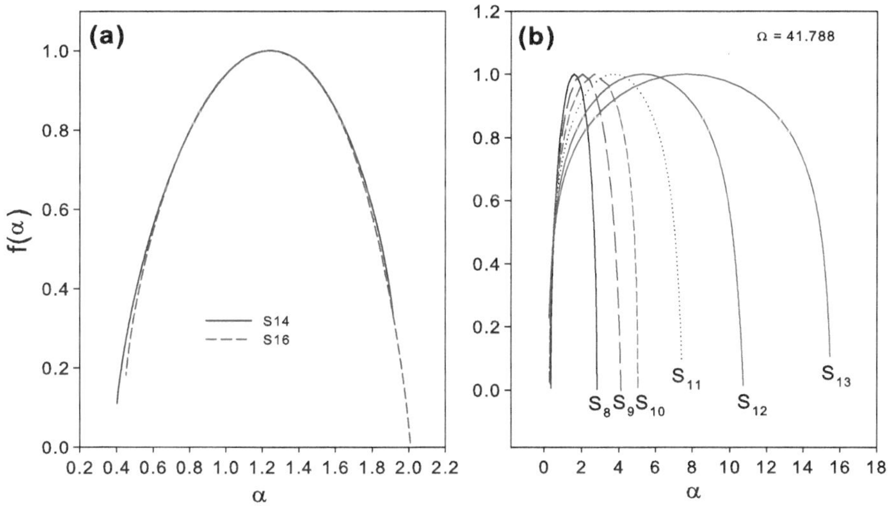

Figure 13.9 (a) The $f(\alpha)$ curve for the distribution of the modulus of the electric field corresponding to the modes labeled 4 (solid line) and 5 (dashed line) in Fig. 13.7. The generations of these modes are the 14th and 16th, respectively. (b) Same as in (a) but for the surface mode 1 at $\Omega = 41.788$.

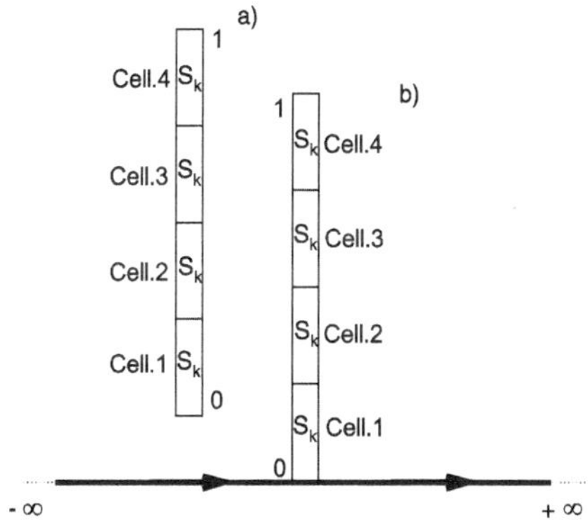

Figure 13.10 (a) Finite superlattice made up of 4 cells, each cell is a Fibonacci generation S_k. (b) Finite superlattice attached vertically to a horizontal guide.

of the interfaces $M = \{0, 1\}$ (Fig. 13.10(a)). Then $g^{-1}(M, M)$ can be written in the form

$$g^{-1}(M, M) = \begin{pmatrix} g^{-1}(0,0) & g^{-1}(0,1) \\ g^{-1}(1,0) & g^{-1}(1,1) \end{pmatrix}. \tag{13.29}$$

If one grafts vertically this structure to two semiinfinite waveguides of impedances Z_s on the site $\{0\}$ (Fig. 13.10(b)), one obtains the Green's function of the final system on this site as follows [57,58]:

$$d(0,0)=\frac{g^{-1}(1,1)}{[g^{-1}(0,0)g^{-1}(1,1)-g^{-1}(0,1)g^{-1}(1,0)]-2jg^{-1}(1,1)/Z_s}. \tag{13.30}$$

The site {0} is the point where the reflection and the transmission of an incident wave launched from the left semiinfinite waveguide $(-\infty,0)$ (Fig. 13.10(b)) is carried out. The transmitted wave in the right semiinfinite waveguide $(0,+\infty)$ is given by

$$t=\frac{-2jg^{-1}(1,1)/Z_s}{[g^{-1}(0,0)g^{-1}(1,1)-g^{-1}(0,1)g^{-1}(1,0)]-2jg^{-1}(1,1)/Z_s}. \tag{13.31}$$

The eigenmodes of the finite structure with $H=0$ surface boundary condition (Fig. 13.10(a)) are given by

$$\det\, g^{-1}(M,M)=g^{-1}(0,0)g^{-1}(1,1)-g^{-1}(0,1)g^{-1}(1,0)=0$$

(Eq. (13.29)). Consequently, the transmission coefficient (Eq. (13.31)) becomes equal to unity. This result shows that all the eigenmodes of the finite structure can be obtained from the maxima of the transmission coefficient of the same structure grafted vertically on an horizontal guide. Some transmission spectra for the generations S_3 and S_4 corresponding to a finite superlattice composed of $N=4$ grafted cells, are presented in Fig. 13.11 for the generations S_3 and S_4. Theoretical results (solid lines) are in good agreement with experimental curves (open circles). As mentioned before, these curves present minima and maxima. The frequencies corresponding to the maxima of these curves give the eigenmodes of the finite structure with $H=0$ surface boundary condition. These modes are shown by open circles in Fig. 13.12(a) for a superlattice composed of $N=4$ cells for each generation. The vertical solid lines represent the bulk bands of the infinite superlattice similarly to Fig. 13.3(a). For $N=4$ cells, one obtains besides the surface modes falling in the gaps, $N-1=3$ modes in each band corresponding to the four cells constituting the finite structure. These results confirm the theorem announced in Chapter 12 about the existence of $N-1$ confined modes in each band and one surface mode in each gap for a finite superlattice composed of N cells [57]. The results in Fig. 13.12(a) show again the fragmentation of the bands when the generation number increases. An analysis of the local density of states as a function of the space position (Fig. 13.12(b)) enables us to distinguish the modes associated with the surrounding surfaces of the finite structure. For example, in the case of four cells, the modes labeled 1 and 2 (Fig. 13.12(a)) are associated with the surface grafted on the guide, while the mode 3 is associated to the other surface

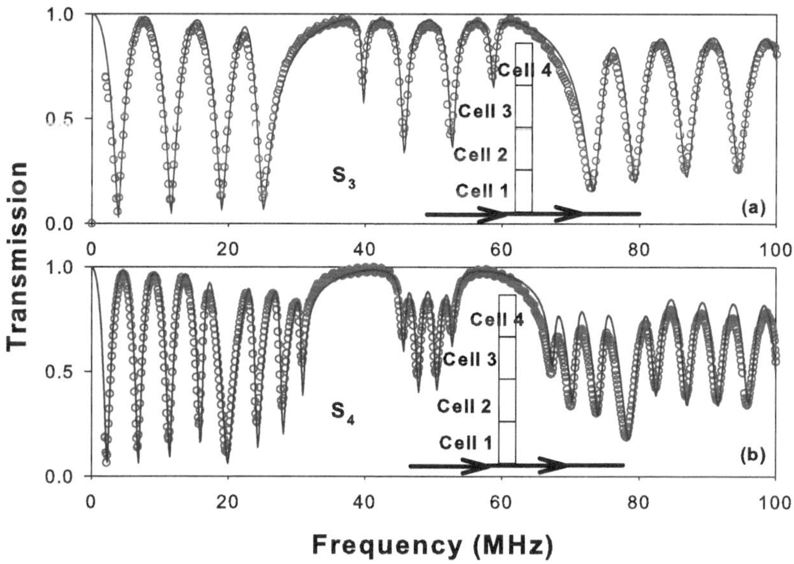

Figure 13.11 Transmission spectrum of a finite periodic structure grafted vertically to a guide. The finite structure is made up of four cells. The blocks A (segment) and B (loop) in the unit cell respect the Fibonacci generation S_k ($k = 3$ (a) and $k = 4$ (b)).

with $H = 0$ boundary condition. Despite the small number of cells (four), the spatial localization of the surface modes is analogous to that of surface modes in semiinfinite superlattices (see Figs. 13.8(a)–(b)).

13.4 Fibonacci sequence

In this section, we study theoretically and experimentally the properties of quasiperiodic one-dimensional serial loop structures made of segments and loops arranged according to one Fibonacci sequence (FS). Two systems are considered: (i) by inserting the FS horizontally between two waveguides, we give an experimental evidence of the scaling behavior of the amplitude and the phase of the transmission coefficient; and (ii) by grafting vertically the FS along a guide, we obtain from the maxima of the transmission coefficient the eigenmodes of the finite structure (assuming the vanishing of the magnetic field at the boundaries of the FS). We show that these two systems (i) and (ii) exhibit the property of self-similarity of order three at certain frequencies where the quasiperiodicity is most effective. In addition, because of the different boundary conditions imposed on the ends of the FS, we show that horizontal and vertical structures give different information on the localization of the different modes inside the FS. Finally, we show that the eigenmodes of the finite FS coincide exactly with the surface modes of two semiinfinite superlattices obtained by the cleavage of

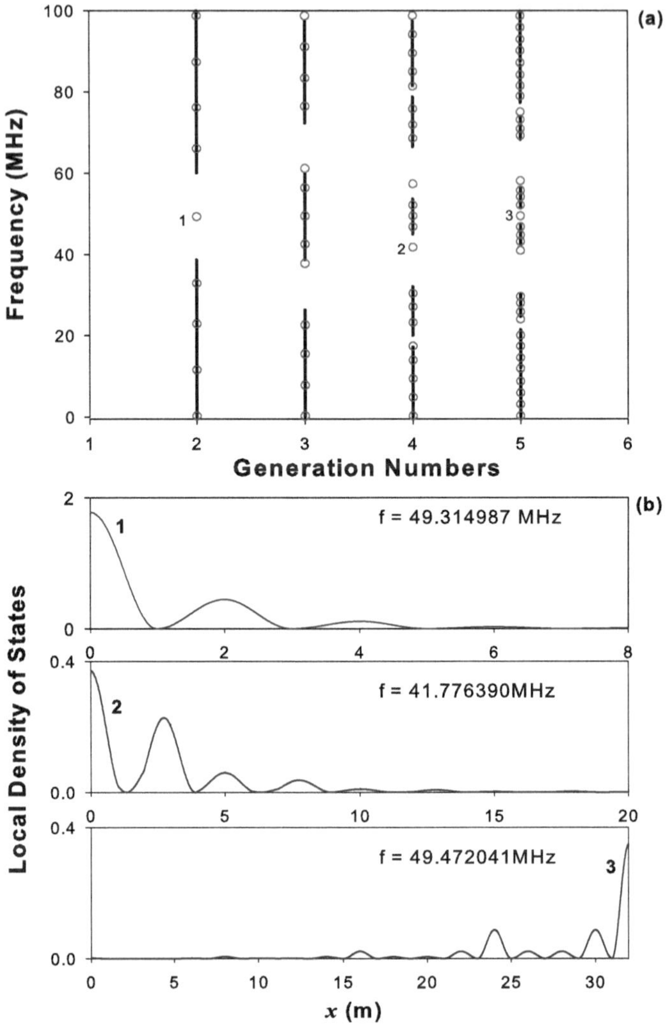

Figure 13.12 (a) Same as in Fig. 13.3(a). The open circles represent the eigenmodes of the finite structure composed of four cells. (b) Analysis of the local density of states as a function of the space position enables to distinguish between the modes associated to the two surfaces of the finite structure. For example, the modes labeled 1 and 2 (Fig. 13.12(a)) are associated with the surface grafted on the guide, while mode 3 is associated with the other surface with $H = 0$ boundary condition.

an infinite superlattice formed by a periodic repetition of a given FS, already discussed in Section 13.3.

We shall focus on the transmission coefficient when a given Fibonacci sequence is either inserted horizontally (Fig. 13.13(a)) between, or grafted vertically (Fig. 13.13(b)) to two semiinfinite waveguides made of cables similar to the one of block A. In particular, we shall show that these two geometric systems may give dif-

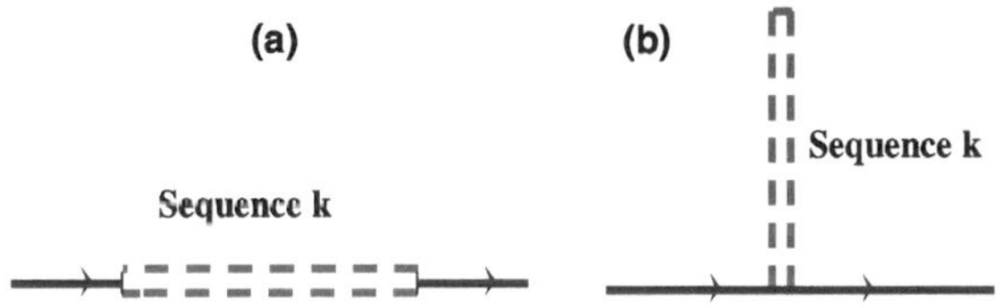

Figure 13.13 (a) Schematic representation of a finite Fibonacci sequence inserted horizontally between two waveguides. (b) Same as (a) but for a sequence grafted vertically along a guide.

ferent information on the propagation and localization of electromagnetic waves in such structures. The theoretical results are the same as in Section 13.2, in which we should take only one sequence instead of a periodic repetition of sequences.

13.4.1 Case of horizontal sequence

In this subsection, we consider a given Fibonacci sequence inserted horizontally between two semiinfinite waveguides (Fig. 13.13(a)). Figs. 13.14(b)–(g) show the transmission coefficient for the generations S_4 (five blocks), S_5 (eight blocks), S_6 (13 blocks), S_7 (21 blocks), S_8 (34 blocks), and S_9 (55 blocks), respectively. The solid (dashed) curves represent the theoretical results with (without) absorption, whereas open circles correspond to the experimental ones. The experimental results are very well fitted by the 1D model using the Green's function method. One can notice in Fig. 13.14 that the attenuation inside the cables induces transmission depletion, especially at high frequencies.

Two regions of frequencies may be distinguished in Figs. 13.14(b)–(g): the regions where the transmission falls down rapidly to zero as the generation number increases, these regions correspond to the forbidden modes (transmission gaps), and the regions where the transmission is more noticeable, these regions correspond to the allowed modes (transmission bands). These results clearly show the existence of more than one gap (transmission dips) in comparison with the transmission spectrum of the periodic case (Fig. 13.14(a)), especially for high generations (Fig. 13.14(g)). This is the consequence of the bandgap fragmentation in quasiperiodic structures. An interesting result in Figs. 13.14(b)–(g) is the existence of new features around the central gap frequency $f_c \simeq 49.34$ MHz of the periodic structure, these resonances present a certain recursive order which is a characteristic of Fibonacci systems. This property, called scaling relation [41,42], has been interpreted as a sign for localization of the waves in Fibonacci systems. Kohmoto et al. [41] have shown that the scaling behavior of the transmission coefficient is characterized by the scaling factor [41,42]

$$F = \sqrt{1 + 4(1 + I)^2} + 2(1 + I), \tag{13.32}$$

where I is an invariant, which remains constant at every step of the recursive procedure. This invariant is given by [41]

$$I = \frac{1}{4}\left(\frac{Z_B}{Z_A} - \frac{Z_A}{Z_B}\right)^2 \sin^2(\omega d\sqrt{\varepsilon}/c) \tag{13.33}$$

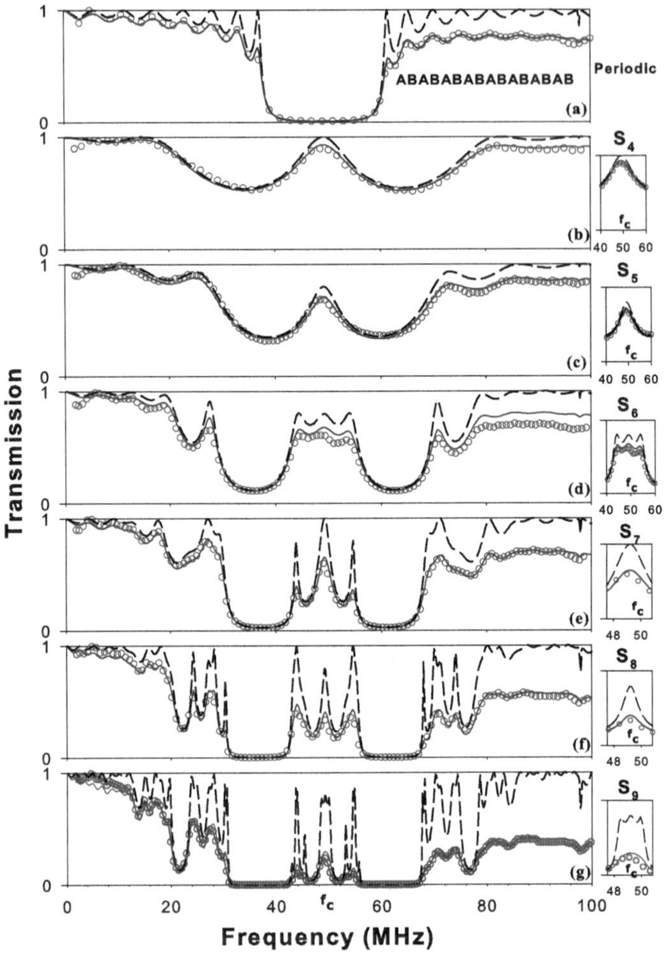

Figure 13.14 Theoretical (solid curves) and experimental (open circles) variations of the transmission coefficient as a function of the frequency for different structures. The block A is a cable of length $d_A = 1$ m and impedance $Z_A = 50\ \Omega$, while the block B is a symmetric loop with two identical cables of length $d_B = 1$ m and $Z_B = 25\ \Omega$: (a) the periodic AB structure, (b)–(g) the generation S_k (k = 4–9) of the Fibonacci structure. The dashed curves in Figs. (a)–(g) are the theoretical curves when the absorption is not taken into account. The blocks A and B in the periodic and Fibonacci structures are the same as in Fig. 13.1. Note the scale change of the frequency axis near the central frequency f_c in the insets at the right-hand side of the figures associated to S_4, S_5, and S_6 compared to S_7, S_8, and S_9, respectively.

where $Z_A = 50\ \Omega$ and $Z_B = 25\ \Omega$ are the impedances of the cables and loops, respectively. Also, it has been demonstrated [41,42] that one can expect scaling around $\Omega = \omega d\sqrt{\varepsilon}/c = (2m+1)\pi/2$ where the quasiperiodicity is most effective (m is an integer). This implies that the transmission coefficient should exhibit a self-similar behavior around the central frequency $f_c = (2m+1)49.34$ MHz with $T_{j+3} = T_j$ (the period of the transmission coefficient is 3, we have 3-recursion) with a scaling factor F. For this frequency, Eqs. (13.32) and (13.33) lead to $I = 0.526$, and thus $F = 6.4061$. This is clearly illustrated in the insets at the right-hand side of Figs. 13.14(b)–(g) near f_c. Note the scale change of the frequency axis in the insets of Figs. 13.14(b), 13.14(c), and 13.14(d) compared to Figs. 13.14(e), 13.14(f), and 13.14(g), respectively. We remark that the S_4 (S_7), S_5 (S_8), and S_6 (S_9) resemble each other near f_c with a periodicity of 3, and a scaling factor F as was found by Gellermann et al. [42] in optical layered media. One can notice, however, that due to the dissipation in the cables the similarities between the different curves around f_c do not show exactly the same features as it should be in the absence of dissipation (see the dashed curves). This is in particular the case for high generations (see the insets of Figs. 13.14(d) and 13.14(g)). However, as it will be shown below, this self-similarity is better illustrated in the transmission phase time.

It is well established that outside the Fibonacci bandgaps, the waves are critically localized [50]. In contrast with the fully disordered (Anderson) localized case, these critically localized modes display weaker than exponential decay, most likely similar to a power law, and have a rich self-similar structure [70]. In order to understand the spatial localization of the different modes in Fig. 13.14, we plotted in Fig. 13.15 the local density of states (LDOS) as a function of the space position x for the modes lying at $f = 98.68$, 49.34, and 68.08 MHz (i.e., for the reduced frequencies $\Omega = \pi, 0.5\pi$, and 0.69π) in Fig. 13.14(f) (the 8th generation). The LDOS reflects the square modulus of the electric field inside the structure. These modes can be classified respectively as: (i) extended modes as it is shown in Fig. 13.15(a) for the completely transparent mode ($\Omega = \pi$) for which the transmission is unity [69]. The wave function, namely the electric field distribution, follows the structure of the Fibonacci sequence (see Fig. 13.15(b)). Similar results are found for electronic and electromagnetic waves in layered media [77,78]; (ii) self-similar modes as it is shown in Fig. 13.15(c) for the mode lying at $\Omega = 0.5\pi$. The corresponding LDOS shows a self-similar behavior [79,80] around the main peak every three generations as it is shown in Fig. 13.15(d) displayed for the 11th generation; and (iii) band-edge modes as it is shown in Fig. 13.15(e) for the mode $\Omega = 0.69\pi$. The LDOS shows a noticeable similarity to the band edge resonances occurring in the periodic structure (see Fig. 13.15(f)) but are less regular. Band edge resonances in photonic periodic crystals are shown to be not localized states since their extension scales linearly with the system size and they do not decay to zero [79,81]. In contrast, the Fibonacci band edge resonances may decay via a power law due to optical path gradient occurring during the growth of the Fibonacci samples [47,82]. As pointed out in [83], these resonances could serve as a new type of complex cavity in 1D multilayer structures that provides the feedback for laser action in the field of random laser systems.

Another important quantity that characterizes the interaction of the incident photon with the different modes in the FS is the transmission phase time τ_φ. This quantity is interpreted as the time needed for a photon to complete the transmission process. The phase time in symmetric loop structures is equivalent to the total density of states [13,26,84]. Fig. 13.16(a) gives the phase time as a function of the frequency for the periodic structure as in Fig. 13.14(a). Similarly to the density of states in 1D structures, the phase time exhibits large values near the band edges, small oscillations

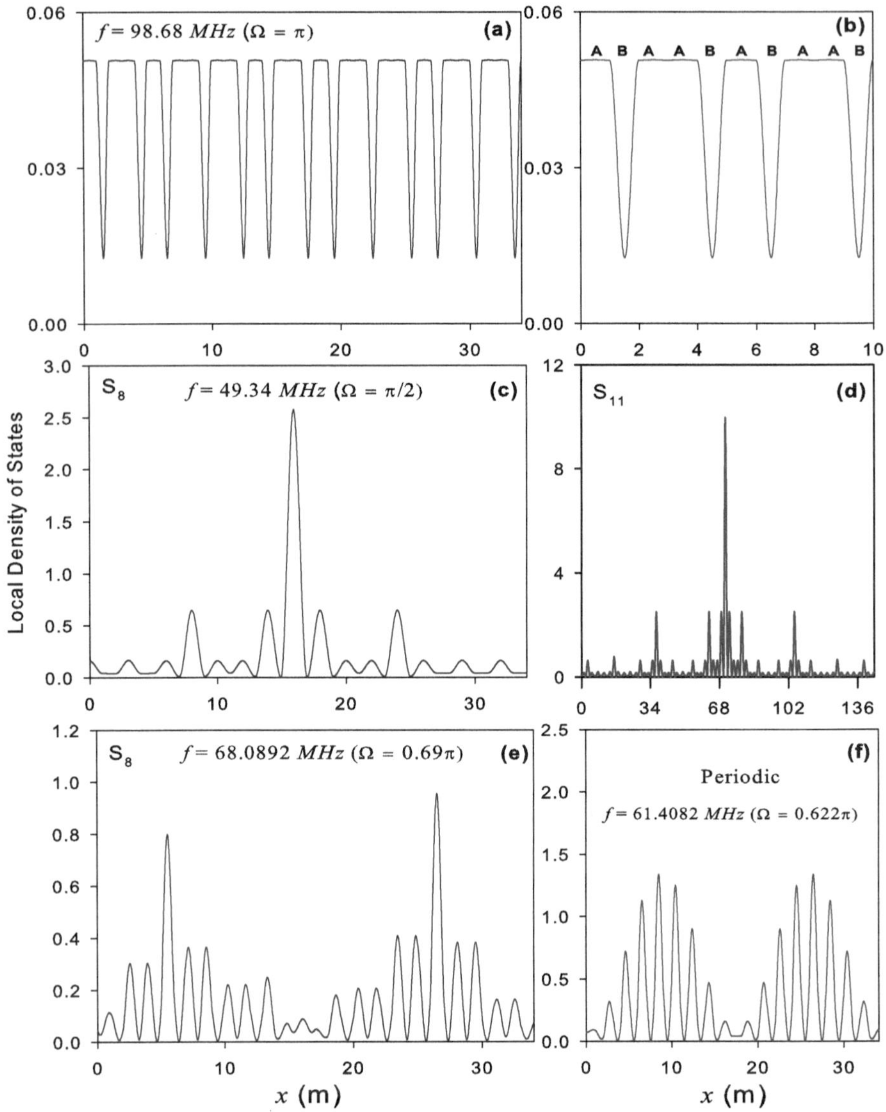

Figure 13.15 The local density of states LDOS (in arbitrary units) as a function of the space position x for three frequencies belonging to Fig. 13.14(f): (a) $\Omega = \pi$ (center of the band), (c) $\Omega = 0.5\pi$ (central frequency), and (e) $\Omega = 0.69\pi$ (band edge). (b) Magnification of (a) in the space domain $0 \leq x \leq 10$ m. (d) LDOS associated to the 11th generation at $\Omega = 0.5\pi$. (f) LDOS associated to the periodic structure at $\Omega = 0.622\pi$.

inside the allowed bands (which are indicative of the discrete modes in the finite size structure) and vanishes inside the forbidden bands. In the frequency domain of the forbidden bands of the periodic structure around f_C, the phase time associated to the FS (Figs. 13.16(b)–(d)) also shows self-similarity (3-recursion). In addition, because

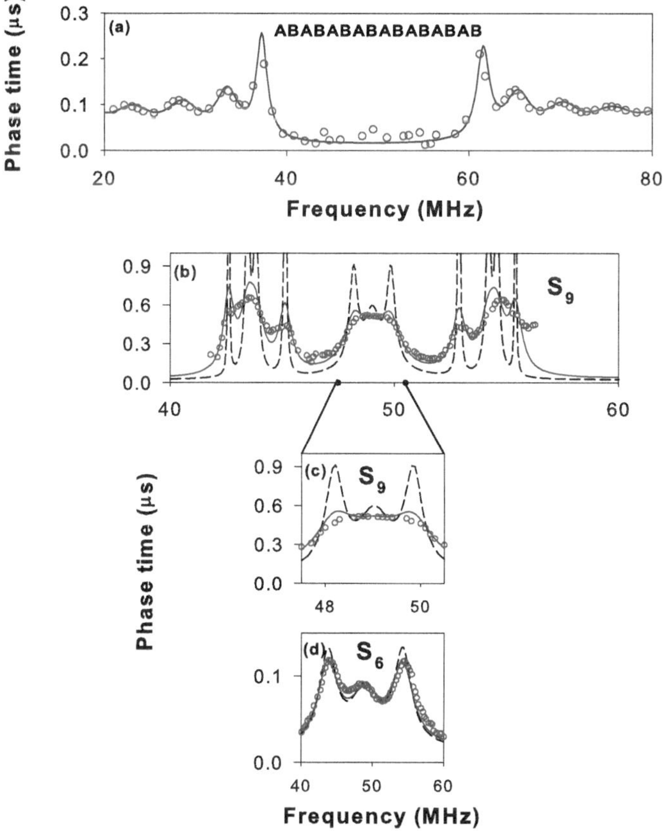

Figure 13.16 (a) Same as in Fig. 13.14(a) but for the phase time. (b)–(d) Same as in Figs. 13.14(d),(g) but for the phase time in the frequency domain [40, 60] MHz.

of the small effect of the attenuation in the cables on the phase, the self-similarity around f_c is better illustrated in the spectra of the phase time than in the spectra of the amplitude as it is shown in Figs. 13.16(b)–(d). To our knowledge, this is the first experimental study on the phase time scaling behavior in FS. It is worth mentioning that these results are different from those we obtained when both blocks A and B were composed of a segment and a loop instead of just one segment [65]. In this case, the periodicity is of order 6 instead of 3 and therefore the scaling factor is about F^2 instead of F.

From the phase time, one can deduce the group velocity defined as $v_g = L/\tau_\varphi$ [85] where L is the total length of the structure, i.e., the sum of the lengths of the A and B blocks constituting the structure (here $L = F_k$ meters). In the case of a periodic structure (Fig. 13.17(a)), an anomalous dispersion occurs inside the gaps and velocities greater than the speed of light are expected [13,14]. Inside the passbands, the group velocity is equal to $0.66c$, which is the normal speed of wave propagation in the cables used in these experiments. Near the band edge a strongly reduced

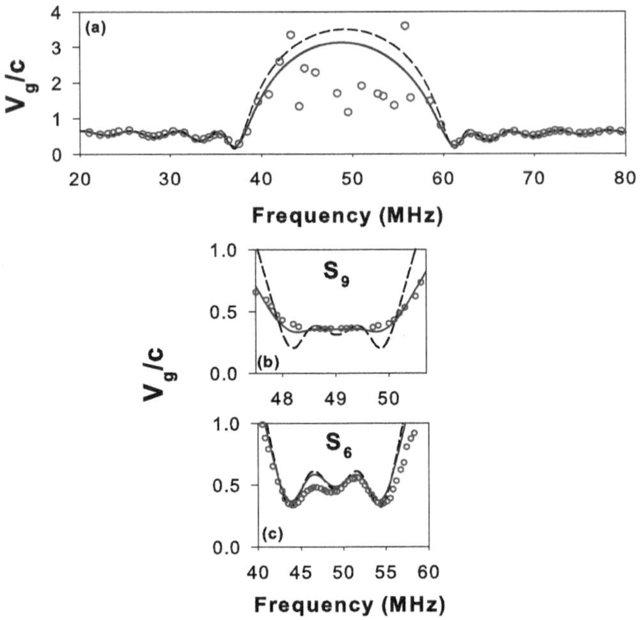

Figure 13.17 (a) Same as in Fig. 13.16 but for the group velocity.

group velocity associated with large delay times, can be observed. In the case of an FS (Figs. 13.17(b)–(c)), which induces large phase times in the frequency domain [40, 60] MHz (Figs. 13.16(b)–(d)), small group velocities $v_g \simeq 0.3c$ lower than the normal speed in the cables (i.e., 0.66c) can be observed. This value is the same as that found by Munday and Robertson [86] in a periodic structure made of two alternating different coaxial cables with an embedded defect cable. Therefore, FS may be used as a tool to reduce the propagation speed of waves in such structures.

13.4.2 Case of vertical sequence

Now, we consider the second structure where the FS is attached vertically to two waveguides (Fig. 13.13(b)). As it was shown in Section 13.3, the maxima of the transmission occur at the eigenfrequencies of the finite FS with vanishing magnetic field on both ends, whereas the minima of the transmission occur at the eigenfrequencies of the finite FS with vanishing magnetic field on one side and electric field on the other. Transmission spectra for the generations S_3–S_8 are presented in Fig. 13.18. The theoretical results (full curves) are in good agreement with experimental results (open circles). In Fig. 13.19(a), we have reported by dots the frequencies corresponding to the maxima of the spectra in Fig. 13.18 for different generations. One can notice that, as predicted, the number of the eigenmodes of the FS increases with increasing generation number. Another interesting point is the accumulation of these modes in certain

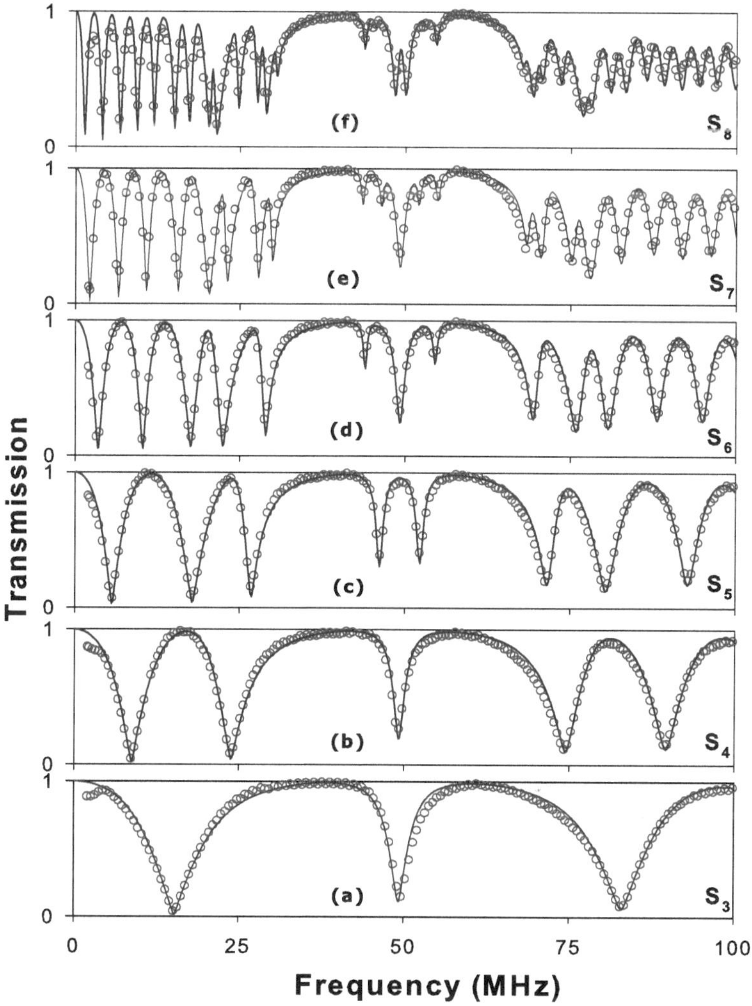

Figure 13.18 Theoretical (solid curves) and experimental (open circles) variations of the transmission coefficient as a function of the frequency for a finite Fibonacci sequence grafted vertically along a guide (Fig. 13.13(b)) for the generations S_3–S_8.

frequency regions and their absence in other regions that coincide with dips in the transmission spectra of Fig. 13.14.

In general, the spatial localization of the eigenmodes of the FS with vanishing magnetic field on both ends presents different behaviors. An example of the frequency dependence of the LDOS for the mode lying exactly at the central frequency $f_c =$ 49.34 MHz is sketched in Fig. 13.20(a). One can notice that the whole spectra of the LDOS associated to the modes lying at $f_c = 49.34$ MHz (see Fig. 13.19(b) which corresponds to Fig. 13.19(a) enlarged in the frequency domain [40, 60] MHz) do not look like self-similar as it is illustrated in the LDOS of S_{11} as compared to S_8 sketched

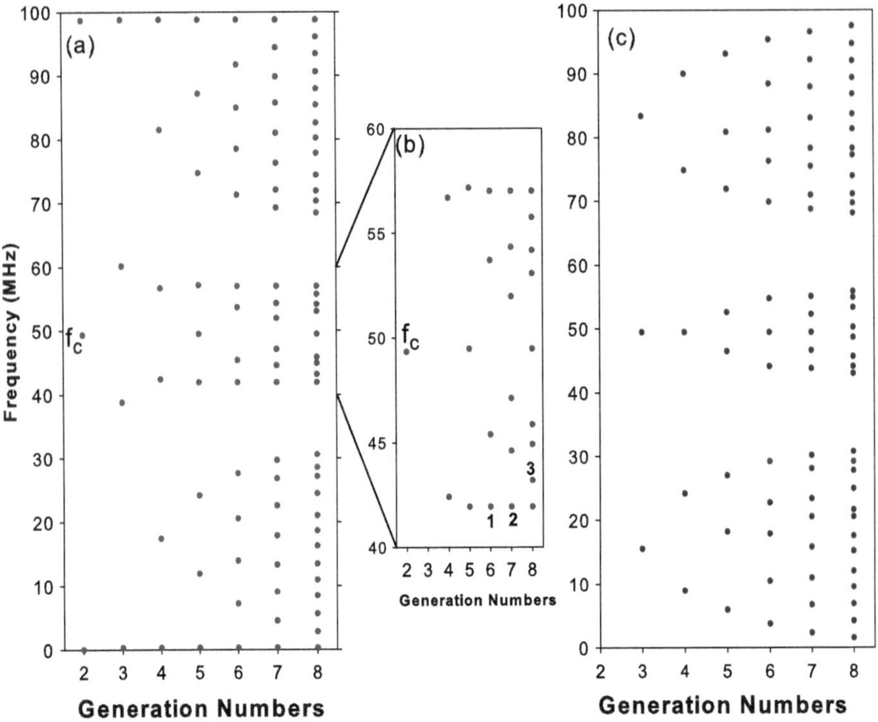

Figure 13.19 (a) Eigenfrequencies (red dots; light gray in print version) of a finite FS with vanishing magnetic field on both sides as a function of the generation numbers. The eigenmodes are obtained from the maxima of the transmission spectra of Fig. 13.18. (b) Same as in (a) but enlarged in the frequency domain [40, 60] MHz. (c) Eigenfrequencies (blue dots; dark gray in print version) of a finite FS with vanishing magnetic field on one side and electric field on the other side as a function of the generation numbers. The eigenmodes are obtained from the minima of the transmission spectra of Fig. 13.18.

in the inset of Fig. 13.20(b). However, a magnification of the spectra associated with S_{11} in the space domain $x \leq 34$ m gives rise to similar spectra as those of S_8 and S_5. This is shown in Figs. 13.20(b) and 13.20(c), where we have magnified the curve of S_{11} (Fig. 13.20(a)) and compared it to the curves of S_8 and S_5 sketched in the insets of Figs. 13.20(b) and 13.20(c), respectively.

For the modes lying at an almost constant frequency ($f = 41.93$ MHz) until the 8th generation (Fig. 13.19(b)), these modes are induced by one of the two surfaces bounding the FS as it is shown in Figs. 13.21(a) and 13.21(b) for the modes labeled 1 and 2 in Fig. 13.19(b), respectively. We observe now a decrease of LDOS from the surface when penetrating into the structure with almost the same localization. This is also the case for the modes lying at $f = 57.01$ MHz (Fig. 13.19(b)). For the modes lying at $f = 98.68$ MHz (i.e., $\Omega = \pi$), one can notice (Fig. 13.21(c)) that these modes show regular oscillations (extended modes). This behavior is different from that obtained for the same mode but in the horizontal structure where the LDOS follows a Fibonacci line shape (see Fig. 13.15(b)). Concerning the other modes of the FS, they exhibit

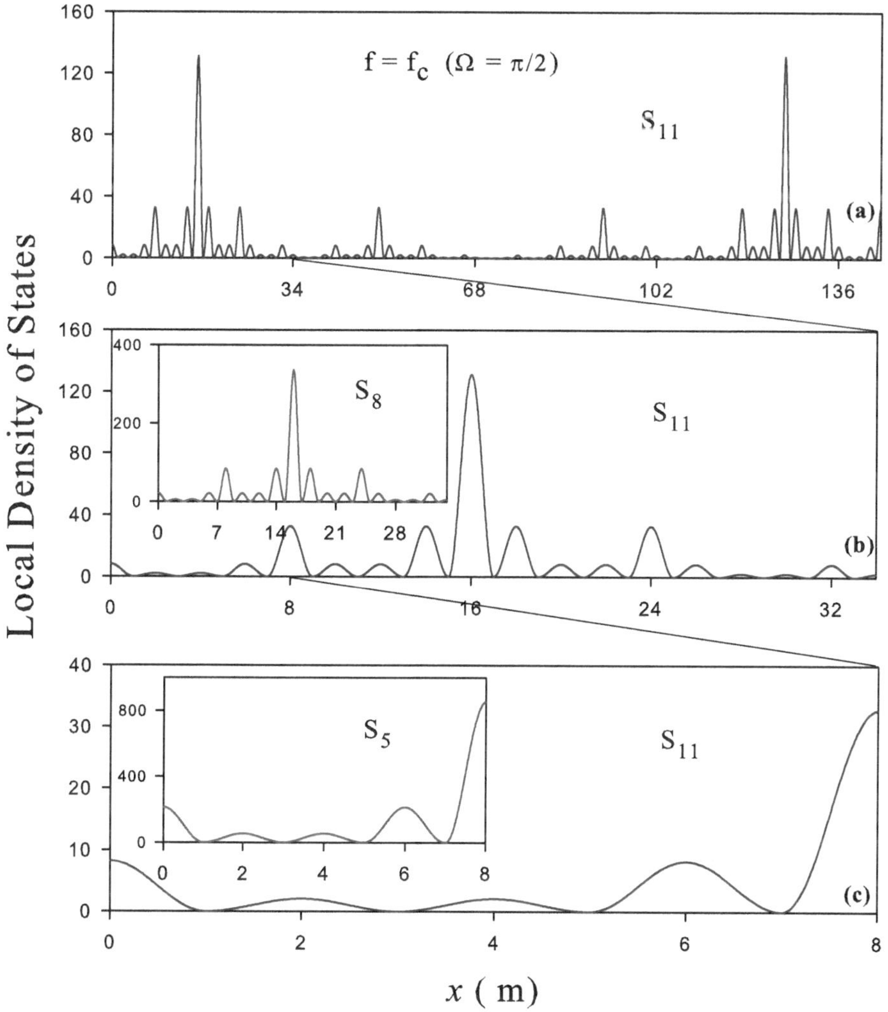

Figure 13.20 LDOS (in arbitrary units) as a function of the space position x for $f = f_c$ and different generations: (a) S_{11}, (b) S_{11} enlarged in the space position $x < 34$ m (the inset represents the LDOS for the generation S_8), and (c) S_{11} enlarged in the space position $x < 8$ m (the inset represents the LDOS for the generation S_5). Notice the high similarity between the enlarged magnification of the results for S_{11} and the results for S_8 and S_5.

an irregular shape inside the structure (see Fig. 13.21(d) associated to the mode labeled 3 in Fig. 13.19(b)). These modes could be qualified as critical modes. It is worth noticing that similar results to those presented here are found theoretically for other excitations in FS such as plasmon-polaritons and magnetostatic excitations [87]. As mentioned above, the minima of the transmission spectra of Fig. 13.18 give the eigenmodes of the FS with vanishing magnetic field on one side and electric field on the other. These results are reported in Fig. 13.19(c). One can notice that the eigenmodes

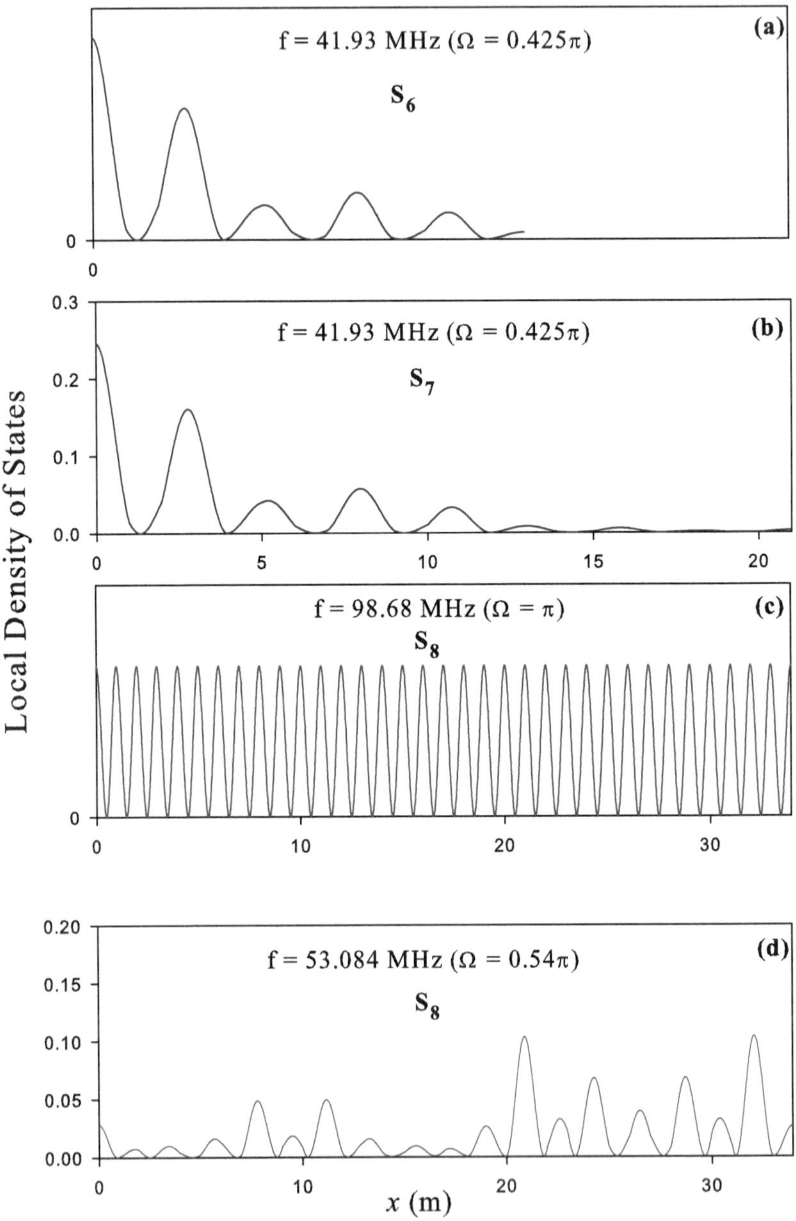

Figure 13.21 (a)–(b) Same as in Fig. 13.20, but for the modes labeled 1 and 2 in Fig. 13.19(b) at the frequency $f = 41.93$ MHz for the generations S_6 and S_7, respectively. (c) LDOS for the mode lying at $\Omega = \pi$ ($f = 98.68$ MHz). (d) LDOS for the mode labeled 3 in Fig. 13.19(b).

of such a structure are quite different from those of the FS with vanishing magnetic field on both ends (Fig. 13.19(a)).

As mentioned in Section 13.3, the eigenfrequencies of the finite FS with vanishing magnetic field on both ends coincide with the surface modes of two complementary superlattices obtained by the cleavage of an infinite superlattice made of a periodic repetition of a given sequence (see Fig. 13.3(a)). Finally, it is worth notincing that similar results to those presented in Section 13.4 for loop structures, have been reported recently for stub structures as well [88].

13.5 Summary and conclusion

In this chapter we have studied electromagnetic waves in quasiperiodic structures of Fibonacci type. Two configurations were considered: Fibonacci superlattices and Fibonacci sequences.

1. In the case of Fibonacci superlattices, we have presented theoretical results of bulk and surface electromagnetic modes in such quasiperiodic systems composed of two blocks A and B made of coaxial cables in the shape of segments and loops, respectively. Different properties related to the bulk bands such as the fragmentation of the frequency spectrum and the power law have been analyzed and discussed. We have generalized a theoretical rule on the existence of surface modes in the quasiperiodic superlattices, namely when one considers two semiinfinite superlattices obtained by the cleavage of an infinite superlattice, one obtains as many surface modes as there are minigaps. Two types of surface modes have been found:

(i) The modes inside the stable gaps. These modes fall at the same frequency for the superlattice having its surface termination unchanged every generation. However, the surface modes of the complementary superlattice fall at two different stable frequencies as their terminations are not the same for odd and even generations. These modes are very localized at the surface and present almost the same localization length. They are related to the periodicity of the structure.

(ii) The modes inside the transient gaps. These modes appear each six generations following a well-defined profile (Fig. 13.7) and their spatial localization shows the properties of self-similarity of order 6 and multifractality which are related to the quasiperiodicity of the structure.

Experimentally, we have shown the existence of the allowed and prohibited bands in Fibonacci superlattices by measuring the transmission through a finite periodic structure embedded horizontally between two semiinfinite waveguides. Moreover, we have proposed an original method to detect the surface modes by measuring the transmission through a guide along which the finite structure is attached vertically. Because of the absorption in the cables, we did not give an experimental verification of all the theoretical results, in particular for high generations. However, for small generations below the 5th generation, the experimental results agree very well with the theoretical predictions. Finally, let us mention that our theoretical model can be applied to the study of other quasiperiodic structures such as Cantor, Thue–Morse, and double period [32]. Even though these structures present different bandgap structures [32], one can expect similar results to those presented here about surface modes, namely the existence of one surface mode per gap for two complementary semi-infinite superlattices, as well as the different localization properties of surface modes in stable and transient gaps.

2. In the case of Fibonacci sequences, we have shown that (i) by grafting the structure horizontally, we can reproduce different results known on layered media as the splitting of the bands and the self-similarity between the transmission spectra with a scaling factor when the generation number increases. In addition, we have shown that besides the transmission amplitude, an analysis of the phase time enables us to determine the density of states, as well as the group velocities in these structures; (ii) by grafting the structure vertically, we have shown that the maxima of the transmission coefficient give the eigenmodes of the FS with vanishing magnetic field at the boundaries, whereas the minima of the transmission coefficient give the eigenmodes of the FS with vanishing magnetic field on one side and electric field on the other. The spatial localization of the modes of the FS is very sensitive to the different boundary conditions on both ends. Some modes of the FS with vanishing magnetic field on both ends fall at an almost constant frequency as a function of the generation number. These modes may exhibit either a strong localization at the surface terminations of the FS, a self-similar behavior of order 3, or just an oscillating behavior inside the FS.

It is worth noticing that in this chapter we disregarded flat bands occurring at the multiple of the frequency 98.68 MHz, above the frequency range we have considered. Indeed, in analogy with Chapter 12 dealing with periodic systems, the loop structures studied here can also present flat bands. These bands appear at the wavelengths $\lambda = 2d_B/n$, where $n = 0, 1, 2, 3, \ldots$ (see Fig. 13.1). They are confined modes within each loop, which can easily be understood as modes for which the length of the loop fits exactly an integer number of wavelengths. Consequently, these states are robust against any disorder in the repetition of the blocks A and B. This opens new perspectives in their utilization for further research and applications of the flat band states.

Acknowledgments

The authors would like to thank Youssef El Hassouani, Hassan Aynaou, and Victor Velasco for their fruitful contribution in the original papers from which this chapter is inspired.

References

[1] E. Yablonovitch, Inhibited spontaneous emission in solid-state physics and electronics, Phys. Rev. Lett. 58 (1987) 2059.

[2] S. John, Strong localization of photons in certain disordered dielectric superlattices, Phys. Rev. Lett. 58 (1987) 2486.

[3] See, for example, J.D. Joannopoulos, R.D. Meade, J.N. Winn, Photonic Crystals, Princeton University Press, Princeton, NJ, 1995.

[4] C.M. Soukoulis (Ed.), Photonic Band Gap and Localization, Plenum, New York, 1993; C.M. Soukoulis (Ed.), Photonic Band Gap Materials, Kluwer, Dordrecht, 1996.

[5] P. Yeh, Optical Waves in Layered Media, John Wiley Sons, New York, 1988.

[6] Y. Fink, J.N. Winn, S. Fan, C. Chen, J. Michel, J.D. Joannopoulos, E.L. Thomas, A dielectric omnidirectional reflector, Science 282 (1998) 1679.

[7] J.P. Dowling, Mirror on the wall: you're omnidirectional after all?, Science 282 (1998) 1841.

[8] E. Yablonovitch, Engineered omnidirectional external-reflectivity spectra from one-dimensional layered interference filters, Opt. Lett. 23 (1998) 1648.

[9] A. Mir, A. Akjouj, E.H. El Boudouti, B. Djafari-Rouhani, L. Dobrzynski, Large omnidirectional band gaps and selective transmission in one-dimensional multilayer photonic structures, Vacuum 63 (2001) 197.

[10] I. Celanovic, F. O'Sullivan, M. Ilak, J. Kassakian, D. Perrault, Design and optimization of one-dimensional photonic crystals for thermophotovoltaic applications, Opt. Lett. 29 (2004) 863;
F. O'Sullivan, I. Celanovic, N. Jovanovic, J. Kassakian, S. Akiyama, K. Wada, Optical characteristics of one-dimensional Si/SiO_2 photonic crystals for thermophotovoltaic applications, J. Appl. Phys. 97 (2005) 033529.

[11] G.J. Schneider, S. Hanna, J.L. Davis, G.H. Watson, Defect modes in coaxial photonic crystals, J. Appl. Phys. 90 (2001) 2642.

[12] A. Haché, A. Slimani, A model coaxial photonic crystal for studying band structures, dispersion, field localization, and superluminal effects, Am. J. Phys. 72 (2004) 916.

[13] E.H. El Boudouti, N. Fettouhi, A. Akjouj, B. Djafari-Rouhani, A. Mir, J.O. Vasseur, L. Dobrzynski, J. Zemmouri, Experimental and theoretical evidence for the existence of photonic bandgaps and selective transmissions in serial loop structures, J. Appl. Phys. 95 (2004) 1102.

[14] A. Haché, L. Poirier, Anomalous dispersion and superluminal group velocity in a coaxial photonic crystal: theory and experiment, Phys. Rev. E 65 (2002) 036608;
J.N. Munday, W. Robertson, Negative group velocity pulse tunneling through a coaxial photonic crystal, Appl. Phys. Lett. 80 (2002) 518.

[15] M. del Mar Sanchez-Lopez, J.A. Davis, K. Crabtree, Coaxial cable analogs of multilayer dielectric optical coatings, Am. J. Phys. 71 (2003) 1314.

[16] L. Poirier, A. Haché, Nonlinear coaxial photonic crystal, Appl. Phys. Lett. 78 (2001) 2626.

[17] J. Vasseur, A. Akjouj, B. Djafari-Rouhani, L. Dobrzynski, E.H. El Boudouti, Photon, electron, magnon, phonon and plasmon mono-mode circuits, Surf. Sci. Rep. 54 (2004) 1.

[18] R.D. Meade, K.D. Brommer, A.M. Rappe, J.D. Joannopoulos, Electromagnetic Bloch waves at the surface of a photonic crystal, Phys. Rev. B 44 (1991) R10961.

[19] W.M. Robertson, G. Arjavalingam, R.D. Meade, K.D. Brommer, A.M. Rappe, J.D. Joannopoulos, Observation of surface photons on periodic dielectric arrays, Opt. Lett. 18 (1993) 528.

[20] F. Ramos-Mendieta, P. Halevi, Surface modes in a 2D array of square dielectric cylinders, Solid State Commun. 100 (1996) 314;
F. Ramos-Mendieta, P. Halevi, Surface electromagnetic waves in two-dimensional photonic crystals: effect of the position of the surface plane, Phys. Rev. B 59 (1999) 15112.

[21] P. Yeh, A. Yariv, C.S. Hong, Electromagnetic propagation in periodic stratified media. I. General theory, J. Opt. Soc. Am. 67 (1977) 423.

[22] N. Ng, P. Yeh, A. Yariv, Optical surface waves in periodic layered medium grown by liquid phase epitaxy, Appl. Phys. Lett. 32 (1978) 370.

[23] A.A. Bulgakov, V.R. Koturn, Surface optical oscillations in a limited stratified-periodic medium, Opt. Spectrosc. 56 (1984) 269.
[24] W.M. Robertson, M.S. May, Surface electromagnetic wave excitation on one-dimensional photonic band-gap arrays, Appl. Phys. Lett. 74 (1999) 1800.
[25] F. Ramos-Mendieta, P. Halevi, Electromagnetic surface modes of a dielectric superlattice: the supercell method, J. Opt. Soc. Am. B 14 (1997) 370.
[26] M.L.H. Lahlaouti, A. Akjouj, B. Djafari-Rouhani, L. Dobrzynski, M. Hammouchi, E.H. El Boudouti, A. Nougaoui, B. Kharbouch, Theoretical analysis of the density of states and phase times: application to resonant electromagnetic modes in finite superlattices, Phys. Rev. B 63 (2001) 035312.
[27] F. Villa, J.A. Gaspar-Armenta, F. Ramos-Mendieta, One-dimensional photonic crystals: equivalent systems to single layers with a classical oscillator like dielectric function, Opt. Commun. 216 (2003) 361.
[28] S. Feng, H-Yi. Sang, Z-Y. Li, B-Y. Cheng, D-Z. Zhang, Sensitivity of surface states to the stack sequence of one-dimensional photonic crystals, J. Opt. A, Pure Appl. Opt. 7 (2005) 374.
[29] D.N. Chigrin, A.V. Lavrinenko, D.A. Yarotsky, V. Gaponenko, All-dielectric one-dimensional periodic structures for total omnidirectional reflection and partial spontaneous emission control, J. Lightwave Technol. 17 (1999) 2018.
[30] T. Okamoto, T. Kamiyama, I. Yamaguchi, All-optical spatial light modulator with surface plasmon resonance, Opt. Lett. 18 (1993) 1570.
[31] W.M. Robertson, Experimental measurement of the effect of termination on surface electromagnetic waves in one-dimensional photonic bandgap arrays, J. Lightwave Technol. 17 (1999) 2013.
[32] E.L. Albuquerque, M.G. Cottam, Theory of elementary excitations in quasiperiodic structures, Phys. Rep. 376 (2003) 225;
E.L. Albuquerque, M.G. Cottam, Polaritons in Periodic and Quasiperiodic Structures, Elsevier, Amsterdam, 2004.
[33] E. Maciá, Aperiodic Structures in Condensed Matter: Fundamentals and Applications, Taylor and Francis Group, CRC Press, 2009.
[34] Luca Dal Negro, Optics of Aperiodic Structures: Fundamentals and Device Applications, Pan Stanford Publishing, CRC Press, 2013.
[35] P.W. Mauriz, M.S. Vasconcelos, E.L. Albuquerque, Optical transmission spectra in symmetrical Fibonacci photonic multilayers, Phys. Lett. A 373 (2009) 496.
[36] L. Dal Negro, J.H. Yi, V. Nguyen, Y. Yi, J. Michel, L.C. Kimerling, Spectrally enhanced light emission from aperiodic photonic structures, Appl. Phys. Lett. 86 (2005) 261905.
[37] S.V. Boriskina, A. Gopinath, L. Dal Negro, Optical gap formation and localization properties of optical modes in deterministic aperiodic photonic structures, Opt. Express 16 (2008) 18813.
[38] M. Hiltunen, L. Dal Negro, F. Ning-Ning, L.C. Kimerling, J. Michel, Modeling of aperiodic fractal waveguide structures for multifrequency light transport, J. Lightwave Technol. 25 (2007) 1841.
[39] N. Zotov, J. Feydt, T. Walther, A. Ludwig Appl, Structure of PtFe/Fe double-period multilayers investigated by X-ray diffraction, reflectivity, diffuse scattering and TEM, Surf. Sci. 253 (2006) 128.
[40] K. Lambropoulos, C. Simserides, Periodic, quasiperiodic, fractal, Kolakoski, and random binary polymers: energy structure and carrier transport, Phys. Rev. E 99 (2019) 032415.
[41] M. Kohmoto, B. Sutherland, K. Iguchi, Localization of optics: quasiperiodic media, Phys. Rev. Lett. 58 (1987) 2436.

[42] W. Gellermann, M. Kohmoto, B. Sutherland, P.C. Taylor, Localization of light waves in Fibonacci dielectric multilayers, Phys. Rev. Lett. 72 (1994) 633.
[43] T. Hattori, N. Tsurumachi, S. Kawato, H. Nakatsuka, Photonic dispersion relation in a one-dimensional quasicrystal, Phys. Rev. B 50 (1994) R4220.
[44] L. Chow, K.H. Guenther, Critical field patterns in optical Fibonacci multilayers, J. Opt. Soc. Am. A 10 (1993) 2231.
[45] R. Pelster, V. Gasparian, G. Nimtz, Propagation of plane waves and of waveguide modes in quasiperiodic dielectric heterostructures, Phys. Rev. E 55 (1997) 7645.
[46] R.W. Peng, Y.M. Liu, X.Q. Huang, F. Qin, M. Wang, A. Hu, S.S. Jiang, D. Feng, L.Z. Ouyang, J. Zou, Dimerlike positional correlation and resonant transmission of electromagnetic waves in aperiodic dielectric multilayers, Phys. Rev. B 69 (2004) 165109.
[47] M. Ghulinyan, C.J. Oton, L. Dal Negro, L. Pavesi, R. Sapienza, M. Colocci, D.S. Wiersma, Light-pulse propagation in Fibonacci quasicrystals, Phys. Rev. B 71 (2005) 094204.
[48] C. Sibilia, I.S. Nefedov, M. Scolara, M. Berlotti, Electromagnetic mode density for finite quasi-periodic structures, J. Opt. Soc. Am. B 15 (1998) 1947.
[49] X. Huang, Y. Wang, C. Gong, Numerical investigation of light-wave localization in optical Fibonacci superlattices with symmetric internal structure, J. Phys. Condens. Matter 11 (1999) 7645.
[50] E. Macia, Exploiting quasiperiodic order in the design of optical devices, Phys. Rev. B 63 (2001) 205421.
[51] N.H. Liu, Propagation of light waves in Thue–Morse dielectric multilayers, Phys. Rev. B 55 (1997) 3543.
[52] A.V. Lavrinenko, S.V. Zhukovsky, K.S. Sandmirski, S.V. Gaponenko, Propagation of classical waves in nonperiodic media: scaling properties of an optical Cantor filter, Phys. Rev. E 65 (2002) 036621.
[53] D. Lusk, I. Abdulhalim, F. Placido, Omnidirectional reflection from Fibonacci quasiperiodic one-dimensional photonic crystal, Opt. Commun. 198 (2001) 273.
[54] R.W. Peng, M. Wang, A. Hu, S.S. Jiang, G.J. Jin, D. Feng, Photonic localization in one-dimensional k-component Fibonacci structures, Phys. Rev. B 57 (1998) 1544.
[55] Z.V. Vardeny, A. Nahata, A. Agrawal, Optics of photonic quasicrystals, Nat. Photonics 7 (2013) 177.
[56] P. Xiao-Ning, D. Jian-Wen, W. He-Zhou, Photonic localization of interface modes at the boundary between metal and Fibonacci quasiperiodic structure, J. Opt. Soc. Am. B 27 (2010) 2009.
[57] E.H. El Boudouti, Y. El Hassouani, H. Aynaou, B. Djafari-Rouhani, Two types of modes in finite size one-dimensional coaxial photonic crystals: general rules and experimental evidence, Phys. Rev. E 76 (2007) 026607.
[58] B. Djafari-Rouhani, E.H. El Boudouti, A. Akjouj, L. Dobrzynski, J.O. Vasseur, A. Mir, N. Fettouhi, J. Zemmouri, Surface states in one-dimensional photonic band gap structures, Vacuum 63 (2001) 177.
[59] El Hassouani, H. Aynaou, E.H. El Boudouti, B. Djafari-Rouhani, A. Akjouj, V.R. Velasco, Surface electromagnetic waves in Fibonacci superlattices: theoretical and experimental results, Phys. Rev. B 74 (2006) 035314.
[60] E.H. El Boudouti, Y. El Hassouani, H. Aynaou, B. Djafari-Rouhani, A. Akjouj, V.R. Velasco, Electromagnetic wave propagation in quasi-periodic photonic circuits, J. Phys. Condens. Matter 19 (2007) 246217.
[61] R. Merlin, K. Bajema, R. Clarke, Quasiperiodic GaAs-AlAs heterostructures, Phys. Rev. Lett. 55 (1985) 1768;
K. Bajema, R. Merlin, Raman scattering by acoustic phonons in Fibonacci GaAs-AlAs superlattices, Phys. Rev. B 36 (1987) 4555.

[62] H. Aynaou, E.H. El Boudouti, B. Djafari-Rouhani, A. Akjouj, V.R. Velasco, Propagation and localization of acoustic waves in Fibonacci phononic circuits, J. Phys. Condens. Matter 17 (2005) 4245.
[63] H. Aynaou, E.H. El Boudouti, Y. El Hassouani, A. Akjouj, B. Djafari-Rouhani, J. Vasseur, A. Benomar, V.R. Velasco, Propagation and localization of electromagnetic waves in quasiperiodic serial loop structures, Phys. Rev. E 72 (2005) 056601.
[64] L. Dobrzynski, Interface response theory of continuous composite systems, Surf. Sci. Rep. 11 (1990) 139.
[65] E.H. El Boudouti, B. Djafari-Rouhani, A. Akjouj, L. Dobrzynski, Theory of surface and interface transverse elastic waves in N-layer superlattices, Phys. Rev. B 54 (1996) 14728.
[66] F. Piechon, M. Benakli, A. Jagannathan, Analytical results for scaling properties of the spectrum of the Fibonacci chain, Phys. Rev. Lett. 74 (1995) 5248.
[67] W. Chen, Y. Lu, H.J. Maris, G. Xiao, Picosecond ultrasonic study of localized phonon surface modes in Al/Ag superlattices, Phys. Rev. B 50 (1994) 14506.
[68] J.W. Feng, G.J. Jin, A. Hu, S.S. Kang, S.S. Jiang, D. Feng, Magnetostatic modes in Fibonacci magnetic and nonmagnetic multilayers, Phys. Rev. B 52 (1995) 15312.
[69] E.S. Zijlstra, A. Fasolino, T. Janssen, Existence and localization of surface states on Fibonacci quasicrystals: a tight-binding study, Phys. Rev. B 59 (1999) 302.
[70] T. Fujiwara, M. Kohmoto, T. Tokihiro, Multifractal wave functions on a Fibonacci lattice, Phys. Rev. B 40 (1989) 7413.
[71] R.W. Peng, G.J. Jin, M. Wang, A. Hu, S.S. Jiang, D. Feng, Electronic transport in k-component Fibonacci quantum waveguides, J. Phys. Condens. Matter 12 (2000) 5701.
[72] For a review, see J.L. McCauley, Introduction to multifractals in dynamical systems theory and fully developed fluid turbulence, Phys. Rep. 189 (1990) 225.
[73] T.C. Halsey, M.H. Jensen, L.P. Kadanoff, I. Procaccia, B.I. Shraiman, Fractal measures and their singularities: the characterization of strange sets, Phys. Rev. A 33 (1986) 1141.
[74] M. Kohmoto, B. Sutherland, C. Tang, Critical wave functions and a Cantor-set spectrum of a one-dimensional quasicrystal model, Phys. Rev. B 35 (1987) 1020.
[75] P. Grassberger, R. Badii, A. Politi, Scaling laws for invariant measures on hyperbolic and nonhyperbolic atractors, J. Stat. Phys. 51 (1988) 135.
[76] A.B. Chhabra, R.V. Jensen, Direct determination of the $f(\alpha)$ singularity spectrum, Phys. Rev. Lett. 62 (1989) 1327.
[77] C.S. Ryu, G.Y. Oh, M.H. Lee, Extended and critical wave functions in a Thue–Morse chain, Phys. Rev. B 46 (1992) 5162;
C.S. Ryu, G.Y. Oh, M.H. Lee, Electronic properties of a tight-binding and a Kronig–Penney model of the Thue–Morse chain, Phys. Rev. B 48 (1993) 132.
[78] M.S. Vasconcelos, E.L. Albuquerque, A.M. Mariz, Optical localization in quasi-periodic multilayers, J. Phys. Condens. Matter 10 (1998) 5839.
[79] M. Kohmoto, J.R. Banvar, Quasiperiodic lattice: electronic properties, phonon properties, and diffusion, Phys. Rev. B 34 (1986) 563.
[80] D. Jin, G. Jin, Matrix maps for substitution sequences in the biquaternion representation, Phys. Rev. B 71 (2005) 014212.
[81] M. Scalora, R.J. Flynn, S.B. Reinhard, R.L. Fork, M.J. Bloemer, M.D. Tocci, C.M. Bowden, H.S. Ledbetter, J.M. Bendickson, J.P. Dowling, Ultrashort pulse propagation at the photonic band edge: large tunable group delay with minimal distortion and loss, Phys. Rev. E 54 (1996) R1078.
[82] L. Dal Negro, C.J. Oton, Z. Gaburro, L. Pavesi, P. Johnson, A. Lagendijk, R. Righini, M. Colocci, D.S. Wiersma, Light transport through the band-edge states of Fibonacci quasicrystals, Phys. Rev. Lett. 90 (2003) 055501.

[83] D. Wiersma, R. Sapienza, S. Mujumdar, M. Colocci, M. Ghulinyan, L. Pavesi, Optics of nanostructured dielectrics, J. Opt. A, Pure Appl. Opt. 7 (2005) S190.
[84] S. Khattou, M. Amrani, A. Mouadili, E.H. El Boudouti, A. Talbi, A. Akjouj, B. Djafari-Rouhani, Comparison of density of states and scattering parameters in coaxial photonic crystals: theory and experiment, Phys. Rev. B 102 (2020) 165310.
[85] J.N. Zhu, N. Liu, H. Zheng, H. Chen, Time delay of light propagation through defect modes of one-dimensional photonic band-gap structures, Opt. Commun. B 174 (2000) 139.
[86] J.N. Munday, W.M. Roberston, Slow electromagnetic pulse propagation through a narrow transmission band in a coaxial photonic crystal, Appl. Phys. Lett. 83 (2003) 1053.
[87] N-h. Liu, Y. He, W-g. Feng, X. Wu, Magnetic-field-induced plasmon polaritons in finite quasiperiodic conducting superlattices, Phys. Rev. B 52 (1995) 11105;
S.S. Kang, Magnetostatic excitations in quasiperiodic antiferromagnetic superlattices, Phys. Rev. B 65 (2002) 064401.
[88] H. Aynaou, A. Mouadili, N. Ouchani, E.H. El Boudouti, A. Akjouj, B. Djafari-Rouhani, Scaling law, confined and surface modes in photonic Fibonacci stub structures: theory and experiment, Appl. Sci. 10 (2020) 7767.

One-dimensional photonic waveguide for filtering and demultiplexing

14

Yan Pennec[a], Bahram Djafari-Rouhani[a], Abdellatif Akjouj[a], El Houssaine El Boudouti[b], Gaëtan Lévêque[a], and Leonard Dobrzyński[a]
[a]Univ. Lille, CNRS, Centrale Lille, Yncréa ISEN, Univ. Polytechnique Hauts-de-France, UMR 8520 - IEMN, Lille, France, [b]Laboratoire de Physique de la Matière et de Rayonnement, Département de Physique, Faculté des Science, Université Mohammed I, Oujda, Morocco

14.1 Introduction

Microresonator devices have received a great deal of attention because of their highly optical circuits integration in relation with relevant applications for optical filtering [1] and switching [2], ultra-small lasers [3], strong light–matter coupling [4], and chemical sensing [5] among others. For signal processing applications, resonators have proven to be one of the most versatile means of obtaining desirable filtering transfer functions. A wide range of functionality has been investigated, both theoretically and experimentally, for future optical communications including channel add/drop filters, and demultiplexers. The challenge has been to produce resonators which offer very good properties such as narrow bandwidth filtering and high-quality factors with dimensions comparable to the wavelength of light. Filtering application are thus being studied in the literature for different physical systems. Among them, one can refer to microring resonators [6,7], cavities in photonic crystal membranes (2D) [8,9], or linear waveguides (1D) [10,11].

During the last decade, the study of 1D photonic waveguide has sparked new interest in optomechanical studies. The simultaneous control of phonons and photons in the same structure, with the aim of enhancing their interaction, has received a great deal of attention during the last few years. Dual phononic–photonic (also called phoxonic) structures hold promises for the simultaneous confinement and tailoring of sound and light waves, with potential applications to acousto-optical devices and highly controllable phonon–photon interactions [12,13]. The existence of dual photonic and phononic bandgaps has been investigated first in 2D infinite crystals [14], while similar demonstrations have been performed closely later in photonic crystal slabs [15,16]. The periodic dual photonic/phononic platform is also used to design linear and cavity defects in which the confinement of both excitations or the existence of slow waves strongly enhance their interaction [17,18]. The enhancement of photon–phonon interaction is currently investigated in the framework of 1D pho-

Photonics. https://doi.org/10.1016/B978-0-12-819388-4.00024-1

tonic waveguides. Eichenfield et al. [19] proposed a 1D patterned optomechanical nanobeam made up of a periodic array of rectangular holes in a straight waveguide. Pennec et al. investigated the simultaneous existence of phoxonic, i.e., dual phononic and photonic, bandgaps in a periodic silicon strip waveguide [20]. The design of corrugated nanobeams displays high acousto-optic coupling rates of a few MHz involving localized phonons in the range of a few GHz and optical cavity modes with high quality factor at the telecommunications frequency [21–23].

The aim of this chapter is to discuss, using numerical simulations, the efficiency of a resonator device constituted by a straight waveguide coupled to a rectangular cavity disposed on each side of the guide, forming a cross. Such a cavity is called a stub. Owing to their very small size, microstubs could serve in future complex and highly integrated structures. We investigate the possibility of achieving sophisticated control of the device wavelength response. In a first part of the chapter (Section 14.2), we define the geometrical parameters of the device and the method of calculation. In the second part (Section 14.3), we show that, in the presence of a stub, the transmission spectrum can display several narrow dips, provided that the stub is covered with a perfect metallic layer [24]. Such a device could easily be used in the far infrared frequency domain (far from the plasma frequency) where the metal behaves as a perfect mirror. The geometrical conditions of an efficient rejective filter are discussed for two different symmetries of the incident waves, namely symmetric and antisymmetric, with respect to the middle line parallel to the waveguide. The third part (Section 14.4) deals with more complex geometries based on the interaction between several stubs [25]. In particular, we show that a periodical set of stubs open a gap by coupling each zero of transmission. Inserting a defect resonator can produce a selective tunneling transmission inside this gap. This structure provides functionalities such as opening a wider gap in which selective transmission can be produced. Then, in Section 14.5, we use those filtering processes to investigate the possibility of a new device for wavelength demultiplexing based on a bent Y-shaped component. In the last Section 14.6, we show that the main features of the transmission spectrum through a waveguide coupled to a lateral stub remain valid if the metal covering the stub is not perfect, but described in the framework of the Drude model [26]. This makes the extension of the above applications to the near optical regime plausible.

14.2 Geometrical parameters and method of calculation

The study is based on a two-dimensional waveguide coupled to a lateral rectangular cavity which geometry is displayed in Fig. 14.1; w and l are respectively the width and the length of the stub. The embedding medium is air. The waveguide and the stub are made of the same material, with an index of refraction $n = 3.2429$, near the index of some material such as InP or InGaAsP [27]. On the other hand, the choice of such a material ensures a large contrast in the optical properties of the waveguide constituent material and the surrounding medium. The incident wave, created at the left of the figure, is adapted to the waveguide through a taper situated at the entrance.

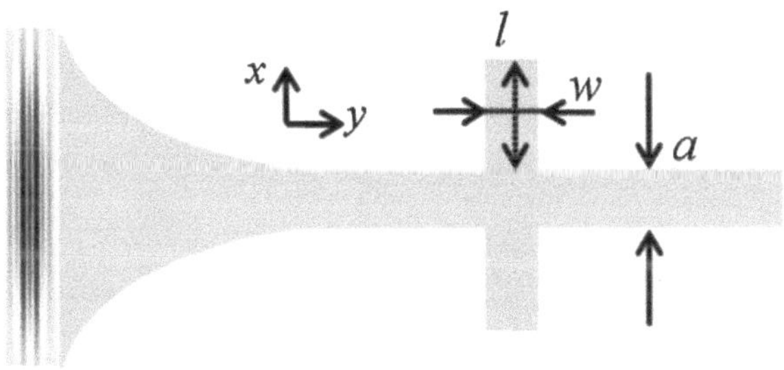

Figure 14.1 Schematic view of a stub coupled to a bus waveguide.

Calculations are performed using Finite Difference Time Domain (FDTD) method which solves the Maxwell's equations [28] for the analysis of high index contrast waveguide coupled microstub resonator. The FDTD method solves the Maxwell electromagnetic equations by discretizing both time and space, and by replacing derivatives by finite differences. Our calculation is performed in a two-dimensional (2D) box (along x and y axes) with Perfect Matching Layer (PML) conditions applied at the boundaries of the box, in order to avoid reflections of outgoing waves [29]. The structure is supposed to be infinite along the z direction. Space is discretized in both x and y directions using a mesh interval equal to $\Delta x = \Delta y = a/30$, where a is the width of the waveguide. The width and length of the box are respectively $7a$ and $25a$. The equations of motion are solved with a time integration step $\Delta t = \Delta x/4$ and a number of time steps equals to 2^{19} or 2^{20}, depending on the size of the structure, which is the necessary tested time for a good convergence of the numerical calculations.

A fundamental mode of the waveguide is launched at the left end of the waveguide, inside the taper region. The incoming wave is a longitudinal pulse, with a Gaussian profile along the x direction, created by an initial current source parallel to the x axis, according to the Maxwell's equations. The transmitted signal, probed at the right end of the waveguide, is recorded as a function of time, integrated over the cross-section of the waveguide, and finally Fourier transformed to obtain the transmission coefficient versus frequency. All the transmission spectra are normalized with respect to the one corresponding to a perfect (without stub) waveguide. The transmission is reported in dB as a function of the dimensionless frequency $\Omega = \omega a/(2\pi c) = a/\lambda_0$ where λ_0 is the wavelength in vacuum and c the velocity of light in vacuum. The polarization of the incident wave is transverse magnetic (TM), meaning that the magnetic field is perpendicular to the incident plane.

According to the symmetry of the perfect waveguide, the propagating waves can be distinguished between symmetric and anti-symmetric modes. Fig. 14.2(a) displays a schematic representation of the current source, as well as the associated electromagnetic field for both types of symmetry of the incident wave. It is quite straightforward to write the solutions of such waves in the perfect waveguide and to obtain their dis-

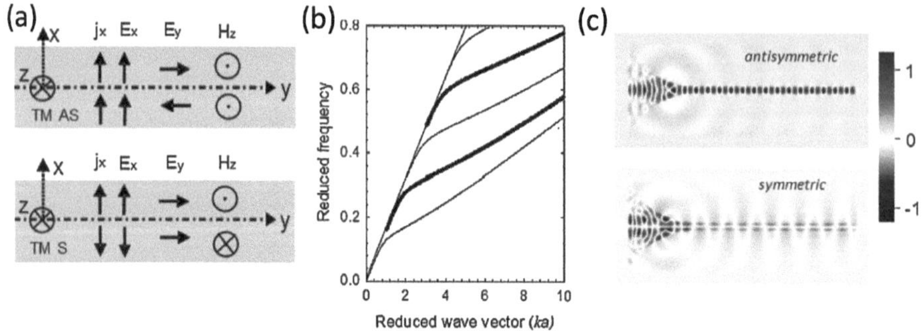

Figure 14.2 (a) Symmetry representation of the current source and of the electromagnetic field inside the waveguide for the antisymmetric and symmetric modes in a transverse magnetic (TM) polarization. (b) Dispersion curves of a 2D waveguide for transverse magnetic polarization: antisymmetric modes (thin solid lines) and symmetric modes (thick solid lines). (c) Maps of the magnetic field for each symmetry of the incident electromagnetic wave at the monochromatic frequency $\Omega = 0.250$. The maps are obtained by averaging the field over half period of oscillation. The red [light gray in print version] (blue [dark gray in print version]) color corresponds to the highest (lowest) values of the magnetic field given in arbitrary units.

persion relations from the boundary conditions:

$$\begin{cases} \text{cotan}\left(\dfrac{\beta a}{2}\right) - \dfrac{n^2\alpha}{\beta} = 0, & \text{antisymmetric mode} \\ \text{cotan}\left(\dfrac{\beta a}{2}\right) + \dfrac{n^2\alpha}{\beta} = 0, & \text{symmetric mode} \end{cases} \tag{14.1}$$

$$\text{with } \alpha^2 = k^2 - \frac{\omega^2}{c^2} \text{ and } \beta^2 = \frac{n^2\omega^2}{c^2} - k^2$$

Here k is the wave vector along the y direction, ω the frequency, and a the width of the waveguide. Fig. 14.2(b) represents the dimensionless frequency Ω versus the wave vector k. Below the vacuum line $\omega = ck$, we find that the dispersion branches are alternatively antisymmetric and symmetric. The waveguide will then be monomodal in the frequency range [0; 0.32] for the antisymmetric mode and [0.16; 0.48] for the symmetric one. As an illustration, we present in Fig. 14.2(c) the map of the magnetic field, obtained from the FDTD calculation, for each symmetry of an incident monochromatic wave at $\Omega = 0.25$. The incident wave is launched at the left of the figures through a taper situated at the entrance of the waveguide. The taper enables adapting the incident wave to the mode of the waveguide. We can notice that the maps of the magnetic field are in agreement with the above symmetry consideration.

14.3 Rejective filter

Fig. 14.3(a) displays, for the antisymmetric waves, the theoretical transmission for a stub grafted on the waveguide with the following parameters: $w = a$ and $l = 1.7a$.

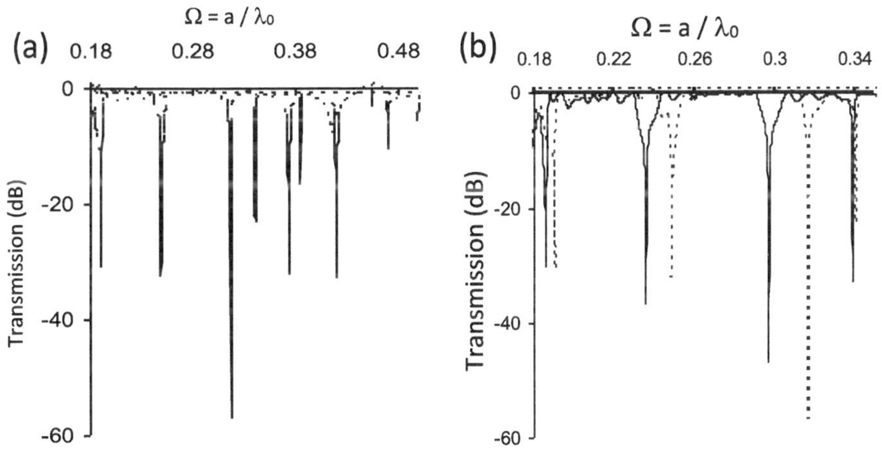

Figure 14.3 (a) Transmission spectra of the waveguide coupled with a stub. The solid and dashed lines correspond to a stub covered or uncovered with a perfectly metallic thin film, respectively. (b) Transmission spectra of the waveguide coupled to a stub covered with a perfectly thin metallic layer: $(w, l) = (a, 1.7a)$ (dashed line) and $(w, l) = (a, 1.9a)$ (solid line).

The results are given for Ω in the range from 0.2 to 0.5. We use two different types of boundary conditions at the limits of the stub. In the first case the uncovered stub is in contact with air, while in the second case the stub is covered with a perfectly metallic thin layer that prevents the radiation to escape outside from the boundaries of the stub. In the first case, the transmission spectrum presented as the dashed line in Fig. 14.3(a) is depressed by an attenuation factor of less than 10 dB at the frequencies $\Omega = 0.243$, $\Omega = 0.318$, and $\Omega = 0.417$. In the second case, presented as a full line in Fig. 14.3(a), there is an increase in the number of extinction peaks in the same frequency range. More importantly, the dips become much narrower with values of attenuation much greater than in the first case. The attenuation is now above 20 dB for several dips and even reaches 57 dB at $\Omega = 0.317$. The presence of the metal has reinforced the interaction between the waveguide and the resonance modes of the stub, leading to a filtering of many narrow frequencies in the transmission spectrum.

In Fig. 14.3(b), we show how the dips in the transmission spectrum evolve as functions of the length l of the stub. As a general trend, the frequencies of these modes decrease by increasing either w or l. This can be seen in Fig. 14.3(b) where the transmission spectra are given for two different lengths $l = 1.7a$ and $l = 1.9a$, for a constant value of w. However, one can notice that the evolution of the frequency and more especially of the attenuation factor may be different from dip to dip when we change the geometrical parameters of the stub.

The above trends can actually be understood in the framework of a simple analytical model considering the electromagnetic modes of a finite rectangular cavity grafted to a line (Fig. 14.4(a)). The interest of this comparison is to yield possible the prediction, at least qualitatively, of the number and frequencies of the dips in the transmission spectrum without having recourse to the numerical calculation. Let us recall

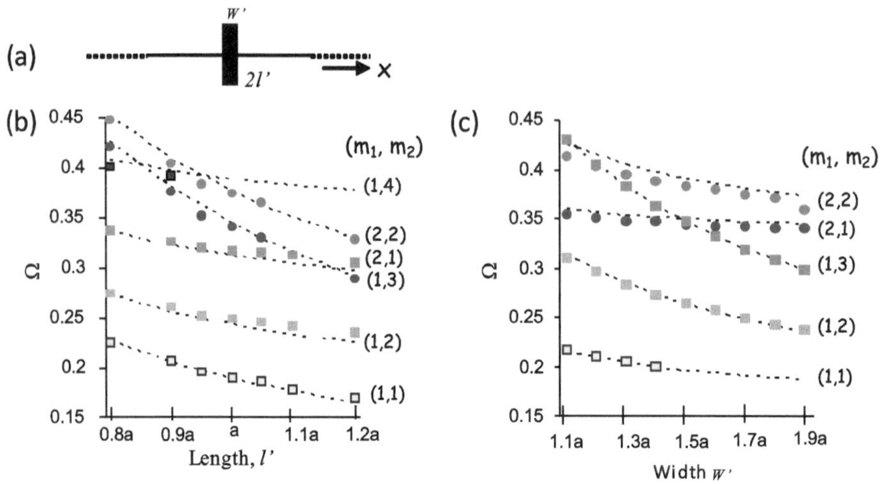

Figure 14.4 (a) Schematic representation of an infinite line coupled to a rectangular cavity. (b)–(c) Dimensionless frequencies Ω of the dips in the transmission coefficients versus: (b) the length l' of the stub for $w' = a$ and (c) the width w' of the stub for $l' = 1.7a$; (m_1, m_2) represents the order of the mode inside the cavity.

the result of a simple analytical model [30] in which we have calculated the transmission coefficient of a waveguide with a grafted side branch by neglecting the transverse dimensions of any wire, which means all the wires are assumed to be infinitely thin. This calculation shows that the zeros of transmission occur at the resonance frequencies of the grafted stub, namely $\omega = \frac{c}{n} m \frac{\pi}{l'}$, n being the index of refraction, m a positive integer, and l' the length of the stub. Now, we replace the wire-shaped stub by a rectangular cavity with dimensions w' (for the width) and l' (for the length), embedded in a perfect metal (see Fig. 14.4(a)). The dimensionless eigenfrequencies Ω of the cavity are given by

$$\Omega = \frac{1}{2n}\sqrt{m_1^2\left(\frac{a}{w'}\right)^2 + m_2^2\left(\frac{a}{l'}\right)^2} \tag{14.2}$$

where m_1 and m_2 are two positive integers.

In Figs. 14.4(b) and 14.4(c), we compare the frequencies of the zeros of transmission obtained from the numerical calculations with the eigenfrequencies of the rectangular cavity (Eq. (14.2)) as a function of the length and the width of the stub. In order to obtain a satisfactory fit between the numerical and analytical results, we have chosen $w' = w$ and $l' = l + 0.2a$. Indeed, it is worth noticing that, in comparison with the actual structure, the length of the stub in the analytical model remains somewhat undefined and can be chosen between $l' = l$ and $l' = l + 0.5a$. Figs. 14.4(b) and 14.4(c) show that, with our choice of $l' = l + 0.2a$, the numerical results are well reproduced for the fundamental or higher-order modes of the dielectric cavity embedded in a perfect metal.

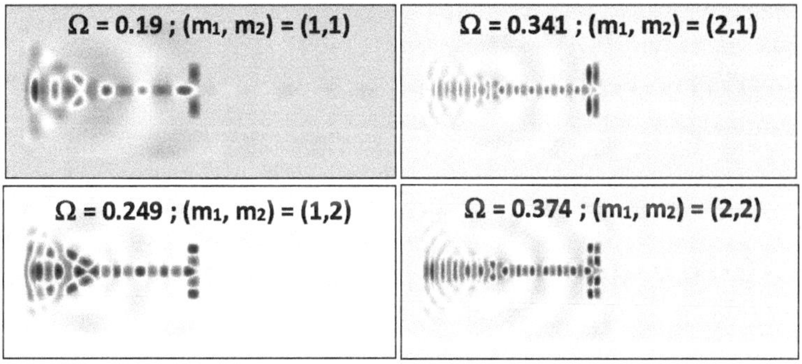

Figure 14.5 Map of the magnetic field for different monochromatic incident radiations at the frequencies of the zeros of transmission. The geometrical parameters of the stub are $w = a$ and $l = 1.7a$. The map is obtained by averaging the field over half period of oscillation. The red [light gray in print version] (blue [dark gray in print version]) color corresponds to the highest (lowest) values of the magnetic field given in arbitrary units. Based on the number of oscillations in the magnetic field inside the stub, an order (m_1, m_2) can be attributed to each mode in relation with the analytical model of Eq. (14.2).

To give a deeper insight about the above comparison, we present in Fig. 14.5 the map of the magnetic field obtained from the FDTD calculation for several monochromatic incident waves corresponding to the frequencies of the zeros of transmission. In this calculation, the geometrical parameters are $w = a$ and $l = 1.7a$. The number of oscillations in the map of the magnetic field inside the stub enables us to attribute an order (m_1, m_2) to each mode, in relation with the analytical result of Eq. (14.2). It is also worth noticing that, for the selected frequencies, the incident field in the waveguide interacts with the stub and is then totally reflected back, the outgoing wave becoming vanishingly small.

Usually, in many filtering devices, traveling waves in a waveguide interact with an adjacent microring or disk by evanescent coupling through a small air gap. As in microring resonators, we examine now the case where the stub is separated from the straight waveguide by such an air gap of thickness g (Fig. 14.6(a)). The thickness of the gap has been modified from 5% to 25% of the width a of the waveguide. Fig. 14.6(b) presents the transmission spectra for three values of g equal to 0, $0.1a$, and $0.2a$. The depression in the transmission coefficient decreases when g increases, which can be explained by a smaller coupling between the waveguide and the resonator. However, the dips become narrower and, then, their quality factor Q (defined as the ration between the frequency and the full width at half maximum) increases. The increase of the quality factor with g is displayed in Fig. 14.6(c).

To demonstrate the extension of the model, we present the transmission of waves through a three-dimensional straight waveguide coupled to a stub with the geometry of the device displayed in Fig. 14.7(a). The waveguide is constituted of an InP/InGaAsP/InP heterostructure where each layer is represented by a square rod with side length a, the total height of the waveguide being $3a$. The stub is constituted by the same materials as the waveguide with the following geometrical parameters: length $l = 1.7a$, width $w = a$, and height $h = 3a$. The refractive indices of InP and InGaAsP

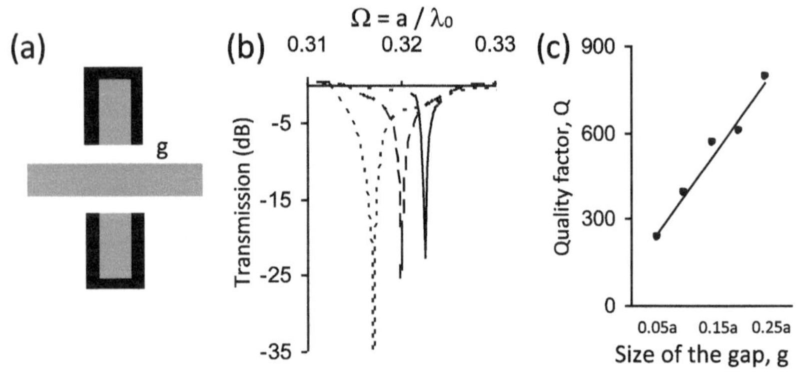

Figure 14.6 (a) Schematic view of a stub covered with a perfectly metallic thin layer, coupled to a bus waveguide via a narrow air gap of thickness g. (b) Transmission spectrum of the waveguide for an air gap of thickness $g = 0$ (dotted lines), $g = 0.1a$ (dashed lines), and $g = 0.2a$ (full lines). (c) Quality factor Q of the dip versus the thickness of the air gap.

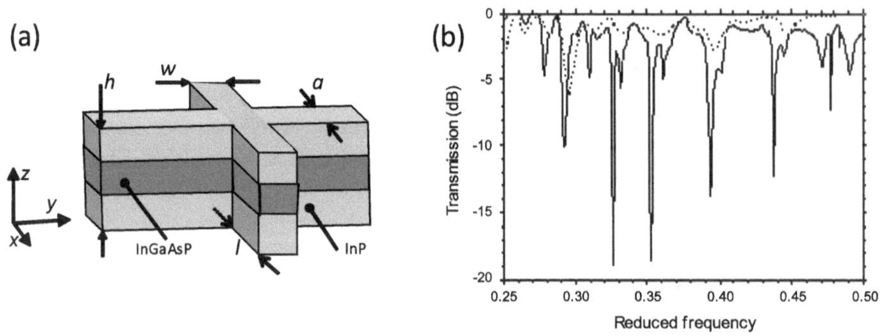

Figure 14.7 (a) Three-dimensional schematic view of a stub coupled to a waveguide. (b) Transmission spectra of the waveguide coupled to the stub. The solid and dotted lines correspond to a stub covered and not covered with a perfect metallic layer, respectively.

are 3.1647 and 3.41, respectively. Since the refractive index of InGaAsP is greater than that of InP, the wave remains preferentially confined in the InGaAsP layer. An incident wave is created by a current source located at the left entrance of the waveguide and adapted to it with the help of a taper. Here, we consider the case of an antisymmetric incident wave, mostly used in laser source excitations. The transmission spectrum (Fig. 14.7(b)) is calculated by assuming that the stub is either covered or uncovered with a thin metallic layer. One can notice that with the uncovered stub, the attenuations at the dips are only of the order of a few dB. However, when the stub is covered with a perfect metallic layer, there is an increase in the number of dips and their attenuations become much stronger than in the previous case. In fact, the uncovered stub behaves like a microantenna, with radiations escaping from the boundaries of the stub. On the contrary, adding a metallic layer at the surface of the stub prevents the waves from escaping out of the boundaries and therefore reinforces the interaction between

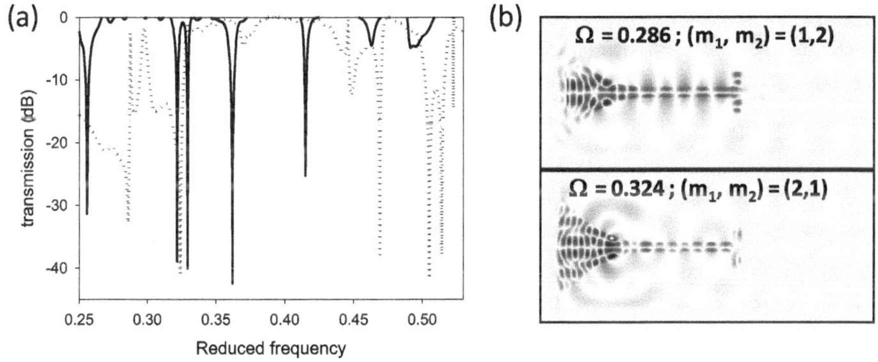

Figure 14.8 (a) Transmission spectra (FDTD model) through the waveguide coupled to a lateral stub covered with a perfect metallic layer for the antisymmetric (solid lines) and symmetric (dashed lines) modes. The geometrical parameters of the stub are $w = a$ for the width and $l = 1.4a$ for the length. (b) Maps of the magnetic field (FDTD model) for two monochromatic incident radiations at frequencies corresponding to the eigenmodes $(1, 2)$ and $(2, 1)$ of the stub for the symmetric modes. The maps are obtained by averaging the field over half period of oscillation. The red [light gray in print version] (blue [dark gray in print version]) color corresponds to the highest (lowest) values of the magnetic field given in arbitrary units.

the stub and the waveguide, leading to a filtering of many narrow frequencies in the transmission spectrum.

Finally, going back to the 2D calculation model, Fig. 14.8(a) presents the numerical transmission through the waveguide coupled to the metallic covered stub when the incoming wave is symmetric. In this calculation, we have changed the geometrical parameters of the stub, namely $w = a$ for the width and $l = 1.4a$ for the length, that ensures four dips lower than 30 dB in the dimensionless frequency range $[0.25, 0.55]$. The frequencies of the dips appear at different values than in the antisymmetric configuration (see the dashed lines in Fig. 14.8(a)).

As before, the frequencies of the dips can be understood and predicted in comparison with the electromagnetic modes of a finite rectangular cavity. Based on the results of the simple analytical model, the eigenfrequencies Ω for the symmetric case are given by

$$\Omega_{sym} = \frac{1}{2n}\sqrt{m_1^2\left(\frac{a}{d_1'}\right)^2 + \left(m_2 + \frac{1}{2}\right)^2\left(\frac{a}{d_2'}\right)^2} \tag{14.3}$$

where n is the refractive index and m_1 and m_2 are two positive integers that define the order of the modes. In Table 14.1, we make a comparison between the frequencies of the zeros of transmission obtained from the numerical calculation (FDTD) and the eigenfrequencies of the rectangular cavity (Eq. (14.3)). As in the antisymmetric case, it is worth noticing that the choice of the length l' in the analytical model remains somewhat undefined because the boundary conditions at the lower face of the cavity can be applied somewhere between the middle and the top of the waveguide; this means that l' can be chosen between $l' = 1.4a + 0.5a$ and $l' = 1.4a$. A good agreement

Table 14.1 Comparison between the reduced resonance frequencies obtained by the FDTD simulation and by the eigenfrequencies of a cavity calculated from Eq. (14.3). The numbers (m_1, m_2) give the orders of the modes.

Ω_{sym} (FDTD Fig. 14.8(a))	Ω_{sym} (Eq. (14.3))	(m_1, m_2)
0.286	0.274	$(1, 2)$
0.324	0.337	$(2, 1)$
0.444	0.436	$(1, 4)$
0.449	0.442	$(2, 3)$
0.469	0.481	$(3, 1)$
0.505	0.511	$(2, 4)$
0.514	0.514	$(3, 2)$
0.523	0.521	$(1, 5)$

can be obtained between the numerical results and the eigenfrequencies of the cavity (see Table 14.1) by choosing $l' = 1.8a$.

For symmetric modes, we plot in Fig. 14.8(b) the maps of the magnetic field obtained from the FDTD calculation for two monochromatic incident waves corresponding respectively to two zeros of transmission, $\Omega = 0.286$ and $\Omega = 0.324$. Each resonance can be labeled by two indices (m_1, m_2) that give the numbers of half wavelength within the stub region, respectively in the y and x directions.

14.4 Selective filter

In this section we present the study of several stubs attached to the waveguide as sketched in Fig. 14.9(a). The dimensions of the stubs are $w = a$ for the width and $l = 1.4a$ for the length, and they are separated from each other by a distance $\delta = 2a$. As seen in Fig. 14.4(c), the transmission curves show a set of 5 dips for a single stub in the dimensionless frequency range $[0.15, 0.45]$. In the rest of the section, we will focus on the zero transmission at $\Omega = 0.256$. The transmission spectrum is presented Fig. 14.9(b). Increasing the number of stubs from one (dashed line) to five (solid line) widen the forbidden band in the range $[0.250, 0.257]$ with an attenuation of the transmission exceeding 20 dB. We now change the geometrical parameters of the third stub in order to make a "defect" inside the periodic chain of stubs (Fig. 14.9(c)). The defect stub could be created by a modification of the width or the length of the stub. For instance, we choose to increase the length of the defect stub from $l = 1.4a$ to $l = 1.5a$ which has the effect of decreasing its resonance frequency with respect to the one associated with the other stubs in the chain, therefore ensuring that the resonance frequency falls inside the gap created in Fig. 14.9(b). The transmission spectrum of such a structure incorporating a defect (solid line in Fig. 14.9(d)) displays one narrow peak centered at frequency $\Omega = 0.2527$ inside the gap. This corresponds to the tunneling of

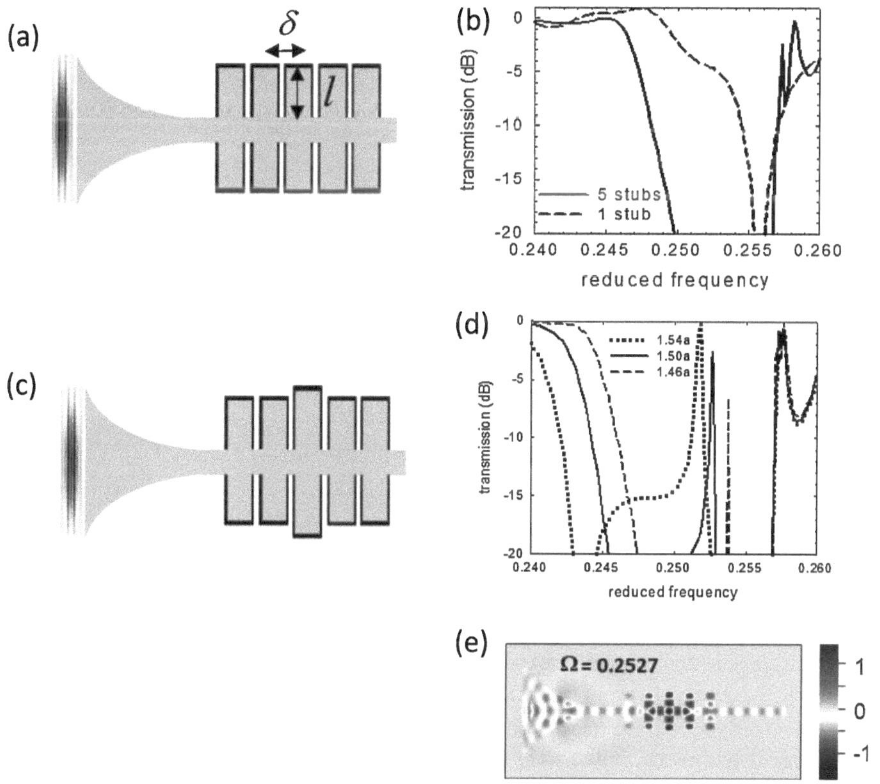

Figure 14.9 (a) Schematic view of a waveguide coupled to a regular periodic arrangement of five stubs with geometrical parameters $(w, l) = (a, 1.4a)$. The distance between two stubs is $\delta = 2a$. (b) Transmission spectrum for one (dashed line) and five stub(s) (solid line) and a separation of $2a$ between the stubs. (c) Schematic representation of the waveguide coupled to 5 stubs where the length of the central stub is increased with respect to the others. (d) Transmission spectra calculated for the above model for three different lengths of the centered stub ($l = 1.46a$, $l = 1.5a$, $l = 1.54a$), showing the shift of a narrow pass band inside the gap with centered frequencies at $\Omega = 0.2538$, $\Omega = 0.2527$, and $\Omega = 0.2518$ respectively. (e) Map of the magnetic field for the above structure at the incident frequency 0.2527.

the wave through the localized mode associated with the defect. The calculated map of the magnetic field at the latter frequency shows a strong enhancement of the field at the defect stub and its nearest neighbors (Fig. 14.9(e)). As a result, such a device could be used as a selective filter. Furthermore, it is also possible to shift the narrow frequency band inside the gap by adjusting the length or the width of the defect stub. In Fig. 14.9(d), the transmission spectrum is presented for three different lengths of the stub, namely $l = 1.46a$, $l = 1.5a$ and $l = 1.54a$. The centered frequency of the narrow pass band shifts to lowest frequencies as soon as the length of the stub increases. We have thus created a tunable selective filter at the respective reduced frequencies $\Omega = 0.2538$, $\Omega = 0.2527$, and $\Omega = 0.2518$.

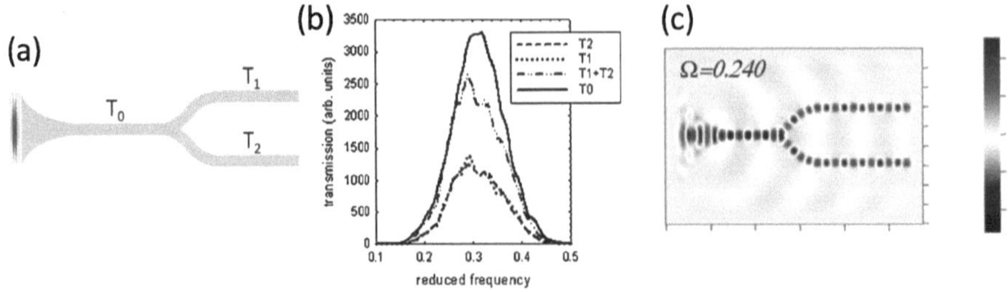

Figure 14.10 (a) Schematic of the Y-shaped waveguide; T_0 is the input signal, T_1 and T_2 the outgoing transmission at the end of the two branches. (b) Representation of the non-normalized transmission for the three ports of the Y-shaped waveguide. (c) Representation of the magnetic field for a Y-structure at the monochromatic reduced frequency ($\Omega = 0.240$). The map is obtained by averaging the field over half period of oscillation. The red [light gray in print version] (blue [dark gray in print version]) color corresponds to the highest (lowest) values of the magnetic field given in arbitrary units.

14.5 Bent Y-branch waveguide

In this section, we shall discuss the division capacity of a Y-shaped waveguide which geometry is displayed in Fig. 14.10(a). This example could easily be extended to a multichannel device. The transmission spectra (Fig. 14.10(b)) are calculated for the entering signal at port T_0 and at the end of the two branches of the device, namely T_1 and T_2. We have also reported on the same scheme, the addition of the two transmissions exiting from the Y-waveguide ($T_1 + T_2$). The Gaussian shape of the input signal is still recovered at the exits. We note that we have an equal partition of the intensity between the two branches ($T_1 = T_2$). Nevertheless, we can see a decrease of 20% between the additional transmission ($T_1 + T_2$) and the fundamental one (T_0). This difference can be attributed to reflections and diffractions during the bending. The calculated map of the magnetic field at the reduced frequency $\Omega = 0.24$ (Fig. 14.10(c)) shows a strong confinement of the wave inside the waveguide with a low diffraction in the air media and a good separation of the input signal between the two branches of the Y-device.

As shown in Fig. 14.11(a), we now insert two different stubs respectively on the two branches of the Y-structure, T_1 and T_2. The parameters of the stub in the T_1-branch are a for the width and $l_1 = 1.4a$ for the length. For the T_2 branch, the parameters are $(w, l_2) = (a, 1.5a)$. The corresponding transmission spectra at port T_1 and T_2 are reported in Fig. 14.11(b) in the reduced frequency range $[0.22, 0.27]$. We note the presence of two deep attenuation peaks at the reduced frequencies $\Omega = 0.2465$ and $\Omega = 0.2557$. These attenuated transmissions appear exactly at the frequencies of the dips calculated in the case of individual stubs coupled to a straight waveguide (Fig. 14.9(d)). The Y-structure has been able to reject independently the frequency of each stub resonator. We can conclude an absence of perturbations coming from the waveguide separation, the bending or an interaction between the stubs. The reflective waves produced by the stubs thus travel towards the entrance of the structure. Such a device is able to reject selective frequencies from a broadband signal, depending on the geometry of the microstub cavity and the choice of the exit port. We present in

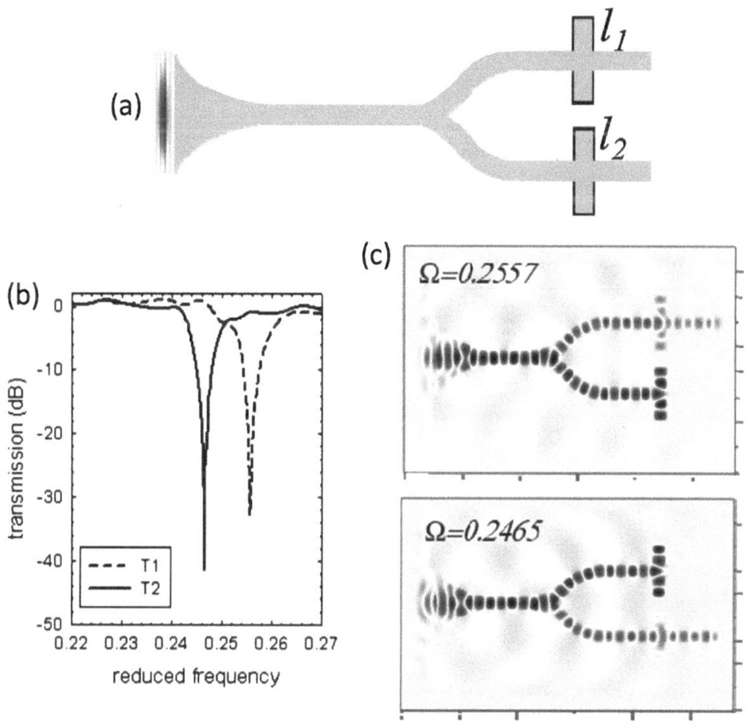

Figure 14.11 (a) 2D schematic view of the Y-shaped waveguide in which the two branches are coupled with two different lengths of stubs ($l_1 = 1.4a$ for T_1 and $l_2 = 1.5a$ for T_2). (b) Transmission spectra when the detector is placed at the exit T_1 (dashed line) and T_2 (solid line). (c) Map of the magnetic field for the two monochromatic incident radiations at the frequencies of the transmission zero $\Omega = 0.2557$ and $\Omega = 0.2465$.

Fig. 14.11(c) the map of the magnetic field obtained from the FDTD calculation for the two monochromatic incident waves corresponding to the two resonant frequencies of the stubs. At the reduced frequency $\Omega = 0.2557$, the incident field in the waveguide interacts with the stub localized in the T_1 branch and then is totally reflected back. At the same frequency, the stub in the T_2 branch is almost not excited and thus let the field travel towards the exit port. We can add a remark on the number of oscillations in the map of the magnetic field inside the stub which enable us to attribute an order to the resonant mode, in relation to the analytical result of Eq. (14.2). Here, we can confirm the order of the mode to be $(m_1, m_2) = (1, 2)$. Similar conclusions can be drawn when the incident wave is exited at the reduced frequency $\Omega = 0.2465$ and the stop frequency is in port T_2 instead of T_1. We have demonstrated the efficiency of the Y-structure bichannel division on a large input signal. Moreover, we could confirm the independence of the two branches when we insert defects as cavities on the branches.

Finally, this proposed structure can be extended towards a multichannel rejective filter which could separate many chosen frequencies inside a large input signal. The purpose is first to realize a Y-demultiplexer based on several microstubs cavities. The

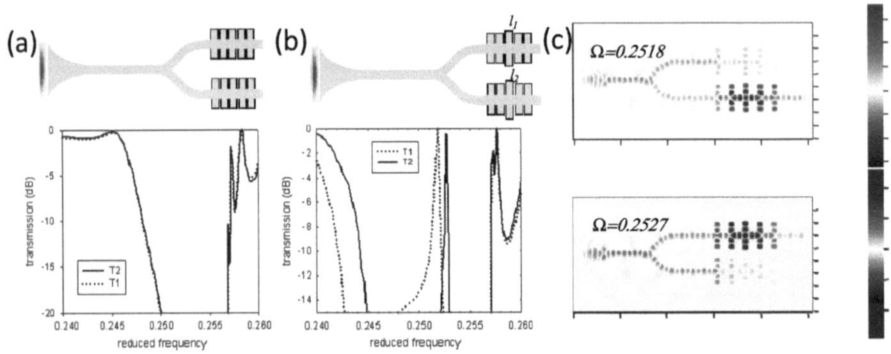

Figure 14.12 (a) 2D schematic of the Y-shaped waveguide in which the two branches are coupled with five similar stubs $(w, l) = (a, 1.4a)$ and FDTD computed transmission spectra of the previous scheme when the detector is positioned at the exit T_1 (dashed line) and T_2 (solid line). (b) Schematic representation of the Y-shaped waveguide in which the branches are coupled to 5 stubs where the length of the central stub is increased with respect to the others: $l_1 = 1.54a$ for T_1 and $l_2 = 1.50a$ for T_2. Transmission spectra when the detector is positioned at the exit T_1 (dashed line) and T_2 (solid line). (c) Map of the magnetic field for the two monochromatic incident radiations at the frequencies of the transmission zero $\Omega = 0.2527$ and $\Omega = 0.2518$.

main goal is to select different frequencies on each branch of the Y-structure from a broad band signal. As explained above (Section 14.4), the first step in the selective filter constitution is to associate microstubs in a periodic structure. The stub periodicity leads to the formation of a gap in the transmission spectrum. Fig. 14.12(a) shows that we can obtain the same forbidden frequency gap on each branch T_1 and T_2, when we place five similar stubs of dimensions $(w, l) = (a, 1.4a)$ in a periodic association. This gap is almost the same as already observed in the case of a straight waveguide (Fig. 14.9(b)). The second step of the selective filter realization is to insert, on each branch, a defect stub in the middle of the periodic stub arrangement as illustrated in Fig. 14.12(b). In this model, we replace on the T_1 (resp. T_2) branch the third stub by a new one with the geometrical parameters $(w, l_1) = (a, 1.5a)$ (resp. $(w, l_2) = (a, 1.54a)$). The transmission spectra calculated from the signals exiting the two branches of the Y-shaped waveguide are displayed below. In those transmission spectra, we recognize the two narrow pass bands we have already met in the straight filters configuration of Fig. 14.9(d). The intensity, the shape and reduced frequencies of the peaks are also the same in the two figures which confirm once again the independence between the two output channels. The Y-demultiplexer is thus able to separate two narrow and near passbands coming from a large input Gaussian signal. Each selected reduced frequency ($\Omega = 0.2518$ and $\Omega = 0.2527$) is available at the end of the corresponding port (T_2 and T_1). The calculated maps of the magnetic field at the latter monochromatic frequencies show a strong enhancement of the field at the defect stub and its nearest neighbors (Fig. 14.12(c)) into the T_1 branch for the reduced frequency $\Omega = 0.2518$ and into the T_2 branch for $\Omega = 0.2527$. At the same frequencies, the regular periodic arrangement of the stubs acts as a stopband and no signal is available at port T_2 for $\Omega = 0.2518$ and at port T_1 for $\Omega = 0.2527$. As a result,

Fig. 14.12 shows clearly that the large band input Gaussian signal T_0 (Fig. 14.10) has been separated and directed towards the two branches of the Y-junction. It is also worth noticing that each branch of the Y-shaped waveguide will select its own narrow passband.

14.6 Real metal

The above applications were discussed by assuming that the stubs are covered with a perfect metal layer. The hypothesis of a perfect metal is well adapted to infrared frequencies but not to optical frequencies where the true behavior of the metal cannot be omitted. Nevertheless, this hypothesis has the advantage of reducing considerably the simulation time, since no particular discretization is required in the metallic part of the structure. In the case of a real metal, the discretization should be performed at the scale of the skin depth, which means a much smaller mesh than in the case of the semiconductor waveguide. To show that the above applications should possibly be also feasible at the telecommunication wavelength, we have studied the transmission through a waveguide coupled to a single stub covered with a real metal whose dielectric function is described by the Drude model

$$\varepsilon(\omega) = \varepsilon_\infty - \frac{\omega_p^2}{\omega(\omega + i.\gamma)}.$$

Here ω_p is the plasma frequency of the metal, γ the absorption coefficient, and ϵ_∞ comes from the contribution of the bound electrons to the polarizability. For metal Ag used here, $\omega_p = 1.36 \times 10^{16}$ rad.s^{-1}, $\gamma = 1.0 \times 10^{14}$ Hz and ($\epsilon_\infty = 4.2$) [31]. In Fig. 14.13(a), we show with solid line the transmission spectrum for a stub with the geometric parameters $(w, l) = (a, a)$. The transmission spectrum displays three dips with attenuation less than 20 dB that confirm the interaction of the waveguide modes with the cavity resonances. The quality factor of the peaks may reach values of the order of 100. Fig. 14.13(b) displays the maps of the magnetic field at the frequencies of the dips, namely $\Omega = 0.2052$, $\Omega = 0.2338$, and $\Omega = 0.3357$. One can notice that the incident field in the waveguide interacts with the stub and then is totally reflected back, the outgoing wave becoming vanishingly small. Depending on the frequency, one can observe inside the stub different number of oscillations in relation with the size of the cavity. In Fig. 14.13(a), we have also investigated the effect of losses on the transmission spectrum by changing the parameter γ in a broad range ($\gamma = 10^{13}$, 10^{14}, and 10^{15} Hz). We find that for γ below or of the order of 10^{14} Hz, there is no significant modification in the results. On the contrary, when γ goes much above 10^{14} Hz, up to 10^{15} Hz, the peaks broaden and become less deep. This means that over a reasonable range of the absorption coefficient ($\gamma \sim 10^{14}$ Hz), the effect of the metal to confine the waves in the stubs can mainly be attributed to the negative value of the real part of its dielectric constant. To give a better insight about the range of validity of the Drude model and its possible application to our structures, we have fitted the experimental values of the real and imaginary parts of the Ag dielectric

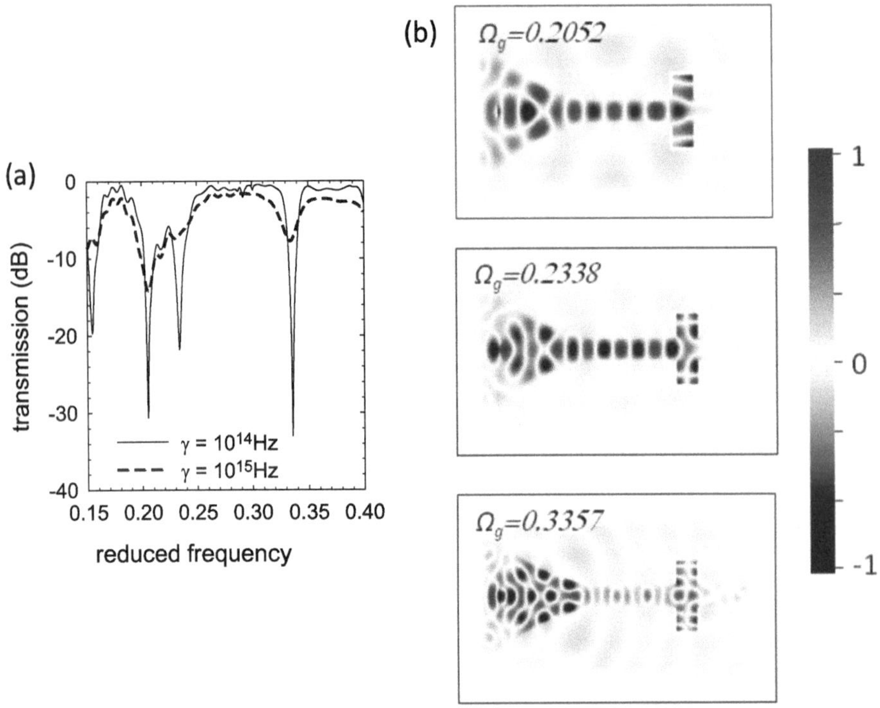

Figure 14.13 (a) Transmission spectrum through the system presented in Fig. 14.1 when the coating material of the stub is a real metal described by the Drude model and the dimensions of the stub are $(w, l) = (a, a)$. The transmission curves are plotted for a value of the absorption coefficient $\gamma = 10^{14}$ Hz (solid line) and $\gamma = 10^{15}$ Hz (dashed line). (b) Maps of the magnetic field (FDTD model) for three monochromatic incident radiations at frequencies corresponding to the deep peaks of Fig. 14.13(a) ($\gamma = 10^{14}$ Hz). The maps are obtained by averaging the field over half period of oscillation. The red [light gray in print version] (blue [dark gray in print version]) color corresponds to the highest (lowest) values of the magnetic field given in arbitrary units.

constant [32] with the Drude model. A good agreement can be obtained for γ ranging from 10^{13} to 10^{14} Hz when the wavelength varies from 2.5 to 0.5 μm, a range which is even broader than the one we have displayed in Fig. 14.13(a). For example, for a common waveguide thickness $a = 0.5$ μm, the wavelength in Fig. 14.13(a) ranges between 3 and 1.25 μm. For wavelengths lower than 0.5 μm, it is no more possible to fit the dielectric constant, even by increasing the absorption coefficient. This is due to the proximity of the plasmon resonance where the Drude model fails to give an appropriate description of the metal. We have done the same analysis for two other metals which are commonly used in the infrared, namely gold and aluminum. To fit the experimental gold data, Drude model with parameters $\omega_p = 1.4 \times 10^{16}$ rad.s^{-1}, $\gamma = 1.0 \times 10^{14}$ Hz, and $\epsilon_\infty = 7.0$ [33] is suitable for wavelengths falling in the range $[0.8, 2.5]$ μm, while it becomes invalid for shorter wavelengths due to the proximity of the resonance. To fit the experimental Al data [34], one can take the following parameters: $\omega_p = 2.3 \times 10^{16}$ rad.s^{-1}, $\epsilon_\infty = 1.0$, and an absorption coefficient $\gamma =$

2.0×10^{14} Hz in the wavelength range [1.6, 2.5] μm. One can notice that this latter material is less adapted for the wavelength range of Fig. 14.13(a) with a waveguide thickness of 0.5 μm.

14.7 Conclusions

In conclusion, we have demonstrated the efficiency of a resonator device constituted by a straight waveguide coupled to a double stub covered with a metal. The insertion of the stub gives rise to a filtering of many narrow frequencies in the transmission spectrum of the waveguide. Using a simple analytical model of a rectangular cavity, all the filtered frequencies have been identified in terms of the fundamental and higher order modes of the stub. Those frequencies can then be chosen by adjusting the length and/or the width of the stub. The quality factor can be improved by the insertion of a small air gap between the stub and the waveguide. We have shown that similar features to those obtained with 2D structures should also be possible in 3D realistic semiconductor waveguides. We have also discussed the occurrence of transmission zeros according to the symmetry of the incident wave. The combination of a set of stubs widens the zeros of transmission into gaps in which a selective tunneling transmission becomes possible through a defect stub. The selective frequency can be adjusted by choosing appropriately the width or the length of the defect stub. Moreover, we have discussed a model of demultiplexer based on a Y-shaped waveguide. This system can be used for separating signals with different frequencies. The validity of such simulations while assuming the metal as a perfect coating conductor is limited to the far-infrared regime, far from the plasma frequency. Nevertheless, the physical trends have been shown to remain valid in the near-optical regime by the modeling of the metallic coating in the framework of the Drude model.

Acknowledgment

The authors would like to thank M. Beaugeois, J.O. Vasseur, J-P Vilcot, M. Bouazaoui, and J-P Vigneron for their participation to the original papers at the basis of this chapter.

References

[1] T.J. Johnson, M. Borselli, O. Painter, Self-induced optical modulation of the transmission through a high-Q silicon microdisk resonator, Opt. Express 14 (2006) 817.

[2] T. Tanabe, M. Notomi, S. Mitsugi, A. Shinya, E. Kuramochi, All-optical switches on a silicon chip realized using photonic crystal nanocavities, Appl. Phys. Lett. 87 (2005) 817.

[3] O. Painter, R.K. Lee, A. Scherer, A. Yariv, J.D. O'Brien, P.D. Dapkus, I. Kim, Two-dimensional photonic band-gap defect mode laser, Science 284 (1999) 1819.

[4] K. Hennessy, A. Badolato, M. Winger, D. Gerace, M. Atature, S. Gulde, S. Falt, E.L. Hu, A. Imamoglu, Quantum nature of a strongly coupled single quantum dot–cavity system, Nature (London) 445 (2007) 896.

[5] D.Q. Yang, B. Duan, X. Liu, A.Q. Wang, X.G. Li, Y.F. Ji, Photonic crystal nanobeam cavities for nanoscale optical sensing: a review, Micromachines 11 (2020) 72.

[6] R. Quidant, J.-C. Weeber, A. Dereux, G. Lévêque, J. Weiner, C. Girard, Addressing and imaging microring resonators with optical evanescent light, Phys. Rev. B 69 (2004) 081402(R).

[7] D. Rafizadeh, J.P. Zhang, S.C. Hagness, A. Taflove, K.A. Stair, S.T. Ho, Waveguide-coupled AlGaAs/GaAs microcavity ring and disk resonators with high finesse and 21.6-nm free spectral range, Opt. Lett. 22 (1997) 1244.

[8] Y. Akahane, T. Asano, B.S. Song, S. Noda, High-Q photonic nanocavity in a two-dimensional photonic crystal, Nature 425 (2003) 944.

[9] C. Meier, K. Hennessy, E.D. Haberer, R. Sharma, Y.S. Choi, K. McGroddy, S. Keller, S.P. DenBaars, S. Nakamura, E.L. Hu, Visible resonant modes in GaN-based photonic crystal membrane cavities, Appl. Phys. Lett. 88 (3) (2006) 031111.

[10] P.B. Deotare, M.W. McCutcheon, I.W. Frank, M. Khan, M. Lončar, High quality factor photonic crystal nanobeam cavities, Appl. Phys. Lett. 94 (2009) 121106.

[11] A.R.M. Zain, N.P. Johnson, M. Sorel, M. Richard, Ultra high quality factor one dimensional photonic crystal/photonic wire micro-cavities in silicon-on-insulator (SOI), Opt. Express 16 (2008) 12084.

[12] M. Aspelmeyer, T.J. Kippenberg, F. Marquardt, Cavity optomechanics, Rev. Mod. Phys. 86 (2014) 1391.

[13] Yan Pennec, Vincent Laude, Nikos Papanikolaou, Bahram Djafari-Rouhani, Mourad Oudich, Said El Jallal, Jean Charles Beugnot, Jose M. Escalante, Alejandro Martínez, Modeling light-sound interaction in nanoscale cavities and waveguides, Nanophotonics 3 (6) (2014) 413–440.

[14] M. Maldovan, E.L. Thomas, Simultaneous complete elastic and electromagnetic band gaps in periodic structures, Appl. Phys. B 83 (2006) 595;
M. Maldovan, E.L. Thomas, Simultaneous localization of photons and phonons in two-dimensional periodic structures, Appl. Phys. Lett. 88 (2006) 251907.

[15] Y. Pennec, B. Djafari-Rouhani, E.H. El Boudouti, C. Li, Y. El Hassouani, J.O. Vasseur, N. Papanikolaou, S. Benchabane, V. Laude, A. Martinez, Simultaneous existence of phononic and photonic band gaps in periodic crystal slabs, Opt. Express 18 (2010) 14301.

[16] S. Mohammadi, A.A. Eftekhar, A. Khelif, A. Adibi, Simultaneous two-dimensional phononic and photonic band gaps in opto-mechanical crystal slabs, Opt. Express 18 (2010) 9164.

[17] Q. Rolland, M. Oudich, S. El-Jallal, S. Dupont, Y. Pennec, J. Gazalet, J.C. Kastelik, G. Lévêque, B. Djafari-Rouhani, Acousto-optic couplings in two-dimensional phoxonic crystal cavities, Appl. Phys. Lett. 101 (2012) 061109.

[18] S. El-Jallal, M. Oudich, Y. Pennec, B. Djafari-Rouhani, V. Laude, J.C. Beugnot, A. Martínez, J.M. Escalante, A. Makhoute, Analysis of optomechanical coupling in two-dimensional square lattice phoxonic crystal slab cavities, Phys. Rev. B 88 (2013) 205410.

[19] M. Eichenfield, J. Chan, R.M. Camacho, K.J. Vahala, O. Painter, Optomechanical crystals, Nature 462 (2009) 78.

[20] Y. Pennec, B. Djafari-Rouhani, C. Li, J.M. Escalante, A. Martinez, S. Benchabane, V. Laude, N. Papanikolaou, Band gaps and cavity modes in dual phononic and photonic strip waveguides, AIP Adv. 1 (2011) 041901.

[21] J. Chan, A.H. Safavi-Naeini, J.T. Hill, S. Meenehan, O. Painter, Optimized optomechanical crystal cavity with acoustic radiation shield, Appl. Phys. Lett. 101 (2012) 081115.

[22] J. Gomis-Bresco, D. Navarro-Urrios, M. Oudich, S. El-Jallal, A. Griol, D. Puerto, E. Chavez, Y. Pennec, B. Djafari-Rouhani, F. Alzina, A. Martínez, C.M. Sotomayor Torres, A one-dimensional optomechanical crystal with a complete phononic band gap, Nat. Commun. 5 (2014) 4452.
[23] M. Oudich, S. El-Jallal, Y. Pennec, B. Djafari-Rouhani, J. Gomis-Bresco, D. Navarro-Urrios, C.M. Sotomayor Torres, A. Martiınez, A. Makhoute, Optomechanic interaction in a corrugated phoxonic nanobeam cavity, Phys. Rev. B 89 (2014) 245122.
[24] Y. Pennec, B. Djafari-Rouhani, A. Akjouj, J.O. Vasseur, L. Dobrzynski, J.P. Vilcot, M. Beaugeois, M. Bouazaoui, R. Fikri, J.P. Vigneron, Transmission filtering of a waveguide coupled to a stub microresonator, Appl. Phys. Lett. 89 (2006) 101113.
[25] Y. Pennec, M. Beaugeois, B. Djafari-Rouhani, A. Akjouj, J.O. Vasseur, L. Dobrzynski, J.P. Vilcot, M. Bouazaoui, J.P. Vigneron, Selective filtering of confined optical waves in a straight waveguide coupled to lateral stubs, J. Opt. A, Pure Appl. Opt. 9 (2007) S431–S436.
[26] Y. Pennec, M. Beaugeois, B. Djafari-Rouhani, R. Sainidou, A. Akjouj, J.O. Vasseur, L. Dobrzynski, E.H. El Boudouti, J.P. Vilcot, M. Bouazaoui, J.P. Vigneron, Microstubs resonators integrated to bent Y-branch waveguide, Photonics Nanostruct. Fundam. Appl. 6 (2008) 26–31.
[27] A. Beaurain, S. Dupont, H.-W. Li, J.-P. Vilcot, C. Legrand, J. Harari, M. Constant, D. Decoster, Characterization and fabrication of InGaAsP/InP deep-etched micro-waveguides, Microw. Opt. Technol. Lett. 40 (2004) 216218.
[28] A. Taflove, Computational Electrodynamics: The Finite-Difference Time-Domain Method, Artech House, Norwood, MA, 1995.
[29] J.P. Berenger, A perfectly matched layer for the absorption of electromagnetic waves, J. Comput. Phys. 114 (1994) 185.
[30] J.O. Vasseur, A. Akjouj, L. Dobrzynski, B. Djafari-Rouhani, E.H. El Boudouti, Photon, electron, magnon, phonon and plasmon mono-mode circuits, Surf. Sci. Rep. 54 (2004) 1.
[31] H. Gao, H. Shi, C. Wang, C. Du, X. Luo, Q. Deng, Y. Lv, X. Lin, H. Yao, Surface plasmon polariton propagation and combination in Y-shaped metallic channels, Opt. Express 13 (2005) 10795.
[32] P.B. Johnson, R.W. Christy, Optical constants of the noble metals, Phys. Rev. B 6 (1972) 4370.
[33] Anuj K. Sharma, B.D. Gupta, Fibre-optic sensor based on surface plasmon resonance with Ag–Au alloy nanoparticle films, Nanotechnology 17 (2005) 124.
[34] E.D. Palik, Handbook of Optical Constants of Solids, Academic Press, 1998.

Silicon nanowires and nanopillars for photovoltaic

15

Y. Pennec, A. Akjouj, G. Lévêque, B. Djafari-Rouhani, and L. Dobrzyński
Institut d'Electronique et de Microélectronique et de Nanotechnologies, IEMN, CNRS, UMR 8520, Groupe de Physique, Cite scientifique, Villeneuve d'Ascq, France

15.1 Introduction

Semiconductor silicon nanostructures as nanowires or nanopillars are currently involved in the fabrication of devices based on the photogeneration of free carriers [1–5], mainly due to their low optical reflectance when used as thin films or their high efficiency to collect carriers in radial junction as their diameter can be lower than the diffusion length of the minority carriers [6]. It has been shown that nanowires with diameters smaller than thin films made up of equivalent materials offer efficient optical properties, thus are expected to have shorter carrier transit times [6–8]. Such a property is crucial for increasing the speed of photodetectors or the photoconductive gain of solar cells. Consequently, nanowire arrays have been studied, and their optical properties have been both experimentally [9–18] and theoretically [19–32] investigated.

To characterize the optical properties of the nanostructures, the so-called ultimate efficiency parameter, α, has been introduced. This parameter encodes the assumption that each absorbed photon with energy greater than the bandgap produces one and only one electron–hole pair. Such a property turns out to be crucial in increasing the speed of photodetectors or the photoconductive gain of solar cells. In addition, it has been demonstrated that arrays of NWs possess appropriate long optical paths for efficient light absorption [33–36]. Therefore, semiconductor NWs offer the prospect of improving the energy conversion efficiency in optoelectronic devices. The efficiency parameter α depends of many parameters, such as the length, the diameter, the filling factor, and periodicity of the array. One key problem is that these parameters are dependent on each other. While generally people arbitrarily assign a value to one parameter (length, diameter, periodicity, or filling factor) and study the optical properties versus the other parameters, a general overview of the role of the geometrical parameters appears necessary.

The present chapter deals with two silicon nanostructures made respectively of two-dimensional nanowires (NWs) and three-dimensional nanopillars (NPs). In both cases, the theoretical results have been supported with experimental measurements. In the first section we address a simulation of the optical absorption of silicon nanowires versus their diameter based on two different theoretical tools, namely the analytical Green's function formalism and the numerical finite difference time domain (FDTD) method. The results of the simulation are compared with photoconductivity measurements performed on n-type doped Si NWs. Using NWs that are slightly tapered, such

Photonics. https://doi.org/10.1016/B978-0-12-819388-4.00025-3

experiments allow verifying the diameter dependence of the optical absorption along the growth direction of single wires, while the structural and compositional of the NWs are kept unchanged, as proved by a complementary electrical analysis of the NWs. In the second part of the chapter, we propose simulations on the optical reflectance and efficiency of nanopillar (NP) arrays [37]. Experimental results are also reported in good agreement with the numerical calculations. The chapter thus gives a general trend on the behavior of the efficiency of the solar cell versus all geometrical parameters.

15.2 Optical absorption of silicon nanowires

15.2.1 Theoretical analysis

15.2.1.1 Formalism

Two theoretical tools have been used to investigate the absorption of light by Si NW. The first method, based on the Green's function formalism, is useful to calculate the scattering and absorption properties of a single or a finite set of NWs. The second simulation tool, based on the Finite Difference Time Domain (FDTD) method [38] is well-adapted to deal with a periodic set of NWs. After a brief presentation of each formalism, we discuss the physical trends of the absorption by NW of various diameters.

We investigated first the single-wire problem with the Green's tensor method. This numerical method is well documented in several articles [39,40] and is particularly convenient for the study of localized objects placed in a multilayered 2D or 3D environment. For convenience, we remind that it relies on the resolution of the Lippmann–Schwinger equation:

$$\boldsymbol{E}(\boldsymbol{r}) = \boldsymbol{E}_0(\boldsymbol{r}) + k_0^2 \int_V d\boldsymbol{r}' \, \boldsymbol{G}(\boldsymbol{r}, \boldsymbol{r}')\Delta(\boldsymbol{r}')\boldsymbol{E}(\boldsymbol{r}') \tag{15.1}$$

where ω is the pulsation, $k_0 = \omega/c$, $\boldsymbol{E}$ is the total electric field at point $\boldsymbol{r}$, $\boldsymbol{E}_0$ is the incident electric field (electric field at $\boldsymbol{r}$ in vacuum), and $\boldsymbol{G}(\boldsymbol{r}, \boldsymbol{r}')$ the Green's function of the multilayered environment between a source point $\boldsymbol{r}'$ and the observation point $\boldsymbol{r}$. The integration is restricted to the volume of the nanowire, which is discretized in several polarizable cells. We have computed the absorption inside the nanowire given by

$$C_{abs}(z, \omega) = \int_V d\boldsymbol{r} \, \varepsilon''(\omega) \, |\boldsymbol{E}(\boldsymbol{r})|^2 \tag{15.2}$$

where $\varepsilon''(\omega)$ is the imaginary part of the nanowire susceptibility at the pulsation ω.

In the FDTD method, the calculations are performed on a periodic array of uniform Si wires that lie in a parallel arrangement at the interface between two dielectric layers. Each array is defined by the diameter d of the Si NWs and the spacing a between two adjacent wires, as shown in Fig. 15.1. The Maxwell's equations are solved by

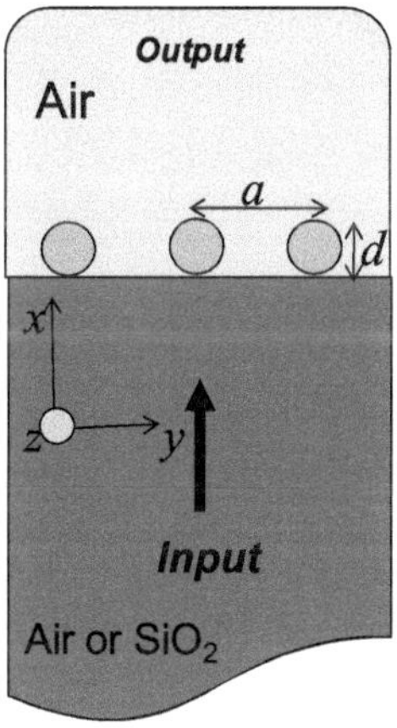

Figure 15.1 Schematic representation of the periodic Si NWs structure. The Si NW diameter is d. The lattice parameter a is defined as the distance between two nearest neighboring Si NWs. The input source is placed in the glass substrate and the detector in air.

discretizing both time and space and by replacing derivatives with finite differences. Our calculation is performed in a 2D box (along x and y axes) with propagation along the y axis. Perfect matching layer conditions are applied at the boundaries y of the box, in order to avoid reflections of outgoing waves [41]. Along the x direction, the unit cell is repeated periodically, and the structure is supposed to be infinite along the z direction. Space is discretized in both x and y directions using a mesh interval equal to $\Delta x = \Delta y = 1$ nm. The equations of motion are solved with a time integration step $\Delta t = \Delta x/4c$, and the number of time steps equals to 2^{20}, which is the necessary tested time for a good convergence of the numerical calculation. The incoming pulse is generated at the bottom part of the unit cell, by a current source parallel to the x axis and having a planar profile along the x direction. The current is generated during a short period of time in such a way as to excite the electromagnetic waves in a broad frequency domain of interest. Then, the reflection and transmission spectra recorded during the time are Fourier-transformed to obtain R and T as functions of frequency and deduce the absorption as $A = 1 - T - R$.

15.2.1.2 Theoretical results

In our calculations, the frequency-dependent complex permittivity of silicon is described by the Lorentz oscillator model [42],

$$\varepsilon(E) = \varepsilon_1(\infty) + \sum_{i=1}^{4} \frac{A_i}{E_i^2 - E^2 + jE\Gamma_i}, \tag{15.3}$$

where $\varepsilon_1(\infty)$ is the high frequency dielectric constant, A_i, Γ_i, and E_i are respectively the amplitude, the damping factor, and the resonant energy of the ith oscillator. The parameters are taken from [43]. They allow fitting the dielectric permittivity of silicon in the range from 0.2 to 4 eV.

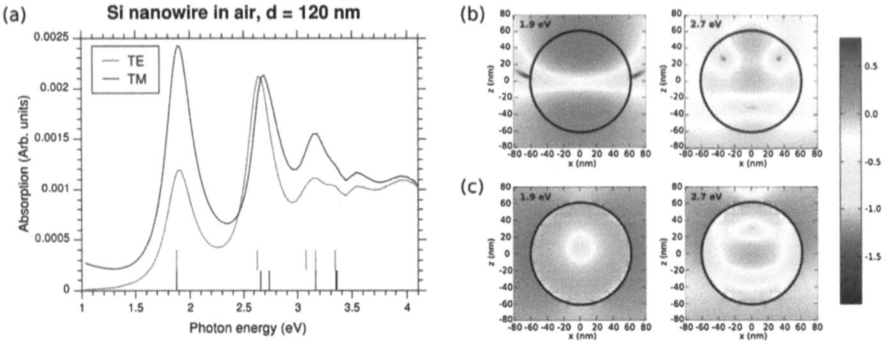

Figure 15.2 (a) Absorption spectra for a silicon NW of diameter $d = 120$ nm, in case of TM (black) and TE (red; light gray in print version) plane wave illumination. Vertical lines indicate the eigenfrequencies of the Si NW surrounded by air: degenerated TE/TM are in magenta (dark gray in print version), TM only in black and TE only in red (light gray in print version); (b)–(c) maps of the electric field intensity for the two main peaks in Fig. 15.2 appearing at 1.9 eV ($\lambda = 650$ nm) and 2.7 eV ($\lambda = 462$ nm): (b) TM illumination and (c) TE illumination.

Fig. 15.2(a) shows the absorption spectra of 120-nm-diameter silicon wire in air, for both TE (transverse electric, magnetic field parallel to the wire axis) and TM (transverse magnetic, electric field parallel to the wire axis) illuminations. Spectra are dominated by two modes at about 1.9 and 2.7 eV, corresponding to the first volume modes of a cylindrical dielectric waveguide, as the dielectric constant is positive in this energy range. The energies of the first TE (red; light gray in print version) and TM (black) eigenmodes are indicated by the ticks at the bottom of the figure. Since the wavelengths λ of the photonic modes in the NWs scale proportionally with their diameter d, one can expect that the position of the absorption peaks will move to lower energies when increasing this diameter. We shall discuss below the relation between the threshold for the absorption of light and the diameter d. For energy larger than 3.0 eV, the absorption increases independently of the wavelength due to the higher losses related to electronic transitions. The maps of the electric field intensity for the main two peaks in Fig. 15.2(a) are plotted in Figs. 15.2(b) and 15.2(c). The dissymmetry along the y-axis is due to the incident illumination along the same direction. Indeed, as the quality factor of the mode around is not very large, the structure of the optical modes is difficult to see because of the superposition of the mode with the incident field. Despite this, we can clearly identify the two modes at 1.9 eV as, respectively, TE01 and TM11 [44].

In Fig. 15.3, we plot the absorption as a function of the photon energy and the diameter d of the wire, averaged on both polarizations of the incident wave. For diameter larger than 20 nm, the successive modes appear, the lowest branch (in cyan; light gray in print version) corresponding to the TM11 and TE01 modes of the cylindrical waveguide. Let us stress that this picture is qualitatively very similar to the results obtained in [43] for the optical absorption of a Ge NW that we can reproduce by our Green's function method.

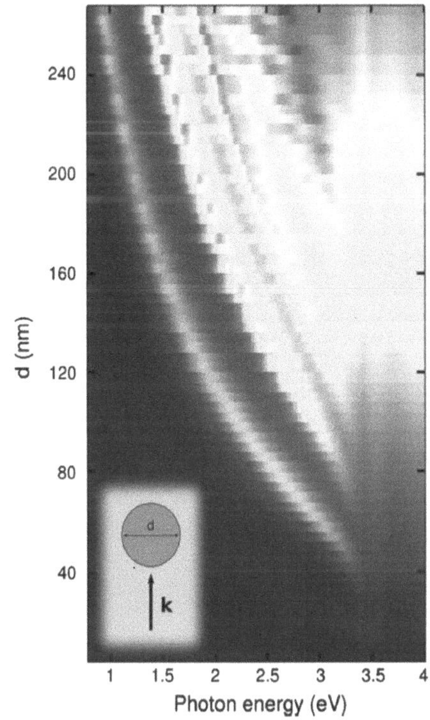

Figure 15.3 Absorption spectra as a function of the NW diameter, for a plane wave in normal incidence, averaged on TE and TM polarizations. The absorption is calculated per volume unit.

Before going to the case of a periodic array of NWs, we also use the Green's function method to calculate the optical absorption of a grating constituted by a finite number N of Si NWs regularly spaced from each other. The results are given in Fig. 15.4 for NWs of diameter 120 nm and spacing 200 nm. In case of TE plane wave illumination, modes are slightly blue-shifted due to the coupling between adjacent NWs, but the spectra remain unchanged when the number of NWs inside the grating is larger than 5: everything happens as if the grating were infinite. For the TM illumination, the situation is more complex, as the lowest energy mode (1.9 eV) splits into several modes for a large number of nanowires. Its average position oscillates and tends to stabilize for a number of wires greater than 9. We note that the Lorentz–Drude model used for the Si dielectric constant is incorrect for photon energy smaller than 1.2 eV, which explains the small increase in absorption on Fig. 15.4(b) around this value.

The above results are completed with the calculation of the optical absorption by a periodic array of Si NW under TM illumination using the FDTD method. Fig. 15.5(a) shows the absorption for various diameters of the NWs ranging from 50 to 160 nm with a period $a = 200$ nm. Overall, the absorption decreases with smaller diameters, which is understandable owing to the smaller volume occupied by the wire. The trends and the position of the absorption peaks are very similar to those obtained above for a

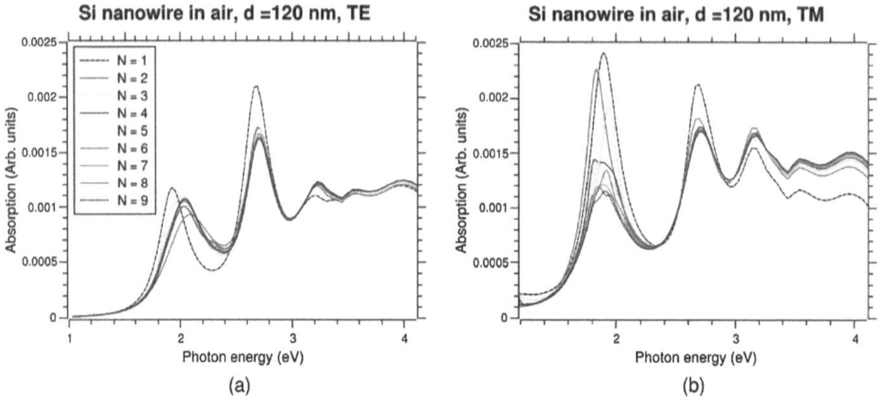

Figure 15.4 Optical absorption for a finite set N of Si NWs of diameter 120 nm regularly spaced by 200 nm, for TE and TM plane wave illumination. The absorption is calculated per volume unit.

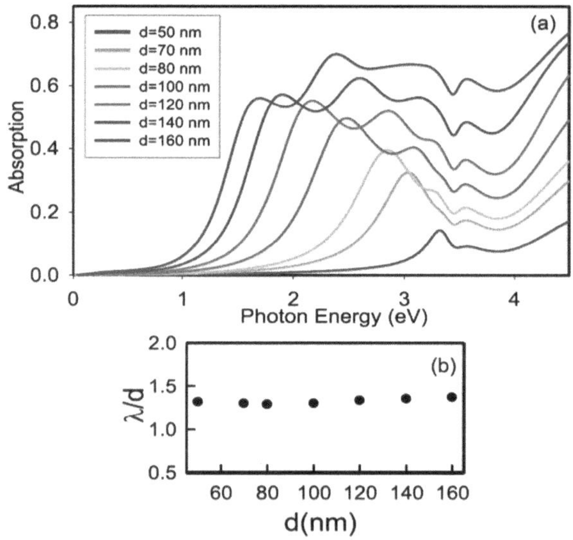

Figure 15.5 (a) Absorption of nanowires with diameter of 50, 70, 80, 100, 120, 140, and 160 nm and period $a = 200$ nm. (b) The evolution of the ratio between the threshold wavelength λ/d versus d.

single NW. Remarkably, the threshold energy for the absorption of light by the NWs increases with decreasing d. For instance, the energy onset starts around 1 eV for an NW diameter of 160 nm and reaches the UV spectral region, showing the first absorption peak at 3.3 eV for wires with a diameter of 50 nm. We believe that this behavior is a geometrical effect displaying an intimate relation between the wavelength λ of light in silicon at the threshold and the diameter d of the NW. In fact, deduced from our simulations, we find that the ratio λ/d versus d is constant as shown in Fig. 15.5(b). In previous works [45], a similar trend was observed when the absorption threshold energy was studied as a function of the filling factor. However, the variation of the

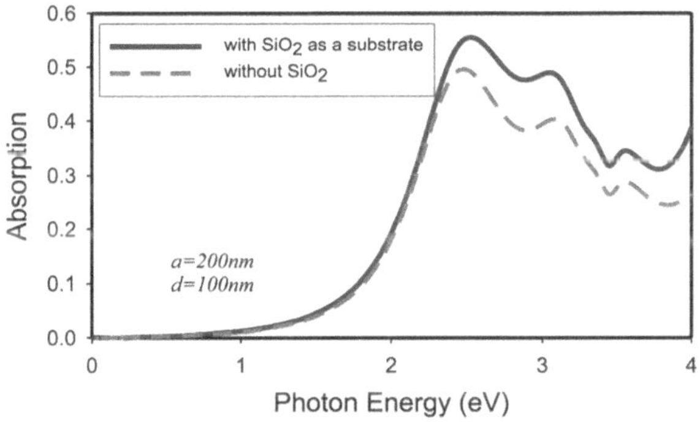

Figure 15.6 Absorption of nanowires with diameter of $d = 100$ nm and period $a = 200$ nm when the nanowires are suspended in air (dashed line) or supported by SiO_2 substrate (heavy line).

filling factor can be realized in two ways, either by changing the diameter d of the nanowire or by changing the period a.

Finally, in Fig. 15.6, we compare the absorption of light by suspended and supported arrays of NWs (see Fig. 15.1). The trends are the same either the incoming medium for light is air or an insulator such as SiO_2. However, one can note that the amplitude of the absorption is increased when the NWs are deposited on SiO_2 substrate.

15.2.2 Experimental results

In order to confirm the dependence of the optical absorption onset on the diameter of a nanowire, n-type Si NWs were synthesized on Si (111) substrate by the vapor–liquid solid mechanism using the chemical vapor deposition (CVD) technique. Gold droplets were first formed by evaporating a 2 nm thick layer of gold on a silicon surface. The Si surface covered with the gold thin film was then annealed at 700°C for 5 min in the CVD reactor prior to NWs growth, in order to produce droplets that were subsequently used as catalysts for the growth of the wires. In the final step, the n-type doped Si NWs were grown at a silane partial pressure of 1 mbar, while the total pressure in the chamber was maintained at 10 mbar with H_2 and the temperature set at 430°C. The silane flow rate was 300 sccm (standard cubic centimeter per minute), whereas the flow rate of phosphine diluted in H_2 was 1 sccm.

Fig. 15.7(a) gives a typical high-angle annular dark field (HAADF) scanning transmission electron microscopy (STEM) image of the end of an n-type doped Si NW grown by CVD. To the right of the gold droplet located at the apex of the NW (left part of the TEM image), the shaft appears speckled over a length of 1 μm. The bright features with sizes of a few nanometers correspond to Au-rich clusters. Such transitional region is attributed to the further growth of the NWs from residual silane gas after the interruption of the gas flow in the growth chamber and also the subsequent diffusion of gold due to the high-temperature inertia of the chamber, after silane has been fully pumped away. Then, in the right-hand side region that corresponds to constant

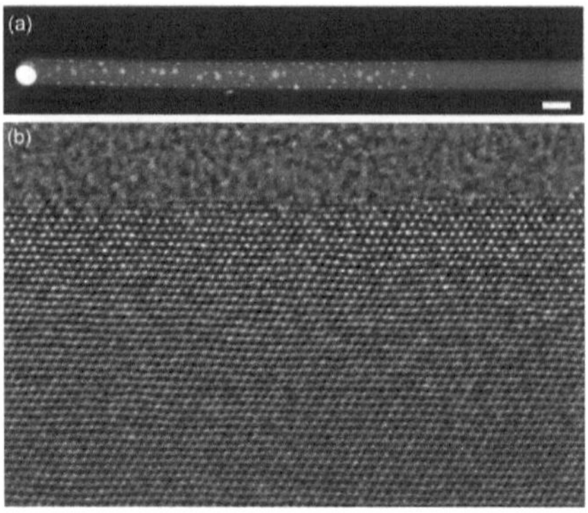

Figure 15.7 (a) Typical TEM image of the apex of an n-type doped Si NW grown by CVD. Scale bar: 80 nm. (b) Lattice-resolved TEM image obtained in the middle of the Si NW to show the straight sidewalls and the absence of Au clusters when the Si NW is grown at high silane partial pressure.

growth conditions, the contrast on the NW is homogeneous and the NW sidewalls are smooth. From the high resolution transmission electron microscopy image acquired in the middle of the NW, a high quality of the crystallographic structure is found and the sidewalls do not appear defective. This result is consistent with the properties obtained in our previous work [46]. From atom probe tomography and secondary ion mass spectrometry, it was found that the concentration of phosphorus impurities establishes at 2.10^{18} at.cm^{-3} in the center of the Si NWs and gradually increases towards the surface (typically, 5.10^{19} at.cm^{-3}), in agreement with several studies performed on Si NWs grown with similar conditions [47,48]. At the surface of the Si NWs, the doping level may be different since the incorporation rate of impurities on the NW sidewall has been previously found to be different when lateral growth occurs [49].

To perform electrical and photoconductivity measurements, the Si NWs were deposited onto an oxidized silicon surface from an isopropyl alcohol (IPA) suspension. Several contacts to single Si NWs were made by defining electrodes with electron-beam lithography and by subsequently evaporating a 50 nm-thick layer of Ti and 200 nm-thick layer of Au. The electrode having the smallest separation with the seed particle was carefully positioned away from the region where Au clusters reside.

To illustrate the reproducibility of the doping among NWs, dark current–voltages (I–V) curves are given in Fig. 15.8. In Fig. 15.8(a), I–V curves have been measured for three different Si NWs through two-terminal electrical contacts. All three Si NWs have a mean diameter of 90 nm between both electrodes. Over a few micron length for each NW, the doping level is homogeneous since the measured resistance (i.e., the slope of the I–V curves) is very similar. Fig. 15.8(b) shows a scanning electron microscopy (SEM) image of Si NW contacted with four electrodes, two at the top of the wire, two at the bottom. We note that the separation between both pairs of electrodes is comparable. The tapered NW clearly appears. When I(V) characteristics are measured between both pairs of electrodes, a slight variation of the resistance is found and varies accordingly to the small variation of distances between both pairs of electrodes. Such a result indicates a homogeneous doping level along the growth axis.

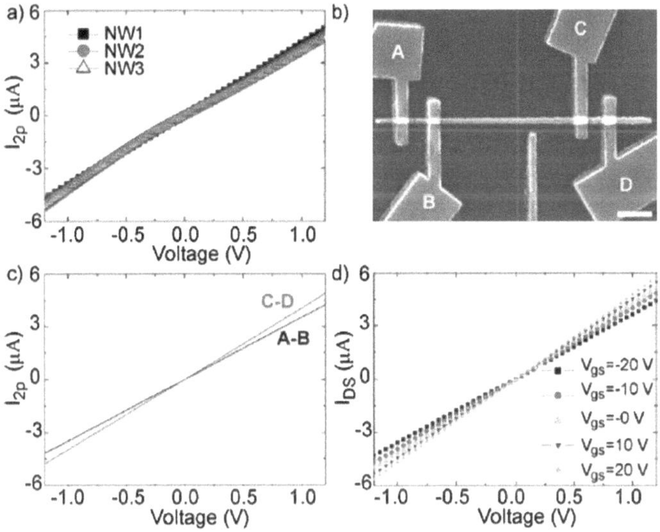

Figure 15.8 (a) I–V curve measured for three different Si NWs through from two-terminal current–voltage characteristics. All three Si NWs have a mean diameter of 90 between both electrodes. (b) SEM image of Si NW in electrical contact with four electrodes. Scale bar: 500 nm. (c) I–V curves measured between the labeled electrodes shown in (b). (d) I–V curves measured between C and D electrodes as a function of gate voltage.

Finally, using a back gate electrode, the I–V curves measured between the electrodes placed at the bottom of the wire as a function of the gate voltage confirm the n-type doping of the wire.

Next, photoconductivity measurements were performed using a monochromator source (ORIEL), with 100 W xenon light, allowing a spectroscopic analysis from 350 to 1100 nm (which means 3.5 to 1.1 eV). The photon flux was calibrated using a thermopile detector with a broad flat spectral response from 200 nm to 50 nm. All the photocurrent curves were therefore normalized. In order to increase the signal/noise ratio, the light flux was chopped at a frequency of 75 Hz and a lock-in detection was used for the photocurrent measurement. Fig. 15.9 shows the photocurrent characteristics measured on a single Si NW. Whatever the wavelength of the illumination is, the characteristics are highly nonlinear. First, the photocurrent increases with the bias up to 1.5 V. For higher voltage values, there is a decrease of the photocurrent up to a voltage of 2 V. The maximum of the photocurrent intensity is observed at the energy where the slope of the dark current significantly changes, as seen in the inset of Fig. 15.9. This behavior is characteristic of a barrier contact (pseudo-Schottky contact with a low barrier) [50] that occurs between the Si NW and one of the electrodes. At low voltages, the photocarriers are accelerated by the local electrical field in the barrier region, whereas at higher voltages, typically equal or greater than the build in potential, the diode turns into flat band conditions. The electric field is lowered and the current is therefore limited by a diffusion mechanism [51,52]. Such observation indicates that there is an electric field associated with the contact between the Si NWs

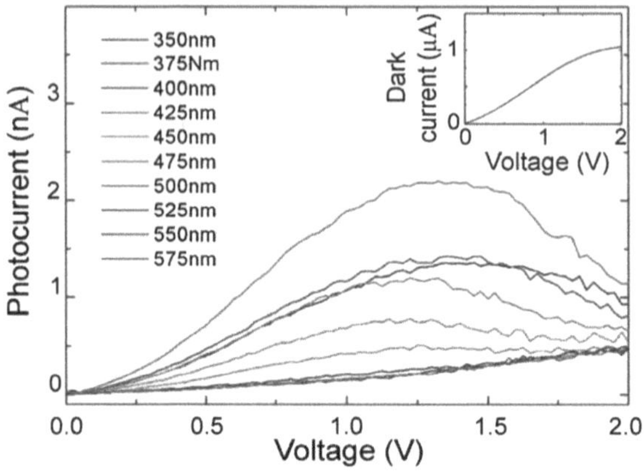

Figure 15.9 Photocurrent–voltage curves as a function of the wavelength of the incident photons on a single NW with a diameter of 100 nm. The wavelength varies from 575 to 350 nm. (Inset) Dark current–voltage curve measured on the same wire.

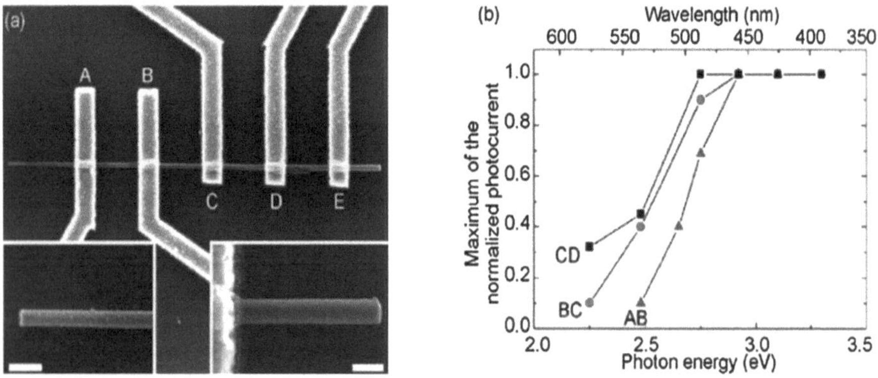

Figure 15.10 (a) SEM view of an isolated Si NW that is connected with five electrodes, labeled A, B, C, D, E, from the top towards the base of the wire. The left and right insets, which correspond to the top part and base of the wire, highlight the tapered shape of the wire. The scale bars correspond to 150 nm. (b) Variation of the maximum of the photocurrent versus the photon energy measured on the portion of the wire inserted between the electrodes.

and the electrode, as in the case, for example, of a Schottky contact or an (n+)–n homojunction. Interestingly, the maximum intensity of the photocurrent increases with decreasing wavelength to reach a saturation regime below 375 nm. By normalizing the peak intensity of the photocurrent with the saturation value of the photocurrent, we can compare the optical absorption of the wire, which is related to the photocurrent, as a function of the photon energy, as shown in Fig. 15.10 for the NW visible in the SEM image. This NW appears slightly tapered with a top part thinner than its base, as it is seen when both insets of Fig. 15.10(a) are compared. Fig. 15.10(b) shows the variation of the maximum of the normalized photocurrent measured in the case of this Si NW for the portions of the wire inserted between electrodes A, B, C, and D.

15.2.3 Discussion

In Fig. 15.10(b), the curves of the maximum of the normalized photocurrent as a function of the wavelength of the incident photons clearly shift depending on the position of the photoconductivity measurements along the wire. The onset energies for the absorption peak are determined by assuming linear onsets in the normalized photocurrent plots. Therefore, by measuring the photon energy, when the onsets reach a normalized photocurrent of 1, we find a shift of about 300 meV towards a lower energy threshold for NW diameter increasing from 70 nm (portion AB) to 85 nm (portion CD). As our electrical and structural characterizations do not show any strong fluctuations of the doping level along the growth direction or a significant presence of other impurities due to Au diffusion from the seed particle for example, such a result is consistent with the simulation performed above. It yields a spectral shift of 312 meV for the same variation in the diameter of the wire. Finally, we would like to note that measurements on NWs with much smaller diameters are quite difficult to perform for two reasons: first, the absorption is very weak, due to a very small volume of silicon; then, it is well known that the ionization energy of the dopants increases with decreasing diameters, leading to a transition towards an insulating material [53]. This effect implies less control on the electrical properties of the contact electrodes and much more fluctuations in the photoconductivity measurements.

15.3 Optical absorption of silicon nanopillars

15.3.1 Theoretical analysis

15.3.1.1 Finite Difference Time Domain (FDTD) method

The calculations have been performed on periodic array of vertical nanopillars (NPs) on Si substrate using the Finite Difference Time Domain method. The structure of the 3D unit cell is shown Fig. 15.11. The incident source is placed in the air medium and propagates along the z-direction, perpendicularly to the periodic structure. Perfect Matching Layers (PML) are applied at the finite boundaries of the box, along z, in order to avoid any reflections of outgoing waves. Along the x and y direction, the unit cell is repeated periodically and the structure is supposed to be finite along the z direction. Periodic boundary conditions (PBC) are applied on both upper and lower sides of the unit cell to build the periodic structure. Fig. 15.11(b) represents a cross-section in the (x, y) plane of the 3D unit through the cylinders. A rectangular unit cell with two cylinders per unit cell has been chosen to respect the square meshing used in the FDTD code. The application of the periodic boundary conditions on each side of the unit cell leads to a hexagonal (triangular) array of pillars. The symmetry of this array insures a quasi-in-plane isotropy of the structure. To calculate the absorption through the Si NPs, a reflection detector is placed in the air medium and a transmission one is placed close to the interface between Si wafer and NPs in the Si medium.

The space is discretized in both x and y directions using a mesh interval equal to $\Delta x = \Delta y = 1$ nm. The equations of motion are solved with a time integration step

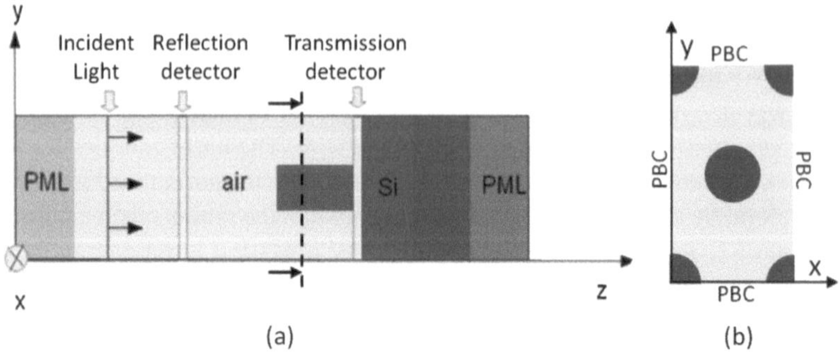

Figure 15.11 Schematic representation of the unit cell used for the calculation. The positions of the detectors for the incident, reflection and transmission calculations are presented (a) in the (y, z) plane and (b) in the (x, y) plane. The boundaries conditions are indicated as PBC (periodic boundary conditions) and PML (perfect matching layer).

$\Delta t = \Delta x/4c$ and a number of time steps equals to 2^{20}, which is the necessary tested time for a good convergence of the numerical calculation.

The incoming pulse is generated at the left part of the unit cell (in the air medium), by a current source perpendicular to the z axis and having a planar profile. Then, the reflection and transmission spectra recorded during the time are Fourier-transformed to obtain R and T as functions of frequency and then deduce the absorption as $A = 1 - T - R$. In order to estimate not only the reflectance or absorption properties of Si NP, we introduce the so-called ultimate efficiency α, given by the equation

$$\alpha = \frac{\int_0^{\lambda_g} \mathrm{I}(\lambda)\mathrm{A}(\lambda)\lambda\, d\lambda}{\lambda_g \int_0^{\infty} \mathrm{I}(\lambda)\, d\lambda} \tag{15.4}$$

where $\mathrm{I}(\lambda)$ is the solar irradiance spectrum, $\mathrm{A}(\lambda)$ is the silicon absorption, and λ_g is the wavelength corresponding to the energy bandgap of the silicon.

This parameter is directly bound to the photocurrent of the solar cell. This expression gives rise to a maximum as we assume that each absorbed photon with energy greater than the bandgap produces one and only one electron–hole pair with energy equal to the gap.

In the next part, we present the general theoretical trends for the absorption $\mathrm{A}(\lambda)$ and the efficient coefficient α as a function of the geometrical parameters of the Si NPs arrays. Many parameters involve in the structure, namely the diameter, the period, and the length of the pillars.

15.3.1.2 Theoretical results

15.3.1.2.1 Effect of the diameter

We first illustrate the properties of absorption for a variation of the diameters of the NPs. Intuitively one can expect that the absorption is linked to the volume of the materials then to the diameter of the NPs.

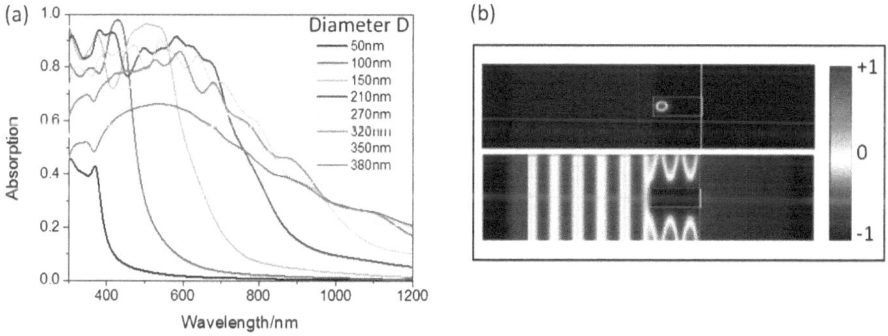

Figure 15.12 (a) Absorption spectra of NPs as a function of the diameter ranging from 50 to 380 nm, with the same length (1000 nm) and period (380 nm) of the NPs; (b) Distribution of the electromagnetic field intensity for a wavelength of 509 nm (upper part) and a wavelength of 1200 nm (lower part).

Fig. 15.12(a) gives the absorption of an Si NPs array of period 380 nm, for a constant length L of the pillars of 1000 nm and a diameter ranging from 50 to 380 nm. We can see that at low wavelengths, the absorption is maximal before drop to almost zero for larger wavelengths. We define the threshold wavelength λ_{th} as the wavelength for which the absorption is equal to half its maximum. Fig. 15.12(b) shows the distribution of the electromagnetic field intensity inside the unit cell, for a diameter of the NP $d = 150$ nm, at two different wavelengths, before (upper part) and after (lower part) the threshold ($\lambda_{th} = 600$ nm), respectively. For a wavelength of 509 nm (Fig. 15.12(b), upper part), the electromagnetic field is localized inside the NP, leading to a maximum of absorption, whereas for an illumination of 1200 nm, the electromagnetic field is outside the NP (Fig. 15.12(b), lower part), and the absorption is minimal. It means that when the diameter is too small compared to the wavelength, the absorption becomes weak.

Fig. 15.13(a) represents the evolution of the threshold wavelength λ_{th} with the diameter, which increases in agreement with previous results [8]. For diameters larger than 300 nm, a phenomenon of saturation is observed due to a close packing of the NPs inside the array. For these large diameters, the benefit of the nanostructuring, which normally increases the photons optical path, is lost. In the simulation, the maximum value of the diameter is 380 nm, which corresponds to the value of the period. It corresponds to a close packed cylinders array, not far from bulk material.

While the optical absorption is a physical key parameter, it is nevertheless not sufficient to master an optimized nanostructuring of the surface. An additional information is the ultimate efficiency, as defined in Eq. (15.4). It gives a property which is not far from the photocurrent, and therefore the yield of a solar cell as the solar irradiance spectrum is taken to be the AM1.5 G spectrum. Fig. 15.13(b) gives the ultimate efficiency calculated versus the diameter, ranging from 50 to 380 nm. The curve has a maximum for a diameter D of the order of 270 nm, i.e., a ratio $D/P = 0.7$. When the diameter tends towards the period (380 nm), the efficiency tends to the value of a bulk layer with a thickness equal to the length of the NPs. It is interesting to note that there is a critical diameter (in that case, 210 nm, $D/P = 0.55$) for which the NPs

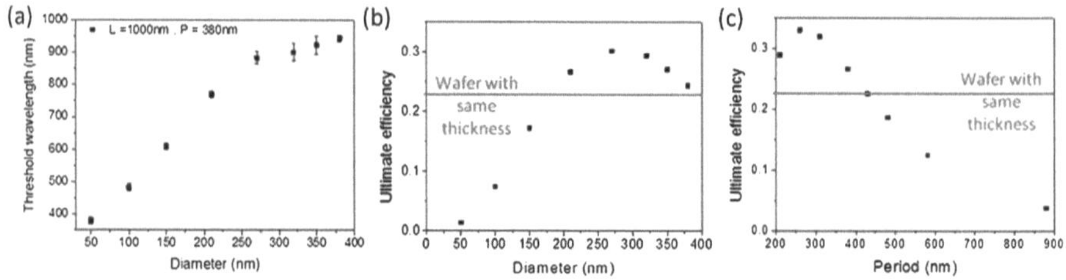

Figure 15.13 (a) Evolution of the threshold wavelength as a function of the diameter, (b) Ultimate efficiency versus the diameter of the NPs. (c) Ultimate efficiency versus the period for a diameter of the NPs of 210 nm and a length of 1000 nm.

array or the bulk wafer with the same thickness has the same efficiency. In order to get a benefit from the nanostructured surface, the NPs should have a diameter greater than the critical diameter. However, this critical value is not independent of the period.

15.3.1.2.2 Effect of the period and the length

Fig. 15.13(c) shows the ultimate efficiency calculated versus the period for a diameter $D = 210$ nm and a length $L = 1000$ nm of the NPs. As for the study versus the diameter, we observe that the efficiency of an array of NPs is better than the associated bulk layer with the same thickness if the period is lower than a critical value. In the case studied in Fig. 15.13, the critical period is equal to 430 nm, i.e., $D/P = 0.5$ and a maximum of efficiency is found for a period of 260 nm, i.e., $D/P = 0.8$. These ratios are very close to those found in the previous section. So, we find that the pertinent key parameter is not the diameter or the period alone but the ratio D/P. Finally, as a function of the length with given diameter and period, the efficiency increases rapidly for small lengths and then saturates for higher ones. The gain obtained with respect to a bulk layer having the same volume is greater for small lengths than for high values of L. In other words, there is a gain associated with the nanostructuring as far as the length L remains relatively small, whereas for larger lengths we detect the bulk effect associated with the volume of NPs.

All these correlated effects are summarized in the following part.

15.3.1.2.3 Efficiency abacus versus the length, the diameter and the period of NPs

Fig. 15.14 summarizes the combined effects of the three parameters: diameter, period, and length on the efficiency. We have chosen three values of the period (380, 600, and 900 nm respectively labeled A, B, and C in Fig. 15.14). For each period, the length varies from 50 to 3000 nm and the diameters vary from 50 nm (part A), 60 nm (part B), and 90 nm (part C) up to its maximum possible value ($D = P$). For the three periods, and *independently* of the length, the best efficiency is obtained for a ratio $D/P = 0.7$–0.8. Of course, for a given D/P ratio, the efficiency varies with the length. The ratio D/P appears definitely as a key parameter: in order to get a maximum of efficiency, the ratio D/P must be ranged between 0.7–0.8 for a given length.

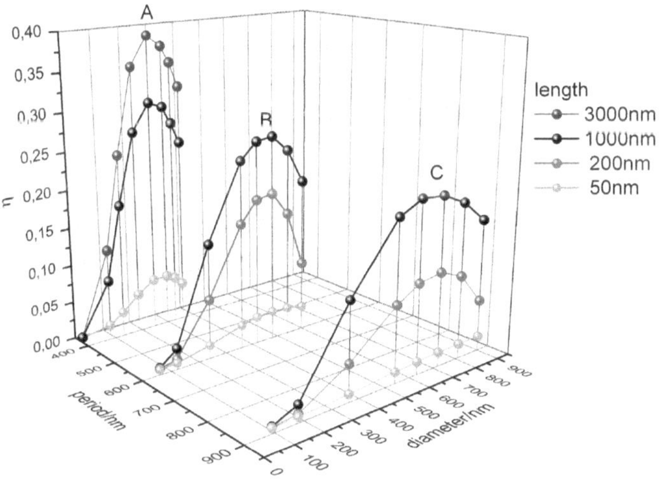

Figure 15.14 Ultimate efficiency versus the period, the diameter and the length of the NPs.

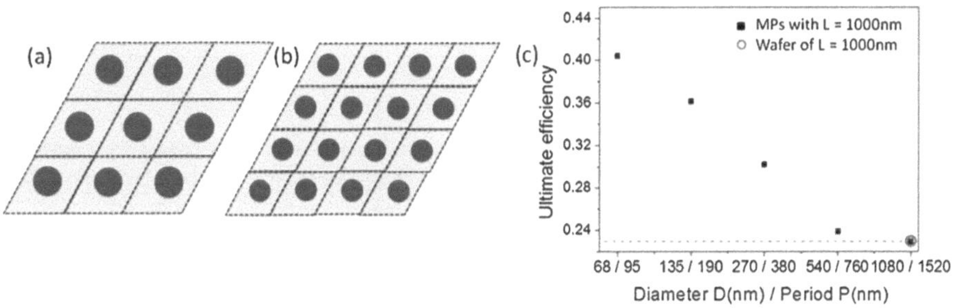

Figure 15.15 Schematic representation of large (a) or small (b) diameters keeping the same ratio $D/P = 0.7$, (c) efficiency versus different diameters with the ratio $D/P = 0.7$ and for a length of 1000 nm. The different diameter/period (D/P) are: (68/95), (135/190), (270/380), (540/760), and (1080/1520) nm. A wafer with the same thickness has the same efficiency as the (1080/1520) nm array.

Another question is to know, for a same filling fraction, whether it is more efficient to cover the surface with bigger or smaller diameter NPs. This is illustrated in the next section.

15.3.1.2.4 Large or small NPs diameters, keeping $D/P = 0.7$–0.8

The configuration "small diameters" or "large diameters", for a given D/P ratio is illustrated Figs. 15.15(a) and 15.15(b), respectively. Keeping $D/P = 0.7$, and a length of 1000 nm, we have calculated the efficiency for different pairs (D, P), from small to large D and P. We chose the following pairs: $(68, 95)$, $(135, 190)$, $(270, 380)$, $(540, 760)$, and $(1080, 1520)$ nm. The associated efficiencies are given in Fig. 15.15(c) where one can easily notice that the best efficiencies are obtained for smaller diameters and, accordingly, periods. For the higher pairs ($D > 540$ nm, $P > 760$ nm), the

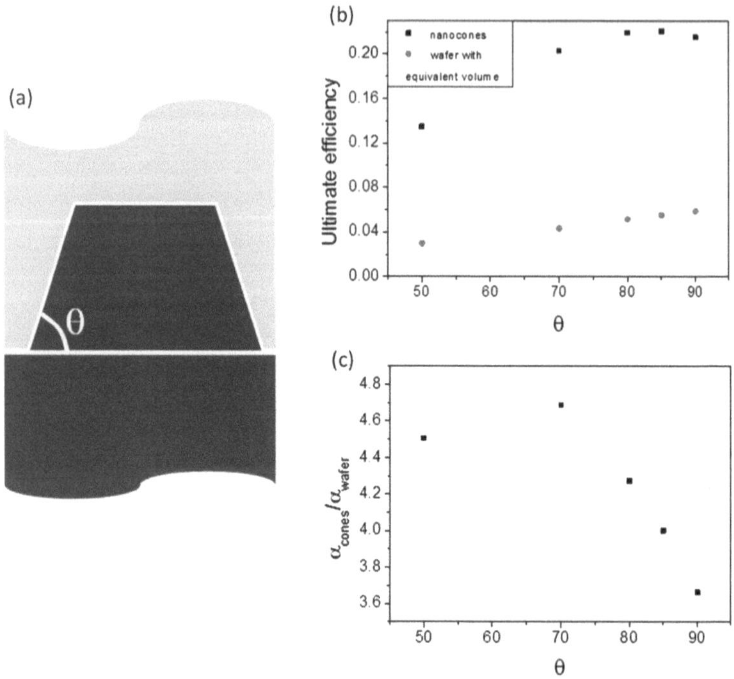

Figure 15.16 Schematic view (a) of NC, (b) efficiency versus the angle θ for NCs (squares dots) and bulk layer (circular dots) with the same volume (c) ratio of the efficiencies of NCs over bulk layer versus the angle θ.

efficiency of the array moves exactly toward the equivalent bulk layer, with the same length.

15.3.1.2.5 From nanopillars to nanocones

Nanopillars are special instances of nanocones, where the top diameter is equal to the bottom diameter. So, the properties of nanopillars can be easily extrapolated to nanocones (NCs). When fixing the bottom diameter and height, nanocones of different shapes are obtained by rotating the lateral sides with the angle θ (see Fig. 15.16(a)). Fig. 15.16(b) gives the efficiency of NCs versus the angle θ (square solid dots). We take the set of parameters for the nanocone as follows: a bottom diameter of $D = 704$ nm, a height of 350 nm, and a period of $P = 880$ nm. At a first glance, NCs display only better efficiencies than NPs (corresponding to θ equal 90°) when the angle θ is close to 90° with the best angle around 85°. However, to explain and understand this behavior, two effects have to be distinguished. First, there is a shape effect associated with the nanocones, namely a gradual change in the effective optical index from top to bottom, resulting in a better adaptation for the absorption of the electromagnetic waves. When θ is greater than 90° (not shown here) the efficiency decreases. In such a case, there is a misadjustment effect from top to bottom of the NCs, compared to NPs. When θ decreases well below 90°, there is also a dramatic decrease of the efficiency. A second effect which produced the decrease of the NC efficiencies, keeping constant the height and bottom diameter of the NC, is the decrease of the volume due to the variation of θ which results in a decrease of the absorption. Therefore, in order to separate the shape and volume effects, we have calculated the efficiency of an equivalent volume of silicon layer (circular dots in Fig. 15.16(b)) with the same volume as the NCs array associated with a given θ angle and we have plotted the ratio $\alpha_{\text{cones}}/\alpha_{\text{wafer}}$ in Fig. 15.16(c), where α_{cones} and α_{wafer} are the efficiencies of the NCs and bulk layers, respectively. This ratio represents the exact gain obtained with NCs array compared to a bulk layer with the same volume. This approach shows that the best efficiency with the NCs is obtained with an angle of 70° instead of 85°.

So, the main trends we found to obtain the best efficiency are (i) a ratio D/P of the order of 0.7–0.8, independently of the length, (ii) the efficiency increases with length, (iii) a given D/P must be used with a small diameter, and (iv) NCs have a better optical behavior than NPs of the same length with a best angle for NCs sidewall with respect to bottom face in the order of 70°, compared to an equivalent bulk layer with the same volume. In the next part, we compare our simulations with experimental measurements on different NPs arrays.

15.3.1.3 Experimental results

15.3.1.3.1 Fabrication of the NPs or the NCs

To fabricate nanopillars (or nanocones), masks with suitable sizes should be deposited on the substrate. In this work, we choose silica balls as mask, and the Langmuir–Blodgett technique is used to deposit modified silica monolayer on the Si substrate, due to its lower-cost compared with photoresist and photolithography. For a desired NPs diameter and period, the initial diameter of the silica balls gives the period, and the diameter after etching gives the NPs diameter. Two types of silica balls with the diameter of 880 and 380 nm, bought from GmbH Microparticles company, are used. To satisfy the needs of LB techniques and obtain amphiphilic silica balls, surface modification is required. The experimental process is as follows: (1) 1 ML of silica ball solution (5% in water) is centrifugalized at 5000 tr/min for 4 minutes. After that, the supernatant is removed and replaced by 1.5 ML of ethanol. This new solution is put in the ultrasonic bath to be well dispersed and then is centrifugalized again to remove the supernatant and replaced by ethanol, (2) 10 μL of aminopropyl-dimethyl-ethoxy silane is put into the solution to react with particles for 24 hours at room temperature, (3) centrifugal and ultrasonic processes, as in step (1), are repeated three times. The modified silica balls are diluted into the mixture of 250 μL ethanol and 250 μL chloroform. Thanks to the amphiphilic nature of modified silica balls, when silica balls spread at the air/water interface, the hydrophilic group is immersed in the water and hydrophobic part is pointing towards air. The monolayer can be compressed by the barrier moving system, meanwhile, the surface pressure is measured by electrobalance.

Silica balls can be transferred into the substrate by successively dipping the substrate down and up through the monolayer under a constant surface pressure (pressure of dip-coating).

Table 15.1 RIE parameter for etching silica balls.

	CHF_3 (sccm)	O_2 (sccm)	Power (W)	Pressure (mTorr)
a	40	8	50	300
b	40	8	70	350
c	40	8	100	550
d	40	8	100	250

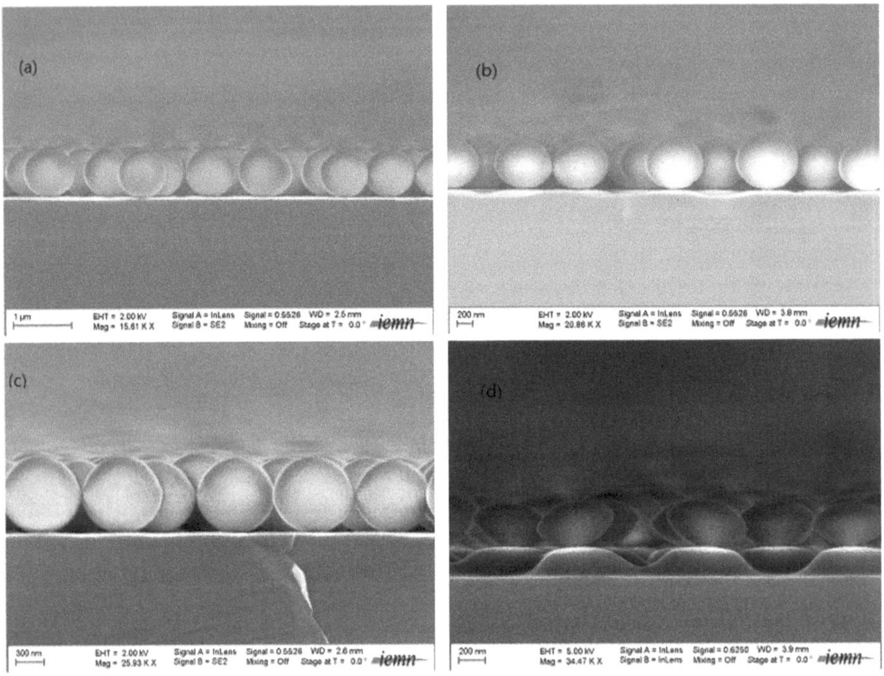

Figure 15.17 SEM images of reduced silica balls with different etching parameter, images (a), (b), (c), and (d) correspond to parameters a, b, c and d in Table 15.1, respectively.

Any contamination will lower surface tension, so the cleaning steps are performed before deposition process: chloroform, isopropanol, and deionized water are used successively to wash the LB trough.

In order to obtain nanopillars (or nanocones) with required diameters, silica balls are reduced by RIE plasma etchings system. In this work, we use CHF_3 and O_2 to etch silica balls, the parameters of the plasma are given in Table 15.1.

Fig. 15.17 shows the morphology of reduced silica balls with different parameters. The variation of power and pressure will result in the configuration change of silica balls. When silica balls have been reduced to a required size, DRIE or RIE etchings are subsequently used to fabricate nanopillars and nanocones, respectively. The main technique of DRIE used in this work is the Bosch process, which is also called time-multiplex etching process, alternating between etching step and passivation step and

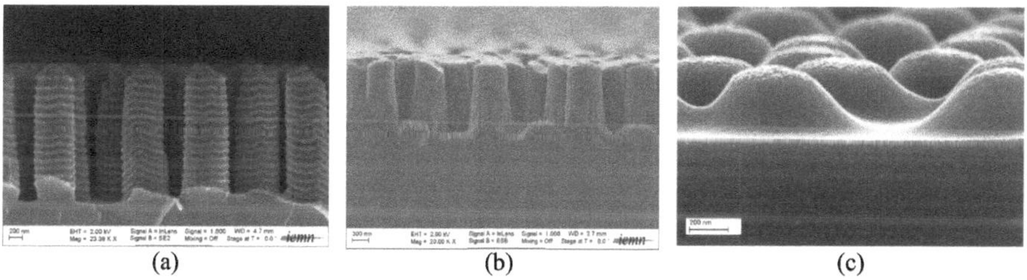

Figure 15.18 SEM images of Si nanopillars made by (a) DRIE etching and (b) ICP etching and (c) Si nanocones made by RIE etching.

Table 15.2 DRIE parameter for fabricating Si nanopillars shown in Fig. 15.18.

Etching time (s)	SF_3 (sccm)	O_2 (sccm)	Etching Power (W)	Passivation time (s)	CHF_3 (sccm)	Passivation Power (W)
2	450	45	2100	2.6	200	2000

then repeating them until the required length is reached. Passivation step is done with a mixture of SF_6 and O_2, this step will deposit polymer on the surface, protecting the sidewall from etching. But the etching step will completely remove polymer on the bottom and partially remove the polymer on the sidewall [54]. This process produces structure with high anisotropy at a higher etch rate, but the apparent drawback is the scallop profiles sidewall, as illustrated in Fig. 15.18(a). The etching parameters are shown in Table 15.2. In fact, NPs with the same length (1000 nm), period (880 nm), and diameter (600 nm), but with smooth or scallop sidewall, are identical because the size of the scallop profile is small compared to the optical wavelength. ICP-RIE system is a RIE system with ICP generator applying a high density of ions which enhances the etching rate, along with RF generator to supply an electrical field near the sample to increase the degree of anisotropy. In this work, Cl_2 is used to fabricate Si nanopillars, as shown in Fig. 15.18(b). The etching parameters are as follows: flux of Cl_2 (30 sccm), pressure of 10 mTorr, and power of 40 W. RIE system is used to fabricate nanocones in this work, the etching parameters are shown in Table 15.2. Fig. 15.18(c) is an illustration of the morphology of nanocones obtained by RIE etching (CHF_3 (40 sccm), O_28 (sccm), power (100 W), and pressure (250 mTorr)). Finally, after etching, we remove the remaining silica balls by immersing them into HF solution (50%HF) for 30 s. Fig. 15.19 is an illustration of Si NPs with different D/P made by ICP etching. The length is 1300 nm, with a period of 380 nm and diameter of (a) 240 nm and (b) 320 nm, respectively, and the length is 1300 nm, period of 880 nm and diameter of (c) 520 nm and (d) 630 nm, respectively.

Fig. 15.20 presents the corresponding reflectance spectra of these Si NPs in the wavelength range of 300–1000 nm. In the range from 700 to 1000 nm, the reflectance spectra of Si NPs with different D/P are similar. However, in the range from 300 to 700 nm, Si NPs with $D/P = 0.6$ show clearly the smallest reflectance. When this ratio increases, the reflectance increases with a best fabricated NPs of around

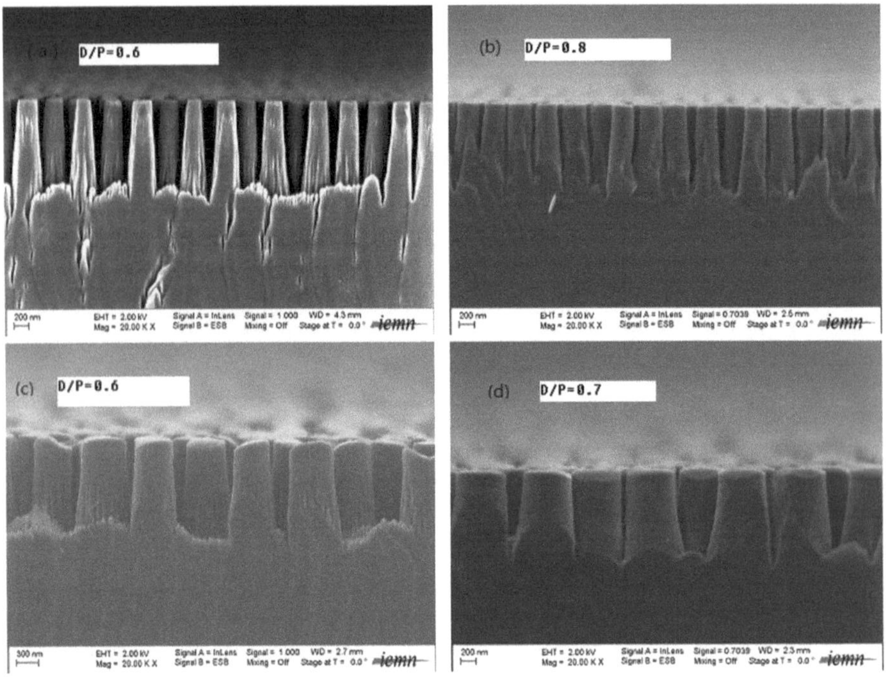

Figure 15.19 Illustration of Si NPs made by ICP etching with different D/P. Fig. 15.10 is an illustration of Si NPs with different D/P made by ICP etching. The length is 1300 nm, with a period of 380 nm and diameter of (a) 240 nm and (b) 320 nm, respectively, and the length is 1300 nm, period of 880 nm and diameter of (c) 520 nm and (d) 630 nm, respectively.

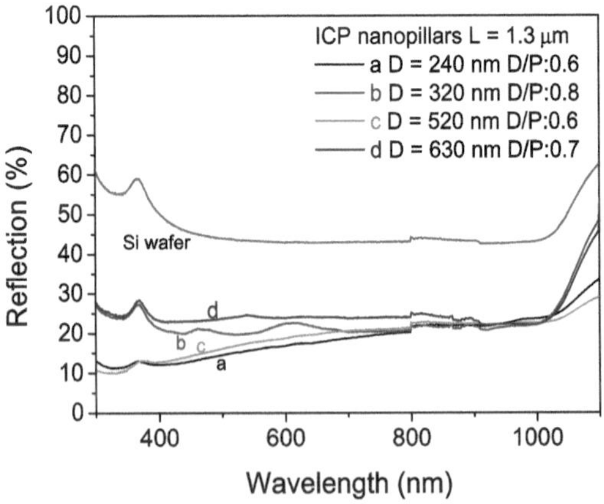

Figure 15.20 Total reflectance spectra for Si NPs film with different D/P, (a), (b), (c), and (d) correspond to the Si NPs in Fig. 15.19.

$D/P = 0.6$–0.7. Comparing the reflectance spectra of NPs with the same $D/P = 0.6$, i.e., diameter of 240 nm with period of 380 nm and diameter of 520 nm with period of 880 nm, we can observe that the NPs with smaller diameter and smaller period present a lower reflectance.

15.3.1.3.2 Comparison between experimental and simulation results

In order to compare easily the measured and calculated reflectances which can exhibit some oscillations due to interference phenomena associated with the NPs array, we have taken the average reflectances from 300 to 1200 nm. The first comparison concerns the reflectance versus the diameter as illustrated in Fig. 15.21(a), with NPs having a length of 750 nm and a period of 380 nm. The experimental and theoretical average reflectance values exhibit the same behavior versus the diameter. With the increase of diameters, the average reflectance first decreases and then increases. The simulated results show that the best NPs have a diameter of 270 nm with $D/P = 0.7$, in agreement with experimental results. In any cases, the experimental reflectance is typically more or less 10% above the theoretical one (and this will appear in the next comparisons as well). In fact, as we use double-faced polished silicon wafer, we believe that part of the light is reflected from the bottom face, and then we measure a reflected signal, a little greater than the theoretical one, for which these experimental characteristics are not taken into account.

Next, we compare the simulated and measured reflectance spectra of NPs with the same length of 1300 nm and different D/P of 0.6, 0.7, and 0.8, as shown in Fig. 15.21(b). As pointed out in the previous paragraph, the calculated results are always smaller than the experimental results, but the average reflectance behavior as a function of D/P is quite similar. It can be seen that the NPs with $D/P = 0.6$ show the lower reflectance, and when we compare the reflectance of NPs with $D/P = 0.6$, the NPs with smaller period of 380 nm and smaller diameter of 240 nm have the lowest reflectance. So, the best NPs have $D/P = 0.6$, diameter of 240 nm, and period of 380 nm, which agrees with the experimental results. Finally, we simulated the reflectance of Si NPs with different lengths of 800, 1300, and 1900 nm. Fig. 15.21(c) gives the average reflectance of simulated results, compared with experimental results. It can be seen that the reflectance first decreases as the length increases (i.e., the efficiency increases with the length) and, for higher length values (typically, in this case 800 nm), the average reflectance becomes a constant. So, the associated variation of the yield compared to bulk wafer decreases, as observed in the previous simulations. We note that for lengths around 2000 nm, there is a good agreement between experimental and simulated reflectance. For this high value of the length, the effect of the reflection of the light on the polished back face has weak importance because the light is scattered and absorbed on an important layer of NPs.

15.4 Conclusion

In conclusion, we have investigated the geometrical influence of diameter, period, and length of nanowires (NWs), nanopillars (NPs), and nanocones (NCs) on their

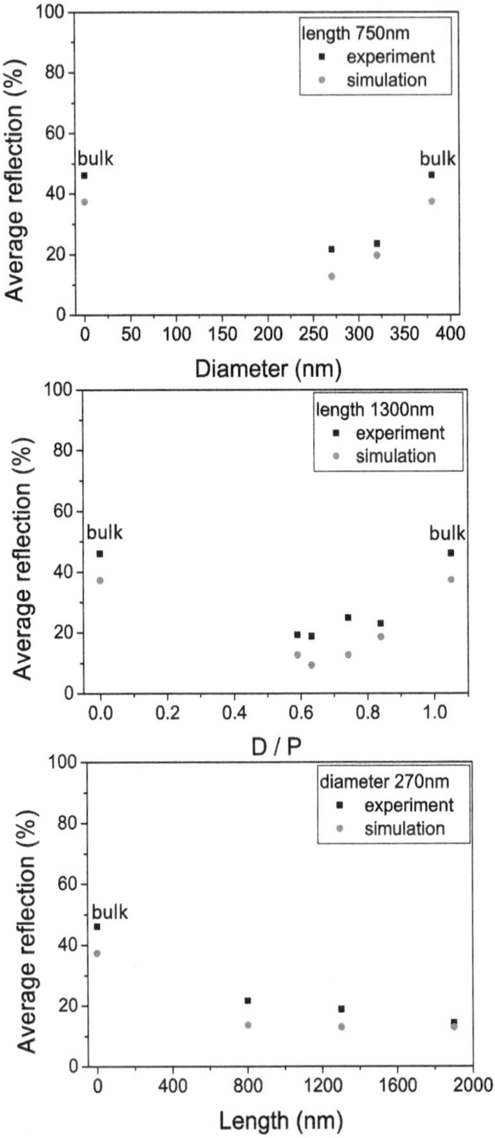

Figure 15.21 Experimental (black squares) and theoretical (red circles; light gray in print version) average reflectance as a function of (a) diameters with same length (750 nm), and period (380 nm); (b) of *D*/*P* with same length (1300 nm); and (c) of length with same diameters (270 nm) and same period (380 nm).

experimental and calculated optical properties for photovoltaic application. More precisely, an increase of the NWs diameter reveals a spectral shift of the photocurrent intensity peak towards lower photon energies, which allows tuning the absorption onset from the ultraviolet radiations to the visible light spectrum. For the NPs, we found that (a) the optimized diameter changes with the variation of period. When $D/P = 0.6–0.8$, the NPs show the highest efficiency, and this ratio is independent of the length; (b) for a given ratio D/P, NPs with smaller diameters and periods exhibited higher efficiency; (c) the longer the length, the higher the efficiency. This enhancement is due to both the increasing of Si volume and the surface structural effect. When the length is high, the gain of structural effect is constant, so the enhancement is consequently due to the increase of Si volume; (d) when the size of the sidewalls is smaller than the measured wavelength, the sidewalls have no effect on the reflectance spectra of NPs; (e) if the bottom diameter, period, and height are constant, NCs with an angle θ of the order of 70° have the highest efficiency. It remains true that the surface nanostructuring is a crucial challenge for the optimization of light trapping in solar cells.

Acknowledgments

The authors would like to thank T. Xu, Y. Lambert, C. Krzeminski, B. Grandidier, Di Zhou, O. Cristini, Y. Deblock, M. Faucher, and D. Stievenard for their fruitful contributions to the original papers from which this chapter is inspired.

References

[1] H. Kind, H. Yan, B. Messer, M. Law, P.D. Yang, Nanowire ultraviolet photodetectors and optical switches, Adv. Mater. 14 (2002) 158.
[2] F. Qian, Y. Li, S. Gradecak, D. Wang, C.J. Barrelet, C.M. Lieber, Gallium nitride-based nanowire radial heterostructures for nanophotonics, Nano Lett. 4 (2004) 1975.
[3] B. Tian, X. Zheng, T.J. Kempa, Y. Fang, N. Yu, G. Yu, J. Huang, C.M. Lieber, Coaxial silicon nanowires as solar cells and nanoelectronic power sources, Nature 449 (2007) 885.
[4] N.A. Sanford, P.T. Blanchard, K.A. Bertness, B. Tian, L. Mansfield, J.B. Schlager, A.W. Sanders, A. Roshko, B.B. Burton, S.M. Geroge, Steady-state and transient photoconductivity in c-axis GaN nanowires grown by nitrogen-plasma-assisted molecular beam epitaxy, Appl. Phys. 107 (2010) 034318.
[5] A. De Luna Bugallo, M. Tchernycheva, G. Jacopin, L. Rigutti, F.H. Julien, S.-T. Chou, Y.-T. Lin, P.-H. Tseng, L.-W. Tu, Visible-blind photodetector based on p–i–n junction GaN nanowire ensembles, Nanotechnology 21 (2010) 315201.
[6] B.M. Kayes, H.A. Atwater, N.S. Lewis, Comparison of the device physics principles of planar and radial junction nanorod solar cells, J. Appl. Phys. 97 (2005) 114302.
[7] T. Xu, Y. Lambert, C. Krzeminski, B. Grandidier, D. Stiévenard, G. Lévêque, A. Akjouj, Y. Pennec, B. Djafari-Rouhani, Optical absorption of silicon nanowires, J. Appl. Phys. 112 (2012) 033506.
[8] G. Brönstrup, N. Jahr, C. Leiterer, A. Csaki, W. Fritzsche, S. Christiansen, Optical properties of individual silicon nanowires for photonic devices, ACS Nano 4 (12) (2010) 7113.
[9] L. Tsakalakos, J. Balch, J. Fronheiser, M.-Y. Shih, S.F. Leboeuf, M. Pietrzykowski, P.J. Codella, B.A. Korevaar, O. Sulima, J. Rand, A. Davuluru, U. Rapol, Strong broadband optical absorption in silicon nanowire films, J. Nanophotonics 1 (2007) 013552.

[10] L. Tsakalakos, Nanostructures for photovoltaics, Mater. Sci. Eng., R Rep. 62 (2008) 175.
[11] M. Nehra, N. Dilbaghi, G. Marrazza, A. Kaushik, R. Abolhassani, Y.K. Mishra, K.H. Kim, S. Kumar, 1D semiconductor nanowires for energy conversion, harvesting and storage applications, Nano Energy 76 (2020) 104991.
[12] J. Zhu, Z. Yu, S. Fan, Y. Cui, Nanostructured photon management for high performance solar cells, Mater. Sci. Eng., R Rep. 70 (2010) 330.
[13] Y. Liu, S.H. Sun, J. Xu, L. Zhao, H.C. Sun, J. Li, W.W. Mu, L. Xu, K.J. Chen, Broadband antireflection and absorption enhancement by forming nano-patterned Si structures for solar cells, Opt. Express 19 (2011) A1051.
[14] K.Q. Peng, S.-T. Lee, Silicon nanowires for photovoltaic solar energy conversion, Adv. Mater. 23 (2011) 198.
[15] M. Farangi, M. Zahedifar, M.R. Mozdianfard, M.H. Pakzamir, Effects of silicon nanowires length on solar cells photovoltaic properties, Appl. Phys. A 109 (2012) 299.
[16] Q. Lin, B. Hua, S.-F. Leung, X. Duan, Z. Fan, Efficient light absorption with integrated nanopillar/nanowell arrays for three-dimensional thin-film photovoltaic applications, ACS Nano 7 (2013) 2725.
[17] J.-Y. Jung, H.-D. Um, S.-W. Jee, K.-T. Park, J.H. Bang, J.-H. Lee, Optimal design for antireflective Si nanowire solar cells, Sol. Energy Mater. Sol. Cells 112 (2013) 84.
[18] H. Tan, E. Psomadaki, O. Isabella, M. Fischer, P. Babal, R. Vasudevan, M. Zeman, A.H.M. Smets, Micro-textures for efficient light trapping and improved electrical performance in thin-film nanocrystalline silicon solar cells, Appl. Phys. Lett. 103 (2013) 173905.
[19] M.A. Elrabiaey, M. Hussein, M.F.O. Hameed, S.S. Obayya, Light absorption enhancement in ultrathin film solar cell with embedded dielectric nanowires, Sci. Rep. 10 (1) (2020) 1–10.
[20] C. Lin, M.L. Povinelli, Optical absorption enhancement in silicon nanowire arrays with a large lattice constant for photovoltaic applications, Opt. Express 17 (2009) 19371.
[21] J. Li, H. Yu, S.M. Wong, X. Li, G. Zhang, P. Guo-Qiang Lo, D.-L. Kwong, Si nanopillar array optimization on Si thin films for solar energy harvesting, Appl. Phys. Lett. 95 (2009) 033102.
[22] J. Li, H. Yu, S.M. Wong, X. Li, G. Zhang, P. Guo-Qiang Lo, D.-L. Kwong, Design guidelines of periodic Si nanowire arrays for solar cell application, Appl. Phys. Lett. 95 (2009) 243113.
[23] J. Zhu, C.-M. Hsu, Z. Yu, S. Fan, Y. Cui, Nanodome solar cells with efficient light management and self-cleaning, Nano Lett. 10 (2010) 1979.
[24] S.E. Han, G. Chen, Optical absorption enhancement in silicon nanohole arrays for solar photovoltaics, Nano Lett. 10 (2010) 1012.
[25] S.E. Han, G. Chen, Toward the Lambertian limit of light trapping in thin nanostructured silicon solar cells, Nano Lett. 10 (2010) 4692.
[26] W.Q. Xie, J.I. Oh, W.Z. Shen, Realization of effective light trapping and omnidirectional antireflection in smooth surface silicon nanowire arrays, Nanotechnology 22 (2011) 6.
[27] Z. Yu, A. Raman, S. Fan, Nanophotonic light-trapping theory for solar cells, Appl. Phys. A 105 (2011) 329.
[28] B. Wang, P.W. Leu, Enhanced absorption in silicon nanocone arrays for photovoltaics, Nanotechnology 23 (2012) 194003.
[29] B.C.P. Sturmberg, K.B. Dossou, L.C. Botten, A.A. Asatryan, C.G. Poulton, R.C. McPhedran, C. Martijn de Sterke, Nanowire array photovoltaics: radial disorder versus design for optimal efficiency, Appl. Phys. Lett. 101 (2012) 173902.
[30] W. Wang, J. Zhang, Z. Xie, G. Qin, Optical absorption enhancement in submicrometre crystalline silicon films with nanotexturing arrays for solar photovoltaic applications, J. Phys. D, Appl. Phys. 46 (2013) 195106.

[31] K.X. Wang, Z. Yu, V. Liu, Y. Cui, S. Fan, Absorption enhancement in ultrathin crystalline silicon solar cells with antireflection and light-trapping nanocone gratings, Nano Lett. 12 (2012) 1616.

[32] M. Boccard, C. Battaglia, F.-J. Haug, M. Despeisse, C. Ballif, Light trapping in solar cells: analytical modeling, Appl. Phys. Lett. 101 (2012) 151105.

[33] E.C. Garnett, P.D. Yang, Silicon nanowire radial p–n junction solar cells, J. Am. Chem. Soc. 130 (2008) 9224.

[34] M.D. Kelzenberg, S.W. Boettcher, J.A. Petykiewicz, D.B. Turner-Evans, M.C. Putnam, E.L. Warren, J.M. Spurgeon, R.M. Briggs, N.S. Lewis, H.A. Atwater, Enhanced absorption and carrier collection in Si wire arrays for photovoltaic applications, Nat. Mater. 9 (2010) 239.

[35] L. Cao, P. Fan, A.P. Vasudev, J.S. White, Z. Yu, W. Cai, J.A. Schuller, S. Fan, M.L. Brongersma, Semiconductor nanowire optical antenna solar absorbers, Nano Lett. 10 (2010) 439.

[36] J. Zhu, Z. Yu, G.D. Burkhard, C.-M. Hsu, S.T. Connor, Y. Xu, Q. Wang, M. McGehee, S. Fan, Y. Cui, Optical absorption enhancement in amorphous silicon nanowire and nanocone arrays, Nano Lett. 9 (2009) 279.

[37] D. Zhou, Y. Pennec, B. Djafari-Rouhani, O. Cristini-Robbe, T. Xu, Y. Lambert, Yves Deblock, Marc Faucher, Didier Stiévenard, Optimization of the optical properties of nanostructured silicon surfaces for solar cell applications, J. Appl. Phys. 115 (13) (2014) 134304.

[38] A. Taflove, Computational Electrodynamics: The Finite-Difference Time-Domain Method, Artech House, Norwood, MA, 1995.

[39] O.J.F. Martin, N.B. Piller, Electromagnetic scattering in polarizable backgrounds, Phys. Rev. E 58 (1998) 3909.

[40] M. Paulus, O.J.F. Martin, Light propagation and scattering in stratified media: a Green's tensor approach, J. Opt. Soc. Am. A 18 (2001) 854.

[41] J.P. Berenger, Three-dimensional perfectly matched layer for the absorption of electromagnetic waves, J. Comput. Phys. 114 (1994) 185.

[42] R.M.A. Azzam, N.M. Basharra, Ellipsometry and Polarized Light, North-Holland, Amsterdam, 1977.

[43] L. Ding, T.P. Chen, Y. Liu, C.Y. Ng, S. Fung, Optical properties of silicon nanocrystals embedded in a matrix, Phys. Rev. B 72 (2005) 125419.

[44] L. Cao, J.S. White, J.-S. Park, J.A. Schuller, B.M. Clemens, M.L. Brongersma, Engineering light absorption in semiconductor nanowire devices, Nat. Mater. 8 (2009) 643.

[45] L. Hu, G. Chen, Analysis of optical absorption in silicon nanowire arrays for photovoltaic applications, Nano Lett. 7 (2007) 3249.

[46] D.H.K. Murthy, T. Xu, W.H. Chen, A.J. Houtepen, T.J. Savenije, L.D.A. Siebbeles, J.P. Nys, C. Krzeminski, B. Grandidier, D. Stiévenard, P. Pareige, F. Jomard, G. Patriarche, O.I. Lebedev, Efficient photogeneration of charge carriers in silicon nanowires with a radial doping gradient, Nanotechnology 22 (2011) 315710.

[47] E.C. Garnett, Y.-C. Tseng, D.R. Khanal, J. Wu, J. Bokor, P. Yang, Dopant profiling and surface analysis of silicon nanowires using capacitance–voltage measurements, Nat. Nanotechnol. 4 (2009) 311.

[48] E. Koren, N. Berkovitch, Y. Rosenwaks, Measurement of active dopant distribution and diffusion in individual silicon nanowires, Nano Lett. 10 (2010) 1163.

[49] D.E. Perea, E.R. Hemesath, E.J. Schwalbach, J.L. Lensch-Falk, P.W. Voorhes, L.J. Lauhon, Direct measurement of dopant distribution in an individual vapour–liquid–solid nanowire, Nat. Nanotechnol. 4 (2009) 315.

[50] Y.H. Ahn, J. Park, Efficient visible light detection using individual germanium nanowire field effect transistors, Appl. Phys. Lett. 91 (2007) 162102.
[51] Y. Gu, J.P. Romankiewicz, J.K. David, J.L. Lensch, L.J. Lauhon, Quantitative measurement of the electron and hole mobility–lifetime products in semiconductor nanowires, Nano Lett. 6 (2006) 948.
[52] R. Graham, C. Miller, E. Oh, D. Yu, Electric field dependent photocurrent decay length in single lead sulfide nanowire field effect transistors, Nano Lett. 11 (2011) 717.
[53] M. Diarra, Y.-M. Niquet, C. Delerue, G. Allan, Ionization energy of donor and acceptor impurities in semiconductor nanowires: importance of dielectric confinement, Phys. Rev. B 75 (2007) 045301.
[54] B. Wu, A. Kumar, S. Pamarthy, High aspect ratio silicon etch: a review, J. Appl. Phys. 108 (2010) 051101.

Transmission line photonic crystals: a comparison of Green's formalism, lumped circuit element model, and finite element method

16

Abdelkrim Talbi[a], *El Houssaine El Boudouti*[b], *Cécile Ghouila-Houri*[a], *Abdellatif Akjouj*[a], *Bahram Djafari-Rouhani*[a], *and Leonard Dobrzyński*[a]
[a]Univ. Lille, CNRS, Centrale Lille, Yncréa ISEN, Univ. Polytechnique Hauts-de-France, UMR 8520 IEMN, Lille, France, [b]LPMR, Department of Physics, Faculty of Sciences, University Mohammed I, Oujda, Morocco

16.1 Introduction

Photonic crystals are periodic structures of dielectric or metallic materials designed to control the propagation of electromagnetic waves in photonic circuits. This periodic media produces an effect on the propagation of an electromagnetic wave similar to that of the periodic potential on the electrons in semiconductor crystals. The periodic potential is responsible for the permitted and prohibited electronic energy bands. This property is the fundamental basis for electron transfer and the electron current in integrated circuits. In analogy to electronic bands, periodic structures of dielectric or metallic materials exhibit permitted and forbidden photonic bands for electromagnetic waves. Indeed, the absence of propagating modes in a range of frequencies or wavelengths for electromagnetic (EM) waves in photonic crystals is also referred to as a "forbidden band". The simplest form of photonic crystal is a one-dimensional periodic structure composed of a multilayer stack also called a "Bragg mirror". It was Lord Rayleigh in 1887 who first showed that a gap or bandgap could be produced in this way. The concept of "photonic crystals" was first proposed in 1987 [1,2]. The first photonic crystal was realized in 1991 [3]. This concept of permitted and forbidden bands in the field of electromagnetic waves offers the possibility of realizing optical or electromagnetic devices capable of storing, filtering, and guiding light or electromagnetic waves at the wavelength scale by using periodic artificial materials. Numerous proposals for photonic crystal devices have been made and sometimes already realized, such as waveguides [4], bends [5,6], selective filters [7], multiplexers, demultiplexers [8], and combiners [9]. For frequencies in the microwave range, these photonic crystals are called electromagnetic bandgap materials. They are often made using periodic metal or metal–dielectric assemblies and are used to make filters, antennas, and various components used in telecommunications.

Photonics. https://doi.org/10.1016/B978-0-12-819388-4.00026-5

One-dimensional (1D) photonic crystals made of coaxial cables have shown several interesting properties and enable mimicking various properties observed in the quantum domain. Among these properties, we mention the existence of bandgaps with or without defect modes in periodic and quasiperiodic structures [10–18]. Based on the property of bandgaps, the design can also be used to mimic other advanced functions in microwave and photonic domain: adding/dropping filters [19], realizing slow and fast light, the so-called superluminal and subluminal, in periodic structures [15–17] and Mach–Zehnder-like interferometers [20,21]. In addition, recent works [22–25] have shown that several properties of quantum graphs could be successfully simulated by microwave networks made of coaxial cables. These structures are attractive since their production is more feasible at any wavelength scale.

Coaxial cable is the most common type of transmission line. This type of transmission line can carry signals at all frequencies, depending on the dimensions. There are two approaches to characterize a transmission line, the use of Maxwell's equations by considering its complete geometry, or the use of lumped circuit elements where the transmission line is represented by equivalent electrical components, such as capacitances and inductances. The lumped elements' approach is of significant importance in the analysis of microwave circuits and devices. The theory starts from a basic circuit principle and considers spatial dependence of voltage and current along the transmission line length. The voltage and current at different positions are not equal, although there is no discontinuity between them. As we will see in this chapter, the phenomenon of wave propagation on transmission lines will be introduced using the extension of the basic circuit theory to derive wave equations in terms of voltage and current and using Maxwell's equations to derive wave equations in term of electrical and magnetic fields. We shall present both viewpoints and show how these approaches enable obtaining results very close to those obtained by Green's function formalism used in Chapters 9, 10, and 11.

For the readers that have basic prerequisite knowledge in microwaves and photonics, practical design skills may be hard to acquire by simply reading a text that presents the Green's function formalism and the explanations of physics of proposed designs in Chapters 9, 10, and 11. Most of the people working in the fields of microwaves and photonics have become familiar with the prevalence of design software employed in microwave circuit and wave optics designs, such as the Advanced Design System (ADS) from Keysight Technologies and Comsol Multiphysics. Advances in custom simulation tools have drastically reduced the need to develop analytical methods and in-house programs for circuit and system design. These tools have made it possible to facilitate exchanges between designers, and consequently to make the design and experimentation of new designs in the field of photonics and microwaves quicker, as shown by the literature that has become abundant on these topics. Our point of view is not to question the analytical developments and in-house programs that have been widely proposed and discussed in this book. However, it seems more than necessary to bring the results presented in these different chapters closer to the end users, which are experimental researchers and engineers.

Advanced Design System (ADS) tool is a simulator like SPICE and Cadence. It focuses on the RF and microwave design, so most of its devices in the library are microwave devices and their models are based on lines theory and lumped elements. Based on Finite Elements method (FEM), Comsol Multiphysics is used to solve the transmission line and electromagnetic wave equations for various photonic configurations. ADS tool enables calculating the complete scattering matrix, while Comsol enables calculating band diagram, scattering matrix, and the electrical field distribution.

This chapter is primarily composed of two parts: The first part focuses on the basic theory for a transmission line based both on lumped elements' circuits and fields' analysis. The basic theory is concisely and clearly explained for the case of a coaxial cable based on the line theory and Maxwell equations' approaches. For these basic concepts, this chapter focuses more on physical concepts and on understanding the meaning of calculated results rather than on exhaustive mathematical calculations. In addition, Comsol and ADS software is used to analyze or verify the basic concepts, enabling readers to achieve a deeper and more thorough understanding of them. The second part highlights the results of numerical simulation obtained using ADS and Comsol for major results presented in Chapter 9 "Electromagnetic Induced Transparency, Induced Absorption and Fano Resonances in Photonic Circuits", Chapter 10 "Photonic Demultiplexers Based on Fano and Induced Transparency Resonances", and Chapter 11 "Photonic Monomode Circuits: Comb Structures".

16.2 The wave equation and method of calculation

A transmission line is a structure that guides electromagnetic waves. Some common examples are coaxial cable, copper wire, and microstrip line. Transmission lines are made of one or more conductors surrounded by insulator. Coaxial cable is the most common type of a transmission line. This type of transmission line can carry signals of all frequencies, depending on the dimensions (Fig. 16.1).

For lines of a few tens of cm in length and at frequencies in the range 100 MHz–1 GHz, the wavelength is of the same order of magnitude as this characteristic dimension. The fundamental phenomena of solid state physics in crystals such as wave diffraction are then transposable to systems whose elementary constituents are made from these transmission lines. Therefore, it is necessary to consider the analysis of such a system from the wave's propagation point of view.

The transmission line is considered to be a system of conductors connected from one point to another and along which electromagnetic energy can be transmitted. Maxwell's equations for the electric field [V/m] and for the magnetic field [A/m] in the harmonic regime (plane wave solution), and for any point in space, in a linear, homogeneous, and isotropic medium $\vec{D} = \varepsilon \vec{E}$, $\vec{B} = \mu \vec{H}$; with losses specified by the conductivity $\sigma \neq 0$ and characterized by a dielectric constant ($\varepsilon = \varepsilon_0 . \varepsilon_r$), have the following forms:

$$\nabla \times \vec{E} = -j\omega\mu \vec{H}, \tag{16.1a}$$

$$\nabla \times \vec{H} = j\omega\varepsilon \vec{E} + \sigma \vec{E}. \tag{16.1b}$$

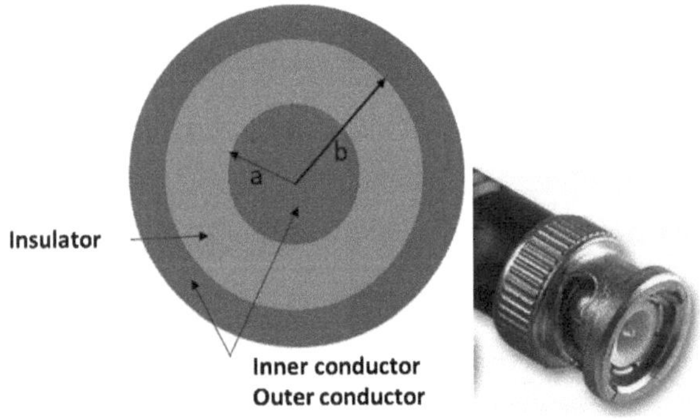

Figure 16.1 Coaxial cable schematic and image.

Taking the curl of Eqs. (16.1a) and (16.1b), we can derive the wave equations for both E and H, called also "Helmholtz equation":

$$\nabla \times \left(\mu_r^{-1}\right) \nabla \times \vec{E} - k^2 \left(\epsilon_r - j\frac{\sigma}{\omega\varepsilon_0}\right) \vec{E} = \vec{0}, \tag{16.2a}$$

$$\nabla \times \left(\mu_r^{-1}\right) \nabla \times \vec{H} - k^2 \left(\epsilon_r + j\frac{\sigma}{\omega\varepsilon_0}\right) \vec{H} = \vec{0}. \tag{16.2b}$$

Using Comsol RF Module: Electromagnetic waves or general form equation in Partial Derivative Equation model (PDE), we can solve these equations for any given geometry. When working at higher frequencies, the wavelength decreases, and the physical dimensions of the system made of transmission lines becomes of the same order of magnitude as the wavelength. In terms of circuit, some phenomena that could be neglected at low frequencies are no longer negligible. The transmission line theory approach allows high frequency circuits to be analyzed in terms familiar to basic circuit analysis: voltage, current, and impedance. At high frequencies, these terms depend on Maxwell's equations, but in a simplified form that enables avoiding the use of electric and magnetic fields and rapid design optimization. Thus in the transverse electric mode, a transmission line can be modeled as a lumped-element circuit as shown in Fig. 16.2, where voltages and currents can vary in magnitude and phase over its length. The piece of line of infinitesimal length "dx" of Fig. 16.2(a) can be represented by an equivalent scheme consisting of the elements *Rdx*, *Ldx*, *Cdx*, and *Gdx*, connected in series and in parallel as shown in Fig. 16.2(b), where R, L, G, and C are per-unit-length quantities defined as follows: R is the series resistance per unit length, for both conductors, in Ω/m; L is the series inductance per unit length, for both conductors, in H/m, G is the shunt conductance per unit length, in S/m, and C is the shunt capacitance per unit length, in F/m. Also R and G represent loss, due respectively to the finite conductivity of the individual conductors and to dielectric loss in the material between the conductors.

From the circuit of Fig. 16.2(b), Kirchhoff's voltage law can be applied and constitutive equations for voltage and current can be derived. In harmonic regime (plane

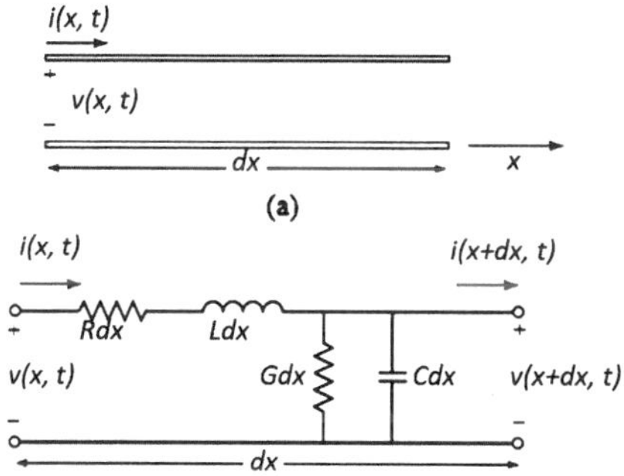

Figure 16.2 Voltage and current definitions and equivalent circuit for an incremental length of transmission line: (a) voltage and current definitions, (b) lumped-element equivalent circuit.

wave solution), these equations can be written as follows:

$$\frac{dV(x)}{dx} = -(R + j\omega L)\,I(x), \tag{16.3a}$$

$$\frac{dI(x)}{dx} = -(G + j\omega C)\,V(x). \tag{16.3b}$$

From the constitutive equations, we can derive the wave propagation equations for voltage and current on a transmission Line, well known as "telegrapher's equations":

$$\frac{d^2V(x)}{dx^2} - \gamma^2 V(x) = 0, \tag{16.4a}$$

$$\frac{d^2I(x)}{dx^2} - \gamma^2 I(x) = 0, \tag{16.4b}$$

where

$$\gamma = \alpha + j\beta = \sqrt{(R + j\omega L)(G + j\omega C)}. \tag{16.4c}$$

The telegrapher's equations can also be derived directly starting from Maxwell's equations. To do that, it is necessary to derive the transmission line parameters (R, L, G, C) as functions of the electric and magnetic fields of the transmission line. In the case of coaxial line, this can be obtained analytically or using FEM method based on integral form to calculate these parameters. We give below only the formulas and final relations, for more details see [26]. The geometrical parameters are shown in Fig. 16.3.

The mentioned formulas are as follows:

$$L = \frac{\mu}{|I_0^2|} \int_S \vec{H}\vec{H}^* dS = \frac{\mu}{2\pi} \ln\left(\frac{b}{a}\right), \tag{16.5a}$$

$$C = \frac{\varepsilon}{|V_0^2|} \int_S \vec{E}\vec{E}^* dS = \frac{2\pi\varepsilon'}{\ln(b/a)}, \tag{16.5b}$$

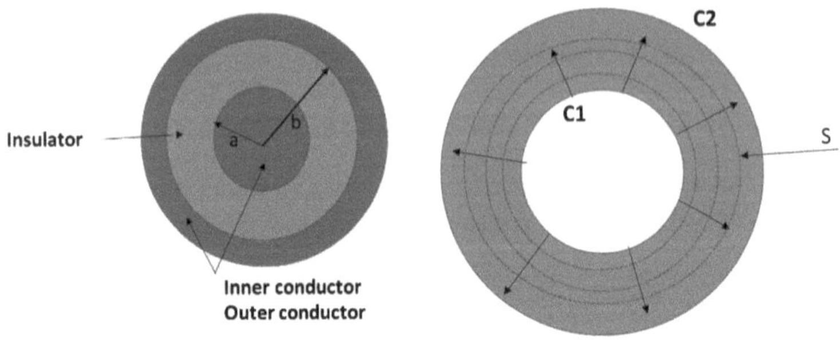

Figure 16.3 Geometry of the coaxial line and field lines of arbitrary TEM lines.

$$R = \frac{R_s}{|I_0^2|} \int_{C_1+C_2} \vec{H}\vec{H}^* dl = \frac{R_s}{2\pi}\left(\frac{1}{a} + \frac{1}{b}\right), \tag{16.5c}$$

$$G = \frac{\omega\varepsilon''}{|V_0^2|} \int_S \vec{E}\vec{E}^* dS = \frac{2\pi\omega\varepsilon''}{\ln\left(\frac{b}{a}\right)}, \tag{16.5d}$$

where $\varepsilon = \varepsilon' - j\varepsilon''$ and $R_s = \frac{1}{\sigma\delta_s}$.

16.3 Method of calculation

16.3.1 *Validation of the basic theory for a transmission line based both on lumped elements' circuits and fields' analysis*

In this section, we present the different models used for calculation: electromagnetic waves and transmission line using Comsol Multiphysics and Radio-Frequency (RF) electrical circuit analysis using ADS software. To validate the models, we consider a coaxial cable of length 10 cm; the geometrical and the physical parameters of the cable are summarized in Table 16.1. Fig. 16.4 shows a screen capture of the different models. Fig. 16.4(a) shows a simple circuit composed of two terminals for both excitation and measurement, we can measure all the coefficients of the matrix scattering. One can modify the geometrical parameters and physical properties directly on the circuit. We can perform parametric analysis and implement algorithm for optimization. Fig. 16.4(b) shows the capture of transmission line Comsol model. In this model, we solve by FEM analysis the telegrapher's equations. The model gives access to all electrical quantities like ADS. In addition, we can access to the distribution of voltage, current, and electrical fields in any point of the circuit. Fig. 16.4(c) shows the electromagnetic waves model considering the complete geometry of the line. In all the studies presented in this chapter, we assume that the inner and outer conductors are perfect. This model is more realistic as it enables to consider the full geometry of the studied system, however it is time consuming.

Table 16.1 Coaxial cable geometrical and physical parameters.

Inner diameter	0.9 mm
Outer diameter	2.95 mm
Relative electrical permittivity of insulator	2.31
Electrical conductivity of conductor	5.96 ∗ 1e7 S/m
Dielectric loss tangent	4 ∗ 1e-4

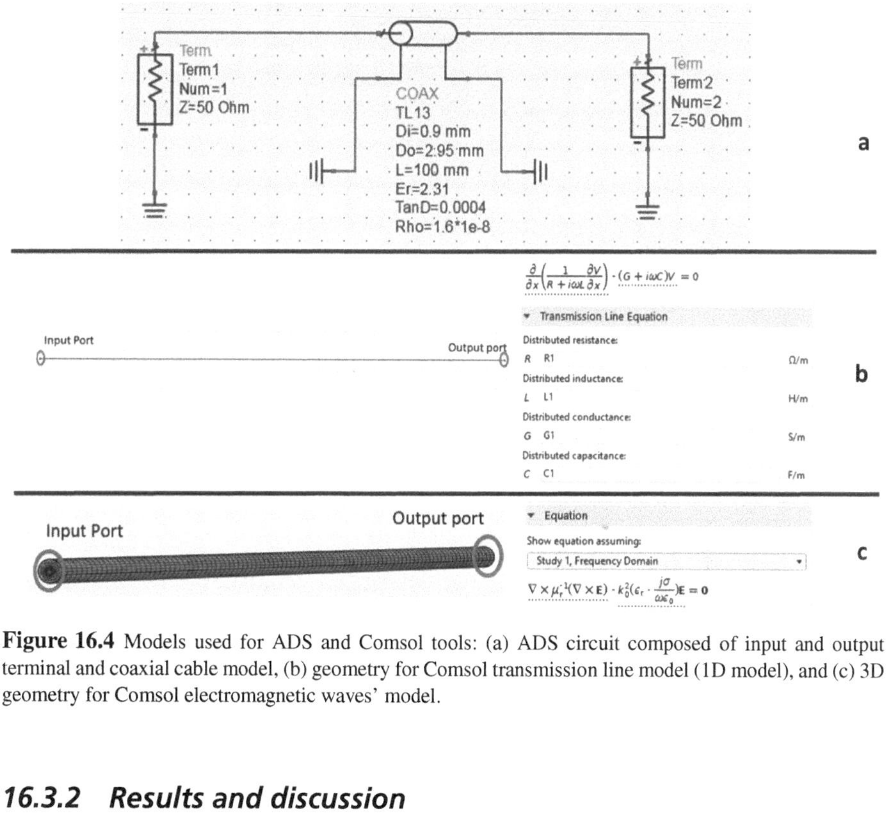

Figure 16.4 Models used for ADS and Comsol tools: (a) ADS circuit composed of input and output terminal and coaxial cable model, (b) geometry for Comsol transmission line model (1D model), and (c) 3D geometry for Comsol electromagnetic waves' model.

16.3.2 Results and discussion

We perform first a frequency analysis for both models. Using Comsol electromagnetic waves model, we calculate the electrical field distribution as shown in Fig. 16.5. We can see clearly that the mode presents TEM polarization, which is in perfect accordance with the theory of electromagnetic wave propagation in a waveguide with two conductors. Secondly, we calculate for the three models the transmission coefficient, both amplitude and phase. The results are displayed in Fig. 16.6, the three models give similar results with errors close to 0.1% for magnitude, and no error was evaluated for the phase. The errors in amplitude can be explained by the effect of mesh for the 3D FEM analysis and also the assumption made on the inner and outer conductors, which are considered as perfect conductors, which is not the case for the two other models.

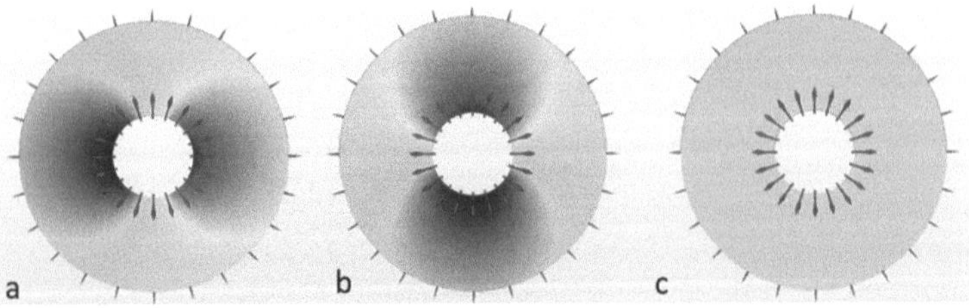

Figure 16.5 Electrical field distribution at 500 MHz in a section perpendicular to the propagation direction: (a) Ey component, (b) Ez component, and (c) Ex component.

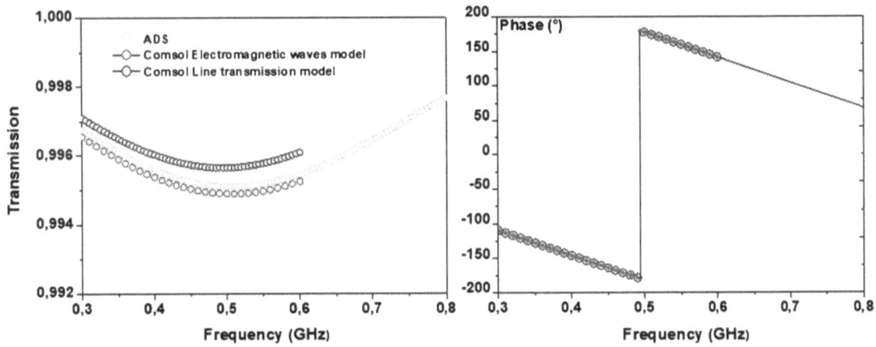

Figure 16.6 Comparison of transmission coefficient (magnitude and phase) calculated using two FEM models and lumped elements circuit.

16.4 Electromagnetic induced transparency, induced absorption and Fano resonances in photonic circuits

As is stated in Chapter 9, scattering of waves in composite materials involves propagation along different paths and, as a consequence, results in interference phenomena, where constructive interference corresponds to resonant enhancement and destructive interference to resonant suppression of the transmission. Fano resonance occurs when a discrete localized state becomes coupled to a continuum of states. Here, for example, we can consider infinite horizontal coaxial cable (continuum) and coaxial stub grafted vertically (parallel stub, localized state (Fig. 16.7)). Fano resonance profile is composed of a peak of transmission followed by a dip of transmission [1] and is characterized by an asymmetric line profile. Electromagnetic induced transparency (EIT) resonance is like Fano resonance which occurs when the peak of transmission falls between two dips of transmission. This kind of resonances, with the more familiar Fabry–Perot and Bragg resonators, are intensively employed as building blocks for sophisticated optical and acoustic devices with unique properties. In recent years, it was shown that Fano resonance and EIT-like behaviors are not restricted to quantum systems and have attracted a lot of attention in photonic, plasmonic, and acoustic systems [27,28], see also references [22–51] cited in Chapter 9. Indeed, almost any local resonant state can be considered as quasidiscrete with a complex frequency that can be described in terms of Fano resonance when interacting with a continuum of states. One major interest of Fano resonances in photonics concerns the design of switching or memory devices that exploit the sharp transmission–reflection curves supplied by this resonance.

In this section, we consider first the two structures considered in Chapter 9 to investigate the existence and properties of Fano, EIT, and EIA phenomena: (i) U-like structure made of two stubs of lengths d_1 and d_2 placed at two different positions and separated by segment of length d_3 (Fig. 16.7), (ii) cross-like structure constituted of two stubs of lengths d_1 and d_2 placed at the same position (Fig. 16.9). For the second instance, we consider the design of the demultiplexers based on Fano and EIT resonances investigated in Chapter 10. The results presented in Chapters 9 and 10 are obtained using the Green's function formalism. Here, we use both Comsol and ADS tools to confirm these results and enable the readers, who have basic prerequisite knowledge, and particularly Master's level graduate students, to reproduce and to understand quickly all the analytical calculations. In this section, we shall focus only on the numerical reproduction of some results in Chapters 9 and 10; however, for an in-depth physics analysis, the readers should refer to these chapters.

16.4.1 EIT-like resonance in a U-structure

Fig. 16.7 shows the schematic of the U-structure used in ADS and Comsol software. We consider the RG-58/U coaxial cable reference with characteristic impedance $Z = 50\ \Omega$. The physical properties are listed in Table 16.1. The upper ends of the stubs are open circuit (i.e., $H = 0$ boundary condition).

The results are displayed in Fig. 16.8. The left panel curves, including transmission, phase, and delay time as functions of frequency, correspond to the results obtained for the U-structure with the following geometrical parameters: $d_1 = 0.44 * 250$ mm, $d_2 = 0.56 * 250$ mm, and $d_3 = 250$ mm. These parameters enable obtaining symmetric Fano-resonance as illustrated in Fig. 9.4 (Chapter 9). The right panel curves correspond to the results obtained with the same structure but the grafted stubs are now of equal length and slightly higher than $0.5 * d_3$ (i.e., $d_1 = 0.56 * 250$ mm, $d_2 = 0.56 * 250$ mm, and $d_3 = 250$ mm). The structure exhibits the asymmetric Fano-resonance as illustrated in Fig. 9.6 (Chapter 9). Both Comsol and ADS models give results in good accordance with those obtained using the Green's function formalism (Sections 9.3 and 9.4 of Chapter 9).

As explained in Chapter 9 (Section 9.3), the above results confirm the slow and fast light behaviors of the EIT resonance observed in this system. In photonics, EIT resonance is the most successful approach used to realize slow light.

16.4.2 EIA-like resonance in a cross structure

In this section, we consider a cross structure made of two stubs of lengths d_1 and d_2 grafted at the same position (Fig. 16.9; see also Fig. 16.10). Here, the objective is to

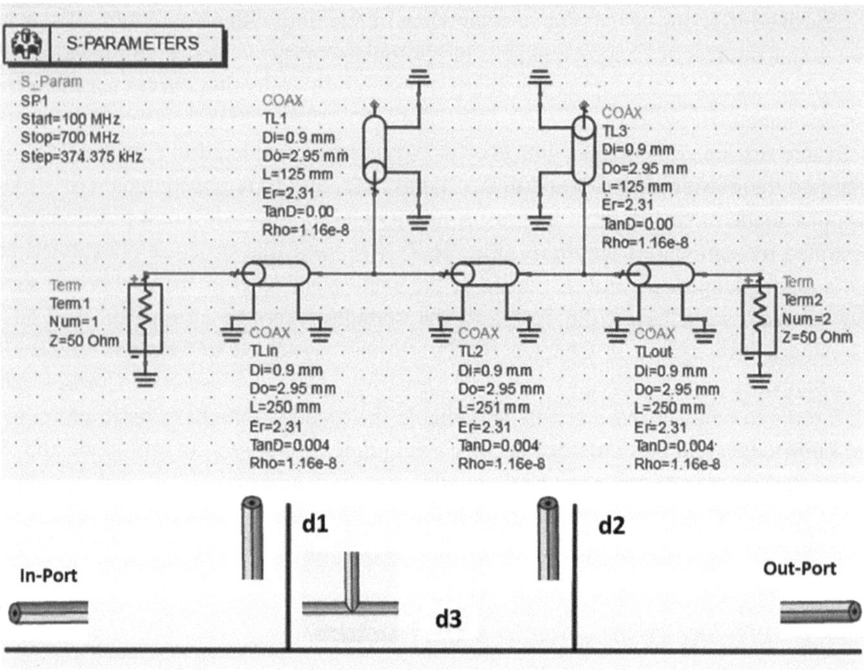

Figure 16.7 Schematic illustration of the studied system (U-structure): (a) ADS lumped circuit elements; (b) Comsol design with horizontal stub of length d_3 connected to input and output ports from each side and grafted parallel stubs (d_1 and d_2).

confirm using ADS and Comsol tools the results obtained in Chapter 9 (Section 9.5) concerning the possibility to realize the classical analogue of electromagnetic induced absorption. The main idea developed in Sections 9.3, 9.4, and 9.5 of Chapter 9 concerns how to clarify resonances as Fano, EIT, and AIT. We recalled in the introduction of this section, the conditions of occurrence of Fano and EIT resonances. In the case of a U-structure, the knowledge of the transmission coefficient magnitude and phase is enough to determine the EIT behavior of the resonance. For a cross structure, this condition is not enough, and we need in this case the knowledge of the reflection coefficient. This will enable determining the absorption, which is essential to guarantee the EIT or EIA behavior of the resonance. The obtained results are in good agreement with those given for the same structure (see Fig. 9.12, Chapter 9).

16.4.3 Photonic demultiplexer based on Fano and induced transparency resonances

In this section, we focus on the simulation of a Y-shaped photonic demultiplexer depicted in Fig. 16.11 using ADS and Comsol. (See also Fig. 16.12.) The study of the design is presented in Chapter 10. The demultiplexer is based on a cross structure, the EIT-type resonance presented by such structures is used to realize perfect demultiplex-

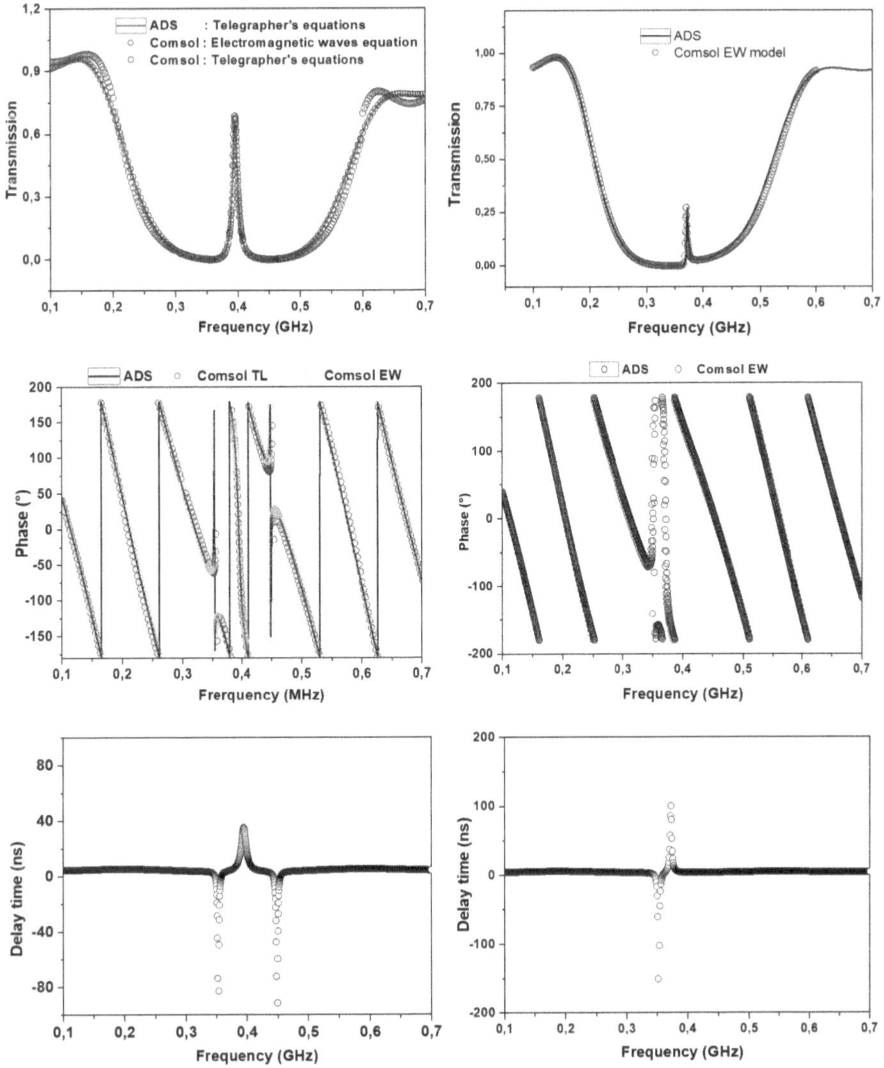

Figure 16.8 Variations of the transmission coefficient (magnitude and phase) and delay time versus the frequency for a U-structure depicted in Fig. 16.7. The lengths of the stubs are chosen such that $d_3 = 250$ mm, $d_1 = 0.44d_3$, $d_2 = 0.56d_3$, and $d_3 = 250$ mm, $d_1 = 0.56d_3$, $d_2 = 0.56d_3$ for the U-structure. Delay times are calculated for both structures using ADS only (Comsol can also be used). Note: the geometrical parameters are not the same as those used in Chapter 9.

ing. It consists on a Y-shaped waveguide with an input line and two output lines. Each line contains two grafted stubs at a given position far from the input line. In Comsol Multiphysics simulation, the transmission line model is used. Both Comsol and ADS models give results in good accordance with those obtained using the Green's function formalism (see Fig. 10.7 of Chapter 10).

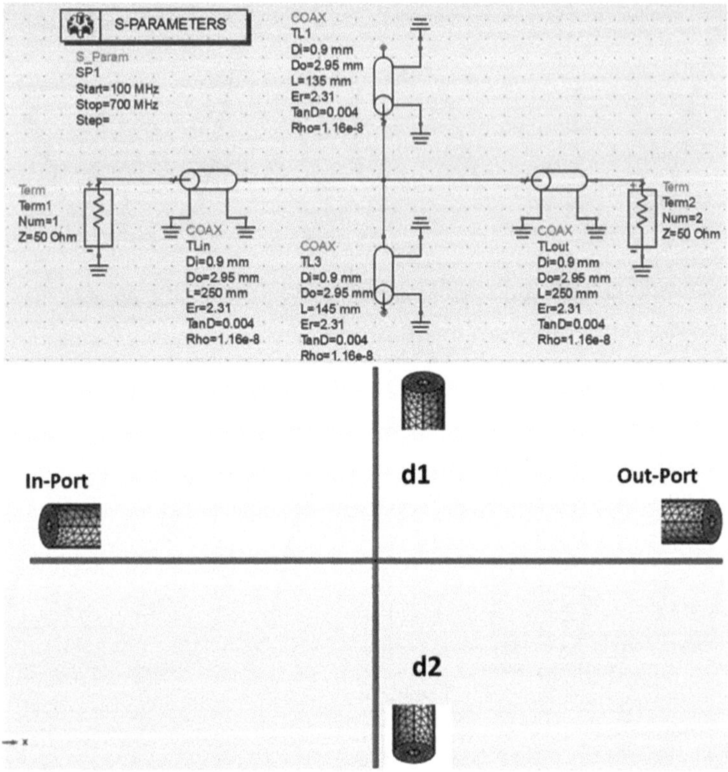

Figure 16.9 Schematic illustration of the studied system (cross structure): (a) ADS lumped circuit elements, (b) Comsol design consisting of one-dimensional electromagnetic waveguide (coaxial cable) with two grafted resonators of lengths d_1 and d_2 at the same point.

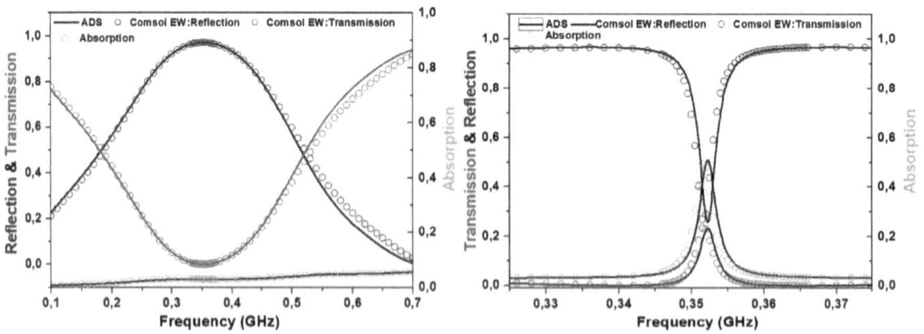

Figure 16.10 Variations of the transmission, reflection, and absorption versus the frequency for the structure depicted in Fig. 16.9: (left) the lengths of the stubs are chosen such that $d_1 = d_2 = 140$ mm, (right) the lengths of the stubs are chosen such that $d_1 = 135$, $d_2 = 145$ mm. Note: the geometrical parameters are not the same as those used in Chapter 9.

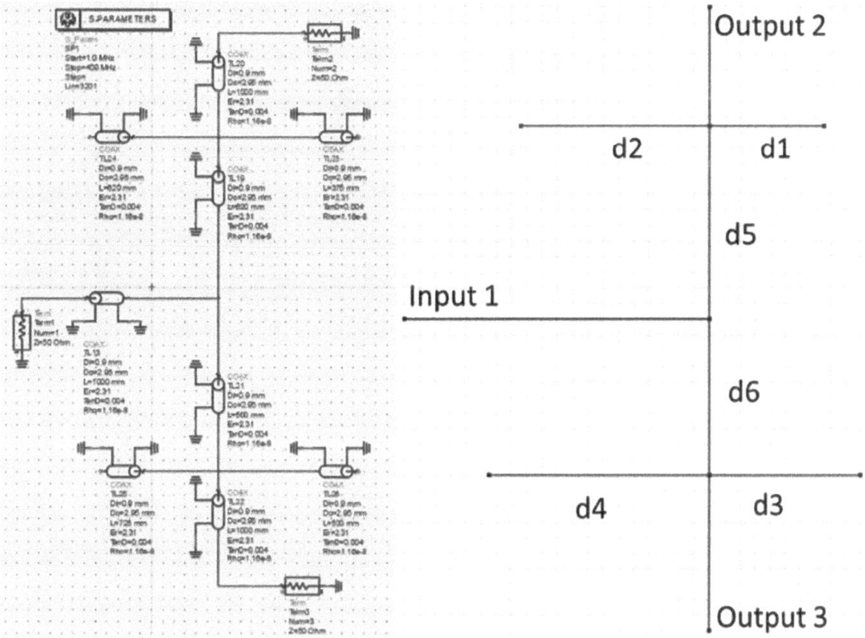

Figure 16.11 Schematic representation of a Y-shaped demultiplexer with one input line and two output lines. Along the first output line, two stubs of lengths d_1 and d_2 are inserted at the same position on the waveguide at a distance d_5 from the input line. Along the second output line, two stubs of lengths d_3 and d_4 are inserted at the same position on the waveguide at a distance d_6 from the input line.

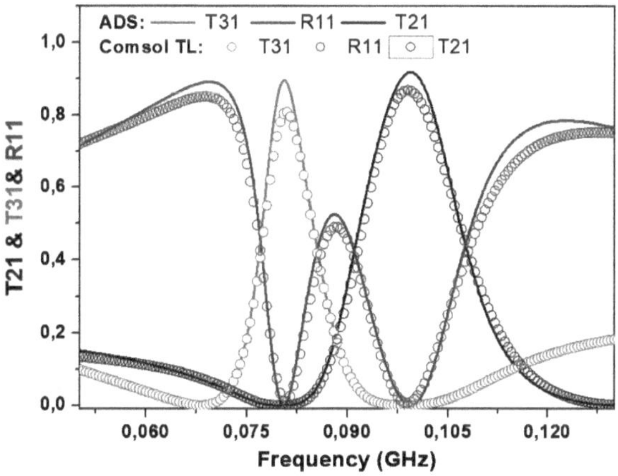

Figure 16.12 Transmission coefficients versus the frequency along ports 2 (black curves) and 3 (red curves; light gray in print version). The reflection coefficient is given by blue curves (dark gray in print version). The lengths of the cables are chosen such that $d_1 = 0.375$ m, $d_2 = d_5 = 0.62$ m, $d_3 = d_6 = 0.5$ m, and $d_4 = 0.725$ m. These parameters are extracted from Chapter 10.

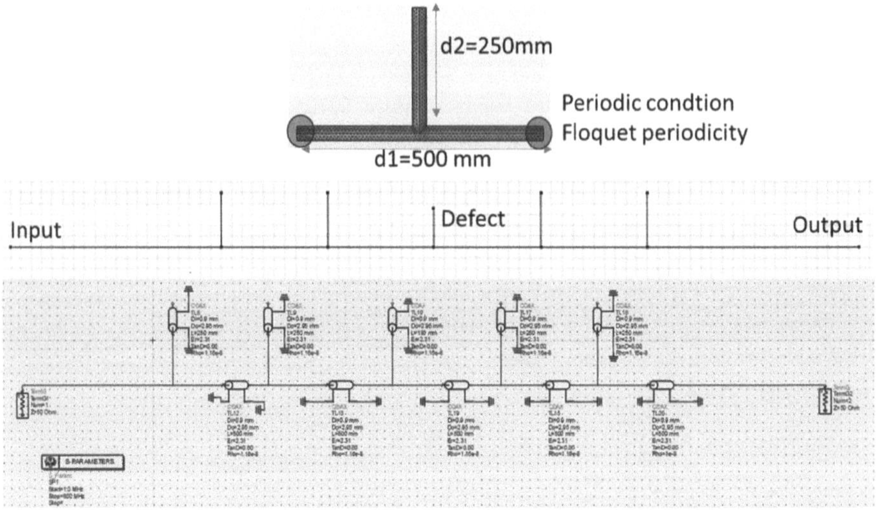

Figure 16.13 Schematic illustration of the studied system (comb structure). The stubs are grafted periodically on a horizontal waveguide: (upper panel) Comsol design for one unit cell for band diagram calculation, (lower panel) ADS lumped circuit elements.

16.5 Photonic monomode circuits: comb structures

In this section, we propose to apply the ADS and Comsol model to study a comb-like structure of one-dimensional waveguides exhibiting stopbands. This structure is composed of a waveguide along which finite side stubs are grafted periodically (Fig. 16.13). First, we calculate the band diagram of the infinite structure using EW Comsol model and the transmission coefficient through a finite structure composed of 6 unit cells using ADS. Secondly, we investigate the presence of a defect stub in the comb structure. The defect is a stub located in the middle of a finite structure composed of 3 unit cells and characterized by a length different from the length of the other grafted stubs. The results are displayed in Fig. 16.14, the band diagram shows the existence of stopbands, and this is in good accordance with the transmission coefficient magnitude. For the structure with a defect, a narrow peak appears in the transmission spectrum. In the case of a system with loss, the transmission is much smaller than one. This is to be noted in contrast to the results presented in Chapter 11 where the system is considered lossless. The localized states are very sensitive to the length of the defects, to their location in the comb, as well as to their number. One more time, the obtained results confirm well those obtained in Chapter 11 using the Green's function formalism.

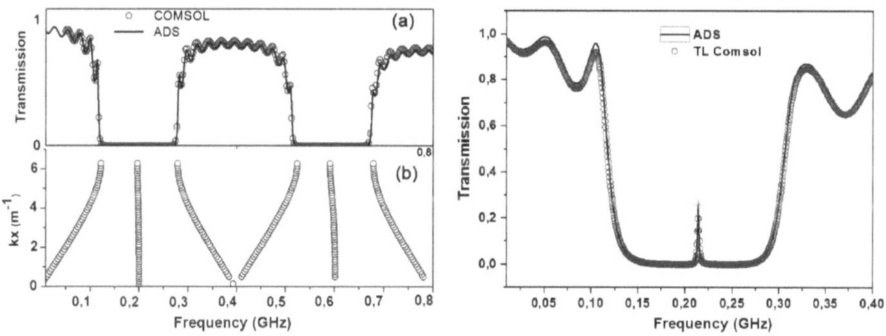

Figure 16.14 (Right) Band diagram of an infinite structure and transmission spectrum for a finite structure with 6 unit cells, (left) transmission spectrum for a finite structure with a defect separating three unit cells from each side. For the infinite structure, $d_1 = 500$ mm and $d_2 = 0.5d_1$. For the structure with a defect, $d_1 = 500$ mm and $d_2 = 0.5d_1$, except for the defect where $d_2 = 200$ mm.

16.6 Conclusions

In conclusion, this chapter compared two different methods for calculating some properties of 1D photonic crystals made of transmission lines. The first method is based on lumped circuit elements using ADS software, while the second was based on solving the electromagnetic wave and telegrapher's equations using Comsol Multiphysics. The obtained results corroborate well with those obtained in Chapters 9, 10, and 11 using the Green's function formalism. As stated in the introduction, these numerical tools will enable: (i) optimizing and experimenting with the different proposed designs in this book and (ii) implementing new teaching pedagogy in solid-state physics based on a digital experiment. As it was shown, these structures can mimic certain phenomena known in quantum systems.

References

[1] E. Yablonovitch, Inhibited spontaneous emission in solid-state physics and electronics, Phys. Rev. Lett. 58 (1987) 2059–2062.

[2] S. John, Strong localization of photons in certain disordered dielectric superlattices, Phys. Rev. Lett. 58 (1987) 2486.

[3] E. Yablonovitch, T.J. Gmitter, K.M. Leung, Photonic band structure: the face-centered-cubic case employing non spherical atoms, Phys. Rev. Lett. 67 (1991) 2295–2298.

[4] S. Olivier, M. Rattier, H. Benisty, C. Weisbuch, C.J.M. Smith, R.M. De La Rue, T.F. Krauss, U. Oesterle, R. Houdre, Mini-stopbands of a one-dimensional system: the channel waveguide in a two-dimensional photonic crystal, Phys. Rev. B 63 (2001) 113311.

[5] S. Olivier, H. Benisty, M. Rattier, C. Weisbuch, M. Qiu, A. Karlsson, C.J.M. Smith, R. Houdré, U. Oesterle, Resonant and nonresonant transmission through waveguide bends in a planar photonic crystal, Appl. Phys. Lett. 79 (2001) 2514–2516.

[6] A. Talneau, Ph. Lalanne, M. Agio, C.M. Soukoulis, Low-reflection photonic-crystal taper for efficient coupling between guide sections of arbitrary widths, Opt. Lett. 27 (2002) 1522–1524.

[7] M. Qiu, M. Mulot, M. Swillo, S. Anand, B. Jaskorzynak, A. Karlsson, M. Kamp, A. Forchel, Photonic crystal optical filter based on contra-directional waveguide coupling, Appl. Phys. Lett. 83 (2003) 5121–5123.

[8] Y. Akahane, M. Mochizuki, T. Asano, Y. Tanaka, S. Noda, Design of a channel drop filter by using a donor-type cavity with high-quality factor in a two-dimensional photonic crystal slab, Appl. Phys. Lett. 82 (2003) 1341–1343.

[9] S. Bocolo, M. Midrio, T.F. Krauss, Y junctions in photonic crystal channel waveguides: high transmission and impedance matching, Opt. Lett. 27 (2002) 1001.

[10] Y. El Hassouani, H. Aynaou, E.H. El Boudouti, B. Djafari-Rouhani, A. Akjouj, V.R. Velasco, Surface electromagnetic waves in Fibonacci superlattices: theoretical and experimental results, Phys. Rev. B 74 (2006) 035314.

[11] B. Djafari-Rouhani, E.H. El Boudouti, A. Akjouj, L. Dobrzynski, J.O. Vasseur, A. Mir, N. Fettouhi, J. Zemmouri, Surface states in one-dimensional photonic band gap structures, Vacuum 63 (2001) 177.

[12] L. Poirier, A. Haché, Nonlinear coaxial photonic crystal, Appl. Phys. Lett. 78 (2001) 2626.

[13] G.J. Schneider, S. Hanna, J.L. Davis, G.H. Watson, Defect modes in coaxial photonic crystals, J. Appl. Phys. 90 (2001) 2642.

[14] J.O. Vasseur, B. Djafari-Rouhani, L. Dobrzynski, A. Akjouj, J. Zemmouri, Defect modes in one-dimensional comb-like photonic waveguides, Phys. Rev. B 59 (1999) 13446.

[15] E.H. El Boudouti, N. Fettouhi, A. Akjouj, B. Djafari-Rouhani, A. Mir, J.O. Vasseur, L. Dobrzynski, J. Zemmouri, Experimental and theoretical evidence for the existence of photonic bandgaps and selective transmissions in serial loop structures, J. Appl. Phys. 95 (2004) 1102.

[16] A. Haché, L. Poirier, Anomalous dispersion and superluminal group velocity in a coaxial photonic crystal: theory and experiment, Phys. Rev. E 65 (2002) 036608;
J.N. Munday, W.M. Robertson, Negative group velocity pulse tunneling through a coaxial photonic crystal, Appl. Phys. Lett. 80 (2002) 518.

[17] H. Aynaou, E.H. El Boudouti, Y. El Hassouani, A. Akjouj, B. Djafari-Rouhani, J. Vasseur, A. Benomar, V.R. Velasco, Propagation and localization of electromagnetic waves in quasiperiodic serial loop structures, Phys. Rev. E 72 (2005) 056601.

[18] E.H. El Boudouti, Y. El Hassouani, B. Djafari-Rouhani, H. Aynaou, Two types of modes in finite size one-dimensional coaxial photonic crystals: general rules and experimental evidence, Phys. Rev. E 76 (2007) 026607.

[19] L. Dobrzynski, B. Djafari-Rouhani, A. Akjouj, J.O. Vasseur, J. Zemmouri, Resonant tunneling between two continua, Phys. Rev. B 60 (1999) 10628.

[20] M.M. Sanchez-Lopez, A. Sanchez-Merono, J. Arias, J.A. Davis, I. Moreno, Observation of superluminal and negative group velocities in a Mach–Zehnder interferometer, Appl. Phys. Lett. 93 (2008) 074102.

[21] J. Arias, A. Sanchez-Merono, M.M. Sanchez-Lopez, I. Moreno, Slow and fast light in three-beam interferometers: theory and experiment, Phys. Rev. A 85 (2012) 033815.

[22] O. Hul, S. Bauch, P. Pakonski, N. Savytskyy, K. Zyczkowski, L. Sirko, Experimental simulation of quantum graphs by microwave networks, Phys. Rev. E 69 (2004) 056205.

[23] M. Lawniczak, O. Hul, S. Bauch, P. Seba, L. Sirko, Experimental and numerical investigation of the reflection coefficient and the distributions of Wigner's reaction matrix for irregular graphs with absorption, Phys. Rev. E 77 (2008) 056210.

[24] M.L. awniczak, S. Bauch, O. Hul, L. Sirko, Experimental investigation of the enhancement factor for microwave irregular networks with preserved and broken time reversal symmetry in the presence of absorption, Phys. Rev. E 81 (2010) 046204.
[25] O. Hul, M.L. awniczak, S. Bauch, A. Sawicki, M. Kuś, L. Sirko, Are scattering properties of graphs uniquely connected to their shapes?, Phys. Rev. Lett. 109 (2012) 040402.
[26] D.M. Pozar, Microwave Engineering, 3rd edition, John Wiley & Sons, Inc., New York, 2005.
[27] M. Limonov, M. Rybin, A. Poddubny, et al., Fano resonances in photonics, Nat. Photonics 11 (2017) 543–554.
[28] Y. Liu, A. Talbi, E.H. El Boudouti, et al., Autler–Townes splitting and acoustically induced transparency based on Love waves interacting with a pillared metasurface, Phys. Rev. Appl. 11 (2019) 064066.

Part Three

Photonic materials

Edited by Abdellatif Akjouj

Interface response function in layered photonic materials

17

Abdellatif Akjouj[a], Leonard Dobrzyński[a], Gaëtan Lévêque[a], and El Houssaine El Boudouti[b]
[a]Institute of Electronic, Microelectronics and Nanotechnology (IEMN), UMR CNRS 8520, Department of Physics, Faculty of Sciences and Technologies, Lille University, Villeneuve d'Ascq Cedex, France, [b]Laboratoire de Physique de la Matière et de Rayonnement, Département de Physique, Faculté des Sciences, Université Mohamed I, Oujda, Morocco

17.1 Introduction

The solution of the Maxwell equations within a bounded homogeneous medium is an old problem. All physics undergraduates know the appropriate boundary conditions the electric and magnetic fields have to obey at the interfaces. This problem was studied before in some cases [1–4] for single interfaces of high symmetry (planes [4], spheres [5–7], cylinders [2], etc.), for a few randomly rough and periodically corrugated interfaces, and for a few layered media (three to four dielectric media separated by planar interfaces, two-layered photonic crystals [4]).

Notice that the aim of this chapter is to show how one can tackle more complex media and solve the Maxwell equations for any composite material as complex as one may wish. For simplicity, we restrict ourselves to nonmagnetic dielectric media, although the present theory may be generalized to media where the complete Maxwell equations have to be considered. The solution of this problem is reached with the help of the interface response theory of composite systems, first proposed for discrete spaces [8,9] and then generalized to continuous ones [10,11].

This theory enables one to calculate the response functions associated with the electromagnetic field for any composite dielectric medium. These functions then enable one to obtain any physical property of these media. In particular, they play a central role in theories of light scattering (both Brillouin and Raman), as well as in a variety of other physical phenomena [12–14]. Much of the analytical results presented in this chapter have been inspired by Bah et al. work published in [15].

New general results are illustrated here by an application to layered isotropic dielectrics, for the interface between two different dielectric media, for one dielectric slab sandwiched between two different semiinfinite dielectrics, and for dielectric photonic crystals.

It is worth mentioning that the interface response theory of electromagnetic continuous media was applied intensively during the last three decades for isotropic multilayered media [15–26], lamellar structures with left-handed materials [27], anisotropic layered photonic crystals [28–31], coaxial cables [32–48], plasmonic waveguides

Photonics. https://doi.org/10.1016/B978-0-12-819388-4.00028-9

[49,50], magnetoplasmons in semiconductor supperlattices [51–53], and nanoplasmonic response in layered composite thin film with nanometallic inclusions [54–58].

This chapter is organized as follows. Section 17.2 gives the response functions for any composite dielectric medium. In Section 17.3, we develop the basic equations for any layered isotropic dielectric media. In Section 17.4 the response functions of the interface between two semiinfinite isotropic dielectrics are given. Section 17.5 gives the analytical results obtained for the response functions in a dielectric slab sandwiched between two semiinfinite isotropic dielectrics and some application for the sandwich electromagnetic guided modes. The response function for a semiinfinite two-layered photonic crystal in contact with another semiinfinite homogeneous dielectric is given in closed form in Section 17.6. This last result is illustrated by some numerical results for dispersion curves and surface electromagnetic modes. Section 17.7 summarizes the principal results of this chapter.

17.2 Response functions for any composite dielectric medium

First the response function for the electromagnetic field will be defined for one homogeneous infinite dielectric medium. Then a dielectric medium limited by a black-box surface will be defined by a completely opaque surface through which the electromagnetic field cannot propagate. Any composite dielectric medium can then be constructed out of pieces of such different dielectrics bounded by black-box surfaces. This way of constructing such a medium enables one to calculate the corresponding response function.

17.2.1 One infinite dielectric medium and one limited by a black-box surface

We begin by defining the response function we are concerned with. We assume that in the infinite physical system under study there is a current density $\boldsymbol{J}$ present. The Maxwell equations for the electromagnetic field in the system in this case are:

$$\nabla \times \boldsymbol{H} = \boldsymbol{J} + \frac{\partial \boldsymbol{D}}{\partial t}, \tag{17.1}$$

$$\nabla \times \boldsymbol{E} = -\frac{\partial \boldsymbol{B}}{\partial t}, \tag{17.2}$$

where $\boldsymbol{E}$ is the macroscopic electric field, $\boldsymbol{D}$ is the displacement vector, $\boldsymbol{H}$ is the magnetic field, and $\boldsymbol{B}$ is the magnetic induction.

For simplicity, we will assume that the medium is nonmagnetic, and will set the magnetic fields $\boldsymbol{H}$ equal to $\boldsymbol{B}/\mu_0$ (μ_0 is the permeability in vacuum). When the effects of spatial dispersion are neglected, the most general linear relation between the displacement vector $\boldsymbol{D}(\boldsymbol{x}, t)$, and the macroscopic electric field $\boldsymbol{E}(\boldsymbol{x}, t)$ in a dielectric medium is given by [59]

$$\boldsymbol{D}(\boldsymbol{x},t)=\int dt'\varepsilon_0\varepsilon_i(t-t')\boldsymbol{E}(\boldsymbol{x},t'), \tag{17.3}$$

where $\varepsilon_i(t-t')$ is the 3×3 relative dielectric tensor (the index i will be necessary when we consider the composite system formed out of different dielectrics) and ε_0 is the permittivity in the vacuum. This tensor is a function of the difference in the time argument, $t-t'$, because we are assuming that the Hamiltonian for the dielectric medium is time-independent [15]. The requirement of causality is that the $\varepsilon_i(t)$ vanish identically for $t<0$.

Create now a dielectric bounded by a black-box surface defined by $x_3=f(x_1,x_2)$. This can be achieved by writing that the speed of light c vanishes for $x_3<f(x_1,x_2)$.

This enables us to rewrite the Maxwell equations for this dielectric limited by a black-box surface as

$$\theta[-f(x_1,x_2)+x_3]\left(\nabla\times\boldsymbol{H}-\boldsymbol{J}\right)-\frac{\partial\boldsymbol{D}}{\partial t}=\boldsymbol{0}, \tag{17.4}$$

$$\theta[-f(x_1,x_2)+x_3]\,\nabla\times\boldsymbol{E}+\frac{\partial\boldsymbol{B}}{\partial t}=\boldsymbol{0}, \tag{17.5}$$

where

$$\theta[-f(x_1,x_2)+x_3]=\begin{cases}1, & \text{for } x_3\geq f(x_1,x_2),\\ 0, & \text{for } x_3<f(x_1,x_2).\end{cases} \tag{17.6}$$

Eliminating $\boldsymbol{H}$ and $\boldsymbol{B}$ between Eqs. (17.4) and (17.5) provides the equation satisfied by the macroscopic electric field. Applying first the operator ∇ to Eq. (17.5), we obtain

$$\nabla\times\theta[-f(x_1,x_2)+x_3]\,\nabla\times\boldsymbol{E}+\frac{\partial}{\partial t}\nabla\times\boldsymbol{B}=\boldsymbol{0}. \tag{17.7}$$

Then using Eq. (17.4) and differentiating provides

$$\begin{aligned}&\theta[-f(x_1,x_2)+x_3]\left(\nabla\times\nabla\times\boldsymbol{E}+\mu_0\frac{\partial^2\boldsymbol{D}}{\partial t^2}\right)\\ &\quad-\delta[-f(x_1,x_2)+x_3]\,\boldsymbol{V}_i(\boldsymbol{x})\boldsymbol{E}=-\mu_0\frac{\partial\boldsymbol{J}}{\partial t},\end{aligned} \tag{17.8}$$

where the black-box cleavage operator $\boldsymbol{V}_i(\boldsymbol{x})$ is defined as a 3×3 matrix. It depends on the subdielectric i only through the form of its black-box surface. Note, however, that $\boldsymbol{V}_i(\boldsymbol{x}_3)$ has the opposite sign if one considers the complementary (in the infinite space) dielectric. Its expression was found to be

$$\boldsymbol{V}_i(\boldsymbol{x})=\begin{pmatrix}-\frac{\partial f(x_1,x_2)}{\partial x_2}\frac{\partial}{\partial x_2}+\frac{\partial}{\partial x_3} & \frac{\partial f(x_1,x_2)}{\partial x_2}\frac{\partial}{\partial x_1} & -\frac{\partial}{\partial x_1}\\ \frac{\partial f(x_1,x_2)}{\partial x_1}\frac{\partial}{\partial x_2} & -\frac{\partial f(x_1,x_2)}{\partial x_1}\frac{\partial}{\partial x_1}+\frac{\partial}{\partial x_3} & -\frac{\partial}{\partial x_2}\\ \frac{\partial f(x_1,x_2)}{\partial x_1}\frac{\partial}{\partial x_3} & \frac{\partial f(x_1,x_2)}{\partial x_2}\frac{\partial}{\partial x_3} & -\frac{\partial f(x_1,x_2)}{\partial x_1}\frac{\partial}{\partial x_1}-\frac{\partial f(x_1,x_2)}{\partial x_2}\frac{\partial}{\partial x_2}\end{pmatrix}. \tag{17.9}$$

We now Fourier transform $\boldsymbol{E}(\boldsymbol{x},t)$, $\boldsymbol{H}(\boldsymbol{x},t)$, $\boldsymbol{J}(\boldsymbol{x},t)$, and $\varepsilon_i(t-t')$ with respect to time:

$$\boldsymbol{E}(\boldsymbol{x},t)=\int\frac{d\omega}{2\pi}\boldsymbol{E}(\boldsymbol{x}/\omega)e^{-i\omega t}, \tag{17.10}$$

$$\boldsymbol{H}(\boldsymbol{x},t)=\int\frac{d\omega}{2\pi}\boldsymbol{H}(\boldsymbol{x}/\omega)e^{-i\omega t}, \tag{17.11}$$

$$\boldsymbol{J}(\boldsymbol{x},t)=\int\frac{d\omega}{2\pi}\boldsymbol{J}(\boldsymbol{x}/\omega)e^{-i\omega t}, \tag{17.12}$$

$$\varepsilon_i(t-t')=\int\frac{d\omega}{2\pi}\varepsilon_i(\omega)e^{-i\omega(t-t')}, \tag{17.13}$$

where $\varepsilon_i(\omega)$, considered as a function of the complex variable ω, can have no poles in the upper half-plane [15].

Remember that $\varepsilon_0\mu_0=1/c^2$, where c is the light speed. When Eqs. (17.10), (17.11), (17.12), and (17.13) are substituted into Eq. (17.8), the latter takes the form

$$\begin{aligned}&\theta[-f(x_1,x_2)+x_3]\left(\frac{\omega^2}{c^2}\varepsilon_i(\omega)-\boldsymbol{H}^0(\boldsymbol{x})\right)\boldsymbol{E}(\boldsymbol{x}/\omega)\\&\quad+\delta[-f(x_1,x_2)+x_3]\,\boldsymbol{V}_i(\boldsymbol{x})\boldsymbol{E}(\boldsymbol{x}/\omega)=-i\omega\mu_0\boldsymbol{J}(\boldsymbol{x}/\omega),\end{aligned} \tag{17.14}$$

where

$$\boldsymbol{H}^0_{\alpha\beta}(\boldsymbol{x})=\frac{\partial^2}{\partial x_\alpha\partial x_\beta}\delta_{\alpha\beta}\boldsymbol{\nabla}^2\quad\text{for }\ \alpha,\beta=1,2,3. \tag{17.15}$$

Define now the response functions $\boldsymbol{G}_i(\omega/\boldsymbol{x},\boldsymbol{x}')$ and $\boldsymbol{g}_{si}(\omega/\boldsymbol{x},\boldsymbol{x}')$, respectively, for the infinite dielectric medium i and the same medium but limited by the black-box surface as

$$\left(\frac{\omega^2}{c^2}\varepsilon_i(\omega)-\boldsymbol{H}^0(\boldsymbol{x})\right)\boldsymbol{G}_i(\omega/\boldsymbol{x},\boldsymbol{x}')=\boldsymbol{I}\delta(\boldsymbol{x}-\boldsymbol{x}') \tag{17.16}$$

and

$$\begin{aligned}&\left[\theta[-f(x_1,x_2)+x_3]\left(\frac{\omega^2}{c^2}\varepsilon_i(\omega)-\boldsymbol{H}^0(\boldsymbol{x})\right)+\delta[-f(x_1,x_2)+x_3]\,\boldsymbol{V}_i(\boldsymbol{x})\right]\\&\quad\times\boldsymbol{g}_{si}(\omega/\boldsymbol{x},\boldsymbol{x}')=\boldsymbol{I}\delta(\boldsymbol{x}-\boldsymbol{x}'),\end{aligned} \tag{17.17}$$

where $\boldsymbol{I}$ is the 3×3 unit matrix.

The interface response theory of continuous media [10,11] provides the following relation between the above response functions [15]:

$$\boldsymbol{g}_{si}(\boldsymbol{x},\boldsymbol{x}')+\int d\boldsymbol{x}''\boldsymbol{g}_{si}(\boldsymbol{x},\boldsymbol{x}'')\boldsymbol{A}_{si}(\boldsymbol{x},\boldsymbol{x}'')=\boldsymbol{G}_i(\boldsymbol{x},\boldsymbol{x}'), \tag{17.18}$$

for $\{\boldsymbol{x}, \boldsymbol{x}'\} \in D_i$ and $\boldsymbol{x}'' \in M_i$, where D_i is the three-dimensional space for which $x_3 \geq f(x_1, x_2)$, and M_i is the surface space defined by $x_3 = f(x_1, x_2)$; $\boldsymbol{A}_{si}$ is the corresponding surface response operator defined by

$$\boldsymbol{A}_{si}(\boldsymbol{x}, \boldsymbol{x}') = \boldsymbol{V}_i(\boldsymbol{x}'')\boldsymbol{G}_i(\boldsymbol{x}, \boldsymbol{x}'')|_{\boldsymbol{x}''=\boldsymbol{x}}, \quad \{\boldsymbol{x}', \boldsymbol{x}''\} \in D_i \text{ and } \boldsymbol{x} \in M_i. \tag{17.19}$$

In Eq. (17.18), and in what follows, we do not write explicitly, for simplicity, the dependence on ω of the response functions.

Once the response function $\boldsymbol{G}_i$ of the infinite dielectric is determined from Eq. (17.16), the surface integral equation (17.18) enables us to calculate the response function $\boldsymbol{g}_{si}$ of the same dielectric but limited by the black-box surface M_i. Of course, the solutions of the integral equation (17.18) have to be obtained in general by numerical methods. One of these numerical methods may be to transform the integral equation (17.18) into a matrix equation by a suitable choice of a discrete number of values for $\boldsymbol{x}, \boldsymbol{x}' \in D_i$. In particular, it is helpful and sufficient to do this inside the surface space M_i. Then one easily shows [8–11] that

$$\boldsymbol{g}_{si}^{-1}(M_i, M_i) = \boldsymbol{\Delta}(M_i, M_i)\boldsymbol{G}_s^{-1}(M_i, M_i) \tag{17.20}$$

where

$$\boldsymbol{\Delta}(M_i, M_i) = \boldsymbol{I}(M_i, M_i) + \boldsymbol{A}_{si}(M_i, M_i). \tag{17.21}$$

All entities in the above two equations have to be understood as matrices defined for a discrete number of space points $\boldsymbol{x}$ within the surface, M_i. In particular $\boldsymbol{g}_{si}^{-1}(M_i, M_i)$ and $\boldsymbol{G}_i^{-1}(M_i, M_i)$ are the inverses of $\boldsymbol{g}_{si}(M_i, M_i)$ and $\boldsymbol{G}_i(M_i, M_i)$ within the space M_i.

Note also that all the above results can be easily generalized [10,11] to one dielectric medium i limited by J disconnected black-box surfaces $M_{ij} \in M_i$, $1 \leqslant j \leqslant J$. One has to calculate $\boldsymbol{V}_i(\boldsymbol{x})$ for each surface M_{ij} and the total response operator $\boldsymbol{A}_{si}$ within the whole surface space M_i.

17.2.2 Any composite dielectric

A composite medium is formed out of N different submedia labeled by the index i and contained in their respective subdomains D_i. The ensemble of these subdomains D_i forms the domain of definition D of the composite medium.

The subdomains are separated from each other by subinterface domains M_{ij}. In general, a domain D_i is bounded by several disconnected subinterface domains M_{ij} and their ensemble is called the interface subdomain M_i. The ensemble of all interface subdomains M_i is the interface domain M of the composite material.

Any composite dielectric medium can in this way be built out of the independent pieces limited by the black-box surfaces, as defined above. In particular, one has to use semiinfinite pieces corresponding to the composite material [15]. Putting all these independent pieces together and allowing them to interact enables one [8–11] to construct the real composite dielectric medium and to calculate its response function $\boldsymbol{g}$ out of the elements of the response functions $\boldsymbol{G}_i$ of the infinite dielectric media.

Once the inverse surface elements $\boldsymbol{g}_{si}^{-1}(M_i, M_i)$ are determined from Eqs. (17.20) and (17.21) for all independent submedia, one gets the inverse interface elements $\boldsymbol{g}^{-1}(M, M)$ of the composite medium in the following manner [10,11]:

$$\boldsymbol{g}^{-1}(M_{ij}, M_{i'j'}) = 0 \qquad M_{i'j'} \notin M_i; \tag{17.22}$$

$$\boldsymbol{g}^{-1}(M_{ij}, M_{ij'}) = \boldsymbol{g}_{si}^{-1}(M_{ij}, M_{ij'}) \quad j' \neq j; \tag{17.23}$$

$$\boldsymbol{g}^{-1}(M_{ij}, M_{ij}) = \sum_{i'} \boldsymbol{g}_{si'}^{-1}(M_{i'j'}, M_{i'j'}) \quad M_{i'j'} \equiv M_{ij}. \tag{17.24}$$

So now we have all the elements of the matrix $\boldsymbol{g}^{-1}(M, M)$, the inverse of the matrix $\boldsymbol{g}(M, M)$. This inverse, as explained above, can be defined in general only by using a finite number of discrete space points in M.

So once the matrix inversion of the black-box tridiagonal matrix $\boldsymbol{g}^{-1}(M, M)$ (Eqs. (17.22), (17.23), and (17.24)) is achieved, one has the matrix elements $\boldsymbol{g}(M, M)$ of the response function $\boldsymbol{g}$ of the composite dielectric medium, within the total interface space M.

All the other matrix elements of $\boldsymbol{g}$ can be obtained from [10,11]:

$$\begin{aligned} \boldsymbol{g}(D, D) &= \boldsymbol{G}(D, D) - \boldsymbol{G}(D, M)\boldsymbol{G}^{-1}(M, M)\boldsymbol{G}(M, D) \\ &\quad + \boldsymbol{G}(D, M)\boldsymbol{G}^{-1}(M, M)\boldsymbol{g}(M, M)\boldsymbol{G}^{-1}(M, M)\boldsymbol{G}(M, D), \end{aligned} \tag{17.25}$$

where $\boldsymbol{G}(D, D)$ is defined by

$$\boldsymbol{G}(x, x') = \begin{cases} \boldsymbol{G}_i(x, x'), & \{x, x'\} \in D_i; \\ 0, & x \in D_i, \ x' \in D_{i'}. \end{cases} \tag{17.26}$$

Further, $\boldsymbol{G}(D, M)$ and $\boldsymbol{G}(M, D)$ are defined in the same manner and $\boldsymbol{G}^{-1}(M, M)$ is the inverse of the block diagonal square matrix $\boldsymbol{G}(M, M)$.

The new polariton states localized at the interfaces M for any composite dielectric are given by

$$\det \left| \boldsymbol{g}^{-1}(M, M) \right| = 0, \tag{17.27}$$

and the total variation of the polariton density of state $\Delta N(\omega^2)$ between the real composite dielectric and the reference one (described by the block-diagonal bulk response function $\boldsymbol{G}$, Eq. (17.26)) is given by [8–11]

$$\Delta N(\omega^2) = \frac{1}{\pi} \frac{d\eta(\omega^2)}{d\omega^2} + \frac{1}{\pi} \sum_{i=1}^{N} \Im \mathrm{Tr} \left[\boldsymbol{G}_i(D_i, M_i)\boldsymbol{G}_i^{-1}(M_i, M_i)\boldsymbol{G}_i(M_i, D_i) \right], \tag{17.28}$$

where the phase shift $\eta(\omega^2)$ is

$$\eta(\omega^2) = -\arg\det \left| \boldsymbol{g}^{-1}(M, M) \right|. \tag{17.29}$$

However, the last term of Eq. (17.28) has no physical meaning, as well as states associated with truncated matrices. So when calculating density of states variations, it is advised to use as a reference system finite elements with free surfaces; see [60]. Both approaches are equivalent as far as the continuous variation of the density of states is concerned. However, only the approach without truncated elements enables checking the number of states conservation rule.

Let us stress finally that if $|U(D)\rangle$ represents an eigenvector of the reference system, the eigenvector $|u(D)\rangle$ of the composite dielectric is given by [61]

$$\begin{aligned}|u(D)\rangle &= |U(D)\rangle - \boldsymbol{G}(D,M)\boldsymbol{G}^{-1}(M,M)|U(M)\rangle \\ &\quad + \boldsymbol{G}(D,M)\boldsymbol{G}^{-1}(M,M)\boldsymbol{g}(M,M)\boldsymbol{G}^{-1}(M,M)|U(M)\rangle.\end{aligned} \tag{17.30}$$

In particular, if $|U(D)\rangle$ represents a bulk wave launched in one homogeneous piece of the composite dielectric, Eq. (17.30) enables all the waves reflected and transmitted by the interfaces to be calculated.

Rather than expanding more on the general applications of the response functions given here for any composite dielectric, let us now illustrate these very general results by applying them to the study of layered isotropic dielectrics.

17.3 Basic equations for layered isotropic dielectrics

An important particular case of the analysis presented above is that of layered isotropic dielectrics. The isotropy of each subdielectric provides

$$\bar{\varepsilon}_i(\omega) = \varepsilon_i(\omega)\boldsymbol{I}, \quad i = 1, 2, \dots, N. \tag{17.31}$$

Further, assuming that the planar interfaces are perpendicular to the x_3 axis enables all the above operators to be Fourier analyzed according to, for example,

$$\boldsymbol{g}(x,x') = \int \frac{d^2k_{\parallel}}{(2\pi)^2}\boldsymbol{g}(k_{\parallel}/x_3, x_3')e^{ik_{\parallel}(x_{\parallel}-x'_{\parallel})}, \tag{17.32}$$

where $k_{\parallel}$ and $x_{\parallel}$ are the propagation and the real space position vectors parallel to the interface, respectively.

The equations of Section 17.2 remain unchanged for these Fourier coefficients. This transformation enables one to obtain simpler expressions for these Fourier coefficients, in particular, from Eqs. (17.15) and (17.9),

$$\boldsymbol{H}^0(k_{\parallel}/x_3) = \begin{pmatrix} k_2^2 - \frac{\partial^2}{\partial x_3^2} & -k_1k_2 & ik_1\frac{\partial}{\partial x_3} \\ -k_1k_2 & k_1^2 - \frac{\partial^2}{\partial x_3^2} & ik_2\frac{\partial}{\partial x_3} \\ ik_1\frac{\partial}{\partial x_3} & ik_2\frac{\partial}{\partial x_3} & k_{\parallel}^2 \end{pmatrix} \tag{17.33}$$

and

$$\boldsymbol{V}_i(k_{||}/x_3) = \begin{pmatrix} \frac{\partial}{\partial x_3} & 0 & -ik_1 \\ 0 & \frac{\partial}{\partial x_3} & -ik_2 \\ 0 & 0 & 0 \end{pmatrix}, \tag{17.34}$$

for $x_3 \geqslant 0$, when we take the planar surface at $x_3 = f(x_1, x_2) = 0$.

It is now convenient to take advantage of the isotropy of our system in the x_1x_2-plane by premultiplying and postmultiplying Eqs. (17.33) and (17.34) by the matrices $\boldsymbol{S}(k_{||})$ and $\boldsymbol{S}^{-1}(k_{||})$, respectively,

$$\boldsymbol{S}(k_{||}) = \frac{1}{k_{||}} \begin{pmatrix} k_1 & k_2 & 0 \\ -k_2 & k_1 & 0 \\ 0 & 0 & k_{||} \end{pmatrix} \tag{17.35}$$

and

$$\boldsymbol{S}(k_{||}) = \frac{1}{k_{||}} \begin{pmatrix} k_1 & -k_2 & 0 \\ k_2 & k_1 & 0 \\ 0 & 0 & k_{||} \end{pmatrix}. \tag{17.36}$$

This transforms $\boldsymbol{H}^0(k_{||}/x_3)$ and $\boldsymbol{V}_i(k_{||}/x_3)$, respectively, into

$$\boldsymbol{H}^0(k_{||}/x_3) = \begin{pmatrix} -\frac{\partial^2}{\partial x_3^2} & 0 & ik_{||}\frac{\partial}{\partial x_3} \\ 0 & k_{||}^2 - \frac{\partial^2}{\partial x_3^2} & 0 \\ ik_{||}\frac{\partial}{\partial x_3} & 0 & k_{||}^2 \end{pmatrix} \tag{17.37}$$

and

$$\boldsymbol{V}_i(k_{||}/x_3) = \begin{pmatrix} \frac{\partial}{\partial x_3} & 0 & -ik_{||} \\ 0 & \frac{\partial}{\partial x_3} & 0 \\ 0 & 0 & 0 \end{pmatrix}, \tag{17.38}$$

for $x_3 \geqslant 0$, which depends now only on the magnitude $k_{||}$ of the propagation vector (rather than k_1 and k_2 in Eqs. (17.33) and (17.34)).

Finally, Eqs. (17.16) and (17.17) become

$$\left(\frac{\omega^2}{c^2}\varepsilon_i(\omega)\boldsymbol{I} - \boldsymbol{H}^0(k_{||}/x_3)\right)\boldsymbol{G}_i(k_{||}, \omega/x_3, x_3') = \boldsymbol{I}\delta(x_3 - x_3') \tag{17.39}$$

and

$$\left[\theta(x_3)\left(\frac{\omega^2}{c^2}\varepsilon_i(\omega)\boldsymbol{I} - \boldsymbol{H}^0(k_{||}/x_3)\right) + \delta(x_3)\boldsymbol{V}_i(k_{||}/x_3)\right]\boldsymbol{g}_{si}(k_{||}, \omega/x_3, x_3') \\ = \boldsymbol{I}\delta(x_3 - x_3'). \tag{17.40}$$

The determination of the matrix elements of $\boldsymbol{G}_i$ is greatly simplified by the fact that $(G_i)_{21}$ and $(G_i)_{23}$ satisfy homogeneous equations, while $(G_i)_{12}$ and $(G_i)_{32}$ satisfy a pair of coupled homogeneous equations. These four matrix elements therefore vanish in $\boldsymbol{G}_i$ and also in $\boldsymbol{g}_{si}$ [12,62], and the 2×2 matrix elements decouple from the others. Eq. (17.39) becomes

$$\left(\frac{\omega^2}{c^2}\varepsilon_i(\omega) - k_{\|}^2 + \frac{\partial^2}{\partial x_3^2}\right)(G_i)_{22} = \delta(x_3 - x_3'), \tag{17.41}$$

and also

$$\begin{pmatrix} \frac{\omega^2}{c^2}\varepsilon_i(\omega) + \frac{\partial^2}{\partial x_3^2} & -ik_{\|}\frac{\partial}{\partial x_3} \\ -ik_{\|}\frac{\partial}{\partial x_3} & \frac{\omega^2}{c^2}\varepsilon_i(\omega) - k_{\|}^2 \end{pmatrix}\begin{pmatrix} (G_i)_{11} & (G_i)_{13} \\ (G_i)_{31} & (G_i)_{33} \end{pmatrix} = \boldsymbol{I}\delta(x_3 - x_3') \tag{17.42}$$

where $\boldsymbol{I}$ is the 2×2 unit matrix, and we omit, from now on, writing explicitly the dependence on $(k_{\|}, \omega/x_3, x_3')$ in the response functions. The same decoupling appears also in Eq. (17.40):

$$\left[\theta(x_3)\left(\frac{\omega^2}{c^2}\varepsilon_i(\omega) - k_{\|}^2 + \frac{\partial^2}{\partial x_3^2}\right) + \delta(x_3)\frac{\partial}{\partial x_3}\right](g_{si})_{22} = \delta(x_3 - x_3') \tag{17.43}$$

and

$$\begin{pmatrix} \frac{\omega^2}{c^2}\varepsilon_i(\omega) + \frac{\partial^2}{\partial x_3^2} + \delta(x_3)\frac{\partial}{\partial x_3} & -ik_{\|}\left(\frac{\partial}{\partial x_3} + \delta(x_3)\right) \\ -ik_{\|}\frac{\partial}{\partial x_3} & \frac{\omega^2}{c^2}\varepsilon_i(\omega) - k_{\|}^2 \end{pmatrix}\begin{pmatrix} (g_{si})_{11} & (g_{si})_{13} \\ (g_{si})_{31} & (g_{si})_{33} \end{pmatrix}$$
$$= \boldsymbol{I}\delta(x_3 - x_3'). \tag{17.44}$$

From Eq. (17.42), we see that $(G_i)_{31}$ and $(G_i)_{33}$ can be obtained simply from $(G_i)_{11}$ and $(G_i)_{13}$, and similarly the following relations exist between the same elements of $\boldsymbol{g}$, the response function of any such layered dielectric:

$$g_{31}(k_{\|}, \omega/x_3, x_3') = -\frac{ik_{\|}}{\alpha_i^2}\frac{\partial}{\partial x_3}g_{11}(k_{\|}, \omega/x_3, x_3'), \quad x_3 \in D_i, \tag{17.45}$$

$$g_{33}(k_{\|}, \omega/x_3, x_3') = -\frac{1}{\alpha_i^2}\left(\delta(x_3 - x_3') + ik_{\|}\frac{\partial}{\partial x_3}g_{13}(k_{\|}, \omega/x_3, x_3')\right), \quad x_3 \in D_i, \tag{17.46}$$

where

$$\alpha_i(k_{\|}, \omega) = \left(k_{\|}^2 - \frac{\omega^2}{c^2}\varepsilon_i(\omega)\right)^{1/2}. \tag{17.47}$$

Eqs. (17.45) and (17.46) hold for the corresponding matrix elements of the response function $\boldsymbol{g}_{si}$ of one dielectric bounded by a black-box surface, as well as for the response function $\boldsymbol{g}$ of any such layered composite material, provided that x_3 lies inside the slab i, as $\boldsymbol{g}$ has to satisfy Eq. (17.42) for $x_3 \in D_i$.

Consequently, we can in what follows regard g_{11} and g_{22} as the primary elements of $\boldsymbol{g}$ and g_{31} and g_{33} as derived elements through Eqs. (17.45) and (17.46). Eq. (17.40), when use is made of the general relations (17.45) and (17.46), can be rewritten as three independent equations for the 22, 11, and 13 elements of $\boldsymbol{g}_{si}$, namely

$$\left[\theta(x_3)\left(\frac{\partial^2}{\partial x_3^2} - \alpha_i^2\right) + \delta(x_3)\frac{\partial}{\partial x_3}\right](g_{si})_{22} = \delta(x_3 - x_3'), \tag{17.48}$$

$$-\frac{\omega^2 \varepsilon_i(\omega)}{\alpha_i^2 c^2}\left[\theta(x_3)\left(\frac{\partial^2}{\partial x_3^2} - \alpha_i^2\right) + \delta(x_3)\frac{\partial}{\partial x_3}\right](g_{si})_{11} = \delta(x_3 - x_3'), \tag{17.49}$$

$$\begin{aligned} &-\frac{i\omega^2 \varepsilon_i(\omega)}{k_{\|} c^2}\theta(x_3)\left(\frac{\partial^2}{\partial x_3^2} - \alpha_i^2\right)(g_{si})_{13} - \delta(x_3) \\ &\times\left(\delta(x_3 - x_3') + \frac{i\omega^2 \varepsilon_i(\omega)}{k_{\|} c^2}\frac{\partial}{\partial x_3}(g_{si})_{13}\right) = \frac{\partial}{\partial x_3}\delta(x_3 - x_3'). \end{aligned} \tag{17.50}$$

Comparing Eqs. (17.49) and (17.50), one sees that

$$(g_{si})_{13}(x_3, x_3') = \frac{ik_{\|}}{\alpha_i^2}\frac{\partial}{\partial x_3'}(g_{si})_{11}(x_3, x_3'), \quad x_3' \in D_i. \tag{17.51}$$

Relation (17.51) is a constitutive one, as well as Eqs. (17.45) and (17.46), and holds also for the g_{13} of any such layered dielectric:

$$g_{13}(x_3, x_3') = \frac{ik_{\|}}{\alpha_{i'}^2}\frac{\partial}{\partial x_3'}g_{11}(x_3, x_3'), \quad x_3' \in D_{i'}, \tag{17.52}$$

and $\alpha_{i'}$ takes the value given by Eq. (17.47) within $D_{i'}$.

So considering Eqs. (17.48), (17.49), and (17.50), one sees that all three are isomorphic to the same basic equation:

$$\frac{F_i}{\alpha_i}\left[\theta(x_3)\left(-\alpha_i^2 + \frac{\partial^2}{\partial x_3^2}\right) + \delta(x_3)\frac{\partial}{\partial x_3}\right]g_{si}(x_3, x_3') = \delta(x_3 - x_3'), \tag{17.53}$$

where

$$F_i = \alpha_i, \qquad \text{for } g_{22}, \tag{17.54}$$

and

$$F_i = -\frac{\omega^2}{c^2}\frac{\varepsilon_i(\omega)}{\alpha_i}, \qquad \text{for } g_{11}. \tag{17.55}$$

So it is sufficient to obtain the basic element of the response function $\boldsymbol{g}$, starting with Eq. (17.53). Then with the help of the relations (17.54) and (17.55) one obtains g_{22} and g_{11}. The g_{13}, g_{31}, and g_{33} are then easily obtained with the help of the constitutive relations (17.52), (17.45), and (17.46). Indeed, the above considerations are based only on the basic form of the bulk differential equations.

The basic response function G_i of an infinite homogeneous dielectric is easily found to be

$$G_i(k_{||}, \omega/x_3, x_3') = -\frac{e^{-\alpha_i|x_3-x_3'|}}{2F_i}. \tag{17.56}$$

Replacing now in Eq. (17.56) F_i by the values given by Eqs. (17.54) and (17.55) and using Eqs. (17.45), (17.46), and (17.52), one obtains the elements of the complete bulk response function $G(k_{||}, \omega/x_3, x_3')$:

$$G_{22}(k_{||}, \omega/x_3, x_3') = -\frac{e^{-\alpha_i|x_3-x_3'|}}{2\alpha_i}, \tag{17.57}$$

$$G_{11}(k_{||}, \omega/x_3, x_3') = \frac{c^2\alpha_i}{2\omega^2\varepsilon_i(\omega)}e^{-\alpha_i|x_3-x_3'|}, \tag{17.58}$$

$$G_{13}(k_{||}, \omega/x_3, x_3') = \frac{ic^2k_{||}}{2\omega^2\varepsilon_i(\omega)}\text{sgn}(x_3 - x_3')e^{-\alpha_i|x_3-x_3'|}, \tag{17.59}$$

$$G_{31}(k_{||}, \omega/x_3, x_3') = \frac{ic^2k_{||}}{2\omega^2\varepsilon_i(\omega)}\text{sgn}(x_3 - x_3')e^{-\alpha_i|x_3-x_3'|}, \tag{17.60}$$

$$G_{33}(k_{||}, \omega/x_3, x_3') = \frac{c^2}{\omega^2\varepsilon_i(\omega)}\delta(x_3 - x_3') - \frac{c^2k_{||}^2}{2\alpha_i\omega^2\varepsilon_i(\omega)}e^{-\alpha_i|x_3-x_3'|}. \tag{17.61}$$

17.4 Interface between two isotropic dielectrics

The elements of the surface response operator for the semiinfinite medium $i = 1$ ($x_3 > 0$) limited by the perfectly reflecting surface ($x_3 = 0$) are easily obtained from Eqs. (17.19) and (17.53) as

$$A_{s1}(0, x_3') = -\frac{1}{2}e^{-\alpha_1 x_3'}. \tag{17.62}$$

The inverse of the surface element $g_{s1}(0, 0)$ is then obtained with the help of Eqs. (17.20) and (17.21) to be

$$g_{s1}^{-1}(0, 0) = -F_1. \tag{17.63}$$

The system studied in this paragraph is constructed by two semiinfinite homogeneous isotropic dielectrics illustrated in Fig. 17.1. The interface element of the basic

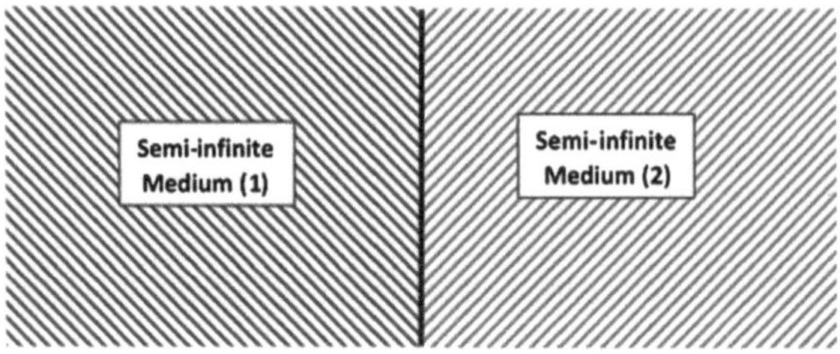

Figure 17.1 Schematic of the interface between two semiinfinite isotropic dielectric media 1 and 2, respectively.

response function is obtained from Eq. (17.24) as

$$g^{-1}(0,0) = g_{s1}^{-1}(0,0) + g_{s2}^{-1}(0,0). \tag{17.64}$$

Using result (17.63) and a similar one for $g_{s2}^{-1}(0,0)$, one obtains

$$g^{-1}(0,0) = -(F_1 + F_2). \tag{17.65}$$

The elements of the basic interface response function are easily calculated from Eqs. (17.25), (17.56), and (17.65):

$$g(k_{||}, \omega/x_3, x_3') = -\frac{1}{2F_2}\left(e^{-\alpha_2|x_3 - x_3'|} - \frac{F_1 - F_2}{F_1 + F_2} e^{-\alpha_2(x_3 + x_3')}\right), \quad x_3 > 0,\ x_3' > 0, \tag{17.66}$$

$$g(k_{||}, \omega/x_3, x_3') = -\frac{1}{F_1 + F_2} e^{\alpha_1 x_3 - \alpha_2 x_3'}, \quad x_3 < 0,\ x_3' > 0, \tag{17.67}$$

$$g(k_{||}, \omega/x_3, x_3') = -\frac{1}{F_1 + F_2} e^{-\alpha_2 x_3 + \alpha_1 x_3'}, \quad x_3 > 0,\ x_3' < 0, \tag{17.68}$$

$$g(k_{||}, \omega/x_3, x_3') = -\frac{1}{2F_1}\left(e^{-\alpha_1|x_3 - x_3'|} + \frac{F_1 - F_2}{F_1 + F_2} e^{\alpha_1(x_3 + x_3')}\right), \quad x_3 < 0,\ x_3' < 0. \tag{17.69}$$

Replacing now in Eqs. (17.66)–(17.69) F_1 and F_2 by their values given by Eqs. (17.54) and (17.55), and using Eqs. (17.45), (17.46), and (17.52), one obtains:

$$g_{22}(k_{||}, \omega/x_3, x_3') = -\frac{1}{2\alpha_2}\left(e^{-\alpha_2|x_3 - x_3'|} - \frac{\alpha_1 - \alpha_2}{\alpha_1 + \alpha_2} e^{-\alpha_2(x_3 + x_3')}\right), \quad x_3 > 0,\ x_3' > 0, \tag{17.70}$$

$$g_{22}(k_{||}, \omega/x_3, x_3') = -\frac{1}{\alpha_1 + \alpha_2} e^{\alpha_1 x_3 - \alpha_2 x_3'}, \quad x_3 < 0,\ x_3' > 0, \tag{17.71}$$

$$g_{11}(k_{||},\omega/x_3,x_3')=\frac{c^2\alpha_2}{2\omega^2\varepsilon_2}\left(e^{-\alpha_2|x_3-x_3'|}-\frac{\alpha_2\varepsilon_1-\alpha_1\varepsilon_2}{\alpha_2\varepsilon_1+\alpha_1\varepsilon_2}e^{-\alpha_2(x_3+x_3')}\right),\quad x_3>0,\ x_3'>0, \tag{17.72}$$

$$g_{11}(k_{||},\omega/x_3,x_3')=\frac{c^2\alpha_1\alpha_2}{\omega^2(\alpha_2\varepsilon_1+\alpha_1\varepsilon_2)}e^{\alpha_1x_3-\alpha_2x_3'},\qquad x_3<0,\ x_3'>0, \tag{17.73}$$

$$g_{13}(k_{||},\omega/x_3,x_3')=\frac{ik_{||}c^2}{2\omega^2\varepsilon_2}\left(e^{-\alpha_2|x_3-x_3'|}\operatorname{sgn}(x_3-x_3')+\frac{\alpha_2\varepsilon_1-\alpha_1\varepsilon_2}{\alpha_2\varepsilon_1+\alpha_1\varepsilon_2}e^{-\alpha_2(x_3+x_3')}\right),\quad x_3>0,\ x_3'>0, \tag{17.74}$$

$$g_{13}(k_{||},\omega/x_3,x_3')=-\frac{ik_{||}c^2\alpha_1}{\omega^2(\alpha_2\varepsilon_1+\alpha_1\varepsilon_2)}e^{\alpha_1x_3-\alpha_2x_3'},\qquad x_3<0,\ x_3'>0, \tag{17.75}$$

$$g_{31}(k_{||},\omega/x_3,x_3')=\frac{ik_{||}c^2}{2\omega^2\varepsilon_2}\left(e^{-\alpha_2|x_3-x_3'|}\operatorname{sgn}(x_3-x_3')-\frac{\alpha_2\varepsilon_1-\alpha_1\varepsilon_2}{\alpha_2\varepsilon_1+\alpha_1\varepsilon_2}e^{-\alpha_2(x_3+x_3')}\right),\quad x_3>0,\ x_3'>0, \tag{17.76}$$

$$g_{31}(k_{||},\omega/x_3,x_3')=-\frac{ik_{||}c^2\alpha_1}{\omega^2(\alpha_2\varepsilon_1+\alpha_1\varepsilon_2)}e^{\alpha_1x_3-\alpha_2x_3'},\qquad x_3<0,\ x_3'>0, \tag{17.77}$$

$$g_{33}(k_{||},\omega/x_3,x_3')=\frac{c^2}{\omega^2\varepsilon_2}\delta(x_3-x_3')-\frac{1}{\alpha_2}\frac{k_{||}^2c^2}{2\omega^2\varepsilon_2}\left(e^{-\alpha_2|x_3-x_3'|}+\frac{\alpha_2\varepsilon_1-\alpha_1\varepsilon_2}{\alpha_2\varepsilon_1+\alpha_1\varepsilon_2}e^{-\alpha_2(x_3+x_3')}\right),\quad x_3>0,\ x_3'>0, \tag{17.78}$$

$$g_{33}(k_{||},\omega/x_3,x_3')=-\frac{k_{||}^2c^2}{\omega^2(\alpha_2\varepsilon_1+\alpha_1\varepsilon_2)}e^{\alpha_1x_3-\alpha_2x_3'},\qquad x_3<0,\ x_3'>0. \tag{17.79}$$

The elements of $g_{\alpha\beta}(k_{||},\omega/x_3,x_3')$ for $x_3<0$ and $x_3'<0$ can be easily deduced from those for $x_3>0$ and $x_3'>0$ by permutations of the indices 2 and 1, together with a change of x_3 to $-x_3'$ and that of x_3' to $-x_3$. The elements of $g_{\alpha\beta}(k_{||},\omega/x_3,x_3')$ for $x_3>0$ and $x_3'<0$ can be deduced from those for $x_3<0$ and $x_3'>0$ by permuting the indices 2 and 1 and by changing x_3 to $-x_3$ and x_3' to $-x_3'$.

We see that the elements g_{11}, g_{13}, g_{31}, and g_{33} all contain terms with a factor $(\alpha_2\varepsilon_1+\alpha_1\varepsilon_2)^{-1}$. The corresponding equation,

$$\varepsilon_1(\omega)\alpha_2(k_{||},\omega)+\varepsilon_2(\omega)\alpha_1(k_{||},\omega)=0, \tag{17.80}$$

is just the well-known [12,62,63] dispersion relation for electromagnetic waves (interface polaritons) localized to the interface $x_3 = 0$.

For the special case of a dielectric/vacuum interface, for which $\varepsilon_2(\omega) = 1$, the results given by Eqs. (17.70)–(17.79) reduce to those obtained before (see [62]) by another method.

The response functions for that case have been used in calculations of Raman scattering of light from metal surfaces [64,65] and of Brillouin scattering from opaque surfaces [66]. They have also been used in the study of nonlinear interactions of surface polaritons [67,68], in calculations of the image force on a moving charge [69], in studies of the interaction of electromagnetic radiation with randomly rough surfaces [62,70], and in the determination of the dispersion relation for surface polaritons [70–73] and of their attenuation by such a surface [74]. This shows a few applications of these response functions.

As another useful application of the above response function, let us indicate here how to obtain the electric and magnetic fields associated with an incident wave reflected and transmitted by the interface. One can use the general equation (17.30) and work out from it the different components of these fields. But as the explicit expressions of the response functions were given above (Eqs. (17.66)–(17.79)), it is now more direct to obtain these electric and magnetic fields in the following manner. Assume the source point x_3' at $+\infty$, so that $x_3 < x_3'$. We can then, in order to get the incident, reflected, and transmitted waves, factorize out, in the useful expressions of $g_{\alpha\beta}(k_{||}, \omega/x_3, x_3')$, a common factor $e^{-\alpha_2 x_3'}$, and the remaining expressions provide us with the components of the electric field. More precisely, we obtain in this way the E_2 components of the s-wave from Eqs. (17.70)–(17.71) and the E_1 and E_3 components of the p-wave from Eqs. (17.72)–(17.73) and (17.76)–(17.77), respectively. Finally, using the Maxwell equation (17.2), one obtains the components of the magnetic fields H_1 and H_3 of the s-wave from:

$$H_1 = \frac{ic}{\omega}\frac{\partial E_2}{\partial x_3}, \tag{17.81}$$

$$H_3 = -\frac{ic}{\omega}\frac{\partial E_2}{\partial x_1}, \tag{17.82}$$

and the component H_2 of the p-wave from,

$$H_2 = -\frac{ic}{\omega}\left(\frac{\partial E_1}{\partial x_3} - \frac{\partial E_3}{\partial x_1}\right), \tag{17.83}$$

remembering that all these waves have the same propagating factor $e^{i(k_{||}x_1 - \omega t)}$ and that in this case $\alpha_2 = -i|\alpha_2|$.

It is then straightforward to obtain also the Poynting vector

$$\boldsymbol{S} = \frac{c}{8\pi}\left[\boldsymbol{E} \times \boldsymbol{H}\right]. \tag{17.84}$$

Finally, let us note also that the above results for $\boldsymbol{E}$ and $\boldsymbol{H}$ can be used for the fields associated with polaritons localized at the interface. In this case, one has to multiply the above values of the fields: for the s-wave by $\alpha_1 + \alpha_2 = 0$ and for the p-wave by $\alpha_1\varepsilon_2 + \alpha_2\varepsilon_1 = 0$. So we are left with exponentially decaying factors on both sides of the interface.

17.5 Dielectric slab bounded by two nonidentical dielectrics

Dielectric slabs are the simplest optical waveguides. Because of their simple geometry, guided and radiation modes of slab waveguides can be described by simple mathematical expressions. The study of slab waveguides and their properties is thus often useful in gaining an understanding of the waveguiding properties of more complicated dielectric waveguides. However, slab waveguides are not only useful as models for more general types of optical waveguides, they are also actually employed for light guidance in integrated optics circuits [75–88]. Planar or slab waveguides restrict light propagation in one dimension (1D waveguide), whilst channel or ridge waveguides confine the light in two dimensions (2D waveguide) [75,76]. The small size of waveguide structures offers high light intensities produced by even very low powers; consequently, the nonlinearities or laser actions in waveguides may be more efficient than those in bulk materials [77]. Moreover, the components based on waveguides may be connected together with optical fibers quite simply because the dimensions of waveguides are directly compatible with fibers used in telecommunications in which waveguides are often used as amplifiers, signal switches, modulators, etc. [78].

There are several techniques for fabricating waveguides in optical materials, which mainly include diffusion of metal ions [79], ion exchange [80,81], sol gel [82], ion implantation [83], and deposition of epitaxial layers (e.g., sputtering [84], molecular beam epitaxy [85], and pulsed laser deposition [86], chemical vapor deposition [87]). Among these techniques, ion implantation possesses one of the most advantageous characteristics, that is, the wide applicability of materials [88,89]. Since the first proton-implanted waveguide in fused silica was reported in 1968 [90], waveguides have so far been fabricated in more than 100 optical materials (including crystals, glasses, and even polymers) by implantation of various ions at the energies of several kilo-electron-volt (keV) up to several mega-electron-volt (MeV) [88–94]. Although the related topics have been studied for more than 30 years, there is no unique explanation for all materials, because of the diversity of configuration and properties, although there are some typical responses where we do understand the processes. However, there is one simple model, which was firstly raised by Townsend in 1986 as a prediction [95], that seems to be a reasonable explanation at least for most crystals. When implantation of light ions at energies of several MeV is performed, a majority of the lattice damage occurs at the end of the ion track inside the substrates, which results in a decrease of physical density by means of volume expansion and hence a reduced refractive index of an optical barrier. Such a barrier confines the light in a narrow layer with relatively high refractive index between itself and the crystal surface, forming an optical waveguide [88]. If the barrier is not adequately high or wide, the so-called tunneling effect may take place, accompanying leakage of light energy [96,97].

In this section, we calculate the response functions for a dielectric slab sandwiched between two different semiinfinite dielectric media. The dispersion relations are given analytically for this asymmetric structure and illustrated by some numerical results for guided electromagnetic waves.

17.5.1 The response function

We consider here a composite system formed of three subsystems, one of which is a slab layer and the two others are two semiinfinite dielectric media. The slab layer being sandwiched between the two other semiinfinite subsystems (see Fig. 17.2). We shall call these media 1, 2, and 3 such that their domains are for $x_3 \leqslant -L$, $-L \leqslant x_3 \leqslant L$, and $x_3 \geqslant L$, respectively.

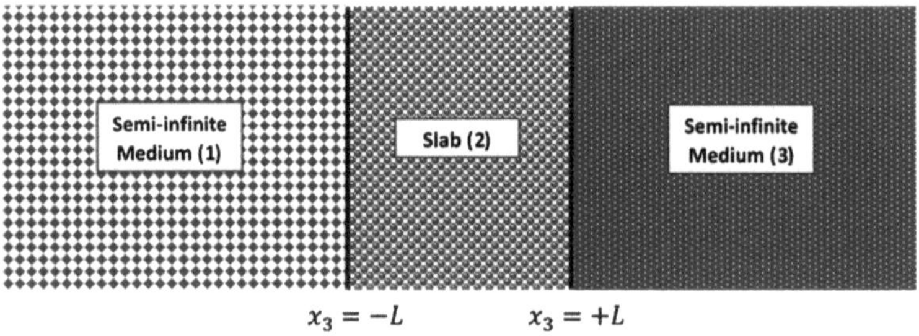

Figure 17.2 "Sandwich" system.

As indicated in Section 17.2 for the calculation of the elements of the response function $\boldsymbol{g}$, it is necessary to know the interface response function $\boldsymbol{g}(M, M)$. In fact, Eq. (17.23) or (17.24) enables us to obtain

$$\boldsymbol{g}^{-1}(M, M) = \begin{pmatrix} g_{s1}^{-1}(-L,-L) + g_{s2}^{-1}(-L,-L) & g_{s2}^{-1}(-L,L) \\ g_{s2}^{-1}(L,-L) & g_{s3}^{-1}(L,L) + g_{s2}^{-1}(L,L) \end{pmatrix}. \tag{17.85}$$

Taking account of the two perfectly reflecting surfaces of the slab at $x_3 = -L$ and $x_3 = +L$ and of the fact that the corresponding cleavage operator can be given by [15],

$$V_2(x_3) = [-\delta(L - x_3) + \delta(L + x_3)]\,\frac{\partial}{\partial x_3}, \tag{17.86}$$

one obtains from Eq. (17.19)

$$\boldsymbol{A}_{s2}(M, M) = -\frac{1}{2}\begin{pmatrix} 1 & e^{-2\alpha_2 L} \\ e^{-2\alpha_2 L} & 1 \end{pmatrix}, \tag{17.87}$$

which provides with the help of Eq. (17.21)

$$\mathbf{\Delta}_2(M,M)=\frac{1}{2}\begin{pmatrix}1 & -e^{-2\alpha_2 L}\\ e^{-2\alpha_2 L} & 1\end{pmatrix}. \tag{17.88}$$

Using Eq. (17.56), the inverse matrix of $\boldsymbol{G}_2(M,M)$ is obtained to be

$$\boldsymbol{G}_2^{-1}(M,M)=-\frac{F_2}{\sinh(2\alpha_2 L)}\begin{pmatrix}e^{2\alpha_2 L} & -1\\ -1 & e^{2\alpha_2 L}\end{pmatrix}. \tag{17.89}$$

Finally, from Eq. (17.20), one obtains

$$\boldsymbol{g}_{s2}^{-1}(M,M)=\frac{F_2}{\sinh(2\alpha_2 L)}\begin{pmatrix}-\cosh(2\alpha_2 L) & 1\\ 1 & -\cosh(2\alpha_2 L)\end{pmatrix}. \tag{17.90}$$

Let us recall from Eq. (17.63) that

$$g_{s3}^{-1}(L,L)=-F_3 \quad \text{and} \quad g_{s1}^{-1}(-L,-L)=-F_1. \tag{17.91}$$

Substituting relations (17.90) and (17.91) in (17.85) and inverting $\boldsymbol{g}^{-1}(M,M)$, one obtains

$$\boldsymbol{g}(M,M)=-\frac{1}{W}\begin{pmatrix}F_3+F_2\coth(2\alpha_2 L) & F_2/\sinh(2\alpha_2 L)\\ F_2/\sinh(2\alpha_2 L) & F_1+F_2\coth(2\alpha_2 L)\end{pmatrix}, \tag{17.92}$$

with

$$W=F_1F_3+F_2(F_1+F_3)\coth(2\alpha_2 L)+F_2^2. \tag{17.93}$$

Let us now derive all the elements of the response function $g(k_{\|},\omega/x_3,x_3')$.

(i) The source and observation points are in dielectric 1
In this case Eq. (17.25) reads

$$\begin{aligned}g(k_{\|},\omega/x_3,x_3')&=G_1(k_{\|},\omega/x_3,x_3')\\ &-G_1(k_{\|},\omega/x_3,-L)G_1^{-1}(k_{\|},\omega/-L,-L)G_1(k_{\|},\omega/-L,x_3')\\ &+G_1(k_{\|},\omega/x_3,-L)G_1^{-1}(k_{\|},\omega/-L,-L)\\ &\times g(k_{\|},\omega/-L,-L)G_1^{-1}(k_{\|},\omega/-L,-L)G_1(k_{\|},\omega/-L,x_3').\end{aligned} \tag{17.94}$$

Substituting Eqs. (17.56) and (17.92) in Eq. (17.94), one obtains for the basic element

$$\begin{aligned}g(k_{\|},\omega/x_3,x_3')&=-\frac{e^{-\alpha_1|x_3-x_3'|}}{2F_1}+\frac{e^{\alpha_1(x_3+x_3'+2L)}}{2F_1}\\ &-\frac{F_3+F_2\cosh(2\alpha_2 L)}{W(k_{\|},\omega)}e^{\alpha_1(x_3+x_3'+2L)}.\end{aligned} \tag{17.95}$$

Consequently, the components of the response function $\boldsymbol{g}$ are calculated with the help of Eqs. (17.54), (17.55), (17.45), (17.46), and (17.52) and have the following expressions:

$$g_{22}(k_{||},\omega/x_3,x_3') = -\frac{1}{2\alpha_1}\left(e^{-\alpha_1|x_3-x_3'|} - e^{\alpha_1(x_3+x_3'+2L)}\right) - \frac{\alpha_3+\alpha_2\cosh(2\alpha_2 L)}{\alpha_1\alpha_3+\alpha_2(\alpha_1+\alpha_3)\coth(2\alpha_2 L)+\alpha_2^2}e^{\alpha_1(x_3+x_3'+2L)}, \tag{17.96}$$

$$g_{11}(k_{||},\omega/x_3,x_3') = -\frac{\alpha_1 c^2}{2\omega^2\varepsilon_1}\left(e^{-\alpha_1|x_3-x_3'|} - e^{\alpha_1(x_3+x_3'+2L)}\right) + \frac{\alpha_1 c^2}{\omega^2}\left(\frac{\alpha_2\varepsilon_3+\alpha_3\varepsilon_2\cosh(2\alpha_2 L)}{\alpha_2\varepsilon_1\varepsilon_3+\varepsilon_2(\alpha_3\varepsilon_1+\alpha_1\varepsilon_3)\coth(2\alpha_2 L)+\varepsilon_2^2\alpha_1\alpha_3/\alpha_2}e^{\alpha_1(x_3+x_3'+2L)}\right), \tag{17.97}$$

$$g_{13}(k_{||},\omega/x_3,x_3') = \frac{ik_{||}c^2}{2\omega^2\varepsilon_1}\left(e^{-\alpha_1|x_3-x_3'|}\mathrm{sgn}(x_3-x_3') - e^{\alpha_1(x_3+x_3'+2L)}\right) + \frac{ik_{||}c^2}{\omega^2}\left(\frac{\alpha_2\varepsilon_3+\alpha_3\varepsilon_2\cosh(2\alpha_2 L)}{\alpha_2\varepsilon_1\varepsilon_3+\varepsilon_2(\alpha_3\varepsilon_1+\alpha_1\varepsilon_3)\coth(2\alpha_2 L)+\varepsilon_2^2\alpha_1\alpha_3/\alpha_2}e^{\alpha_1(x_3+x_3'+2L)}\right), \tag{17.98}$$

$$g_{31}(k_{||},\omega/x_3,x_3') = \frac{ik_{||}c^2}{2\omega^2\varepsilon_1}\left(e^{-\alpha_1|x_3-x_3'|}\mathrm{sgn}(x_3-x_3') + e^{\alpha_1(x_3+x_3'+2L)}\right) - \frac{ik_{||}c^2}{\omega^2}\left(\frac{\alpha_2\varepsilon_3+\alpha_3\varepsilon_2\cosh(2\alpha_2 L)}{\alpha_2\varepsilon_1\varepsilon_3+\varepsilon_2(\alpha_3\varepsilon_1+\alpha_1\varepsilon_3)\coth(2\alpha_2 L)+\varepsilon_2^2\alpha_1\alpha_3/\alpha_2}e^{\alpha_1(x_3+x_3'+2L)}\right), \tag{17.99}$$

$$g_{33}(k_{||},\omega/x_3,x_3') = \frac{c^2}{\omega^2\varepsilon_1}\delta(x_3-x_3') - \frac{k_{||}^2c^2}{2\alpha_1\omega^2\varepsilon_1}\left(e^{-\alpha_1|x_3-x_3'|} + e^{\alpha_1(x_3+x_3'+2L)}\right) + \frac{k_{||}^2c^2}{\alpha_1\omega^2}\left(\frac{\alpha_2\varepsilon_3+\alpha_3\varepsilon_2\cosh(2\alpha_2 L)}{\alpha_2\varepsilon_1\varepsilon_3+\varepsilon_2(\alpha_3\varepsilon_1+\alpha_1\varepsilon_3)\coth(2\alpha_2 L)+\varepsilon_2^2\alpha_1\alpha_3/\alpha_2}e^{\alpha_1(x_3+x_3'+2L)}\right). \tag{17.100}$$

In the case when the source and observation points are in dielectric 3, the elements of $g(k_{||},\omega/x_3,x_3')$ can be easily deduced from the above expressions (17.95)–(17.100) by obvious permutations of the indices 3 and 1, together with a change of x_3 to $-x_3'$ and that of x_3' to $-x_3$.

(ii) The source point is in dielectric 3 and the observation point in dielectric 1

We examine presently the case where the excitation is in the third medium and the response is observed in the first medium. In this case, Eq. (17.25) has the following form:

$$g(k_{||}, \omega/x_3, x_3') = G_3(k_{||}, \omega/x_3, L)G_3^{-1}(k_{||}, \omega/L, L)g(k_{||}, \omega/L, -L) \times G_1^{-1}(k_{||}, \omega/-L, -L)G_1(k_{||}, \omega/-L, x_3'). \tag{17.101}$$

Using the above results, one obtains

$$g(k_{||}, \omega/x_3, x_3') = -\frac{F_2}{W(k_{||}, \omega)\sinh(2\alpha_2 L)} e^{-\alpha_3(x_3-L)} e^{\alpha_1(L+x_3')}, \quad x_3 \geqslant L, \; x_3' \leqslant -L. \tag{17.102}$$

From this, one obtains as indicated above the components of $g(k_{||}, \omega/x_3, x_3')$ as

$$g_{22}(k_{||}, \omega/x_3, x_3') = -\frac{\alpha_2 e^{-\alpha_3(x_3-L)} e^{\alpha_1(L+x_3')}}{\left[\alpha_1\alpha_3 + \alpha_2(\alpha_1+\alpha_3)\coth(2\alpha_2 L) + \alpha_2^2\right]\sinh(2\alpha_2 L)}, \tag{17.103}$$

$$g_{11}(k_{||}, \omega/x_3, x_3') = \frac{c^2}{\omega^2 \sinh(2\alpha_2 L)} \times \frac{\alpha_1\alpha_3\varepsilon_2 e^{-\alpha_3(x_3-L)} e^{\alpha_1(L+x_3')}}{\alpha_2\varepsilon_1\varepsilon_3 + \varepsilon_2(\alpha_3\varepsilon_1 + \alpha_1\varepsilon_3)\coth(2\alpha_2 L) + \varepsilon_2^2\alpha_1\alpha_3/\alpha_2}, \tag{17.104}$$

$$g_{13}(k_{||}, \omega/x_3, x_3') = \frac{ik_{||}c^2}{\omega^2 \sinh(2\alpha_2 L)} \times \frac{\alpha_3\varepsilon_2 e^{-\alpha_3(x_3-L)} e^{\alpha_1(L+x_3')}}{\alpha_2\varepsilon_1\varepsilon_3 + \varepsilon_2(\alpha_3\varepsilon_1 + \alpha_1\varepsilon_3)\coth(2\alpha_2 L) + \varepsilon_2^2\alpha_1\alpha_3/\alpha_2}, \tag{17.105}$$

$$g_{31}(k_{||}, \omega/x_3, x_3') = \frac{ik_{||}c^2}{\omega^2 \sinh(2\alpha_2 L)} \times \frac{\alpha_1\varepsilon_2 e^{-\alpha_3(x_3-L)} e^{\alpha_1(L+x_3')}}{\alpha_2\varepsilon_1\varepsilon_3 + \varepsilon_2(\alpha_3\varepsilon_1 + \alpha_1\varepsilon_3)\coth(2\alpha_2 L) + \varepsilon_2^2\alpha_1\alpha_3/\alpha_2}, \tag{17.106}$$

$$g_{33}(k_{||}, \omega/x_3, x_3') = -\frac{1}{\alpha_3^2}\delta(x_3 - x_3') - \frac{k_{||}^2 c^2}{\omega^2 \sinh(2\alpha_2 L)} \times \frac{\varepsilon_2 e^{-\alpha_3(x_3-L)} e^{\alpha_1(L+x_3')}}{\alpha_2\varepsilon_1\varepsilon_3 + \varepsilon_2(\alpha_3\varepsilon_1 + \alpha_1\varepsilon_3)\coth(2\alpha_2 L) + \varepsilon_2^2\alpha_1\alpha_3/\alpha_2}. \tag{17.107}$$

When the source point is in dielectric 1 and the observation point in dielectric 3 ($x_3 \leqslant -L$ and $x_3' \geqslant L$), the $g_{\alpha\beta}(k_{||}, \omega/x_3, x_3')$ are obtained from Eqs. (17.102)–(17.107) just by permuting the indices 1 and 3 and by changing x_3 to $-x_3$ and x_3' to $-x_3'$.

(iii) The source point is in dielectric 1 and the observation point in dielectric 2

We consider the case where one realizes the excitation in the first medium and observes the response in the slab layer, and conversely. In this case, from Eq. (17.25), we obtain for the basic element

$$
\begin{aligned}
g(k_{||},\omega/x_3,x_3') &= G_1(k_{||},\omega/x_3,-L)G_1^{-1}(k_{||},\omega/-L,-L)\\
&\times \big(g(k_{||},\omega/-L,-L),\, g(k_{||},\omega/-L,L)\big)\\
&\times \begin{pmatrix} G_2(k_{||},\omega/-L,-L) & G_2(k_{||},\omega/-L,L)\\ G_2(k_{||},\omega/L,-L) & G_2(k_{||},\omega/L,L)\end{pmatrix}\begin{pmatrix} G_2(k_{||},\omega/-L,x_3')\\ G_2(k_{||},\omega/L,x_3')\end{pmatrix}.
\end{aligned}
\tag{17.108}
$$

When developing these products and using the values given by Eqs. (17.56) and (17.92), one obtains

$$
\begin{aligned}
g(k_{||},\omega/x_3,x_3') &= -\frac{e^{\alpha_1(x_3+L)}}{W(k_{||},\omega)\sinh(2\alpha_2 L)}\\
&\times\left([F_3+F_2\coth(2\alpha_2 L)]\sinh[\alpha_2(L-x_3')] + \frac{F_2}{\sinh(2\alpha_2 L)}\sinh[\alpha_2(L+x_3')]\right),\\
&x_3\leqslant -L,\quad -L\leqslant x_3'\leqslant L.
\end{aligned}
\tag{17.109}
$$

When the source point is in dielectric 2 and the observation point in dielectric 1, similarly we obtain

$$
\begin{aligned}
g(k_{||},\omega/x_3,x_3') &= -\frac{e^{\alpha_1(L+x_3')}}{W(k_{||},\omega)\sinh(2\alpha_2 L)}\\
&\times\left([F_3+F_2\coth(2\alpha_2 L)]\sinh[\alpha_2(L-x_3)] + \frac{F_2}{\sinh(2\alpha_2 L)}\sinh[\alpha_2(L+x_3)]\right),\\
&x_3'\leqslant -L,\quad -L\leqslant x_3\leqslant L.
\end{aligned}
\tag{17.110}
$$

Calculation of the components g_{22}, g_{11}, g_{13}, g_{31}, and g_{33} of $g(k_{||},\omega/x_3,x_3')$ gives the following results:

$$
\begin{aligned}
g_{22}(k_{||},\omega/x_3,x_3') &= -\frac{e^{\alpha_1(x_3+L)}}{\left[\alpha_1\alpha_3+\alpha_2(\alpha_1+\alpha_3)\coth(2\alpha_2 L)+\alpha_2^2\right]\sinh(2\alpha_2 L)}\\
&\times\left([\alpha_3+\alpha_2\coth(2\alpha_2 L)]\sinh[\alpha_2(L-x_3')] + \frac{\alpha_2}{\sinh(2\alpha_2 L)}\sinh[\alpha_2(L+x_3')]\right),\\
&x_3\leqslant -L,\quad -L\leqslant x_3'\leqslant L,
\end{aligned}
\tag{17.111}
$$

$$g_{22}(k_{||},\omega/x_3,x_3') = -\frac{e^{\alpha_1(L+x_3')}}{\left[\alpha_1\alpha_3+\alpha_2(\alpha_1+\alpha_3)\coth(2\alpha_2 L)+\alpha_2^2\right]\sinh(2\alpha_2 L)}$$
$$\times\left([\alpha_3+\alpha_2\coth(2\alpha_2 L)]\sinh[\alpha_2(L-x_3)]+\frac{\alpha_2}{\sinh(2\alpha_2 L)}\sinh[\alpha_2(L+x_3)]\right),$$
$$x_3'\leqslant -L,\quad -L\leqslant x_3\leqslant L, \tag{17.112}$$

$$g_{11}(k_{||},\omega/x_3,x_3') = \frac{e^{\alpha_1(L+x_3)}}{\left[\alpha_2\varepsilon_1\varepsilon_3+\varepsilon_2(\alpha_1\varepsilon_3+\alpha_3\varepsilon_1)\coth(2\alpha_2 L)+(\alpha_1\alpha_3/\alpha_2)\varepsilon_2^2\right]}$$
$$\times\frac{c^2}{\omega^2\sinh(2\alpha_2 L)}\left([\alpha_2\varepsilon_3+\alpha_3\varepsilon_2\coth(2\alpha_2 L)]\sinh[\alpha_2(L-x_3')]\right.$$
$$\left.+\frac{\alpha_3\varepsilon_2}{\sinh(2\alpha_2 L)}\sinh[\alpha_2(L+x_3')]\right),$$
$$x_3\leqslant -L,\quad -L\leqslant x_3'\leqslant L, \tag{17.113}$$

$$g_{11}(k_{||},\omega/x_3,x_3') = \frac{e^{\alpha_1(L+x_3')}}{\left[\alpha_2\varepsilon_1\varepsilon_3+\varepsilon_2(\alpha_1\varepsilon_3+\alpha_3\varepsilon_1)\coth(2\alpha_2 L)+(\alpha_1\alpha_3/\alpha_2)\varepsilon_2^2\right]}$$
$$\times\frac{c^2}{\omega^2\sinh(2\alpha_2 L)}\left([\alpha_2\varepsilon_3+\alpha_3\varepsilon_2\coth(2\alpha_2 L)]\sinh[\alpha_2(L-x_3)]\right.$$
$$\left.+\frac{\alpha_3\varepsilon_2}{\sinh(2\alpha_2 L)}\sinh[\alpha_2(L+x_3)]\right),$$
$$x_3'\leqslant -L,\quad -L\leqslant x_3\leqslant L, \tag{17.114}$$

$$g_{13}(k_{||},\omega/x_3,x_3') = -\frac{e^{\alpha_1(L+x_3)}}{\left[\alpha_2\varepsilon_1\varepsilon_3+\varepsilon_2(\alpha_1\varepsilon_3+\alpha_3\varepsilon_1)\coth(2\alpha_2 L)+(\alpha_1\alpha_3/\alpha_2)\varepsilon_2^2\right]}$$
$$\times\frac{ik_{||}c^2}{\omega^2\alpha_2\sinh(2\alpha_2 L)}\left([\alpha_2\varepsilon_3+\alpha_3\varepsilon_2\coth(2\alpha_2 L)]\cosh[\alpha_2(L-x_3')]\right.$$
$$\left.-\frac{\alpha_3\varepsilon_2}{\sinh(2\alpha_2 L)}\cosh[\alpha_2(L+x_3')]\right),$$
$$x_3\leqslant -L,\quad -L\leqslant x_3'\leqslant L, \tag{17.115}$$

$$g_{13}(k_{||},\omega/x_3,x_3') = \frac{e^{\alpha_1(L+x_3')}}{\left[\alpha_2\varepsilon_1\varepsilon_3+\varepsilon_2(\alpha_1\varepsilon_3+\alpha_3\varepsilon_1)\coth(2\alpha_2 L)+(\alpha_1\alpha_3/\alpha_2)\varepsilon_2^2\right]}$$
$$\times\frac{ik_{||}c^2}{\omega^2\sinh(2\alpha_2 L)}\left([\alpha_2\varepsilon_3+\alpha_3\varepsilon_2\coth(2\alpha_2 L)]\sinh[\alpha_2(L-x_3)]\right.$$

$$+\frac{\alpha_3\varepsilon_2}{\sinh(2\alpha_2 L)}\sinh[\alpha_2(L+x_3)]\Bigg),$$

$$x_3'\leqslant -L,\quad -L\leqslant x_3\leqslant L, \tag{17.116}$$

$$g_{31}(k_{\|},\omega/x_3,x_3')=-\frac{e^{\alpha_1(L+x_3)}}{[\alpha_2\varepsilon_1\varepsilon_3+\varepsilon_2(\alpha_1\varepsilon_3+\alpha_3\varepsilon_1)\coth(2\alpha_2 L)+(\alpha_1\alpha_3/\alpha_2)\varepsilon_2^2]}$$

$$\times\frac{ik_{\|}c^2}{\omega^2\sinh(2\alpha_2 L)}\Bigg([\alpha_2\varepsilon_3+\alpha_3\varepsilon_2\coth(2\alpha_2 L)]\sinh[\alpha_2(L-x_3')]$$

$$+\frac{\alpha_3\varepsilon_2}{\sinh(2\alpha_2 L)}\sinh[\alpha_2(L+x_3')]\Bigg),$$

$$x_3\leqslant -L,\quad -L\leqslant x_3'\leqslant L, \tag{17.117}$$

$$g_{13}(k_{\|},\omega/x_3,x_3')=\frac{e^{\alpha_1(L+x_3')}}{[\alpha_2\varepsilon_1\varepsilon_3+\varepsilon_2(\alpha_1\varepsilon_3+\alpha_3\varepsilon_1)\coth(2\alpha_2 L)+(\alpha_1\alpha_3/\alpha_2)\varepsilon_2^2]}$$

$$\times\frac{ik_{\|}c^2}{\omega^2\alpha_2\sinh(2\alpha_2 L)}\Bigg([\alpha_2\varepsilon_3+\alpha_3\varepsilon_2\coth(2\alpha_2 L)]\cosh[\alpha_2(L-x_3)]$$

$$-\frac{\alpha_3\varepsilon_2}{\sinh(2\alpha_2 L)}\cosh[\alpha_2(L+x_3)]\Bigg),$$

$$x_3'\leqslant -L,\quad -L\leqslant x_3\leqslant L, \tag{17.118}$$

$$g_{33}(k_{\|},\omega/x_3,x_3')=-\frac{1}{\alpha_1^2}\delta(x_3-x_3')-\frac{k_{\|}^2c^2}{\omega^2\alpha_2\sinh(2\alpha_2 L)}$$

$$\times\frac{e^{\alpha_1(L+x_3)}}{[\alpha_2\varepsilon_1\varepsilon_3+\varepsilon_2(\alpha_1\varepsilon_3+\alpha_3\varepsilon_1)\coth(2\alpha_2 L)+(\alpha_1\alpha_3/\alpha_2)\varepsilon_2^2]}$$

$$\times\Bigg([\alpha_2\varepsilon_3+\alpha_3\varepsilon_2\coth(2\alpha_2 L)]\cosh[\alpha_2(L-x_3')] \tag{17.119}$$

$$-\frac{\alpha_3\varepsilon_2}{\sinh(2\alpha_2 L)}\cosh[\alpha_2(L+x_3')]\Bigg),$$

$$x_3\leqslant -L,\quad -L\leqslant x_3'\leqslant L,$$

$$g_{33}(k_{\|},\omega/x_3,x_3')=-\frac{1}{\alpha_2^2}\delta(x_3-x_3')-\frac{k_{\|}^2c^2}{\omega^2\alpha_2\sinh(2\alpha_2 L)}$$

$$\times\frac{e^{\alpha_1(L+x_3')}}{[\alpha_2\varepsilon_1\varepsilon_3+\varepsilon_2(\alpha_1\varepsilon_3+\alpha_3\varepsilon_1)\coth(2\alpha_2 L)+(\alpha_1\alpha_3/\alpha_2)\varepsilon_2^2]}$$

$$\times \left([\alpha_2\varepsilon_3 + \alpha_3\varepsilon_2 \coth(2\alpha_2 L)] \cosh[\alpha_2(L - x_3)] - \frac{\alpha_3\varepsilon_2}{\sinh(2\alpha_2 L)} \cosh[\alpha_2(L + x_3)] \right),$$
$$x_3' \leqslant -L, \quad -L \leqslant x_3 \leqslant L. \tag{17.120}$$

When the source point is in dielectric 3 and the observation point is in dielectric 2, the $g_{\alpha\beta}(k_{\|}, \omega/x_3, x_3')$ can be obtained from Eqs. (17.109)–(17.120) by interchanging the indices 1 and 3 and by replacing x_3 by $-x_3'$ and x_3' by $-x_3$. The same transformation can also be used on Eqs. (17.109)–(17.120) when the source point is in dielectric 2 and the observation point in dielectric 3.

(iv) The source and the observation points are in dielectric 2

In this case, Eq. (17.25) becomes

$$\begin{aligned}
g(k_{\|}, \omega/x_3, x_3') &= G_2(k_{\|}, \omega/x_3, x_3') - \big(G_2(k_{\|}, \omega/x_3, -L), G_2(k_{\|}, \omega/x_3, L)\big) \\
&\times \begin{pmatrix} G_2(k_{\|}, \omega/-L, -L) & G_2(k_{\|}, \omega/-L, L) \\ G_2(k_{\|}, \omega/L, -L) & G_2(k_{\|}, \omega/L, L) \end{pmatrix}^{-1} \begin{pmatrix} G_2(k_{\|}, \omega/-L, x_3') \\ G_2(k_{\|}, \omega/L, x_3') \end{pmatrix} \\
&+ \big(G_2(k_{\|}, \omega/x_3, -L), G_2(k_{\|}, \omega/x_3, L)\big) \\
&\times \begin{pmatrix} G_2(k_{\|}, \omega/-L, -L) & G_2(k_{\|}, \omega/-L, L) \\ G_2(k_{\|}, \omega/L, -L) & G_2(k_{\|}, \omega/L, L) \end{pmatrix}^{-1} \\
&\times \begin{pmatrix} g(k_{\|}, \omega/-L, -L) & g(k_{\|}, \omega/-L, L) \\ g(k_{\|}, \omega/L, -L) & g(k_{\|}, \omega/L, L) \end{pmatrix} \\
&\times \begin{pmatrix} G_2(k_{\|}, \omega/-L, -L) & G_2(k_{\|}, \omega/-L, L) \\ G_2(k_{\|}, \omega/L, -L) & G_2(k_{\|}, \omega/L, L) \end{pmatrix}^{-1} \begin{pmatrix} G_2(k_{\|}, \omega/-L, x_3') \\ G_2(k_{\|}, \omega/L, x_3') \end{pmatrix}
\end{aligned} \tag{17.121}$$

or

$$\begin{aligned}
g(k_{\|}, \omega/x_3, x_3') &= \frac{e^{-\alpha_2|x_3 - x_3'|}}{2F_2} + \frac{1}{2F_2 \sinh(2\alpha_2 L)} \\
&\times \left(\cosh[\alpha_2(x_3 + x_3')] - e^{-2\alpha_2 L} \cosh[\alpha_2(x_3 - x_3')]\right) \\
&- \frac{1}{2W \sinh^2(2\alpha_2 L)} \Big([F_3 + F_2 \cosh(2\alpha_2 L)] \\
&\times \big[\cosh[\alpha_2(x_3 + x_3' - 2L)] - \cosh[\alpha_2(x_3 - x_3')]\big] \\
&+ [F_1 + F_2 \cosh(2\alpha_2 L)]\big[\cosh[\alpha_2(x_3 + x_3' + 2L)] - \cosh[\alpha_2(x_3 - x_3')]\big] \\
&+ \frac{F_2}{\sinh(2\alpha_2 L)} \big[\cosh[\alpha_2(x_3 - x_3' - 2L)] \\
&- 2\cosh[\alpha_2(x_3 + x_3')] + \cosh[\alpha_2(x_3 - x_3' + 2L)]\big] \Big).
\end{aligned} \tag{17.122}$$

Calculation of the components $g_{22}(k_{||}, \omega/x_3, x_3')$, $g_{11}(k_{||}, \omega/x_3, x_3')$, $g_{13}(k_{||}, \omega/x_3, x_3')$, $g_{31}(k_{||}, \omega/x_3, x_3')$, and $g_{33}(k_{||}, \omega/x_3, x_3')$ of $g(k_{||}, \omega/x_3, x_3')$ gives the following results:

$$\begin{aligned}
g_{22}(k_{||}, \omega/x_3, x_3') = &-\frac{1}{2\alpha_2}\Bigg(e^{-\alpha_2|x_3-x_3'|} - \frac{1}{\sinh(2\alpha_2 L)}\big(\cosh[\alpha_2(x_3+x_3')] \\
&- e^{-2\alpha_2 L}\cosh[\alpha_2(x_3-x_3')]\big)\Bigg) \\
&- \frac{1}{2\left[\alpha_1\alpha_3 + \alpha_2(\alpha_1+\alpha_3)\coth(2\alpha_2 L) + \alpha_2^2\right]\sinh^2(2\alpha_2 L)} \\
&\times \Bigg([\alpha_3 + \alpha_2\cosh(2\alpha_2 L)]\big[\cosh[\alpha_2(x_3+x_3'-2L)] - \cosh[\alpha_2(x_3-x_3')]\big] \\
&+ [\alpha_1 + \alpha_2\cosh(2\alpha_2 L)]\big[\cosh[\alpha_2(x_3+x_3'+2L)] - \cosh[\alpha_2(x_3-x_3')]\big] \\
&+ \frac{\alpha_2}{\sinh(2\alpha_2 L)}\big[\cosh[\alpha_2(x_3-x_3'-2L)] - 2\cosh[\alpha_2(x_3+x_3')] \\
&+ \cosh[\alpha_2(x_3-x_3'+2L)]\big]\Bigg),
\end{aligned} \tag{17.123}$$

$$\begin{aligned}
g_{11}(k_{||}, \omega/x_3, x_3') = &\frac{\alpha_2 c^2}{2\omega^2\varepsilon_2}\Bigg(e^{-\alpha_2|x_3-x_3'|} - \frac{1}{\sinh(2\alpha_2 L)}\big(\cosh[\alpha_2(x_3+x_3')] \\
&- e^{-2\alpha_2 L}\cosh[\alpha_2(x_3-x_3')]\big)\Bigg) \\
&+ \frac{\alpha_1\alpha_3 c^2}{2\omega^2\sinh^2(2\alpha_2 L)\left[\alpha_2\varepsilon_1\varepsilon_3 + \varepsilon_2(\alpha_1\varepsilon_3 + \alpha_3\varepsilon_1)\coth(2\alpha_2 L) + (\alpha_1\alpha_3/\alpha_2)\varepsilon_2^2\right]} \\
&\times \Bigg(\frac{\alpha_2\varepsilon_3 + \alpha_3\varepsilon_2\cosh(2\alpha_2 L)}{\alpha_3}\big[\cosh[\alpha_2(x_3+x_3'-2L)] - \cosh[\alpha_2(x_3-x_3')]\big] \\
&+ \frac{\alpha_2\varepsilon_1 + \alpha_1\varepsilon_2\cosh(2\alpha_2 L)}{\alpha_1}\big[\cosh[\alpha_2(x_3+x_3'+2L)] - \cosh[\alpha_2(x_3-x_3')]\big] \\
&+ \frac{\varepsilon_2}{\sinh(2\alpha_2 L)}\big[\cosh[\alpha_2(x_3-x_3'-2L)] - 2\cosh[\alpha_2(x_3+x_3')] \\
&+ \cosh[\alpha_2(x_3-x_3'+2L)]\big]\Bigg),
\end{aligned} \tag{17.124}$$

$$\begin{aligned}
g_{13}(k_{||}, \omega/x_3, x_3') = &\frac{ik_{||}c^2}{2\omega^2\varepsilon_2}\Bigg(\mathrm{sgn}(x_3-x_3')e^{-\alpha_2|x_3-x_3'|} \\
&- \frac{1}{\sinh(2\alpha_2 L)}\big(\sinh[\alpha_2(x_3+x_3')] + e^{-2\alpha_2 L}\sinh[\alpha_2(x_3-x_3')]\big)\Bigg)
\end{aligned}$$

$$+\frac{ik_{||}\alpha_1\alpha_3c^2}{2\omega^2\alpha_2\sinh^2(2\alpha_2L)\left[\alpha_2\varepsilon_1\varepsilon_3+\varepsilon_2(\alpha_1\varepsilon_3+\alpha_3\varepsilon_1)\coth(2\alpha_2L)+(\alpha_1\alpha_3/\alpha_2)\varepsilon_2^2\right]}$$
$$\times\left(\frac{\alpha_2\varepsilon_3+\alpha_3\varepsilon_2\cosh(2\alpha_2L)}{\alpha_3}\left[\sinh[\alpha_2(x_3+x_3'-2L)]+\sinh[\alpha_2(x_3-x_3')]\right]\right.$$
$$+\frac{\alpha_2\varepsilon_1+\alpha_1\varepsilon_2\cosh(2\alpha_2L)}{\alpha_1}\left[\sinh[\alpha_2(x_3+x_3'+2L)]+\sinh[\alpha_2(x_3-x_3')]\right]$$
$$-\frac{\varepsilon_2}{\sinh(2\alpha_2L)}\left[\sinh[\alpha_2(x_3-x_3'-2L)]+2\sinh[\alpha_2(x_3+x_3')]\right.$$
$$\left.\left.+\sinh[\alpha_2(x_3-x_3'+2L)]\right]\right),\tag{17.125}$$

$$g_{31}(k_{||},\omega/x_3,x_3')=\frac{ik_{||}c^2}{2\omega^2\varepsilon_2}\left(\mathrm{sgn}(x_3-x_3')e^{-\alpha_2|x_3-x_3'|}\right.$$
$$\left.+\frac{1}{\sinh(2\alpha_2L)}\left(\sinh[\alpha_2(x_3+x_3')]-e^{-2\alpha_2L}\sinh[\alpha_2(x_3-x_3')]\right)\right)$$
$$-\frac{ik_{||}\alpha_1\alpha_3c^2}{2\omega^2\alpha_2\sinh^2(2\alpha_2L)\left[\alpha_2\varepsilon_1\varepsilon_3+\varepsilon_2(\alpha_1\varepsilon_3+\alpha_3\varepsilon_1)\coth(2\alpha_2L)+(\alpha_1\alpha_3/\alpha_2)\varepsilon_2^2\right]}$$
$$\times\left(\frac{\alpha_2\varepsilon_3+\alpha_3\varepsilon_2\cosh(2\alpha_2L)}{\alpha_3}\left[\sinh[\alpha_2(x_3+x_3'-2L)]-\sinh[\alpha_2(x_3-x_3')]\right]\right.$$
$$+\frac{\alpha_2\varepsilon_1+\alpha_1\varepsilon_2\cosh(2\alpha_2L)}{\alpha_1}\left[\sinh[\alpha_2(x_3+x_3'+2L)]-\sinh[\alpha_2(x_3-x_3')]\right]$$
$$+\frac{\varepsilon_2}{\sinh(2\alpha_2L)}\left[\sinh[\alpha_2(x_3-x_3'-2L)]-2\sinh[\alpha_2(x_3+x_3')]\right.$$
$$\left.\left.+\sinh[\alpha_2(x_3-x_3'+2L)]\right]\right),\tag{17.126}$$

$$g_{33}(k_{||},\omega/x_3,x_3')=\frac{c^2}{\omega^2\varepsilon_2}\delta(x_3-x_3')-\frac{k_{||}^2c^2}{2\omega^2\alpha_2\varepsilon_2}\left(e^{-\alpha_2|x_3-x_3'|}+\frac{1}{\sinh(2\alpha_2L)}\right.$$
$$\left.\times\left(\cosh[\alpha_2(x_3+x_3')]+e^{-2\alpha_2L}\cosh[\alpha_2(x_3-x_3')]\right)\right)$$
$$+\frac{k_{||}c^2\alpha_1\alpha_3}{2\omega^2\alpha_2^2\sinh^2(2\alpha_2L)\left[\alpha_2\varepsilon_1\varepsilon_3+\varepsilon_2(\alpha_1\varepsilon_3+\alpha_3\varepsilon_1)\coth(2\alpha_2L)+(\alpha_1\alpha_3/\alpha_2)\varepsilon_2^2\right]}$$
$$\times\left(\frac{\alpha_2\varepsilon_3+\alpha_3\varepsilon_2\cosh(2\alpha_2L)}{\alpha_3}\left[\cosh[\alpha_2(x_3+x_3'-2L)]+\cosh[\alpha_2(x_3-x_3')]\right]\right.$$
$$+\frac{\alpha_2\varepsilon_1+\alpha_1\varepsilon_2\cosh(2\alpha_2L)}{\alpha_1}\left[\cosh[\alpha_2(x_3+x_3'+2L)]+\cosh[\alpha_2(x_3-x_3')]\right]$$

$$- \frac{\varepsilon_2}{\sinh(2\alpha_2 L)}\big[\cosh[\alpha_2(x_3 - x_3' - 2L)] + 2\cosh[\alpha_2(x_3 + x_3')]$$
$$+ \cosh[\alpha_2(x_3 - x_3' + 2L)]\big]\Bigg). \tag{17.127}$$

17.5.2 The confined photons

The knowledge of the above response functions enables us to calculate for the systems described above all their optical properties. However, here we shall only illustrate their usefulness by a study of confined photons.

In the denominator of all the elements of this function, the term $[F_1 F_3 + F_2(F_1 + F_3)\coth(2\alpha_2 L) + F_2^2]$ appears. So the photons confined in slab 2 can be obtained when the two different values (the F_i take two different values, see Eqs. (17.54) and (17.55)) of $[F_1 F_3 + F_2(F_1 + F_3)\coth(2\alpha_2 L) + F_2^2]$ vanish. More precisely, from g_{22} one obtains the condition for possible interface polaritons to be

$$\begin{aligned}\alpha_1(k_{||},\omega)\alpha_3(k_{||},\omega) + \alpha_2(k_{||},\omega)\big[\alpha_1(k_{||},\omega) + \alpha_3(k_{||},\omega)\big]\\ \times \coth[2\alpha_2(k_{||},\omega)L] + \alpha_2^2(k_{||},\omega) = 0.\end{aligned} \tag{17.128}$$

From the four other nonzero elements of $\boldsymbol{g}$, one has

$$\begin{aligned}\frac{\varepsilon_1}{\alpha_1(k_{||},\omega)}\frac{\varepsilon_3}{\alpha_3(k_{||},\omega)} + \frac{\varepsilon_2}{\alpha_2(k_{||},\omega)}\left(\frac{\varepsilon_1}{\alpha_1(k_{||},\omega)} + \frac{\varepsilon_3}{\alpha_3(k_{||},\omega)}\right)\\ \times \coth[2\alpha_2(k_{||},\omega)L] + \frac{\varepsilon_2^2}{\alpha_2^2(k_{||},\omega)} = 0.\end{aligned} \tag{17.129}$$

This result (Eq. (17.129)) corresponds to the confined slab photons, for which the electric field has x_1 and x_3 components (transverse magnetic (TM) modes or p-mode geometry). The solutions of Eq. (17.128) would correspond to confined polaritons for which the electric field has only one x_2 component (transverse electric (TE) modes or s-mode geometry).

Thus, the confined photons will reveal themselves as peaks in the spectral density defined by Eqs. (17.107) and (17.108), and the knowledge of the eigenvectors defined by Eq. (17.109) enables one to study the localization of these confined photons [12, 19–21,98].

Fig. 17.3 shows the existence of localized electromagnetic modes in a slab inserted between two semiinfinite media (see Fig. 17.2). Each medium is characterized by its dielectric constant ε_1 for the semiinfinite medium 1, ε_2 for the slab of thickness $2L$, and ε_3 for the semiinfinite medium 3. In all three figures (Figs. 17.3(a), 17.3(b), and 17.3(c)), the shaded areas represent the bulk bands of the two semiinfinite media and the dielectric constant of the slab is $\varepsilon_2 = 16$. The drawn lines indicate the light line in the slab. Dashed–dotted lines indicate the light line in the semiinfinite medium 1 and

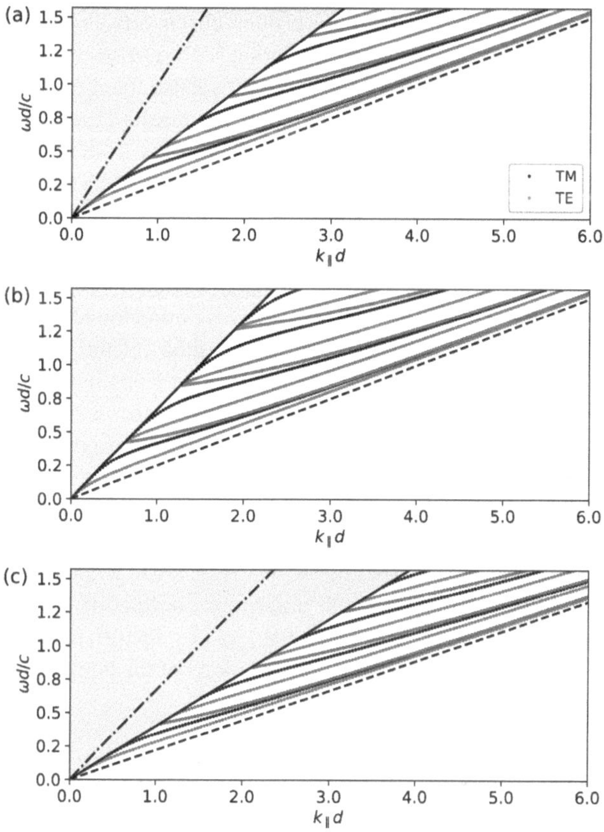

Figure 17.3 Dispersion of localized TE modes (red dotted lines, light gray in print version) and TM modes (black dotted lines) induced by a slab, sandwiched between two semiinfinite media 1 and 3. The dielectric constant of the slab is $\varepsilon_2 = 16$. The drawn lines indicate the light line in the slab. Dashed–dotted lines indicate the light line in the semiinfinite medium 1, and the bold lines indicate the light lines in the semiinfinite medium 3. The shaded areas correspond to the bulk bands of the media 1 and 3: (a) $\varepsilon_1 = 1$ and $\varepsilon_3 = 4$, (b) $\varepsilon_1 = \varepsilon_3 = 2.25$, and (c) $\varepsilon_1 = 2.25$ and $\varepsilon_3 = 6.25$.

the bold lines indicate the light lines in the semiinfinite medium 3. The localized electromagnetic modes in the dielectric slab presented by dotted lines are situated between the light lines of the substrates and the light line of the slab. Localized modes shown in red (light gray in print version) correspond to the transverse electric (TE) modes. The black dotted lines represent transverse magnetic (TM) localized modes. Figs. 17.3(a) and 17.3(b) show localized electromagnetic modes of two asymmetrical structures ($\varepsilon_1 = 1$ and $\varepsilon_3 = 4$ for Fig. 17.3(a) and $\varepsilon_1 = 2.25$ and $\varepsilon_3 = 6.25$ for Fig. 17.3(c)). Fig. 17.3(b) shows the localized modes for a symmetrical structure characterized by $\varepsilon_1 = \varepsilon_3 = 2.25$.

The above response functions can be used in calculations of Raman scattering of light from a metal slab and of Brillouin scattering from opaque slabs. They could also prove to be helpful in the studies of nonlinear interactions between confined polaritons and of the interaction of electromagnetic radiation with randomly rough slabs.

As in the above section, we can also use these response functions in order to obtain the electric and magnetic fields associated with the reflected and transmitted waves associated with an incident wave $e^{i|\alpha_1|x_3}e^{i(k_{\|}x_1-\omega t)}$ coming from $x_3=-\infty$ in the following manner. Assuming the source point x_3' at $-\infty$, so that $x_3' < x_3$, this enables us to factorize out in the useful expressions of $g_{\alpha\beta}(k_{\|},\omega/x_3,x_3')$ a common factor $e^{\alpha_1 x_3'}$ and the remaining expressions provide us with the components of the electric field. More precisely, we obtain in this way the E_2 components of the s-wave: for the incident and reflected wave from Eq. (17.96), for the wave transmitted inside the slab from Eq. (17.111), and for the wave transmitted into medium 3 from Eq. (17.103). In the same manner the E_1 and E_3 components of the p-wave are obtained: from Eqs. (17.97) and (17.99) in medium 1, from Eqs. (17.114) and (17.118) in medium 2, and from Eqs. (17.104) and (17.106) in medium 3. Finally, using the Maxwell equation (17.2), one obtains the components of the magnetic field in the same manner (Eqs. (17.81), (17.82), and (17.83)) as explained above with the example of the interface. When the Poynting vector is needed, it can also be obtained from Eq. (17.84).

Finally, let us stress that here also the above results for $\boldsymbol{E}$ and $\boldsymbol{H}$ can be used in order to derive the fields associated with polaritons confined in the slab and having exponentially decaying factors outside the slab, just by factorizing out the vanishing denominators, which provides the dispersion relations of the confined polaritons.

17.6 Dielectric layered photonic crystals

The photonic crystals are periodic arrays of dielectrics and/or conductors used to scatter light [99,100]. In a similar manner to how semiconductors control the passage of electrons, photonic crystals possess passbands which allow photons in certain frequency ranges to propagate through the crystal and photonic bandgaps which inhibit photon flow, producing regions of suppressed transmission. The existence of these pass- and stopbands is governed by Floquet–Bloch's theorem. These materials have many potential applications in optoelectronics and optical communications, such as cavities [101], waveguides [102], light-emitting diodes [99], superfast optical switches and detectors [103], etc. In addition, they make it possible to solve some fundamental problems, that is, problems related to controlling spontaneous light emission from atoms and molecules and localizing and channeling light [99,100,104–106]. In particular, it is very important that the concepts of photonic bandgaps continue to offer us new ideas for opto-electronic integrated circuits, as well as for the design of high-performance optoelectronic devices.

Layered photonic crystal devices comprise multiple periodic components that can produce transmission properties and field localization not seen in isolated crystals [107,108]. Heterostructures with a single photonic crystal interface have been extensively studied. Examples of localized behavior are the surface or interface modes, also known as optical Tamm [109] states. These modes can exist at a boundary only if their field amplitudes decay away as the distance from the boundary increases in either direction. This means that the wave vectors must be imaginary. In the case of a photonic crystal, this occurs if the mode is trying to travel through a photonic bandgap. These modes have been found in a variety of photonic structures, including 1D [110–113] and 2D [114] photonic crystal interfaces, air-photonic crystal surfaces at oblique angles [115], and photonic crystals bordering media with a graded refractive index [116]. Tamm states have also been investigated in various systems containing a photonic crystal with a tunable cap layer adjacent to a uniform medium. Examples include PCs containing superconducting layers [29], systems containing metamaterials, both the layered photonic crystals [117,118] and the uniform medium [119], and systems with liquid crystal [120], and chiral [121] cap layers. Note that in [115], despite the photonic crystal being adjacent to a uniform medium with positive dielectric constant, surface modes can still form due to total internal reflection. The component of the wave vector parallel to the interface, $k_{\|}$, is large enough to cause the normal component, $k_{\perp}$ $[k_{\perp} = (k^2 - k_{\|}^2)^{1/2}]$ to become imaginary.

In this section, we calculate the response functions for an infinite layered photonic crystal and a semiinfinite layered photonic crystal in contact with another semiinfinite homogeneous dielectric. The dispersion relations are given analytically for these two systems and illustrated by some numerical results for dispersion curves and surface electromagnetic modes.

17.6.1 Infinite isotropic dielectric photonic crystal

The layered photonic crystal is formed out of an infinite repetition of two different dielectric slabs labeled by the unit cell index n. Each slab of width $2L_i = d_i$ is labeled by the index i (1 or 2) within the unit cell designated by an index n. All the interfaces are taken to be parallel to the $(\boldsymbol{x}_1, \boldsymbol{x}_2)$-plane. This infers that the $\boldsymbol{x}_3$ axis is the photonic crystal axis, which observes the periodicity (with a period $d = d_1 + d_2$) of the system (see Fig. 17.4). We replace the x_3 coordinate by two variables (m, z) such that

$$m \equiv (n, i) = 2n + i, \quad -\infty < n < +\infty, \tag{17.130}$$

$$-1 \le x_3/L_i = z \le 1, \quad \text{within the slab } (n, i). \tag{17.131}$$

Also, there are two different ways to label one and the same interface:

$$(m, -1) \equiv (n, i, -1) \equiv (m-1, 1).$$

Let us first rewrite Eq. (17.91) for a black-box slab in a form which will be convenient for further formulation. We rename Eq. (17.91) as $\boldsymbol{g}_{sm}^{-1}$ and rewrite it as

$$\boldsymbol{g}_{sm}^{-1}(M, M) = \begin{pmatrix} g_{sm}^{-1}(-1,-1) & g_{sm}^{-1}(-1,1) \\ g_{sm}^{-1}(1,-1) & g_{sm}^{-1}(1,1) \end{pmatrix} = -\frac{F_m}{S_m}\begin{pmatrix} C_m & -1 \\ -1 & C_m \end{pmatrix}, \tag{17.132}$$

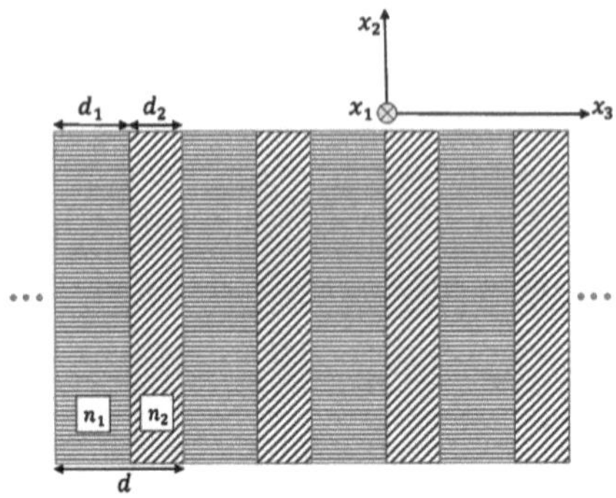

Figure 17.4 Schematic of the one dimensional photonic crystal showing the layer parameters (n_i and d_i are the index of refraction and thickness of layer i, respectively); d is the period of the structure.

with

$$C_m = \cosh(\alpha_m d_m), \tag{17.133}$$

$$S_m = \sinh(\alpha_m d_m). \tag{17.134}$$

Let us call $g(k_{||}, \omega/m, z; m', z')$ the basic element of the response function of an infinite photonic crystal. Out of it, one can construct easily, as seen above, all the elements of the response function $\boldsymbol{g}(k_{||}, \omega/m, z; m', z')$ of an infinite photonic crystal [122,123]. First the interface elements of $\boldsymbol{g}(k_{||}, \omega/M_m, M_{m'})$ are derived, where the interface spaces M_m and $M_{m'}$ are respectively defined as

$$M_m = (m, z = \pm 1), \tag{17.135}$$

$$M_{m'} = (m', z' = \pm 1). \tag{17.136}$$

In fact, as we will see later, only these interface elements are needed in order to construct the basic element $d(k_{||}, \omega/m, z; m', z')$ and then the full $\boldsymbol{g}(k_{||}, \omega/m, z; m', z')$ of a semiinfinite photonic crystal in contact with another semiinfinite homogeneous dielectric. However, the full response function $\boldsymbol{g}(k_{||}, \omega/m, z; m', z')$ is also of interest for the study of the bulk properties of dielectric photonic crystals.

The inverse of the interface response function (within the discrete interface space M) of the two-layered photonic crystal is a block-tridiagonal matrix whose elements are [122,123]

$$g^{-1}(m, -1; m', -1) = -\delta_{m,m'} \left(F_{m-1} \frac{C_{m-1}}{S_{m-1}} + F_m \frac{C_m}{S_m} \right), \tag{17.137}$$

$$g^{-1}(m, -1; m', 1) = -\delta_{m,m'} \frac{F_m}{S_m} = g^{-1}(m, 1; m', -1), \tag{17.138}$$

$$g^{-1}(m, 1; m', 1) = -\delta_{m,m'} \left(F_m \frac{C_m}{S_m} + F_{m+1} \frac{C_{m+1}}{S_{m+1}} \right). \tag{17.139}$$

The block-tridiagonal matrix $\boldsymbol{g}^{-1}(M, M)$ can be inverted with the mathematical methods as pointed out before [122,123]. Since $\boldsymbol{g}^{-1}(M, M)\boldsymbol{g}(M, M) = \boldsymbol{I}$, one can write

$$\boldsymbol{K}(m)\begin{pmatrix} g(m, 1; m', 1) \\ g(m+1, 1; m', 1) \end{pmatrix} + \boldsymbol{H}(m)\begin{pmatrix} g(m, -1; m', 1) \\ g(m, 1; m', 1) \end{pmatrix} = \delta_{mm'}\begin{pmatrix} 1 \\ 0 \end{pmatrix}, \tag{17.140}$$

where

$$\boldsymbol{K}(m) = \begin{pmatrix} g^{-1}(m+1, -1; m+1, -1) & g^{-1}(m+1, -1; m+1, 1) \\ -1 & 0 \end{pmatrix} \tag{17.141}$$

and

$$\boldsymbol{H}(m) = \begin{pmatrix} g^{-1}(m, 1; m, -1) & 0 \\ 0 & 1 \end{pmatrix}. \tag{17.142}$$

It is easy task to calculate $\boldsymbol{K}(M)$ and $\boldsymbol{H}(M)$. The result is:

$$\boldsymbol{K}(m) = \begin{pmatrix} -F_m \dfrac{C_m}{S_m} - F_{m+1}\dfrac{C_{m+1}}{S_{m+1}} & \dfrac{F_{m+1}}{S_{m+1}} \\ -1 & 0 \end{pmatrix}, \tag{17.143}$$

$$\boldsymbol{H}(m) = \begin{pmatrix} F_m/S_m & 0 \\ 0 & 1 \end{pmatrix}, \tag{17.144}$$

$$\boldsymbol{P}(m) = -\boldsymbol{K}^{-1}(m)\boldsymbol{H}(m) = \begin{pmatrix} 0 & 1 \\ -\dfrac{F_m}{F_{m+1}}\dfrac{S_{m+1}}{S_m} & C_{m+1} + S_{m+1}\dfrac{F_m}{F_{m+1}}\dfrac{C_m}{S_m} \end{pmatrix}. \tag{17.145}$$

Now the transfer matrix, which is defined as [122,123]

$$\boldsymbol{R}(m, m') = \prod_{m''=m'+1}^{m} \boldsymbol{P}(m''), \ m \geqslant m'', \tag{17.146}$$

where the symbol $\boldsymbol{\Pi}$ means the matrix product, gives

$$\boldsymbol{R}(m, m-N) = \boldsymbol{P}(m)\boldsymbol{P}(m-1)\cdots\boldsymbol{P}(m-N+1). \tag{17.147}$$

This leads us to infer that, for an $N(=2)$-layered photonic crystal,

$$\boldsymbol{R}(2, 1) = \boldsymbol{P}(2) \text{ and } \boldsymbol{R}(2, 0) = \boldsymbol{P}(2)\boldsymbol{P}(1). \tag{17.148}$$

Consequently,

$$\boldsymbol{R}(2, 0) = \begin{pmatrix} -\frac{F_1 S_2}{F_2 S_1} & \frac{F_1 S_2}{F_2 S_1}C_1 + C_2 \\ -C_2 - \frac{F_1 S_2}{F_2 S_1}C_1 & 2C_1C_2 + \frac{F_1 S_2}{F_2 S_1}C_1^2 + \frac{F_2}{F_1}S_1S_2 \end{pmatrix}. \tag{17.149}$$

It is worthwhile mentioning that the collective (bulk) excitations of the polaritons in infinite dielectric photonic crystals at hand are given by

$$\det \left| \boldsymbol{R}(2,0) - e^{ik_3 d} \boldsymbol{I} \right| = 0. \tag{17.150}$$

Now let us define $\boldsymbol{R}(2,0)$ as the following 2×2 matrix:

$$\boldsymbol{R}(2,0) = \begin{pmatrix} R_{11} & R_{12} \\ R_{21} & R_{22} \end{pmatrix}. \tag{17.151}$$

Then, premultiplying Eq. (17.140) by $\boldsymbol{K}^{-1}(m)$ and using it N times gives

$$\begin{pmatrix} g(n+1,1,-1;n',i',1) \\ g(n+1,1,1;n',i',1) \end{pmatrix} = \boldsymbol{R}(2,0) \begin{pmatrix} g(n,1,-1;n',i',1) \\ g(n,1,1;n',i',1) \end{pmatrix} + \delta_{nn'} \boldsymbol{B}(i'), \tag{17.152}$$

where $\boldsymbol{B}(i')$ is a 2×1 matrix defined by

$$\boldsymbol{B}(i') \equiv \begin{pmatrix} B_1(i') \\ B_2(i') \end{pmatrix} = \boldsymbol{R}(2,i') \boldsymbol{K}^{-1}(i') \begin{pmatrix} 1 \\ 0 \end{pmatrix}, \quad i' = 1, 2. \tag{17.153}$$

It is an easy task to calculate $\boldsymbol{B}(1)$ and $\boldsymbol{B}(2)$. The result is:

$$\boldsymbol{B}(1) = \begin{pmatrix} S_2/F_2 \\ S_1 C_2/F_1 + C_1 S_2/F_2 \end{pmatrix} \tag{17.154}$$

and

$$\boldsymbol{B}(2) = \begin{pmatrix} 0 \\ S_1/F_1 \end{pmatrix}. \tag{17.155}$$

Using the above results (17.141)–(17.155) in Eq. (17.152) gives the interface elements of $g(k_{\|}, \omega/M_m, M_{m'})$ we seek, namely

$$g(n,i,1;n',i,1) = \left(\frac{S_1 C_2}{F_1} + \frac{S_2 C_1}{F_2} \right) \frac{t^{|n-n'|+1}}{t^2 - 1}, \quad i = 1, 2, \tag{17.156}$$

$$g(n,1,1;n',2,1) = \frac{t}{t^2-1} \left(\frac{S_1}{F_1} t^{|n-n'|} + \frac{S_2}{F_2} t^{|n-n'-1|} \right), \tag{17.157}$$

$$g(n,2,1;n',1,1) = \frac{t}{t^2-1} \left(\frac{S_1}{F_1} t^{|n-n'|} + \frac{S_2}{F_2} t^{|n-n'+1|} \right), \tag{17.158}$$

where

$$t = \begin{cases} \eta - (\eta^2 - 1)^{1/2}, & \eta > 1, \\ \eta + i(1-\eta^2)^{1/2}, & -1 < \eta < 1, \\ \eta + (\eta^2 - 1)^{1/2}, & \eta < -1, \end{cases} \tag{17.159}$$

and

$$\eta = C_1 C_2 + \frac{1}{2}\left(\frac{F_1}{F_2} + \frac{F_2}{F_1}\right) S_1 S_2. \tag{17.160}$$

In these results and from now on, we no longer explicitly write the dependence on $(k_{||}, \omega)$, bearing in mind that all the response functions given in this paper are functions of $k_{||}$ and ω. Note that the dispersion relation obtained from Eq. (17.150) is identical to that obtained from

$$2\cos(k_3 d) = R_{11} + R_{22} = 2\eta, \tag{17.161}$$

where $d = d_1 + d_2$ is the photonic crystal period.

We write the following 2×2 matrices $\boldsymbol{g}(M_m, M_{m'})$ which have four different values depending upon the four possible combinations of the layer indices $i(\equiv 1, 2)$ and $i'(\equiv 1, 2)$ in $m(\equiv ni)$ and $i(\equiv n', i')$, namely

$$g(M_{n1}, M_{n'1}) = \frac{t}{t^2-1}\begin{pmatrix} \left(\frac{S_1C_2}{F_1} + \frac{S_2C_1}{F_2}\right) t^{|n-n'|} & \frac{S_1}{F_1} t^{|n-n'-1|} + \frac{S_2}{F_2} t^{|n-n'|} \\ \frac{S_1}{F_1} t^{|n-n'+1|} + \frac{S_2}{F_2} t^{|n-n'|} & \left(\frac{S_1C_2}{F_1} + \frac{S_2C_1}{F_2}\right) t^{|n-n'|} \end{pmatrix}, \tag{17.162}$$

$$g(M_{n1}, M_{n'2}) = \frac{t}{t^2-1}\begin{pmatrix} \frac{S_1}{F_1} t^{|n-n'-1|} + \frac{S_2}{F_2} t^{|n-n'|} & \left(\frac{S_1C_2}{F_1} + \frac{S_2C_1}{F_2}\right) t^{|n-n'-1|} \\ \left(\frac{S_1C_2}{F_1} + \frac{S_2C_1}{F_2}\right) t^{|n-n'|} & \frac{S_1}{F_1} t^{|n-n'|} + \frac{S_2}{F_2} t^{|n-n'-1|} \end{pmatrix}, \tag{17.163}$$

$$g(M_{n2}, M_{n'1}) = \frac{t}{t^2-1}\begin{pmatrix} \frac{S_1}{F_1} t^{|n-n'+1|} + \frac{S_2}{F_2} t^{|n-n'|} & \left(\frac{S_1C_2}{F_1} + \frac{S_2C_1}{F_2}\right) t^{|n-n'|} \\ \left(\frac{S_1C_2}{F_1} + \frac{S_2C_1}{F_2}\right) t^{|n-n'+1|} & \frac{S_1}{F_1} t^{|n-n'|} + \frac{S_2}{F_2} t^{|n-n'+1|} \end{pmatrix}, \tag{17.164}$$

$$g(M_{n2}, M_{n'2}) = \frac{t}{t^2-1}\begin{pmatrix} \left(\frac{S_1C_2}{F_1} + \frac{S_2C_1}{F_2}\right) t^{|n-n'|} & \frac{S_1}{F_1} t^{|n-n'|} + \frac{S_2}{F_2} t^{|n-n'-1|} \\ \frac{S_1}{F_1} t^{|n-n'|} + \frac{S_2}{F_2} t^{|n-n'+1|} & \left(\frac{S_1C_2}{F_1} + \frac{S_2C_1}{F_2}\right) t^{|n-n'|} \end{pmatrix}. \tag{17.165}$$

The above 2×2 matrices have their first-line elements taken between $(n, i, -1)$ and respectively $(n', i', -1)$ and $(n', i', 1)$, in this order. The second line is between $(n, i, 1)$ and respectively $(n', i', -1)$ and $(n', i', 1)$.

Now it is necessary to rewrite the bulk response function, Eq. (17.56), according to the notations used in the photonic crystal system.

This requires replacing x_3, x'_3 by zL_i and $z'L_i$. Thus

$$G_i(z, z') = -\frac{e^{-\alpha_i L_i |z-z'|}}{2F_i}. \tag{17.166}$$

This means

$$\boldsymbol{G}_i(M_i, M_i) = -\frac{1}{2F_i}\begin{pmatrix} 1 & e^{-2\alpha_i L_i} \\ e^{-2\alpha_i L_i} & 1 \end{pmatrix}, \tag{17.167}$$

$$\boldsymbol{G}_i^{-1}(M_i, M_i) = -\frac{F_i}{\sinh(2\alpha_i L_i)}\begin{pmatrix} e^{2\alpha_i L_i} & 1 \\ 1 & e^{2\alpha_i L_i} \end{pmatrix}, \tag{17.168}$$

$$G_i(z, 1) = -\frac{e^{-\alpha_i(1-z)L_i}}{2F_i}, \tag{17.169}$$

$$G_i(z, -1) = -\frac{e^{-\alpha_i(1+z)L_i}}{2F_i}, \tag{17.170}$$

$$G_i(1, z') = -\frac{e^{-\alpha_i(1-z')L_i}}{2F_i}, \tag{17.171}$$

$$G_i(-1, z') = -\frac{e^{-\alpha_i(1+z')L_i}}{2F_i}. \tag{17.172}$$

We now have everything at hand to write the complete expression for the response function of the two-layered infinite photonic crystal $g(mz, m'z')$ given by

$$\begin{aligned} g(mz, m'z') &= \delta_{mm'}\Bigg[G_m(z, z') - \Big(G_m(z, -1), G_m(z, 1)\Big) \\ &\times \boldsymbol{G}_m^{-1}(M_m, M_m)\begin{pmatrix} G_m(-1, z') \\ G_m(1, z') \end{pmatrix}\Bigg] \\ &+ \Big(G_m(z, -1), G_m(z, 1)\Big)\boldsymbol{G}_m^{-1}(M_m, M_m)\boldsymbol{g}(M_m, M_{m'}) \\ &\times \boldsymbol{G}_m^{-1}(M_{m'}, M_{m'})\begin{pmatrix} G_{m'}(-1, z') \\ G_{m'}(1, z') \end{pmatrix}. \end{aligned} \tag{17.173}$$

The result is

$$\begin{aligned} g(mz, m'z') &= \delta_{mm'} U_m(z, z') \\ &+ \frac{1}{S_m S_{m'}}\Big(\sinh\big[\alpha_m(1-z)L_m\big], \sinh\big[\alpha_m(1+z)L_m\big]\Big) \\ &\times \boldsymbol{g}(M_m, M_{m'})\begin{pmatrix} \sinh\big[\alpha_{m'}(1-z)L_{m'}\big] \\ \sinh\big[\alpha_{m'}(1+z)L_{m'}\big] \end{pmatrix}, \end{aligned} \tag{17.174}$$

where

$$U_m(z,z') = -\frac{e^{-\alpha_m L_m |z-z'|}}{2\Gamma_m} + \frac{1}{2\Gamma_m S_m}\bigg(\sinh\big[\alpha_m(1-z)L_m\big]e^{-\alpha_m(1-z')L_m}$$
$$+\sinh\big[\alpha_m(1+z)L_m\big]e^{-\alpha_m(1+z')L_m}\bigg), \tag{17.175}$$

and the 2 × 2 matrix $\boldsymbol{g}(M_m, M_{m'})$ is given in Eqs. (17.162)–(17.165) for four different possible combinations of i and i' ($m \equiv ni$ and $m' \equiv n'i'$).

Finally, all the five nonzero elements of the response function $\boldsymbol{g}(mz; m'z')$ of an infinite dielectric photonic crystal are easily written down from Eq. (17.173) using the constituent relations (17.45)–(17.46), (17.51), and (17.54)–(17.55).

The photonic bulk band structure for a typical Bragg reflector is shown in Fig. 17.5(a) for transverse magnetic (TM, left panel) and transverse electric (TE, right panel) electromagnetic waves in a polystyrene/tellurium infinite photonic crystal. The filling fraction d_1/d in the photonic crystal is assumed to be equal to 0.25, where d_1 is the thickness of the tellurium layer, and the dielectric constants are: $\varepsilon_2 = 2.25$ for polystyrene and $\varepsilon_1 = 16$ for tellurium, respectively. In Fig. 17.5(a), we report the reduced frequency $\omega d/2\pi c$ versus the parallel reduced wave vector $k_{||}d/2\pi$. The thick dashed straight lines indicate the light lines of the vacuum. The shaded areas correspond to passbands, where photons are allowed to propagate in the photonic crystal; these shaded areas are separated by minigaps, where photons are prohibited to propagate. Due to the large difference between the dielectric constant values for the polystyrene and the tellurium, these gaps are rather large, in contrast to the case of other systems like GaAs/AlAs photonic crystal's. The band structures are calculated using Eqs. (17.54), (17.55), and (17.160).

As follows from Fig. 17.5(a), any wave launched from the ambient medium within the frequency range $0.21 < \omega d/2\pi c < 0.305$ is prohibited to propagate in the photonic crystal; and therefore will be subject to the phenomenon of total reflection (and hence, there will be no transmission). This means that in the above frequency range, the photonic crystal acts as an omnidirectional reflector. To be more specific, let us first notice that, due to the translational symmetry in the (x_1, x_2)-plane (see Fig. 17.2), the wave vector $k_{||} = \omega \sin\theta/c$ is conserved for a TE (TM) wave launched from the vacuum with an incident angle θ. A wave launched from the ambient medium has a velocity c and therefore its frequency lies above the line $\omega = ck_{||}$. As a consequence, for frequencies ranging from 0.21 to 0.305, any wave propagating in the ambient medium (i.e., with an arbitrary angle of incidence and any polarization) is forbidden to propagate in the SL and will be reflected back into the vacuum. On the contrary, in other frequency domains, part of the incident power will be transmitted into the photonic crystal. The necessary and sufficient criterion for omnidirectional reflectivity at a given frequency is that no transmitting state of the structure exists inside the vacuum cone defined by the velocity of vacuum. In Fig. 17.3, this criterion is that there exists a

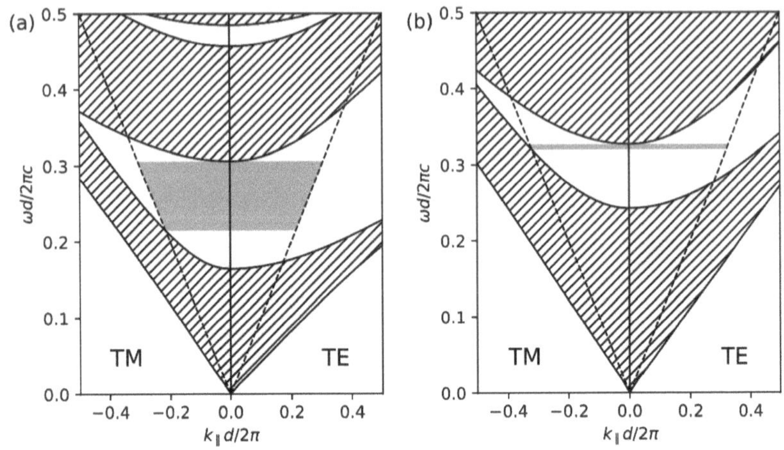

Figure 17.5 Photonic band structure of a typical 1D photonic crystal for TM (left panel) and TE (right panel) polarizations. Shaded area corresponds to the photonic crystal bulk bands. The heavy dashed lines indicate the light line of the vacuum. The horizontal shaded area delimit the edges of the omnidirectional band gap: (a) the stack is assumed to have dielectric constants $\varepsilon_1 = 16$, $\varepsilon_2 = 2.25$ and a filling fraction $d_1/d = 0.25$, (b) the stack is assumed to have dielectric constants $\varepsilon_1 = 6.25$, $\varepsilon_2 = 2.25$ and a filling fraction $d_1/d = 0.25$.

frequency zone in which the projected bands of the photonic crystal have no states for $\omega > ck_{||}$. Let us notice that this first forbidden band exists in different frequency domains of the electromagnetic spectrum for different values of the period d of the structure. More precisely, if one considers a structure with $d = 150$ nm, the omnidirectional total reflection is expected within the visible spectral band $\Delta\lambda = 455$–770 nm. Fig. 17.5 shows that, under certain conditions, such a 1D photonic structure can exhibit large reflection bandwidth for all incident angles and polarizations.

Fig. 17.5(b) shows the photonic bulk band structure for another parameters of the dielectric media. In this case the dielectric constants are: $\varepsilon_2 = 2.25$ for medium 2 and $\varepsilon_1 = 6.25$ for medium 1, respectively. We obtained also in this case the minipassband separated by the minigaps, but the bandgaps is narrower than in Fig. 17.5. Note also that the absolute gap, represented by the shaded area in the figure, is narrower. This result is explained by the low contrast between the two dielectric media, $\varepsilon_1/\varepsilon_2 = 2.97$ (for Fig. 17.5(a), $\varepsilon_1/\varepsilon_2 = 7.11$).

Finally, it should be noted that the widths of the minipassbands and minigaps of the photonic crystal depend on the geometrical parameters of the two materials constituting the photonic crystal, i.e., d_1 and d_2 or the filling fraction d_1/d, where $d = d_1 + d_2$. They also depend on the nature of the dielectric materials used in the photonic crystal.

Rather than expanding on this simple rewriting of the above Eqs. (17.162)–(17.175), let us now turn to another important physical problem, namely that of a semiinfinite dielectric photonic crystal in contact with another semiinfinite homogeneous dielectric.

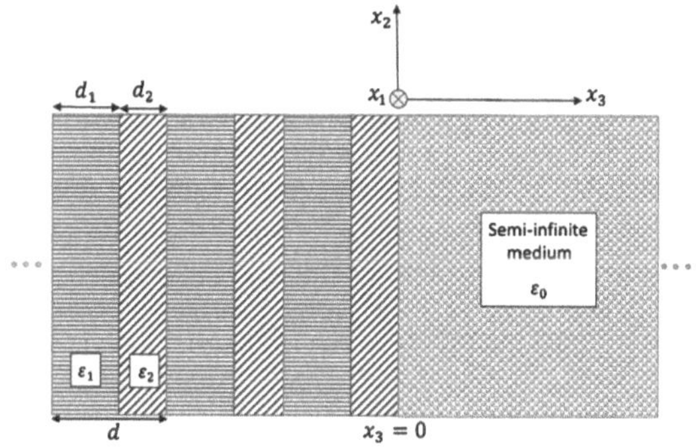

Figure 17.6 Schematic of the interface between one-dimensional photonic crystal and semi-infinite isotropic dielectric medium.

17.6.2 Semiinfinite photonic crystal bounded by another homogeneous dielectric

We consider an infinite layered photonic crystal truncated at an interface $x_3 = 0$ such that we have a semiinfinite photonic crystal with a different medium in the region $-\infty < x_3 < 0$. Let us assign a suffix "0" to the quantities in this medium (see Fig. 17.6). The surface $x_3 = 0$ (or $(0, 1, \bar{1}) \equiv (\bar{1}, 2, 1)$) perturbs the otherwise periodic photonic crystal system. The perturbation is accounted for by a cleavage operator which removes the slab m $(\equiv (ni) \equiv (\bar{1}, 2))$ and a "filling-operator" which fills in the region $-\infty < x_3 < 0$, and is given by

$$\boldsymbol{V}_f(M_{\bar{1}2}, M_{\bar{1}2}) = \boldsymbol{V}_{\boldsymbol{PI}} - \begin{pmatrix} g_s^{-1}(\bar{1}, 2, \bar{1}; \bar{1}, 2, \bar{1}) & g_s^{-1}(\bar{1}, 2, \bar{1}; \bar{1}, 2, 1) \\ g_s^{-1}(\bar{1}, 2, 1; \bar{1}, 2, \bar{1}) & g_s^{-1}(\bar{1}, 2, 1; \bar{1}, 2, 1) \end{pmatrix}, \tag{17.176}$$

where the filling operator

$$\boldsymbol{V}_{\boldsymbol{PI}} = \begin{pmatrix} 0 & 0 \\ 0 & -F_0 \end{pmatrix}. \tag{17.177}$$

Furthermore, F_0 is defined by Eqs. (17.54)–(17.55) for a semiinfinite homogeneous medium labeled by the index 0.

As such, we obtain from Eq. (17.176)

$$\boldsymbol{V}_f(M_{\bar{1}2}, M_{\bar{1}2}) = -\begin{pmatrix} g_{s2}^{-1}(\bar{1}, \bar{1}) & g_{s2}^{-1}(\bar{1}, 1) \\ g_{s2}^{-1}(1, \bar{1}) & g_{s2}^{-1}(1, 1) + F_0 \end{pmatrix}. \tag{17.178}$$

Therefore, the interface response operator $\boldsymbol{A}_f$ for this problem is

$$\begin{aligned} A_f(0,1,\bar{1};M_{m'}) &= -g_s^{-1}(\bar{1},2,1;\bar{1},2,\bar{1})g(\bar{1},2,\bar{1};M_{m'}) \\ &\quad - g_s^{-1}(\bar{1},2,1;\bar{1},2,1)g(\bar{1},2,1;M_{m'}) \\ &= \frac{F_2C_2}{S_2}g(\bar{1},2,\bar{1};M_{m'}) - \left(\frac{F_2}{S_2}+F_0\right)g(\bar{1},2,1;M_{m'}). \end{aligned} \tag{17.179}$$

With the help of Eqs. (17.162)–(17.165), Eq. (17.179) assumes the form

$$A_f(0,1,\bar{1};n',i',1) = \frac{t^{n'+1}}{t^2-1}Y_{i'}, \quad i'=1,2; \;\; n' \geqslant 0, \tag{17.180}$$

where

$$Y_1 = C_1C_2 + \frac{F_2}{F_1}S_1S_2 - t - F_0\left(\frac{S_1C_2}{F_1} + \frac{S_2C_1}{F_2}\right) \tag{17.181}$$

and

$$Y_2 = C_2 - C_1t - F_0\left(\frac{S_1}{F_1}t + \frac{S_2}{F_2}\right). \tag{17.182}$$

Define then

$$\Delta_f(0,1,\bar{1};0,1,\bar{1}) = 1 + A_f(0,1,\bar{1};0,1,\bar{1}), \tag{17.183}$$

which using Eqs. (17.180)–(17.182) has the following value:

$$\Delta_f(0,1,\bar{1};0,1,\bar{1}) = \frac{t}{t^2-1}\frac{F_2W}{Y_1}\left(\frac{S_1C_2}{F_1} + \frac{S_2C_1}{F_2}\right), \tag{17.184}$$

where

$$W = \left(\frac{F_0^2}{F_1F_2} - \frac{F_1}{F_2}\right)S_1C_2 + \left(\frac{F_0}{F_2^2} - 1\right)S_2C_1 + \left(\frac{F_0F_1}{F_2^2} - \frac{F_0}{F_1}\right)S_1S_2. \tag{17.185}$$

The polariton modes localized at the interface between the semiinfinite photonic crystal and the semiinfinite homogeneous medium can be obtained from

$$W = 0. \tag{17.186}$$

Their explicit expressions will be given at the end of this section. Let us first write down the basic response function $\boldsymbol{d}$ associated with this system. Its elements within the interface space M_m can be obtained from the discrete interface response theory to be

$$\boldsymbol{d}(M_m, M_{m'}) = \boldsymbol{g}(M_m, M_{m'}) - \boldsymbol{g}(M_m, M_s)\boldsymbol{\Delta}^{-1}(M_s, M_s)\boldsymbol{A}_f(M_s, M_{m'}). \tag{17.187}$$

The use of this equation, together with Eqs. (17.156)–(17.158), (17.180)–(17.182) and (17.183)–(17.184), gives us all the interface elements of the basic response function $\boldsymbol{d}$ of the semiinfinite photonic crystal bounded by a semiinfinite homogeneous dielectric. The results for $n, n' \geqslant 0$ are:

$$d(n1\bar{1}; n'1\bar{1}) = \frac{t}{t^2-1}\left[\left(\frac{S_1C_2}{F_1} + \frac{S_2C_1}{F_2}\right)t^{|n-n'|} - \frac{Y_1^2}{F_2W}t^{n+n'}\right], \tag{17.188}$$

$$d(n2\bar{1}; n'2\bar{1}) = \frac{t}{t^2-1}\left[\left(\frac{S_1C_2}{F_1} + \frac{S_2C_1}{F_2}\right)t^{|n-n'|} - \frac{Y_2^2}{F_2W}t^{n+n'}\right], \tag{17.189}$$

$$d(n1\bar{1}; n'2\bar{1}) = \frac{t}{t^2-1}\left[\frac{S_1}{F_1}t^{|n-n'-1|} + \frac{S_2}{F_2}t^{|n-n'|} - \frac{Y_1Y_2}{F_2W}t^{n+n'}\right], \tag{17.190}$$

$$d(n2\bar{1}; n'1\bar{1}) = \frac{t}{t^2-1}\left[\frac{S_1}{F_1}t^{|n-n'+1|} + \frac{S_2}{F_2}t^{|n-n'|} - \frac{Y_1Y_2}{F_2W}t^{n+n'}\right]. \tag{17.191}$$

In order to calculate the complete basic response function $\boldsymbol{d}$, we need to write down the following 2×2 matrices $\boldsymbol{d}(M_m, M_{m'})$, as done already before for the infinite photonic crystal. Here we also have four different 2×2 matrices depending upon the four possible combinations of i and i', namely, for n and $n' \geqslant 0$:

$$\boldsymbol{d}(M_{n1}, M_{n'1}) = \begin{pmatrix} d(n,1,\bar{1};n',1,\bar{1}) & d(n,1,\bar{1};n',2,\bar{1}) \\ d(n,2,\bar{1};n',1,\bar{1}) & d(n,2,\bar{1};n',2,\bar{1}) \end{pmatrix}, \tag{17.192}$$

$$\boldsymbol{d}(M_{n2}, M_{n'2}) = \begin{pmatrix} d(n,2,\bar{1};n',2,\bar{1}) & d(n,2,\bar{1};n'+1,1,\bar{1}) \\ d(n+1,1,\bar{1};n',2,\bar{1}) & d(n+1,1,\bar{1};n'+1,1,\bar{1}) \end{pmatrix}, \tag{17.193}$$

$$\boldsymbol{d}(M_{n1}, M_{n'2}) = \begin{pmatrix} d(n,1,\bar{1};n',2,\bar{1}) & d(n,1,\bar{1};n'+1,1,\bar{1}) \\ d(n,2,\bar{1};n',2,\bar{1}) & d(n,2,\bar{1};n'+1,1,\bar{1}) \end{pmatrix}, \tag{17.194}$$

$$\boldsymbol{d}(M_{n2}, M_{n'1}) = \begin{pmatrix} d(n,2,\bar{1};n',1,\bar{1}) & d(n,2,\bar{1};n',2,\bar{1}) \\ d(n+1,1,\bar{1};n',1,\bar{1}) & d(n+1,1,\bar{1};n',2,\bar{1}) \end{pmatrix}, \tag{17.195}$$

where all elements are given by Eqs. (17.188)–(17.191). With these results in hand, the application of the general equation provides us in the same manner as for the infinite photonic crystal (Eq. (17.174)) within the semiinfinite photonic crystal with

$$\begin{aligned}\boldsymbol{d}(mz,m'z') &= \delta_{mm'}U_m(z,z')\\ &+\frac{1}{S_mS_{m'}}\Big(\sinh\big[\alpha_m(1-z)L_m\big],\sinh\big[\alpha_m(1+z)L_m\big]\Big)\\ &\times \boldsymbol{d}(M_m,M_{m'})\begin{pmatrix}\sinh\big[\alpha_{m'}(1-z)L_{m'}\big]\\ \sinh\big[\alpha_{m'}(1+z)L_{m'}\big]\end{pmatrix},\end{aligned}\tag{17.196}$$

where $m\equiv(ni)$, $i=1,2$ and $-1\leqslant z,z'\leqslant+1$. In this result, we use also the S_m and $U_m(z,z')$ defined by Eqs. (17.134) and (17.175).

To have the complete basic response function $\boldsymbol{d}$, we have still to write its elements between two points of the homogeneous semiinfinite dielectric and between one point situated in this homogeneous medium and another situated in the semiinfinite photonic crystal. These results are also easily obtained with the help of the general equation (17.25), together with the bulk response functions given by Eq. (17.56) and the interface elements of the response function $\boldsymbol{d}$ given by Eqs. (17.188)–(17.191).

Inside the semiinfinite homogeneous dielectric, we also use dimensionless positions z such that $z=x_3/L_0$, where L_0 is an arbitrary unit length. Inside this medium $-\infty<z<0$, and the interface with the photonic crystal is at $z=0\equiv(m=1,z=\bar{1})$. With these notations, the results inside the homogeneous medium are:

(a) for z and $z'<0$,

$$\begin{aligned}d(z,z') &= -\frac{e^{-\alpha_0L_0|z-z'|}}{2F_i}\\ &\quad+e^{-\alpha_0L_0(z+z')}\left[\frac{1}{2F_0}+\frac{t}{t^2-1}\left(\frac{S_1C_2}{F_1}+\frac{S_2C_1}{F_2}-\frac{Y_1^2}{F_2W}\right)\right].\end{aligned}\tag{17.197}$$

When one point is inside the homogeneous medium and the other inside the photonic crystal, one has

(b) for $z<0$, $m'\geqslant1$ and $-1<z<+1$,

$$\begin{aligned}d(z;m'z') &= \frac{e^{\alpha_0L_0z}}{S_{m'}}\Big\{d(1,\bar{1};m',\bar{1})\sinh\big[\alpha_{m'}L_{m'}(1-z')\big]\\ &\quad+d(1,\bar{1};m',1)\sinh\big[\alpha_{m'}L_{m'}(1+z')\big]\Big\};\end{aligned}\tag{17.198}$$

(c) for $m\geqslant1$, $-1<z<+1$ and $z'<0$,

$$\begin{aligned}d(mz;z') &= \frac{e^{-\alpha_0L_0z'}}{S_m}\Big\{d(m,\bar{1};1,\bar{1})\sinh\big[\alpha_mL_m(1-z)\big]\\ &\quad+d(m,1;1,\bar{1})\sinh\big[\alpha_mL_m(1+z)\big]\Big\},\end{aligned}\tag{17.199}$$

where $d(m, \pm 1; m', \pm 1) \equiv d(n, i, \pm 1; n', i', \pm 1)$ is given by Eqs. (17.188)–(17.191).

Finally, all the five nonzero elements of the response function $\boldsymbol{d}$ of this dielectric system (semiinfinite photonic crystal bounded by a semiinfinite homogeneous medium) are easily written down from Eqs. (17.197)–(17.199) using the constituent relations (17.45), (17.46), (17.52), (17.54), and (17.55).

Without trying to give here the numerous physical results possible to obtain with the above response function, let us stress only that in the denominator of all the elements of this function a new term W appears, whose value is given by Eq. (17.185). So the interface polaritons associated with this system are obtained when the two different values (the F_i take two different values, see Eqs. (17.54)–(17.55)) of $W(k_{||}, \omega)$ vanish. More precisely, from g_{22} one obtains a condition for possible interface localized states to be

$$\alpha_2(\varepsilon_0 - \varepsilon_1)S_1C_2 + \alpha_1(\varepsilon_0 - \varepsilon_2)S_2C_1 + \alpha_0(\varepsilon_1 - \varepsilon_2)S_1S_2 = 0, \tag{17.200}$$

together with the condition

$$\left| C_1C_2 + \frac{\alpha_2}{\alpha_1}S_1S_2 - \alpha_0\left(\frac{C_1S_2}{\alpha_2} + \frac{C_2S_1}{\alpha_1}\right)\right| > 1. \tag{17.201}$$

And from the four other nonzero elements of $\boldsymbol{g}$, one has

$$\begin{aligned}&\left(\frac{\varepsilon_0^2}{\varepsilon_1\varepsilon_2}\frac{\alpha_1\alpha_2}{\alpha_0^2} - \frac{\varepsilon_1}{\varepsilon_2}\frac{\alpha_2}{\alpha_1}\right)S_1C_2 + \left(\frac{\varepsilon_0^2}{\varepsilon_2^2}\frac{\alpha_2^2}{\alpha_0^2} - 1\right)C_1S_2\\&+\frac{\varepsilon_0}{\varepsilon_2}\frac{\alpha_2}{\alpha_0}\left(\frac{\varepsilon_1}{\varepsilon_2}\frac{\alpha_2}{\alpha_1} - \frac{\varepsilon_2}{\varepsilon_1}\frac{\alpha_1}{\alpha_2}\right)S_1S_2 = 0,\end{aligned} \tag{17.202}$$

together with the condition

$$\left| C_1C_2 + \frac{\varepsilon_2\alpha_1}{\varepsilon_1\alpha_2}S_1S_2 - \frac{\varepsilon_0}{\alpha_0}\left(\frac{C_1S_2\alpha_2}{\varepsilon_2} + \frac{C_2S_1\alpha_1}{\varepsilon_1}\right)\right| > 1. \tag{17.203}$$

The conditions given by Eqs. (17.201) and (17.203) ensure that the optical interface localized modes decay far from the surface of the layered photonic crystal.

This latter result was obtained with other approaches as well, with retardation [124] as here and also without [125] ($c \to \infty$). This result corresponds to interface polaritons for which the electric field has x_1 and x_3 components (p-mode geometry). The eventual interface modes, solution of Eq. (17.200) would correspond to interface polaritons for which the electric field would have only one x_2 component (s-mode geometry). Let us also remind the readers that Yariv and Yeh [126,127] seem to be the first to study surface polaritons in semiinfinite photonic crystals, although they did not derive the analytical expressions given by Eqs. (17.200) and (17.202).

Fig. 17.7 gives the dispersion of bulk bands and surface electromagnetic modes as a function of reduced wave vector. The slab at the surface of the photonic crystal is either 1 or 2 with the same thickness as in the bulk. The thicknesses of the dielectric layers are taken such that $d_1 = 0.25d$ and $d_2 = 0.75d$, where $d = d_1 + d_2$ is the period of

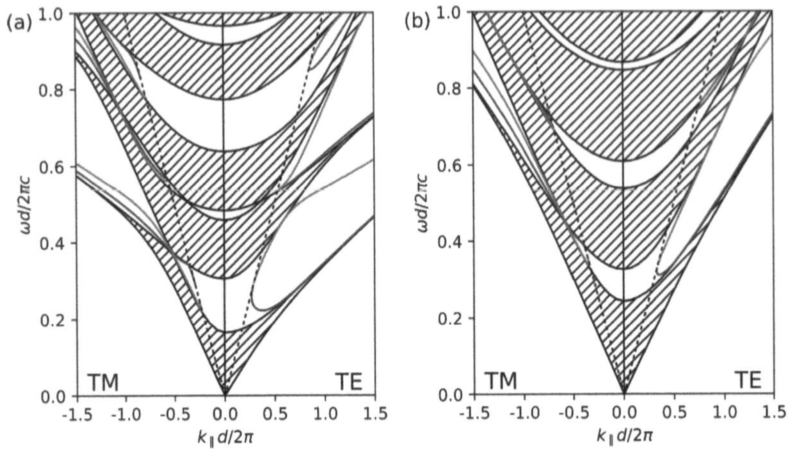

Figure 17.7 Bulk and surface TE (right panel) and TM (left panel) optical waves in a photonic crystal made of two dielectric materials. The curves give $\omega d/2\pi c$ as a function of $k_{||}d/2\pi$, where ω is the frequency, c the velocity of light, $k_{||}$ the propagation vector parallel to the interfaces, and $d = d_1 + d_2$ the period of the photonic crystal. The shaded areas represent the bulk bands. The thicknesses of the layers are taken such that $d_1 = 0.25d$ and $d_2 = 0.75d$. The red solid lines (light gray in print version) represent the surface electric waves for the semiinfinite photonic crystal terminated by the slab of material 1. The blue solid lines (dark gray in print version) represent the surface electric waves for the complementary photonic crystal terminated by the slab of material 2. The dashed lines indicate the light line of vacuum: (a) $\varepsilon_1 = 16$ and $\varepsilon_2 = 2.25$, (b) $\varepsilon_1 = 6.25$ and $\varepsilon_2 = 2.25$.

the photonic crystal. In each figure (Fig. 17.7(a) or Fig. 17.7(b)), the bulk and surface electromagnetic TM modes (left panel) and TE modes (right panel) are presented. The curves give $\omega d/2\pi c$ as a function of $k_{||}d/2\pi$, where ω is the frequency, c the velocity of light, and $k_{||}$ the propagation vector parallel to the interfaces. The dashed lines indicate the light line of vacuum. The shaded areas are the bulk bands separated by gaps where surface electromagnetic modes are represented for the two complementary semiinfinite photonic crystals obtained by cleaving the infinite photonic crystal at the interface dielectric 1/dielectric 2. The red lines (light gray in print version) represent the surface electromagnetic modes for the semiinfinite photonic crystal terminated by a dielectric 1 layer of thickness $d_1 = 0.25d$. The blue lines (dark gray in print version) represent surface electromagnetic modes for the complementary photonic crystal terminated by a dielectric 2 layer of thickness $d_2 = 0.75d$. One can observe that these electromagnetic surface modes are very dependent on the type of material which is at the surface. Fig. 17.7(a) presents the result for the photonic crystal characterized by dielectric constant $\varepsilon_1 = 16$ for the medium 1 and $\varepsilon_2 = 2.25$ for the medium 2. Note that in this case, the first and third minigaps are wide because the contrast between the two dielectric media is important ($\varepsilon_1/\varepsilon_2 = 7.1$). Fig. 17.7(b) presents the result for the photonic crystal characterized by dielectric constant $\varepsilon_1 = 6.25$ for the medium 1 and $\varepsilon_2 = 2.25$ for the medium 2. These electromagnetic surface modes can become resonant modes inside the bulk band of the vacuum (i.e., for $\omega < ck_{||}$) or/and inside the minibands of the photonic crystal.

Here also in the same manner as done above for the interface and sandwich systems we can obtain from the basic response function $\boldsymbol{d}$ given in this section all the $d_{ij}(k_{\|}, \omega)$ $(i, j = 1, 2$ or $3)$ elements and then the electric and magnetic fields, as well as the Poynting vector.

All these results should prove to be useful in particular in calculations of Raman scattering of light from metal surfaces and of Brillouin scattering from opaque surfaces. Other potential applications are: calculations of polariton densities of states [18], studies of nonlinear interactions of interface or confined polaritons [1], calculations of the image force of a moving charge, studies of the interaction of electromagnetic radiation with randomly rough photonic crystal surfaces [2], determinations of the dispersion relations for surface polaritons, and studies of the attenuation of such polaritons [16].

Let us mention also another extension of the theory presented here. In the presence of a magnetic field, the dielectric tensor is no longer diagonal as considered here. Depending on the orientation of this magnetic field within the photonic crystal, three different geometries were considered: the Voigt [51], the perpendicular [52], and the Faraday [53] geometry. In these three cases, the theory was developed in order to obtain explicit expressions for the photonic crystal and heterostructure magnetoplasmons and the response functions.

17.7 Conclusion and perspectives

The interface response theory of electromagnetism is established for any composite dielectric. Its main general result given by Eqs. (17.22), (17.23), and (17.24) takes into account automatically all the boundary conditions of the electromagnetic fields.

We have applied this theory to the determination of the complete response functions of the interface between two dielectrics, of one dielectric slab sandwiched between two different semiinfinite dielectrics, and for dielectric photonic crystals.

These results are far from exhausting the numerous interesting physical problems of electrodynamics one can study easily with interface response theory. Without trying to be complete let us just list a few of them: N-layered photonic crystals, effects of an adsorbed layer on a semiinfinite photonic crystal, effects of different width of the surface layers, applications to different kinds of polaritons of the above results, etc. In particular, a closed-form expression was obtained [128] for the dispersion relation of polaritons in N-layered dielectric photonic crystals.

We outlined here in closed form the response function of an infinite dielectric photonic crystal and those of a semiinfinite one bounded by another semiinfinite homogeneous dielectric. It is indeed the basis for many other forthcoming investigations.

However, if one wants to calculate electromagnetic properties when nonlocal effects must be considered and additional boundary conditions must be imposed, one has to define the corresponding modifications in the interface space and calculate the appropriate response function. The macroscopic electrodynamic theory contained in Section 17.6 is also not directly applicable to photonic crystals with quantum-well structures. For such a photonic crystal, one would also have to recalculate the appropriate response functions.

Some similarity can be found between the present theory and that on "the Green's function of a macroscopic electromagnetic field in a dispersive plane-stratified anisotropic medium" [129]. Both are after the corresponding response or Green's function, by different mathematical approaches.

The complete response function can be used to solve many kinds of problems which, for a given physical system, can range from the calculation of light scattering spectra of interface polaritons, through calculation of cross-sections for the scattering of interface waves, and of volume waves, by surface irregularities (roughness) or point defects in the vicinity of the surface, and for nonlinear generation of mixing of surface polaritons, to the study of the interaction of external probes (e.g., charges, point defects) with surfaces [12].

This, of course, does not exhaustively treat all the possible applications one can think of. If any of the readers of this chapter is stimulated by them and contributes to the further development of this approach to the solutions of problems of composite material science, then that will be a source of considerable satisfaction to its authors.

Acknowledgments

The authors would like to thank Mamadou Lamarana BAH for his fruitful contribution in the original papers from which this chapter is inspired.

References

[1] See, for example, V.M. Agranovich, D.L. Mills (Eds.), Surface Polaritons, Modern Problems in Condensed Matter Sciences, vol. 1, North-Holland, Amsterdam, 1982.
[2] A.D. Bordman (Ed.), Electromagnetic Surface Modes, Wiley, New York, 1982.
[3] R.F. Wallis, G.I. Stegeman (Eds.), Electromagnetic Surface Excitations, Wave Phenomena, vol. 3, Springer, Berlin, 1986.
[4] Pochi Yeh, Optical Waves in Layered Media, John Wiley and Sons, New York, 1988.
[5] Gustav Mie, Beitäge zur Optik trüber Medien, speziell kolloidaler Metallösungen, Ann. Phys. 3 (1908) 377.
[6] H.C. Van de Hulst, Light Scattering by Small Particles, Wiley, New York, 1957.
[7] M. Kerker, The Scattering of Light and Other Electromagnetic Radiation, Academic, New York, 1969.
[8] L. Dobrzynski, Interface response theory of discrete composite systems, Surf. Sci. Rep. 6 (1986) 118.
[9] L. Dobrzynski, Interface response and scattering matrix theories for composite systems, Surf. Sci. 175 (1986) 1.
[10] L. Dobrzynski, Interface response theory of continuous composite materials, Surf. Sci. 180 (1987) 489.
[11] L. Dobrzynski, Interface response theory of continuous composite systems, Surf. Sci. Rep. 11 (1990) 139.

[12] M.G. Cottam, A.A. Maradudin, in: Surface Excitations, in: Modern Problems in Condensed Matter Sciences, vol. 9, North-Holland, Amsterdam, 1986.
[13] D. Schneider, F. Liaqat, E.H. El Boudouti, Y. El Hassouani, B. Djafari-Rouhani, W. Tremel, H-J. Butt, G. Fytas, Engineering the hypersonic phononic band gap of hybrid Bragg stacks, Nano Lett. 12 (2012) 3101.
[14] D. Schneider, F. Liaqat, O. El Abouti, E.H. El Boudouti, W. Tremel, H-J. Butt, B. Djafari-Rouhani, G. Fytas, Defect-controlled hypersound propagation in hybrid superlattices, Phys. Rev. Lett. 111 (2013) 164301.
[15] M.L. Bah, A. Akjouj, L. Dobrzynski, Response functions in layered dielectric media, Surf. Sci. Rep. 16 (1992) 95.
[16] M.L. Bah, A. Akjouj, E.H. El Boudouti, B. Djafari-Rouhani, L. Dobrzynski, The influence of capping layers on surface phonon polaritons in superlattices, J. Phys. Condens. Matter 7 (1995) 3445.
[17] A. Akjouj, M.L. Bah, E.H. El Boudouti, B. Djafari-Rouhani, L. Dobrzynski, Surface and interface s-polarized optical waves in superlattices, Vacuum 46 (1995) 621.
[18] M.L. Bah, A. Akjouj, E.H. El Boudouti, B. Djafari-Rouhani, L. Dobrzynski, Surface and interface optical waves in superlattices: transverse electric localized and resonant modes, J. Phys. Condens. Matter 8 (1996) 4171.
[19] M.L.H. Lahlaouti, A. Akjouj, B. Djafari-Rouhani, L. Dobrzynski, M. Hammouchi, E.H. El Boudouti, A. Nougaoui, Electromagnetic waves in finite superlattices with buffer and cap layers, J. Opt. Soc. Am. A 16 (1999) 1703.
[20] M.L.H. Lahlaouti, A. Akjouj, B. Djafari-Rouhani, L. Dobrzynski, Resonant and localized electromagnetic modes in finite superlattices, Phys. Rev. B 61 (2000) 2059.
[21] M.L.H. Lahlaouti, A. Akjouj, B. Djafari-Rouhani, L. Dobrzynski, M. Hammouchi, E.H. El Boudouti, A. Nougaoui, B. Kharbouch, Theoretical analysis of the density of states and phase times: application to resonant electromagnetic modes in finite superlattices, Phys. Rev. B 63 (2001) 035312.
[22] A. Mir, A. Akjouj, E.H. El Boudouti, B. Djafari-Rouhani, L. Dobrzynski, Large omnidirectional band gaps and selective transmission in one-dimensional multilayer photonic structures, Vacuum 63 (2001) 197.
[23] D. Bria, B. Djafari-Rouhani, E.H. El Boudouti, A. Mir, A. Akjouj, A. Nougaoui, Omnidirectional optical mirror in a cladded-superlattice structure, J. Appl. Phys. 91 (2002) 2569.
[24] Y. El Hassouani, E.H. El Boudouti, H. Aynaou, B. Djafari-Rouhani, V.R. Velasco, Comment on "Sensitivity of surface states to the stack sequence of one-dimensional photonic crystals", J. Opt. A, Pure Appl. Opt. 9 (2007) 308.
[25] A. Bouzidi, D. Bria, A. Akjouj, Y. Pennec, B. Djafari-Rouhani, A tiny gas-sensor system based on 1D photonic crystal, J. Phys. D, Appl. Phys. 48 (2015) 495102.
[26] O. El Abouti, E.H. El Boudouti, Y. El Hassouani, A. Noual, B. Djafari-Rouhani, Optical Tamm states in one-dimensional superconducting photonic crystal, Phys. Plasmas 23 (2016) 082115.
[27] D. Bria, B. Djafari-Rouhani, A. Akjouj, L. Dobrzynski, J.P. Vigneron, E.H. El Boudouti, A. Nougaoui, Band structure and omnidirectional photonic band gap in lamellar structures with left-handed materials, Phys. Rev. B 69 (2004) 66613.
[28] N. Ouchani, D. Bria, B. Djafari-Rouhani, A. Nougaoui, Transverse-electric/transverse-magnetic polarization converter using 1D finite biaxial photonic crystal, J. Opt. Soc. Am. A 24 (2007) 2710–2718.
[29] N. Ouchani, D. Bria, B. Djafari-Rouhani, A. Nougaoui, Defect modes in one-dimensional anisotropic photonic crystal, J. Appl. Phys. 106 (2009) 113107.

[30] N. Ouchani, A. El Moussaouy, H. Aynaou, Y. El Hassouani, E.H. El Boudouti, B. Djafari-Rouhani, Superluminal and negative delay times in isotropic-anisotropic one-dimensional photonic crystal, J. Appl. Phys. 122 (2017) 183106.
[31] N. Ouchani, A. El Moussaouy, H. Aynaou, Y. El Hassouani, E.H. El Boudouti, B. Djafari-Rouhani, Optical transmission properties of an anisotropic defect cavity in one-dimensional photonic crystal, Phys. Lett. A 382 (2018) 231.
[32] L. Dobrzynski, A. Akjouj, B. Djafari-Rouhani, J.O. Vasseur, J. Zemmouri, Giant gaps in photonic band structures, Phys. Rev. B 57 (1998) R9388.
[33] J.O. Vasseur, B. Djafari-Rouhani, L. Dobrzynski, A. Akjouj, J. Zemmouri, Defect modes in one-dimensional comblike photonic waveguides, Phys. Rev. B 59 (1999) 13446.
[34] L. Dobrzynski, B. Djafari-Rouhani, A. Akjouj, J.O. Vasseur, J. Zemmouri, Photonic tunneling between two wires, Prog. Surf. Sci. 67 (2001) 347.
[35] B. Djafari-Rouhani, E.H. El Boudouti, A. Akjouj, L. Dobrzynski, J.O. Vasseur, A. Mir, N. Fettouhi, J. Zemmouri, Surface states in one-dimensional photonic band gap structures, Vacuum 63 (2001) 177.
[36] A. Mir, A. Akjouj, J.O. Vasseur, B. Djafari-Rouhani, N. Fettouhi, E.H. El Boudouti, L. Dobrzynski, J. Zemmouri, Observation of large photonic band gaps and defect modes in one-dimensional networked waveguides, J. Phys. Condens. Matter 15 (2003) 1593.
[37] B. Djafari-Rouhani, E.H. El Boudouti, A. Akjouj, J.O. Vasseur, L. Dobrzynski, Transmission and filtering in photonic circuits: effects of absorption and amplification, Prog. Surf. Sci. 74 (2003) 389.
[38] E.H. El Boudouti, N. Fettouhi, A. Akjouj, B. Djafari-Rouhani, A. Mir, J.O. Vasseur, L. Dobrzynski, J. Zemmouri, Experimental and theoretical evidence for the existence of photonic bandgaps and selective transmissions in serial loop structures, J. Appl. Phys. 95 (2004) 1102.
[39] J.O. Vasseur, A. Akjouj, L. Dobrzynski, B. Djafari-Rouhani, E.H. El Boudouti, Photon, electron, magnon, phonon and plasmon mono-mode circuits, Surf. Sci. Rep. 54 (2004) 1.
[40] H. Aynaou, E.H. El Boudouti, Y. El Hassouani, A. Akjouj, B. Djafari-Rouhani, J.O. Vasseur, L. Dobrzynski, V.R. Velasco, Propagation of electromagnetic waves in periodic and Fibonacci photonic loop structures, Physica A 358 (2005) 68.
[41] H. Aynaou, E.H. El Boudouti, Y. El Hassouani, A. Akjouj, B. Djafari-Rouhani, J.O. Vasseur, A. Benomar, V.R. Velasco, Propagation and localization of electromagnetic waves in quasiperiodic serial loop structures, Phys. Rev. E 72 (2005) 056601.
[42] Y. El Hassouani, H. Aynaou, E.H. El Boudouti, B. Djafari-Rouhani, A. Akjouj, V.R. Velasco, Surface electromagnetic waves in Fibonacci superlattices: theoretical and experimental results, Phys. Rev. B 74 (2006) 035314.
[43] G.H. Cocoletzi, L. Dobrzynski, B. Djafari-Rouhani, H. Al-Wahsh, D. Bria, Electromagnetic wave propagation in quasi-one-dimensional comb-like structures made up of dissipative negative-phase-velocity materials, J. Phys. Condens. Matter 18 (2006) 3683.
[44] E.H. El Boudouti, Y. El Hassouani, H. Aynaou, B. Djafari-Rouhani, A. Akjouj, Electromagnetic wave propagation in quasi-periodic photonic circuits, J. Phys. Condens. Matter 19 (2007) 246217.
[45] E.H. El Boudouti, Y. El Hassouani, B. Djafari-Rouhani, H. Aynaou, Two types of modes in finite size one-dimensional coaxial photonic crystals: general rules and experimental evidence, Phys. Rev. E 76 (2007) 026607.
[46] A. Mouadili, E.H. El Boudouti, A. Soltani, A. Talbi, A. Akjouj, B. Djafari-Rouhani, Theoretical and experimental evidence of Fano-like resonances in simple monomode photonic circuits, J. Appl. Phys. 113 (2013) 164101.

[47] A. Mouadili, E.H. El Boudouti, A. Soltani, A. Talbi, A. Akjouj, B. Djafari-Rouhani, Electromagnetically induced absorption in detuned stub waveguides: a simple analytical and experimental model, J. Phys. Condens. Matter 26 (2014) 505901.

[48] A. Mouadili, E.H. El Boudouti, A. Soltani, A. Talbi, K. Haddadi, A. Akjouj, B. Djafari-Rouhani, Photonic demultiplexer based on electromagnetically induced transparency resonances, J. Phys. D, Appl. Phys. 52 (2019) 075101.

[49] Qiong-gan Zhu, Zhi-guo Wang, The Green's function method for metal–dielectric–metal SPP waveguide network, Europhys. Lett. 103 (2013) 17004.

[50] Qiong-gan Zhu, Zhi-guo Wang, Analytical method for metal–insulator–metal surface plasmon polaritons waveguide networks, Opt. Express 27 (2019) 303.

[51] M.S. Kushwaha, B. Djafari-Rouhani, Theory of magnetoplasmons in semiconductor superlattices in the Voigt geometry: a Green function approach, Phys. Rev. B 43 (1991) 9021.

[52] M.S. Kushwaha, B. Djafari-Rouhani, Theory of magnetoplasmons in semiconductor superlattices in the perpendicular geometry: a Green-function approach, Surf. Sci. 244 (1991) 336.

[53] M.S. Kushwaha, B. Djafari-Rouhani, Theory of magnetoplasmons in semiconductor superlattices in the Faraday geometry: a Green-function approach, Surf. Sci. 268 (1992) 457.

[54] I. Baryakhtar, Y. Demidenko, V. Lozovski, Interaction between localized-on-nanoparticles plasmonpolaritons and surface plasmon polaritons, J. Opt. Soc. Am. B 30 (2013) 1022.

[55] V.Z. Lozovski, M. Razumova, G. Strilchuk, Self-consistent approach to calculation of the optical response and absorption profiles of thin nanocomposite films, Plasmonics 10 (2015) 1779.

[56] V.Z. Lozovski, M.A. Razumova, Optical response of thin nanocomposite films with transverse inhomogeneity, Physica B, Condens. Matter 545 (2018) 420.

[57] V.Z. Lozovski, M.A. Razumova, T.A. Vasiliev, Electrodynamic configurational resonances in nanocomposite thin films, Plasmonics 13 (2018) 545.

[58] V.Z. Lozovski, C. Lienau, G.G. Tarasov, T.A. Vasyliev, Z.Ya. Zhuchenko, Configurational resonances in absorption of metal nanoparticles seeded onto a semiconductor surface, Results Phys. 12 (2019) 1197.

[59] L. Dobrzynski, Interface response theory of electromagnetism in composite dielectric materials, Surf. Sci. 180 (1987) 505.

[60] L. Dobrzynski, A. Akjouj, E.H. El Boudouti, et al., Interface response theory, in: Phononics, in: Interface Transmission Tutorial Book Series, Elsevier, 2017, p. 1.

[61] L. Dobrzynski, H. Puszkarski, Eigenvectors of composite systems. I. General theory, J. Phys. Condens. Matter 1 (1989) 1239.

[62] A.A. Maradudin, D.L. Mills, Scattering and absorption of electromagnetic radiation by a semi-infinite medium in the presence of surface roughness, Phys. Rev. B 11 (1975) 1392.

[63] D.L. Mills, E. Burstein, Polaritons: the electromagnetic modes of media, Rep. Prog. Phys. 37 (1974) 817.

[64] D.L. Mills, A.A. Maradudin, E. Burstein, Theory of the Raman effect in metals, Phys. Rev. Lett. 21 (1968) 1178.

[65] D.L. Mills, A.A. Maradudin, E. Burstein, in: G.B. Wright (Ed.), Light Scattering Spectra of Solids, Springer, New York, 1969, p. 399.

[66] B.I. Bennett, A.A. Maradudin, L.R. Swanson, A theory of the Brillouin scattering of light by acoustic phonons in a metal, Ann. Phys. (NY) 71 (1972) 357.

[67] D.L. Mills, Theory of second harmonic generation by surface polaritons on metals, Solid State Commun. 24 (1977) 669.

[68] L. Bonsall, A.A. Maradudin, Nonlinear interaction of surface polaritons, J. Appl. Phys. 49 (1978) 253.
[69] D.L. Mills, Image force on a moving charge, Phys. Rev. B 15 (1977) 763.
[70] A.A. Maradudin, W. Zierau, Effects of surface roughness on the surface-polariton dispersion relation, Phys. Rev. B 14 (1976) 484.
[71] A.A. Maradudin, Surface wave, in: I.J. Lalov (Ed.), Modern Problems of Surface Physics, Bulgarian Academy of Sciences, Sofia, 1981, p. 11.
[72] A.A. Maradudin, in: Surface Polaritons, in: Modern Problems in Condensed Matter Sciences, vol. 1, North-Holland, Amsterdam, 1982.
[73] M.G. Cottam, D.R. Tilley, in: Introduction to Surface and Photonic Crystal Excitations, Cambridge University Press, Cambridge, 1989.
[74] D.L. Mills, Attenuation of surface polaritons by surface roughness, Phys. Rev. B 12 (1975) 4036.
[75] G. Lifante, Integrated Photonics: Fundamentals, John Wiley and Sons Ltd., Atrium, England, 2003.
[76] R.G. Hunsperger, Integrated Optics: Theory and Technology, Springer-Verlag, Berlin, 2002.
[77] G.I. Stegeman, C.T. Seaton, Nonlinear integrated optics, J. Appl. Phys. 58 (1985) R57.
[78] E.J. Murphy, Integrated Optical Circuits and Components: Design and Applications, Marcel Dekker Inc., New York, 1999.
[79] J. Hukriede, D. Kip, E. Kratzig, Permanent narrow-band reflection holograms for infrared light recorded in $LiNbO_3$:Ti:Cu channel waveguides, Appl. Phys. B 72 (2001) 749.
[80] R. Ramponi, M. Malangoni, R. Osellame, Dispersion of the ordinary refractive-index change in a proton-exchanged $LiNbO_3$ waveguide, Appl. Phys. Lett. 78 (2001) 2098.
[81] G.L. Yip, J. Albert, Characterization of planar optical waveguides by K^+-ion exchange in glass, Opt. Lett. 10 (1985) 151.
[82] S. Iraj Najafi, T. Touam, R. Sara, M.P. Andrews, M.A. Fardad, Sol–gel glass waveguide and grating on silicon, J. Lightwave Technol. 16 (1998) 1640.
[83] P.D. Townsend, Development of ion implantation for optical applications, Vacuum 51 (1998) 301.
[84] Z.Y. Yin, B.K. Garside, Low-loss GeO_2 optical waveguide fabrication using low deposition rate rf sputtering, Appl. Opt. 21 (1982) 4324.
[85] T. Yamashiki, K. Tsuda, Low-loss waveguides of benzylidene-aniline derivatives by organic molecular beam heteroepitaxy, Opt. Lett. 28 (2003) 316.
[86] K.M. Wang, B.R. Shi, N. Cue, Y.Y. Zhu, F.R. Xiao, F. Lu, W. Li, Y.G. Liu, Waveguide laser film in erbium-doped $KTiOPO_4$ by pulsed laser deposition, Appl. Phys. Lett. 73 (1998) 1020.
[87] F. Bruno, M. Guidice, R. Recca, F. Testa, Plasma-enhanced chemical vapor deposition of low-loss SiON optical waveguides at 1.5-μm wavelength, Appl. Opt. 30 (1991) 4560.
[88] P.D. Townsend, P.J. Chandler, L. Zhang, Optical Effects of Ion Implantation, Cambridge University Press, Cambridge, 2006.
[89] Ch. Buchal, Ion implantation for optical applications, Nucl. Instrum. Methods B 96 (1995) 370.
[90] E.R. Schineller, R.P. Flam, D.W. Wilmot, Optical waveguides formed by proton irradiation of fused silica, J. Opt. Soc. Am. 58 (1968) 1171.
[91] C. Meneghini, K.L. Foulgoc, E.J. Knystautas, A. Villeneuve, T. Cardinal, K.A. Richardson, Proc. SPIE 3413 (1998) 46.
[92] J.R. Kulish, H. Franke, A. Singh, R.A. Lessard, E.J. Knystautas, Ion implantation, a method for fabricating light guides in polymers, J. Appl. Phys. 63 (1988) 2517.

[93] R.G. Wilson, D.A. Betts, D.K. Sadana, J.M. Zavada, R.G. Hunsperger, Proton, deuteron, and helium implantation into GaAs and $LiNbO_3$ for waveguide fabrication, J. Appl. Phys. 57 (1985) 5006.
[94] U.G. Akano, I.V. Mitchell, F.R. Shepherd, C.J. Miner, R. Rousina, Implant-damage isolation of InP and InGaAsP, J. Vac. Sci. Technol. A 11 (1993) 1016.
[95] P.D. Townsend, Ion implantation—an introduction, Contemp. Phys. 27 (1986) 241.
[96] S.P. Pogossian, H.L. Gall, Modeling planar leaky optical waveguides, J. Appl. Phys. 93 (2003) 2337.
[97] F. Chen, H. Hu, X.L. Wang, F. Lu, K.M. Wang, Property investigation of C^+-ion-implanted $LiNbO_3$ planar optical waveguides, J. Appl. Phys. 98 (2005) 044507.
[98] B. Sylla, L. Dobrzynski, H. Puszkarski, Eigenvectors of composite systems. II. Phonon eigenvectors in some layered materials, J. Phys. Condens. Matter 1 (1989) 1247.
[99] E. Yablonovitch, Inhibited spontaneous emission in solid-state physics and electronics, Phys. Rev. Lett. 58 (1987) 2059.
[100] S. John, Strong localization of photons in certain disordered dielectric super lattices, Phys. Rev. Lett. 58 (1987) 2486.
[101] K.M. Chen, A.W. Sparks, H.C. Luan, D.R. Lim, K. Wada, L.C. Kimerling, SiO_2/TiO_2 omnidirectional reflector and microcavity resonator via the sol–gel method, Appl. Phys. Lett. 75 (1999) 3805.
[102] B. Temelkuran, E. Ozbay, Experimental demonstration of photonic crystal based waveguides, Appl. Phys. Lett. 74 (1999) 486.
[103] M. Scalora, J.P. Dowling, C.M. Bowden, M.J. Bloemer, Optical limiting and switching of ultrashort pulses in nonlinear photonic band gap materials, Phys. Rev. Lett. 73 (1994) 1368.
[104] J.C. Knight, J. Broeng, T.A. Birks, P.St.J. Russell, Photonic band gap guidance in optical fibers, Science 282 (1998) 1476.
[105] Y. Fink, J.N. Winn, S. Fan, C. Chen, J. Michel, J.D. Joannopoulos, E.L. Thomas, A dielectric omnidirectional reflector, Science 282 (1998) 1679.
[106] J. Joannopoulos, R. Meade, J. Winn, Photonic Crystals, Princeton University Press, Princeton, 1995.
[107] E. Istrate, M. Charbonneau-Lefort, E.H. Sargent, Theory of photonic crystal heterostructures, Phys. Rev. B 66 (2002) 075121.
[108] E. Istrate, E.H. Sargent, Photonic crystal heterostructures and interfaces, Rev. Mod. Phys. 78 (2006) 455.
[109] I. Tamm, Über eine mögliche Art der Elektronenbindung an Kristalloberflächen, Z. Phys. 76 (1932) 849.
[110] A.P. Vinogradov, A.V. Dorofeenko, S.G. Erokhin, M. Inoue, A.A. Lisyansky, A.M. Merzlikin, A.B. Granovsky, Surface state peculiarities in one-dimensional photonic crystal interfaces, Phys. Rev. B 74 (2006) 045128.
[111] A.P. Vinogradov, A.V. Dorofeenko, A.M. Merzlikin, A.A. Lisyansky, Surface states in photonic crystals, Sov. Phys. Usp. 53 (2010) 243.
[112] A.V. Kavokin, I.A. Shelykh, G. Malpuech, Lossless interface modes at the boundary between two periodic dielectric structures, Phys. Rev. B 72 (2005) 233102.
[113] D. Gao, W. Mao, R. Zhang, J. Liu, Q. Zhao, W.Y. Tam, X. Wang, Tunable interface state in one dimensional composite photonic structure, Opt. Commun. 453 (2019) 124324.
[114] L.L. Lin, Z.Y. Li, Interface states in photonic crystal heterostructures, Phys. Rev. B 63 (2001) 033310.
[115] S. Feng, H.Y. Sang, Z.Y. Li, B.Y. Cheng, D.Z. Zhang, Sensitivity of surface states to the stack sequence of one-dimensional photonic crystals, J. Opt. A, Pure Appl. Opt. 7 (2005) 374.

[116] Y. Zheng, Y. Wang, J. Lou, P. Xu, Optical Tamm states in photonic structures made of inhomogeneous material, Opt. Commun. 406 (2018) 103.

[117] T.B. Wang, C.P. Yin, W.Y. Liang, J.W. Dong, H.Z. Wang, Electromagnetic surface modes in one-dimensional photonic crystals with dispersive metamaterials, J. Opt. Soc. Am. B 26 (2009) 1635.

[118] J. Barvestani, M. Kala, A. Soltani-Vala, A. Namdar, Backward surface electromagnetic waves in semi-infinite one-dimensional photonic crystals containing left-handed materials, Phys. Rev. A 77 (2008) 013805.

[119] A. Namdar, I.V. Shadrivov, Y.S. Kivshar, Backward Tamm states in left-handed metamaterials, Appl. Phys. Lett. 89 (2006) 114104.

[120] H. Hajian, B. Rezaei, A.S. Vala, M. Kala, Tuned switching of surface waves by a liquid crystal cap layer in one-dimensional photonic crystals, Appl. Opt. 51 (2012) 2909.

[121] J. Bashiri, B. Rezael, J. Barvestani, C.J. Zapata-Rodriguez, Bloch surface waves engineering in one-dimensional photonic crystals with a chiral cap layer, J. Opt. Soc. Am. B 36 (2019) 2106.

[122] L. Dobrzynski, Interface response theory of N-layered continuous superlattices, Surf. Sci. 182 (1987) 362.

[123] L. Dobrzynski, Interface response theory of electromagnetism in dielectric superlattices, Phys. Rev. B 37 (1988) 8027.

[124] G.E. Giuliani, J.J. Quinn, R.F. Wallis, Surface and bulk plasmon-polaritons in periodic metallic heterostructures, J. Phys. (Paris) Colloq. 45 (1984) C5-285.

[125] R.E. Camley, D.L. Mills, Collective excitations of semi-infinite superlattice structures: surface plasmons, bulk plasmons, and the electron-energy-loss spectrum, Phys. Rev. B 29 (1984) 1695.

[126] A. Yariv, P. Yeh, in: Optical Waves in Crystals, Wiley, New York, 1984.

[127] A. Yariv, P. Yeh, Electromagnetic propagation in periodic stratified media. II. Birefringence, phase matching, and X-ray lasers, J. Opt. Soc. Am. 67 (1977) 423.

[128] B. Djafari-Rouhani, L. Dobrzynski, Simple excitations in N-layered superlattices, Solid State Commun. 62 (1987) 609.

[129] Ya.A. Iosilevskii, The Green function of a macroscopic electromagnetic field in a dispersive plane-stratified anisotropic medium, J. Phys. (Paris) Colloq. Suppl. 4 (45) (1984) C5-305.

Optical Tamm states in semiinfinite layered photonic crystals

18

Abdellatif Akjouj[a], *El Houssaine El Boudouti*[b], *Leonard Dobrzyński*[a], *Bahram Djafari-Rouhani*[a], *and Gaëtan Lévêque*[a]
[a]Institute of Electronic, Microelectronics and Nanotechnology (IEMN), UMR CNRS 8520, Department of Physics, Faculty of Sciences and Technologies, Lille University, Villeneuve d'Ascq Cedex, France, [b]Laboratoire de Physique de la Matière et de Rayonnement, Département de Physique, Faculté des Sciences, Université Mohamed I, Oujda, Morocco

18.1 Introduction

Photonic crystals (PCs), the photonic analogues of electronic crystals, are a novel class of electromagnetic media characterized by natural or artificial periodic modulation of materials with different electromagnetic impedances [1,2]. In analogy with crystallographic structures in solid state physics, PC materials can be divided into three categories, namely one-dimensional (1D), two-dimensional (2D), and three-dimensional (3D) structures. These heterostructures have been the subject of intensive study, both theoretically and experimentally; this is due to the exceptional physical properties exhibited by these systems in comparison with bulk materials [3–6]. For instance, they show the capacity to control and engineer light propagation in the same manner as periodic electronic potentials do for electronic waves in semiconductors. These properties are rendered possible as a result of the existence of bandgaps induced by periodic refractive indices to which photons are submitted. Among the different existing PCs, the 1D systems, called dielectric Bragg mirrors or superlattices, have been extensively studied due to their relative ease of fabrication and the simple analytical models required [4].

In addition to the conventional dielectric 1D photonic crystals (PCs), the plasma PCs composed of alternating layers of dielectric and plasma layers have been intensively studied [7–9]. The effects of incidence angles, plasma density, plasma layer width, and the external magnetized field on the band structures have been demonstrated. Also it has been shown that the presence of a defect such as magnetized plasma layers [10], breaking the periodicity of the structure, modifies the band structure of PCs where new electromagnetic modes are created within the bandgaps as evanescent modes. The presence of surfaces in a PC plays also the same role as defects; these localized modes at the extremities of the structure have been studied in many papers as we will shortly review below. In this chapter we shall be interested in the surface localized modes of three types of PCs: (i) a new type of PC constituted by alternate layers

Photonics. https://doi.org/10.1016/B978-0-12-819388-4.00029-0

of a dielectric and a superconductor, the interest in which has been recognized only recently, especially in view of the sensitivity of their band structure to the temperature and hence their usefulness for the sensing applications [11–15], (ii) a conventional PC constituted by alternating two different dielectric layers, and (iii) the influence of capping InP layer on surface phonon–polaritons in GaAs–AlAs PC.

The surface localized modes were theoretically predicted in dielectric PCs by Yeh et al. in 1977 [16]. Experimental evidence of the existence of these modes has been conducted on PCs consisting of GaAs and $Al_{0.2}Ga_{0.8}As$ bilayers [17,18], and on PCs made of ZnS/Cryolite bilayers [19]. Since then, the existence of surface modes in 1D-multilayered structures composed out of different materials has been the subject of a great interest [20–22]. Many years ago, some of us have determined analytically the dispersion relation of surface waves in semiinfinite PCs with a cap layer [23]. Different methods have also been proven to highlight the presence of surface modes in PCs as in [24]. Lahlaouti et al. [25] have studied electromagnetic surface modes in a finite PC deposited on a substrate with an embedded buffer layer, and ended with a thin cap layer. Surface modes occurring in the junction of two different 1D PCs placed in series were envisaged for the design of narrow-bandpass filters [26]. Surface waves properties have been analyzed in different PCs such as in semiconductor and dielectric layers [27], quasicrystals [28], dielectric Bragg reflectors [29], metals and metamaterials [30,31].

Transverse electric surface modes characteristics in semiinfinite 1D plasma-based PCs were studied recently [32]. Particular surface states called optical Tamm states (OTSs) have attracted a lot of attention due to their huge potential for the fabrication of polaritons lasers [33,34], optical data storage, and applications in analytical chemistry and medicine [11,35,36]. Originally, Tamm states are electronic surface energies that are localized on crystals surface that were predicted by Tamm in 1932 [37]. Analogously, OTSs are surface modes that could be generated by perturbing properly periodic optical impedances at the surface of PCs [38,39]. Unlike surface plasmon modes that are confined along the interface between a dielectric and a conductor, OTSs have a smaller wavevector in magnitude (parallel to the interface) compared with that of the light wave in vacuum. This enables a direct optical excitation of such modes both for transverse magnetic (TM) and transverse electric (TE) incident polarizations [40–44]. Kaliteevski et al. [29] have investigated theoretically OTS modes at the interface of a gold metal and a PC constituted of GaAs and AlAs layers for TM and TE polarizations. First experimental evidence of direct optical excitation of OTS modes at the interface of GaAs/AlAs based 1D periodic structure with a gold metal was reported by Sasin et al. [40] Further experiments have been realized to show the existence of OTSs in 1D periodic structures [12,45], photonic heterostructures based on microstrip transmission lines [39], and at the interface between 1D magnetophotonic and nonmagnetic PCs [46].

Coupling of OTSs at the surface of a metal with cavity polaritons has been proven to produce the formation of hybrid plasmon–exciton–polariton modes in a semiconductor based 1D PC on which a thin gold layer is deposited [47]. Bruckner et al. [45] have proposed an analytical model with an experimental setup to observe and describe the interaction of OTSs and cavity photons modes. TM hybrid modes formed due to a coupling between Tamm plasmon polariton modes and surface plasmon polariton (SPP) modes have been employed to propose a highly sensitive sensor $\geq$ 900 nm/RIU [12]. Other contributions have shown the possibility of experimentally observing surface modes at the 1D PC metal interface [48–51].

Some years ago [52–56], the existence of bandgap structures, as well as omnidirectional gaps, in new superconductor–dielectric based PCs was intensively investigated. In comparison with metal materials, superconductors for which the dielectric function depends on temperature T have shown the possibility to manipulate bandgaps as a function of T. However, despite of the attractive optical phenomenon exhibited by superconductor–dielectric PCs, the existence of surface (or OTS) modes has not been investigated in such materials. Therefore, it may be of great interest to examine the possibility of existence, as well as the properties, of OTS modes at the interface between a semiinfinite superconductor–dielectric PC and vacuum, which can be useful for sensing applications [11–15].

In this chapter, we investigate different possible surface modes that could be supported by a thin layer deposited on a semiinfinite PC made out of a periodic repetition of superconductor and dielectric materials [57]. The cap layer ending the whole structure can be a superconductor or a dielectric and is in contact with vacuum. The calculated band structures and surface modes are analyzed for incident waves of both TE and TM polarizations. Furthermore, the temperature effect on bulk and surface modes is also investigated. We show that the TE modes give similar results as in dielectric–dielectric PCs; however, TM modes exhibit new SPP modes without analogue in dielectric–dielectric PCs. Also, we recall at the end of this chapter, the results concerning the case of OTS modes induced by a dielectric cap layer at the surface of a dielectric–dielectric PC, as well as the case of the influence of InP cap layer on surface phonon–polaritons in GaAs–InAs PCs.

This chapter is organized as follows. Section 18.2 gives the analytical results obtained for the densities of states in semiinfinite PC with a cap layer and those of the dispersion relation of bulk and surface modes. These expressions are obtained using the Green's function method [58,59]. In Section 18.3, we show the existence of OTS modes in a PC constituted out of (i) superconductor–dielectric layers and terminated with a dielectric or superconductor thin layer on its top (Section 18.3.1), (ii) dielectric–dielectric layers terminated with a dielectric thin layer (Section 18.3.2), and (iii) the influence of capping InP layer on surface phonon–polaritons in GaAs–InAs PC (Section 18.3.3). Section 18.3.1 is devoted to both TE and TM polarizations, Section 18.3.2 for TE polarization only, and Section 18.3.3 for TM polarization only. Section 18.4 summarizes the principal results of this chapter.

18.2 Theoretical formalism

The PC is formed out of an infinite repetition of two different slabs, labeled by the unit-cell index n. Each of these slabs of width d_i is labeled by the index $i = 1$ or 2, within the unit cell n. All the interfaces are taken to be parallel to the (x_1, x_2)-plane.

A space position along the x_3 axis in medium i belonging to the unit cell n is indicated by (n, i, x_3), where $-d_i/2 < x_3 < d_i/2$. The period of the photonic crystal is called $D = d_1 + d_2$.

It was shown [58,59] that all the elements of the response function of layered isotropic dielectrics, and in particular those of a superlattice, $\boldsymbol{g}(k_\parallel, \omega; x_3, x_3')$, can be easily deduced from one basic element $g(k_\parallel, \omega; x_3, x_3')$ where $k_\parallel$ is the modulus of the propagation vector parallel to the interfaces and ω the frequency. The bulk basic equation for one homogeneous infinite medium i can be written as

$$\left(\frac{F_i}{\alpha_i}\frac{\partial^2}{\partial x_3^2} - \alpha_i^2\right) G_i(k_\parallel, \omega; x_3, x_3') = \delta(x_3 - x_3'), \quad -\infty < x_3, x_3' < +\infty, \tag{18.1}$$

with

$$\alpha_i(k_\parallel, \omega) = \left(k_\parallel^2 - \frac{\omega^2}{c^2}\varepsilon_i(\omega)\right)^{1/2}, \tag{18.2}$$

where c is the speed of light, $\varepsilon_i(\omega)$ the dielectric function, $G_i(k_\parallel, \omega; x_3, x_3')$ the bulk response function and

$$F_i = \alpha_i, \qquad \text{for TE polarization modes,} \tag{18.3}$$

and

$$F_i = -\frac{\omega^2}{c^2}\frac{\varepsilon_i(\omega)}{\alpha_i}, \qquad \text{for TM polarization modes.} \tag{18.4}$$

The implicit expression giving the bulk dispersion relation for such an infinite superlattice is

$$\cos(k_3 D) = C_1 C_2 + \frac{1}{2}\left(\frac{F_1}{F_2} + \frac{F_2}{F_1}\right) S_1 S_2, \tag{18.5}$$

where

$$C_i = \cosh(\alpha_i d_i), \tag{18.6}$$

$$S_i = \sinh(\alpha_i d_i), \tag{18.7}$$

and k_3 is the component perpendicular to the slabs of the propagation vector $\boldsymbol{k} \equiv (\boldsymbol{k}_\parallel, k_3)$.

18.2.1 Green's function for a semiinfinite photonic crystal–cap layer–vacuum system

Knowing the Green's function for an infinite homogeneous medium given by

$$G_i(x_3, x_3') = -\frac{1}{2F_i} e^{-\alpha_i |x_3 - x_3'|}, \tag{18.8}$$

we obtained the following results with the help of the interface response theory [59].

18.2.1.1 For the infinite layered photonic crystal

(a) The elements $g(m, m')$, where $m \equiv (n, i, \pm d_i/2)$ of the response function g between the different interface planes, as functions of C_i, S_i, F_i, t, and η (Eqs. (18.3), (18.4), (18.6), and (18.7)), are

$$g\left(n, 1, -\frac{d_1}{2}; n', 1, -\frac{d_1}{2}\right) = \left(\frac{S_1 C_2}{F_1} + \frac{S_2 C_1}{F_2}\right) \frac{t^{|n-n'|+1}}{t^2 - 1}, \tag{18.9}$$

$$g\left(n, 1, -\frac{d_1}{2}; n', 1, +\frac{d_1}{2}\right) = \frac{t}{t^2 - 1} \left(\frac{S_2}{F_2} t^{|n-n'|} + \frac{S_1}{F_1} t^{|n-n'-1|}\right), \tag{18.10}$$

$$g\left(n, 1, +\frac{d_1}{2}; n', 1, -\frac{d_1}{2}\right) = \frac{t}{t^2 - 1} \left(\frac{S_2}{F_2} t^{|n-n'|} + \frac{S_1}{F_1} t^{|n-n'+1|}\right), \tag{18.11}$$

$$g\left(n, 1, +\frac{d_1}{2}; n', 1, +\frac{d_1}{2}\right) = \left(\frac{S_1 C_2}{F_1} + \frac{S_2 C_1}{F_2}\right) \frac{t^{|n-n'|+1}}{t^2 - 1}, \tag{18.12}$$

where

$$t = \begin{cases} \eta - (\eta^2 - 1)^{1/2}, & \eta > 1, \\ \eta + i(1 - \eta^2)^{1/2}, & -1 < \eta < 1, \\ \eta + (\eta^2 - 1)^{1/2}, & \eta < -1, \end{cases} \tag{18.13}$$

and

$$\eta = C_1 C_2 + \frac{1}{2}\left(\frac{F_1}{F_2} + \frac{F_2}{F_1}\right) S_1 S_2. \tag{18.14}$$

(b) The element of this response function between any two points of the infinite layered photonic crystal was found to be

$$g(n,i,x_3;n',i',x_3')=\delta_{nn'}\delta_{ii'}U_i(x_3,x_3')+\frac{1}{S_iS_{i'}}\left\{\sinh\left[\alpha_i\left(\frac{d_i}{2}-x_3\right)\right];\right.$$
$$\left.\sinh\left[\alpha_i\left(\frac{d_i}{2}+x_3\right)\right]\right\}g(M_m,M_{m'})\begin{pmatrix}\sinh\left[\alpha_{i'}\left(\frac{d_{i'}}{2}-x_3'\right)\right]\\ \sinh\left[\alpha_{i'}\left(\frac{d_{i'}}{2}+x_3'\right)\right]\end{pmatrix},\tag{18.15}$$

where

$$U_i(x_3,x_3')=-\frac{e^{-\alpha_i|x_3-x_3'|}}{2F_i}+\frac{1}{2F_iS_i}\left\{\sinh\left[\alpha_i\left(\frac{d_i}{2}-x_3'\right)\right]\exp\left[-\alpha_i\left(\frac{d_i}{2}+x_3\right)\right]\right.\tag{18.16}$$
$$\left.+\sinh\left[\alpha_i\left(\frac{d_i}{2}+x_3'\right)\right]\exp\left[-\alpha_i\left(\frac{d_i}{2}-x_3\right)\right]\right\}.$$

In Eq. (18.15) the last three terms are the product of a 1×2 matrix, the 2×2 matrix $g(M_m,M_{m'})$, and a 2×1 matrix; $\mathbf{g}(M_m,M_{m'})$ is the 2×2 matrix formed out of the elements given by Eqs. (18.9)–(18.12), for $m=(n,1,\pm_1/2)$ and $m'=(n',1,\pm d_1/2)$.

18.2.1.2 For the semiinfinite photonic crystal with a surface cap layer

The semiinfinite superlattice with a surface cap layer under consideration here is terminated by the unit cell $n=0$ formed of a surface layer $i=c$ of width d_c deposited on the $i=1$ layer of the semiinfinite superlattice. The underneath unit cell $n=-1$ is formed out of the $i=2$ and then the $i=1$ layers of the superlattice, and so on (see Fig. 18.1).

(a) In this chapter, we need the following elements of the response function d between different interface planes:

$$d\left(0,c,-\frac{d_c}{2};0,c,\frac{d_c}{2}\right)=d\left(0,c,\frac{d_c}{2};0,c,-\frac{d_c}{2}\right)=\frac{C_1S_2/F_2+C_2S_1/F_1}{(1+F_vS_c/F_cC_c)C_c\Delta},\tag{18.17}$$

$$d\left(0,c,\frac{d_c}{2};0,c,\frac{d_c}{2}\right)=\frac{C_1S_2/F_2+C_2S_1/F_1+(S_c/F_cC_c)[C_1C_2+(F_1/F_2)S_1S_2-t]}{(1+F_vS_c/F_cC_c)\Delta},\tag{18.18}$$

$$d\left(0, c, -\frac{d_c}{2}; 0, c, -\frac{d_c}{2}\right) = \frac{1}{\Delta}\left(\frac{C_1 S_2}{F_2} + \frac{C_2 S_1}{F_1}\right), \tag{18.19}$$

and for $n, n' \leq 0$ and $i \neq c$,

$$d\left(n, 1, -\frac{d_1}{2}; n', 1, -\frac{d_1}{2}\right) = \frac{t}{t^2 - 1}\left\{\left(\frac{C_1 S_2}{F_2} + \frac{C_2 S_1}{F_1}\right) t^{|n-n'|} - t^{-n-n'}\left(\frac{S_1 t}{F_1} + \frac{S_2}{F_2}\right)\frac{Y}{\Delta}\right\}, \tag{18.20}$$

$$d\left(n, 1, -\frac{d_1}{2}; n', 1, +\frac{d_1}{2}\right) = \frac{t}{t^2 - 1}\left\{\frac{S_2 t^{|n-n'|}}{F_2} + \frac{S_1 t^{|n-n'-1|}}{F_1} - t^{-n-n'}\left(\frac{C_1 S_2}{F_2} + \frac{C_2 S_1}{F_1}\right)\frac{Y}{\Delta}\right\}, \tag{18.21}$$

$$d\left(n, 1, +\frac{d_1}{2}; n', 1, -\frac{d_1}{2}\right) = \frac{t}{t^2 - 1}\left\{\frac{S_2 t^{|n-n'|}}{F_2} + \frac{S_1 t^{|n-n'+1|}}{F_1} - t^{-n-n'}\left(\frac{C_1 S_2}{F_2} + \frac{C_2 S_1}{F_1}\right)\frac{Y}{\Delta}\right\}, \tag{18.22}$$

$$d\left(n, 1, +\frac{d_1}{2}; n', 1, +\frac{d_1}{2}\right) = \frac{t}{t^2 - 1}\left\{\left(\frac{C_1 S_2}{F_2} + \frac{C_2 S_1}{F_1}\right) t^{|n-n'|} - t^{-n-n'}\left(\frac{S_1}{F_1} + \frac{S_2 t}{F_2}\right)\frac{Y}{\Delta}\right\}, \tag{18.23}$$

where

$$\Delta = C_1 C_2 + \frac{F_2}{F_1} S_1 S_2 - \frac{1}{t} - R F_v \left(\frac{C_1 S_2}{F_2} + \frac{C_2 S_1}{F_1}\right), \tag{18.24}$$

$$R = \frac{1 + F_c S_c / F_v C_c}{1 + F_v S_c / F_c C_c}, \tag{18.25}$$

$$Y = C_2 - C_1 t - R F_v \left(\frac{S_1}{F_1} t + \frac{S_2}{F_2}\right). \tag{18.26}$$

(b) The elements of this response function between any two points of this heterostructure can also be obtained in closed form. As in the present study we need only the trace of this response function, we give here only those expressions for two points belonging both to the layered photonic crystal or both to the surface cap layer, or both to vacuum.

(i) When the two points are inside the superlattice, $d(n, i, x_3; n', i', x_3')$ is given by Eq. (18.15) in which one has to replace $\boldsymbol{g}(M_m, M_{m'})$ by $\boldsymbol{d}(M_m, M_{m'})$ given by Eqs. (18.20)–(18.23).

(ii) When the two points are inside the surface cap layer,

$$\begin{aligned} d(0, c, x_3; 0, c, x_3') &= U_c(x_3, x_3') \\ &+ \frac{1}{S_c^2}\left\{\sinh\left[\alpha_c\left(\frac{d_c}{2} - x_3\right)\right]; \sinh\left[\alpha_c\left(\frac{d_c}{2} + x_3\right)\right]\right\} \\ &\times \boldsymbol{d}(M_c, M_c)\begin{pmatrix} \sinh\left[\alpha_c\left(\frac{d_c}{2} - x_3'\right)\right] \\ \sinh\left[\alpha_c\left(\frac{d_c}{2} + x_3'\right)\right] \end{pmatrix}, \end{aligned} \tag{18.27}$$

where

$$\begin{aligned} U_c(x_3, x_3') &= -\frac{e^{-\alpha_c|x_3 - x_3'|}}{2F_c} \\ &+ \frac{1}{2F_cS_c}\left\{\sinh\left[\alpha_c\left(\frac{d_c}{2} - x_3'\right)\right]\exp\left[-\alpha_c\left(\frac{d_c}{2} + x_3\right)\right]\right. \\ &\left. + \sinh\left[\alpha_c\left(\frac{d_c}{2} + x_3'\right)\right]\exp\left[-\alpha_c\left(\frac{d_c}{2} - x_3\right)\right]\right\}, \end{aligned} \tag{18.28}$$

and $\boldsymbol{d}(M_c, M_c)$ is the 2×2 matrix formed by the elements given by Eqs. (18.17)–(18.19) for $M_c = (0, c, \pm d_c/2)$.

(iii) When the two points are inside vacuum,

$$\begin{aligned} d(x_3, x_3') = \frac{-1}{2F_v} + \left(\frac{1}{2F_v} + \left[\frac{C_1S_2}{F_2} + \frac{C_2S_1}{F_1} + \frac{S_c}{F_cC_c}\left(C_1C_2 + \frac{F_1}{F_2}S_1S_2 - t\right)\right]\right. \\ \left. \times\left\{\left(1 + \frac{F_vS_c}{F_cC_c}\right)\Delta\right\}^{-1}\right)\exp\left[-\alpha_v(x_3 + x_3')\right]. \end{aligned} \tag{18.29}$$

18.2.2 Density of states

Knowing the response functions of an infinite layered PC given in Section 18.2.1, one obtains for a given value of $k_\parallel$ the local and total density of states for a semiinfinite

PC with a surface cap layer. We shall indicate at the end of this section how one can obtain from these quantities similar results for a limiting case, namely the case of the interface between a photonic crystal and a homogeneous substrate.

18.2.2.1 The local densities of states

The local densities of states on the plane (n, i, x_3) are given by

$$n(\omega^2, k_\parallel; n, i, x_3) = -\frac{\varepsilon_i}{\pi c^2}\text{Im}\, d^+(\omega^2, k_\parallel; n, i, x_3; n, i, x_3), \tag{18.30}$$

where

$$d^+(\omega^2) = \lim_{\Gamma\to 0} d(\omega^2 + i\Gamma) \tag{18.31}$$

and $d(\omega^2)$ is the response function whose elements are given in [23]. The density of states can also be given as a function of ω, instead of ω^2 using the well known relation $n(\omega) = 2\omega n(\omega^2)$. From the elements of the response function given in [23], we obtained the following explicit expressions for the local densities of states on the surface of the semiinfinite PC with a cap layer $(n = 0, i = c)$ of width d_c (see Fig. 18.1):

$$\begin{aligned} n_s\left(\omega^2, k_\parallel; 0, c, \frac{d_c}{2}\right) = &-\frac{1}{\pi}\text{Im}\left[\frac{C_1 S_2}{F_2} + \frac{C_2 S_1}{F_1}\right. \\ &\left. + \frac{S_c}{F_c C_c}\left(C_1 C_2 + \frac{F_1}{F_2} S_1 S_2 - t\right)\right] \\ &\times \frac{1}{(1 + F_v S_c / F_c C_c)\Delta}, \end{aligned} \tag{18.32}$$

where C_c, S_c, F_c have the same definitions as C_i, S_i, F_i given by Eqs. (18.3)–(18.4) and (18.6)–(18.7) for $i = c$.

In the same manner, the local density of states at the interface between the cap layer and the semiinfinite photonic crystal was found to be

$$n_i\left(\omega^2, k_\parallel; 0, c, -\frac{d_c}{2}\right) = -\frac{1}{\pi}\text{Im}\left(\frac{C_1 S_2}{F_2} + \frac{C_2 S_1}{F_1}\right)\Delta^{-1}. \tag{18.33}$$

18.2.2.2 The total density of states

The total density of states for a given value of $k_\parallel$ is obtained by integrating over x_3 and summing over n and i the local density $n(\omega^2, k_\parallel; n, i, x_3)$. A particularly interesting quantity is the variation of the total density of states between the semiinfinite PC with the cap layer $n = c$ and the infinite PC having the same number of slabs as

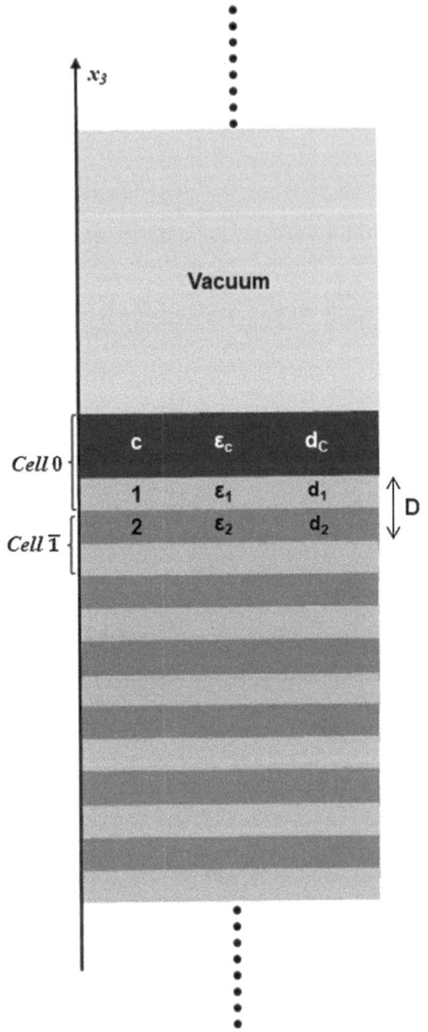

Figure 18.1 Schematic representation of a semiinfinite PC made of two layers ($i = 1, 2$) with a cap layer ($n = 0; i = c$) in contact with vacuum; d_c, d_1, and d_2, respectively, are the thicknesses of the cap layer and of the two different slabs out of which the semiinfinite photonic crystal is built; D is the period of the superlattice.

the semiinfinite photonic crystal without the cap layer. The variation $\Delta n(\omega^2)$ can be written as the sum of the variations $\Delta n_1(\omega^2)$ and $\Delta n_2(\omega^2)$ of the density of states in slabs 1 and 2, the density of states $n_0(\omega^2)$ inside the cap layer, and the variation $\Delta n_v(\omega^2)$ in the density of states in vacuum:

$$\Delta n(\omega^2) = \Delta n_1(\omega^2) + \Delta n_2(\omega^2) + \Delta n_v(\omega^2) + n_c(\omega^2), \tag{18.34}$$

where

$$\Delta n_1(\omega^2) = -\frac{\varepsilon_1}{\pi c^2} \sum_{n=-\infty}^{0} \mathrm{Im}\left(\int_{-\frac{d_1}{2}}^{+\frac{d_1}{2}} \left[d(n,1,x_3;n,1,x_3) - g(n,1,x_3;n,1,x_3) \right] dx_3 \right), \tag{18.35}$$

$$\Delta n_2(\omega^2) = -\frac{\varepsilon_2}{\pi c^2} \sum_{n=-\infty}^{-1} \mathrm{Im}\left(\int_{-\frac{d_2}{2}}^{+\frac{d_2}{2}} \left[d(n,2,x_3;n,2,x_3) - g(n,2,x_3;n,2,x_3) \right] dx_3 \right), \tag{18.36}$$

$$n_c(\omega^2) = -\frac{\varepsilon_c}{\pi c^2} \mathrm{Im} \int_{-\frac{d_c}{2}}^{+\frac{d_c}{2}} d(0,c,x_3;0,c,x_3) dx_3, \tag{18.37}$$

$$\Delta n_v(\omega^2) = -\frac{\varepsilon_v}{\pi c^2} \mathrm{Im}\left(\int_0^{+\infty} \left[d(x_3,x_3) - G_v(x_3,x_3) \right] dx_3 \right). \tag{18.38}$$

Here d and g are the response functions of the semiinfinite photonic crystal with the cap layer and of the infinite photonic crystal, respectively; G_v represents the response function of the infinite homogeneous vacuum. With the help of the explicit expressions of the response functions given in [23], we obtain the following expressions:

$$\Delta n_1(\omega^2) = -\frac{\varepsilon_1}{\pi c^2} \mathrm{Im}\left(\frac{t}{(t^2-1)} \left\{ \frac{S_1}{\alpha_1 F_1} \left[C_2 S_1 + \frac{1}{2} C_1 S_2 \left(\frac{F_1}{F_2} + \frac{F_2}{F_1} \right) \right] + \frac{d_1 S_2}{2F_2} \left(1 - \frac{F_2^2}{F_1^2} \right) \right\} \frac{Y}{\Delta} \right), \tag{18.39}$$

$$\Delta n_2(\omega^2) = -\frac{\varepsilon_2}{\pi c^2} \mathrm{Im}\left(\frac{t}{(t^2-1)} \left\{ \frac{S_2}{\alpha_2 F_2} \left[C_1 S_2 + \frac{1}{2} C_2 S_1 \left(\frac{F_1}{F_2} + \frac{F_2}{F_1} \right) \right] + \frac{d_2 S_1}{2F_1} \left(1 - \frac{F_1^2}{F_2^2} \right) \right\} \frac{Y}{\Delta} \right), \tag{18.40}$$

$$n_c(\omega^2) = -\frac{\varepsilon_c}{2\pi c^2}\mathrm{Im}\left\{\frac{S_c}{\alpha_c C_c}\left[\left(\frac{C_1S_2}{F_2}+\frac{C_2S_1}{F_1}\right)\right.\right.$$
$$\left.-\frac{F_v}{F_c^2}\left(C_1C_2+\frac{F_1}{F_2}S_1S_2-t\right)\right]\frac{1}{1+F_vS_c/F_cC_c}$$
$$\left.+d_c\left[\left(\frac{C_1S_2}{F_2}+\frac{C_2S_1}{F_1}\right)+R\frac{F_v}{F_c^2}\left(C_1C_2+\frac{F_1}{F_2}S_1S_2-t\right)\right]\right\}\frac{1}{\Delta}, \tag{18.41}$$

and

$$\Delta n_v(\omega^2) = -\frac{\varepsilon_v}{\pi c^2}\mathrm{Im}\left[\frac{1}{2\alpha_v}\left(\frac{1}{2F_v}+\left\{\left(\frac{C_1S_2}{F_2}+\frac{C_2S_1}{F_1}\right)\right.\right.\right.$$
$$\left.\left.\left.+\frac{S_c}{F_cC_c}\left(C_1C_2+\frac{F_1}{F_2}S_1S_2-t\right)\right\}\frac{1}{(1+F_vS_c/F_cC_c)\Delta}\right)\right]. \tag{18.42}$$

At the limits of the bulk bands of the PC given by $t(\omega_0)=\pm 1$, an expansion to first order in $\omega-\omega_0$ provides

$$\frac{t}{(t^2-1)^2} = \frac{1}{8}\left\{\left(\frac{d\eta}{d\omega}\right)_{\omega_0}\right\}^{-1}\left\{P\left(\frac{1}{\omega-\omega_0}\right)-i\pi\delta(\omega-\omega_0)\right\}, \tag{18.43}$$

and then

$$\Delta n_1(\omega^2)+\Delta n_2(\omega^2) = -\frac{1}{4}\delta(\omega-\omega_0). \tag{18.44}$$

So, the creation of a semiinfinite photonic crystal from an infinite one gives rise to δ peaks of weight $-1/4$ in the density of states at the edges of the photonic crystal bulk bands.

18.2.3 Localized states

When the denominator of $\Delta n(\omega^2)$ vanishes for a frequency lying inside the gaps of the infinite PC, one obtains localized states within the cap layer which decay exponentially inside the bulk of the PC. The explicit expression giving these localized states is given by

$$C_1S_2\left(\frac{F_2}{RF_v}-\frac{RF_v}{F_2}\right)+S_1S_2\left(\frac{F_2}{F_1}-\frac{F_1}{F_2}\right)+C_2S_1\left(\frac{F_1}{RF_v}-\frac{RF_v}{F_1}\right)=0, \tag{18.45}$$

together with the condition

$$\left| C_1C_2 + \frac{F_2}{F_1} S_1 S_2 - RF_v \left(\frac{C_1 S_2}{F_2} + \frac{C_2 S_1}{F_1} \right) \right| > 1. \tag{18.46}$$

The latter condition ensures that the OTS modes decay far from the surface of the PC.

18.2.4 The case of a semiinfinite photonic crystal without a cap layer

This system is obtained when the thickness d_c of the cap layer is going to zero; then $S_c \to 0$ and the above results (18.32), (18.39), (18.40), (18.42), (18.45), and (18.46) remain valid for a semiinfinite superlattice ending with a complete $i = 1$ surface layer in contact with vacuum (Fig. 18.1). Let us note that in this limit $n_c(\omega)$ (Eq. (18.41)) vanishes.

In the limit where the cap layer $i = c$ is of the same nature as the $i = 2$ superlattice layer and $d_c = d_s < d_2$, the same results provide the localized modes for a semiinfinite superlattice ending with an incomplete $i = 2$ surface layer (see Fig. 18.1). In this case, we can calculate the variation in the density of states between such a semiinfinite photonic crystal and the same amount of the bulk photonic crystal, using in Eq. (18.40) $\Delta n_2(\omega^2)$ integrated to $d_s/2$ rather than to $d_2/2$ in the last layer, and taking $n_c(\omega^2) = 0$.

A particularly interesting result is obtained when considering two PCs obtained by cleaving an infinite PC along the space position $x_3 = d_s$ in such a way that one obtains one PC with an incomplete surface layer d_s and its complementary PC with a surface layer of width $d_2 - d_s$. The variation in the $\boldsymbol{k}_{\parallel}$-resolved density of states when creating two complementary semiinfinite PCs from an infinite PC is equal to zero [60] for frequencies ω which belong simultaneously to a PC bulk band and are above the light line in vacuum.

18.2.5 The case of an interface between a semiinfinite photonic crystal and a homogeneous substrate

This system is obtained by replacing the vacuum by an homogeneous substrate. All the above expressions (18.32)–(18.46) and limits (Section 18.2.4) remain valid upon replacing the index v by s for the substrate.

18.3 Numerical applications and discussion of the results

18.3.1 Optical Tamm states in 1D superconductor–dielectric photonic crystal

In Fig. 18.1, we replace the dielectric 1 of the semiinfinite PC by superconductor characterized by a thickness $d_1 \equiv d_S$ and a permittivity $\varepsilon_1 \equiv \varepsilon_S$, and by dielectric medium characterized by a thickness $d_2 \equiv d_D$ and a permittivity $\varepsilon_2 \equiv \varepsilon_D$. The superconductor dielectric function ε_S could be determined analytically based on two-fluid model [61] as

$$\varepsilon_S = 1 - \frac{\omega_p^2}{\omega^2}, \tag{18.47}$$

where $\omega_p^2 = c^2/\lambda_L^2$ is the threshold frequency of the superconductor and λ_L is the London penetration depth given by $\lambda_L = \lambda_0/\sqrt{1-(T/T_C)^4}$; T_C represents the critical temperature of the superconductor and λ_0 is the London penetration depth at $T = 0$ K; c is the velocity of light in vacuum.

18.3.1.1 *Semiinfinite photonic crystal with a dielectric cap layer*

18.3.1.1.1 Surface modes of TE polarization

The numerical results presented here are based on the numerical resolution of the dispersion relations for bulk (Eq. (18.12)) and surface modes (Eqs. (18.45) and (18.46)). We chose niobium (Nb) as a superconductor material S characterized by a dielectric function ε_S (Eq. (18.47)) and a dielectric material having a dielectric constant $\varepsilon_d = 10$ (such as gallium phosphide, GaP). The critical temperature of the superconductor S is $T_C = 9.2$ K and $\lambda_0 = 40$ nm. Fig. 18.2(a) gives the dispersion curves, namely the reduced frequency $\omega D/c$ versus the reduced wave vector $k_\parallel D$, where $k_\parallel$ is the wave vector component parallel to the PC interfaces. The calculations have been performed for $d_S = d_d = 50$ nm, and such that the temperature is fixed at $T = 4.2$ K. In addition, the threshold frequency for the bulk superconductor is calculated to be $\omega_p = 7.33 \times 10^{15}$ rad/s and does not play a significant role for TE waves in comparison with TM waves (see below). The cap layer is made out of the same dielectric as in the PC and its thickness is kept constant to $d_c = 75$ nm. The gray areas in Fig. 18.2(a) correspond to the bulk bands where electromagnetic waves can propagate through the PC. These bands are separated by frequency gaps (white regions) where the propagation is not allowed. The dotted curves inside the gaps are associated with surface modes induced by the dielectric cap layer. These are evidently caused by the symmetry breaking of the SL periodicity. The straight and dashed lines represent the velocities of light in vacuum and dielectric media, respectively.

One can distinguish two types of surface modes: the modes lying between the light lines of the cap layer and vacuum (blue dotted curves, dark gray in print version); these modes represent localized surface modes confined within the dielectric cap layer. It is worth noting the existence of one surface mode within each gap. The extension of these modes inside the vacuum bulk bands represents resonant (or leaky) modes (red dotted curves, light gray in print version) in Fig. 18.2(a). One can notice a discontinuity of the surface modes across the light line. This can be understood by the fact that below the light line the surface modes are evanescent in vacuum, which plays the role of barrier for photons and the interaction with the latter is weak. However, above the light line the modes propagate inside the vacuum and interact strongly with the latter, giving rise to resonant or pseudosurface modes. Therefore, the modes exhibit a slight red shift across the light line.

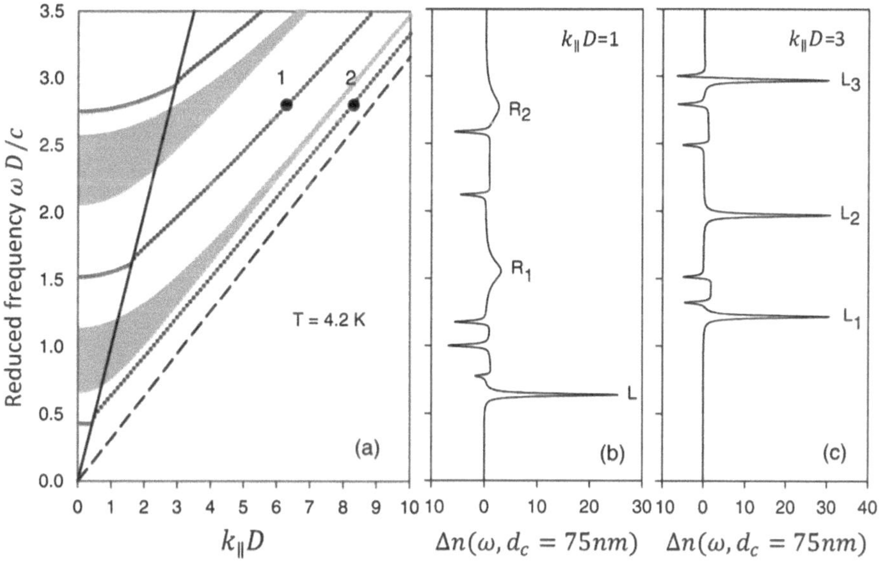

Figure 18.2 (a) Dispersion curves (reduced frequency $\omega D/c$ versus reduced wave vector $k_{\parallel} D$) of bulk (gray regions) and surface modes (blue [dark gray in print version] and red [light gray in print version] dotted curves) of a semiinfinite SL made out of a superconductor (niobium) and a dielectric ($\varepsilon_d = 10$) layers materials with thicknesses $d_S = d_D = 50$ nm. A thin dielectric layer ($d_c = 75$ nm) is coating the SL top surface. Straight and dashed lines refer to light lines in vacuum and in the dielectric material of the PC, respectively. (b) Variation of total density of states $\Delta_n(\omega)$ (in units of D/c) versus the reduced frequency $\omega D/c$ for $k_{\parallel} D = 1$; L_i and R_i indicate localized and resonant surface modes, respectively. (c) Same as in (b) but for $k_{\parallel} D = 3$.

All these modes appear as peaks in the variation of total density of states ($\Delta_n(\omega)$) as it is shown in Fig. 18.2(b) for $k_{\parallel} D = 1$ and in Fig. 18.2(c) for $k_{\parallel} D = 3$. Let us recall that $\Delta_n(\omega)$ represents the total DOS of the semiinfinite PC in contact with vacuum from which we have subtracted the DOS of the infinite PC and infinite vacuum (Eq. (18.34)). The localized surface modes appear as poles in the Green's functions and therefore give rise to delta peaks labeled L_i in $\Delta_n(\omega)$; these peaks are enlarged artificially by adding a small imaginary part Γ to the angular frequency ($\omega \to \omega + i\Gamma$). However, the resonant modes labeled R_i emerge as large peaks, they are intrinsically enlarged due to their interaction with the continuum modes in vacuum. The negative peaks in the VDOS (Figs. 18.2(b) and 18.2(c)) correspond to delta peaks of weight $-1/4$; these modes fall at the vicinities of the SL and vacuum bulk bands (Eq. (18.44)).

In order to show the spatial localization of these modes, we have plotted in Fig. 18.3 the square modulus of the electric field versus the reduced space position z/D for the modes labeled L, R_1, and R_2 in Fig. 18.2(b) at $k_{\parallel} D = 1$ (Figs. 18.3(a)–(c)) and for the modes labeled L_1, L_2, and L_3 in Fig. 18.2(c) at $k_{\parallel} D = 3$ (Figs. 18.3(d)–(f)). These results clearly show that the localized modes L, L_1, L_2, and L_3 are strongly confined in the top surface layer and vanish exponentially in the surrounding media, namely the PC and vacuum. Furthermore, one can notice the increasing number of

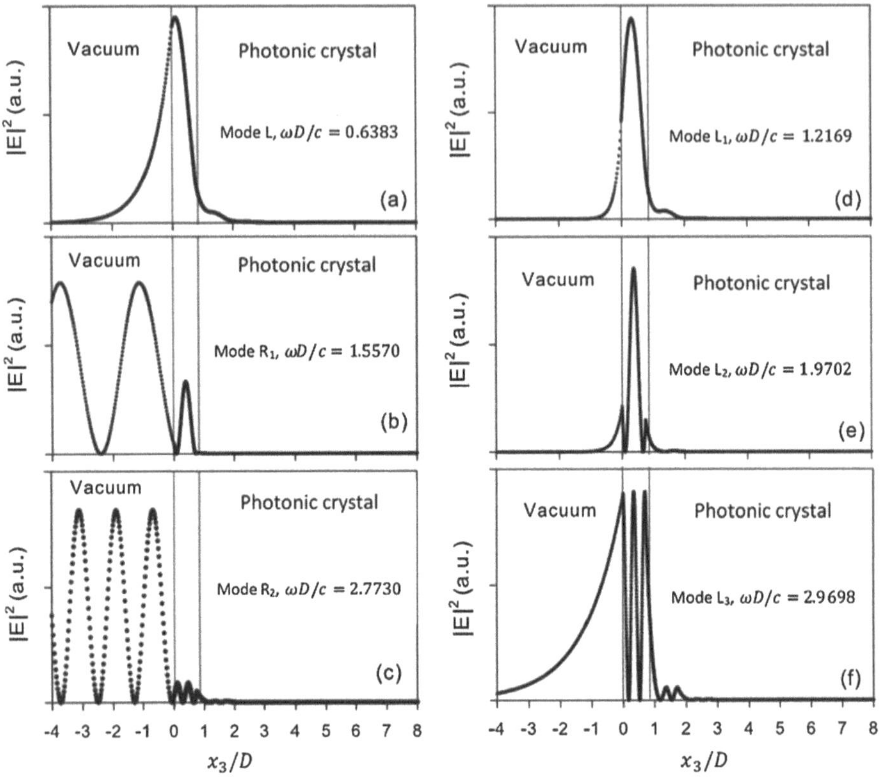

Figure 18.3 Square modulus of the electric field versus the reduced space position z/D for the modes labeled L ($\omega D/c = 0.638$) (a), R_1 ($\omega D/c = 1.557$) (b), R_2 ($\omega D/c = 2.7730$) (c) at $k_{\|} D = 1$, and the modes labeled L_1 ($\omega D/c = 1.2169$) (d), L_2 ($\omega D/c = 1.9702$) (e), and L_3 ($\omega D/c = 2.9698$) (f) at $k_{\|} D = 3$. The interface spaces separating vacuum, cap layer, and the SL are indicated by vertical lines.

oscillations for high frequency modes (L_3 mode exhibits more oscillations compared with L_1 mode, for instance). On the other side, resonant modes R_1 and R_2 show a nonnegligible confinement (though lower as compared to L_i modes) of light inside the top surface layer, with a vanishing behavior in the interior of the PC and a notable propagation in vacuum as the corresponding frequencies fall above the vacuum light line (Fig. 18.2(a)).

Another method of calculation that enables revealing the surface modes existence in a PC is the so-called Kretschmann configuration based on attenuated total internal reflection of an incident light [13]. This method was often used experimentally to excite surface plasmon–polariton modes on metal surfaces. This method has been used to excite surface electromagnetic waves in PCs [62,63]. Fig. 18.4 provides an example of the reflection coefficient versus the incidence angle for a given value of the reduced frequency in the case of a superconductor–dielectric PC terminated with a dielectric cap layer in contact with vacuum. In this configuration, we considered a finite PC composed of only two periods and deposited on a dielectric substrate made out of the

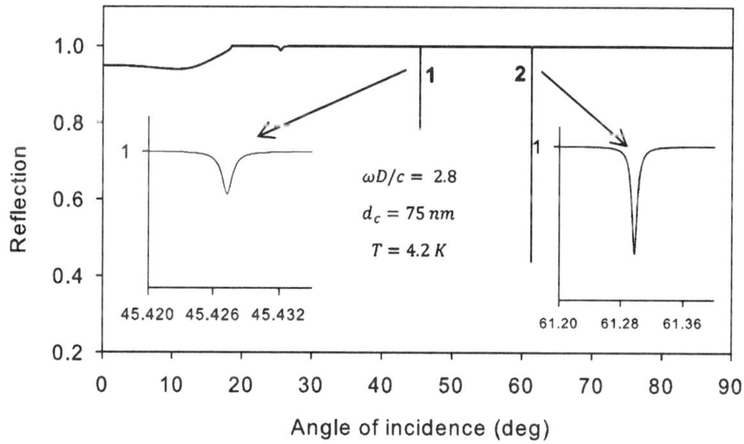

Figure 18.4 Reflection coefficient versus the incidence angle for the reduced frequency $\omega D/c = 2.8$. (Insert) Zoom around the reflection dips situated about $\theta = 45.4°$ and $\theta = 61.3°$.

same dielectric as those of the PC. The incident light interacts with surface modes in the SL, giving rise to well defined dips in the reflection amplitude versus the incidence angle. Indeed, this coupling occurs when the incidence light momentum matches the surface mode wave vector. In Fig. 18.4, the reflection spectra sketched at $\omega D/c = 2.8$ indicates the existence of two surface modes labeled 1 and 2 around $\theta = 45.4°$ and $\theta = 61.3°$, respectively. These two modes correspond to the localized modes indicated by two empty circles 1 and 2 in Fig. 18.2(a). However, the resonant mode lying above the vacuum light line at $\omega D/c = 2.8$ (i.e., $\theta < 20°$ in Fig. 18.4) does not give any signature in the reflection spectra (Fig. 18.4). The small peak around $\theta = 27°$ corresponds to an allowed band.

To examine the temperature effect on band structures and surface modes of the PC, we report in Fig. 18.5, the reduced frequency $\omega D/c$ versus temperature for a given value of the wave vector $k_{\parallel} D = 3$. The bulk bands (gray areas) and surface modes (blue dotted curves; dark gray in print version) undergo a small variation versus temperature especially when the latter is smaller than $T_C = 9.2$ K. Let us notice that for $T = T_C$, the dielectric constant of the superconductor layers in the PC becomes the same as in vacuum.

Also, we have studied the effect of the thickness of the surface cap layer on the localized surface modes. Fig. 18.6 gives the variation of the reduced frequency $\omega D/c$ of surface modes versus the reduced thickness d_c/D for $k_{\parallel} D = 3$. When d_c increases, the frequencies of the surface modes decrease until the corresponding branches merge into the bulk bands as resonant modes, at the same time new localized modes are extracted from the bulk bands. For any frequency $\omega D/c$ in Fig. 18.6, there is a periodic repetition of the modes as a function of d_c. Also, when d_c increases, all the surface modes tend asymptotically towards the light line in the dielectric cap layer (horizontal dashed line).

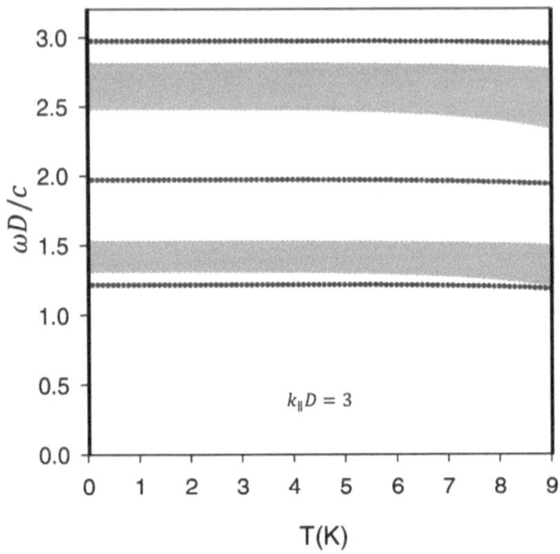

Figure 18.5 Variation of the reduced frequency of bulk and surface modes in Fig. 18.2(a) versus temperature T for $k_{\|}D = 3$. Gray areas represent bulk bands and dotted blue curves (dark gray in print version) correspond to surface modes.

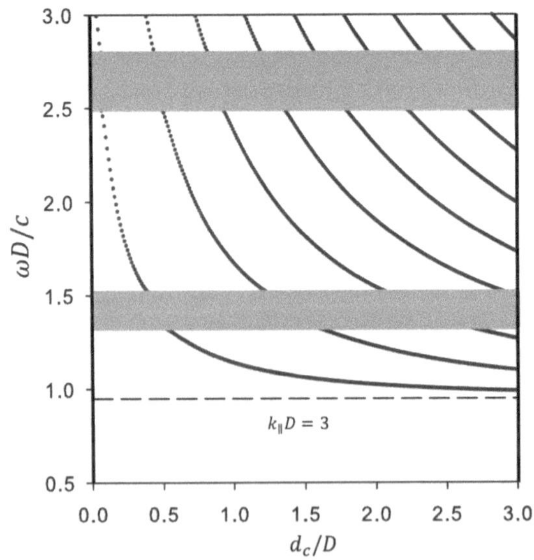

Figure 18.6 Variation of the reduced frequency of bulk and surface modes in Fig. 18.2(a) versus temperature T for $k_{\|}D = 3$. Gray areas represent bulk bands and dotted blue curves (dark gray in print version) correspond to surface modes. The horizontal dashed line corresponds to the cap layer light line.

18.3.1.1.2 Surface modes of TM polarization

In this subsection, we are interested in the bulk bands and surface modes of the PC when the incident light waves are of TM polarization. Figs. 18.7(a)–(c) give the disper-

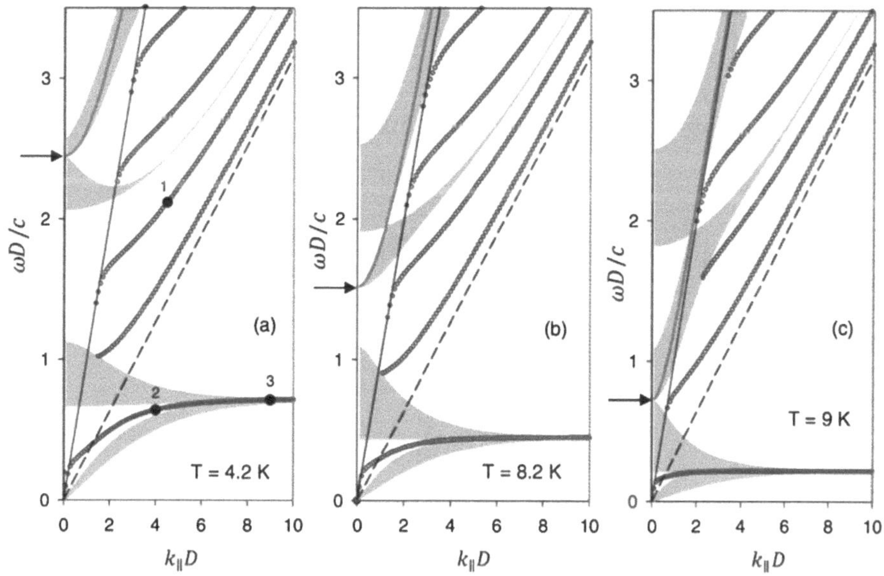

Figure 18.7 Dispersion curves of bulk bands (gray areas) and surface modes (blue dotted curves, dark gray in print version) for TM waves in a semiinfinite PC composed out of superconductor (niobium) and dielectric ($\varepsilon_d = 10$) bilayers and terminated by a superconductor cap layer of thickness $d_c = 125$ nm. The curves ($\omega D/c$ versus $k_{\parallel} D$) are plotted for three values of T, namely $T = 4.2$ K (a), $T = 8.2$ K (b), and $T = 9$ K (c). Full and dashed lines correspond to light lines in vacuum and in the dielectric material, respectively. The red curves (light gray in print version) show the light line in the superconductor material. The horizontal arrows indicate the plasma frequencies for each dispersion curve. The spatial localization of the modes 1, 2, and 3 in (a) will be discussed in Fig. 18.8.

sion curves of bulk bands (gray areas) and localized surface modes (blue dotted curves, dark gray in print version) for three different temperatures $T = 4.2$, 8.2, and 9 K, respectively. All the parameters of the PC are kept the same as in the previous study, while the thickness of the cap layer is chosen equal to $d_c = 125$ nm in order to get several surface guided modes induced by the cap layer. In comparison with the band structure of TE polarizations at $T = 4.2$ K (Fig. 18.2(a)), TM waves clearly show the existence of a large absolute gap (Fig. 18.7(a)). This gap extends form $\omega D/c = 1.15$ to $\omega D/c = 2.07$ and was discussed to some extent by Lin et al. [52]. Nevertheless, the gap width decreases drastically up to zero when T increases from 4.2 to 8.75 K. When $k_{\parallel} D$ increases, the two first bulk bands approach the surface plasmon–polariton (SPP) mode between the dielectric and the superconductor, namely $\omega = \omega_p/\sqrt{1+\varepsilon_d}$. These two bands fall below the light lines in the dielectric material (dashed line) and in the superconductor (red curve, light gray in print version). They are induced by interface modes between the dielectric and the superconductor.

The localized modes within the surface cap layer (blue dotted curves, dark gray in print version) can be divided into two types: (i) surface modes lying between the light lines in the dielectric material and in vacuum; these modes are strongly confined in the surface layer and rapidly vanish inside the PC and in vacuum as it is clearly shown in

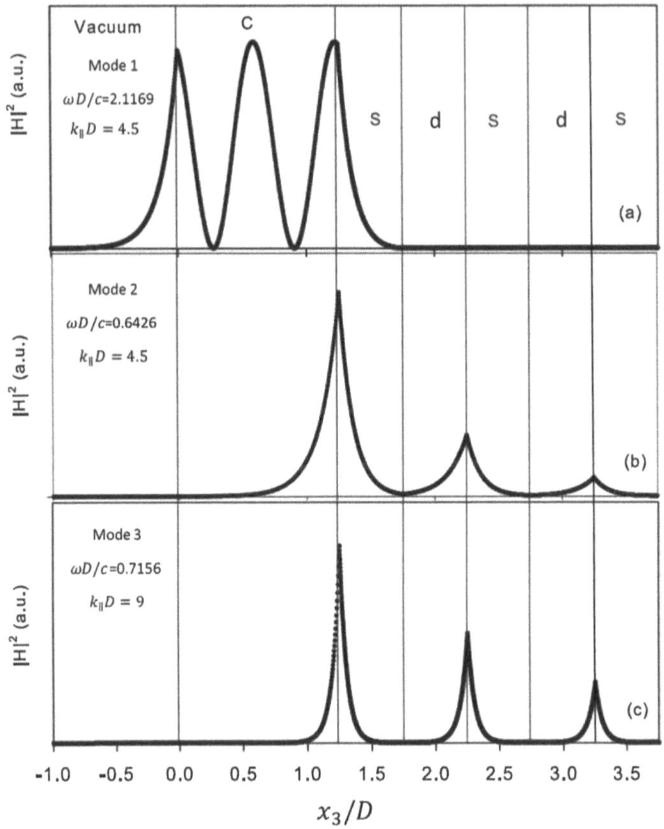

Figure 18.8 Square modulus of the magnetic field profile as a function of the reduced space position x_3/D for the modes labeled 1 ($\omega D/c = 2.1169$) (a), 2 ($\omega D/c = 0.6426$) (b), and 3 ($\omega D/c = 0.7156$) (c) in Fig. 18.7(a). The vertical lines indicate the interfaces of the structure. S, d, and c refer to superconductor, dielectric, and cap layers, respectively.

Fig. 18.8(a), where we have plotted the profile of the magnetic field versus x_3/D for the mode labeled 1 in Fig. 18.7(a); (ii) surface modes lying at lower frequencies below the light line in vacuum. The first part of this branch lies between the light lines in vacuum and the dielectric and when $k_{\|}D$ increases, this branch tends asymptotically to a SPP mode induced by the interface between the dielectric cap layer and the superconductor layer beneath it as it is clearly shown in Figs. 18.8(b)–(c) where we have plotted the profile of the magnetic field versus x_3/D for the modes labeled 2 and 3 belonging to the SPP branch in Fig. 18.7(a). Indeed, for large values of $k_{\|}D$, the limit of the SPP branch is the interface mode between the dielectric and the superconductor given by $\omega = \omega_p/\sqrt{1+\varepsilon_d}$ where the dimensionless plasma frequency is mentioned by the horizontal arrows in Fig. 18.7.

As mentioned above, Figs. 18.7(a)–(c) show that the bulk bands are very dependent on temperature. However, confined surface modes inside the cap layer indicate almost

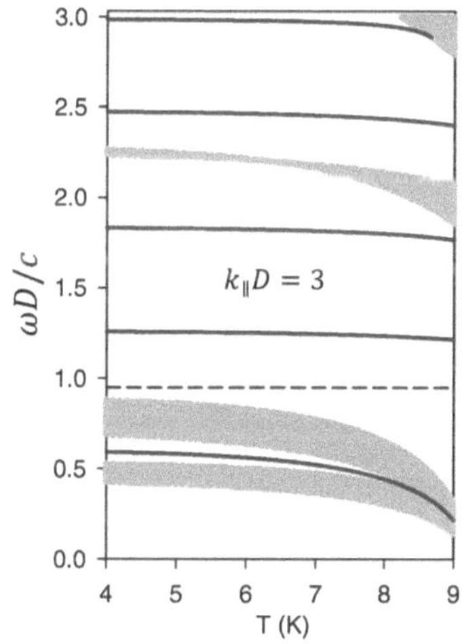

Figure 18.9 Reduced frequency versus temperature for $k_{\parallel}D = 3$. Gray areas represent bulk bands and blue dotted curves (dark gray in print version) indicate surface modes. The horizontal dashed line represents light line in the dielectric material.

no sensitivity towards T. On the other hand, the SPP mode appears to be significantly affected by temperature as it is shown in Fig. 18.9 for $k_{\parallel}D = 3$. Indeed, for $T \geq 7$ K, the SPP at lower frequencies decreases significantly and reaches roughly the half of its initial magnitude when T approaches T_C.

The modes induced by the cap layer are very sensitive to its thickness. An example is given in Fig. 18.10 where we have plotted the reduced frequency of bulk (gray areas) and surface modes (blue dotted curves, dark gray in print version) versus the cap layer thickness d_c/D at $k_{\parallel}D = 3$. Apart from the lowest branch that starts at $\omega D/c = 1.5$ for $d_c = 0$ nm, crosses the second bulk band, and goes asymptotically to the SPP mode at $\omega D/c = 0.59$ for large d_c, the other higher branches approach asymptotically the light-line in the surface cap layer. These modes exhibit a periodic repetition as a function of d_c for any frequency $\omega D/c$.

18.3.1.2 Semiinfinite photonic crystal with a superconductor cap layer

In the following, we shall discuss the surface modes in the case when the PC is capped with a thin superconductor material layer. The surface modes are given by Eqs. (18.3)–(18.4) and (18.45)–(18.47), where one only needs to exchange the superconductor and dielectric layers forming the PC. For TE waves, we have found that no localized surface modes exist as all the modes induced by the superconductor cap layer fall above the vacuum light line.

However, TM localized surface modes do exist. Fig. 18.11(a) gives the dispersion curves of these modes along with bulk bands (gray areas) for four values of the superconductor cap layer thickness: $d_c = 0$ nm, $d_c = 1$ nm, $d_c = 5$ nm, and $d_c = 100$ nm. The reason for the choices of small thicknesses will appear below. The tempera-

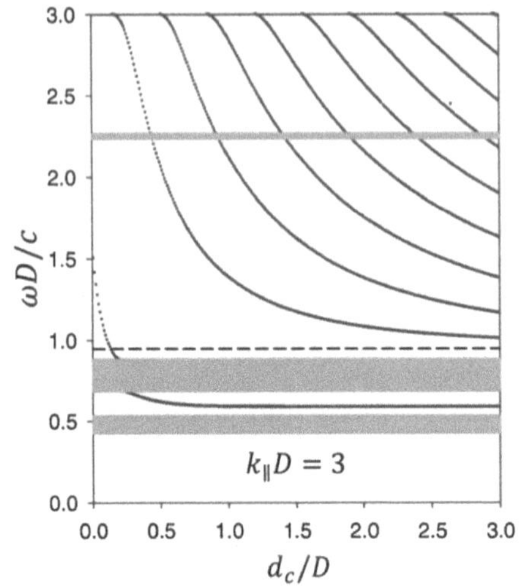

Figure 18.10 Variation of the reduced frequency as a function of d_c/D for $k_{\|}D = 3$ where d_c is the cap layer thickness. The gray areas indicate the first three bulk bands of the PC and the blue dotted curves (dark gray in print version) indicate surface modes. The horizontal dashed line represents light line in the dielectric material.

ture being fixed to $T = 4.2$ K. For $d_c = 0$ nm (i.e., when the PC terminates by a full dielectric layer of thickness 50 nm), one obtains the dispersion curves indicated by black dots. For a very thin superconductor layer such as $d_c = 1$ nm (blue dots, dark gray in print version), one obtains an anticrossing between a photonic branch induced by the dielectric surface layer and an almost horizontal plasmonic branch induced by the thin superconductor cap layer around the plasma frequency. This coupling occurs at $\omega_p D/c = 2.45$ where $\omega_p = 7.33 \times 10^{15}$ rad/s and $k_{\|}D = 4.5$ (see also Fig. 18.11(b) where we highlight the frequency range around $\omega D/c = 2$). When d_c increases ($d_c = 5$ nm, red dots [mid gray in print version]), the two branches repel each other and for large values of d_c ($d_c = 100$ nm, pink dots [light gray in print version]), the lower branch tends to the SPP mode at the interface between the superconductor cap layer and vacuum (around the typical SPP mode $\omega_p/\sqrt{2}$, indicated by the high arrow on the right hand side in Fig. 18.11(a)), while the higher branch penetrates inside the third bulk band (see below). Also, one can notice the existence of two surface branches, one below the bulk bands and one inside the first gap. The behavior of these modes versus d_c (cap layer thickness) is quite striking as they literally disappear when $d_c = 20$ nm. This is clearly shown in Fig. 18.12 where we have plotted their reduced frequencies versus the cap layer thickness d_c/D for $k_{\|}D = 3$. Except the surface mode situated in the second gap that tends to the SPP mode around $\omega D/c = 1.7$, all the other branches merge with the bulk bands around $d_c/D \sim 0.2$.

There exist also two surface branches at lower frequencies, below the light lines in vacuum and dielectric in Fig. 18.11. The lowest branch corresponds to modes which are localized at the interface between the thin superconductor cap layer and the first dielectric layer of the PC and decay both in the PC and into vacuum as shown in

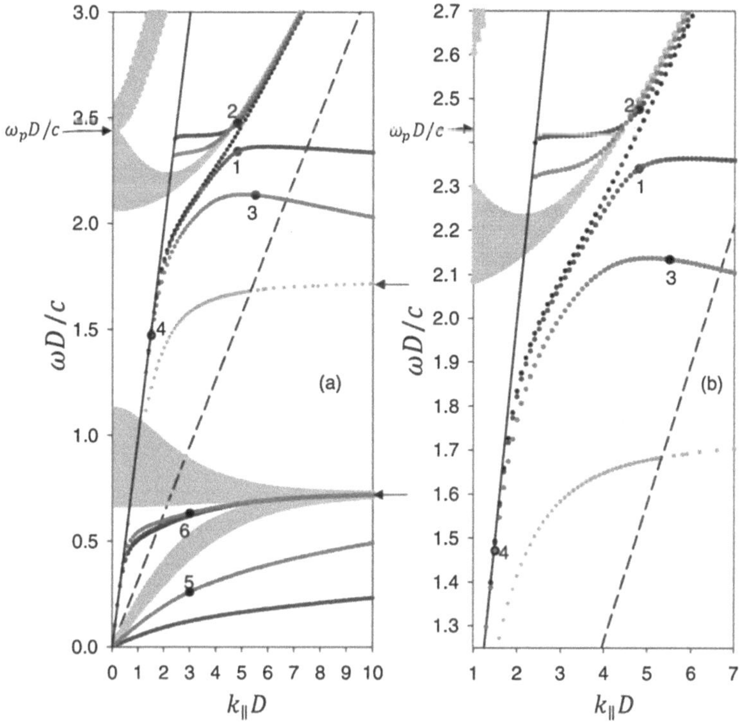

Figure 18.11 (a) Dispersion curves of bulk (gray areas) and surface (dotted curves) TM modes of a semi-infinite PC with a superconductor cap layer. The surface modes are sketched for $d_c = 0$ nm (black dots), $d_c = 1$ nm (blue dots, dark gray in print version), $d_c = 5$ nm (red dots, mid gray in print version), and $d_c = 100$ nm (pink dots, light gray in print version). (b) Zoom around the anticrossing points of surface modes around $\omega D/c = 2.45$ and $k_{\parallel} D = 4.5$ in (a).

Fig. 18.13(e) for the mode labeled 5 in Fig. 18.11. The second surface branch lying in the first gap corresponds to SPP modes. These modes are mainly localized at the interface between the dielectric layer just beneath the cap layer and the first superconductor layer in the PC. They exhibit a fast decay on both sides of this interface, as illustrated in Fig. 18.13(f) for the mode labeled 6 in Fig. 18.11. These modes are induced by the interface between the dielectric and the superconductor constituting the PC (indicated by the lower arrow on the right hand side in Fig. 18.11(a)).

In order to check the spatial localization of the surface modes around the anticrossing points in Fig. 18.11, we have presented in Figs. 18.13(a)–(b) the square modulus of the magnetic field as a function of the space position x_3/D for the modes labeled 1 and 2 in Fig. 18.11 for $d_c = 1$ nm. Even though these modes are very close to each other, mode 1 is strongly localized inside the superconductor thin cap layer (Fig. 18.13(a)), whereas mode 2 is well confined within the dielectric layer just below the cap layer (Fig. 18.13(b)). Similar trends can be seen in Figs. 18.13(c)–(d) concerning the modes labeled respectively 3 and 4 in Fig. 18.11 for $d_c = 5$ nm. These modes belong to the SPP branch induced by the interface between the superconductor cap layer and vac-

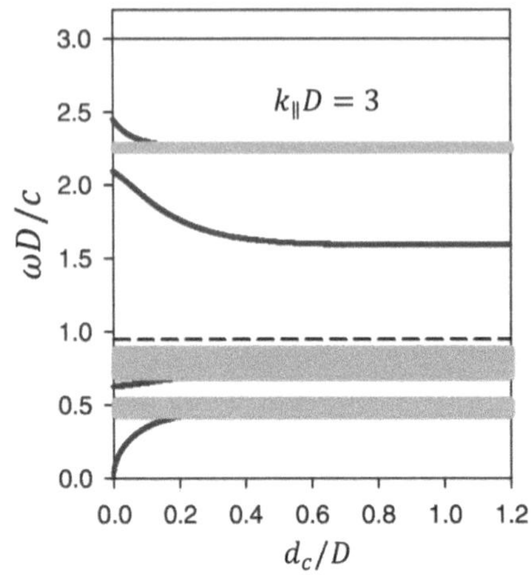

Figure 18.12 Reduced frequency versus reduced thickness d_C/D of the superconductor cap layer for $k_{\|}D = 3$. Straight and dashed horizontal lines refer to light lines in vacuum and in the dielectric material of the PC, respectively.

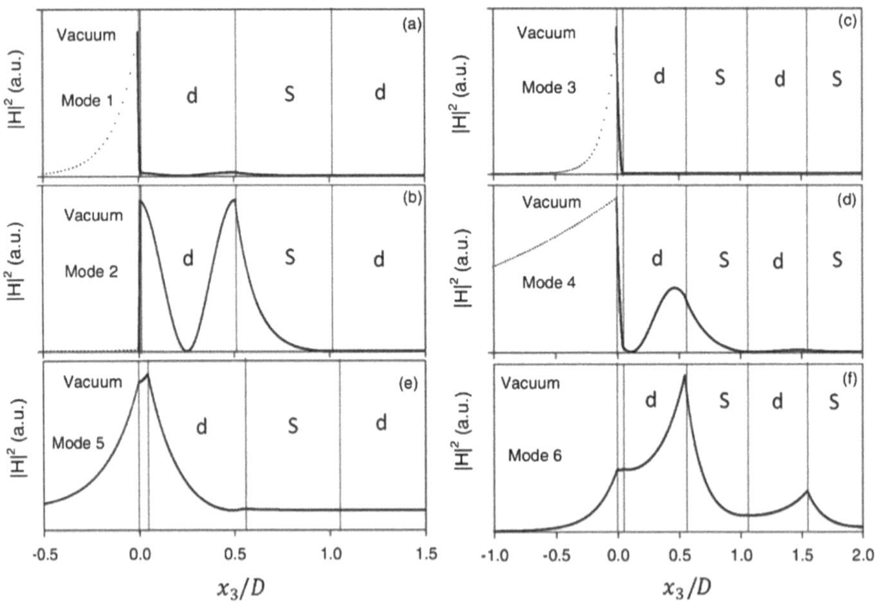

Figure 18.13 Profile of the square modulus of the magnetic field associated with surface modes labeled 1 ($\omega D/c = 2.3407$ and $k_{\|}D = 4.8$) (a), 2 ($\omega D/c = 2.4757$ and $k_{\|}D = 4.8$) (b) for $d_c = 1$ nm, 3 ($\omega D/c = 2.1340$ and $k_{\|}D = 5.5$) (c), 4 ($\omega D/c = 1.4707$ and $k_{\|}D = 1.5$) (d), 5 ($\omega D/c = 0.2586$ and $k_{\|}D = 3$) (e), and 6 ($\omega D/c = 0.6316$ and $k_{\|}D = 3$) (f) for $d_c = 5$ nm in Fig. 18.11. Vertical solid lines represent different interfaces of the structure.

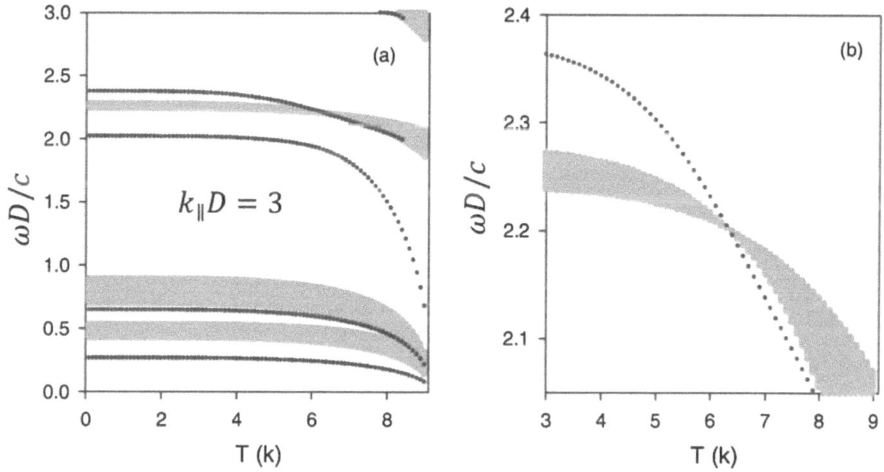

Figure 18.14 (a) Variation of the reduced frequency as a function of temperature T for $k_{\parallel}D = 3$ and $d_c = 5$ nm. The gray areas correspond to the first three bulk bands of the PC. Blue dotted curves (dark gray in print version) refer to surface modes. (b) Zoom around the crossing point between a surface branch and a bulk band around $\omega D/c = 2.23$ and $T \sim 6.3$ K.

uum. Nevertheless, mode 4 is less localized in vacuum than mode 3 as its frequency falls near the vacuum light line (Fig. 18.11).

Surface modes seem to be deeply sensitive to temperature when the cap layer is made of a superconductor. This is shown in Fig. 18.14 where we have presented the reduced frequency of bulk and surface modes versus T for $k_{\parallel}D = 3$ and $d_c = 5$ nm. It is clear that for $T \geq 6$ K surface modes drop down to lower frequencies; this is especially the case for the branch lying inside the second bandgap. Also, one can notice that the third allowed band closes around $T \sim 6.3$ K where it is crossed by the fourth surface branch. This is highlighted in Fig. 18.14(b) showing a zoom of Fig. 18.14(a) around $\omega D/c = 2.2$ and $T = 6.5$ K.

18.3.2 Optical Tamm states in 1D dielectric–dielectrique photonic crystal

In what follows, specific results will be given for photonic crystals (PCs) consisting of alternate layers of dielectric materials 1 and 2 with a cap layer at the surface (denoted by subscript c), and also for such PCs in contact with a substrate. The constituents are assumed to be characterized by their dielectric constants denoted ε_1, ε_2, and ε_c. The corresponding layer thicknesses will be denoted d_1, d_2, and d_c, and the period of the PC is $D = d_1 + d_2$. The calculations are performed for $\varepsilon_1 = 3$, $\varepsilon_2 = 10$, and $\varepsilon_c = 2$, and layer thicknesses $d_1 = 2d_2 = 2D/3$ (see Figs. 18.1 and 18.16(a)).

We shall first consider a semiinfinite PC with a surface cap layer (Section 18.3.2.1) and then a semiinfinite PC in contact with a semiinfinite homogeneous substrate (Section 18.3.2.2).

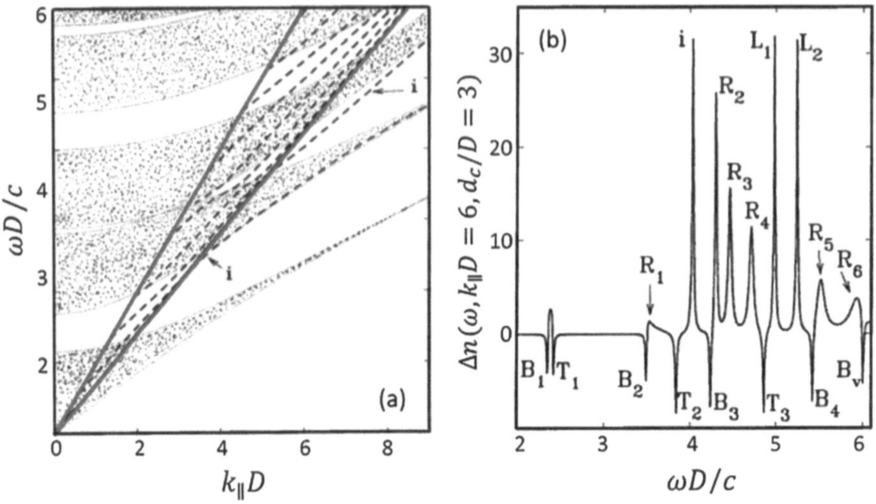

Figure 18.15 (a) Dispersion of localized and resonant modes (broken lines) induced by a cap layer of material c of thickness $d_c = 3D$ and dielectric constant $\varepsilon_c = 2$, deposited on the top of the superlattice formed of dielectrics 1 and 2 and terminated by a layer of dielectric 1. The shaded areas are the superlattice bulk bands. The red bold lines (mid gray in print version) indicate the light line in vacuum and in material c, respectively. The branches labeled (i) correspond to modes localized at the superlattice–cap layer interface. (b) Variation in the density of states (in units of D/c) corresponding to the case depicted in Fig. 18.15(a), for $k_\parallel D = 6$ and $d_c = 3D$; B_i and T_i refer to δ peaks of weight $-1/4$ at the edges of the bulk bands, L_i indicates the modes localized at the surface, and B_v refers to the δ peak of weight $-1/4$ situated at the light line of vacuum; R_i and i refer to the resonant modes and localized mode at the photonic crystal–adlayer interface, respectively.

18.3.2.1 Semiinfinite photonic crystal with a surface cap layer

The cap layer is in contact with material 1 of the PC (see Fig. 18.1). Fig. 18.15(a) gives the dispersion of localized and resonant electromagnetic modes induced by the cap layer of relative width $d_c/D = 3$. They are obtained as well-defined peaks in the variation $\Delta n_c(\omega)$ in the total density of states between the coupled (PC-cap layer-vacuum) and the uncoupled (infinite PC and vacuum alone) systems. An example of $\Delta n(\omega)$ is plotted in Fig. 18.15(b), for $k_\parallel D = 6$. These localized and resonant modes are very sensitive to the thickness and to the nature of the cap layer [64,65]. Most of these can be understood as modified (shifted and broadened) confined modes of a free slab and are therefore different from the surface modes of a PC without a cap layer (labeled i in Fig. 18.15). One can, however, see the existence of interface modes localized at the photonic crystal–cap layer interface. Among the resonances appearing in Figs. 18.15(a)–(b), the most intense resonance R_2 is the lowest, situated just above the cap layer light line. The next resonances are less intense, especially at higher frequencies where the separation between the successive branches increases.

With the help of Eqs. (18.32) and (18.33), we also studied local densities of states. We found that they change with the position x_3 where they are calculated. In particular, we found that the local density of states at the surface of the cap layer shows the same resonances as the total density of states illustrated in Fig. 18.15(b). On the contrary, in the local density of states at the superlattice–adlayer interface the resonances are shifted with respect to those in Fig. 18.15(b). These behaviors can be understood by the very different boundary conditions existing on these two planes.

Depending on their frequencies, the modes induced by the cap layer may propagate along the direction perpendicular to the interfaces in both the superlattice and the cap layer, or they may propagate in one and decay in the other, or finally, they may decay on both sides of the photonic crystal–adlayer interface. The latter, which corresponds to an interface localized mode, is labeled by the index i in Figs. 18.15(a)–(b). To illustrate these three types of behavior, we have plotted in Figs. 18.16(b)–(d) the square modulus of the electric field $|E|^2$ as a function of the space position x_3 for a given wavevector $k_{\|}D = 6$ and for different reduced frequencies $\omega D/c = 4.458$, $\omega D/c = 4.981$, and $\omega D/c = 4.027$ corresponding to a resonant mode R_3, a localized mode L_1, and an interface mode i, respectively, in Fig. 18.15(b).

In the first case (Fig. 18.16(b)), the reduced frequency ($\omega D/c = 4.458$) falls inside the bulk band of the PC. Consequently, $|E|^2$ corresponding to this resonant mode shows an oscillatory behavior both in the space of the PC and in the space of the cap layer and decays in vacuum. However, the local density of states is on average more important inside the cap layer than in the PC. One can also observe that, in the space occupied by the PC, this state is mostly concentrated in the layer made of dielectric 1. A similar behavior was found for electronic states in multiquantum wells where some resonant states are mostly concentrated in the barriers rather than in wells [66]. In the second case (Fig. 18.16(c)), the reduced frequency ($\omega D/c = 4.981$) falls inside the gap of the PC. Now, the corresponding local density of states shows an oscillatory behavior in the space occupied by the cap layer and a decaying behavior either in the PC or in vacuum. In the third case (Fig. 18.16(d)) corresponding to the reduced frequency of a localized state ($\omega D/c = 4.027$) at the PC–adlayer interface, $|E|^2$ decays on both sides of this interface.

The frequencies of the localized and resonant electromagnetic modes vary with the thickness d_c of the cap layer. Fig. 18.17 presents this variation for $k_{\|}D = 6$. The lowest branch (chain lines) corresponds to interface modes. The next branches (broken lines) become closer to each other when dc increases, and as a consequence the intensities of the corresponding resonances increase. However, the intensity of the resonant states decreases or may even vanish, in particular when d_c is small or the frequency is high. Moreover, one can observe in Fig. 18.17 that the branch corresponding to interface modes remains almost flat (independent of d_c) even for small values of d_c. Let us mention here too that, for any given frequency in Fig. 18.17, there is a periodic repetition of the modes as a function of d_c.

Finally, when the cap layer is deposited on a superlattice terminated with a full layer of dielectric 2 instead of 1, localized and resonant modes become quite different from those discussed in Figs. 18.15, 18.16, and 18.17.

18.3.2.2 Semiinfinite layered photonic crystal in contact with a substrate

To show the interface localized and resonant modes associated with the deposition of a semiinfinite PC on a semiinfinite substrate, we have chosen the same PC as before

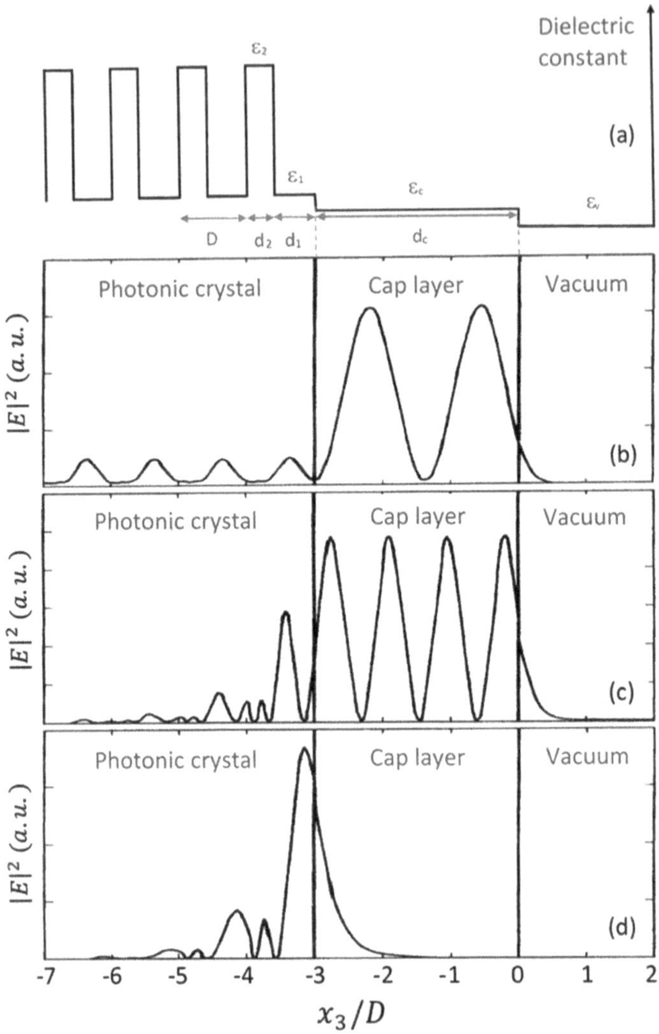

Figure 18.16 (a) A schematic representation of the dielectric constants corresponding to the different constituents of the coupled (PC-cap layer-vacuum) system. (b) Spatial representation of $|E|^2$ (in arbitrary units) at $\omega D/c = 4.458$ and $k_{\parallel} D = 6$. (c) Same as in (b) but at $\omega D/c = 4.981$ and $k_{\parallel} D = 6$. (d) Same as in (b) but at $\omega D/c = 4.027$ and $k_{\parallel} D = 6$.

deposited on a substrate with a dielectric constant $\varepsilon_s = 2$. Fig. 18.18(a) gives the localized and resonant interface modes for two complementary PCs in which the substrate is in contact with either a full layer of dielectric 2 or a full layer of dielectric 1. In the former case, the two solid lines in the minigaps of the PC are localized interface modes which continue (broken lines) as resonances inside the bulk band of the substrate. However, the resonance appearing in the first minigap is rather wide and not

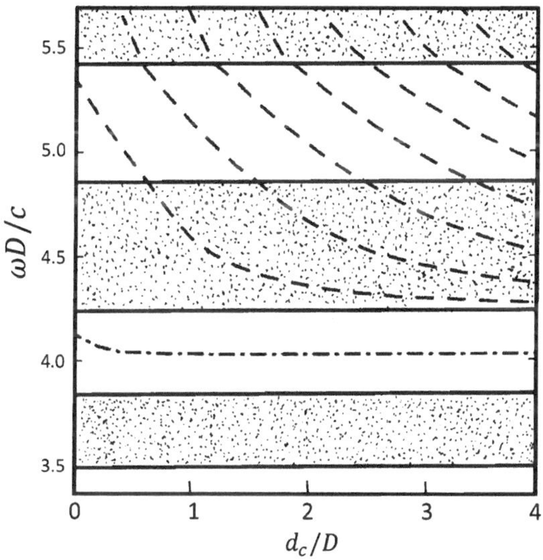

Figure 18.17 Variation in the dimensionless frequencies of localized and resonant modes induced by a cap layer of width d_c deposited on the surface of the semiinfinite photonic crystal of Fig. 18.15 for $k_\parallel D = 6$. The lowest branch (chain lines) corresponds to the localized state located at the interface between the photonic crystal and the cap layer.

a very well defined feature. Now, if the substrate is in contact with dielectric 2, one obtains the dotted branches near the top of the first two bulk bands of the PC.

When one creates the two complementary PCs used in Fig. 18.18(a) from the infinite PC and the infinite substrate, the variation $\Delta n_{Ic}(\omega)$ in the density of states is shown to be equal to zero [60] for frequencies ω belonging at the same time to the bulk bands of the substrate and the PC. Bearing in mind the loss of the $-1/2$ state at the limits of any bulk band and the conservation of the total number of states, we are led to the necessary existence of resonant modes lying at the same time inside the gaps of the PC and in the bulk band of the substrate. We have presented in Fig. 18.18(b) an example of this variation in the density of states for $k_\parallel D = 1$; the loss of states due to the δ peaks of weight $-1/2$ at every edge of the bulk bands is mostly compensated by the appearance of peaks of the density of states labeled R_1, R_2, R_3, R_4, and R_5. Every resonance, of weight almost equal to 1, is distributed differently over each of the two complementary PCs. One can also check the validity of the statement presented above, namely that $\Delta n_{Ic}(\omega)$ is equal to zero for ω belonging to the bulk bands of the substrate and the PC at the same time.

One can also remark that the position of the interface modes in Figs. 18.15(a) and 18.18(a) are almost the same even though in the former case the substrate is replaced by a cap layer of finite thickness $d_c/D = 3$; moreover, the localization of the interface modes is similar in both cases.

The frequencies of the interface modes are very sensitive to the nature of the substrate in contact with the PC. We have presented in Fig. 18.19 the dispersion of interface modes for different values of the parameter ε_s, when $\varepsilon_1 < \varepsilon_s < \varepsilon_2$. In fact,

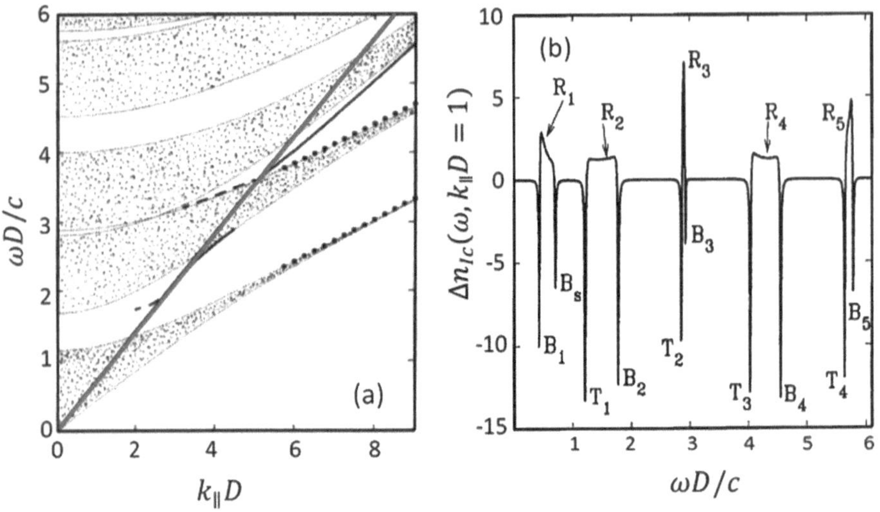

Figure 18.18 (a) Interface localized and resonant modes associated with the two complementary photonic crystals in which the substrate is either in contact with a full layer of dielectric 1 or 2. The shaded areas are the bulk bands of the superlattice. The red bold straight line (mid gray in print version) indicates the bottom of the substrate bulk band. The dielectric constant of the substrate is taken to be $\varepsilon_s = 2$. When the photonic crystal terminates with a layer 1, the localized (resonant) modes are presented by the full (broken) lines. The dotted lines are interface branches associated with termination of the photonic crystal by dielectric 2. (b) Variation in the density of states (in units of D/c), at $k_{\|}D = 1$, for the two complementary photonic crystals of Fig. 18.18(a) created from the infinite PC and the infinite substrate; B_i and T_i are δ peaks of weight $-1/2$ appearing at the edges of the PC bulk bands, B_s refers to a δ peak of weight $-1/2$ situated at the bottom of the substrate bulk band. The loss of states due to the δ peaks of weight $-1/2$ at every edge of the bulk bands is compensated by the peaks of the density of states labeled R_i.

when ε_s increases, the frequencies of the interface modes shift downwards, in particular the branch which appears above the first bulk band of the PC for $\varepsilon_v = 1$ (vacuum) and $\varepsilon_s = 2$ (see Fig. 18.18(a)) merges into this band for $\varepsilon_s = 3$; when $\varepsilon_s > 3$, an interface branch appears below the first bulk band as illustrated in Fig. 18.19 for $\varepsilon_s = 4$, $\varepsilon_s = 6$, and $\varepsilon_s = 8$. (Note that, for clarity, we have plotted in this figure the reduced velocity $\omega/ck_{\|}$ instead of the reduced frequency $\omega D/c$.)

18.3.3 Phonon–polariton Tamm states in 1D semiconductor–semiconductor photonic crystal

The coupling between lattice vibrations and electromagnetic waves in an ionic crystal results in infrared absorption and polariton excitation, which causes the opening of the polaritonic bandgap in which the propagation of EM waves is forbidden. The phenomenon in the coupling of phonons and photons has been found in both bulk and superlattice systems [69].

Phonon polaritons in layered photonic crystals have been studied before theoretically by Yeh et al. [16], Giuliani et al. [70], Lambin et al. [67,68], Haupt and Wendler

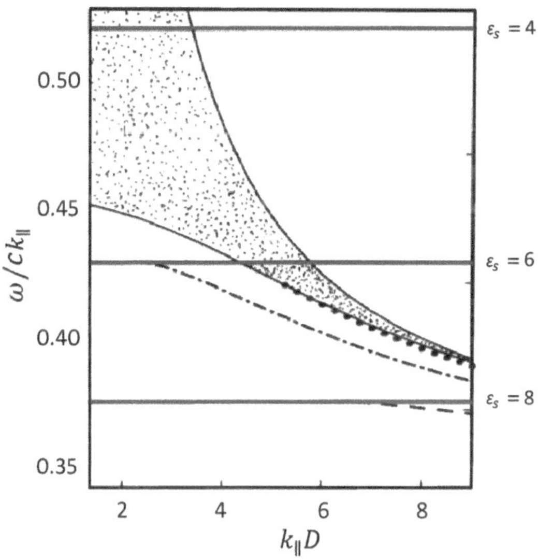

Figure 18.19 Dispersion of the lowest bulk band and interface modes for $\varepsilon_s = 4$ (dotted line), $\varepsilon_s = 6$ (chain line), and $\varepsilon_s = 8$ (broken line). The horizontal red bold lines (mid gray in print version) indicate the bottom of the substrate bulk bands characterized by their dielectric constants cited above. In order to obtain good separation between the interface modes, we have represented the reduced velocity $\omega/ck_{\parallel}$ instead of the reduced frequency $\omega D/c$.

[71], Djafari-Rouhani and Dobrzynski [72], Dereux et al. [73,74], Mendialdua et al. [75], Bah et al. [76], and references therein. Corresponding experimental investigations appeared also; see, for example, Yeh et al. [4,17], Lambin et al. [77], Dumelow et al. [78], and Zhang et al. [79]. The surface phonon polaritons can be studied by electron energy loss spectroscopy, infrared reflectivity measurement, and attenuated total reflection. Recently, employing the transfer matrix method, Wang et al. investigated the property of the interface phonon polariton modes in a two-layer finite layered photonic crystal with a cap layer and a structural defect layer [80].

In this Section 18.3.3, we present some results and discussions of the effects of capping layers on surface phonon–polaritons in layered photonic crystals. The specific results will be given for TM polarized nonretarded polaritons in GaAs–InAs photonic crystals with or without an InP surface layer. We characterize each material by the usual dielectric constant

$$\varepsilon_i(\omega) = \varepsilon_{\infty i}\frac{\omega_{Li}^2 - \omega^2}{\omega_{Ti}^2 - \omega^2}, \tag{18.48}$$

where ω_{Li} and ω_{Ti} are the frequencies of the longitudinal and transverse optical phonons and $\varepsilon_{\infty i}$ the high-frequency dielectric constant in material i. The numerical values of $\varepsilon_{\infty i}$, ω_{Li}, and ω_{Ti} are given in Table 18.1.

We shall first consider (Section 18.3.3.1) a semiinfinite PCs in contact with a vacuum, then (Section 18.3.3.2) semiinfinite PCs with a surface cap layer in contact with

Table 18.1 Transverse and longitudinal frequencies and high-frequency dielectric constants of GaAs, InAs, and InP.

	ω_L (cm^{-1})	ω_T (cm^{-1})	ε_∞
GaAs	297	273	10.9
InAs	243	218	12.3
InP	351	307	9.6

a vacuum. All the illustrative curves presented here are given for transverse magnetic (TM) polarization nonretarded polaritons for which F_i takes the value given by Eq. (18.4) and the speed of light $c \to \infty$.

18.3.3.1 GaAs–InAs photonic crystal in contact with a vacuum

The applications presented here are for a GaAs–InAs photonic crystal with $d_1 = 2d_2$ and the period $D = d_1 + d_2 = 30$ nm. Fig. 18.20(a) gives the dispersion of bulk bands

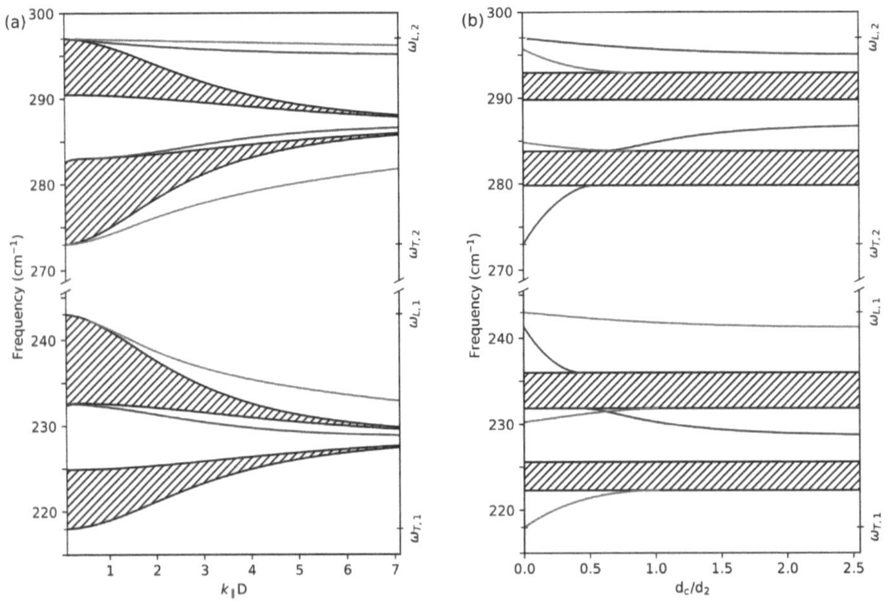

Figure 18.20 (a) The unretarded bulk and surface phonon polaritons with p polarization for a GaAs–InAs PC. The curves give ω (cm^{-1}) as a function of $k_\parallel D$, where ω is the frequency, $k_\parallel$ the propagation vector parallel to the interface, and $D = d_1 + d_2 = 30$ nm the period of the PC with $d_1 = 2d_2$. The shaded areas represent the bulk bands. The blue full lines (dark gray in print version) represent the surface polaritons for the semiinfinite PC terminated by a GaAs layer of thickness $d_c = 0.8d_2$. The red full lines (light gray in print version) represent the surface polaritons for the complementary PC terminated by a GaAs layer of thickness $d_c = 0.2d_2$. (b) Variation of the frequencies of the surface localized modes of a semiinfinite GaAs–InAs PC as a function of d_c/D and $k_\parallel D = 2.5$. The surface layer may be GaAs (blue lines, dark gray in print version) or InAs (red lines, light gray in print version).

and surface phonon–polariton modes as a function of $k_{\parallel} D$ where $k_{\parallel}$ is the propagation vector parallel to the interface. We have presented the surface phonon–polariton modes of two complementary semiinfinite PCs obtained by cleaving an infinite GaAs–InAs PC within one GaAs slab, such that the thickness of the remaining surface GaAs layer is respectively $d_c = 0.2d_2$ (red full lines, light gray in print version) and $d_c = 0.8d_2$ (blue full lines, dark gray in print version) in each semiinfinite part. One can observe that the frequencies of the surface phonon–polariton modes are very dependent on the thickness of the last surface layer of GaAs.

Having seen that the frequencies of the surface states are very sensitive to the width d_c of the last surface layer, we present in Fig. 18.20(b) the variation of these frequencies, for $k_{\parallel} D = 2.5$, as a function of d_c/d_2, both for a surface GaAs layer (blue lines, dark gray in print version) and for a surface InAs layer (red lines, light gray in print version). All the surface branches that cannot reach the value $d_c = 0$ actually stop at the frequencies $\omega = \omega_{Li}$ or ω_{Ti} $(i = 1, 2)$. Let us note that when d_c increases, the frequencies of the existing surface modes may either increase or decrease, before going to an asymptotic limit for $d_c/d_2 > 3$. In both cases, for $d_c/d_2 = 0.5$, we observe only one branch of the surface mode, all the other branches become resonant waves (or leaky waves) by merging into the bulk bands of the superlattice.

18.3.3.2 GaAs–InAs photonic crystal with InP cap layer

Now, we assume that a cap layer of InP, of thickness d_c, is deposited on top of the GaAs–InAs photonic crystal terminated by a full GaAs layer. The dispersions of surface and interface localized modes are presented in Fig. 18.21(a) for $d_1/d_2 = 0.5$ and $d_c/D = 1$. An important variation of the frequencies of the localized modes is obtained with the presence of the cap layer. Here the second and third surface modes counted from the bottom of Fig. 18.21(a) are respectively below the third and fourth bulk bands; they become resonant modes by merging into the bulk band at $k_{\parallel} D \simeq 1$. Without a cap layer, Lambin et al. [67,68] found these modes above the same bands. Note that different localized modes appear when the InP cap layer is deposited on top of an InAs layer of the photonic crystal rather than a GaAs layer.

The frequencies of the localized modes show significant variations with the thickness d_c of the cap layer, as shown in Fig. 18.21(b) for $k_{\parallel} D = 2.5$. These variations are faster when d_c is smaller or of the order of D. For $d_c/D < 0.25$, the second and the third localized branches become resonant by merging into the bulk bands. Finally, let us mention that for $d_c/D > 1$, the frequencies of the localized branches reach their asymptotic limits, which correspond, except for the highest branch, to the frequencies of the interface modes between a semiinfinite GaAs–InAs PC and a semiinfinite homogeneous InP substrate. In this case, the highest branch corresponds to a localized mode at the interface between the InP cap layer and the vacuum above. These asymptotic limits can be found from Eqs. (18.24), (18.25), (18.26), (18.45), and (18.46) by taking the limit of $d_c \to \infty$.

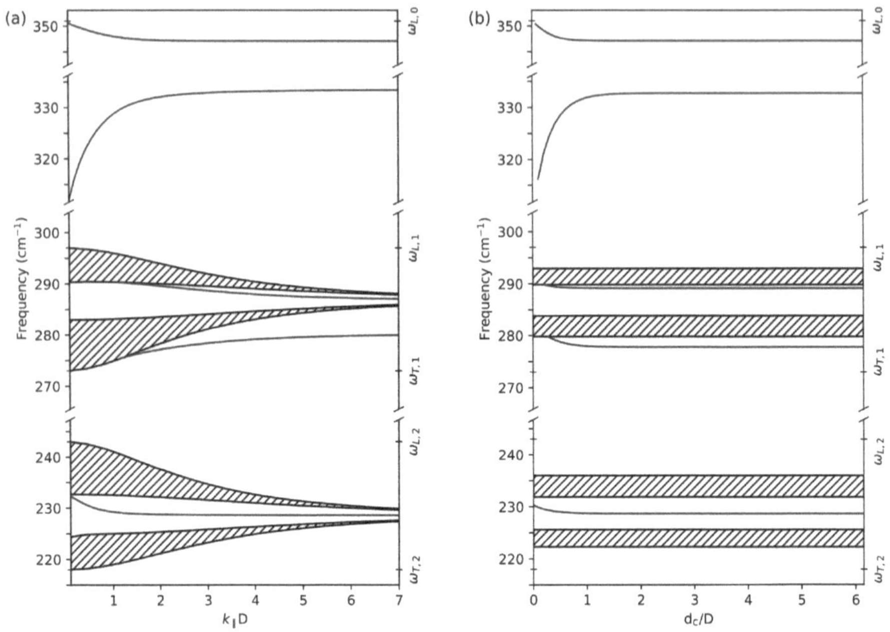

Figure 18.21 (a) The dispersion of localized phonon polaritons of TM polarization for an InP cap layer of thickness $d_c = D$, deposited on top of a GaAs–InAs PC terminated by a full GaAs layer with $d_1 = 0.5d_2$. The shaded areas represent the bulk bands of the PC. (b) The variation of the frequencies of the interface and surface localized modes of an InP cap layer of thickness d_c, deposited on top of the GaAs–InAs PC terminated by a full GaAs layer with $d_1 = 0.5d_2$ as a function of d_c/D for $k_\parallel D = 2.5$.

18.4 Summary and perspectives

In this chapter we have presented an analytical calculation of the response function for electromagnetic waves in a semiinfinite PC, with or without a cap layer. These results have been restricted to isotropic materials and are applicable to any PC system for which the dielectric functions of the component materials can be specified. These complete response functions can be used to study any physical property of the PC systems [76,81]. These include the calculation of light scattering spectra of interface electromagnetic modes, the calculation of the electric and magnetic fields associated with the reflected and transmitted waves, the determination of the dispersion relations for surface (or interface) modes and studies of the attenuation of such modes, and the calculation of the density of states.

Using the interface response theory, we have deduced the closed-form relations of the density of states for electromagnetic waves in a semiinfinite PC, with or without a cap layer or in contact with a substrate. These closed-form expressions enabled us to derive the dispersion of both localized and resonant states for these heterostructures and their spatial distribution. Particular attention was devoted to resonances (also called leaky waves) appearing in such heterostructures and to their relations with the localized modes. In particular, we discussed different kinds of localized and resonant mode induced by a cap layer deposited on top of a semiinfinite PC. These surface dispersion curves may serve as a tool for the characterization of the material at the surface of the PC.

We have demonstrated the possibility of existence of optical surface modes in a semiinfinite 1D PC constituted by superconductor/dielectric bilayers and terminated by a thin surface cap layer. The cap layer can be made out of a superconductor or a dielectric and the electromagnetic waves can be of TE or TM polarization. The dielectric constant of the superconductor is temperature dependent; therefore both bulk and surface modes can be tuned. When the cap layer is a dielectric material, two types of surface modes can exist for TE polarization: (i) true guided modes which are confined within the surface layer, and (ii) resonant (or pseudosurface) guided modes that are localized in the surface layer, but leaking energy in vacuum. For TM waves, we have shown that in addition to the guided modes analogous to TE waves, there exists SPP mode confined at the interface between the superconductor and dielectric layers of the SL. This mode falls below the light lines in dielectric and superconductor materials and inside the PC bandgap. In the case where a superconductor is capping the PC surface, we have shown that no surface modes exist for TE waves. However, in the case of TM modes and for very thin surface layer ($\sim$a few nm), we have shown the existence of an anticrossing between a photonic branch induced by the dielectric layer just beneath the superconductor cap layer and a plasmonic branch induced by the former layer. These plasmonic modes are found to be very dependent on temperature and on the thickness of the superconductor surface layer. This property can make these structures very useful for realizing highly sensitive PC sensors [11–15].

It worth mentioning that in the case of dielectric–superconductor PC, we have assumed that the permittivity of the dielectric layer is constant ($\varepsilon = 10$). If we take due account the dispersion of the dielectric constant of GaP material by using the expression given in the literature [82], namely $\varepsilon(\lambda) = 2.102 + 3.689\lambda^2/(\lambda^2 - 432^2)$ for $\lambda > 432$ nm, we found that the dispersion curves as well as the surface modes for TE and TM modes show the same behavior as those presented here but with a slight blue shift. However, in the case of TE modes, the dispersion curves become different from those presented here when the wavelength approaches the threshold value of 432 nm.

Acknowledgments

The authors would like to thank Mamadou Lamarana Bah, Oussama El Abouti, Youssef El Hassouani, and Adnane Noual for their fruitful contribution in the original papers from which this chapter is inspired.

References

[1] E. Yablonovitch, Inhibited spontaneous emission in solid-state physics and electronics, Phys. Rev. Lett. 58 (1987) 2059.

[2] S. John, Strong localization of photons in certain disordered dielectric superlattices, Phys. Rev. Lett. 58 (1987) 2486.
[3] Y. Fink, J.N. Winn, S. Fan, C. Chen, J. Michel, J.D. Joannopoulos, E.L. Thomas, A dielectric omnidirectional reflector, Science 282 (1998) 1679.
[4] P. Yeh, Optical Wave in Layered Media, Wiley, New York, 1988.
[5] C.H. Raymond Ooi, T.C. Au Yeung, Polariton gap in a superconductor–dielectric superlattice, Phys. Lett. A 259 (1999) 413.
[6] J.C. Knight, J. Broeng, T.A. Birks, P.St.J. Russell, Photonic band gap guidance in optical fibers, Science 282 (1998) 1476.
[7] B. Guo, Photonic band gap structures of obliquely incident electromagnetic wave propagation in a one-dimension absorptive plasma photonic crystal, Phys. Plasmas 16 (2009) 043508.
[8] L.M. Qi, Z.Q. Yang, F. Lan, X. Gao, Z.J. Shi, Properties of obliquely incident electromagnetic wave in one-dimensional magnetized plasma photonic crystals, Phys. Plasmas 17 (2010) 042501.
[9] B. Guo, M.Q. Xie, L. Peng, Photonic band structures of one-dimensional photonic crystals doped with plasma, Phys. Plasmas 19 (2012) 072111.
[10] Xiang-kun Kong, Shao-bin Liu, Hai-feng Zhang, Chun-zao Li, A novel tunable filter featuring defect mode of the TE wave from one-dimensional photonic crystals doped by magnetized plasma, Phys. Plasmas 17 (2010) 103506.
[11] J. Homola, S.S. Yee, G. Gauglitz, Surface plasmon resonance sensors, Sens. Actuators B 54 (1999) 3.
[12] R. Das, T. Srivastava, R. Jha, Tamm-plasmon and surface-plasmon hybrid-mode based refractometry in photonic bandgap structures, Opt. Lett. 39 (2014) 15.
[13] F. Villa, V.F. Regalado, L.E. Ramos-Mendieta, F. Gaspar-Armenta, J.T. Lopez-Rios, Photonic crystal sensor based on surface waves for thin-film characterization, Opt. Lett. 27 (2002) 646.
[14] A. Farmer, A.C. Friedli, S.M. Wright, W.M. Robertson, Biosensing using surface electromagnetic waves in photonic band gap multilayers, Sens. Actuators B 173 (2012) 79.
[15] R. Das, T. Srivastava, R. Jha, On the performance of Tamm-plasmon and surface-plasmon hybrid-mode refractive-index sensor in metallo-dielectric heterostructure configuration, Sens. Actuators B 206 (2015) 443.
[16] P. Yeh, A. Yariv, C.S. Hong, Electromagnetic propagation in periodic stratified media: I. General theory, J. Opt. Soc. Am. 67 (1977) 423.
[17] P. Yeh, A. Yariv, F. Villa, A.Y. Cho, Optical surface waves in periodic layered media, Appl. Phys. Lett. 32 (1978) 104.
[18] W. Ng, P. Yeh, P.C. Chen, A. Yariv, Optical surface waves in periodic layered medium grown by liquid phase epitaxy, Appl. Phys. Lett. 32 (1978) 370.
[19] A.A. Bulgakov, V.R. Koturn, Surface optical oscillations in a limited stratified-periodic medium, Opt. Spectrosc. 56 (1984) 269.
[20] R.D. Meade, K.D. Brommer, A.M. Rappe, J.D. Joannopoulos, Electromagnetic Bloch waves at the surface of a photonic crystal, Phys. Rev. B 44 (1991) 10961.
[21] S.Y. Vetrov, A.V. Shabanov, Surface electromagnetic waves at the interface of an isotropic medium and a superlattice, Sov. Phys. JETP 74 (1992) 719.
[22] W.M. Robertson, G. Arjavalingam, R.D. Meade, K.D. Brommer, A.M. Rappe, J.D. Joannopoulos, Observation of surface photons on periodic dielectric arrays, Opt. Lett. 18 (1993) 528.
[23] M.L. Bah, A. Akjouj, E.H. El Boudouti, B. Djafari-Rouhani, L. Dobrzynski, Surface and interface optical waves in superlattices: transverse electric localized and resonant modes, J. Phys. Condens. Matter 8 (1996) 4171.

[24] F. Ramos-Mendieta, P. Halevi, Electromagnetic surface modes of a dielectric superlattice: the supercell method, J. Opt. Soc. Am. B 14 (1997) 370.

[25] M.L.H. Lahlaouti, A. Akjouj, B. Djafari-Rouhani, L. Dobrzynski, M. Hammouchi, E.H. El Boudouti, A. Nougaoui, B. Kharbouch, Theoretical analysis of the density of states and phase times: application to resonant electromagnetic modes in finite superlattices, Phys. Rev. B 63 (2001) 035312.

[26] F. Villa, J.A. Gaspar-Armenta, Electromagnetic surface waves: photonic crystal–photonic crystal interface, Opt. Commun. 223 (2003) 109.

[27] X.-X. Liu, C.-F. Tsai, R.-L. Chern, D.-P. Tsai, Dispersion mechanism of surface magnetoplasmons in periodic layered structures, Appl. Opt. 48 (2009) 3102.

[28] Y. El Hassouani, H. Aynaou, E.H. El Boudouti, B. Djafari-Rouhani, A. Akjouj, V.R. Velasco, Surface electromagnetic waves in Fibonacci superlattices: theoretical and experimental results, Phys. Rev. B 74 (2006) 035314.

[29] M. Kaliteevski, I. Iorsh, S. Brand, R.A. Abram, J.M. Chamberlain, A.V. Kavokin, I.A. Shelykh, Tamm plasmon-polaritons: possible electromagnetic states at the interface of a metal and a dielectric Bragg mirror, Phys. Rev. B 76 (2007) 165415.

[30] J. Gaspar-Armenta, F. Villa, Photonic surface-wave excitation: photonic crystal–metal interface, J. Opt. Soc. Am. B 20 (2003) 2349.

[31] T.-B. Wang, C.-P. Yin, W.-Y. Liang, J.-W. Dong, H.-Z. Wang, Electromagnetic surface modes in one-dimensional photonic crystals with dispersive metamaterials, J. Opt. Soc. Am. B 26 (2009) 1635.

[32] S. Shukla, S. Prasad, V. Singh, Properties of surface modes in one dimensional plasma photonic crystals, Phys. Plasmas 22 (2015) 022122.

[33] A.V. Kavokin, I.A. Shelykh, G. Malpuech, Lossless interface modes at the boundary between two periodic dielectric structures, Phys. Rev. B 72 (2005) 233102.

[34] A. Kavokin, I. Shelykh, G. Malpuech, Optical Tamm states for the fabrication of polariton lasers, Appl. Phys. Lett. 87 (2005) 261105.

[35] K. Kneipp, H. Kneipp, I. Itzkan, R.R. Dasari, M.S. Feld, Surface-enhanced Raman scattering and biophysics, J. Phys. Condens. Matter 14 (2002) R597.

[36] A.V. Krasavin, A.V. Zayats, N.I. Zheludev, Active control of surface plasmon–polariton waves, J. Opt. A, Pure Appl. Opt. 7 (2005) S85.

[37] I. Tamm, A possible binding of the electrons on a crystal surface, Zh. Eksp. Teor. Fiz. 3 (1933) 34.

[38] N. Malkova, C.Z. Ning, Interplay between Tamm-like and Shockley-like surface states in photonic crystals, Phys. Rev. B 76 (2007) 045305.

[39] J. Guo, Y. Sun, Y. Zhang, H. Li, H. Jiang, H. Chen, Experimental investigation of interface states in photonic crystal heterostructures, Phys. Rev. E 78 (2008) 026607.

[40] M.E. Sasin, R.P. Seisyan, M.A. Kalitteevski, S. Brand, R.A. Abram, J.M. Chamberlain, A.Yu. Egorov, A.P. Vasilev, V.S. Mikhrin, A.V. Kavokin, Tamm plasmon polaritons: slow and spatially compact light, Appl. Phys. Lett. 92 (2008) 251112.

[41] C. Symonds, A. Lemaître, E. Homeyer, J.C. Plenet, J. Bellessa, Emission of Tamm plasmon/exciton polaritons, Appl. Phys. Lett. 95 (2009) 151114.

[42] H.X. Da, Z. Huang, Z.Y. Li, Electrically controlled optical Tamm states in magnetophotonic crystal based on nematic liquid crystals, Opt. Lett. 34 (2009) 1693.

[43] Y. Zheng, Y. Wang, J. Lou, P. Xu, Optical Tamm states in photonic structures made of inhomogeneous material, Opt. Commun. 406 (2018) 103.

[44] D. Gao, W. Mao, R. Zhang, J. Liu, Q. Zhao, W.Y. Tam, X. Wang, Tunable interface state in one-dimensional composite photonic structure, Opt. Commun. 453 (2019) 124324.

[45] R. Brückner, M. Sudzius, S.I. Hintschich, H. Frob, V.G. Lyssenko, M.A. Kaliteevski, I. Iorsh, R.A. Abram, A.V. Kavokin, K. Leo, Parabolic polarization splitting of Tamm states in a metal–organic microcavity, Appl. Phys. Lett. 100 (2012) 062101.
[46] T. Goto, A.V. Dorofeenko, A.M. Merzlikin, A.V. Baryshev, A.P. Vinogradov, M. Inoue, A.A. Lisyansky, A.B. Granovsky, Optical Tamm states in one-dimensional magnetophotonic structures, Phys. Rev. Lett. 101 (2008) 113902.
[47] M. Kaliteevski, S. Brand, R.A. Abram, I. Iorsh, A.V. Kavokin, I.A. Shelykh, Hybrid states of Tamm plasmons and exciton polaritons, Appl. Phys. Lett. 95 (2009) 251108.
[48] B.I. Afinogenov, V.O. Bessonov, A.A. Nikulin, A.A. Fedyaninb, Observation of hybrid state of Tamm and surface plasmon-polaritons in one-dimensional photonic crystals, Appl. Phys. Lett. 103 (2013) 061112.
[49] M. Lopez-Garcia, Y.-L.D. Ho, M.P.C. Taverne, L.-F. Chen, M.M. Murshidy, A.P. Edwards, M.Y. Serry, A.M. Adawi, J.G. Rarity, R. Oulton, Efficient out-coupling and beaming of Tamm optical states via surface plasmon polariton excitation, Appl. Phys. Lett. 104 (2014) 231116.
[50] A. Ramirez-Duverger, J. Gaspar-Armenta, R. Garcia Llamas, Experimental determination of a surface wave at the one-dimensional photonic crystal–metal interface, J. Opt. Soc. Am. B 25 (2008) 1016.
[51] M.E. Sasin, R.P. Seisyan, M.A. Kaliteevski, S. Brand, R.A. Abram, J.M. Chamberlain, I.V. Shelykh, A.Y. Egorov, A.P. Vasilev, V.S. Mikhrin, A.V. Kavokin, Tamm plasmon-polaritons: first experimental observation, Superlattices Microstruct. 47 (2010) 44.
[52] W.-H. Lin, C.-J. Wu, T.-J. Yang, S.J. Chang, Terahertz multichanneled filter in a superconducting photonic crystal, Opt. Express 18 (2010) 27155.
[53] H.-M. Lee, J.-C. Wu, Transmittance spectra in one-dimensional superconductor-dielectric photonic crystal, J. Appl. Phys. 107 (2010) 09E149.
[54] C.-J. Wu, C.-L. Liu, T.-J. Yang, Investigation of photonic band structure in a one-dimensional superconducting photonic crystal, J. Opt. Soc. Am. B 26 (2009) 2089.
[55] C.-J. Wu, Transmission and reflection in a periodic superconductor/dielectric film multilayer structure, J. Electromagn. Waves Appl. 19 (2005) 1991.
[56] O.L. Berman, Y.E. Lozovik, S.L. Eiderman, R.D. Coalson, Superconducting photonic crystals: numerical calculations of the band structure, Phys. Rev. B 74 (2006) 092505.
[57] O. El Abouti, E.H. El Boudouti, Y. El Hassouani, A. Noual, B. Djafari-Rouhani, Optical Tamm states in one-dimensional superconducting photonic crystal, Phys. Plasmas 23 (2016) 082115.
[58] L. Dobrzynski, Interface response theory of discrete composite systems, Surf. Sci. Rep. 6 (1986) 119.
[59] L. Dobrzynski, Interface response theory of continuous composite systems, Surf. Sci. Rep. 11 (1990) 139.
[60] E.H. El Boudouti, B. Djafari-Rouhani, E.M. Khourdifi, L. Dobrzynski, Surface and interface elastic waves in superlattices: transverse localized and resonant modes, Phys. Rev. B 48 (1993) 10987.
[61] T. Van Duzer, C.W. Turner, Principles of Superconductive Devices and Circuits, Edward Arnold, London, 1981, Chapter 3.
[62] W.M. Robertson, M.S. May, Surface electromagnetic wave excitation on one-dimensional photonic band-gap arrays, Appl. Phys. Lett. 74 (1999) 1800.
[63] W.M. Robertson, Experimental measurement of the effect of termination on surface electromagnetic waves in one-dimensional photonic bandgap arrays, J. Lightwave Technol. 17 (1999) 2013.
[64] A. Akjouj, M.L. Bah, E.H. El Boudouti, B. Djafari-Rouhani, L. Dobrzynski, Surface and interface s-polarized optical waves in superlattices, Vacuum 46 (1995) 621.

[65] A. Akjouj, M.L. Bah, E.H. El Boudouti, B. Djafari-Rouhani, L. Dobrzynski, Localised and resonant electromagnetic waves in superlattices, Ann. Chim. Fr. 19 (1994) 465.
[66] M. Steslicka, R. Kucharczyk, E.H. El Boudouti, B. Djafari-Rouhani, M.L. Bah, A. Akjouj, L. Dobrzynski, Electronic surface states in GaAs/$Ga_{1-x}Al_xAs$ superlattice: effect of surface location, Vacuum 46 (1995) 459.
[67] Ph. Lambin, J.P. Vigneron, A.A. Lucas, Theory of electron energy loss spectroscopy at the surface of a planar stratified medium, application to semiconductor superlattices, Solid State Commun. 54 (1985) 257.
[68] Ph. Lambin, J.P. Vigneron, A.A. Lucas, Electron-energy-loss spectroscopy of multilayered materials: theoretical aspects and study of interface optical phonons in semiconductor superlattices, Phys. Rev. B 32 (1985) 8203.
[69] M.L. Bah, A. Akjouj, E.H. El Boudouti, B. Djafari-Rouhani, L. Dobrzynski, The influence of capping layers on surface phonon polaritons in superlattices, J. Phys. Condens. Matter 7 (1995) 3445.
[70] G.E. Giuliani, J.J. Quinn, Charge-density excitations at the surface of a semiconductor superlattice: a new type of surface polariton, Phys. Rev. Lett. 51 (1983) 919.
[71] R. Haupt, L. Wendler, Damping of polaritons in finite semiconductor superlattices, Solid State Commun. 5 (1987) 341.
[72] B. Djafari-Rouhani, L. Dobrzynski, Simple excitations in N-layered superlattices, Solid State Commun. 62 (1987) 609.
[73] A. Dereux, J.P. Vigneron, Ph. Lambin, A.A. Lucas, Polaritons in semiconductor multilayered materials, Phys. Rev. B 38 (1988) 5438.
[74] A. Dereux, J.P. Vigneron, Ph. Lambin, A.A. Lucas, Radiative and non-radiative polariton structure of superlattices, Phys. Scr. 38 (1988) 462.
[75] J. Mendialdua, A. Rodriguez, M. More, A. Akjouj, L. Dobrzynski, Bulk and surface phonon polaritons in three-layer superlattices, Phys. Rev. B 50 (1994) 14605.
[76] M.L. Bah, A. Akjouj, L. Dobrzynski, Response functions in layered dielectric media, Surf. Sci. Rep. 16 (1992) 95.
[77] Ph. Lambin, J.P. Vigneron, A.A. Lucas, P.A. Thiry, M. Liehr, J.J. Pireaux, R. Caudano, T.J. Kuech, Observation of long-wavelength interface phonons in a GaAs/AlGaAs superlattice, Phys. Rev. Lett. 56 (1986) 1842.
[78] T. Dumelow, T.J. Parker, S.R.P. Smith, D.R. Tilley, Far-infrared spectroscopy of phonons and plasmons in semiconductor superlattices, Surf. Sci. Rep. 17 (1993) 151.
[79] X.J. Zhang, R.Q. Zhu, J. Zhao, Y.F. Chen, Y.Y. Zhu, Phonon-polariton dispersion and the polariton-based photonic band gap in piezoelectric superlattices, Phys. Rev. B 69 (2004) 085118.
[80] X.J. Wang, J.F. Liu, Y.F. Luo, S. Li, The influence of cap and defect layer on interface optical-phonon modes in finite superlattices, Chin. Phys. Lett. 27 (2010) 016301.
[81] M.G. Cottam, A.A. Maradudin, Surface Excitations, Mod. Problems Cond. Matter Sci., vol. 9, North-Holland, Amsterdam, 1986.
[82] N. Pimpabute, T. Burinprakhon, W. Somkhunthot, Determination of optical constants and thickness of amorphous GaP thin film, Opt. Appl. 41 (2011) 257.

Optical waves in finite layered photonic crystals

19

Abdellatif Akjouj[a], El Houssaine El Boudouti[b], Leonard Dobrzyński[a], and Bahram Djafari-Rouhani[a]
[a]Institute of Electronic, Microelectronics and Nanotechnology (IEMN), UMR CNRS 8520, Department of Physics, Faculty of Sciences and Technologies, Lille University, Villeneuve d'Ascq Cedex, France, [b]Laboratoire de Physique de la Matière et de Rayonnement, Département de Physique, Faculté des Sciences, Université Mohamed I, Oujda, Morocco

19.1 Introduction

Progress in research on artificial periodic structures has produced exciting advances in fundamental and technological knowledge. One of the most prominent examples is semiconductor multiquantum wells, or photonic crystals (PCs) [1]. The PCs, which consist of alternating layers of two different materials with layer thicknesses comparable to atomic dimensions, offer an enormous field of research for optical and optoelectronic devices [1–3]. Such devices include Bragg-wave-type couplers, bandpass filters, multiquantum-well lasers, and optical waveguides. With the progresses in crystal-growth technologies such as molecular beam epitaxy and metal–organic and chemical-vapor phase deposition, artificial microstructures of high quality are available [4,5].

These structures exhibit photonic bandgaps in the dispersion of the electromagnetic fields [2,3,6,7]. One of the most important properties of the photonic bandgap structures is the emergence of localized modes within the bandgaps associated with the perturbation of the infinite layered photonic crystals. An understanding of the interface properties in a PC is of particular importance because, by choosing a proper termination of the structure, it is possible, for example, to eliminate the losses associated with radiation into interface modes [8].

The existence of electromagnetic modes localized at the surface of a layered photonic crystal (PC) was predicted by Yeh et al. [9]. Subsequently, experiments in a periodic-layer structure that consisted of 12 pairs of alternating layers of GaAs and $Al_{0.2}Ga_{0.8}As$ on a GaAs substrate [10,11] and in ZnS/cryolite PCs [12] proved their existence. On the theoretical side, it is also known that the presence of a PC–vacuum interface [2,3,13–16] and a PC–substrate interface [15] or of a defect layer [17–19] gives rise to localized states inside the photonic bandgaps. These properties have also been extended to two- and three-dimensional photonic crystals [20–22]. Surface modes occurring at the junction of two different 1D PCs placed in series were envisaged for the design of narrow-bandpass filters [23]. Surface wave properties have been analyzed in different kinds of PCs such as in semiconductor and dielectric layers

Photonics. https://doi.org/10.1016/B978-0-12-819388-4.00030-7

[24], quasicrystals [25], dielectric Bragg reflectors [26,27], metals and metamaterials [28–30]. These localized modes in the forbidden band can also be utilized to manufacture semiconductor microlasers with almost perfect quantum efficiency and extremely low threshold [31,32]. Maas et al. present an experimental realization of a metamaterial design in which the dispersion is described by spherical rather than hyperbolic isofrequency wavevector contours, emulating an isotropic refractive index $n = -1$. They fabricate a flat lens that operates in the UV at $\lambda = 364$ nm using a silver/titanium dioxide photonic crystal [33]. Lang et al. show that one-dimensional photonic crystal can exhibit edge states and topological phases which are generally believed to appear in two-dimensional systems [34].

The first object of this chapter is to investigate the electromagnetic waves in the realistic geometry of a finite layered photonic crystals bounded by two semiinfinite media, a substrate and vacuum, and separated respectively by a buffer and a cap layer (see Fig. 19.1). This kind of geometry is often used experimentally [10,11,35] and has been proposed theoretically for analyzing the guiding properties of a multilayer planar waveguide [36,37]. However, in such studies the effect of the buffer and cap layers on electromagnetic waves in finite PCs is usually ignored. On the other hand, only a few experimental studies have dealt with surface and interface optical waves in layered photonic crystals—without, however, studying in detail the effect of buffer and cap layers, and especially the behavior of surface waves as functions of layer thicknesses. In this research we study localized and resonant electromagnetic modes (the latter are also called leaky waves) in this system, through an investigation of the local and the total density of states (DOS) in which these modes appear as well-defined peaks. One aim of this work is to study the existence and behavior of the localized and resonant modes induced by the buffer and cap layers when a finite PC is deposited on a substrate. We also calculate the transmission coefficient through this structure and compare its behavior with that of the DOS.

The wave dispersion, the local and total DOSs, and the transmission coefficients are given and discussed here for such a system as a function of (1) the wavevector $k_{\parallel}$ (parallel to the layers), (2) the number N of the periods in the PC, and (3) the width and the nature of the buffer and cap layers. We show that the positions of the peaks obtained from the DOS coincide with those given by the transmission coefficient. We also discuss the position and penetration of the localized modes inside the PC and their effect on the transmission coefficient. The method of calculation used in this paper is the Green's function (response) method. This method derives a response function that contains all the physical properties of the composite system under study. In particular, this response function gives directly the electric field at any point of the system in response to a unit input excitation introduced at any other point. This method is equivalent to the usual calculation of the eigenvalues and eigenvectors of the composite system by, for example, the transfer matrix method. Its advantages lie in a more compact treatment of the interface boundary conditions and in the fact that this response function enables us to make direct calculations of all the physical properties, in particular those connected with defects and scattering problems. These Green's functions enable us to obtain analytic expressions of local and total DOSs, as well as the reflection and transmission coefficients through the structure. The complete response functions can also be used to derive all eigenvectors [38,39] in the finite PCs. These quantities are, for instance, involved in the study of the power spectrum of spontaneous emission [35]. Although our results are in general obtained for transverse electric (TE) waves (for which the electric field is in the plane of the layers and perpendicular to $k_{\|}$), they can also be used for transverse magnetic (TM) polarization.

Several studies addressing the problem of propagation of electromagnetic waves through layered photonic crystal focus on the tunneling of optical pulses through a finite PC. It has been shown theoretically [40–43] and experimentally [44,45] that the time needed for a particle to tunnel through an opaque barrier reaches asymptotically a limit independent of the barrier thickness. It has also been shown experimentally [44,45] that the best measure of the delay time of a wave packet is the phase time.

The second aim of the present chapter is to investigate the exact relations between the DOS and the reflection and transmission phase times of scalar modes for any lamellar photonic crystal sandwiched between two different homogeneous semiinfinite media (see Fig. 19.2). Using the Green's function method, we derive in closed form the relations between the DOS and the reflection (τ_R) and transmission (τ_T) phase times. Avishai and Band [46] have found that the density of states is directly proportional to the phase time of the transmission amplitude. In some of our works [47,48], we have noticed similar qualitative behaviors between the DOS and reflection time. In this chapter, we establish that these two entities are exactly proportional when the incident wave is entirely reflected. We also demonstrate here that, apart from the delta peaks that can be induced by the reflection coefficient, $\tau_R = \tau_T$ for a symmetric composite system. We illustrate these analytical results on the example of a buffer layer-finite layered photonic crystal sandwiched between substrate and air (Fig. 19.11(a)). Different localized and resonant discrete modes due to the interaction between the different constituents of this system are also investigated. Particular attention is concentrated on the influence of buffer layers. In particular, we show the possibility of finding well-defined guided resonances in the PC, as a consequence of its separation from the substrate by the buffer layer. This phenomenon results from the fact that some of the discrete modes of the PC, even if they are in resonance with the substrate bulk modes, cannot propagate in the buffer layer and therefore remain well-defined guided waves in the finite PC. On the other hand, we illustrate numerically the equivalence between the reflection phase time and the DOS for different frequencies. Our phase time results include the special limit studied before [44,45], i.e., in the absence of a buffer layer. We found that the presence of a buffer layer strongly increases the magnitude of the time delay for reflection or transmission waves through the finite size PC.

The chapter is organized as follows. Section 19.2 gives (i) the analytical expressions obtained for the Green's function in the substrate–buffer layer–finite photonic crystal–cap layer–vacuum system (see Fig. 19.1), (ii) the analytical results obtained for the density of states and the transmission and reflection coefficients in the above described composite system, and (iii) the demonstration of the analytical relations between the densities of states and the reflection and transmission phase times in the structure illustrated in Fig. 19.2. Section 19.3 contains the numerical results about the dispersion curves, densities of states, and transmission spectrum for a layered photonic crystal depicted in Fig. 19.1. In Section 19.4, we present analytical and numerical results about delay times and density of states in GaAs substrate–SiO_2 buffer layer–GaAs/AlAs PC–air system (see Fig. 19.11(a)). Summary and perspectives are presented in Section 19.5.

19.2 Model and method of calculation

In a previous study [15], some of the present authors derived closed-form expressions of the Green's function and the DOS for a semiinfinite PC (with or without a cap layer) or for a PC in contact with a substrate. The purpose of such a study was the investigation of the semiinfinite PC–cap layer–vacuum interaction in relation to experimental results [10,11] of these systems. In this chapter, we discuss a more general structure, often used experimentally, that consists of a substrate–buffer layer (a)–finite PC–cap layer (b)–vacuum system (see Fig. 19.1). The finite PC has a total of N unit cells, where a unit cell is composed of two different slabs denoted by the unit-cell index n. Each of these slabs is characterized by its dielectric function ε_i and thickness d_i ($i = 1, 2$) within the unit cell n. The repetition period is called $D = d_1 + d_2$. Similarly, the buffer layer, cap layer, vacuum, and substrate involve the parameters (ε_a, d_a), (ε_b, d_b), ε_v, and ε_s, respectively. A space position along the x_3 axis in medium i belonging to the cell n is indicated by (n, i, x_3), where $-d_i/2 < x_3 < d_i/2$ ($i = 1, 2$).

The Green's function (from which the total DOS, the local DOS, and the reflection and transmission coefficients are deduced) is calculated by using the interface response theory [38,39] in composite materials in which the solution is first sought in the restricted space of the interfaces before being extended to the whole material. The complete Green's functions can also be used to derive any other physical property of the system at hand. They play a central role in the theories of light scattering (both Brillouin and Raman), as well as in various other physical phenomena [49,50].

19.2.1 Green's function for a substrate–buffer layer–finite photonic crystal–cap layer–vacuum system

Let us define the following quantities in each material $i = 1$, 2, b, c, s, and v:

$$\begin{aligned} C_i &= \cosh(\alpha_i d_i), \\ S_i &= \sinh(\alpha_i d_i), \end{aligned} \tag{19.1}$$

$$R_s = \frac{1 + F_b S_b / F_s C_b}{1 + F_s S_b / F_b C_b}, \tag{19.2}$$

$$R_v = \frac{1 + F_c S_c / F_v C_c}{1 + F_v S_c / F_c C_c}, \tag{19.3}$$

$$t = \exp(j k_3 D), \tag{19.4}$$

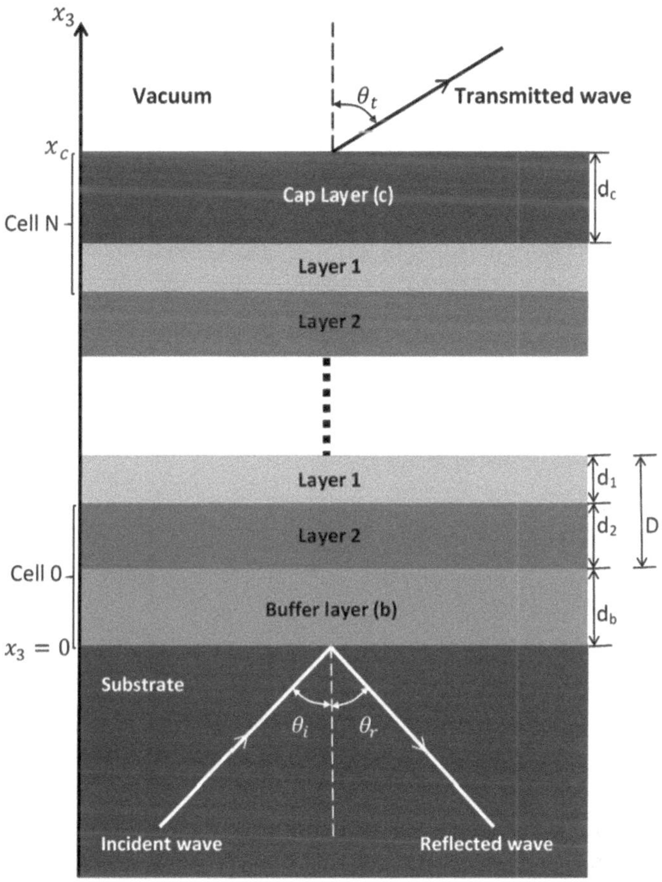

Figure 19.1 Schematic representation of a finite PC ($i = 1, 2$) with a buffer layer ($n = 0, i = b$) and a cap layer ($n = N, i = c$) and sandwiched between a semiinfinite substrate (S) and vacuum (v); d_b, d_c, d_1, and d_2 are respectively the thicknesses of the buffer layer b, the cap layer c, and the two different slabs out of which the finite layered photonic crystal is built; D is the period of the photonic crystal. The reflection and transmission of TE electromagnetic waves are also represented; θ_i, θ_r, θ_t are the incident, reflection, and transmission angles, respectively.

where k_3 is a wavevector along the axis of the layered photonic crystal and $j = \sqrt{-1}$.

Knowing the Green's function for an infinite homogeneous medium given by

$$G_i(x_3, x_3') = -\frac{1}{2F_i} e^{-\alpha_i |x_3 - x_3'|}, \tag{19.5}$$

where

$$\alpha_i^2(k_\parallel, \omega) = k_\parallel^2 - \frac{\omega^2}{c^2}\varepsilon_i, \tag{19.6}$$

$$F_i = \alpha_i, \qquad \text{for TE polarization,} \tag{19.7}$$

$$F_i = -\frac{\omega^2}{c^2}\frac{\varepsilon_i}{\alpha_i}, \qquad \text{for TM polarization,} \tag{19.8}$$

and for an infinite photonic crystal given in Chapter 16, we obtained, with the help of the interface response theory [39], the Green's function for the substrate–buffer layer (b)–finite layered photonic crystal–cap layer (c)–vacuum system (see Fig. 19.1).

1. In this chapter we need the following elements of the response function d between different interface planes:

(i) For $n = 0$ and $i = b$,

$$d\left(0, b, -\frac{d_b}{2}; 0, b, -\frac{d_b}{2}\right) = -\frac{\left(Y + Z\frac{S_b}{F_bC_b}\right) B_0 t - t^{2N}\left(Y + Z_0\frac{S_b}{F_bC_b}\right) Bt}{(1 + F_sS_b/F_bC_b)\,\Delta_-}, \tag{19.9}$$

$$\begin{aligned} d\left(0, b, -\frac{d_b}{2}; 0, b, \frac{d_b}{2}\right) &= d\left(0, b, \frac{d_b}{2}; 0, b, -\frac{d_b}{2}\right) \\ &= -Yt\frac{B_0 - t^{2N}B}{C_b[(1 + F_sS_b/F_bC_b)\,\Delta_-]}, \end{aligned} \tag{19.10}$$

$$d\left(0, b, \frac{d_b}{2}; 0, b, \frac{d_b}{2}\right) = -Yt\frac{B_0 - t^{2N}B}{\Delta_-}; \tag{19.11}$$

(ii) For $n = N$ and $i = c$,

$$d\left(N, c, -\frac{d_c}{2}; N, c, -\frac{d_c}{2}\right) = -\frac{XA_0 - t^{2N}X_0A}{\Delta_-}, \tag{19.12}$$

$$\begin{aligned} d\left(N, c, -\frac{d_c}{2}; N, c, \frac{d_c}{2}\right) &= d\left(N, c, \frac{d_c}{2}; N, c, -\frac{d_c}{2}\right) \\ &= -\frac{XA_0 - t^{2N}X_0A}{C_c[(1 + F_sS_c/F_cC_c)\,\Delta_-]}, \end{aligned} \tag{19.13}$$

$$d\left(N, c, \frac{d_c}{2}; N, c, \frac{d_c}{2}\right) = \frac{\left(-X + K\frac{S_c}{F_cC_c}\right) A_0 - t^{2N}\left(-X_0 + K_0\frac{S_c}{F_cC_c}\right) A}{(1 + F_vS_c/F_cC_c)\,\Delta_-}; \tag{19.14}$$

(ii) For $0 \leq n, n' \leq N$ and $i \neq b, c$,

$$d\left(n, 1, -\frac{d_1}{2}; n', 1, -\frac{d_1}{2}\right) = \frac{t}{t^2-1}\left\{Yt^{|n-n'|} - \frac{1}{\Delta_-}\left[XAB_0t^{n+n'-1}\right.\right.$$
$$\left.\left. - X_0ABt^{2N}(t^{n-n'}+t^{n'-n}) + X_0A_0Bt^{2N+1-n-n'}\right]\right\}, \tag{19.15}$$

$$d\left(n, 1, -\frac{d_1}{2}; n', 1, \frac{d_1}{2}\right) = \frac{t}{t^2-1}\left\{\frac{S_2}{F_2}t^{|n-n'|} + \frac{S_1}{F_1}t^{|n-n'-1|} - \frac{1}{\Delta_-}\left[YAB_0t^{n+n'}\right.\right.$$
$$\left.\left. - YABt^{2N+1}\left(t^{n-n'-1} + \frac{X_0}{X}t^{n'-n}\right) + YA_0Bt^{2N+1-n-n'}\right]\right\}, \tag{19.16}$$

$$d\left(n, 1, \frac{d_1}{2}; n', 1, -\frac{d_1}{2}\right) = d\left(n', 1, -\frac{d_1}{2}; n, 1, \frac{d_1}{2}\right), \tag{19.17}$$

$$d\left(n, 1, \frac{d_1}{2}; n', 1, \frac{d_1}{2}\right) = \frac{t}{t^2-1}\left\{Yt^{|n-n'|} - \frac{1}{\Delta_-}\left[X_0AB_0t^{n+n'}\right.\right.$$
$$\left.\left. -X_0ABt^{2N}(t^{n-n'}+t^{n'-n}) + XA_0Bt^{2N-n-n'}\right]\right\}. \tag{19.18}$$

2. The elements of this response function between any two points of this heterostructure can also be obtained in closed form. Since in the present study we need only the trace of this response function, we give here only those expressions for two points both of which belong to the substrate, both to cap layer (b) or (c), both to the photonic crystal, or both to vacuum.

(i) When the two points are inside the substrate,

$$d(x_3, x_3') = -\frac{e^{-\alpha_s|x_3-x_3'|}}{2F_s}$$
$$+\left[\frac{1}{2F_s} - \frac{\left(Y + Z\frac{S_b}{F_bC_b}\right)B_0t - t^{2N}\left(Y + Z_0\frac{S_b}{F_bC_b}\right)Bt}{(1+F_sS_b/F_bC_b)\,\Delta_-}\right]e^{-\alpha_s(x_3-x_3')}. \tag{19.19}$$

(ii) When the two points are inside buffer layer b,

$$d(0,b,x_3;0,b,x_3') = U_b(x_3,x_3') + \frac{1}{S_b^2}\left(\sinh\left[\alpha_b\left(\frac{d_b}{2}-x_3\right)\right], \sinh\left[\alpha_b\left(\frac{d_b}{2}+x_3\right)\right]\right) \times \boldsymbol{d}(M_b,M_b)\begin{pmatrix}\sinh\left[\alpha_b\left(\frac{d_b}{2}-x_3'\right)\right]\\ \sinh\left[\alpha_b\left(\frac{d_b}{2}+x_3'\right)\right]\end{pmatrix}, \tag{19.20}$$

where

$$U_b(x_3,x_3') = -\frac{e^{-\alpha_b|x_3-x_3'|}}{2F_b} + \frac{1}{2F_bS_b}\left\{\sinh\left[\alpha_b\left(\frac{d_b}{2}-x_3'\right)\right]\exp\left[-\alpha_b\left(\frac{d_b}{2}+x_3\right)\right] + \sinh\left[\alpha_b\left(\frac{d_b}{2}+x_3'\right)\right]\exp\left[-\alpha_b\left(\frac{d_b}{2}-x_3\right)\right]\right\}, \tag{19.21}$$

and $d(M_b, M_b)$ is the 2×2 matrix formed by the elements given by Eqs. (19.9)–(19.11) for $M_b = (0, b, \pm d_b/2)$.

(iii) When the two points are inside the photonic crystal,

$$d(n,i,x_3;n',i',x_3') = \delta_{nn'}\delta_{ii'}U_i(x_3,x_3') + \frac{1}{S_iS_{i'}}\left(\sinh\left[\alpha_i\left(\frac{d_i}{2}-x_3\right)\right], \sinh\left[\alpha_i\left(\frac{d_i}{2}+x_3\right)\right]\right) \times \boldsymbol{d}(M_m,M_{m'})\begin{pmatrix}\sinh\left[\alpha_{i'}\left(\frac{d_{i'}}{2}-x_3'\right)\right]\\ \sinh\left[\alpha_{i'}\left(\frac{d_{i'}}{2}+x_3'\right)\right]\end{pmatrix}, \tag{19.22}$$

where

$$U_i(x_3,x_3') = -\frac{e^{-\alpha_i|x_3-x_3'|}}{2F_i} + \frac{1}{2F_iS_i}\left\{\sinh\left[\alpha_i\left(\frac{d_i}{2}-x_3'\right)\right]\exp\left[-\alpha_i\left(\frac{d_i}{2}+x_3\right)\right] + \sinh\left[\alpha_i\left(\frac{d_i}{2}+x_3'\right)\right]\exp\left[-\alpha_i\left(\frac{d_i}{2}-x_3\right)\right]\right\}. \tag{19.23}$$

(iv) When the points are inside the surface cap layer c,

$$
\begin{aligned}
d(N, c, x_3; N, c, x_3') = {} & U_c(x_3, x_3') \\
& + \frac{1}{S_c^2}\left(\sinh\left[\alpha_c\left(\frac{d_c}{2} - x_3\right)\right], \sinh\left[\alpha_c\left(\frac{d_c}{2} + x_3\right)\right]\right) \\
& \times \boldsymbol{d}(M_c, M_c)\begin{pmatrix} \sinh\left[\alpha_c\left(\frac{d_c}{2} - x_3'\right)\right] \\ \sinh\left[\alpha_c\left(\frac{d_c}{2} + x_3'\right)\right] \end{pmatrix},
\end{aligned}
\tag{19.24}
$$

where

$$
\begin{aligned}
U_c(x_3, x_3') = {} & -\frac{e^{-\alpha_c|x_3 - x_3'|}}{2F_c} \\
& + \frac{1}{2F_c S_c}\left\{\sinh\left[\alpha_c\left(\frac{d_c}{2} - x_3'\right)\right]\exp\left[-\alpha_c\left(\frac{d_c}{2} + x_3\right)\right]\right. \\
& \left. + \sinh\left[\alpha_c\left(\frac{d_c}{2} + x_3'\right)\right]\exp\left[-\alpha_c\left(\frac{d_c}{2} - x_3\right)\right]\right\},
\end{aligned}
\tag{19.25}
$$

and $d(M_c, M_c)$ is the 2×2 matrix formed by the elements given by Eqs. (19.12)–(19.14) for $M_c = (N, c, \pm d_c/2)$.

(v) When the two points are inside the vacuum,

$$
\begin{aligned}
d(x_3, x_3') = {} & -\frac{e^{-\alpha_s|x_3 - x_3'|}}{2F_v} \\
& + \left[\frac{1}{2F_v} + \frac{\left(-X + K\frac{S_c}{F_c C_c}\right)A_0 - t^{2N}\left(-X_0 + K_0\frac{S_c}{F_c C_c}\right)A}{(1 + F_v S_c/F_c C_c)\,\Delta_-}\right] \\
& \times e^{-\alpha_v(x_3 + x_3' - 2x_c)},
\end{aligned}
\tag{19.26}
$$

where $x_c = d_b + ND + d_c$,

$$
\Delta_- = Yt\left(\frac{A_0 B_0}{X_0} - t^{2N}\frac{AB}{X}\right), \tag{19.27}
$$

$$
A_0 = F_s R_s X_0 + K_0, \quad A = F_s R_s X + K, \tag{19.28}
$$

$$
B_0 = F_v R_v y - Z_0, \quad B = F_v R_v Y - Z, \tag{19.29}
$$

$$X_0 = \frac{S_1}{F_1} t + \frac{S_2}{F_2}, \quad X = \frac{S_1}{F_1} + \frac{S_2}{F_2} t, \tag{19.30}$$

$$K_0 = C_2 - C_1 t, \quad K = C_2 t - C_1, \tag{19.31}$$

and

$$Y = \frac{C_1 S_2}{F_2} + \frac{C_2 S_1}{F_1}. \tag{19.32}$$

The localized waves are given by the poles of the Green's function, namely,

$$\Delta_- = 0. \tag{19.33}$$

The general expressions admit the following particular cases:

– A finite layered photonic crystal sandwiched between two semiinfinite substrates (i.e., $d_b = d_1$ and $d_c = d_2$).
– A finite layered photonic crystal deposited on the substrate (s) with only a cap layer (i.e., $d_b = 0$ and $F_v = 0$).
– A finite layered photonic crystal deposited on the substrate (s) with only a buffer layer (i.e., $d_c = 0$ and $F_v = 0$).

We can also deduce the limiting case $N \to \infty$ that corresponds to a semiinfinite photonic crystal with or without a cap layer or in contact with a substrate.

19.2.2 Density of states

We limit ourselves to the simplest case of TE polarization. Knowing the Green's function for the system, one obtains for a given value of $k_\parallel$ the local and total DOS for a substrate–buffer layer (b)–finite phononic crystal–cap layer (c)–semiinfinite medium composite system (see Fig. 19.1).

19.2.2.1 Local density of states

The local densities of states on the plane (n, i, x_3) are given by

$$n(\omega^2, k_\parallel; n, i, x_3) = -\frac{\varepsilon_i}{\pi c^2} \mathrm{Im} d^+(\omega^2, k_\parallel; n, i, x_3; n, i, x_3), \tag{19.34}$$

where

$$d^+(\omega^2) = \lim_{\Gamma \to 0} d(\omega^2 + i\Gamma) \tag{19.35}$$

and $d(\omega^2)$ is the response function whose elements are given in Section 19.2.1. As mentioned before, the density of states can also be given as a function of ω, instead of ω^2 using the well known relation $n(\omega) = 2\omega n(\omega^2)$.

19.2.2.2 Total density of states

The total density of states for a given value of $k_{\|}$ is obtained by integrating over x_3 and summing over n and i the local density $n(\omega^2, k_{\|}; n, i, x_3)$. In particular, we are interested in this total density of states from which the substrate and semiinfinite medium contributions have been subtracted. The expression of the variation of the density of states, when the buffer layer (b) and the cap layer (c) are in contact with materials 2 and 1, respectively, of the photonic crystal, can be written as the sum of six contributions,

$$n(\omega^2) = n_1(\omega^2) + n_2(\omega^2) + n_b(\omega^2) + n_c(\omega^2) + \Delta n_s(\omega^2) + \Delta n_v(\omega^2), \tag{19.36}$$

where $n_1(\omega^2)$ and $n_2(\omega^2)$ are the contributions of layers 1 and 2, respectively, of the photonic crystal; $n_b(\omega^2)$ and $n_c(\omega^2)$ are the contributions of buffer layer (b) and cap layer (c); and $\Delta n_s(\omega^2)$ and $\Delta n_v(\omega^2)$ come from the substrate and semiinfinite medium, respectively. Actually, in the latter terms we subtract the bulk contributions of the substrate and the semiinfinite medium.

Then the six quantities in Eq. (19.36) are given by

$$n_1(\omega^2) = -\frac{\varepsilon_1}{\pi c^2}\sum_{n=1}^{N} \text{Im} \int_{-\frac{d_1}{2}}^{+\frac{d_1}{2}} d(n, 1, x_3; n, 1, x_3) dx_3, \tag{19.37}$$

$$n_2(\omega^2) = -\frac{\varepsilon_2}{\pi c^2}\sum_{n=0}^{N-1} \text{Im} \int_{-\frac{d_2}{2}}^{+\frac{d_2}{2}} d(n, 2, x_3; n, 2, x_3) dx_3, \tag{19.38}$$

$$n_b(\omega^2) = -\frac{\varepsilon_b}{\pi c^2}\text{Im} \int_{-\frac{d_b}{2}}^{+\frac{d_b}{2}} d(0, b, x_3; 0, b, x_3) dx_3, \tag{19.39}$$

$$n_c(\omega^2) = -\frac{\varepsilon_c}{\pi c^2}\text{Im} \int_{-\frac{d_c}{2}}^{+\frac{d_c}{2}} d(N, c, x_3; N, c, x_3) dx_3, \tag{19.40}$$

$$\Delta_s n(\omega^2) = -\frac{\varepsilon_s}{\pi c^2}\text{Im} \int_{-\infty}^{0} [d(x_3, x_3) - G_s(x_3, x_3)] dx_3, \tag{19.41}$$

$$\Delta_v n(\omega^2) = -\frac{\varepsilon_v}{\pi c^2}\text{Im} \int_{x_c}^{+\infty} [d(x_3, x_3) - G_v(x_3, x_3)] dx_3, \tag{19.42}$$

where

$$x_c = d_b + ND + d_c \tag{19.43}$$

and N is the number of the finite layered photonic crystal cells.

The quantities d, G_s, and G_v are respectively the Green's functions of (1) the coupled substrate–buffer layer (b) finite layered photonic crystal–cap layer (c) semiinfinite medium system, (2) the infinite homogeneous substrate (s), and (3) the infinite vacuum (v).

The wavevector k_3, defined along the axis of the layered photonic crystal, satisfies the following dispersion relation in the infinite layered photonic crystal:

$$\cos(k_3 D) = C_1 C_2 + \frac{1}{2}\left(\frac{F_1}{F_2} + \frac{F_2}{F_1}\right) S_1 S_2. \tag{19.44}$$

Therefore, with the help of the explicit expressions of the Green's functions given in Section 19.2.1 and Eqs. (19.37)–(19.43), we obtain

$$\begin{aligned} n_1(\omega^2) = &-\frac{\varepsilon_1}{\pi c^2}\mathrm{Im}\frac{t}{(t^2-1)\Delta_-}\left(t\frac{1-t^{2N}}{t^2-1}(AB_0 t + A_0 B)\right. \\ &\times\left\{\frac{S_1}{\alpha_1 F_1}\left[C_2 S_1 + \frac{1}{2}C_1 S_2\left(\frac{F_1}{F_2}+\frac{F_2}{F_1}\right)\right] + \frac{S_2 d_1}{2F_2}\left(1-\frac{F_2^2}{F_1^2}\right)\right\} \\ &\left.+ N\Delta_+\left\{\frac{d_1}{F_1}\left[C_2 S_1 + \frac{1}{2}C_1 S_2\left(\frac{F_1}{F_2}+\frac{F_2}{F_1}\right)\right] + \frac{S_2 S_1}{2\alpha_1 F_2}\left(1-\frac{F_2^2}{F_1^2}\right)\right\}\right), \end{aligned} \tag{19.45}$$

$$\begin{aligned} n_2(\omega^2) = &-\frac{\varepsilon_2}{\pi c^2}\mathrm{Im}\frac{t}{(t^2-1)\Delta_-}\left(t\frac{1-t^{2N}}{t^2-1}(AB_0 + A_0 B t)\right. \\ &\times\left\{\frac{S_2}{\alpha_2 F_2}\left[C_1 S_2 + \frac{1}{2}C_2 S_1\left(\frac{F_1}{F_2}+\frac{F_2}{F_1}\right)\right] + \frac{S_1 d_2}{2F_1}\left(1-\frac{F_1^2}{F_2^2}\right)\right\} \\ &\left.+ N\Delta_+\left\{\frac{d_2}{F_2}\left[C_1 S_2 + \frac{1}{2}C_2 S_1\left(\frac{F_1}{F_2}+\frac{F_2}{F_1}\right)\right] + \frac{S_2 S_1}{2\alpha_2 F_1}\left(1-\frac{F_1^2}{F_2^2}\right)\right\}\right), \end{aligned} \tag{19.46}$$

$$\begin{aligned} n_b(\omega^2) = &-\frac{\varepsilon_b}{\pi c^2}\mathrm{Im}\frac{1}{\left(1+\frac{F_s S_b}{F_b C_b}\right)\Delta_-} \\ &\times\left(\left[Y\left(\frac{S_b}{\alpha_b C_b}+d_b\right) + \frac{d_b S_b}{F_b C_b}(Z+YF_s) - \frac{ZF_s}{F_b^2}\left(\frac{S_b}{\alpha_b C_b}-d_b\right)\right]B_0 t\right. \\ &\left.- t^{2N}\left[Y\left(\frac{S_b}{\alpha_b C_b}+d_b\right) + \frac{d_b S_b}{F_b C_b}(Z_0+YF_s) - \frac{Z_0 F_s}{F_b^2}\left(\frac{S_b}{\alpha_b C_b}-d_b\right)\right]Bt\right), \end{aligned} \tag{19.47}$$

$$\begin{aligned} n_c(\omega^2) = &-\frac{\varepsilon_c}{\pi c^2}\mathrm{Im}\frac{1}{\left(1+\frac{F_v S_c}{F_c C_c}\right)\Delta_-} \\ &\times\left(\left[X\left(\frac{S_c}{\alpha_c C_c}+d_c\right) - \frac{d_c S_c}{F_c C_c}(K-XF_v) + \frac{KF_v}{F_c^2}\left(\frac{S_c}{\alpha_c C_c}-d_c\right)\right]A_0\right. \end{aligned}$$

$$-t^{2N}\left[X_0\left(\frac{S_c}{\alpha_c C_c}+d_c\right)-\frac{d_c S_c}{F_c C_c}(K_0-X_0F_v)+\frac{K_0F_v}{F_c^2}\left(\frac{S_c}{\alpha_c C_c}-d_c\right)\right]A\Bigg), \tag{19.48}$$

$$\Delta_s n(\omega^2)=-\frac{\varepsilon_s}{\pi c^2}\times \mathrm{Im}\frac{1}{2\alpha_s}\left[\frac{1}{2F_s}+\frac{\left(Y+Z\frac{S_b}{F_bC_b}\right)B_0t-t^{2N}\left(Y+Z_0\frac{S_b}{F_bC_b}\right)Bt}{\left(1+\frac{F_sS_b}{F_bC_b}\right)\Delta_-}\right], \tag{19.49}$$

$$\Delta_v n(\omega^2)=-\frac{\varepsilon_v}{\pi c^2}\times \mathrm{Im}\frac{1}{2\alpha_v}\left[\frac{1}{2F_v}+\frac{\left(-X+K\frac{S_c}{F_cC_c}\right)A_0-t^{2N}\left(-X_0+K_0\frac{S_c}{F_cC_c}\right)A}{\left(1+\frac{F_vS_c}{F_cC_c}\right)\Delta_-}\right], \tag{19.50}$$

where

$$Z_0=C_1C_2+\frac{F_1}{F_2}S_1S_2-\frac{1}{t},\quad Z=C_1C_2+\frac{F_1}{F_2}S_1S_2-t, \tag{19.51}$$

and

$$\Delta_+=Yt\left(\frac{A_0B_0}{X_0}+t^{2N}\frac{AB}{X}\right). \tag{19.52}$$

The general expressions admit the following particular cases:

(i) A finite layered photonic crystal sandwiched between two semiinfinite substrates (i.e., $d_b=d_1$ and $d_c=d_2$);

(ii) A finite layered photonic crystal deposited on the substrate (s) with only a cap layer (i.e., $d_b=0$ and $F_v=0$);

(iii) A finite layered photonic crystal deposited on the substrate (s) with only a buffer layer (i.e., $d_c=0$ and $F_v=0$).

We can also deduce the limiting case $N\to\infty$ that corresponds to a semiinfinite photonic crystal with or without a cap layer or in contact with a substrate.

19.2.3 Reflection and transmission coefficients

For a composite material contained in its space of definition D, the ensemble of all the interface spaces bounding the homogeneous pieces between themselves is called the interface space M. The Green's function $g(D,D)$ of the composite material can be obtained from the knowledge of a bulk reference Green's function $G(D,D)$ and of the interface elements $g(M,M)$ of $g(D,D)$.

The inverse of the matrix $g(M, M)$ can be obtained [39] by linear superposition of the inverses of the $g_i(M, M)$ of all the homogeneous pieces i bounded by perfectly free interfaces with appropriate boundary conditions.

Moreover, an incident propagating vector $U(D)$ of the reference system depicted by the above $G(D, D)$ generates [56] in the composite the following vector $u(D)$:

$$u(D) = U(D) + G(D, M)\Big\{[G(M, M)]^{-1} g(M, M)[G(M, M)]^{-1} - [G(M, M)]^{-1}\Big\}U(M). \tag{19.53}$$

Consider an incident electromagnetic wave in the substrate represented by the plane wave $\exp(-\alpha_s x_3)$ of unit amplitude, propagating from $x_3 = -\infty$ towards the photonic crystal. The incident waves are scattered from the interfaces between dissimilar layers constituting the system. With the help of Eq. (19.53) and of the interface response theory [39], the reflected and transmitted waves can be respectively written as follows:

$$r = -1 - 2F_s d\left(0, b, -\frac{d_b}{2}; 0, b, -\frac{d_b}{2}\right) \tag{19.54}$$

and

$$t = -2F_s d\left(0, b, -\frac{d_b}{2}; N, c, \frac{d_c}{2}\right), \tag{19.55}$$

where $d\left(0, b, -\frac{d_b}{2}; 0, b, -\frac{d_b}{2}\right)$ is given by Eq. (19.9) and

$$d\left(0, b, -\frac{d_b}{2}; N, c, \frac{d_c}{2}\right) = \frac{-Yt^N(t^2-1)}{C_bC_c(1 + F_sS_b/F_bC_b)(1 + F_vS_c/F_cC_c)\Delta_-}. \tag{19.56}$$

The square moduli of the wave function of the reflected waves (denoted $R = |r|^2$) in the substrate and the transmitted waves in semiinfinite medium v (denoted $T = (F_v/F_s)|t|^2$) are expressed as

$$R = \left|\left(\frac{1 - F_sS_b/F_bC_b}{1 + F_sS_b/F_bC_b}\right)\frac{Yt\left(\frac{B_0A_{0-}}{X_0} - t^{2N}\frac{BA_-}{X}\right)}{\Delta_-}\right|^2 \tag{19.57}$$

and

$$T = \frac{F_v}{F_s}\left|\frac{2YF_st^N(t^2-1)}{C_bC_c(1 + F_sS_b/F_bC_b)(1 + F_vS_c/F_cC_c)\Delta_-}\right|^2, \tag{19.58}$$

respectively, where

$$A_{0-} = -F_sX_0R_{s-} + K_0, \quad A_- = -F_sXR_{s-} + K, \tag{19.59}$$

and

$$R_{s-}=\frac{1-F_bS_b/F_sC_b}{1-F_sS_b/F_bC_b}. \tag{19.60}$$

In the particular case corresponding to a finite layered PC with N periods sandwiched between two semiinfinite identical media, Eq. (19.58) is the same as Eq. (19) given by Bendickson et al. in [51]. It has also been derived by Yariv and Yeh in the special case where the unit cell is composed of a two-layer step-index profile [2].

19.2.4 Relations between densities of states and phase times

Besides the information obtained from the imaginary part of the system Green's function (namely the density of states) giving the frequencies of the different modes of the system [54], it is possible to obtain additional information by using the transmission and reflection coefficients and the corresponding phase times [55]. The aim of this Section 19.2.4 is to investigate the exact relations between the DOS and the reflection and transmission phase times of scalar modes for any layered photonic crystal sandwiched between two different homogeneous semiinfinite media. Using the Green's function method, we derive in closed form the relations between the DOS and the reflection (τ_R) and transmission (τ_T) phase times. Avishai and Band [46] have found that the density of states is directly proportional to the phase time of the transmission amplitude. However, to the best of our knowledge, any exact relation between the DOS and the reflection phase time has not been demonstrated, yet. In recent works [47,48], we have noticed similar qualitative behaviors of these two quantities. In this chapter, we establish that these two entities are exactly proportional when the incident wave is entirely reflected. We also demonstrate here that $\tau_R=\tau_T$ for a symmetric composite system.

In the case of nonmagnetic dielectric media and TE modes, the well-known [57] Green's function for an infinite medium i is

$$G_i(k_\parallel;x_3,x_3')=-\frac{\exp(-j\alpha_i|x_3-x_3'|)}{2j\alpha_i}, \tag{19.61}$$

where $k_\parallel$ is the two-dimensional wavevector parallel to the x_1Ox_2 interfaces, $j^2=-1$,

$$\alpha_i=\left(\frac{\omega^2}{c^2}\varepsilon_i(\omega)-k_\parallel^2\right)^{1/2}, \tag{19.62}$$

ω the angular frequency, c the speed of light in vacuum, and $\varepsilon_i(\omega)$ the relative permittivity.

Let us now consider (Fig. 19.2) a finite layered photonic crystal ($i=2$) sandwiched between the substrate and one homogeneous semiinfinite medium ($i=1$ and 3). The interface space M is formed here by the two planes separating these three media. For each scalar TE mode, the above defined $[g(M,M)]^{-1}$ can be obtained from the

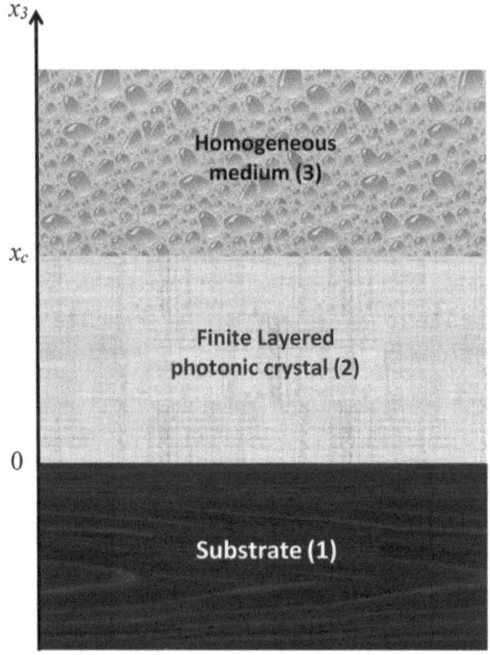

Figure 19.2 The general system under consideration constructed by a finite layered photonic crystal (2) sandwiched between a substrate (1) and a semiinfinite homogeneous medium (3).

surface $[g_i(M, M)]^{-1}$ of these three media, namely [39]

$$[g_i(0,0)]^{-1} = -j\alpha_i, \quad \text{for } i = 1 \text{ and } 3, \tag{19.63}$$

and

$$[g_2(M, M)]^{-1} = \begin{pmatrix} A_1 & B \\ B & A_2 \end{pmatrix}, \tag{19.64}$$

the detailed expressions for A_1, A_2, and B will be given below for the model structure depicted in Fig. 19.1 with $d_c = 0$, but the important point to notice is that these three quantities are purely real functions for scalar modes in a finite system. Then

$$[g(M, M)]^{-1} = \begin{pmatrix} -j\alpha_1 + A_1 & B \\ B & -j\alpha_3 + A_2 \end{pmatrix}. \tag{19.65}$$

Let us consider an incident wave $U(x_3) = \exp(-j\alpha_1 x_3)$. With the help of Eqs. (19.53) and (19.62), one easily finds the transmitted wave in medium 3 to be

$$u_T(x_3) = -2j\alpha_1 g(1, 2)\exp(-j\alpha_3 x_3), \tag{19.66}$$

where $g(0, x_c)$ is the element of g between the two interfaces labeled by $M \equiv (0, x_c)$, and the reflected one is

$$u_R(x_3) = -[1 + 2j\alpha_1 g(0, 0)] \exp(j\alpha_1 x_3). \tag{19.67}$$

By inversion of Eq. (19.65), let us write

$$g(0, x_c) = -B \det[g(M, M)] \tag{19.68}$$

and

$$g(0, 0) = (-j\alpha_3 + A_2) \det[g(M, M)], \tag{19.69}$$

where

$$\det[g(M, M)] = [-\alpha_1\alpha_3 + A_1 A_2 - B^2 - j(\alpha_1 A_2 + \alpha_3 A_1)]^{-1}. \tag{19.70}$$

Thus, Eqs. (19.66) and (19.67) can be written as

$$u_T(x_3) = C_T \exp(-j\alpha_3 x_3) \tag{19.71}$$

and

$$u_R(x_3) = C_R \exp(j\alpha_1 x_3), \tag{19.72}$$

where C_T and C_R are the transmission and reflection amplitudes given respectively by

$$C_T = 2j\alpha_1 B \det[g(M, M)] \tag{19.73}$$

and

$$C_R = -[\alpha_1\alpha_3 + A_1 A_2 - B^2 + j(\alpha_1 A_2 - \alpha_3 A_1)] \det[g(M, M)]. \tag{19.74}$$

From Eqs. (19.73)–(19.74), one can obtain the phases θ_T and θ_R of the transmission and reflection coefficients. Of more interest are the derivatives of these phases with respect to the frequency that are indicative of the times needed by a wave packet to complete the transmission or reflection processes. These quantities, usually called phase times [40–43], are defined by

$$\tau_T = \frac{d\theta_T}{d\omega} \tag{19.75}$$

and

$$\tau_R = \frac{d\theta_R}{d\omega}. \tag{19.76}$$

Let us mention that other characteristic times have also been defined in the literature [58–60] which are not considered here. For instance, the "*dwell time*" [42] is defined as the average time spent in a given region by all incoming particles.

From Eqs. (19.73) and (19.74) one easily obtains

$$\tau_T = \frac{d}{d\omega}\mathrm{Arg}\,\det[g(M,M)]. \tag{19.77}$$

Although the general expressions for τ_R can also be derived from Eq. (19.74), we limit ourselves to two particular cases when τ_R can be related to τ_T as well as to the DOS. When the composite system is symmetric (i.e., $\alpha_1 = \alpha_3$ and $A_1 = A_2$) and apart from the delta peaks that can be induced by the term $\alpha_1^2 + A_1^2 - B^2$ (see Eq. (19.74)), we obtain

$$\tau_R = \tau_T = \frac{d}{d\omega}\mathrm{Arg}\,\det[g(M,M)]. \tag{19.78}$$

It is worth mentioning that in the presence of loss, the above equation is no longer valid as the delta peaks induced by the term $\alpha_1^2 + A_1^2 - B^2$ in the reflection spectra broaden and therefore τ_R and τ_T become different.

When α_3 becomes pure imaginary (total reflection), the transmission vanishes and Eqs. (19.70) and (19.74) provide us with another exact result, namely

$$\tau_R = 2\frac{d}{d\omega}\mathrm{Arg}\,\det[g(M,M)] = 2\tau_T. \tag{19.79}$$

Let us now recall [61] that the difference of the DOS between the present composite system and a reference system formed out of two semiinfinite media 1 and 3 and the finite medium 2 can be obtained from

$$\Delta' n(\omega) = -\frac{1}{\pi}\frac{d}{d\omega}\mathrm{Arg}\,\det\left(\frac{g_2(M,M)g_{s1}(0,0)g_{s3}(0,0)}{g(M,M)}\right), \tag{19.80}$$

where $g_{s1}(0,0)$ and $g_{s3}(0,0)$ are the $(0,0)$ elements of the free surface response functions of the semiinfinite media 1 and 3.

The relation between the surface and bulk response functions is given in Chapter 1 of this Photonics book. It is shown there also that cleaving at $x_3 = 0$ an infinite medium into two semiinfinite ones produces a $\pi/2$ phase shift. This shift is due to the shift, due to the cleavage, of all antisymmetric (through $x_3 = 0$) states $\sin(\alpha_i x_3)$, with $i = 1, 3$ of the infinite medium into $\cos(\alpha_i x_3)$ states within each of the two semiinfinite media, with conservation of the number of states.

If one subtracts the discrete states of the finite medium 2, then the variation of the DOS between the composite system and the two semiinfinite media 1 and 3 is given by

$$\Delta n(\omega) = \frac{1}{\pi}\frac{d}{d\omega}\mathrm{Arg}\,\det\left(\frac{g(M,M)}{\alpha_1\alpha_3}\right). \tag{19.81}$$

This relation is equivalent to that obtained by using a direct calculation of the imaginary part of the trace of the Green's function [39,47]. Then for all frequencies, but $\omega \neq ck_{\|}/\sqrt{\varepsilon_1}$ and $\omega \neq ck_{\|}/\sqrt{\varepsilon_3}$, one obtains

$$\tau_T = \pi \Delta n(\omega), \tag{19.82}$$

which is equivalent to the result of [45]. Moreover,

$$\tau_R = \tau_T = \pi \Delta n(\omega) \tag{19.83}$$

when the composite system is symmetric, and

$$\tau_R = 2\pi \Delta n(\omega) \tag{19.84}$$

when one has total reflection.

Here also we should mention that in the presence of loss the above equation (19.83) remains valid for τ_T, whereas it ceases to be valid for τ_R because of additional delta peaks existing in the reflection spectra. Therefore, τ_R and DOS become quite different [62].

19.3 Dispersion curves and densities of states

We limit ourselves to the case of transverse electric (TE) polarization modes where the electric field has only one transverse component along the x_2 axis and the wave vector $k_{\|}$ (parallel to the interfaces) is directed parallel to the x_1 axis. Our study is performed for isotropic dielectric photonic crystal slabs of thicknesses comparable to the wavelength of the electromagnetic radiation. The retardation is fully taken into account.

This section contains a few illustrations of local and total DOS, dispersion curves, and transmission spectra for TE polarization modes in a substrate–buffer layer (b)–finite PC–cap layer (c)–vacuum system (see Fig. 19.1). The numerical calculations are performed for $\varepsilon_1 = 10$, $\varepsilon_2 = 3$, $\varepsilon_b = 5$, $\varepsilon_c = 7$, and $\varepsilon_s = 2$, and layer thicknesses $d_2 = d_c = 2d_1 = 2d_b = 2D/3$ where D is the period of the layered PC. For the sake of clarity, we assume that the layered PC contains $N = 5$ periods of 1–2 dielectrics. The buffer layer (b) and cap layer (c) are in contact with the materials 2 and 1, respectively, of the layered PC.

Fig. 19.3 gives the dispersion curves (dimensionless frequency $\omega D/c$ versus the reduced wavevector $k_{\|}D$) for TE polarization waves in the structure depicted in the paragraph above. The localized and resonant discrete modes are induced by the buffer layer (b)–finite PC–cap layer (c) interactions, and they are very sensitive to the thicknesses and nature of the buffer, cap layers, and of the PC layers. They are obtained from the maxima of the peaks in the total DOS (see Fig. 19.4(a)). In Fig. 19.3 we plot two light lines: the vacuum light line (blue heavy solid line, dark gray in print version), given by the equation $k_{\|}D = \omega D\sqrt{\varepsilon_v}/c$, and the substrate light line (blue heavy

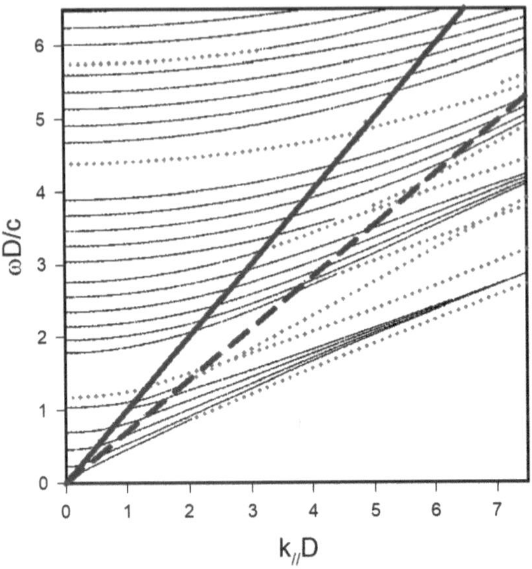

Figure 19.3 Dispersion of localized and resonant modes (red dotted lines, light gray in print version) induced by a finite PC and buffer and cap layers (see Fig. 19.1). The PC contains $N = 5$ layer pairs of 2–1 dielectrics. The blue heavy solid and the blue heavy dashed lines (dark gray in print version) are the vacuum light line and the substrate light line, respectively. The red dotted lines (light gray in print version) below the substrate light lines are the localized modes, and those above the substrate light line are the resonant modes. The thin solid lines below the substrate light line are the confined modes of the PC, and those above the substrate light line are resonant modes. The numerical calculations are performed for $\varepsilon_1 = 10$, $\varepsilon_2 = 3$, $\varepsilon_b = 5$, $\varepsilon_c = 7$, and $\varepsilon_s = 2$, and layer thicknesses $d_2 = d_c = 2d_1 = 2d_b = 2D/3$ where D is the period of the layered PC.

dashed line, dark gray in print version), given by the equation $k_{\|} D = \omega D\sqrt{\varepsilon_s}/c$. These lines delimit the oscillatory and the decaying behavior of the field in vacuum or in the substrate. The branches situated below the substrate bulk band are the resonant modes associated with the deposition of the finite layered PC on the top of the substrate and decay exponentially when penetrating in the latter.

The branches represented by the thin solid lines represent the confined modes of the finite layered PC; their number increases with the number N of periods in the PC, leading to the bulk bands of the infinite layered PC in the limit $N \to \infty$. The branches represented by the red dotted lines (light gray in print version) and falling in the minigaps of the layered PC are localized and resonant modes induced by the buffer or the cap layer; they are localized mainly at the boundaries of the PC or are guided waves of the buffer or the cap layer. In the following we shall concentrate on the discussion of the latter modes, which are very sensitive to the physical parameters of the cap and buffer layers.

In Fig. 19.3, there are six branches (red dotted lines, light gray in print version) situated below the substrate bulk band and falling inside the gaps of the PC; these branches correspond to the localized modes induced by the buffer or the cap layer. The localization properties of these modes will be discussed below. The lowest lo-

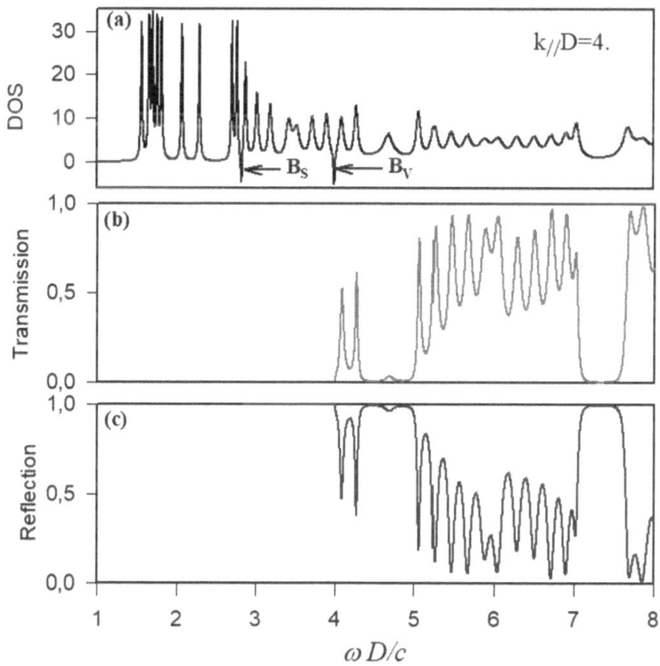

Figure 19.4 Frequency dependence of (a) the DOS, (b) the transmission coefficient, and (c) the reflection coefficient for $k_{\parallel}D = 4$. The geometrical parameters are the same as in Fig. 19.3. The bulk contribution of the substrate and vacuum to the DOS are subtracted; $\mathbf{B}_s$ and $\mathbf{B}_v$ are δ functions of weight $-1/4$ appearing at the limits of light cones of the substrate and the vacuum bulk band, respectively, within an approach [55] using a reference system made out of the response functions $g_2(M, M)$ and truncated G_1, G_3.

calized branch emerges from the first miniband of the PC at $k_{\parallel}D \approx 2$. Similarly, the lowest branch of the second PC miniband becomes a localized mode when $k_{\parallel}D \geq 4.3$. Notice that one of the three localized modes inside the first gap of the PC is peeled off from the top branch of the first miniband of the PC for $k_{\parallel}D = \omega D\sqrt{\varepsilon_s}/c$. This mode becomes a localized mode for $k_{\parallel}D \geq \omega D\sqrt{\varepsilon_s}/c$ and sweeps almost the entire first bandgap after its interaction with the two other localized modes. The branches falling above the substrate light line correspond to resonant modes, as they can propagate in the substrate. Notice the mixing of the two resonant modes situated in the third gap, around $k_{\parallel}D \approx 6.75$. It is worth noting that for other buffer and cap layers, the positions of these modes will be quite different.

An example of the behavior of localized and resonant modes in the variation of the DOS is illustrated in Fig. 19.4(a). Note that the delta peaks lying below B_s are enlarged by adding a small imaginary part to the frequency ω. Indeed, the denominator of the Green's function goes to zero for the frequencies corresponding to localized modes, which leads to infinite δ peaks in the DOS. Consequently, to enlarge these δ peaks, we introduced an absorption factor μ as a small imaginary part added to the frequency ω ($\mu = 10^{-2}c/D$). The peaks that lie below B_s represent confined modes inside the layered PC or inside the cap or the buffer layer, and they decay on both sides of these two materials. Those lying between B_s and B_v may propagate in both the PC and the substrate, and those falling above B_v propagate in the whole composite system.

Next we compare the DOS and the transmission spectrum. Fig. 19.4(b) presents the frequency dependence of the transmission coefficient through the buffer layer–finite PC–cap layer for $k_{\parallel} D = 4$. The TE polarization wave incident from the substrate is represented by a plane wave. The incident photons are scattered from the interfaces between the dissimilar layers that constitute the system. The transmission factor (Fig. 19.4(b)) shows sharp peaks with frequency positions corresponding to the peaks in Fig. 19.4(a). In the interior of the first PC minigap, a broad resonance appears in the curve of the total DOS (Fig. 19.4(a)) at $\omega D/c = 4.68$. This resonance also appears with a very broad intensity in the transmission spectrum. Apart from the frequency associated with this resonance, no transmission arises in the frequency gaps of the PC. Note that the reflection coefficient R is equal to $1 - T$ for any value of the frequency (see Figs. 19.4(b) and 19.4(c)), as it should be in the absence of any dissipation.

The localized and resonant modes can be divided into different groups according to the behavior of the corresponding eigenstates along the x_3 axis of the PC; the modes may propagate in both the PC and the buffer and cap layers, or they may propagate in one and decay in the other or decay on both sides of the layered PC–buffer layer or the layered PC–cap layer interfaces. To illustrate these different types of behaviors, we have plotted in Fig. 19.5 the local DOS as a function of the space position x_3 for a given reduced wavevector $k_{\parallel} D = 7.5$ and for different reduced frequencies $\omega D/c = 2.726$, $\omega D/c = 3.194$, $\omega D/c = 3.791$, $\omega D/c = 3.899$, $\omega D/c = 4.222$, $\omega D/c = 4.438$, $\omega D/c = 4.887$, $\omega D/c = 5.451$, and $\omega D/c = 5.607$ corresponding to different modes in Fig. 19.3. This local DOS reflects the spatial behavior of the square modulus of the electric field. Indeed, the DOS is obtained from the imaginary part of the Green's function. Therefore it can be expressed [52] as the sum of delta functions at the eigenfrequencies of the system, each having a weight equal to the square modulus of the wave function, i.e., the electric field in our case.

Case (a) ($\omega D/c = 2.726$) corresponds to a mode localized at the photonic crystal–cap layer interface, as the local DOS decays exponentially on both sides of this interface. Cases (b) and (c) ($\omega D/c = 3.194$ and $\omega D/c = 3.791$) reflect guided modes of the cap layer that decay outside this layer, with a small penetration into the photonic crystal mainly in the first layer in contact with the cap. This situation also occurs in Case (f) ($\omega D/c = 4.438$), where the guided mode penetrates more deeply into the layered photonic crystal. Case (d) ($\omega D/c = 3.899$) illustrates a guided mode of the buffer layer, whereas case (g) ($\omega D/c = 4.887$) refers to a mode localized at the photonic crystal–buffer layer interface. In case (e) ($\omega D/c = 4.222$) the frequency falls inside a miniband of the PC while being below the substrate bulk band; therefore this corresponds to a confined mode inside the finite PC. Finally, the last two cases, (h) and (i) ($\omega D/c = 5.451$ and $\omega D/c = 5.607$), are associated with resonant modes that can propagate inside the substrate, as their frequencies are situated above the substrate light line. Both modes have a decaying behavior inside the PC because their frequencies fall inside a minigap of the PC. However, one can notice that the resonant mode presented in (h) is in essence induced by the cap layer and has the character-

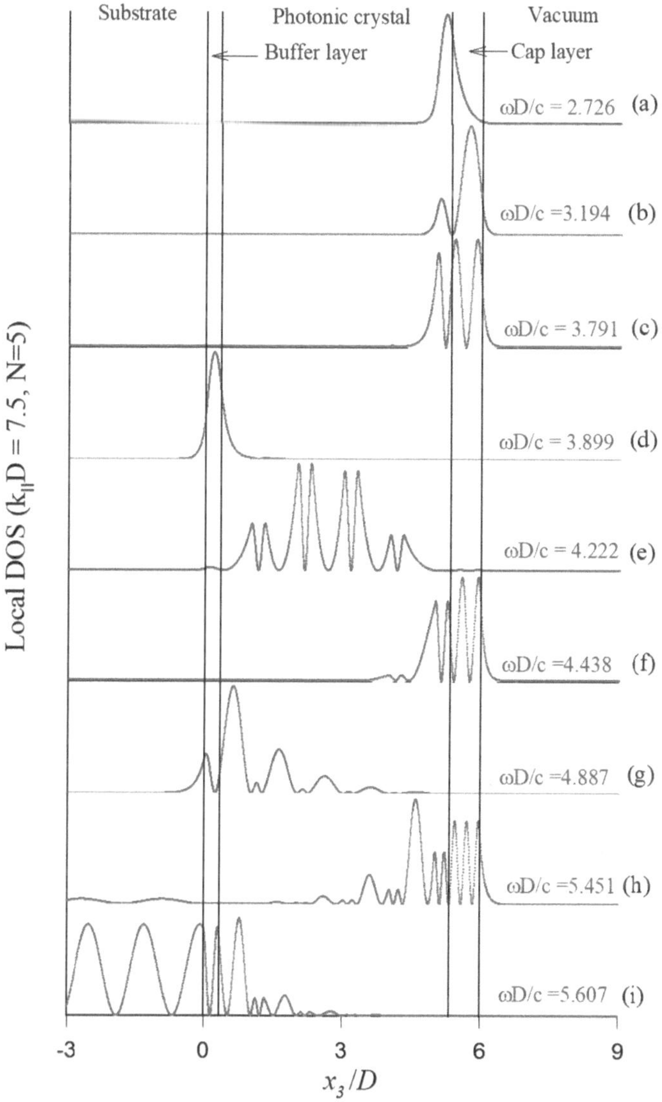

Figure 19.5 Spatial representation of the local DOS corresponding to the case depicted in Fig. 19.3 for (a) $k_{\|} D = 7.5$ and $\omega D/c = 2.726$, (b) $\omega D/c = 3.194$, (c) $\omega D/c = 3.791$, (d) $\omega D/c = 3.899$, (e) $\omega D/c = 4.222$, (f) $\omega D/c = 4.438$, (g) $\omega D/c = 4.887$, (h) $\omega D/c = 5.451$, and (i) $\omega D/c = 5.607$. The positions of the different interfaces are indicated by vertical lines.

istic of a guided wave in the vicinity of this layer, even though it can propagate in the substrate; this behavior results from the fact that the cap layer and the substrate are separated by the PC, where this mode cannot penetrate. On the other hand, the resonant mode displayed in (i) is induced by the presence of the substrate and remains rather weak in the DOS.

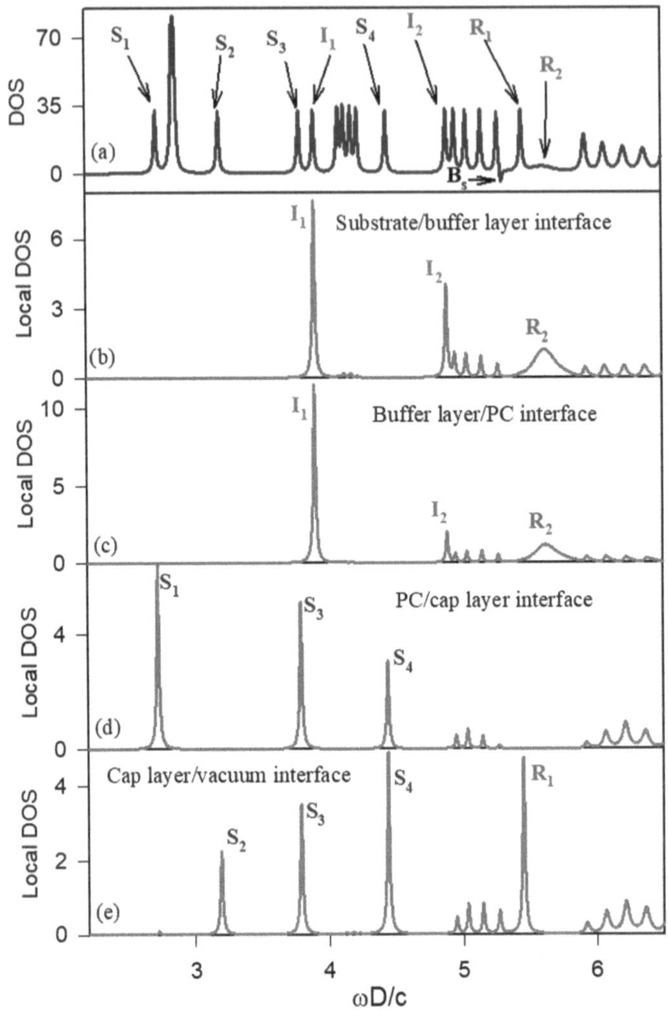

Figure 19.6 (a) DOS versus $\omega D/c$ for $k_{\|} D = 7.5$. Curves (b), (c), (d), and (e) represent the local densities of states versus $\omega D/c$ and for $k_{\|} D = 7.5$ calculated at the following interfaces: substrate–buffer layer, buffer layer–photonic crystal, photonic crystal–cap layer, and cap layer–vacuum, respectively.

In the following, we shall denote by **I** (interface) the modes localized inside the buffer layer or at the buffer layer–photonic crystal interface. We also denote by **S** (surface) the modes localized inside the cap layer or at the cap layer–PC interface. Finally, the modes situated above the substrate light line and falling inside the PC minigaps are denoted **R** (resonant). With these denotations, we give in Fig. 19.6 another view of the spatial distribution of the localized and resonant modes induced by the cap and buffer layers and that appear as well-defined peaks of the DOS. This figure displays, for $k_{\|} D = 7.5$, the total DOS (Fig. 19.6(a)) together with the local DOS at different interface planes, namely, at the substrate–buffer layer (Fig. 19.6(b)), buffer layer–PC

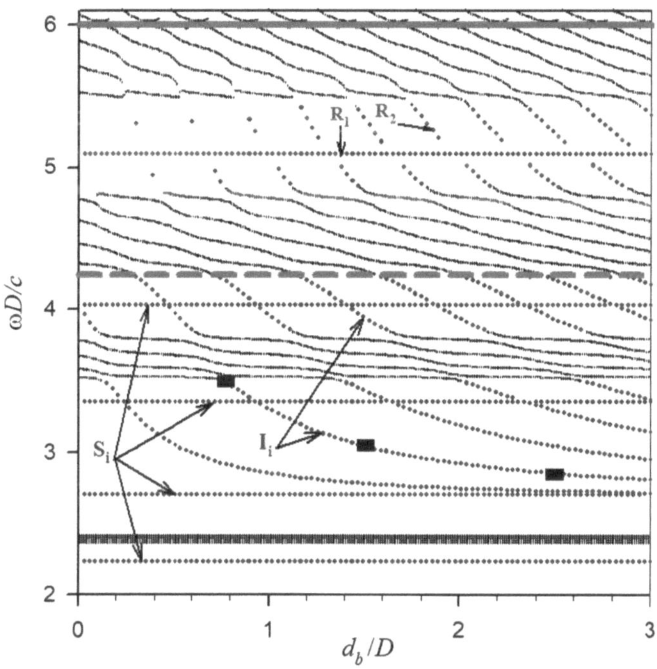

Figure 19.7 Variation of the frequencies of the localized and the resonant modes as a function of the width d_b of the buffer layer, for the system depicted in Fig. 19.1, for $k_{\|}D = 6$. Symbols I_i $(i = 1, 2)$, S_i $(i = 1, 2, 3, 4)$, and R_i $(i = 1, 2)$ (blue dotted lines, dark gray in print version) correspond to the localized modes at the PC–cap layer interface or inside the cap layer, the localized modes in the buffer layer–PC interface or inside the buffer layer, and the resonant modes, respectively. The thin solid lines represent the modes of the PC. The solid rectangles highlight the modes whose local DOS are presented in Fig. 19.8. The upper red heavy solid line (light gray in print version) represents the vacuum light line, and the red heavy dashed line (light gray in print version) represents the substrate light line.

(Fig. 19.6(c)), PC–cap layer (Fig. 19.6(d)), and cap layer–vacuum (Fig. 19.6(e)) interfaces. As one can expect, the modes labeled S_i $(i = 1, 2, 3, 4)$ are localized mainly inside or in the vicinity of the cap layer; however, one can notice that the spatial distributions of the modes labeled $\mathbf{S_1}$ and $\mathbf{S_2}$ are quite different from those corresponding to the modes $\mathbf{S_3}$ and $\mathbf{S_4}$. The modes noted $\mathbf{I_1}$ and $\mathbf{I_2}$ are localized in the buffer layer and do not contribute to the local DOS in the cap layer; in the same manner the resonant mode $\mathbf{R_2}$ may propagate in both the buffer layer and the substrate. Finally, at the frequencies of the resonant mode labeled $\mathbf{R_1}$, the main contribution to the LDOS comes from the cap layer; as indicated above in relation to Fig. 19.5(h), these modes have the characteristic of guided waves in the vicinity of the cap layer.

Next we investigate how the thickness d_b of the buffer layer affects the dispersion curves of the resonant and localized modes. We show in Fig. 19.7 the frequencies of these modes versus the thickness of the buffer layer for $k_{\|}D = 7.5$. Let us first observe that for two kinds of modes the frequencies remain independent of the width d_b, that is, for the modes localized at the PC–cap layer interface ($\mathbf{S_1}$) or confined in the cap

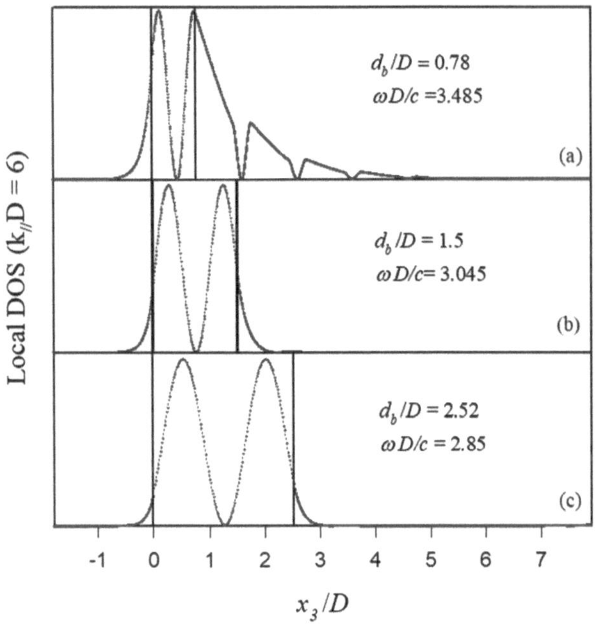

Figure 19.8 Spatial representation of the LDOS for the three modes (belonging to the branch $\mathbf{S}_1$) indicated by the solid rectangles in Fig. 19.7.

layer, ($\mathbf{S}_2$), ($\mathbf{S}_3$), and ($\mathbf{S}_4$), and for the resonant modes ($\mathbf{R}_1$) that decay into the PC from the cap layer and propagate inside the substrate. The localized modes confined in the first PC minigap are peeled off from the bottom of the second PC miniband when the width d_b increases. They become localized modes confined in the buffer layer and sweep almost all of the first bandgap. They become closer to one another when d_b increases and converge asymptotically to the bottom of the buffer layer bulk band given by $\omega D\sqrt{\varepsilon_b}/c$.

If one examines the localization properties of the localized modes along any branch (at the points indicated by solid rectangles in Fig. 19.7), one can see that the local DOS given in Fig. 19.8 shows for the top of this branch that the mode is propagating in the buffer layer and is penetrating into the PC with a decaying behavior. At the bottom of this branch the mode lies only in the buffer layer. Therefore we can conclude that these modes become confined in the buffer layer without penetrating into the PC when the thickness d_b becomes larger than D (here D is the period of the layered PC). In the second bandgap, when the thickness of the buffer layer increases, the modes are extracted from the third PC miniband and become localized modes characterized by an oscillatory behavior in the buffer layer and an exponentially decaying behavior into the PC. Notice that the same behavior occurs in the third gap, where the modes extracted from the fourth PC miniband become resonant modes characterized by an oscillatory behavior in the substrate and in the buffer layer, and they decay in the photonic crystal. The frequencies of these modes decrease until they merge with the lower bulk band. We note that when a resonant mode of the third gap disappears for

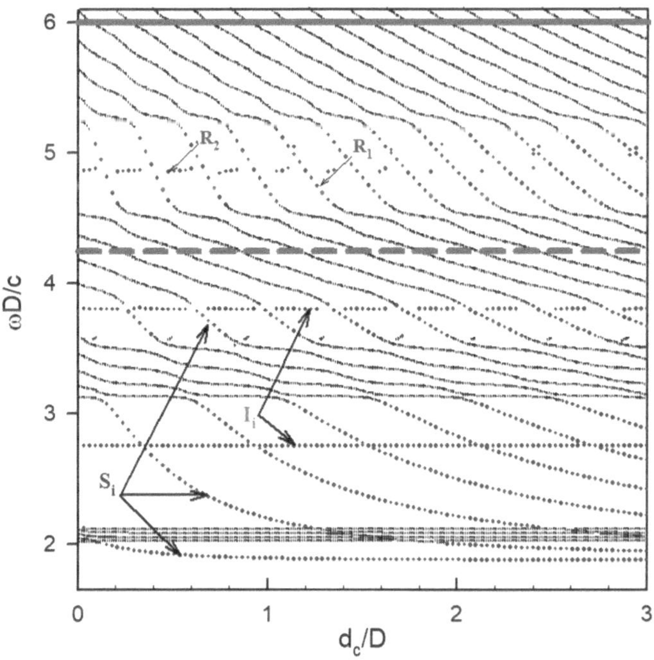

Figure 19.9 Same as in Fig. 19.7, but the variation of the frequencies is given versus d_c/D.

a given value of d_b, one mode is extracted from the bottom of the third PC miniband, giving rise to a localized mode in the buffer layer. As d_b increases, this mode sweeps through the second and first bandgaps until it reaches the light line of the buffer layer, acquiring the confined mode character in the buffer layer. Finally, we mention that for any given frequency ω in Fig. 19.7, there is a periodic repetition of the modes as a function of d_b.

As we have seen in Fig. 19.7, the frequencies of the resonant and localized modes are very sensitive to the width of the buffer layer. In this case, only the frequencies of the buffer layer–PC interface modes and those confined inside the buffer layer are affected. Now we investigate how the thickness of the cap layer affects the eigenfrequencies of the localized and resonant modes. Fig. 19.9 shows the variation of the frequency of these modes as a function of the thickness ratio d_c/D for a given value of the reduced wave vector $k_{\|}D$ such that $k_{\|}D = 6$. We can notice that contrary to the previous case (Fig. 19.7), only the surface resonant mode $\mathbf{R_2}$ (characterized by an oscillatory behavior in the substrate and in the cap layer) and the surface localized modes $\mathbf{S_1}$, $\mathbf{S_2}$, and $\mathbf{S_3}$ (confined in the cap layer or localized at the PC–cap layer interface) are affected.

As a point of additional interest, the transmission rate gives us information on the interaction of an incident photon with resonant modes confined in the buffer layer–PC–cap layer system. We investigate how the localized and resonant modes appearing in the bandgaps of the PC affect the transmission of waves that propagate through

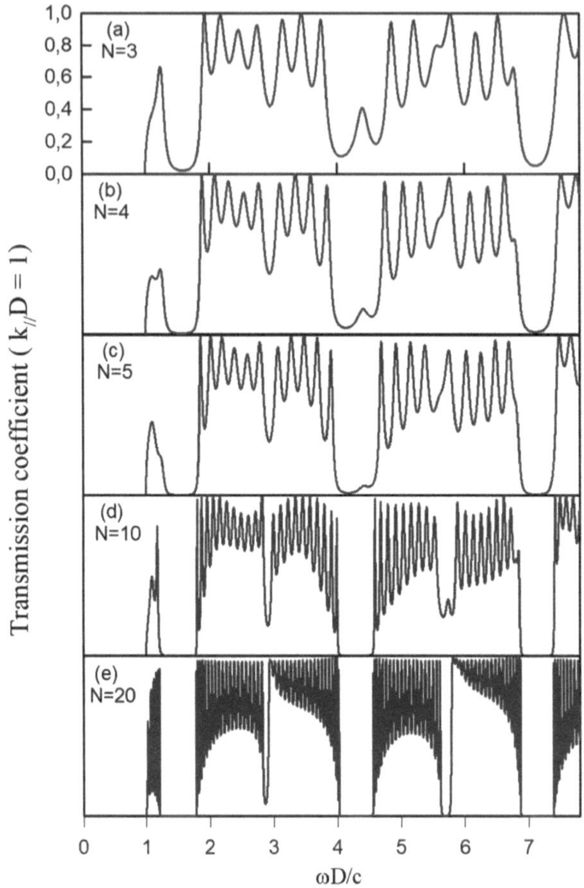

Figure 19.10 Evolution of the transmission coefficient as a function of the number of cells N, for $k_{\|} D = 1$. The parameters (both material and geometrical) are the same as in Fig. 19.3.

the photonic crystal. We consider an incident plane wave coming from the substrate and calculate the transmission spectrum through the composite system. For instance, Fig. 19.10 shows the variation of the transmission factor versus reduced frequencies for $k_{\|} D = 1$ and different number of periods in the layered photonic crystal, namely, $N = 3, 4, 5, 10$, and 20. The transmission coefficient displays a number of oscillations in each miniband of the photonic crystal in relation to the number N of periods. The edges of the minibands become more sharply defined with increasing N. In the second and third minigaps, the transmission coefficient reveals the existence of a resonant mode, in accordance with our results on the DOS. However, the intensity of this mode in the transmission spectrum decreases when the number N of periods in the photonic crystal is increased because the eigenfunction of this mode is exponentially decaying into the photonic crystal. For instance, the peaks associated with these resonant modes are visible in Fig. 19.10(a), 19.10(b), and 19.10(c), but are totally depressed in Figs. 19.10(d) and 19.10(e).

19.4 Delay times and density of states

In order to illustrate the general results given in Section 19.2.4, we present here a simple and novel application for transverse electromagnetic (TE) modes in the special case of a buffer layer–finite PC sandwiched between the substrate and air (when we take $d_c = 0$ in Fig. 19.1). Different localized and resonant discrete modes due to the interaction between the different constituents of this system are also investigated. Particular attention is focused on the influence of buffer layers. In particular, we show the possibility of finding well-defined guided resonances in the layered photonic crystal, as a consequence of its separation from the substrate by the buffer layer. This phenomenon results from the fact that some of the discrete modes of the PC, even if they are in resonance with the substrate bulk modes, cannot propagate in the buffer layer and therefore remain well-defined guided waves in the finite layered PC. On the other hand, we illustrate numerically the equivalence between the reflection phase time and the DOS for different frequencies. Our phase time results include the special limit studied before [44,45], i.e., in the absence of a buffer layer. We found that the presence of a buffer layer strongly increases the magnitude of the time delay for reflection or transmission waves through the finite size PC.

19.4.1 Analytical results

With the help of the interface response theory [39], it is straightforward to obtain analytical expressions for the entities defined by Eq. (19.64), namely

$$A_1 = \frac{g(x_c, x_c)}{g(0,0)g(x_c, x_c) - g(0, x_c)^2}, \tag{19.85}$$

$$A_2 = \frac{g(0,0)}{g(0,0)g(x_c, x_c) - g(0, x_c)^2}, \tag{19.86}$$

and

$$B = -\frac{g(0, x_c)}{g(0,0)g(x_c, x_c) - g(0, x_c)^2}, \tag{19.87}$$

where

$$g(0,0) = \frac{1}{\Delta^-}\left\{\left(\frac{t^2-1}{t} + Y_1^S\right)\left[\left(\frac{C_1 S_2}{F_2} + \frac{C_2 S_1}{F_1}\right) - \frac{S_b}{F_b C_b}\left(\frac{t^2-1}{t} + Y_1^S\right)\right]\right.$$
$$\left. -t^{2N} Y_2^S\left[\frac{S_1}{F_1} + \frac{S_2}{F_2}t - \frac{S_b}{F_b C_b} Y_2^S\right]\right\}, \tag{19.88}$$

$$g(0,x_c)=\frac{t^N}{C_b\Delta^-}\frac{t^2-1}{t}\left(\frac{S_1}{F_1}+\frac{S_2}{F_2}t\right), \tag{19.89}$$

$$g(x_c,x_c)=\frac{1}{\Delta^-}\left\{\left(\frac{C_1S_2}{F_2}+\frac{C_2S_1}{F_1}\right)\left(\frac{t^2-1}{t}+Y_1^b\right)-t^{2N}Y_2^b\left(\frac{S_1}{F_1}+\frac{S_2}{F_2}t\right)\right\}, \tag{19.90}$$

with

$$\Delta^-=\left(\frac{t^2-1}{t}+Y_1^S\right)\left(\frac{t^2-1}{t}+Y_1^b\right)-t^{2N}Y_2^SY_2^b, \tag{19.91}$$

$$Y_1^S=C_1C_2+\frac{F_1}{F_2}S_1S_2-t, \tag{19.92}$$

$$Y_2^S=C_1-C_2t, \tag{19.93}$$

$$Y_1^b=Y_1^S-\frac{F_bS_b}{C_b}\left(\frac{C_1S_2}{F_2}+\frac{C_2S_1}{F_1}\right), \tag{19.94}$$

and

$$Y_2^b=Y_2^S-\frac{F_bS_b}{C_b}\left(\frac{S_1}{F_1}+\frac{S_2}{F_2}t\right). \tag{19.95}$$

19.4.2 Numerical results and discussion

This section contains new results of local and total densities of states, dispersion curves and phase times for electromagnetic modes in a finite GaAs–AlAs PC deposited on a GaAs substrate with a buffer layer of SiO_2 embedded between these two materials. The numerical calculations are performed for $\varepsilon(\text{GaAs})=12.39$, $\varepsilon(\text{AlAs})=8.41$, $\varepsilon(SiO_2)=2.13$ and $\varepsilon(\text{air})=1$ and layer thicknesses $d_1(\text{GaAs})=d_2(\text{AlAs})/2=D/3$ where $D=d_1+d_2$ is the period of the PC. The velocity of the electromagnetic waves in the buffer layer is higher than those in the surrounding media (i.e., the substrate and the PC); therefore, the buffer layer plays the role of a barrier for photons between the PC and the substrate.

Fig. 19.11(b) shows the dispersion curves (reduced frequency versus the reduced wave vector) for electromagnetic waves in the structure depicted in Fig. 19.11(a). The thickness d_b of the SiO_2 buffer layer is such that $d_b=2D$. These modes are obtained from the maxima of the DOS, shown in Fig. 19.12 for a few values of the reduced wavevector $k_\parallel D$. Let us notice that the δ peaks lying below B_b are enlarged by adding a small imaginary part to the frequency ω. For the sake of clarity, in Fig. 19.11(b), the PC contains $N=5$ bilayers of GaAs/AlAs. All the branches in Fig. 19.11(b) represent resonant modes (also called leaky waves) induced by the buffer/PC multilayers in the continuum of the substrate bulk band, which means above the light line of the GaAs substrate (heavy straight line). There are three categories of normal modes in

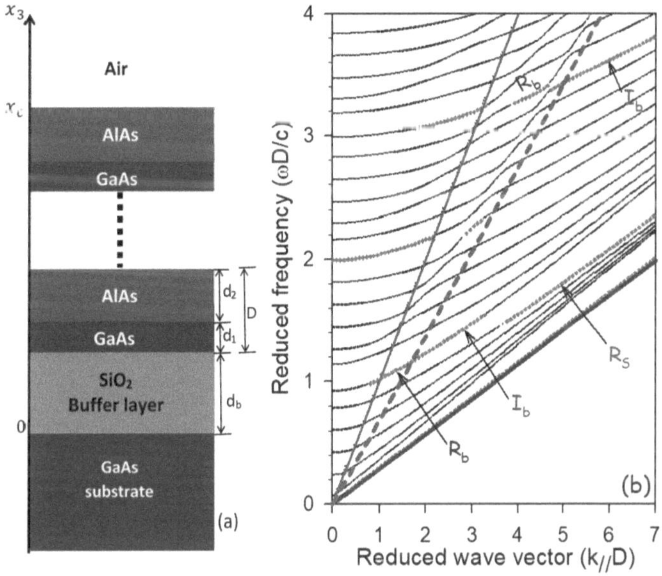

Figure 19.11 (a) Schematic representation of a finite GaAs–AlAs PC deposited on a GaAs substrate with an embedded SiO_2 buffer layer; d_b, d_1, and d_2, respectively, are the thicknesses of the SiO_2 buffer layer and of the two different slabs out of which the unit cell of the PC is built; D is the period of the layered PC. (b) Dispersion of electromagnetic waves of the structure presented in Fig. 19.11(a) with $d_b = 2D$. The PC contains $N = 5$ bilayers of GaAs/AlAs. The blue heavy line (dark gray in print version) and green heavy-dashed line (mid gray in print version), respectively, correspond to the light lines of the substrate and of the SiO_2 buffer layer. Full lines correspond to extended states belonging to the PC minibands. Red dotted lines refer to surface modes (labeled R_s), interface modes (labeled I_b) and buffer layer modes (labeled R_b) falling in the PC minigaps.

the present structure according to their propagating (full lines) or localized (red dotted lines, light gray in print version) character in the PC. The former are extended (or bulk) states, while the latter are either decaying from the free surface (red dotted lines [light gray in print version], labeled R_s) or confined in the buffer layer (red dotted lines [light gray in print version], labeled I_b and R_b). The localization properties of these different modes will be shown below. The number of branches corresponding to extended states in the PC increases with increasing number N of periods, leading to the bulk bands of an infinite PC in the limit $N \to \infty$.

On the other hand, the positions of the surface and buffer resonant modes, which are already distinguishable in Fig. 19.11(b), even for a small number of periods as $N = 5$, shift only slightly with N when going to the limit of a semiinfinite layered PC. Among these resonant modes, one can distinguish those falling below and above the buffer light line (green dashed straight line, mid gray in print version). We shall focus on the branches situated below the buffer light line. They appear as well-defined peaks in the DOS (see Fig. 19.12), even though they are in resonance with the bulk states of the GaAs substrate. One can, however, notice that the widths of these peaks become large when either $k_{\parallel} D$ or the thickness d_b of the buffer layer decreases (see below).

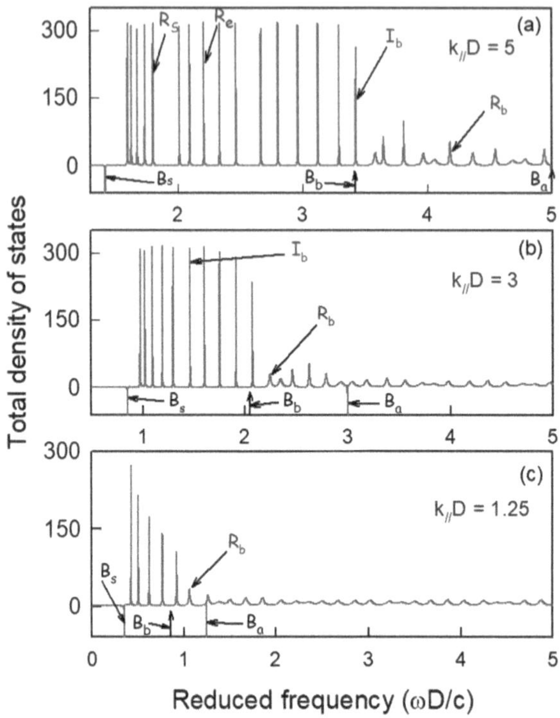

Figure 19.12 Density of states $n(\omega, k_\parallel)$, in units of D/c, depicted for (a) $k_\parallel D = 5$, (b) $k_\parallel D = 3$ and (c) $k_\parallel D = 1.25$ in Fig. 19.11(b). The symbols R_s, I_b, and R_b have the same meanings as in Fig. 19.11(b); B_s and B_a are delta functions of weight $-1/4$ appearing at the bottom of the substrate and air bulk bands; B_b indicates the limit of the SiO_2 bulk band within an approach [55] using a reference system made out of the response functions $g_2(M, M)$ and truncated G_1, G_3.

The presence of the buffer layer strongly increases the intensity of the peaks in the DOS (Fig. 19.12), lying below the buffer layer light line and appearing as δ peaks for high values of $k_\parallel D$ (Fig. 19.12(a)). Therefore, such confined modes in the PC could be observed easily. One can notice also in Fig. 19.11(b) a branch labeled I_b below the buffer light line which correspond to localized modes at the buffer–PC interface; the continuation of this branch above the buffer light line corresponds to modes confined in the buffer layer (branch labeled R_b in Fig. 19.11(b)).

As mentioned above, the resonant modes can be divided in different groups according to the behavior of the corresponding eigenstates along the axis x_3 of the PC. To illustrate these different types of behavior, we have plotted in Fig. 19.13 the local DOS as a function of the space position x_3, for different modes labeled R_e, R_s, I_b, and R_b corresponding to extended, surface, interface, and buffer modes, respectively, in Fig. 19.12(a). This local DOS reflects the spatial behavior of the squared modulus of the electric field. Indeed, the density of states is obtained from the imaginary part of the Green's function. Therefore, it can be expressed as a sum of delta functions at the

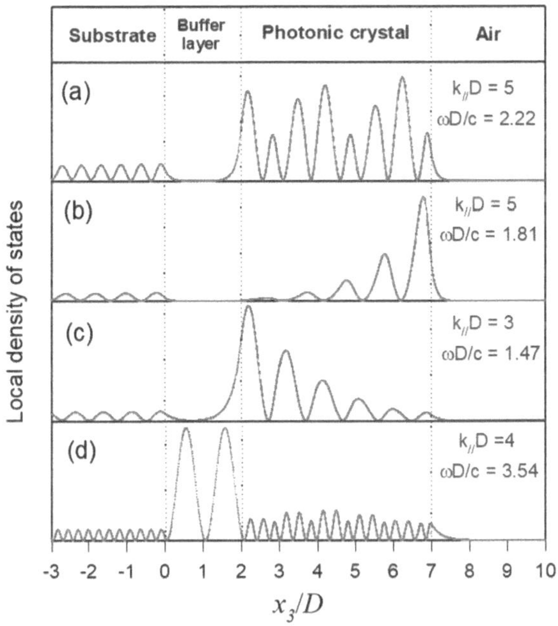

Figure 19.13 Spatial representation of the local DOS corresponding to the case depicted in Fig. 19.11(b) for (a) $k_{\|} D = 5$ and $\omega D/c = 2.22$, (b) $k_{\|} D = 5$ and $\omega D/c = 1.81$, (c) $k_{\|} D = 3$ and $\omega D/c = 1.47$, and (d) $k_{\|} D = 5$ and $\omega D/c = 3.54$. The space positions of the different interfaces are marked by vertical dotted lines.

eigenfrequencies of the system, each having a weight equal to the squared modulus of the wave function, i.e., the electric field in our case.

Fig. 19.13(a) shows that the extended mode R_e is confined in the finite PC and do not propagate into the SiO_2 buffer layer. Consequently, it remains a well-defined guided wave of the finite GaAs–AlAs PC. On the contrary, the surface R_s and interface I_b modes in Fig. 19.12 reflect a strong localization at the surface of the PC and at the PC/buffer layer interface, respectively. Above the buffer light line, the resonant modes are due to the interaction between the layered PC/buffer layer modes and the substrate modes. Now, an analysis of the local DOS as a function of the space position shows a propagating behavior in the substrate/buffer layer/PC system (see Fig. 19.13(d) related the mode labeled R_b in Fig. 19.12(a)) with a pronounced amplitude in the SiO_2 buffer layer.

The resonant modes induced by the PC/buffer are dependent on the width of the buffer layer. Fig. 19.14 illustrates the variation of the frequency of the resonant modes as a function of the thickness ratio d_b/D, for a given value of the reduced wavevector $k_{\|} D$ such that $k_{\|} D = 5.5$. Let us recall that apart from the surface and interface curves falling, respectively, in the first and third PC minigaps and situated below the buffer light line (dashed line), the other curves represent sharp resonant guided waves in the PC; the corresponding frequencies of all these modes present a very small variation with d_b. Moreover, the intensity of the resonances in this region of frequen-

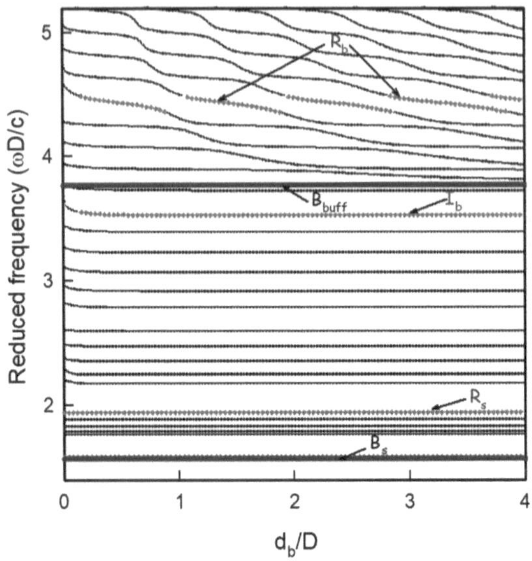

Figure 19.14 Variations of the frequencies of the localized and resonant modes versus the thickness d_b of the buffer layer for the PC depicted in Fig. 19.11(a). The wavevector $k_\parallel$ is chosen such that $k_\parallel D = 5.5$. The notations have the same meaning as in Fig. 19.12.

cies increases by increasing the thickness d_b of the SiO_2 buffer layer (Figs. 19.15(a) and 19.15(b)), and for large values of d_b (Fig. 19.15(b)) the resonances appear as delta functions of weight 1 as the interaction between the GaAs–AlAs PC and the GaAs substrate becomes weak. Concerning the curves lying above the buffer light line (Fig. 19.14), they present a noticeable variation with varying d_b/D and tend asymptotically to the buffer bulk band when $d_b \to \infty$. These resonances correspond to waves with predominant amplitudes in the SiO_2 buffer layer (see Fig. 19.13(d)) even though they are propagating in the whole GaAs/SiO_2/GaAs–AlAs (PC) system. The corresponding peaks in the DOS (or in the phase time) present a noticeable intensity only for large value of d_b.

The reflection from the buffer/PC of an incident photon originated in the GaAs substrate is explained quantitatively in terms of photon trapping times. Because of the free surface of the PC, the incident photons with frequencies lying between the air and the GaAs substrate light lines are entirely reflected. Therefore, the reflection rate is one and only the phase time (Eq. (19.76)) gives us information about the interaction of an incident photon with resonant modes confined in the layered PC/buffer system. This time is interpreted as the time needed for a photon to complete the reflection process. In this part of work, we study the interaction of photons with quasilocalized modes in the layered PC. In addition, we give a comparison between the DOS and phase time as a function of the frequency. In Fig. 19.15, we give a comparison between the DOS and phase time for two different values of the buffer layer thickness: $d_b = 0$ (Figs. 19.15(a) and 19.15(b)) and $d_b = 2D$ (Figs. 19.15(c) and 19.15(d)). In Fig. 19.15(b), the resonances are not very sharp; one can distinguish, in particular,

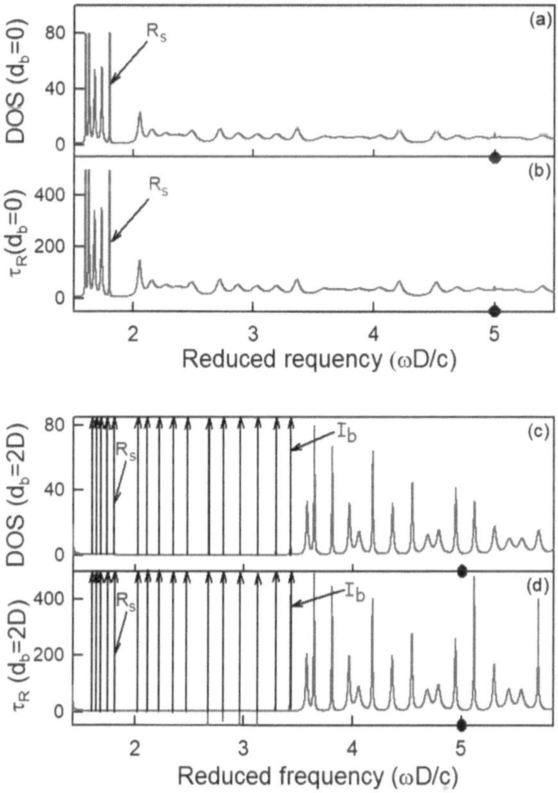

Figure 19.15 Frequency dependence of the DOS and reflection phase times (in units of D/c) for $d_b = 0$ (top panel) and $d_b = 2D$ (bottom panel) in Fig. 19.5. The wavevector $k_{\parallel}$ is chosen such that $k_{\parallel} D = 5$. The filled squares (dots) on the frequency axis indicate the positions of the buffer (air) light line.

the resonance mode (R_s) which exhibits a large time delay. Indeed, the magnitude of the peak associated with this surface resonance is about 224 (in units of D/c). This means that for a typical value of the period D used in experiments [63,64], for example, $D = 200$ nm, the magnitude of the time delay is of the order of 150 fs, which is in the range detectable by a photon experiment using laser techniques [44,45]. The half-widths of the peaks in this quantity is related to the lifetimes of the resonances. The intensity of the peaks in the phase time describes the time needed for photons to complete the reflection process. For a large value of d_b (Figs. 19.15(c) and 19.15(d)) and for frequencies lying below the buffer light line, the resonant modes become quasilocalized modes and give rise to extremely sharp resonances in the phase time (indicated by arrows in Fig. 19.15(c) and 19.15(d)) as the interaction between the layered PC and the substrate modes is rather weak. Indeed, by using a very fine mesh in the numerical calculation, the magnitude of the sharp resonances in the phase time (the bottom panel of Fig. 19.15) reaches 2.10^4 (in units of D/c), i.e., about 13.3 ps for a period $D = 200$ nm. Of course, to observe these resonances, a very extended wave packet (of spectral width smaller than that of the resonance) would have to be used. On the other hand, one can notice that in addition to the surface mode R_s in Figs. 19.15(c) and 19.15(d), an interface mode I_b associated with the buffer–PC interface appears also in the frequency gap. Moreover, for the frequency range situated above the buffer light line, the intensity of the resonances becomes important with respect to the case $d_b = 0$. On the other hand, one can clearly see that below the air light line (indicated by dots in the frequency axis), the reflection phase time τ_R is equal to 2π times the DOS (Figs. 19.15(c) and 19.15(d)) in accordance with the exact result of Eq. (19.80). On the contrary, for frequencies lying above the air light line, there is no proportionality relation between the DOS and the reflection phase time whereas Δn remains equal to τ_T/π as demonstrated in Eq. (19.82).

19.5 Summary and perspectives

In this chapter we have presented an analytical calculation of the response function (Green's function) for electromagnetic waves in a substrate–buffer layer–PCs–cap layer–vacuum system. We have deduced the closed-form relations of the density of states for electromagnetic waves in the system described above, as well as the reflection and transmission coefficients. We have focused our attention on the derivation of closed-form expressions for the local and total densities of states and the dispersion of both localized and resonant modes for these heterostructures and their spatial distribution. This Green's function can also be used to study any physical property of the composite systems [49,50].

Our general results are illustrated by a few applications to a photonic crystal (PC) that is composed of $N = 5$ cells and capped on both sides by two materials different from the PC components and that is in contact with a substrate on one side and vacuum on the other side.

We have shown that the deposition of the photonic crystal onto a substrate and the introduction of the buffer and cap layers induce localized and resonant modes. These modes appear as well-defined peaks of the total DOS, their localization properties are investigated by means of local DOS, and they also affect the transmission coefficient. One part of these modes is confined inside the bulk band of the PC and is characterized by an oscillatory behavior in the space occupied by the PC. The other part is confined in the buffer or the cap layer and decays within the PC. The penetration depth of the modes localized in the PC gaps varies with their positions among the dispersion curves of Fig. 19.3. It has been found that the occurrence, the position, and the localization properties of interface modes inside photonic bandgaps are very sensitive to the buffer and cap layers inserted inside the substrate–PC and the PC–vacuum interface, respectively. When the thickness of the buffer layer and/or the cap layer varies, the electromagnetic modes undergo modifications by moving across the PC gaps. Their frequencies are also lowered with increasing the thickness of the buffer and/or the cap layers.

Knowing that the layered photonic crystal containing a defect layer may be more useful and have more applications than the perfect one [19,20,53], we can study in the same theoretical framework the effect of a defect layer introduced into the photonic bandgap material.

We have also derived exact relations between the density of states and the reflection and transmission phase times of a scalar mode in a finite lamellar composite system sandwiched between two different homogeneous semiinfinite media.

We have also investigated the effect of the buffer layer on electromagnetic modes in a finite layered PC deposited on a substrate. The closed-form expressions obtained for the local and total DOS enabled us to study the dispersion of resonant modes for the substrate–buffer layer–PC structure, and their spatial distribution. Particular attention was devoted to the effect of the buffer layer when its dielectric permittivity is lower than those in the PC and the substrate. In this case, the buffer layer plays the role of a barrier for photons between the PC and the substrate. The localized and resonant modes associated with these structures appear as well-defined DOS peaks. The intensity of these peaks was found to be very dependent on the wavevector $k_{\|}$ and the buffer layer thickness as well as on the parameters of the different constituent materials.

We have also calculated the photon phase time expressions and shown that they have the same behavior as the DOS versus the frequency. Our phase time results include a special limit treated in an earlier worker, i.e., in the absence of buffer layer. The presence of the buffer layer (Fig. 19.15) was found to increase strongly the magnitude of the time delay of reflected photons. The experimental observation of the localized and resonant modes predicted here in such finite PCs can be possible with different laser techniques.

It is worthwhile to mention that the exact relation between the DOS and the transmission and reflection phase times obtained here can also be transposed straightforwardly to other excitations of scalar nature such as transverse acoustic phonons or free electrons.

The results presented in this chapter and those of Chapter 6 enable addressing also the creation of new path states due to the adsorption of stubs along robust zero lines of interface localized eigenfunctions. Such states may create invisibility cloaks within very large wavelength bands.

Acknowledgments

The authors would like to thank Mohamed El Hassane Lahlaouti, Mokhtar Hammouchi, and Abdelkrim Nouhaoui for their fruitful contribution in the original papers from which this chapter is inspired.

References

[1] L. Esaki, Implications of Semiconductor Superlattice Research, Plenum, New York, 1991.
[2] A. Yariv, P. Yeh, Optical Waves in Crystals, Wiley, New York, 1984, pp. 155–219.

[3] P. Yeh, Optical Waves in Layered Media, Wiley, New York, 1988.
[4] L.H. Qin, Y.D. Zheng, R. Zhang, Study of Ge_xSi_{1-x}/Si superlattices by ellipsometry, Appl. Phys. A 55 (1992) 297.
[5] T. Hattori, N. Tsurumachi, S. Kawato, H. Nakatsuka, Photonic dispersion relation in a one-dimensional quasicrystal, Phys. Rev. B 50 (1994) 4220.
[6] D.R. Smith, R. Dalichaouch, N. Kroll, S. Schultz, S.L. Mc-Call, P.M. Platzman, Photonic band structure and defects in one and two dimensions, J. Opt. Soc. Am. B 10 (1993) 314.
[7] J.P. Dowling, C.M. Bowden, Anomalous index of refraction in photonic bandgap materials, J. Mod. Opt. 41 (1994) 345.
[8] R.D. Meade, K.D. Brommer, A.M. Rapper, J.D. Joannopoulos, Electromagnetic Bloch waves at the surface of a photonic crystal, Phys. Rev. B 44 (1991) 10961.
[9] P. Yeh, A. Yariv, C.S. Hong, Electromagnetic propagation in periodic stratified media. I. General theory, J. Opt. Soc. Am. 67 (1977) 423.
[10] P. Yeh, A. Yariv, A.Y. Cho, Optical surface waves in periodic layered media, Appl. Phys. Lett. 32 (1978) 104.
[11] W. Ng, P. Yeh, P.C. Chen, A. Yariv, Optical surface waves in periodic layered medium grown by liquid phase epitaxy, Appl. Phys. Lett. 32 (1978) 370.
[12] A.A. Bulgakov, V.R. Kovtun, Surface optical oscillations in a limited stratified-periodic medium, Opt. Spectrosc. 56 (1984) 269.
[13] A.A. Bulgakov, V.R. Kovtun, Study of surface optical oscillations in periodical multi-layer media, Solid State Commun. 56 (1985) 781.
[14] X.I. Saldana, G. Gonzalez de la Cruz, Electromagnetic surface waves in semi-infinite superlattices, J. Opt. Soc. Am. A 8 (1991) 36.
[15] M.L. Bah, A. Akjouj, E.H. El Boudouti, B. Djafari-Rouhani, L. Dobrzynski, Surface and interface optical waves in superlattices: transverse electric localized and resonant modes, J. Phys. Condens. Matter 8 (1996) 4171.
[16] F. Ramos-Mendieta, P. Halevi, Electromagnetic surface modes of a dielectric superlattice: the supercell method, J. Opt. Soc. Am. B 14 (1997) 370.
[17] Y.F. Li, J.W.Y. Lit, General formulas for the guiding properties of a multilayer slab waveguide, J. Opt. Soc. Am. A 4 (1987) 671.
[18] T. Hattori, N. Tsurumachi, H. Nakatsuka, Analysis of optical nonlinearity by defect states in one-dimensional photonic crystals, J. Opt. Soc. Am. B 14 (1997) 348.
[19] R. Wang, J. Dong, D.Y. Xing, Defect studies in a one-dimensional photonic band gap structure, Phys. Status Solidi B 200 (1997) 529.
[20] E. Yablonovitch, T.J. Gmitter, R.D. Meade, A.M. Rappe, K.D. Brommer, J.D. Joannopoulos, Donor and acceptor modes in photonic band structure, Phys. Rev. Lett. 67 (1991) 3380.
[21] M. Sigalas, C.M. Soukoulis, E.N. Economou, C.T. Chan, K.M. Ho, Photonic band gaps and defects in two dimensions: studies of the transmission coefficient, Phys. Rev. B 48 (1993) 14121.
[22] A. Mekis, J.C. Chen, I. Kurland, S. Fan, P.R. Villeneuve, J.D. Joannopoulos, High transmission through sharp bends in photonic crystal waveguides, Phys. Rev. Lett. 77 (1996) 3787.
[23] F. Villa, J.A. Gaspar-Armenta, Electromagnetic surface waves: photonic crystal–photonic crystal interface, Opt. Commun. 223 (2003) 109.
[24] X.-X. Liu, C.-F. Tsai, R.-L. Chern, D.-P. Tsai, Dispersion mechanism of surface magnetoplasmons in periodic layered structures, Appl. Opt. 48 (2009) 3102.
[25] Y. El Hassouani, H. Aynaou, E.H. El Boudouti, B. Djafari-Rouhani, A. Akjouj, V.R. Velasco, Surface electromagnetic waves in Fibonacci superlattices: theoretical and experimental results, Phys. Rev. B 74 (2006) 035314.

[26] M. Kaliteevski, I. Iorsh, S. Brand, R.A. Abram, J.M. Chamberlain, A.V. Kavokin, I.A. Shelykh, Tamm plasmon-polaritons: possible electromagnetic states at the interface of a metal and a dielectric Bragg mirror, Phys. Rev. B 76 (2007) 165415.
[27] W. Shen-Yun, L. Shao-Bin, L.-W. Joshua Li, Finite-difference time-domain studies of low-frequency stop band in superconductor–dielectric superlattice, Chin. Phys. B 19 (2010) 084101.
[28] J. Gaspar-Armenta, F. Villa, Photonic surface-wave excitation: photonic crystal–metal interface, J. Opt. Soc. Am. B 20 (2003) 2349.
[29] T.-B. Wang, C.-P. Yin, W.-Y. Liang, J.-W. Dong, H.-Z. Wang, Electromagnetic surface modes in one-dimensional photonic crystals with dispersive metamaterials, J. Opt. Soc. Am. B 26 (2009) 1635.
[30] T. Xu, A. Agrawal, M. Abashin, K.J. Chau, H.J. Lezec, All-angle negative refraction and active flat lensing of ultraviolet light, Nature 497 (2013) 470.
[31] E. Yablonovitch, Inhibited spontaneous emission in solid-state physics and electronics, Phys. Rev. Lett. 58 (1987) 2059.
[32] A. Szameit, S. Nolte, Discrete optics in femtosecond-laser-written photonic structures, J. Phys. B, At. Mol. Opt. Phys. 43 (2010) 163001.
[33] R. Maas, J. Van De Groep, A. Polman, Planar metal/dielectric single-periodic multilayer ultraviolet flat lens, Optica 3 (2016) 592.
[34] L.-J. Lang, X. Cai, S. Chen, Edge states and topological phases in one-dimensional optical superlattices, Phys. Rev. Lett. 108 (2012) 220401.
[35] M.D. Tocci, M. Scalora, M.J. Bloemer, J.P. Dowling, C.M. Bowden, Measurement of spontaneous-emission enhancement near the one-dimensional photonic band edge of semiconductor heterostructures, Phys. Rev. A 53 (1996) 2799.
[36] Y.F. Li, K. Iizuka, J.W.Y. Lit, Periodic stratified structure in a multilayer planar optical waveguide, J. Opt. Soc. Am. A 9 (1992) 559.
[37] M.L.H. Lahlaouti, A. Akjouj, B. Djafari-Rouhani, L. Dobrzynski, M. Hammouchi, E.H. El Boudouti, A. Nougaoui, Electromagnetic waves in finite superlattices with buffer and cap layers, J. Opt. Soc. Am. A 16 (1999) 1703.
[38] L. Dobrzynski, Interface response theory of discrete composite systems, Surf. Sci. Rep. 6 (1986) 119.
[39] L. Dobrzynski, Interface response theory of continuous composite systems, Surf. Sci. Rep. 11 (1990) 139.
[40] M. Buttiker, R. Landauer, Traversal time for tunneling, Phys. Rev. Lett. 49 (1982) 1739.
[41] M. Buttiker, Larmor precession and the traversal time for tunneling, Phys. Rev. B 27 (1983) 6178.
[42] E.H. Hauge, J.A. Stovneng, Tunneling times: a critical review, Rev. Mod. Phys. 61 (1989) 917.
[43] V. Laude, P. Tournois, Superluminal asymptotic tunneling times through one-dimensional photonic bandgaps in quarter-wave-stack dielectric mirrors, J. Opt. Soc. Am. B 16 (1999) 194.
[44] A.M. Steinberg, P.G. Kwiat, R.Y. Chiao, Measurement of the single-photon tunneling time, Phys. Rev. Lett. 71 (1993) 708.
[45] Ch. Spielmann, R. Szpocs, A. Stingl, F. Krausz, Tunneling of optical pulses through photonic band gaps, Phys. Rev. Lett. 73 (1994) 2308.
[46] Y. Avishai, Y.B. Band, One-dimensional density of states and the phase of the transmission amplitude, Phys. Rev. B 32 (1985) 2674.
[47] M. Hammouchi, E.H. El Boudouti, A. Nougaoui, B. Djafari-Rouhani, M.L.H. Lahlaouti, A. Akjouj, L. Dobrzynski, Acoustic waves in finite superlattices: influence of buffer layers, Phys. Rev. B 59 (1999) 1999.

[48] M.L.H. Lahlaouti, A. Akjouj, B. Djafari-Rouhani, L. Dobrzynski, Resonant and localized electromagnetic modes in finite superlattices, Phys. Rev. B 61 (2000) 2059.
[49] M.G. Cottam, A.A. Maradudin, Surface linear response functions, in: V.M. Agranovich, R. Loudon (Eds.), Surface Excitations, in: Mod. Probl. Condens. Matter Sci., vol. 9, North-Holland, Amsterdam, 1986, pp. 5–20.
[50] M.L. Bah, A. Akjouj, L. Dobrzynski, Response functions in layered dielectric media, Surf. Sci. Rep. 16 (1992) 95.
[51] J.M. Bendickson, J.P. Dowling, M. Scalora, Analytic expressions for the electromagnetic mode density in finite, one-dimensional, photonic band-gap structures, Phys. Rev. E 53 (1996) 4107.
[52] E.N. Economou, Green's Functions in Quantum Physics, Spring-Verlag, Heidelberg, 1990.
[53] R.D. Meade, K.D. Brommer, A.M. Rappe, Photonic bound states in periodic dielectric materials, Phys. Rev. B 44 (1991) 13772.
[54] E.H. El Boudouti, B. Djafari-Rouhani, A. Akjouj, L. Dobrzynski, Acoustic waves in solid and fluid layered materials, Surf. Sci. Rep. 64 (2009) 471.
[55] M.L.H. Lahlaouti, A. Akjouj, B. Djafari-Rouhani, L. Dobrzynski, M. Hammouchi, E.H. El Boudouti, A. Nougaoui, B. Kharbouch, Theoretical analysis of the density of states and phase times: application to resonant electromagnetic modes in finite superlattices, Phys. Rev. B 63 (2001) 035312.
[56] L. Dobrzynski, H. Puszkarski, Eigenvectors of composite systems. I. General theory, J. Phys. Condens. Matter 1 (1989) 1239.
[57] P.M. Morse, H. Feshbach, Methods of Theoretical Physics, McGraw-Hill, New York, 1953, Chap. 9.
[58] R. Landauer, Barrier traversal time, Nature (London) 341 (1989) 567.
[59] C.R. Leavens, G.C. Aers, Dwell time and phase times for transmission and reflection, Phys. Rev. B 39 (1989) 1201.
[60] G. Iannaccone, B. Pellegrini, Characteristic times in the motion of a particle, Phys. Rev. B 49 (1994) 16548.
[61] B. Djafari-Rouhani, L. Dobrzynski, Acoustic resonances of adsorbed wires and channels, J. Phys. Condens. Matter 5 (1993) 139.
[62] S. Khattou, M. Amrani, A. Mouadili, E.H. El Boudouti, A. Talbi, A. Akjouj, B. Djafari-Rouhani, Comparison of density of states and scattering parameters in coaxial photonic crystals: analytical and experimental study, Phys. Rev. B 102 (2020) 165310.
[63] Y. Fink, J.N. Winn, S. Fan, C. Chen, J. Michel, J.D. Joannopoulos, E.L. Thomas, A dielectric omnidirectional reflector, Science 282 (1998) 1679.
[64] D.N. Chigrin, A.V. Lavrinenko, D.A. Yarotsky, S.V. Gaponenko, Observation of total omnidirectional reflection from a one-dimensional dielectric lattice, Appl. Phys. A 68 (1999) 25.

Omnidirectional bandgaps and selective transmission in layered photonic crystals

20

Abdellatif Akjouj[a], El Houssaine El Boudouti[b], Leonard Dobrzyński[a], and Bahram Djafari-Rouhani[a]
[a]Institute of Electronic, Microelectronics and Nanotechnology (IEMN), UMR CNRS 8520, Department of Physics, Faculty of Sciences and Technologies, Lille University, Villeneuve d'Ascq Cedex, France, [b]Laboratoire de Physique de la Matière et de Rayonnement, Département de Physique, Faculté des Sciences, Université Mohamed I, Oujda, Morocco

20.1 Introduction

Photonic crystals [1,2] made of a periodic repetition of dielectric (or metallic) structures have attracted much attention during the last three decades, in relation to their ability of controlling the propagation of light and the possibility of many new optical devices [3–5]. In particular, three-dimensional (3D) photonic crystals may exhibit absolute photonic bandgaps in which the propagation of light is prohibited in any direction of the space. This means that any incident light with a frequency falling in the gap of the photonic crystal will be totally reflected independently of the incidence angle and the polarization of the wave, i.e., the photonic crystal behaves like an omnidirectional mirror in the corresponding frequency range. In other words, such structures play the role of a perfect mirror for light in the frequency window of the bandgaps, where electromagnetic waves, spontaneous emission, and zero point fluctuations are all absent. These properties are more pronounced when the bandgap is larger.

Many applications require the control of laser beams, in which the beams are very close to plane waves. In this case, one does not need 2D or 3D photonic crystals, but prefers the one-dimensional (1D) photonic crystal, realized by dielectric multilayers. Obviously, 1D photonic bandgap (PBG) systems such as layered photonic crystals (LPCs) have several advantages. First, in contrast to the 3D case, the 1D photonic crystals are attractive since their production is more feasible at any wavelength scale. Second, the required analytical and numerical calculations are simpler. It has been shown that 1D periodic structures such as layered photonic crystals can also exhibit the property of omnidirectional reflection [6–15] provided the optical indices of the LPC constituents are chosen appropriately. Different designs of 1D photonic structures have been investigated due to their ability to control the propagation of light [16–20]. In particular, dielectric reflectors are one of the most widely studied optical devices due to low absorption of the light as compared to the metallic mirrors [6,21].

Photonics. https://doi.org/10.1016/B978-0-12-819388-4.00031-9

As it was reported by Fink et al. [6], the well-known thin-film dielectric mirror can totally reflect an arbitrarily polarized light coming from outside at all angles, within a certain frequency range. Fink et al. [6] have demonstrated this property at infrared frequencies, and Chigrin et al. [8] succeeded in demonstrating it at optical frequencies. The mirror designed in [6] consists of 9 alternating layers of tellurium and polystyrene polymer and totally reflects light at all angles in the infrared (10–15 micron wavelength) region. Made of 19 alternating layers of Na_3AlF_6 (low-refractive index material) and ZnSe (high-reflective index material), the mirror designed in [8] totally and omnidirectionally reflects visible light (in the 604.3–638.4 nm range) at all angles. Many other different materials have been used to fabricate high-quality 1D photonic structures, e.g., SiO_2/ZrO_2 [22], TiO_2/SiO_2 [23], AlGaAs–GaAs [24], B_4C/Si [25], porous silicon (PSi) [26], etc. In particular, a wide variety of photonic structures fabricated with PSi have been reported [27,29–31,33–35] due to its ease of fabrication [36] and relatively economical preparation method [37]. Being nearly free of dissipative losses up to the optical frequencies, thin-film all-dielectric omnidirectional mirrors open up a cheap and effective possibility to control the propagation of light. At optical frequencies, the mirrors can be used as enclosure coatings for high finesse microcavities, where lossy metallic mirrors cannot be used [6,8]. Broadband Bragg mirrors [26,38], high quality microcavities [27], transmission and energy-saving filters [28], sharp rugate filters [29], omnidirectional mirrors [30], waveguides antenna substrates, and enclosure coatings for waveguides [31,32], etc., have been fabricated in the last decade for their possible applications as photonic devices, filters, and sensors [39]. Moreover, in order to obtain a multiple-peak optical response from the same photonic structure, the fabrication of composite structures has been studied through sequential [40] and superposition [41] of refractive index profile functions.

Meade et al. [41] demonstrated the spectrally encoded structures through superposition of sinusoidal current-time wave forms. The superposition addition method has been used to prepare optically encoded composite structures, with the similar reflectance peak corresponding to each independent rugate. The possibility to construct an encoded library with PSi photonic structures has also been investigated. Some of the possible applications of such photonic structures have been oriented towards molecular sensing field [42], in which the physical thickness of the structures becomes a limiting factor due to the limited availability of reagents and the penetration inhomogeneity along the depth. Hence, a reduction in physical thickness of the filter amounts to its greater applicability in the sensing field. In 2009, Sailor's research group [43] used the lithographic procedure to isolate similar spectrally encoded PSi structures in the form of microparticles for the simultaneous detection of multiple analytes. Superposition structures (double/triple rugate filters) have been found useful for the gas sensing applications due to faster recovery after vapor exposure as compared to sequential structures [44–46]. In 2014, Dyachenko et al. reported on ceramic photonic glasses, new disordered materials obtained by sedimentation of zirconia microparticles [47]. This novel material can be used for a wide range of applications, such as thermal radiation barrier coatings, high temperature structural coloration, broadband visible reflectors, resonant random lasers, and photonic Anderson localization. A reflectivity up to 93% in the visible range was achieved for 40 μm film thickness. Qian et al. have demonstrated that the metallic grating structure with gradient index materials can be used to realize nearly total omnidirectional reflection, regardless of the polarizations of the incident wave [48]. The presence of the Brewster point in layered photonic crystals restricts the possibility of achieving full omnidirectional reflection at large incident wave vectors. Al-Juboori and Reece have shown experimentally that it is possible to achieve omnidirectional reflection in both rarefied and refracting incident media using periodic structures consisting of three or more distinct layers with differing refractive indices [49].

This chapter is divided into two parts: in the first part, we show that large omnidirectional reflection of electromagnetic waves can be obtained by associating two or several LPCs in tandem. We discuss the existence conditions for these gaps as compared to the usual case of a binary layered photonic crystal. We emphasize that, under certain conditions, such a 1D photonic structure can exhibit large reflection bandwidth for all incident angles and polarizations. Indeed, with an appropriate choice of the dielectric parameters in the LPCs and in the ambient medium, there may exist a frequency range in which any incident wave launched from the ambient medium falls inside a minigap of the LPC, and therefore undergoes a total reflection. By introducing a defect layer in the finite layered structure, one can obtain localized modes inside these omnidirectional bandgaps, giving rise to the selective transmission through the structure. These modes appear as peaks in the transmission spectrum with a very high quality factor, and therefore may have useful applications in the frame of photonic bandgap materials.

The LPC can exhibit the property of omnidirectional reflection under certain conditions [6–15]. However, since the photonic band structure of a layered photonic crystal never contains an absolute bandgap (i.e., a gap for any value of the wavevector), the property of omnidirectional reflection requires that the incident light is launched from vacuum, or from a substrate in which the refraction index is relatively lower than the refractive indices of the materials that compose the LPC. In other words, the ability of the LPC to behave like a perfect mirror imposes a limitation on the choice of the substrate. The second object of this chapter is to show and discuss to which extent this limitation can be removed or at least relaxed. We propose to combine the LPC with a cladding layer constituted by a material with a low optical index (or high velocity) that acts like a barrier for the propagation of light generated in the substrate. We demonstrate that such a cladded LPC can be used as a tool for a practical realization of an omnidirectional reflector, independently of the nature of the substrate.

20.2 Theoretical model

The LPC considered here is made of an array of alternating polystyrene and tellurium layers coupled to a homogeneous medium, characterized by n_0 (such as the vacuum with $n_0 = 1$) at the interface. Let us mention that the polystyrene/tellurium materials were recently used to demonstrate the existence of the omnidirectional reflection over the wavelength range from 10 to 15 μm. Electromagnetic waves are incident

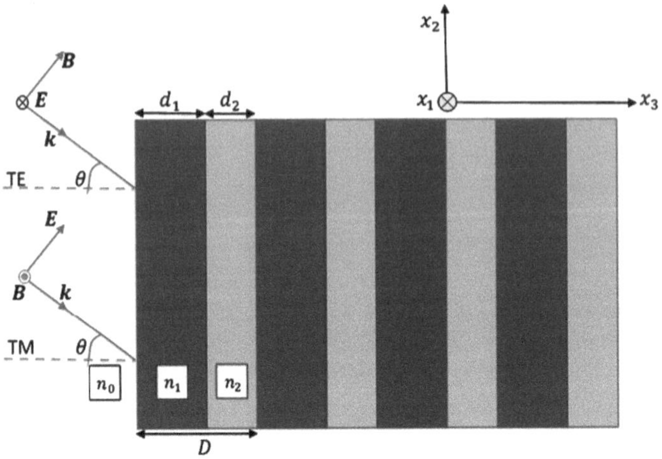

Figure 20.1 Schematic of the multilayer system showing the layer parameters (n_i and d_i are the index of refraction and thickness of layer i, respectively), the incident wave vector $\boldsymbol{k}$, and the electromagnetic mode convention (transverse electric (TE) and transverse magnetic (TM)); $\boldsymbol{E}$ and $\boldsymbol{B}$ are the electric and magnetic fields, respectively.

upon the multilayer film from the homogeneous medium. Although such a system has been extensively analyzed in the literature [28,50,51], the possibility of total omnidirectional reflectivity was not recognized until the year 1998 [6–8,11,52–55]. The generic system is described in Fig. 20.1, where d_1 and d_2 are the layer thickness and n_1 and n_2 are the indices of refraction of the respective layers. The incident wave has a wavevector $\boldsymbol{k} = k_2\boldsymbol{i}_2 + k_3\boldsymbol{i}_3$ and a frequency $\omega = c|\boldsymbol{k}|/n_0$, where c is the speed of light in vacuum, while $\boldsymbol{i}_2$ and $\boldsymbol{i}_3$ are the unit vectors in the x_2 and x_3 directions, respectively. The wavevector, together with the normal to the periodic structure, defines a symmetry mirror plane that allows us to distinguish between two independent electromagnetic modes: transverse electric (TE) and transverse magnetic (TM) modes. For the TE (TM) mode, the electric (magnetic) field is perpendicular to the plane. All the interfaces of the LPC are taken to be parallel to the (x_1, x_2)-plane (see Fig. 20.1).

In the following, the calculation of the dispersion curves, as well as the transmittance through finite layered photonic crystals, is performed in the framework of a Green's function formalism [56,58]. We shall avoid giving these expressions which are detailed in Chapter 18 of this book.

20.3 Onmidirectional bandgaps

For an arbitrary direction of propagation, it is convenient to examine the so-called projected band structure, i.e., the frequency ω versus the wavevector $k_{\parallel}$ (parallel to the interfaces). The photonic bulk band structure for a typical Bragg reflector is shown in Fig. 20.2 for TM (left panel) and TE (right panel) electromagnetic waves in a polystyrene/tellurium semiinfinite LPC in contact with the vacuum. The filling

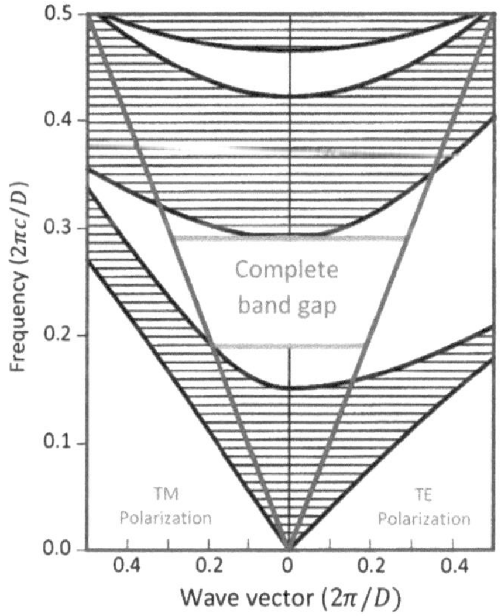

Figure 20.2 Photonic band structure of a typical 1D photonic crystal for TM (left panel) and TE (right panel) polarizations. The stack is assumed to have refractive indices $n_1 = 4.6$, $n_2 = 1.6$ and a filling fraction $d_1/D = 0.22$. Shaded areas correspond to the LPC bulk bands. The heavy green solid lines (light gray in print version) indicate the light line of vacuum. The horizontal dashed lines delimit the edges of the omnidirectional bandgap.

fraction d_1/D in the LPC is assumed to be equal to 0.22, where d_1 is the thickness of the tellurium layer, and the indices of refraction are 1.6 for polystyrene and 4.6 for tellurium, respectively. In Fig. 20.2, we report the frequency versus the parallel wave vector. The thick solid straight lines indicate the light lines of the ambient media (vacuum). The shaded areas correspond to minibands, where photons are allowed to propagate in the LPC; these shaded areas are separated by minigaps, where photons are prohibited to propagate. Due to the large difference between the refractive index values for the polystyrene and tellurium, these gaps are rather large, in contrast to the case of other systems like GaAs–AlAs LPCs. The band structure and the transmission coefficients discussed below are calculated using the Green's function method [56].

As follows from Fig. 20.2, any wave launched from the ambient medium within the frequency range $0.185 < \omega D/2\pi c < 0.29$ is prohibited to propagate in the LPC; and therefore will be subject to the phenomenon of total reflection (and hence, there will be no transmission). This means that in the above frequency range, the LPC acts as an omnidirectional reflector. To be more specific, let us first notice that, due to the translational symmetry in the (x_1, x_2)-plane (see Fig. 20.1), the wave vector $k_\parallel = \omega \sin(\theta)/c$ is conserved for a TE (TM) wave launched from the vacuum with an incident angle θ. A wave launched from the ambient medium has a velocity c, and therefore its frequency lies above the lines $\omega = ck_\parallel$. As a consequence, for frequencies ranging from 0.185 to 0.29, any wave propagating in the ambient medium (i.e., with an arbitrary angle of incidence and any polarization) is forbidden to propagate in the LPC and will be reflected back into the ambient medium. On the contrary, in

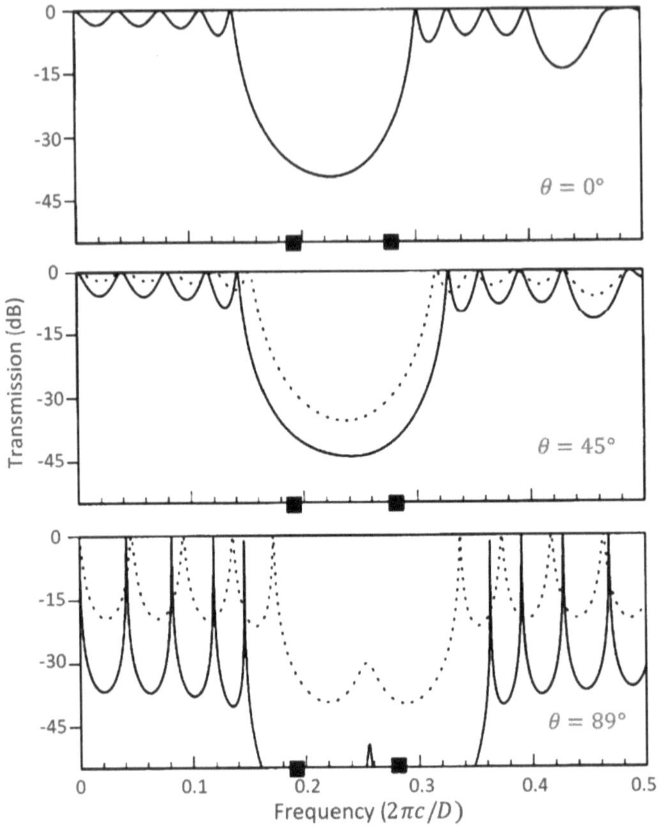

Figure 20.3 Transmission factor for a polystyrene–tellurium LPC in the ambient vacuum with the same parameters as in Fig. 20.2. The curves are plotted for three angles of incidence and TE (solid lines) and TM (dotted lines) polarizations. The filled squares on the frequency axis indicate the edges of the common omnidirectional stopband.

other frequency domains, part of the incident power will be transmitted into the LPC. The necessary and sufficient criterion for omnidirectional reflectivity at a given frequency is that no transmitting state of the structure exists inside the ambient medium cone defined by the velocity of the ambient medium. In Fig. 20.2, this criterion is that there exists a frequency zone in which the projected bands of the photonic crystal have no states for $\omega > ck_{\parallel}$. Let us notice that this first forbidden band exists in different frequency domains of the electromagnetic spectrum for different values of the period D of the structure. More precisely, if one considers a structure with $D = 150$ nm, the omnidirectional total reflection is expected within the visible spectral band $\Delta\lambda = 517$–810 nm.

To check the efficiency of the LPC for preventing the propagation of light, we have calculated the transmission coefficients through a finite LPC constituted of 5 layers of polystyrene and 5 layers of tellurium, sandwiched between two ambient media

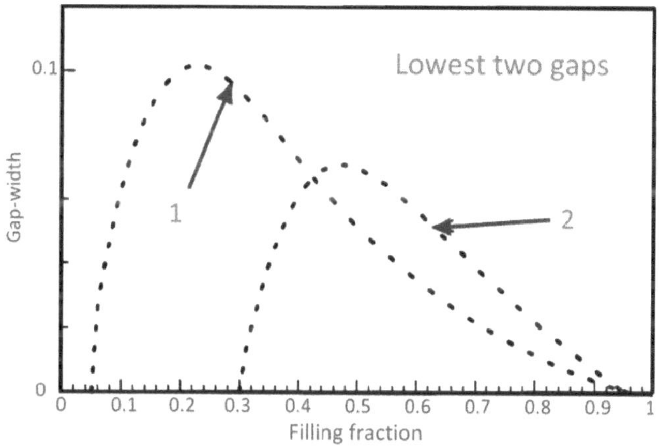

Figure 20.4 Gap width of the first two complete band-gaps versus filling fraction for a polystyrene–tellurium LPC. It is evident that there is no electromagnetic bandgap for $d_1/D \le 5\%$. The numbers 1 and 2 refer to the lowest and the second lowest complete bandgaps.

(Fig. 20.3). The power transmission is calculated for three angles of incidence and different polarizations of the incident wave. The top, middle, and bottom panels in Fig. 20.3 display the frequency dependence of the transmission for $\theta = 0°$, $\theta = 45°$, and $\theta = 89°$, respectively. Inside the gaps, the transmission decreases exponentially with the thickness of the structure. For any given incident angle, the transmission spectrum shows a stopband for both the TE (solid lines) and TM (dotted lines) polarizations. This large common gap associated with the omnidirectional reflection extends from 0.185 to 0.29. (This is indicated by filled squares on the frequency axis in Fig. 20.3. The filled squares are obtained by cutting the transmission coefficient at −30 dB.) One can conclude that a thin LPC composed of a few periods may be sufficient to prevent the propagation of electromagnetic waves inside the omnidirectional PBG.

One of parameters optimizing the width of the omnidirectional bandgaps is the filling fraction d_1/D. Fig. 20.4 gives the evolution of gap widths of the two existing complete stop bands in the 1D photonic band structure of Fig. 20.2. The gap width on the y-axis represents just a difference of the top and bottom frequency of the omnidirectional stop band, for a given filling fraction. As can be seen in Fig. 20.4, the filing fraction must exceed a certain minimum value, $(d_1/D)_{\min}$, for a complete gap to be opened. This leads us to infer that there is no bandgap for $d_1/D \le 0.05$. The first gap opens for a very large range of filling fraction ($0.05 \le d_1/D \le 0.95$) and the second gap opens between 0.3 and 0.95. We notice that the largest first (second) gap is obtained when this ratio is around 22% (48%). It is interesting to note that at $d_1/D = 95\%$ there are no gaps opening up in the system, just as for $d_1/D \le 5\%$. As a general trend, a large PBG requires a huge contrast between the reflective indices of the constituent materials in the LPC [6,8].

Next, we propose the fabrication of a multiperiodic system in tandem that could be designed so as to give rise to wider bandgaps in the desired frequency range. The

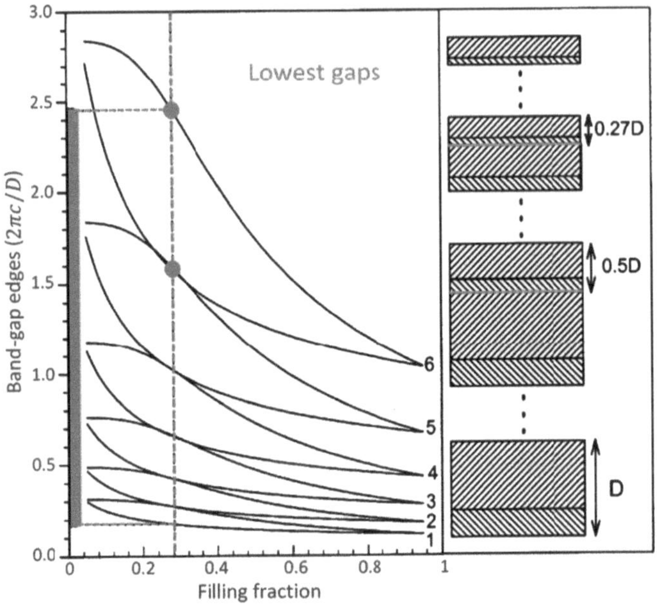

Figure 20.5 Design of an ultrawideband filter corresponding to the lowest stop band in Fig. 20.2. The bandgap edges from the bottom to top correspond respectively to the periods of $1D$, $0.64D$, $0.41D$, $0.265D$, $0.16D$, and $0.11D$; D is the period of the lowest polystyrene–tellurium LPC. Each of the six LPCs produces a complete stopband, whose upper and lower edges are plotted as a function of the filling fraction d_1/D. For $d_1/D = 0.24$ (vertical dotted line), the six stop bands join perfectly so as to form a "superomnidirectional stopband" within the frequency range between 0.185 and 2.45 (see the bold vertical frequency bar between the two horizontal dotted lines). The right panel shows a schematic representation of three layered photonic crystals in tandem.

bandgap edges of the lowest, which is also the widest, complete stopband are computed as a function of the filling fraction for a large number of systems with different periods. The numerical results of such investigations are illustrated in Fig. 20.5. The "wedges", labeled 1–6, correspond to different periods (in the decreasing order from bottom to top) and are based on the numerous band structure calculations-one for every value of the filling fraction d_1/D. These are, in fact, eigenvalue problems for the frequency as a function of the wave vector scanned in all directions and for both the TE and TM polarizations. It is important to note here that the eigenfrequency is inversely proportional to the period of the system. That means that, given the specific medium in the background, the frequencies of a "wedge" for a period of 100 nm will be 10 times higher than those of a "wedge" corresponding to a period of 1000 nm. Consider two dots on "wedge" number 6 for a filling fraction $d_1/D = 0.24$. The dots mark the upper and lower edges of the complete stopband in the band structure and the vertical distance between them is the width of the complete stopband. The upper edge of the next "wedge" number 2 crosses the lower edge of "wedge" number 1, and so on. The six bandgap edges join perfectly so as to form a "superomnidirectional bandgap" within the reduced frequency range between 0.185 and 2.45. The range of the "superomnidirectional bandgap" is highlighted by a bold vertical frequency bar in Fig. 20.5. Let us notice that if one takes $D = 500$ nm, the "superomnidirectional total reflection" is expected within the spectral band $\Delta\Lambda = 204$–2700 nm. This covers the whole range of optical frequencies.

20.4 Selective transmission

In addition to the perfect mirror behavior shown by the LPC in the omnidirectional PBG frequency region, the LPC may also help build a very selective electromagnetic filter when a defect layer is introduced inside the LPC [57]. We shall focus only on the omnidirectional PBG frequency region and report (Figs. 20.6(a)–(c)) the transmission spectra as a function of frequency for three angles of incidence θ and when one replaces the thickness of the tellurium layer ($d_1 = 0.22D$) in the middle of the finite size LPC structure by $d_0 = 0.5D$. The top, middle, and bottom panels display the frequency dependence of the transmission for $\theta = 0°$, $\theta = 20°$, and $\theta = 40°$, respectively. When the frequency of incident photons coincides with one of the discrete frequency levels of the defect layer, the photons are completely transmitted through the structure even if they are in resonance with the ambient medium states. The quality factors of

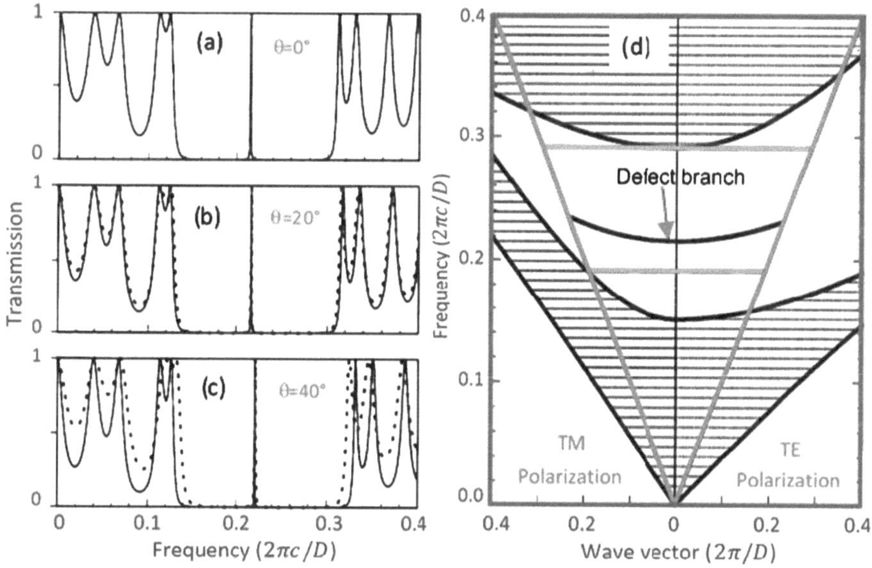

Figure 20.6 Transmission spectra for a finite LPC with a tellurium defect layer of thickness $d_0 = 0.5D$. The finite LPC contains 4 layers of polystyrene and 4 layers of tellurium: (a) $\theta = 0°$, (b) $\theta = 20°$, and (c) $\theta = 40°$. The defect layer of tellurium is inserted in the middle of the LPC. A sharp resonance associated with a defect layer appears in the first gap of the LPC. The curves are plotted for three angles of incidence and the TE (solid lines) and TM (dotted lines) polarizations. (d) Same as in Fig. 20.2 but with a tellurium defect layer of thickness $d_0/D = 0.5$. A defect branch appears inside the omnidirectional bandgap.

the sharp peaks, defined as the ratio between the central frequency and the full width at half maximum, are of the order of 2000. The quality factor becomes more important when the complete bandgap is larger. Let us notice that with our system, we can easily answer the telecommunication demands, where the signal wavelengths are of the order of 1550 nm and have a width of the order of 0.5 nm.

The dispersion curve of the mode induced by the tellurium defect layer in the omnidirectional PBG is shown in Fig. 20.6(d). This curve is represented by a branch inside the PBG for both TE and TM polarizations. These modes are obtained from the maxima of the transmission spectrum (see Figs. 20.6(a)–(c)).

20.5 Optical mirror in a cladded–layered photonic crystal

In the following, the calculation of the dispersion curves, as well as the transmittance through finite layered photonic crystals, is performed in the framework of a Green's function formalism [58] (see Chapter 18).

Fig. 20.7 displays the photonic band structure of a superlattice composed of two materials with optical indices $n_1 = 2.36$ and $n_2 = 4.6$, and thicknesses d_1 and d_2 such that $d_1 = 2D/3$ and $d_2 = D/3$, $D = d_1 + d_2$ being the period of the layered photonic crystal (LPC). We have used a dimensionless frequency $\Omega = \omega D/(2\pi c)$, $k_\parallel$ is the wavevector parallel to the layers, and c is the speed of light in vacuum. For every value of $k_\parallel$, the shaded and white areas in the projected band structure respectively correspond to the minibands and to the minigaps of the LPC where the propagation of light is allowed or forbidden. The left and right panels respectively give the band structure for transverse magnetic (TM) and transverse electric (TE) modes. One can notice that the band structure shown in Fig. 20.7 does not contain any absolute gap, i.e., a minigap existing for every value of $k_\parallel$. However, the LPC can display an omnidirecitional reflection gap around the frequency $\Omega = 0.18$, provided the optical index of the substrate, from which the incident light is launched, remains less than 1.85. For instance, if the incident light is launched onto the LPC from vacuum, the omnidirectional gap for total reflection extends from $\Omega = 0.138$ to $\Omega = 0.194$. On the contrary, if the incident wave is coming from a substrate with a high optical index, let us say $n_s = 4$ (see the heavy straight lines in Fig. 20.7), part of the incident wave can be transmitted through the superlattice depending upon the wavevector $k_\parallel$ or the incidence angle θ given by $\sin\theta = ck_\parallel/n_s\omega$. Therefore, the superlattice does not display the property of omnidirectional reflection.

To study the transmission of light through a finite structure, let us now consider a finite LPC cladded on one side by a layer of thickness d_0 and optical index n_0, and bounded on both sides by a substrate of optical index n_s. Figs. 20.8 compares the transmitted intensity through the superlattice with or without the clad layer, for two different values of the incidence angle. In this example, we have chosen $n_s = 4$, $n_0 = 1.34$ (see the heavy dashed lines in Fig. 20.7), $d_0 = 8.5D$, and the LPC contains $N = 10$ layers of material 2 and $N + 1 = 11$ layers of material 1. Due to the finiteness of the LPC structure, the transmission never becomes exactly equal to zero, but can

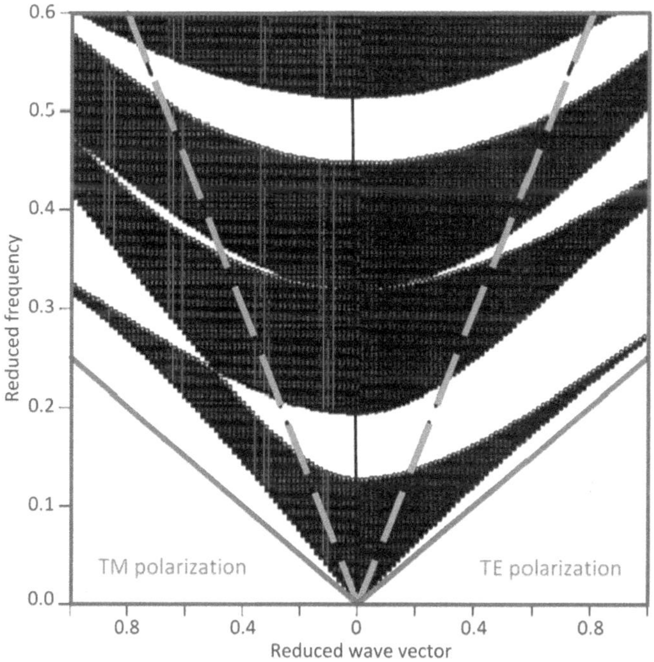

Figure 20.7 Projected band structure of the layered photonic crystal for TE (right panel) and TM (left panel) modes. The reduced frequency $\Omega = \omega D/2\pi c$ is presented as a function of the reduced wavevector $k_{\parallel} D$. The shaded and white areas respectively correspond to minibands and minigaps of the layered photonic crystal. The heavy straight lines show the light lines of a substrate with $n_s = 4$ (full red lines; mid gray in print version) and of a clad layer with $n_0 = 1.34$ (dashed green lines; light gray in print version).

decrease to almost vanishing values in some frequency range. At normal incidence $\theta = 0°$ or $k_{\parallel} = 0$ (Fig. 20.8(b)), one can observe that the clad layer does not affect the bandgap which totally coincides with the minigap of the LPC (see Fig. 20.7). However, at $\theta = 32°$ (Figs. 20.8(c)–(d)), corresponding approximately to the Brewster angle in the layered photonic crystal, the decrease in the transmittance in the frequency range from $\Omega = 0.156$ to 0.192 is only due to the presence of the clad layer: indeed, the propagation of light is prohibited for those frequencies that lie below the light line of the clad layer, even though they fall inside a miniband of the superlattice, as the clad layer play the role of a barrier between photons in the superlattice and the substrate.

Let us define the omnidirectional gap as a frequency interval in which the transmission intensity falls below a certain threshold whatever the incidence angle or the polarization of the wave. By fixing the threshold to 10^{-3} (respectively 10^{-2}), we have checked that with the above structure the omnidirectional gap extends from $\Omega = 0.161$ to 0.187 (respectively from $\Omega = 0.156$ to 0.192). To emphasize more the formation of the omnidirectional gap, we present in Fig. 20.9 the frequency domain in which the transmission through the cladded superlattice remains below a threshold of 10^{-3}. In the same figure, we have sketched the dispersion curves of the finite cladded LPC ob-

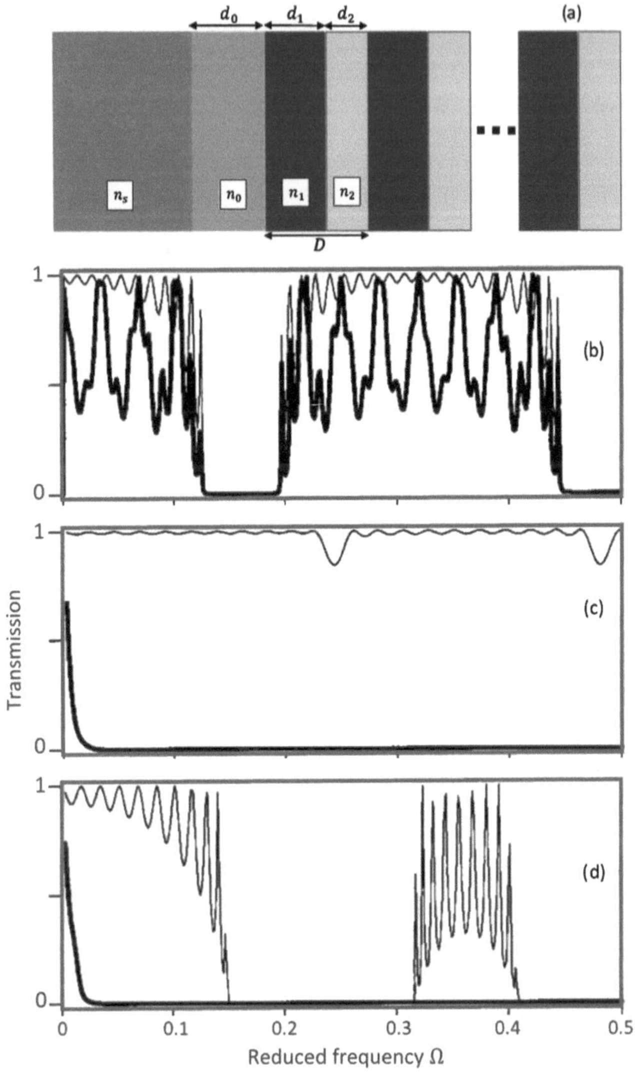

Figure 20.8 (a) Schematic of the cladded–layered photonic crystal system. Comparison of the transmission rate through a finite layered photonic crystal with (heavy lines) and without (thin lines) a clad layer; (b) Normal incidence; (c)–(d) Incidence at the Brewster angle ($\theta = 32°$) for TM and TE modes, respectively.

tained from a calculation of the density of electromagnetic modes [58]. Although the existence of the gap is due to the introduction of the clad, one can notice that the limitation on the lower edge of the gap comes nevertheless from the transmittance through the modes induced by the clad layer, more precisely those modes that lie very near to the light line of the clad. Let us notice that the modes induced by the cladded LPC appear as well defined peaks in the density of electromagnetic modes, but contribute slightly in the transmission spectrum.

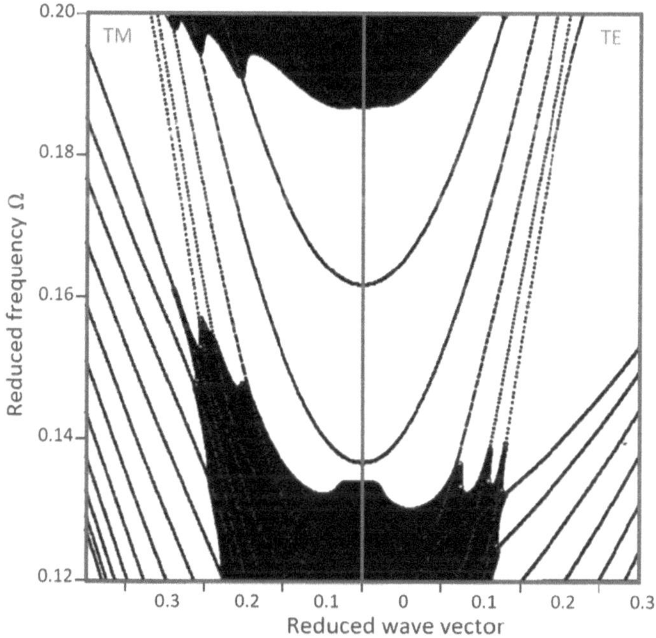

Figure 20.9 Omnidirectional gap of a finite cladded layered photonic crystal when the threshold for the maximum allowed transmission is fixed to 10^{-3}. The dispersion curves are the eigenmodes (resonant states) of the finite structure obtained from the maxima in the density of electromagnetic modes. The left and right panels respectively refer to TM and TE modes.

We can now briefly discuss how the existence and behavior of the omnidirectional gap are affected by the choice of the different parameters in our structure. Fig. 20.10(a) shows the frequency range of the gap as a function of the thickness d_0 of the clad layer, the other parameters being kept the same as in the above illustrations ($n_s = 4$, $n_0 = 1.34$, $n_1 = 2.36$, $d_1 = 0.66D$, $n_2 = 4.6$, and $d_2 = 0.34D$). The gap opens when d_0 exceeds a certain threshold (approximately $6D$), widens for increasing the thickness d_0, and finally saturates for d_0 greater than $12D$. In Fig. 20.10(b), the omnidirectional gap is sketched as a function of the refractive index n_0 of the clad layer of thickness $d_0 = 8.5D$. Fig. 20.10(b) shows that the largest gap is obtained if the clad layer is made of an air gap, the width of the gap decreases with increasing n_0, and finally the gap closes for n_0 of the order of 1.6. In Fig. 20.10(c), we display the gap as a function of the number of unit cells in the layered photonic crystal, assuming that the thickness d_0 of the clad layer is fixed to $d_0 = 8.5D$ and $n_0 = 1.34$. The existence of the gap requires that the superlattice contains a minimum number N of unit cells, around 8 in our illustration.

Instead of cladding the superlattice, one may think that the layer of refractive index n_0 can also be inserted into the superlattice as a defect layer. However, we have checked that the best solution for lowering the transmission is to add this layer at the boundary of the superlattice rather than at another place between two interior layers.

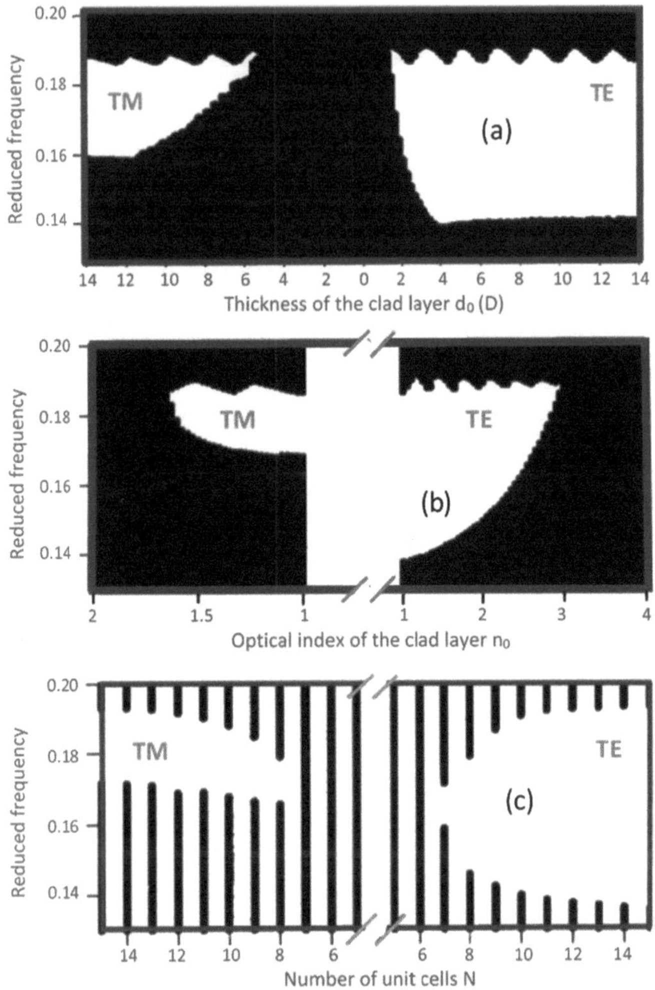

Figure 20.10 Dependence of the omnidirectional gap on different parameters of the structure. (a) Dependence with the thickness d_0 of the clad layer ($n_0 = 1.34$). (b) Dependence with the refractive index n_0 of the clad layer of thickness $d_0 = 8.5D$. (c) Dependence with the number of unit cells in the layered photonic crystal with $n_0 = 1.34$ and $d_0 = 8.5D$. In (a), (b), and (c), the other parameters are $n_s = 4$, $n_1 = 2.36$, $d_1 = 0.66D$, $n_2 = 4.6$, and $d_2 = 0.34D$. The left and right panels respectively refer to TM and TE modes.

20.6 Summary and perspectives

In the first part of this chapter, we have reviewed the existence of large complete omnidirectional PBGs in one-dimensional periodic structures built out of two alternating isotropic dielectric layered materials. These materials exhibit total reflection for all angles of incidence and both TE and TM polarizations. For the polystyrene-tellurium LPC, it is observed that a minimal filling fraction $d_1/D \approx 5\%$ is needed for the appearance of the lowest forbidden frequency band. The 1D tandem structures proposed here could be designed to achieve large "superomnidirectional stopbands" within the desired frequency range. The frequency range of such "superomnidirectional stopbands" can be raised (lowered) by decreasing (increasing) the period of the constituent systems. Inclusion of a defect layer in the structure gives rise to localized states in the complete stopbands; these frequencies lead to the enhancement of the transmission of an incident photon in the ambient medium. The omnidirectional PBG studied here may have potential applications, such as mirrors electromagnetic filters allowing only selective frequencies to pass through. Owing to the advances in epitaxial crystal growth techniques, we can modulate the photon frequency of the forbidden gap and selective modes by changing the thickness of the constituent materials.

In the second part of this chapter, we have shown that an omnidirectional mirror can be realized with a layered photonic crystal even if the substrate is made of a high refractive index (or low velocity) material; this requires adding a clad layer of low refractive index on one side of the layered photonic crystal. We have discussed the behavior of this gap as a function of the physical parameters involved in the structure. Let us stress that an alternative solution to remove the limitation about the choice of the substrate would be to combine in tandem two layered photonic crystals in such a way that the band structure of the coupled structure displays an absolute bandgap, this means that for a certain frequency range a miniband of one layered photonic crystal overlaps with a minigap of the other layered photonic crystal, and vice versa. We have found that such a possibility exists when, for instance, the layered photonic crystal described above is combined with a second layered photonic crystal of the same period D but the two constituting materials are of optical index and thickness: $n'_1 = 1$, $d'_1 = 0.573D$ and $n'_2 = 1.34$, $d'_2 = 0.427D$, respectively. Finally, we have proposed similar solutions for the realization of an omnidirectional acoustic mirror [59,60].

Acknowledgments

The authors would like to thank Abdellah Mir, Driss Bria, and Abelkarim Nougaoui for their fruitful contributions to the original paper from which this chapter is inspired.

References

[1] E. Yablonovitch, Inhibited spontaneous emission in solid-state physics and electronics, Phys. Rev. Lett. 58 (1987) 2059.
[2] S. John, Strong localization of photons in certain disordered dielectric superlattices, Phys. Rev. Lett. 58 (1987) 2486.
[3] C.M. Soukoulis (Ed.), Photonic Band Gaps and Localization, Plenum, New York, 1993.
[4] J.D. Joannopoulos, R.D. Meade, J.N. Winn, Photonic Crystals, Princeton University Press, Princeton, 1995.
[5] C.M. Soukoulis (Ed.), Photonic Band Gap Materials, Kluwer, Dordrecht, 1996.

[6] Y. Fink, J.N. Winn, S. Fan, C. Chen, J. Michel, J.D. Joannopoulos, E.L. Thomas, A dielectric omnidirectional reflector, Science 282 (1998) 1679.
[7] J.P. Dowling, Mirror on the wall: you're omnidirectional after all?, Science 282 (1998) 1841.
[8] D.N. Chigrin, A.V. Lavrinenko, D.A. Yarotsky, S.V. Gaponenko, Observation of total omnidirectional reflection from a one-dimensional dielectric lattice, Appl. Phys. A 68 (1999) 25.
[9] K.M. Chen, A.W. Sparks, H. Luan, D.R. Lim, K. Wada, L.C. Kimerling, SiO_2/TiO_2 omnidirectional reflector and microcavity resonator via the sol–gel method, Appl. Phys. Lett. 75 (1999) 3805.
[10] C. Hooijer, D. Lenstra, A. Lagendijk, Mode density inside an omnidirectional mirror is heavily directional but not small, Opt. Lett. 25 (2000) 1666.
[11] W.H. Southwell, Omnidirectional mirror design with quarter-wave dielectric stacks, Appl. Opt. 38 (1999) 5464.
[12] J. Lekner, Omnidirectional reflection by multilayer dielectric mirrors, J. Opt. A, Pure Appl. Opt. 2 (2000) 349.
[13] E. Cojocaru, Omnidirectional reflection from Šolc-type anisotropic periodic dielectric structures, Appl. Opt. 39 (2000) 6441.
[14] I. Abdulhalim, Omnidirectional reflection from anisotropic periodic dielectric stack, Opt. Commun. 174 (2000) 43.
[15] A. Mir, A. Akjouj, E.H. El Boudouti, B. Djafari Rouhani, L. Dobrzynski, Large omnidirectional band gaps and selective transmission in one-dimensional multilayer photonic structures, Vacuum 63 (2001) 197.
[16] M. Malak, N. Gaber, F. Marty, N. Pavy, E. Richalo, T. Bourouina, Analysis of Fabry–Perot optical microcavities based on coating-free all-Silicon cylindrical Bragg reflectors, Opt. Express 21 (2013) 2378.
[17] D. Bria, B. Djafari-Rouhani, E.H. El Boudouti, A. Mir, A. Akjouj, A. Nougaoui, Omnidirectional optical mirror in a cladded-superlattice structure, J. Appl. Phys. 91 (2002) 2569.
[18] A. Mekis, J.C. Chen, I. Kurland, S. Fan, P.R. Villeneuve, J.D. Joannopoulos, High transmission through sharp bends in photonic crystal waveguides, Phys. Rev. Lett. 77 (1996) 3787.
[19] S. Wanga, X. Yanga, C.T. Liu, Omnidirectional reflection in one-dimensional ternary photonic crystals and photonic heterostructures, Phys. Lett. A 378 (2014) 1332.
[20] J.P. Pandey, Enlargement of omnidirectional reflection range using cascaded photonic crystals, Int J. Pure Appl. Phys 13 (2017) 167.
[21] A. Hosseini, Y. Massoud, A low-loss metal–insulator–metal plasmonic Bragg reflector, Opt. Express 14 (2006) 11318.
[22] S.F. Chichibu, T. Ohmori, N. Shibata, T. Koyama, Dielectric SiO_2/ZrO_2 distributed Bragg reflectors for ZnO microcavities prepared by the reactive helicon-wave-excited-plasma sputtering method, Appl. Phys. Lett. 88 (2006) 161914.
[23] Q. Qin, H. Lu, S.N. Zhu, C.S. Yuan, Y.Y. Zhu, N.B. Ming, Resonance transmission modes in dual-periodical dielectric multilayer films, Appl. Phys. Lett. 82 (2003) 004654.
[24] K. Kojima, S. Noda, K. Mitsunaga, K. Kyuma, K. Hamanaka, Continuous wave operation of a surface emitting AlGaAs/GaAs multiquantum well distributed Bragg reflector laser, Appl. Phys. Lett. 50 (1987) 001705.
[25] V.V. Medvedev, A.E. Yakshin, R.W.E. van de Kruijs, V.M. Krivtsun, A.M. Yakunin, K.N. Koshelev, F. Bijkerk, Infrared antireflective filtering for extreme ultraviolet multilayer Bragg reflectors, Opt. Lett. 37 (2012) 1169.

[26] V. Agarwal, J.A. del Rio, Tailoring the photonic band gap of a porous silicon dielectric mirror, Appl. Phys. Lett. 82 (2003) 001512.
[27] P.J. Reece, G. Lerondel, W.H. Zheng, M. Gal, Optical microcavities with subnanometer linewidths based on porous silicon, Appl. Phys. Lett. 81 (2002) 004895.
[28] P. Yeh, Optical Waves in Layered Media, Wiley, New York, 1988.
[29] S. Ilyas, T. Bocking, K. Kilian, P.J. Reece, J. Gooding, K. Gaus, M. Gal, Porous silicon based narrow linewidth rugate filters, Opt. Mater. 29 (2007) 619–622.
[30] D. Ariza-Flores, L.M. Gaggero-Sager, V. Agarwal, White metal-like omnidirectional mirror from porous silicon dielectric multilayers, Appl. Phys. Lett. 101 (2012) 031119.
[31] P. Apiratikul, A.M. Rossi, T.E. Murphy, Nonlinearities in porous silicon optical waveguides at 1550 nm, Opt. Express 17 (2009) 3396.
[32] M.L.H. Lahlaouti, A. Akjouj, B. Djafari Rouhani, L. Dobrzynski, Resonant and localized electromagnetic modes in finite superlattices, Phys. Rev. B 61 (2000) 2059.
[33] K.S. Perez, J.O. Estevez, A. Mendez-Blas, J. Arriaga, G. Palestino, M.E. Mora-Ramos, Tunable resonance transmission modes in hybrid heterostructures based on porous silicon, Nanoscale Res. Lett. 7 (2012) 000392.
[34] M.Y. Chen, S.O. Meade, M.J. Sailor, Preparation and analysis of porous silicon multilayers for spectral encoding applications, Phys. Status Solidi C 6 (2009) 1610.
[35] S.O. Meade, M.J. Sailor, Microfabrication of freestanding porous silicon particles containing spectral barcodes, Phys. Status Solidi RRL 1 (2007) R71.
[36] L. Pavesi, R. Turan, Silicon Nanocrystals, Wiley-VCH, 2010.
[37] M.J. Sailor, S.O. Meade, Method for forming optically encoded thin films and particles with grey scale spectra, U.S. Patent 8 (2012) 308.
[38] C. Mazzoleni, L. Pavesi, Application to optical components of dielectric porous silicon multilayers, Appl. Phys. Lett. 67 (1995) 002983.
[39] P.A. Snow, E.K. Squire, P.St.J. Russell, L.T. Canham, Vapor sensing using the optical properties of porous silicon Bragg mirrors, J. Appl. Phys. 86 (1999) 1781.
[40] E. Xifre-Perez, L.F. Marsal, J. Ferre-Borrull, J. Pallares, Low refractive index contrast porous silicon omnidirectional reflectors, Appl. Phys. B, Lasers Opt. 95 (2009) 169–172.
[41] S.O. Meade, M.S. Yoon, K.H. Ahn, M.J. Sailor, Porous silicon photonic crystals as encoded microcarriers, Adv. Mater. 16 (2004) 1811.
[42] T. Jalkanen, V. Torres-Costa, J. Salonen, M. Bjorkqvist, E. Makila, J.M. Martınez-Duart, V.P. Lehto, Optical gas sensing properties of thermally hydrocarbonized porous silicon Bragg reflectors, Opt. Express 17 (2009) 5446.
[43] S.O. Meade, M.Y. Chen, M.J. Sailor, G.M. Miskelly, Multiplexed DNA detection using spectrally encoded porous SiO_2 photonic crystal particles, Anal. Chem. 81 (2009) 2618.
[44] S. Li, D. Hu, J. Huang, L. Cai, Optical sensing nanostructures for porous silicon rugate filters, Nanoscale Res. Lett. 7 (2012) 79.
[45] T. Jalkanen, J. Salonen, V. Torres-Costa, K. Fukami, T. Sakka, Y.H. Ogata, Structural considerations on multistopband mesoporous silicon rugate filters prepared for gas sensing purposes, Opt. Express 19 (2011) 13291.
[46] A. Mukherjee, A.D. Ariza-Flores, R.F. Balderas-Valadez, V. Agarwal, Controlling the optical properties of composite multilayered photonic structures: effect of superposition, Opt. Express 21 (2013) 17324.
[47] P.N. Dyachenko, J.J. do Rosario, E.W. Leib, A.Yu. Petrov, R. Kubrin, G.A. Schneider, H. Weller, T. Vossmeyer, M. Eich, Ceramic photonic glass for broadband omnidirectional reflection, ACS Photonics 1 (2014) 1127.
[48] E. Qian, Y. Fu, Y. Xu, H. Chen, Total omnidirectional reflection by sub-wavelength gradient metallic gratings, Europhys. Lett. 114 (2016) 34003.

[49] A. Al-Juboori, P.J. Reece, Extending omnidirectional reflection bands in one-dimensional photonic crystals, J. Phys. Commun. 2 (2018) 055003.
[50] F. Abeles, Optical properties of thin absorbing films, J. Opt. Soc. Am. 47 (1957) 473.
[51] M. Born, E. Wolf, Principles of Optics, 6th ed., Pergamon, 1980, p. 67.
[52] J.N. Winn, Y. Fink, S. Fan, J.D. Joannopoulos, Omnidirectional reflection from a one-dimensional photonic crystal, Opt. Lett. 23 (1998) 1573.
[53] E. Yablonovitch, Engineered omnidirectional external-reflectivity spectra from one-dimensional layered interference filters, Opt. Lett. 23 (1998) 1648.
[54] P.S.J. Russell, S. Tredwell, P.J. Roberts, Full photonic bandgaps and spontaneous emission control in 1D multilayer dielectric structures, Opt. Commun. 160 (1999) 66.
[55] I. Abdulhalim, Reflective phase-only modulation using one-dimensional photonic crystals, J. Opt. A, Pure Appl. Opt. 2 (2000) 9.
[56] M.L.H. Lahlaouti, A. Akjouj, B. Djafari-Rouhani, L. Dobrzynski, M. Hammouchi, E.H. El Boudouti, A. Nougaoui, Electromagnetic waves in finite superlattices with buffer and cap layers, J. Opt. Soc. Am. A 16 (1999) 1703.
[57] R. Wang, J. Dong, D.Y. Xing, Defect studies in a one-dimensional photonic band gap structure, Phys. Status Solidi B 200 (1997) 529.
[58] M.L.H. Lahlaouti, A. Akjouj, B. Djafari Rouhani, L. Dobrzynski, M. Hammouchi, E.H. El Boudouti, A. Nougaoui, Theoretical analysis of the density of states and phase times: application to resonant electromagnetic modes in finite superlattices, Phys. Rev. B 63 (2001) 035312.
[59] D. Bria, B. Djafari-Rouhani, A. Bousfia, E.H. El Boudouti, A. Nougaoui, Absolute acoustic band gap in coupled multilayer structures, Europhys. Lett. 55 (2001) 841.
[60] A. Bousfia, E.H. El Boudouti, B. Djafari-Rouhani, D. Bria, A. Nougaoui, V.R. Velasco, Omnidirectional phononic reflection and selective transmission in one-dimensional acoustic layered structures, Surf. Sci. 482–485 (2001) 1175.

Layered photonic crystals with left-handed materials

21

Abdellatif Akjouj[a], El Houssaine El Boudouti[b], Leonard Dobrzyński[a], and Bahram Djafari-Rouhani[a]
[a]Institute of Electronic, Microelectronics and Nanotechnology (IEMN), UMR CNRS 8520, Department of Physics, Faculty of Sciences and Technologies, Lille University, Villeneuve d'Ascq Cedex, France, [b]Laboratoire de Physique de la Matière et de Rayonnement, Département de Physique, Faculté des Sciences, Université Mohamed I, Oujda, Morocco

21.1 Introduction

About 50 years ago, a Russian scientist, Victor Veselago, had the idea that materials with negative refractive indices can bend light into the backward direction and behave in many other counterintuitive ways [1]. However, due to the absence of such materials in nature, this idea has been silent for four decades, until 20 years ago, when Pendry proposed artificial designs that can achieve negative permeability [2]. For the first time, this brought Veselago's idea into reality and opened up the new field of left-handed materials (LHMs) as compared to right-handed materials (RHMs).

Left-handed materials, in which the dielectric permittivity ε and magnetic permeability μ are simultaneously negative, have received a great deal of attention during the last 20 years [3–6]. This is due to the unusual physical properties of these materials that have raised strong theoretical interest and may lead to potential applications in optical devices. Some peculiar properties of LHMs have already been discussed some 50 years ago by Veselago [1], for instance, a Poynting vector being directed opposite to the propagation wavevector k, as well as the reversal of Doppler and Cerenkov effects. Because of the absence of naturally existing LHMs, the experimental realization of an artificial heterogeneous medium exhibiting both negative $\varepsilon(\omega)$ and $\mu(\omega)$ was performed by Schelby et al. [7,8]. The realization of such media [7–10] is based on the proposition of Pendry et al. for specific structures [11,12]. Great interest in these metamaterials has been directed towards the theoretical and experimental study of Snell's law of refraction at the boundary with an LHM [13–17], the focusing and imaging properties of a metamaterial lens [2,18,19], the tunneling in the presence of an LHM layer [20], the emission in an LHM metamaterial [21], etc.

A left-handed material is an excellent candidate to study the Goos–Hänchen (G–H) shifts [22]. This phenomenon has been analyzed theoretically [23–27] and experimentally [22,28–32] in several papers. Some papers analyzed the G–H shift associated with left- and right-handed media [33–35]. Qing and Chen showed that for two media of the same (opposite) handedness, the G–H phase shift opposes (reinforces) the phase variation associated with propagation through the incident medium. The G–H lateral shift is in the same (opposite) direction as (of) the direction of the horizontal compo-

Photonics. https://doi.org/10.1016/B978-0-12-819388-4.00032-0

nent of the incident energy flux [33]. Shadrivov et al. have studied the beam reflection from a layered structure with an LHM [36]. They have predicted a giant lateral G–H shift and splitting of the beam due to the resonant excitation of surface polaritons with a vortex like energy flow between the right- and left-handed media. Some other papers have theoretically investigated the G–H effect in the LHM slab inserted between two semiinfinite right-handed media [35,37,38]. Wang and Zhu showed that the G–H shifts can be large and positive near the resonance condition if the absorption of the medium is sufficiently weak [37]. The G–H shift characteristics for the uniaxially anisotropic LHM film having two interfaces and for the interface between semiinfinite LHM and RHM media with an arbitrary angle have been investigated for transverse electric (TE) and transverse magnetic (TM) polarizations [39–41].

Assuming the possibility of realizing such an LHM in the form of layered media, a few works have investigated the photonic band structure of one-dimensional layered structures constituted by a periodical repetition of right- and left-handed materials (RHMs and LHMs) [42–44]. Some peculiar properties related to the presence of LHM layers have been underlined, for instance, the possibility of gap widening with respect to usual photonic crystals constituted only by RHM [42], the theoretical and experimental investigation of a new type of gap when the average index of refraction in the photonic crystal vanishes [43], and the possibility of discrete and photon tunneling modes [44,45]. These works have mainly concentrated on the propagation along the axis of the layered photonic crystal, i.e., normal incidence.

Recently it has been shown that the interface separating a layered photonic crystal and a homogeneous left-handed material can support backward surface modes [46]. And also it has been shown that electromagnetic surface waves can exist in the interface separating a conventional dielectric and one-dimensional photonic crystal made of alternate left-handed metamaterial and right-handed material [47–49]. The existence of these modes depends only on the presence of homogeneous left-handed metamaterial. Rostami et al. studied the properties of surface modes at an interface between a semiinfinite LHM and a semiinfinite LHM–RHM photonic crystal. They have shown that in the presence of the left-handed material, the surface modes can be either forward or backward depending on the cap layer width of LHM–RHM photonic crystal [50]. Shokri et al. studied the optical dispersion and transmittance spectrum in a finite layered PC composed of a material with negative refractive index and the $Al_xGa_{1-x}As$ with positive refractive index [51]. They showed a defect mode in the optical dispersion spectrum when a defect layer with negative refractive index is added to the photonic crystal. This mode was shown to be not sensitive to the incident angle and the type of polarization. But by adding a defect layer with positive refractive index to the PC, the defect mode shifts slowly toward the higher frequencies when the incident angle increases [51].

It should also be noted that quasiperiodic layered structures made of RHM, especially for Fibonacci sequence, have received a great deal of interest [52–54]. Light transmission and reflection from Fibonacci photonic crystals differ from those in binary ones. Moreover, omnidirectional reflection and bandgaps of Fibonacci photonic crystals containing traditional or metamaterials with a frequency-independent refractive index have been investigated [55–58]. According to measured data from the experiment, the electric permittivity and magnetic permeability of metamaterials are usually frequency-dependent [2,7,43,59–64]. Using frequency-independent parameters, the bandgaps and omnidirectional reflection of Fibonacci photonic crystals have been presented [65,66]. Most of the studies address the effect induced by different generation orders for a set of widths of a refractive index of traditional and metamaterial layers. Also, a zero-$\bar{n}$ gap is found in the photonic spectra of such quasiperiodic structures. The self-similar and non-self-similar wavefunctions are exhibited at the frequencies close to the zero-$\bar{n}$ gap. It was also shown that the zero-$\bar{n}$ gap is less sensitive to the incident angle and the polarization in contrast to the Bragg gap. The effects of layer widths and materials on zero-$\bar{n}$ gaps and omnidirectional reflection for TE and TM polarizations in Fibonacci basis photonic crystal containing frequency dependent metamaterial have been proposed by Hsueh et al. [67,68].

In 2013, Jose investigated large enhancement of the nonlinear wave-mixing process (third-harmonic generation) in LHM–RHM photonic bandgap materials [69]. He showed that when incorporating a metamaterial, electromagnetically induced transparency (EIT)-like resonance in the nonlinear layer leads to unprecedented line narrowing of the generated third harmonic. He also showed that this scheme permits the mapping of the narrow EIT resonance at the fundamental frequency to the generated third-harmonic frequency with even narrower line shape [69]. In another study, Jose presented bistable switching operations of an LHM–RHM photonic crystal with a center nonlinear defect medium [70]. The inclusion of a metamaterial-based electromagnetic induced transparency (EIT)-like response for the positive medium splits the defect mode with a sharp EIT transmission peak at the resonant frequency. Jose obtained bistable response near the EIT resonance that is found to be extremely sensitive to the operating frequency about the EIT peak [70].

The object of this chapter is to theoretically present a detailed study of the dispersion relation and photonic band structure in photonic crystals constituted by alternate layers of LHMs and RHMs, with the aim of giving the different trends that can occur [71]. We present and discuss the band structure with various physical parameters and different ratios of the LHM to RHM layer thicknesses. In these calculations, the dielectric permittivity ε and magnetic permeability μ are, in general, assumed to take constant values. Although these parameters in LHM are in general frequency-dependent, our results can be used to design specific metamaterials that would lead to a typical behavior around a given frequency. We also illustrate the photonic band structure in a case with frequency-dependent parameters. We discuss, in particular, the photonic bands of the photonic crystal originating from the interface modes at the boundary between an RHM and LHM, and those bands that are confined in one type of layer in the photonic crystal. When the permittivity ε and permeability μ are constant parameters, one of the passbands can decrease down to the static limit of the frequency $\omega = 0$ at a value of the wavevector $k_{||}$, parallel to the layers, that is different from zero.

For fixed values of $k_{||}$, we also investigate the dispersion curves ω versus k_3, where k_3 is the Bloch wavevector along the axis of the photonic crystal which is limited to the reduced Brillouin zone $-\pi/D < k_3 < \pi/D$ (D being the period of the layered photonic crystal). In contrast to the case of usual photonic crystals where ω displays a monotonic variation with k_3, while k_3 goes from 0 to π/D and vice versa, in the LHM–RHM composite photonic crystals some of the dispersion curves may exist only in a limited part of the Brillouin zone and even display a zigzag behavior.

We also show that for some particular choices of the physical and geometrical parameters, the RHM–LHM photonic crystal can exhibit an absolute (or omnidirectional) bandgap, for either TE or TM polarization of the electromagnetic field. This situation is without analog in the case of usual photonic crystals. Thus, a combination in tandem of two LHM–RHM photonic crystals enables us to propose an omnidirectional reflector structure for both polarizations of the light. Let us notice that the search of omnidirectional reflection gaps has been the object of several works [42,72–77]. In particular, the possibility of an omnidirectional reflection gap in a lamellar structure containing left handed media has been mentioned in [42]. Most of the results presented in this chapter have been published by our group in [71].

The chapter is organized as follows. In Section 21.2 we give the details of the Green's function of any composite electromagnetic medium, and in Section 21.3 we present the basic equations for layered isotropic media. Then, for the clarity of the discussion, we briefly present in Section 21.4 the interface modes localized at an LHM–RHM boundary, and in Section 21.5 the confined modes of an LHM layer embedded between two semiinfinite RHM media. Section 21.6 is devoted to the presentation of our main results concerning the photonic band structure of LHM–RHM photonic crystals. Finally, some conclusions are drawn in Section 21.7.

Let us notice that the derivation of the single interface mode and the modes in a slab of LHM have been considered in two papers by Ruppin [78,79]. Therefore, the object of Section 21.4 is mainly to emphasize the physical behaviors in these two problems for a clear understanding of the results presented in Section 21.6.

21.2 Response functions for any composite electromagnetic medium

First the response function for the electromagnetic field will be defined for one homogeneous infinite electromagnetic medium. Then an electromagnetic medium limited by a black-box surface will be defined by a completely opaque surface through which the electromagnetic field cannot propagate. Any composite electromagnetic medium can then be constructed out of pieces of such different media bounded by black-box surfaces. This way of constructing such a medium enables one to calculate the corresponding response function.

21.2.1 *One infinite electromagnetic medium and one limited by a black-box surface*

We begin by defining the response function we are concerned with. We assume that in the infinite physical system under study here is present a current density $\boldsymbol{J}$ that could be due, e.g., to the passage of a charged particle through it. The Maxwell equations for the electromagnetic field in the system in this case are:

$$\nabla \times \boldsymbol{H} = \boldsymbol{J} + \frac{\partial \boldsymbol{D}}{\partial t}, \tag{21.1}$$

$$\nabla \times \boldsymbol{E} = -\frac{\partial \boldsymbol{B}}{\partial t}, \tag{21.2}$$

where $\boldsymbol{E}$ is the macroscopic electric field, $\boldsymbol{D}$ is the displacement vector, $\boldsymbol{H}$ is the magnetic field, and $\boldsymbol{B}$ is the magnetic induction.

When the effects of spatial dispersion are neglected, the most general linear relation between the displacement vector $\boldsymbol{D}(\boldsymbol{x}, t)$, and the macroscopic electric field $\boldsymbol{E}(\boldsymbol{x}, t)$ in an electromagnetic medium is given by [80]

$$\boldsymbol{D}(\boldsymbol{x}, t) = \int dt' \varepsilon_0 \varepsilon_i (t - t') \boldsymbol{E}(\boldsymbol{x}, t'), \tag{21.3}$$

where $\varepsilon_i(t - t')$ is the 3×3 relative dielectric tensor (the index i will be necessary when we consider the composite system formed out of different media) and ε_0 is the permittivity in the vacuum. This tensor is a function of the difference in the time argument, $t - t'$, because we are assuming that the Hamiltonian for the electromagnetic medium is time-independent. The requirement of causality is that the $\varepsilon_i(t)$ vanish identically for $t < 0$.

In the same way we define the general linear relation between the magnetic induction $\boldsymbol{B}(\boldsymbol{x}, t)$, and the macroscopic magnetic field $\boldsymbol{H}(\boldsymbol{x}, t)$ in an electromagnetic medium is given by

$$\boldsymbol{B}(\boldsymbol{x}, t) = \int dt' \mu_0 \mu_i (t - t') \boldsymbol{H}(\boldsymbol{x}, t'), \tag{21.4}$$

where $\mu_i(t - t')$ is the 3×3 relative magnetic tensor (the index i will be necessary when we consider the composite system formed out of different media) and μ_0 is the permeability in the vacuum.

Create now a medium bounded by a black-box surface defined by $x_3 = f(x_1, x_2)$. This can be achieved by writing that the speed of light c vanishes for $x_3 < f(x_1, x_2)$.

This enables us to rewrite the Maxwell equations for this medium limited by a black-box surface as:

$$\theta[-f(x_1, x_2) + x_3]\left(\nabla \times \boldsymbol{H} - \boldsymbol{J}\right) - \frac{\partial \boldsymbol{D}}{\partial t} = \boldsymbol{0}, \tag{21.5}$$

$$\theta[-f(x_1, x_2) + x_3]\nabla \times \boldsymbol{E} + \frac{\partial \boldsymbol{B}}{\partial t} = \boldsymbol{0}, \tag{21.6}$$

where

$$\theta[-f(x_1, x_2) + x_3] = \begin{cases} 1, & \text{for } x_3 \geq f(x_1, x_2), \\ 0, & \text{for } x_3 < f(x_1, x_2). \end{cases} \tag{21.7}$$

Eliminating $\boldsymbol{H}$ between Eqs. (21.5) and (21.6) provides the equation satisfied by the macroscopic electric field. Applying first the operator $\boldsymbol{\nabla}$ to Eq. (21.6), we obtain:

$$\boldsymbol{\nabla}\times\theta[-f(x_1,x_2)+x_3]\,\boldsymbol{\nabla}\times\boldsymbol{E}+\frac{\partial}{\partial t}\boldsymbol{\nabla}\times\boldsymbol{B}=\boldsymbol{0}. \tag{21.8}$$

Then using Eq. (21.5) and differentiating provides

$$\begin{aligned}\theta[-f(x_1,x_2)+x_3]&\left(\boldsymbol{\nabla}\times\boldsymbol{\nabla}\times\boldsymbol{E}+\mu_i\frac{\partial^2\boldsymbol{D}}{\partial t^2}\right)\\ -\,\delta[-f(x_1,x_2)+x_3]\,&\boldsymbol{V}_i(\boldsymbol{x})\boldsymbol{E}=-\mu_i\frac{\partial \boldsymbol{J}}{\partial t},\end{aligned} \tag{21.9}$$

where the black-box cleavage operator $\boldsymbol{V}_i(\boldsymbol{x})$ is defined as a 3×3 matrix. It depends on the submedium i only through the form of its black-box surface. Note, however, that $\boldsymbol{V}_i(\boldsymbol{x}_3)$ has the opposite sign, if one considers the complementary (in the infinite space) medium. Its expression was found to be

$$\boldsymbol{V}_i(\boldsymbol{x})=\begin{pmatrix}-\frac{\partial f(x_1,x_2)}{\partial x_2}\frac{\partial}{\partial x_2}+\frac{\partial}{\partial x_3} & \frac{\partial f(x_1,x_2)}{\partial x_2}\frac{\partial}{\partial x_1} & -\frac{\partial}{\partial x_1}\\ \frac{\partial f(x_1,x_2)}{\partial x_1}\frac{\partial}{\partial x_2} & -\frac{\partial f(x_1,x_2)}{\partial x_1}\frac{\partial}{\partial x_1}+\frac{\partial}{\partial x_3} & -\frac{\partial}{\partial x_2}\\ \frac{\partial f(x_1,x_2)}{\partial x_1}\frac{\partial}{\partial x_3} & \frac{\partial f(x_1,x_2)}{\partial x_2}\frac{\partial}{\partial x_3} & -\frac{\partial f(x_1,x_2)}{\partial x_1}\frac{\partial}{\partial x_1}-\frac{\partial f(x_1,x_2)}{\partial x_2}\frac{\partial}{\partial x_2}\end{pmatrix}. \tag{21.10}$$

Now Fourier transform $\boldsymbol{E}(\boldsymbol{x},t)$, $\boldsymbol{H}(\boldsymbol{x},t)$, $\boldsymbol{J}(\boldsymbol{x},t)$, $\varepsilon_i(t-t')$, and $\mu_i(t-t')$ with respect to time:

$$\boldsymbol{E}(\boldsymbol{x},t)=\int\frac{d\omega}{2\pi}\boldsymbol{E}(\boldsymbol{x}/\omega)e^{-i\omega t}, \tag{21.11}$$

$$\boldsymbol{H}(\boldsymbol{x},t)=\int\frac{d\omega}{2\pi}\boldsymbol{H}(\boldsymbol{x}/\omega)e^{-i\omega t}, \tag{21.12}$$

$$\boldsymbol{J}(\boldsymbol{x},t)=\int\frac{d\omega}{2\pi}\boldsymbol{J}(\boldsymbol{x}/\omega)e^{-i\omega t}, \tag{21.13}$$

$$\varepsilon_i(t-t')=\int\frac{d\omega}{2\pi}\varepsilon_i(\omega)e^{-i\omega(t-t')}, \tag{21.14}$$

$$\mu_i(t-t')=\int\frac{d\omega}{2\pi}\mu_i(\omega)e^{-i\omega(t-t')}, \tag{21.15}$$

where $\varepsilon_i(\omega)$ and $\mu_i(\omega)$ considered as functions of the complex variable ω can have no poles in the upper half-plane.

Remember that $\varepsilon_0\mu_0=1/c^2$, where c is the light speed. When Eqs. (21.11), (21.12), (21.13), (21.14), and (21.15) are substituted into Eq. (21.9), the latter takes the form

$$\theta[-f(x_1,x_2)+x_3]\frac{1}{\mu_i(\omega)}\left(\frac{\omega^2}{c^2}\varepsilon_i(\omega)\mu_i(\omega)-\boldsymbol{H}^0(\boldsymbol{x})\right)\boldsymbol{E}(\boldsymbol{x}/\omega) + \frac{1}{\mu_i(\omega)}\delta[-f(x_1,x_2)+x_3]\,\boldsymbol{V}_i(\boldsymbol{x})\boldsymbol{E}(\boldsymbol{x}/\omega) = -i\omega\boldsymbol{J}(\boldsymbol{x}/\omega), \tag{21.16}$$

where

$$H^0_{\alpha\beta}(\boldsymbol{x}) = \frac{\partial^2}{\partial x_\alpha \partial x_\beta}\delta_{\alpha\beta}\nabla^2 \quad \text{for } \alpha,\beta = 1,2,3. \tag{21.17}$$

Define now the response functions $\boldsymbol{G}_i(\omega/\boldsymbol{x},\boldsymbol{x}')$ and $\boldsymbol{g}_{si}(\omega/\boldsymbol{x},\boldsymbol{x}')$, respectively, for the infinite electromagnetic medium i and the same semiinfinite medium limited by the black-box surface as

$$\frac{1}{\mu_i(\omega)}\left(\frac{\omega^2}{c^2}\varepsilon_i(\omega)\mu_i(\omega)-\boldsymbol{H}^0(\boldsymbol{x})\right)\boldsymbol{G}_i(\omega/\boldsymbol{x},\boldsymbol{x}') = \boldsymbol{I}\delta(\boldsymbol{x}-\boldsymbol{x}') \tag{21.18}$$

and

$$\frac{1}{\mu_i(\omega)}\left[\theta[-f(x_1,x_2)+x_3]\left(\frac{\omega^2}{c^2}\varepsilon_i(\omega)\mu_i(\omega)-\boldsymbol{H}^0(\boldsymbol{x})\right) + \delta[-f(x_1,x_2)+x_3]\,\boldsymbol{V}_i(\boldsymbol{x})\right]\boldsymbol{g}_{si}(\omega/\boldsymbol{x},\boldsymbol{x}') = \boldsymbol{I}\delta(\boldsymbol{x}-\boldsymbol{x}'), \tag{21.19}$$

where $\boldsymbol{I}$ is the 3×3 unit matrix.

The interface response theory of continuous media provides [83,84] the following relation between the above response functions:

$$\boldsymbol{g}_{si}(\boldsymbol{x},\boldsymbol{x}') + \int d\boldsymbol{x}''\,\boldsymbol{g}_{si}(\boldsymbol{x},\boldsymbol{x}'')\boldsymbol{A}_{si}(\boldsymbol{x},\boldsymbol{x}'') = \boldsymbol{G}_i(\boldsymbol{x},\boldsymbol{x}'), \tag{21.20}$$

for $\{\boldsymbol{x},\boldsymbol{x}'\}\in D_i$ and $\boldsymbol{x}''\in M_i$, where D_i is the three-dimensional space for which $x_3 \geq f(x_1,x_2)$ and M_i is the surface space defined by $x_3 = f(x_1,x_2)$; $\boldsymbol{A}_{si}$ is the corresponding surface response operator defined by

$$\boldsymbol{A}_{si}(\boldsymbol{x},\boldsymbol{x}') = \boldsymbol{V}_i(\boldsymbol{x}'')\boldsymbol{G}_i(\boldsymbol{x},\boldsymbol{x}'')|_{\boldsymbol{x}''=\boldsymbol{x}}, \quad \{\boldsymbol{x}',\boldsymbol{x}''\}\in D_i \text{ and } \boldsymbol{x}\in M_i. \tag{21.21}$$

In Eq. (21.20), and in what follows, we do not write explicitly, for simplicity, the dependence on ω of the response functions.

So once the response function $\boldsymbol{G}_i$ of the infinite medium is determined from Eq. (21.18), the surface integral equation (21.20) enables us to calculate the response function $\boldsymbol{g}_{si}$ of the same medium but limited by the black-box surface M_i. Of course, the solutions of the integral Eq. (21.20) have to be obtained in general by numerical methods. One of these numerical methods may be to transform the integral equation (21.20) into a matrix equation by a suitable choice of a discrete number of values for

$\boldsymbol{x}, \boldsymbol{x}' \in D_i$. In particular, it is helpful and sufficient to do this inside the surface space M_i. Then one easily shows [81–84] that

$$\boldsymbol{g}_{si}^{-1}(M_i, M_i) = \boldsymbol{\Delta}(M_i, M_i)\boldsymbol{G}_s^{-1}(M_i, M_i), \tag{21.22}$$

where

$$\boldsymbol{\Delta}(M_i, M_i) = \boldsymbol{I}(M_i, M_i) + \boldsymbol{A}_{si}(M_i, M_i). \tag{21.23}$$

All entities in the above two equations have to be understood as matrices defined for a discrete number of space points $\boldsymbol{x}$ within the surface M_i. In particular, $\boldsymbol{g}_{si}^{-1}(M_i, M_i)$ and $\boldsymbol{G}_i^{-1}(M_i, M_i)$ are the inverses of $\boldsymbol{g}_{si}(M_i, M_i)$ and $\boldsymbol{G}_i(M_i, M_i)$ within the space M_i.

Note also that all the above results can be easily generalized [83,84] to one electromagnetic medium i limited by J disconnected black-box surfaces $M_{ij} \in M_i$, $1 \leqslant j \leqslant J$. One has to calculate $\boldsymbol{V}_i(\boldsymbol{x})$ for each surface M_{ij} and the total response operator $\boldsymbol{A}_{si}$ within the whole surface space M_i.

21.2.2 Any composite electromagnetic medium

A composite medium is formed out of N different submedia labeled by the index i and contained in their respective subdomains D_i. The ensemble of these subdomains D_i forms the domain of definition D of the composite medium.

The subdomains are separated from each other by subinterface domains M_{ij}. In general, a domain D_i is bounded by several disconnected subinterface domains M_{ij} and their ensemble is called the interface subdomain M_i. The ensemble of all interface subdomains M_i is the interface domain M of the composite material.

Any composite electromagnetic medium can in this way be built out of the independent pieces limited by the black-box surfaces, as defined above. In particular, one has to use semiinfinite pieces corresponding to the composite material. Putting all these independent pieces together and allowing them to interact enables one [81–84] to construct the real composite electromagnetic medium and to calculate its response function $\boldsymbol{g}$ out of the elements of the response functions $\boldsymbol{G}_i$ of the infinite electromagnetic media.

Once the inverse surface elements $\boldsymbol{g}_{si}^{-1}(M_i, M_i)$ are determined from Eqs. (21.22) and (21.23) for all independent submedia, one can obtain the inverse interface elements $\boldsymbol{g}^{-1}(M, M)$ of the composite medium in the following manner [83,84]:

$$\boldsymbol{g}^{-1}(M_{ij}, M_{i'j'}) = 0, \qquad M_{i'j'} \notin M_i; \tag{21.24}$$

$$\boldsymbol{g}^{-1}(M_{ij}, M_{ij'}) = \boldsymbol{g}_{si}^{-1}(M_{ij} M_{ij'}), \quad j' \neq j; \tag{21.25}$$

$$\boldsymbol{g}^{-1}(M_{ij}, M_{ij}) = \sum_{i'} \boldsymbol{g}_{si'}^{-1}(M_{i'j'}, M_{i'j'}), \quad M_{i'j'} \equiv M_{ij}. \tag{21.26}$$

So now we have all the elements of the matrix $g^{-1}(M, M)$, the inverse of the matrix $g(M, M)$. This inverse, as explained above, can be defined in general only by using a finite number of discrete space points in M.

Once the matrix inversion of the black-box tridiagonal matrix $g^{-1}(M, M)$ (Eqs. (21.24), (21.25), and (21.26)) is obtained, one has the matrix elements $g(M, M)$ of the response function g of the composite electromagnetic medium, within the total interface space M.

All the other matrix elements of g can be obtained from [83,84]:

$$\begin{aligned} g(D, D) = {} & G(D, D) - G(D, M)G^{-1}(M, M)G(M, D) \\ & + G(D, M)G^{-1}(M, M)g(M, M)G^{-1}(M, M)G(M, D), \end{aligned} \tag{21.27}$$

where $G(D, D)$ is defined by

$$G(x, x') = \begin{cases} G_i(x, x'), & \{x, x'\} \in D_i; \\ 0, & x \in D_i;\ x' \in D_{i'}. \end{cases} \tag{21.28}$$

Also $G(D, M)$ and $G(M, D)$ are defined in the same manner, and $G^{-1}(M, M)$ is the inverse of the block diagonal square matrix $G(M, M)$.

The new polariton states localized at the interfaces M for any composite electromagnetic medium are given by

$$\det\left[g^{-1}(M, M)\right] = 0. \tag{21.29}$$

21.3 Basic equations for layered isotropic media

An important particular case of the analysis presented above is that of layered isotropic media. The isotropy of each submedium provides:

$$\bar{\bar{\varepsilon}}_i(\omega) = \varepsilon_i(\omega)I, \quad i = 1, 2, \ldots, N. \tag{21.30}$$

$$\bar{\bar{\mu}}_i(\omega) = \mu_i(\omega)I, \quad i = 1, 2, \ldots, N, \tag{21.31}$$

where $\varepsilon_i(\omega)$ and $\mu_i(\omega)$ are the relative dielectric permittivity and magnetic permeability.

Further, assuming that the planar interfaces are perpendicular to the x_3-axis enables all the above operators to be Fourier analyzed according to, for example,

$$g(x, x') = \int \frac{d^2k_{||}}{(2\pi)^2} g(k_{||}/x_3, x_3')e^{ik_{||}(x_{||}-x_{||}')}, \tag{21.32}$$

where $k_{||}$ and $x_{||}$ are the propagation and the real space position vectors parallel to the interface, respectively.

The equations of Section 21.2 remain unchanged for these Fourier coefficients. This transformation enables one to obtain simpler expressions for these Fourier coefficients, in particular, from Eqs. (21.17) and (21.10) we get

$$\boldsymbol{H}^0(k_{||}/x_3) = \begin{pmatrix} k_2^2 - \frac{\partial^2}{\partial x_3^2} & -k_1k_2 & ik_1\frac{\partial}{\partial x_3} \\ -k_1k_2 & k_1^2 - \frac{\partial^2}{\partial x_3^2} & ik_2\frac{\partial}{\partial x_3} \\ ik_1\frac{\partial}{\partial x_3} & ik_2\frac{\partial}{\partial x_3} & k_{||}^2 \end{pmatrix} \tag{21.33}$$

and

$$\boldsymbol{V}_i(k_{||}/x_3) = \begin{pmatrix} \frac{\partial}{\partial x_3} & 0 & -ik_1 \\ 0 & \frac{\partial}{\partial x_3} & -ik_2 \\ 0 & 0 & 0 \end{pmatrix}, \tag{21.34}$$

for $x_3 \geqslant 0$, when we take the planar surface at $x_3 = f(x_1, x_2) = 0$.

It is now convenient to take advantage of the isotropy of our system in the (x_1, x_2)-plane by premultiplying and postmultiplying Eqs. (21.33) and (21.34) by the matrices $\boldsymbol{S}(k_{||})$ and $\boldsymbol{S}^{-1}(k_{||})$, respectively,

$$\boldsymbol{S}(k_{||}) = \frac{1}{k_{||}} \begin{pmatrix} k_1 & k_2 & 0 \\ -k_2 & k_1 & 0 \\ 0 & 0 & k_{||} \end{pmatrix} \tag{21.35}$$

and

$$\boldsymbol{S}^{-1}(k_{||}) = \frac{1}{k_{||}} \begin{pmatrix} k_1 & -k_2 & 0 \\ k_2 & k_1 & 0 \\ 0 & 0 & k_{||} \end{pmatrix}. \tag{21.36}$$

This transforms $\boldsymbol{H}^0(k_{||}/x_3)$ and $\boldsymbol{V}_i(k_{||}/x_3)$, respectively, into

$$\boldsymbol{H}^0(k_{||}/x_3) = \begin{pmatrix} -\frac{\partial^2}{\partial x_3^2} & 0 & ik_{||}\frac{\partial}{\partial x_3} \\ 0 & k_{||}^2 - \frac{\partial^2}{\partial x_3^2} & 0 \\ ik_{||}\frac{\partial}{\partial x_3} & 0 & k_{||}^2 \end{pmatrix} \tag{21.37}$$

and

$$\boldsymbol{V}_i(k_{||}/x_3) = \begin{pmatrix} \frac{\partial}{\partial x_3} & 0 & -ik_{||} \\ 0 & \frac{\partial}{\partial x_3} & 0 \\ 0 & 0 & 0 \end{pmatrix}, \tag{21.38}$$

for $x_3 \geqslant 0$, which depends now only on the magnitude $k_{||}$ of the propagation vector (rather than k_1 and k_2 in Eqs. (21.33) and (21.34)).

Finally, Eqs. (21.18) and (21.19) become, when one takes due account of $\mu_i \neq 1$,

$$\frac{1}{\mu_i(\omega)}\left(\frac{\omega^2}{c^2}\varepsilon_i(\omega)\mu_i(\omega)\boldsymbol{I} - \boldsymbol{H}^0(k_{||}/x_3)\right)\boldsymbol{G}_i(k_{||},\omega/x_3,x_3') = \boldsymbol{I}\delta(x_3 - x_3') \tag{21.39}$$

and

$$\frac{1}{\mu_i(\omega)}\left[\theta(x_3)\left(\frac{\omega^2}{c^2}\varepsilon_i(\omega)\mu_i(\omega)\boldsymbol{I} - \boldsymbol{H}^0(k_{||}/x_3)\right) + \delta(x_3)\boldsymbol{V}_i(k_{||}/x_3)\right]\boldsymbol{g}_{si}(k_{||},\omega/x_3,x_3') = \boldsymbol{I}\delta(x_3 - x_3'). \tag{21.40}$$

The determination of the matrix elements of $\boldsymbol{G}_i$ is greatly simplified by the fact that $(G_i)_{21}$ and $(G_i)_{23}$ satisfy homogeneous equations, while $(G_i)_{12}$ and $(G_i)_{32}$ satisfy a pair of coupled homogeneous equations. These four matrix elements therefore vanish in $\boldsymbol{G}_i$ and also in $\boldsymbol{g}_{si}$ [14,85,86], and the 2×2 matrix elements decouple from the others. Eq. (21.39) becomes

$$\frac{1}{\mu_i(\omega)}\left(\frac{\omega^2}{c^2}\varepsilon_i(\omega)\mu_i(\omega) - k_{||}^2 + \frac{\partial^2}{\partial x_3^2}\right)(G_i)_{22} = \delta(x_3 - x_3') \tag{21.41}$$

and

$$\frac{1}{\mu_i(\omega)}\begin{pmatrix} \frac{\omega^2}{c^2}\varepsilon_i(\omega)\mu_i(\omega) + \frac{\partial^2}{\partial x_3^2} & -ik_{||}\frac{\partial}{\partial x_3} \\ -ik_{||}\frac{\partial}{\partial x_3} & \frac{\omega^2}{c^2}\varepsilon_i(\omega)\mu_i(\omega) - k_{||}^2 \end{pmatrix}\begin{pmatrix} (G_i)_{11} & (G_i)_{13} \\ (G_i)_{31} & (G_i)_{33} \end{pmatrix} = \boldsymbol{I}\delta(x_3 - x_3'), \tag{21.42}$$

where $\boldsymbol{I}$ is the 2×2 unit matrix, and we omit, from now on, to write explicitly the dependence on $(k_{||}, \omega/x_3, x_3')$ in the response functions. The same decoupling appears also in Eq. (21.40):

$$\frac{1}{\mu_i(\omega)}\left[\theta(x_3)\left(\frac{\omega^2}{c^2}\varepsilon_i(\omega)\mu_i(\omega) - k_{||}^2 + \frac{\partial^2}{\partial x_3^2}\right) + \delta(x_3)\frac{\partial}{\partial x_3}\right](g_{si})_{22} = \delta(x_3 - x_3') \tag{21.43}$$

and

$$\frac{1}{\mu_i(\omega)}\begin{pmatrix} \frac{\omega^2}{c^2}\varepsilon_i(\omega)\mu_i(\omega)+\frac{\partial^2}{\partial x_3^2}+\delta(x_3)\frac{\partial}{\partial x_3} & -ik_{\parallel}\left(d\frac{\partial}{\partial x_3}+\delta(x_3)\right) \\ -ik_{\parallel}\frac{\partial}{\partial x_3} & \frac{\omega^2}{c^2}\varepsilon_i(\omega)\mu_i(\omega)-k_{\parallel}^2 \end{pmatrix} \times \begin{pmatrix} (g_{si})_{11} & (g_{si})_{13} \\ (g_{si})_{31} & (g_{si})_{33} \end{pmatrix} = \boldsymbol{I}\delta(x_3-x_3'). \tag{21.44}$$

From Eq. (21.42) we see that $(G_i)_{31}$ and $(G_i)_{33}$ can be obtained simply from $(G_i)_{11}$ and $(G_i)_{13}$, and similarly the following relations exist between the same elements of $\boldsymbol{g}$, the response function of any such layered medium:

$$g_{31}(k_{\parallel},\omega/x_3,x_3') = -\frac{ik_{\parallel}}{\alpha_i^2}\frac{\partial}{\partial x_3}g_{11}(k_{\parallel},\omega/x_3,x_3'), \quad x_3 \in D_i, \tag{21.45}$$

$$g_{33}(k_{\parallel},\omega/x_3,x_3') = -\frac{1}{\alpha_i^2}\left(\delta(x_3-x_3')+ik_{\parallel}\frac{\partial}{\partial x_3}g_{13}(k_{\parallel},\omega/x_3,x_3')\right), \quad x_3 \in D_i, \tag{21.46}$$

where

$$\alpha_i(k_{\parallel},\omega) = \left(k_{\parallel}^2-\frac{\omega^2}{c^2}\varepsilon_i(\omega)\mu_i(\omega)\right)^{1/2}. \tag{21.47}$$

Eqs. (21.45) and (21.46) hold for the corresponding matrix elements of the response function $\boldsymbol{g}_{si}$ of one medium bounded by a black-box surface, as well as for the response function $\boldsymbol{g}$ of any such layered composite material, provided that x_3 lies inside the slab i, as $\boldsymbol{g}$ has to satisfy Eq. (21.42) for $x_3 \in D_i$.

Consequently, we can regard in what follows g_{11} and g_{22} as the primary elements of $\boldsymbol{g}$ and g_{31} and g_{33} as derived elements through Eqs. (21.45) and (21.46). Eq. (21.40), when use is made of the general relations (21.45) and (21.46), can be rewritten as three independent equations for the 22, 11, and 13 elements of $\boldsymbol{g}_{si}$, namely:

$$\frac{1}{\mu_i(\omega)}\left[\theta(x_3)\left(\frac{\partial^2}{\partial x_3^2}-\alpha_i^2\right)+\delta(x_3)\frac{\partial}{\partial x_3}\right](g_{si})_{22} = \delta(x_3-x_3'), \tag{21.48}$$

$$-\frac{\omega^2\varepsilon_i(\omega)}{\alpha_i^2c^2}\left[\theta(x_3)\left(\frac{\partial^2}{\partial x_3^2}-\alpha_i^2\right)+\delta(x_3)\frac{\partial}{\partial x_3}\right](g_{si})_{11} = \delta(x_3-x_3'), \tag{21.49}$$

$$-\frac{i\omega^2\varepsilon_i(\omega)}{k_{\parallel}c^2}\theta(x_3)\left(\frac{\partial^2}{\partial x_3^2}-\alpha_i^2\right)(g_{si})_{13} - \delta(x_3) \times \left(\delta(x_3-x_3')+\frac{i\omega^2\varepsilon_i(\omega)}{k_{\parallel}c^2}\frac{\partial}{\partial x_3}(g_{si})_{13}\right) = \frac{\partial}{\partial x_3}\delta(x_3-x_3'). \tag{21.50}$$

Comparing Eqs. (21.49) and (21.50), one sees that

$$(g_{si})_{13}(x_3, x_3') = \frac{ik_{||}}{\alpha_i^2}\frac{\partial}{\partial x_3'}(g_{si})_{11}(x_3, x_3'), \quad x_3' \in D_i. \tag{21.51}$$

Relation (21.51) is a constitutive one, as well as Eqs. (21.45) and (21.46), and holds also for the g_{13} of any such layered medium:

$$g_{13}(x_3, x_3') = \frac{ik_{||}}{\alpha_{i'}^2}\frac{\partial}{\partial x_3'}g_{11}(x_3, x_3'), \quad x_3' \in D_{i'}, \tag{21.52}$$

and $\alpha_{i'}$ takes the value given by Eq. (21.47) within $D_{i'}$.

So considering Eqs. (21.48), (21.49), and (21.50), one sees that all three are isomorphic to the same basic equation:

$$\frac{F_i}{\alpha_i}\left[\theta(x_3)\left(-\alpha_i^2 + \frac{\partial^2}{\partial x_3^2}\right) + \delta(x_3)\frac{\partial}{\partial x_3}\right] g_{si}(x_3, x_3') = \delta(x_3 - x_3'), \tag{21.53}$$

where

$$F_i = \frac{\alpha_i}{\mu_i(\omega)}, \qquad \text{for } g_{22}, \tag{21.54}$$

$$F_i = -\frac{\omega^2}{c^2}\frac{\varepsilon_i(\omega)}{\alpha_i}, \qquad \text{for } g_{11}. \tag{21.55}$$

So it is sufficient to obtain the basic element of the response function $\boldsymbol{g}$, starting with Eq. (21.53). Then with the help of the relations (21.54) and (21.55), one obtains g_{22} and g_{11}. The elements g_{13}, g_{31}, and g_{33} are then easily obtained with the help of the constitutive relations (21.52), (21.45), and (21.46). Indeed, the above considerations are based only on the basic form of the bulk differential equations.

The basic response function G_i of an infinite homogeneous medium is easily found to be

$$G_i(k_{||}, \omega/x_3, x_3') = -\frac{e^{-\alpha_i|x_3 - x_3'|}}{2F_i}. \tag{21.56}$$

Replacing now in Eq. (21.56) F_i by the values given by Eqs. (21.54) and (21.55) and using Eqs. (21.45), (21.46), and (21.52), one obtains the elements of the complete bulk response function $G(k_{||}, \omega/x_3, x_3')$:

$$G_{22}(k_{||}, \omega/x_3, x_3') = -\mu_i(\omega)\frac{e^{-\alpha_i|x_3 - x_3'|}}{2\alpha_i}, \tag{21.57}$$

$$G_{11}(k_{||}, \omega/x_3, x_3') = \frac{c^2\alpha_i}{2\omega^2\varepsilon_i(\omega)}e^{-\alpha_i|x_3 - x_3'|}, \tag{21.58}$$

$$G_{13}(k_{||},\omega/x_3,x_3')=\frac{ic^2k_{||}}{2\omega^2\varepsilon_i(\omega)}\mathrm{sgn}(x_3-x_3')e^{-\alpha_i|x_3-x_3'|}, \tag{21.59}$$

$$G_{31}(k_{||},\omega/x_3,x_3')=\frac{ic^2k_{||}}{2\omega^2\varepsilon_i(\omega)}\mathrm{sgn}(x_3-x_3')e^{-\alpha_i|x_3-x_3'|}, \tag{21.60}$$

$$G_{33}(k_{||},\omega/x_3,x_3')=\frac{c^2}{\omega^2\varepsilon_i(\omega)}\delta(x_3-x_3')-\frac{c^2k_{||}^2}{2\alpha_i\omega^2\varepsilon_i(\omega)}e^{-\alpha_i|x_3-x_3'|}. \tag{21.61}$$

21.4 Interface between two isotropic media

In the following, we assume that the x_3-axis is along the normal to the interfaces and the wavevector component parallel to the layers $k_{||}$ is along the x_1-axis. The elements of the surface response operator for the semiinfinite medium $i=1$ ($x_3>0$) limited by the perfectly reflecting surface ($x_3=0$) are easily obtained from Eqs. (21.21) and (21.53) as

$$A_{s1}(0,x'3)=-\frac{1}{2}e^{-\alpha_1x_3'}. \tag{21.62}$$

The inverse of the surface element $g_{s1}(0,0)$ is then obtained with the help of Eqs. (21.22) and (21.23) to be

$$g_{s1}^{-1}(0,0)=-F_1. \tag{21.63}$$

The system studied in this paragraph is constructed by two semiinfinite homogeneous isotropic media illustrated in Fig. 21.1. The interface element of the basic response function is obtained from Eq. (21.26) as

$$g^{-1}(0,0)=g_{s1}^{-1}(0,0)+g_{s2}^{-1}(0,0). \tag{21.64}$$

Using result (21.63) and a similar one for $g_{s2}^{-1}(0,0)$, one obtains

$$g^{-1}(0,0)=-(F_1+F_2). \tag{21.65}$$

In this section, we are interested in the interface modes localized at the boundary $x_3=0$ between an LHM and RHM presented in Fig. 21.1 (see also [78]). The interface mode should be exponentially decaying on both sides of the interface and, therefore, its frequency lies below the light lines of both media. The equation giving the interface modes is given by $g^{-1}(0,0)=0$, namely,

$$F_1+F_2=0, \tag{21.66}$$

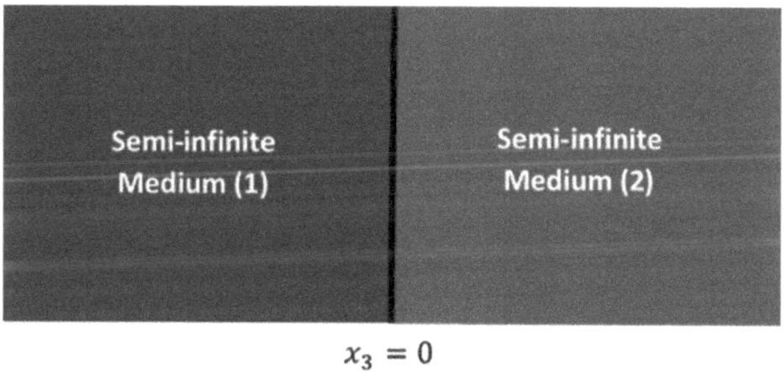

Figure 21.1 Schematic representation of the interface between two semiinfinite media.

where

$$F_i = \frac{\alpha_i(k_\parallel, \omega)}{\mu_i(\omega)}, \qquad \text{for TE polarization modes,} \tag{21.67}$$

$$F_i = -\frac{\varepsilon_i(\omega)}{\alpha_i(k_\parallel, \omega)}\frac{\omega^2}{c^2}, \qquad \text{for TM polarization modes,} \tag{21.68}$$

and

$$\alpha_i(k_\parallel, \omega) = \sqrt{k_\parallel^2 - \varepsilon_i(\omega)\mu_i(\omega)\frac{\omega^2}{c^2}} \tag{21.69}$$

where $i = 1, 2$; c is the speed of light in vacuum, ε and μ are the relative dielectric permittivity and magnetic permeability of the material, and the index of refraction is defined by $n_i = \sqrt{\varepsilon_i(\omega)\mu_i(\omega)}$ with the plus and minus sign being used for RHM and LHM, respectively. The parameters F_i are proportional to the electromagnetic admittance of the corresponding materials for each polarization. By taking the squares in these equations, it is possible to solve for the frequency ω and obtain

$$\frac{c^2 k_\parallel^2}{\omega^2} = \frac{\dfrac{\varepsilon_1}{\mu_1} - \dfrac{\varepsilon_2}{\mu_2}}{\dfrac{1}{\mu_1^2} - \dfrac{1}{\mu_2^2}} \tag{21.70}$$

for TE modes, and

$$\frac{c^2 k_\parallel^2}{\omega^2} = \frac{\dfrac{\varepsilon_1}{\mu_1} - \dfrac{\varepsilon_2}{\mu_2}}{\dfrac{1}{\varepsilon_1^2} - \dfrac{1}{\varepsilon_2^2}} \tag{21.71}$$

for TM modes. However, it should be pointed out that these solutions are valid provided the slope $\omega/k_\parallel$ of the corresponding lines remains below the velocities of light in both media 1 and 2. Thus, one can easily derive the condition for the existence of interface modes as follows: for TE modes, either $\varepsilon_2\mu_2 < \varepsilon_1\mu_1$ and $\mu_2^2 > \mu_1^2$ or $\varepsilon_2\mu_2 > \varepsilon_1\mu_1$ and $\mu_2^2 < \mu_1^2$; for TM modes, either $\varepsilon_2\mu_2 < \varepsilon_1\mu_1$ and $\varepsilon_2^2 > \varepsilon_1^2$ or $\varepsilon_2\mu_2 > \varepsilon_1\mu_1$ and $\varepsilon_2^2 < \varepsilon_1^2$.

Unlike the case of an interface between two RHM, the RHM–LHM interface can support a localized mode of TE polarization. However, one can notice that the TE and TM interface localized modes can never exist simultaneously, i.e., the interface supports at most one localized mode of either TE or TM polarization.

21.5 Confined modes of a layer

In this section, we are interested in the confined modes of an LHM layer of medium 2 embedded between two semiinfinite RHM made of materials 1 and 3 presented in Fig. 21.2 (see also [79]). We consider here a composite system formed of three subsystems, one of which is an LHM slab layer and the others are two semiinfinite RHM media; the LHM slab layer being placed between the other two subsystems (see Fig. 21.2). We shall call these media 1, 2, and 3 such that their domains are for $x_3 \leqslant 0$, $0 \leqslant x_3 \leqslant d$, and $x_3 \geqslant d$, respectively.

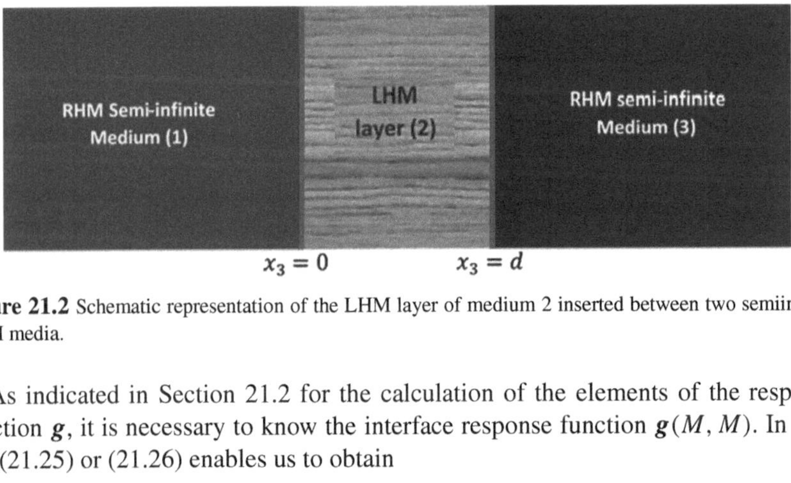

Figure 21.2 Schematic representation of the LHM layer of medium 2 inserted between two semiinfinite RHM media.

As indicated in Section 21.2 for the calculation of the elements of the response function $\boldsymbol{g}$, it is necessary to know the interface response function $\boldsymbol{g}(M,M)$. In fact, Eq. (21.25) or (21.26) enables us to obtain

$$\boldsymbol{g}^{-1}(M,M) = \begin{pmatrix} g_{s1}^{-1}(0,0) + g_{s2}^{-1}(0,0) & g_{s2}^{-1}(0,d) \\ g_{s2}^{-1}(d,0) & g_{s3}^{-1}(d,d) + g_{s2}^{-1}(d,d) \end{pmatrix}. \tag{21.72}$$

Taking account of the two perfectly reflecting surfaces of the slab at $x_3 = 0$ and $x_3 = d$, and of the fact that the corresponding cleavage operator can be given by [87]

$$V_2(x_3) = [-\delta(d - x_3) + \delta(x_3)]\frac{\partial}{\partial x_3}, \tag{21.73}$$

one obtains from Eq. (21.21)

$$A_{s2}(M,M) = -\frac{1}{2}\begin{pmatrix} 1 & e^{-\alpha_2 d} \\ e^{-\alpha_2 d} & 1 \end{pmatrix}, \tag{21.74}$$

which provides, with the help of Eq. (21.23),

$$\Delta_2(M,M) = \frac{1}{2}\begin{pmatrix} 1 & -e^{-\alpha_2 d} \\ e^{-\alpha_2 d} & 1 \end{pmatrix}. \tag{21.75}$$

Using Eq. (21.56), the inverse matrix of $G_2(M,M)$ is obtained to be

$$G_2^{-1}(M,M) = -\frac{F_2}{\sinh(\alpha_2 d)}\begin{pmatrix} e^{\alpha_2 d} & -1 \\ -1 & e^{\alpha_2 d} \end{pmatrix}. \tag{21.76}$$

Finally, from Eq. (21.22), one obtains

$$g_{s2}^{-1}(M,M) = \frac{F_2}{\sinh(\alpha_2 d)}\begin{pmatrix} -\cosh(\alpha_2 d) & 1 \\ 1 & -\cosh(\alpha_2 d) \end{pmatrix}. \tag{21.77}$$

Let us recall from Eq. (21.63) that

$$g_{s3}^{-1}(d,d) = -F_3 \quad \text{and} \quad g_{s1}^{-1}(0,0) = -F_1. \tag{21.78}$$

Substituting relations (21.77) and (21.78) in (21.72) and inverting $g^{-1}(M,M)$, one obtains

$$g(M,M) = -\frac{1}{W}\begin{pmatrix} F_3 + F_2\coth(\alpha_2 d) & F_2/\sinh(\alpha_2 d) \\ F_2/\sinh(\alpha_2 d) & F_1 + F_2\coth(\alpha_2 d) \end{pmatrix}, \tag{21.79}$$

with

$$W = F_1 F_3 + F_2(F_1 + F_3)\coth(\alpha_2 d) + F_2^2. \tag{21.80}$$

Now, we are interested in the confined modes of an LHM layer of medium 2 embedded between two semiinfinite RHMs made of material 1 ($F_1 = F_3$) (see also [79]). The photons confined in LHM slab 2 can be obtained when the two different values (the F_i take two different values, see Eqs. (21.54) and (21.55)) of $[(F_1^2 + F_2^2)\sinh(\alpha_2 d) + 2F_1F_2\cosh(\alpha_2 d)]$ vanish. More precisely, for TE polarization modes,

$$\left[\left(\frac{\alpha_1}{\mu_1}\right)^2 + \left(\frac{\alpha_2}{\mu_2}\right)^2\right]\sinh(\alpha_2 d) + 2\frac{\alpha_1\alpha_2}{\mu_1\mu_2}\cosh(\alpha_2 d) = 0, \tag{21.81}$$

and for TM polarization modes,

$$\left[\left(\frac{\varepsilon_1}{\alpha_1}\right)^2 + \left(\frac{\varepsilon_2}{\alpha_2}\right)^2\right]\sinh(\alpha_2 d) + 2\frac{\varepsilon_1\varepsilon_2}{\alpha_1\alpha_2}\cosh(\alpha_2 d) = 0. \tag{21.82}$$

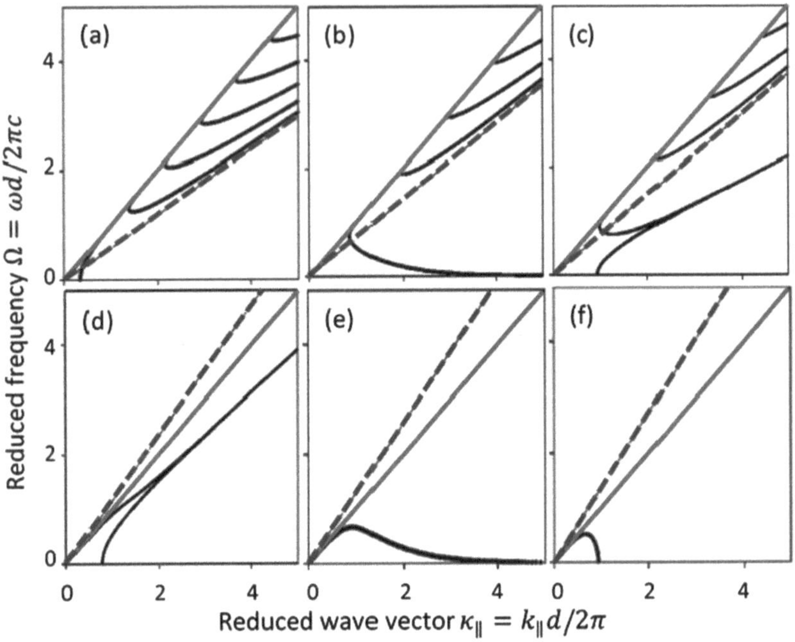

Figure 21.3 Dispersion curves of confined TE optical modes in LHM layer of thickness d sandwiched between vacuum. The parameters [ε; μ] of LHM are taken to be (a) [−1.4; −2], (b) [−2; −1], (c) [−2; −0.9], (d) [−0.6; −1.2], (e) [−0.6; −1], and (f) [−0.6; −0.9]. The reduced frequency $\Omega = \omega d/2\pi c$ is presented as a function of the reduced wavevector $\kappa_{||} = k_{||}d/2\pi$. The straight lines show the light lines of vacuum (red full line; light gray in print version) and of the LHM layer (blue dashed line; dark gray in print version).

In Fig. 21.3, we show all possible behaviors of the TE modes dispersion curves by choosing different parameters for the LHM layer, the RHM medium being vacuum. Let us emphasize that we obtain similar results for TM modes since the results for one polarization can be transposed to the other by interchanging the parameters ε_i and μ_i. Also, exactly the same behaviors are observed if a finite RHM layer is embedded between two semiinfinite LHMs, so for the sake of briefness we omit these results. In Fig. 21.3, we have presented the confined modes of the embedded LHM layer, so all the dispersion curves lie below the light line of the external medium, i.e., vacuum in our case. The panels (a), (b), (c) in the left column ((d), (e), (f) in the right column) refer to an LHM material with an index of refraction $|n_2|$ greater (smaller) than $n_1 = 1$. The panels in the upper, middle, and lower rows, respectively, correspond to an LHM material with $|\mu_2|$ greater than, equal to, or smaller than μ_1.

In the example of panel (a) where $|\mu_2| > 1$, most of the confined modes resemble those of the usual RHM. In particular, for increasing wavevector $k_{||}$, the slopes of the dispersion curves go to the velocity of light in medium 2. Still, one can observe a nonmonotonic behavior of the dispersion curves around their crossing points with vacuum light line. However, the main novelty in this figure is the existence of the lower branch that starts at $\omega = 0$ for a wavevector $k_{||}$ different from zero. In the limit of

$\omega = 0$, the solution for the electromagnetic field reduces to a static magnetic field while the electric field should be equal to zero to prevent the divergence of the magnetic field. In panel (c), the parameters are chosen in such a way that the RHM–LHM interface can support a localized mode ($|\mu_2| < 1$). Thus, in addition to the confined modes of the layer, one can observe two degenerate interface modes in the limit of large $k_{||}d$ (see the two lowest dispersion curves); this degeneracy is lifted at lower values of $k_{||}d$ where the corresponding modes can interact together due to the proximity of the interfaces. Thus, for small values of $k_{||}d$, these modes become more spread over the entire layer and less localized around the interfaces $x_3 = 0$ and $x_3 = d$. Panel (b) corresponds to an intermediate case between the examples of panels (a) and (c), namely, $\mu_2 = -1$. In this case, the interface mode ceases to exist, but there is a dispersion curve which goes asymptotically to $\omega = 0$.

The examples sketched in panels (d), (e), and (f) (right column of Fig. 21.3) correspond to an index of refraction $|n_2|$ in the LHM lower than 1, so the light line of medium 2 is above the vacuum light line. Consequently, the dispersion curves of the confined modes can be searched for only below the latter line. In case (d) where $|\mu_2| > 1$, the dispersion curves display two branches that become the localized interface modes at the LHM–RHM boundaries in the limit of large $k_{||}d$; for decreasing $k_{||}d$, the two interface modes interact more strongly and their degeneracy is lifted. In case (f) where $|\mu_2| < 1$, the RHM–LHM interface does not support any localized mode, so a dispersion curve appears only in a limited range of the wave vector $k_{||}$. Finally, case (e) where $\mu_2 = -1$ is intermediate between cases (d) and (f), i.e., there is one dispersion curve going asymptotically to $\omega = 0$.

In the next section, we shall study the photonic band structure of a periodic stack of LHM–RHM layers. The bands that will appear below the light lines of either media 1 or 2 are those resulting from the interaction of the confined modes discussed above. Therefore, some new behaviors are expected with respect to usual RHM–RHM photonic crystals in relation with the dispersion curves sketched in Fig. 21.3.

21.6 Photonic band structure of LHM–RHM photonic crystal

We shall call d_1 and d_2 the thicknesses of layers 1 and 2, respectively, with $D = d_1 + d_2$ being the period of the layered photonic crystal (see Fig. 21.4). In the examples presented in this section, medium 1 is vacuum whereas medium 2 is an LHM with different dielectric permittivity and magnetic permeability.

The derivation of the dispersion relation of the layered photonic crystal is quite simple and well known. Then, we use the periodicity of the system to introduce a Bloch wavevector k_3 along the axis of the layered photonic crystal that relates the field in two consecutive unit cells by the factor e^{ik_3D} (remember that k_3 is limited to the first Brillouin zone $-\pi/D < k_3 < \pi/D$).

The inverse of the Green's function in the interface space $M=\{\frac{-d_1}{2},\frac{+d_1}{2}\}$ of a medium 1 of thickness d_1 is given by a 2×2 matrix under the following form [87]:

$$g_1^{-1}(M,M)=\begin{pmatrix} \frac{F_1C_1}{S_1} & \frac{F_1}{S_1} \\ \frac{F_1}{S_1} & \frac{F_1C_1}{S_1} \end{pmatrix} \tag{21.83}$$

where $C_1=\cosh(\alpha_1 d_1)$ and $S_1=\sinh(\alpha_1 d_1)$.

Similarly, the inverse of the Green's function of a medium 2 of thickness d_2 in the interface space $M=\{\frac{-d_2}{2},\frac{+d_2}{2}\}$ is given by

$$g_2^{-1}(M,M)=\begin{pmatrix} \frac{F_2C_2}{S_2} & \frac{F_2}{S_2} \\ \frac{F_2}{S_2} & \frac{F_2C_2}{S_2} \end{pmatrix} \tag{21.84}$$

where $C_2=\cosh(\alpha_2 d_2)$ and $S_2=\sinh(\alpha_2 d_2)$.

The Green's function of the infinite structure (Fig. 21.4) is obtained by a linear superposition of the 2×2 matrices (Eqs. (21.83) and (21.84)) at different interfaces, leading to an infinite tridiagonal matrix

$$g^{-1}(M,M)=\begin{pmatrix} \ddots & \ddots & \ddots & & & \\ & \frac{F_1}{S_1} & \frac{F_2C_2}{S_2}+\frac{F_1C_1}{S_1} & \frac{F_2}{S_2} & & \\ & & \frac{F_2}{S_2} & \frac{F_2C_2}{S_2}+\frac{F_1C_1}{S_1} & \frac{F_1}{S_1} & \\ & & & \frac{F_1}{S_1} & \frac{F_2C_2}{S_2}+\frac{F_2C_1}{S_1} & \frac{F_2}{S_2} \\ & & & & \ddots & \ddots & \ddots \end{pmatrix}. \tag{21.85}$$

After a Fourier transform of the above tridiagonal matrix within one unit cell, one obtains

$$g^{-1}(k_3,M,M)=\begin{pmatrix} \frac{F_2C_2}{S_2}+\frac{F_1C_1}{S_1} & \frac{F_2}{S_2}+\frac{F_1}{S_1}e^{-ik_3D} \\ \frac{F_2}{S_2}+\frac{F_1}{S_1}e^{ik_3D} & \frac{F_2C_2}{S_2}+\frac{F_1C_1}{S_1} \end{pmatrix}. \tag{21.86}$$

Here $i=\sqrt{-1}$ and k_3 is the Bloch wavevector of the infinite periodic structure (note that k_3 is real inside the passbands and complex inside the gaps of the infinite system). From the previous matrix, one can deduce the dispersion relation of the infinite photonic crystal (Fig. 21.4(a)) from the equation $\det[g^{-1}(M,M)]=0$, namely,

$$\cos(k_3D)=C_1C_2+\frac{1}{2}\left(\frac{F_1}{F_2}+\frac{F_2}{F_1}\right)S_1S_2. \tag{21.87}$$

The expressions of F_i for both TE and TM polarizations were given in Section 21.2. The dispersion relation (21.87) can be computed in the following way. The right-hand side of Eq. (21.87) is evaluated for any values of ω and $k_{||}D$. If the result is smaller than 1 in absolute value, one can obtain a real solution for k_3, i.e., the corresponding

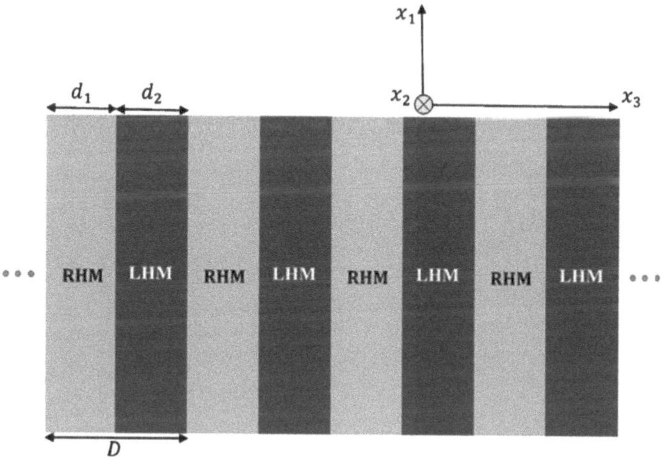

Figure 21.4 Schematic of the RHM–LHM photonic crystal (PC); d_1 and d_2 are the thicknesses of layers 1 (RHM) and 2 (LHM); D is the period of the layered PC.

wave propagates along the axis of the layered photonic crystal and ω belongs to a passband for the chosen value of $k_{||}D$. Otherwise, k_3 becomes a complex number, the wave cannot propagate, and ω belongs to a gap of the photonic crystal. The dispersion curves can be sketched in two different ways. One way is to fix the wave vector component $k_{||}D$ and give the frequency ω as a function of the Bloch wavevector k_3. In the other way, the so-called projected band structure is presented in which all the passbands and minigaps are displayed as a function of $k_{||}D$. In Figs. 21.5 and 21.6, we give two illustrations in a few cases that cover most of the possible behaviors for the RHM–LHM photonic crystal photonic band structure. In these figures, the parameters of material 2 are taken to be $\varepsilon_2 = -0.6$ and $\mu_2 = -1$, whereas a few values are given to the volume fraction d_2/D of this medium. The results are presented as a function of dimensionless frequency $\Omega = \omega D/2\pi c$ and dimensionless wavevectors $\kappa_{||} = k_{||}D/2\pi$ and $\kappa_3 = k_3 D/2\pi$. The results presented in Figs. 21.5 and 21.6 are complementary; nevertheless, we have given the band structures of Fig. 21.6 for both TE (right side of Fig. 21.6) and TM (left side of Fig. 21.6) polarizations whereas, for the sake of briefness, the dispersion curves of Fig. 21.5 are presented only for TM modes.

Fig. 21.5 shows the dispersion curves Ω versus k_3 for three different values of $\kappa_{||}$ (namely, $\kappa_{||} = 0.5$ (left column), $\kappa_{||} = 0.7$ (middle column), and $\kappa_{||} = 2.5$ (right column)) and for three values of the filling fraction d_1/D (namely, $d_1/D = 0.3$ (upper row), $d_1/D = 0.43649$ (middle row), and $d_1/D = 0.5$ (lower row)). The value $d_1/D = 0.43649$ of the filling fraction is that for which the average index of refraction in the photonic crystal $\langle n \rangle = (d_1 n_1 + d_2 n_2)/D$ becomes equal to zero; it leads to peculiar behaviors of the dispersion curves (see below) that have also been discussed in [42–44], mainly at $\kappa_{||} = 0$.

In panel (a) of Fig. 21.5, the dispersion curves behave similarly to the usual case of photonic crystals made only of RHM. By changing slightly $\kappa_{||}$, a new dispersion curve starts to emerge around $\Omega = 0$ and $\kappa_3 = 0$ (see the lowest branch in panel (b)), that

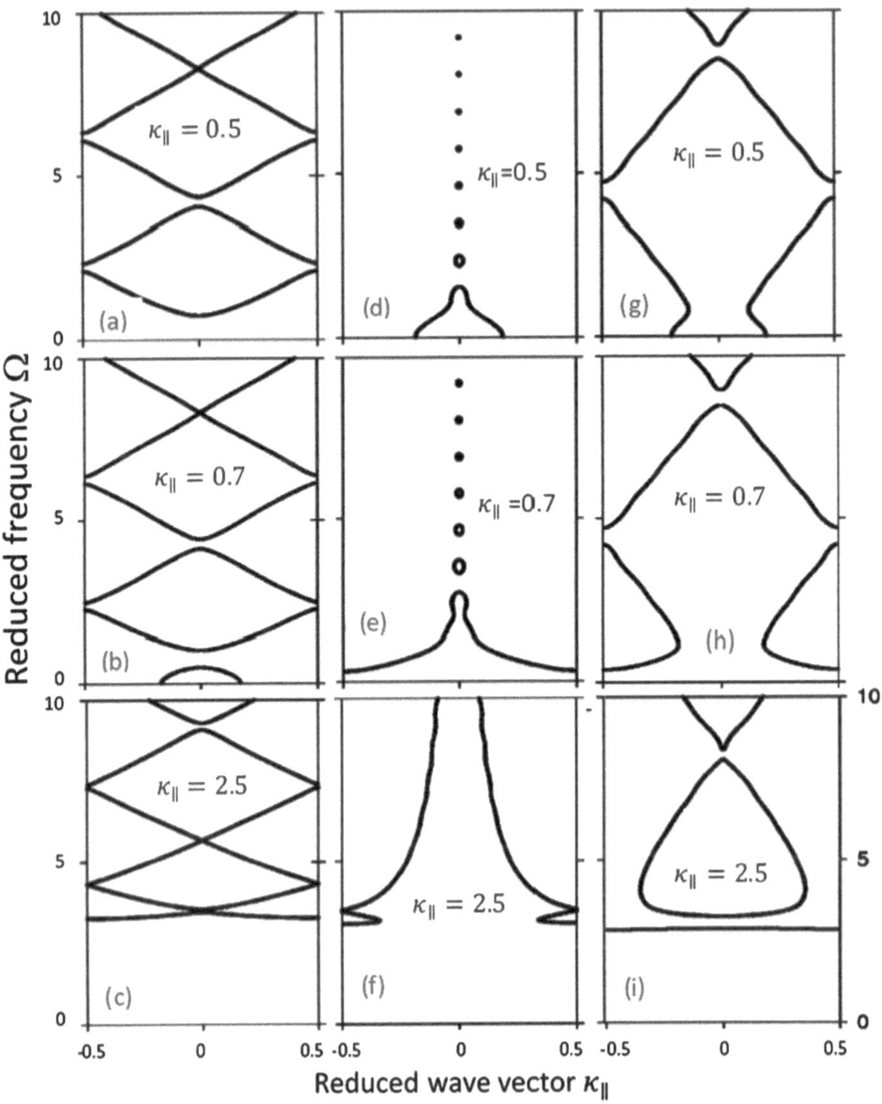

Figure 21.5 Dispersion curves of a vacuum–LHM photonic crystal for different values of the wave vector component $k_{||}$, parallel to the layers, and different volume fraction of vacuum d_1/D. The parameters of LHM are $\varepsilon_2 = -0.6$, $\mu_2 = -1$. The reduced frequency $\Omega = \omega D/2\pi c$ is presented as function of the reduced wave vector $\kappa_3 = k_3 D/2\pi$. The panels in the upper, middle, and lower rows, respectively, correspond to $d_1/D = 0.3$, 0.43649, and 0.5. The panels in the left, middle, and right columns refer, respectively, to the dimensionless wavevector parallel to the layers $\kappa_{||} = k_{||}D/2\pi = 0.5$, 0.7, and 2.5.

extends only over a limited range of the reduced Brillouin zone. Increasing $\kappa_{||}$ further, the latter branch covers the whole range of the Brillouin zone and even (see panel (c)) a cut-off frequency appears. Let us notice that this branch is situated below the light lines of the media constituting the photonic crystal in panel (b) but moves above these lines in panel (c). A global view of these results can be seen in the projected band structure displayed in Fig. 21.6(a) (left side) where the shaded area corresponds to the passbands and the white area to the gaps. The novel behavior resulting from the presence of the LHM is the existence of a passband that falls below the light lines of both media in the photonic crystal and reaches the frequency $\omega = 0$ for nonvanishing values of the wave vector $k_{||}$. This band results from the interaction between the confined modes (see the lowest branch in Fig. 21.3(e)) in different layers of the photonic crystal. This behavior is without analog in the case of usual photonic crystals made of RHM.

Another type of behavior for the dispersion curves is presented in the second row of Fig. 21.5 (see also Fig. 21.6(c)), corresponding to a filling fraction $d_1/D = 0.43649$ such that the average index of refraction in the photonic crystal vanishes, $d_1n_1 + d_2n_2 = 0$. It has been argued [43] that for this choice of the filling fraction, the propagation along the axis of the photonic crystal, i.e., for $\kappa_{||} = 0$, becomes prohibited except at some discrete values of the frequency (see also [42,44]). This can be clearly seen in the projected band structure of Fig. 21.6(c), where the consecutive gaps join together to constitute a very large gap. Now, when $\kappa_{||}$ departs from zero, the gaps become separated by very narrow bands, i.e., the discrete modes transform into narrow bands (see panels (d) and (e) in Fig. 21.5) in which the wavevector κ_3 describes a small region around $\kappa_3 = 0$, whereas the lowest dispersion curve extends over a more or less spread domain of the Brillouin zone. For higher values of κ_3 (see panel (f) of Fig. 21.5 and also Fig. 21.6(c)), the passbands widen and some of them join together to constitute a continuous band. Therefore the occurrence of narrow bands and discrete modes is related to the fact that some of the bandgaps join together and the passbands that separate them close.

A third example of the dispersion curves is presented for the filling fraction $d_1/D = 0.5$ (third row of Fig. 21.5 and Fig. 21.6(d)). At $\kappa_{||} = 0$, these curves are again similar to those of usual photonic crystals. However, by increasing $\kappa_{||}$ (see panels (g) and (h) of Fig. 21.5), the lowest curve covers only a limited part of the Brillouin zone; in particular, for increasing Ω, this curve starts from $\kappa_3 = 0.5$ and returns back to the same wave vector, displaying a zigzag behavior. The width of the corresponding passband that extends over the frequency range $0.4 < \Omega < 4.2$ at $\kappa_{||} = 0.7$ (panel (h)) can be reduced very much when $\kappa_{||}$ increases up to 1.8 where the band approaches the limit of a discrete mode. Panel (i), corresponding to $\kappa_{||} = 2.5$, also shows a zigzag type of behavior, but now for a dispersion curve starting from, and ending at, $\kappa_3 = 0$. A global view of these results is presented in the projected band structure of Fig. 21.6(d) (left side). Here again, one can recognize a passband situated below the light lines of the constituting materials in the photonic crystal. The occurrence of very narrow bands and discrete modes mentioned above can be more clearly seen in the TE band structure of Fig. 21.6(d) (right side). For instance, one can notice that at $\kappa_{||} \approx 1.84$ two bandgaps, delimited by loop shaped curves, join together around the frequency $\Omega \approx 3.52$. Therefore around these values of $\kappa_{||}$ and Ω, there is a very narrow band that separates the two gaps; this passband goes to the limit of a discrete mode at the particular value of $\kappa_{||}$ where the gaps join together and the width of the

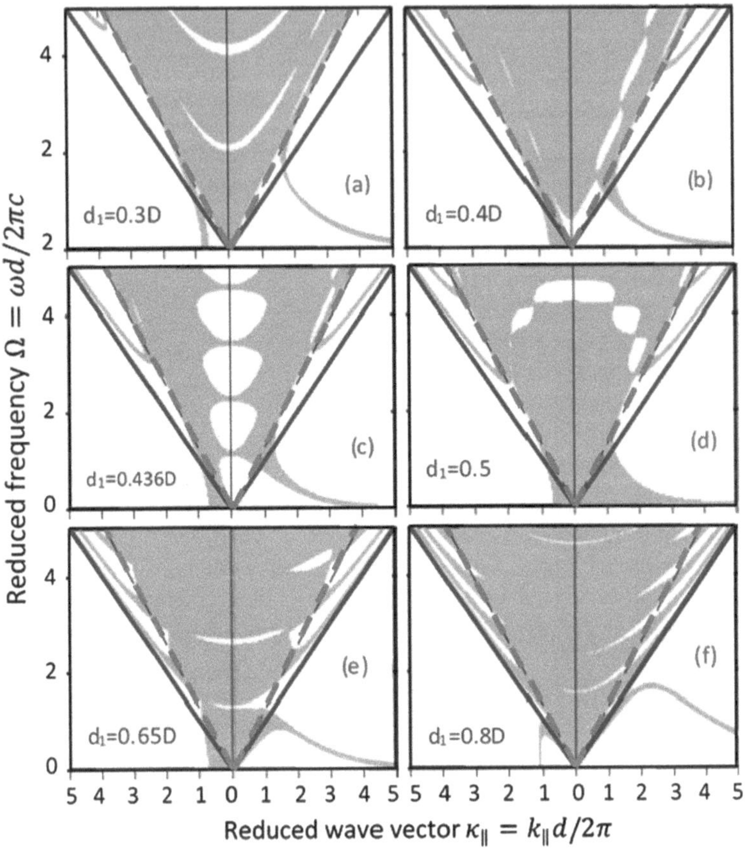

Figure 21.6 Projected photonic band structure of a vacuum–LHM photonic crystal for different values of the volume fraction of vacuum d_1/D. The reduced frequency $\Omega = \omega D/2\pi c$ is presented as a function of the reduced wave vector $\kappa_{||} = k_{||}D/2\pi$. The shaded and white areas respectively correspond to the passbands and to the gaps of the photonic crystal. The left and right sides of the figures respectively give the band structures of the TM and TE modes. The parameters of LHM are $\varepsilon = -0.6$, $\mu = -1$. The filling fraction d_1/D takes the following values in the different panels: (a) 0.3, (b) 0.4, (c) 0.43649 (corresponding to $\langle n \rangle = 0$), (d) 0.5, (e) 0.65, and (f) 0.8. The straight lines show the light lines of vacuum (red full line; dark gray in print version) and of LHM (blue dashed line; mid gray in print version).

passband vanishes. However, looking to the dispersion curves Ω versus κ_3, the narrow band extends only over a very limited range of the Brillouin zone, namely, κ_3 starts and ends at $\kappa_3 = 0.5$ while remaining always in the near vicinity of the Brillouin zone edge.

Now, let us make a few complementary comments about the projected photonic band structures which is displayed in Fig. 21.6 for several values of the filling fraction, namely, $d_1/D = 0.3$ (a), $d_1/D = 0.4$ (b), $d_1/D = 0.43649$ (c), $d_1/D = 0.5$ (d), $d_1/D = 0.65$ (e), and $d_1/D = 0.8$ (f). First, one can notice a number of narrow bands situated between the light lines of vacuum and the LHM. These bands result from the interaction of confined modes of vacuum layer embedded between two LHMs (we remember that in our example the vacuum light line is below the LHM light line). Thus, the number of these bands increases with the filling fraction d_1/D of vacuum layers (from panel (a) to panel (f)). The other point to notice is about the different trends that can be observed around the frequency zero. Referring to TM polarization (left side of the figure), in panel (a) there is a band reaching $\Omega = 0$ over a range of the wave vector $\kappa_{||}$ outside zero. A triangular gap separates this band from the next band that starts at $\kappa_{||} = 0$. Increasing the filling fraction (panel (b)), the former band reaches $\kappa_{||} = 0$ while the latter have moved to a nonvanishing value of the frequency Ω. Increasing further the filling fraction, both bands merge together (panels (d) and (e)). For even higher filling fraction, we find the situation of panel (f) where the lowest band bends downwards to reach $\Omega = 0$ over a range of the wave vector $\kappa_{||}$ outside zero.

The above discussion can be illustrated both analytically or numerically by looking to the trends of the dispersion relation (21.87) either along the Ω-axis or along the $\kappa_{||}$-axis. To have an analytical insight about these behaviors, let us first assume that both $\kappa_{||}$ and Ω are much smaller than 1. Then, we can make a Taylor expansion of Eq. (21.87) that yields

$$\cos(k_3 D) \cong 1 - 4\pi^2\langle\mu\rangle\langle\varepsilon\rangle\Omega^2 + 4\pi^2\left\langle\frac{1}{\varepsilon}\right\rangle\langle\varepsilon\rangle\kappa_{||}^2, \quad \text{for TM modes,} \qquad (21.88)$$

and

$$\cos(k_3 D) \cong 1 - 4\pi^2\langle\mu\rangle\langle\varepsilon\rangle\Omega^2 + 4\pi^2\left\langle\frac{1}{\mu}\right\rangle\langle\mu\rangle\kappa_{||}^2, \quad \text{for TE modes.} \qquad (21.89)$$

In these equations the symbol $\langle A\rangle$ means the average of the quantity A, i.e., $\langle A\rangle = (d_1 A_1 + d_2 A_2)/D$. From the above Taylor expansions, it is easy to analyze the existence of gaps or passbands around Ω and $\kappa_{||} \sim 0$. Referring to TM polarization, for instance, and choosing $\Omega \equiv 0$, one can see from the right-hand side of Eq. (21.88) that a passband (a gap) exists along the $\kappa_{||}$-axis if $\left\langle\frac{1}{\varepsilon}\right\rangle\langle\varepsilon\rangle$ is negative (positive). Similarly, by choosing $\kappa_{||} \equiv 0$, one finds that a passband (gap) occurs along the Ω axis if $\langle\mu\rangle\langle\varepsilon\rangle$ is positive (negative). We also illustrate these behaviors numerically in Fig. 21.7 by choosing either $\kappa_{||} \equiv 0$ or $\Omega \equiv 0$. In Fig. 21.7(a), we present the photonic bands of the photonic crystal at $\kappa_{||} \equiv 0$ as a function of the filling fraction d_1/D. Notice that at $\kappa_{||} \equiv 0$ there is no distinction between TE and TM modes. In this figure, one can see that a frequency cut-off exists for a range of the filling fraction extending from 0.375 (where $\langle\varepsilon\rangle = 0$) to 0.5 (where $\langle\mu\rangle = 0$). In particular, at the filling fraction 0.43649, where $\langle n\rangle$ vanishes, the propagation becomes prohibited except at some discrete frequencies. In Fig. 21.7(b), we choose $\Omega \equiv 0$ and represent the TM photonic bands as a function of $\kappa_{||}$. One can see that the passband, which reaches $\Omega = 0$, extends over a limited range of $\kappa_{||}$ that either includes $\kappa_{||} = 0$ (for $0.375 < d_1/D < 0.625$) or excludes it.

Although the above discussions of Figs. 21.5–21.7 were mainly concentrated on modes of TM polarization, it should be emphasized that the TE photonic band structure displays qualitatively similar results to those of TM modes. The fact that in

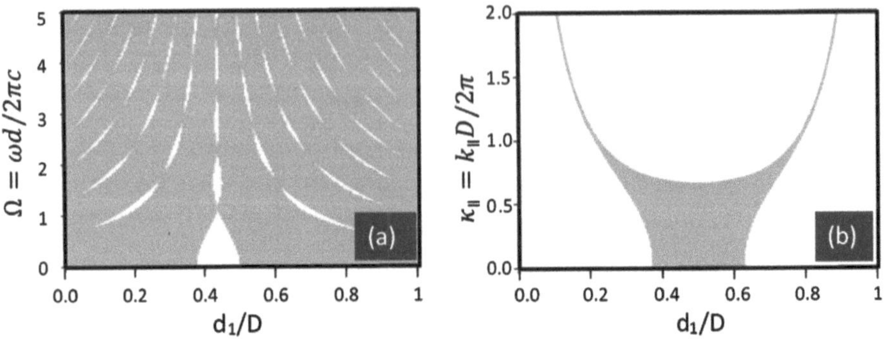

Figure 21.7 (a) Frequencies of photonic bands at $\kappa_{||} \equiv 0$ as a function of the filling fraction d_1/D in a vacuum–LHM photonic crystal. The parameters of LHM are $\varepsilon = -0.6$, $\mu = -1$. The shaded and white areas, respectively, correspond to the passbands and to the gaps of the photonic crystal. (b) Values of $\kappa_{||}$ for which a photonic band exists at $\Omega = 0$ versus d_1/D.

Fig. 21.6 the lowest TE band asymptotically goes to zero, instead of cutting the $\kappa_{||}$ axis, is due to the choice of $\mu_2 = -1$ in our example. This is related to the behavior of the confined modes of a layer (see Fig. 21.3(e)) when $\mu_2 = -\mu_1$.

An interesting and unexpected result which is due to the presence of LHM layers is the existence in Fig. 21.6(e) of an absolute (or omnidirectional) bandgap of TE polarization in the frequency range $1.22 < \Omega < 1.4$. Indeed, this frequency interval is free of TE modes for any value of the wave vector $\kappa_{||}$. Consequently, a wave launched from any substrate with an arbitrary angle of incidence is prohibited from propagation and will be reflected back. The photonic crystal becomes a perfect mirror for the TE modes, or a filter for TM modes, in this frequency range. This situation is without analogue in usual RHM photonic crystals where the photonic band structure never contains an absolute bandgap and the property of omnidirectional reflection requires that the incident light is launched from a substrate in which the index of refraction is relatively lower than those of the materials composing the photonic crystal [75].

Let us mention that the authors of [42] have also reported the possibility of an omnidirectional reflection gap in a lamellar structure containing left-handed media. The case studied in this paper should be similar to that reported in our Fig. 21.6(c): in this figure, a given frequency situated in the lowest gap is omnidirectional reflective for TE polarization, provided the incident light is launched from a substrate in which the index of refraction does not exceed a certain limit. However, in contrast to the case of Fig. 21.6(e) (see also Fig. 21.9(b)), Fig. 21.6(c) does not display any absolute gap, this means that the band structure does not contain any frequency range which remains free of a mode for any value of the wave vector $\kappa_{||}$.

Another illustration of the projected photonic band structures is given in Fig. 21.8 for a photonic crystal in which the parameters of LHM are $\varepsilon_2 = -1.4$ and $\mu_2 = -2$ and the RHM is still vacuum. Although the velocity of light in the LHM is here lower than c, the results are overall qualitatively similar to those of Fig. 21.6. The narrow bands between the light lines of the constituting materials are now originating from the confined modes of the LHM layers embedded between vacuum. One can also

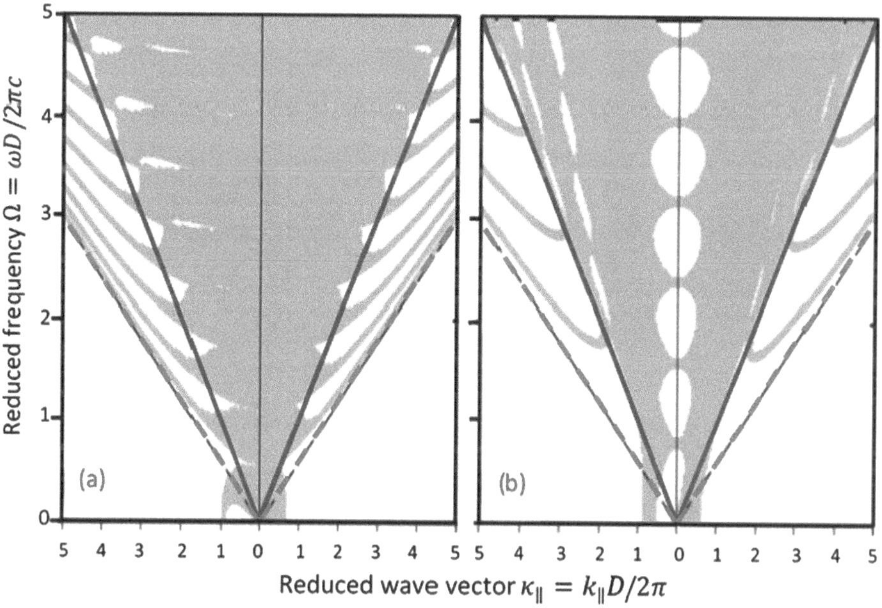

Figure 21.8 Same as in Fig. 21.6 for the following parameters of LHM: $\varepsilon = -1.4$, $\mu = -2$. In the different panels, the layer thicknesses are defined as $d_1 = 0.3D$ (a) and $d_1 = 0.626D$ (b) (corresponding to $\langle n \rangle = 0$).

notice the existence of a band that reaches $\Omega = 0$ for $\kappa_{||}$ different from zero, as well as the peculiar behavior of the band structure at a filling fraction such that $\langle n \rangle = 0$ (panel (b)). An interesting result of Fig. 21.8(a) is the existence of an absolute bandgap of TM polarization that extends in the frequency interval $0.54 < \Omega < 0.61$ while such an omnidirectional gap cannot exist in usual RHM photonic crystals.

In the previous examples of Figs. 21.6 and 21.8, we have shown that for an appropriate choice of the material parameters and their volume fraction, the LHM–RHM can display an omnidirectional gap for either TE or TM polarization. Now, by making a combination in tandem of two such multilayer structures [75–77], it should be possible to realize in a certain frequency range an omnidirectional reflector of light for both polarizations. This operation is sketched in Fig. 21.9. The upper panel is just a magnification of the TM absolute gap of Fig. 21.8(a). The middle panel is a magnification of the TE absolute gap of Fig. 21.6(e); however, the layer thicknesses d_1 and d_2 and the period D of the photonic crystal are multiplied here by a factor of 2.3 in order to obtain the coincidence of the TE and TM gaps. The lower panel in Fig. 21.9 shows the frequency domain in which the propagation is prohibited for both polarizations when the bandgaps of panels (a) and (b) are superimposed. Of course, in a real structure the number of periods in each photonic crystal is finite and a small part of an incident signal will be transmitted. It would be necessary to investigate the decaying of the transmitted wave as a function of the total thicknesses of the multilayer structures, as we did in our previous works dealing with RHM materials [75] or with acoustic waves [76,77].

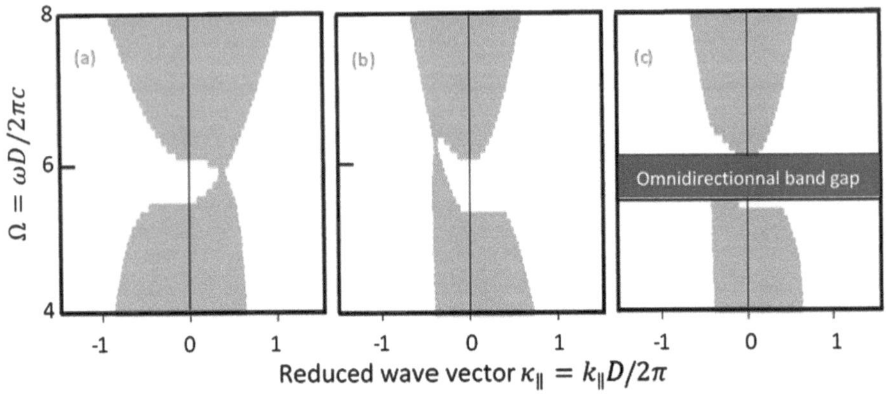

Figure 21.9 (a) Magnification of the projected band structure of Fig. 21.8(a) at the frequency of the TM absolute band gap ($\Omega \approx 0.55$). (b) Magnification of the projected band structure of Fig. 21.6(e) in the vicinity of the TE absolute bandgap. All the thicknesses are scaled by a factor of 2.3 with respect to those of Fig. 21.6(e) in order to make the TE and TM absolute gaps coincide. (c) Overlap of the bandgaps displayed in panels (a) and (b) to show the frequency range in which the propagation will be prohibited through a structure resulting from a combination in tandem of two multilayers.

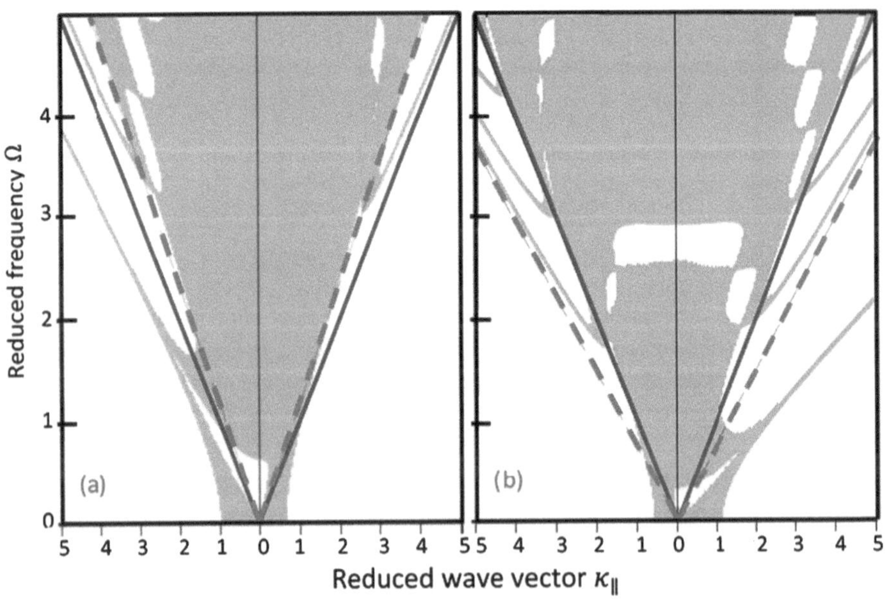

Figure 21.10 Same as in Fig. 21.6 for the following parameters of LHM: (a) $\varepsilon = -1.2$, $\mu = -0.6$ and (b) $\varepsilon = -2$, $\mu = -0.9$. In both cases, the thickness d_2 of the LHM layers is equal to $0.5D$.

A third example of the projected photonic band structure is given in Fig. 21.10 with the following parameters of the RHM: (a) $\varepsilon_2 = -1.2$, $\mu_2 = -0.6$ and (b) $\varepsilon_2 = -2$, $\mu_2 = -0.9$. The filling fraction is taken to be 0.5. The novelty in this case with respect to the previous illustrations is the existence of a narrow band that originates from

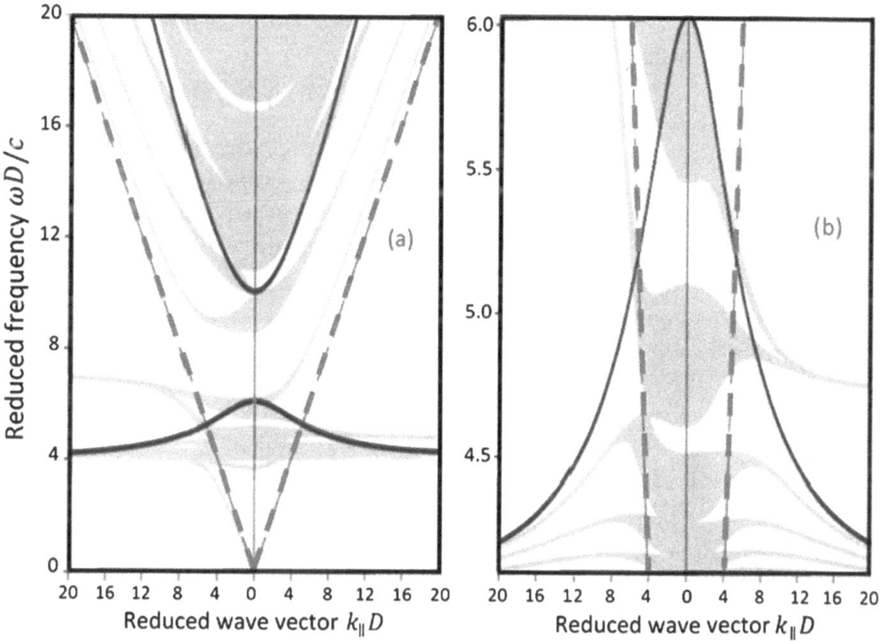

Figure 21.11 (a) Same as in Fig. 21.5 when ε_2 and μ_2 are defined by Eq. (21.90). (b) Magnification of Fig. 21.11(a) in the frequency range $\omega D/c \sim 4$ to 6 where both ε_2 and μ_2 are negative. The straight blue dashed line (mid gray in print version) is the vacuum light line. The black solid lines, defined by the equation $\alpha_2 = 0$, separate the regions of propagative and evanescent waves in LHM.

the interface mode at the RHM–LHM boundary. This band, which is either of TM (panel (a)) or TE (panel (b)) polarization, is situated below the light lines of both materials and becomes similar to a straight line in the limit of high $\kappa_{||}$. For small $\kappa_{||}$, it widens and can divide into two different bands. From the discussion of Section 21.4, let us remember that it is not possible to obtain the interface band simultaneously for TE and TM polarizations.

Finally, in Fig. 21.11 we illustrate the photonic band structure for a photonic crystal in which the parameters ε_2 and μ_2 of the LHM are frequency-dependent and take the following forms [43,78,79]:

$$\varepsilon_2(\omega) = 1 - \frac{\omega_p^2}{\omega^2},$$
$$\mu_2(\omega) = 1 - \frac{F\omega^2}{\omega^2 - \omega_0^2}. \tag{21.90}$$

In Fig. 21.11, we have chosen $\omega_p D/c = 10$, $\omega_0 D/c = 4$, $F = 0.56$, and $d_1 = d_2 = D/2$. The heavy solid lines are obtained by putting $\alpha_2 = 0$ in LHM, i.e., these curves separate the regions of propagating and evanescent waves in LHM. The evanescent waves appear in the region between the two curves. The blue dashed straight line (mid gray in print version) is the vacuum light line. The band structure is magnified in the frequency range $4 \leqslant \omega D/c \leqslant 6$ (Fig. 21.11(b)) where both ε_2 and μ_2 are negative. The following points should be emphasized. For $k_{||}D$ going to infinity, there are two bands of TE polarization around $\omega D/c \simeq 4.75$ (see Fig. 21.11(b)) which are separated from each other by a very small gap (not visible at the scale of the figure). These bands originate from the interface mode at the LHM–vacuum boundary; they broaden when $k_{||}D$ decreases due to the interaction between the different interface modes, before penetrating in the regions where the waves become propagative in vacuum and/or in LHM. One can also notice the existence of several branches of both TE and TM polarizations in the frequency range $\omega D/c \sim 4$ to 4.2 which are essentially slab modes of the LHM layers, because they fall in the regions where the modes are propagating in LHM but evanescent in vacuum. Similarly, the narrow bands that become asymptotic to the light line of vacuum when $k_{||}D$ increases are essentially slab modes of the vacuum layers. Finally, one can recognize around $\omega D/c \simeq 4.5$ (see Fig. 21.11(b)) the existence of an omnidirectional bandgap for TE polarization of the electromagnetic field. These general trends parallel those mentioned in the previous discussions. It is also interesting to mention that an omnidirectional bandgap occurs in the TM band structure around $\omega D/c \simeq 7$ (see Fig. 21.11(a)); however, in this frequency range only the dielectric permittivity ε_2 is negative while the magnetic permittivity μ_2 is positive.

21.7 Summary and perspectives

In this chapter, the fundamental properties of the left-handed materials are examined from a systematic electromagnetic wave theory viewpoint. We reviewed recent works in the left-handed materials based on layered media. We have also reviewed the interface response theory of electromagnetism for any composite medium characterized by the permittivity $\varepsilon(\omega)$ and the permeability $\mu(\omega)$. Its main general results given by Eqs. (21.24), (21.25), and (21.26) take into account automatically all the boundary conditions of the electromagnetic fields. We have applied this theory to the determination of the complete response functions of the infinite homogeneous medium and the dispersion relation for the interface between two media, of one slab sandwiched between two different semiinfinite media, and for layered photonic crystal.

We have also reported a detailed study of the photonic band structure of one-dimensional photonic crystals composed of alternate layers of right- and left-handed materials. The different possible behaviors have been illustrated by varying the physical parameters and the volume fraction of LHM materials. In particular, the formation of the bands below or between the light lines of the constituting materials can be understood on the basis of the localized interface modes at an LHM–RHM boundary and the confined modes of an LHM (RHM) layer sandwiched between two semiinfinite RHMs (LHMs). While the RHM–LHM layer can, in principle, support both TE and TM interface modes, only one of these modes can exist. Thus, in the photonic crystal band structure, there can be at most one band originating from the localized interface mode. Between the light lines of the constituting materials, there are also narrow bands originating from the confined modes of a layer, their number increasing with the volume fraction of the corresponding material in the photonic crystal. Among other peculiar behaviors associated with the presence of the LHM with fixed parameters ε_2 and μ_2, one can notice the behavior of the band structure in the vicinity of $\omega = 0$ and, in particular, the existence of a band that lies below the light lines of the constituting materials and can reach a vanishing frequency at nonvanishing values of the wave vector $k_{||}$. Also, the dispersion curves ω versus k_3, calculated for a given value of $k_{||}$, do not behave always monotonically as in usual photonic crystals, but may describe only a limited range of the Brillouin zone and display a zigzag behavior. For specific values of the parameters, such bands may become even very narrow and appear as a discrete mode ω at $k_3 = 0$ or π/D. This happens, in particular, at $k_{||} \cong 0$ when the average index of refraction in the photonic crystal is equal to zero, but such a behavior is not limited only to the case $\langle n \rangle = 0$.

Finally, a new phenomenon associated with the presence of the LHM is the possibility of an absolute bandgap, of either TE or TM polarization, when the material parameters are chosen appropriately. This enables us to propose an application of our structure for realizing an omnidirectional optical mirror that prevents propagation of optical waves of TE or TM polarization for a given frequency range. Combination in tandem of two such multilayers can yield an omnidirectional reflector of light for both polarizations.

Acknowledgments

The authors would like to thank Driss Bria, Jean Pol Vigneron, and Abelkarim Nougaoui for their fruitful contributions to the original paper from which this chapter is inspired.

References

[1] V.G. Veselago, The electrodynamics of substances with simultaneously negative values of permittivity and permeability, Sov. Phys. Usp. 10 (1968) 509.
[2] J.B. Pendry, Negative refraction makes a perfect lens, Phys. Rev. Lett. 85 (2000) 3966.
[3] J.B. Pendry, in: C.M. Soukoulis (Ed.), Photonic Crystals and Light Localization in the 21st Century, in: NATO Science, Series C, vol. 563, Kluwer, Dordrecht, 2002, p. 329.
[4] D.R. Smith, W.J. Padilla, D.C. Vier, R. Shelby, S.C. Nemat-Nasser, N. Kroll, S. Schultz, in: C.M. Soukoulis (Ed.), Photonic Crystals and Light Localization in the 21st Century, in: NATO Science, Series C, vol. 563, Kluwer, Dordrecht, 2002, p. 351.
[5] E. Ozbay, G. Ozkan, K. Aydin, Left-handed metamaterials — a review, in: A.V. Narlikar, Y.Y. Fu (Eds.), Oxford Handbook of Nanoscience and Technology: Basic Aspects, Oxford Handbooks Online, 2010.
[6] W. Cai, V. Shalaev, Optical Metamaterials: Fundamentals and Applications, Springer Science+Business Media, LLC, 2010.
[7] R.A. Shelby, D.R. Smith, S. Schultz, Experimental verification of a negative index of refraction, Science 292 (2001) 77.
[8] R.A. Shelby, D.R. Smith, S.C. Nemat-Nasser, S. Schultz, Microwave transmission through a two-dimensional, isotropic, left-handed metamaterial, Appl. Phys. Lett. 78 (2001) 489.

[9] R. Marqués, J. Martel, F. Mesa, F. Medina, Left-handed-media simulation and transmission of EM waves in subwavelength split-ring-resonator-loaded metallic waveguides, Phys. Rev. Lett. 89 (2002) 183901.
[10] P. Gay-Balmaz, O.J.F. Martin, Electromagnetic resonances in individual and coupled split-ring resonators, J. Appl. Phys. 92 (2002) 2929.
[11] J.B. Pendry, A.J. Holden, D.J. Robbins, W.J. Steweat, Magnetism from conductors and enhanced nonlinear phenomena, IEEE Trans. Microw. Theory Tech. 47 (1999) 2075.
[12] J.M. Pitarke, F.J. Garcia-Vidal, J.B. Pendry, Effective electronic response of a system of metallic cylinders, Phys. Rev. B 57 (15) (1998) 261.
[13] S. Foteinopoulou, E.N. Economou, C.M. Soukoulis, Refraction in media with a negative refractive index, Phys. Rev. Lett. 90 (2003) 107402.
[14] A.A. Houck, J.B. Brock, I.L. Chuang, Experimental observations of a left-handed material that obeys Snell's law, Phys. Rev. Lett. 90 (2003) 137401.
[15] C.G. Parazzoli, R.B. Greegor, K. Li, B.E.C. Koltenbah, M. Tanielian, Experimental verification and simulation of negative index of refraction using Snell's law, Phys. Rev. Lett. 90 (2003) 107401.
[16] R.W. Ziolkowski, E. Heyman, Wave propagation in media having negative permittivity and permeability, Phys. Rev. E 64 (2001) 056625.
[17] D.R. Smith, D. Schurig, J.B. Pendry, Negative refraction of modulated electromagnetic waves, Appl. Phys. Lett. 81 (2002) 2713.
[18] M.W. Feise, P.J. Bevelacqua, J.B. Schneider, Effects of surface waves on the behavior of perfect lenses, Phys. Rev. B 66 (2002) 035113.
[19] N. Fang, X. Zhang, Imaging properties of a metamaterial superlens, Appl. Phys. Lett. 82 (2003) 161.
[20] Z.M. Zhang, C.J. Fu, Unusual photon tunneling in the presence of a layer with a negative refractive index, Appl. Phys. Lett. 80 (2002) 1097.
[21] S. Enoch, G. Tayeb, P. Sabouroux, N. Guérin, P. Vincent, A metamaterial for directive emission, Phys. Rev. Lett. 89 (2002) 213902.
[22] F. Goos, H. Hänchen, A new and fundamental experiment on total reflection, Ann. Phys. (Leipz.) 1 (1947) 333.
[23] K.V. Artmann, Berechnung der Seitenversetzung des totalreflektierten Strahles, Ann. Phys. (Leipz.) 2 (1948) 87.
[24] W.J. Wild, C.L. Giles, Goos–Hänchen shifts from absorbing media, Phys. Rev. A 25 (1982) 2099.
[25] H.M. Lai, S.W. Chan, Large and negative Goos–Hänchen shift near the Brewster dip on reflection from weakly absorbing media, Opt. Lett. 27 (2002) 680.
[26] C.-F. Li, Negative lateral shift of a light beam transmitted through a dielectric slab and interaction of boundary effects, Phys. Rev. Lett. 91 (2003) 133903.
[27] L.-G. Wang, H. Chen, S.-Y. Zhu, Large negative Goos–Hänchen shift from a weakly absorbing dielectric slab, Opt. Lett. 30 (2005) 2936.
[28] J.J. Cowan, B. Anicin, Longitudinal and transverse displacements of a bounded microwave beam at total internal reflection, J. Opt. Soc. Am. 67 (1977) 1307.
[29] F. Bretenaker, A. Le Floch, L. Dutriaux, Direct measurement of the optical Goos–Hänchen effect in lasers, Phys. Rev. Lett. 68 (1992) 931.
[30] E. Pfeghaar, A. Marseille, A. Wei, Quantitative investigation of the effect of resonant absorbers on the Goos–Hänchen shift, Phys. Rev. Lett. 70 (1993) 2281.
[31] B.M. Jost, A.-A.R. Al-Rashed, B.E.A. Saleh, Observation of the Goos–Hänchen effect in a phase-conjugate mirror, Phys. Rev. Lett. 81 (1998) 2233.
[32] H. Gilles, S. Girard, J. Hamel, Simple technique for measuring Goos–Hänchen effect with polarization modulation and a position-sensitive detector, Opt. Lett. 27 (2002) 1421.

[33] D.-K. Qing, G. Chen, Goos–Hänchen shifts at the interfaces between left- and right-handed media, Opt. Lett. 29 (2004) 872.
[34] P.R. Berman, Goos–Hänchen shift in negatively refractive media, Phys. Rev. E 66 (2002) 067603.
[35] S.A. Ramakrishna, Physics of negative refractive index materials, Rep. Prog. Phys. 68 (2005) 449.
[36] I.V. Shadrivov, A.A. Zharov, Y.S. Kivshar, Giant Goos–Hänchen effect at the reflection from left-handed metamaterials, Appl. Phys. Lett. 83 (2003) 2713.
[37] L.G. Wang, S.-Y. Zhu, Large positive and negative Goos–Hänchen shifts from a weakly absorbing left-handed slab, J. Appl. Phys. 98 (2005) 043522.
[38] H. Chen, B.-I. Wu, J.A. Kong, Review of electromagnetic theory in left handed materials, J. Electromagn. Waves Appl. 20 (2006) 2137.
[39] Z.P. Wang, C. Wang, Z.H. Zhang, Goos–Hänchen shift of the uniaxially anisotropic left-handed material film with an arbitrary angle between the optical axis and the interface, Opt. Commun. 281 (2008) 3019.
[40] W. Ding, L. Chen, C.-H. Liang, Numerical study of Goos–Hänchen shift on the surface of anisotropic left-handed materials, Prog. Electromagn. Res. 2 (2008) 151.
[41] T.M. Grzegorczyk, X. Chen, J. Pacheco Jr., J. Chen, B.-I. Wu, J.A. Kong, Reflection coefficients and Goos–Hänchen shifts in anisotropic and bianisotropic left handed materials, Prog. Electromagn. Res. 51 (2005) 83.
[42] I.S. Nefedov, S.A. Tretyakov, Photonic band gap structure containing metamaterial with negative permittivity and permeability, Phys. Rev. E 66 (2002) 036611.
[43] J. Li, L. Zhou, C.T. Chan, P. Sheng, Photonic band gap from a stack of positive and negative index materials, Phys. Rev. Lett. 90 (2003) 083901.
[44] L. Wu, S. He, L. Shen, Band structure for a one-dimensional photonic crystal containing left-handed materials, Phys. Rev. B 67 (2003) 235103.
[45] T. Pan, G. Xu, T. Zang, L. Gao, Goos–Hänchen shift in one-dimensional photonic crystals containing uniaxial indefinite medium, Phys. Status Solidi B 246 (2009) 1088.
[46] A. Namdar, I. Shadrivov, Y. Kivshar, Backward Tamm states in left-handed metamaterials, Appl. Phys. Lett. 89 (2006).
[47] J. Barvestani, M. Kalafi, A. Soltani-vala, A. Namdar, Backward surface electromagnetic waves in semi-infinite one-dimensional photonic crystals containing left-handed materials, Phys. Rev. A 77 (2008).
[48] F. Villa-Villa, J.A. Gaspar-Armenta, A. Mendoza-Suarez, Surface modes in one dimensional photonic crystals that include left handed materials, J. Electromagn. Waves Appl. 21 (2007) 485.
[49] Y. Xiang, X. Dai, S. Wen, Omnidirectional gaps and one-dimensional photonic crystals containing indefinite metamaterials, J. Opt. Soc. Am. B 24 (2007) 2033.
[50] A. Rostami, M. Khezri, S. Golmohammadi, Investigation of optical properties of Tamm states in higher bands of multilayer stacks including right and left handed materials, Optik 123 (2012) 847.
[51] A.A. Shokri, R. Jamshidi, Merging of defect modes in a superlattice of one-dimensional metamaterials photonic crystals, AIP Adv. 9 (2019) 055318.
[52] M. Kohmoto, B. Sutherland, K. Iguchi, Localization of optics: quasiperiodic media, Phys. Rev. Lett. 58 (1987) 2436.
[53] W. Gellermann, M. Kohmoto, B. Sutherland, P.C. Taylor, Localization of light waves in Fibonacci dielectric multilayers, Phys. Rev. Lett. 72 (1994) 633.
[54] S. Golmohammadi, M.K. Moravvej-Farshi, A. Rostami, A. Zarifkar, Narrowband DWDM filters based on Fibonacci-class quasi-periodic structures, Opt. Express 15 (2007) 10520.

[55] R. Pelster, V. Gasparian, G. Nimtz, Propagation of plane waves and of waveguide modes in quasiperiodic dielectric heterostructures, Phys. Rev. E 55 (1997) 7645.
[56] A.G. Barriuso, J.J. Monzón, L.L. Sánchez-Soto, A. Felipe, Comparing omnidirectional reflection from periodic and quasiperiodic one-dimensional photonic crystals, Opt. Express 13 (2005) 3913.
[57] K. Ben Abdelaziz, J. Zaghdoudi, M. Kanzari, B. Rezig, A broad omnidirectional reflection band obtained from deformed Fibonacci quasi-periodic one dimensional photonic crystals, J. Opt. A, Pure Appl. Opt. 7 (2005) 544.
[58] D. Lusk, I. Abdulhalim, F. Placido, Omnidirectional reflection from Fibonacci quasi-periodic one-dimensional photonic crystal, Opt. Commun. 198 (2001) 273.
[59] D.R. Smith, J. Pendry, M. Wiltshire, Metamaterials and negative refractive index, Science 305 (2004) 788.
[60] D.R. Smith, N. Kroll, Negative refractive index in left-handed materials, Phys. Rev. Lett. 85 (2000) 2933.
[61] J. Pacheco Jr., T.M. Grzegorczyk, B.-I. Wu, Y. Zhang, J.A. Kong, Power propagation in homogeneous isotropic frequency-dispersive left-handed media, Phys. Rev. Lett. 89 (2002) 257401.
[62] T. Koschny, M. Kafesaki, E.N. Economou, C.M. Soukolis, Effective medium theory of left-handed materials, Phys. Rev. Lett. 93 (2004) 107402.
[63] A.N. Grigorenko, A.K. Geim, H.F. Gleeson, Y. Zhang, A.A. Firsov, I.Y. Khrushchev, J. Petrovic, Nanofabricated media with negative permeability at visible frequencies, Nature (London) 438 (2005) 335.
[64] V.M. Shalaev, Optical negative-index metamaterials, Nat. Photonics 1 (2007) 41.
[65] J. Li, D. Zhao, A. Liu, Zero-n photonic band gap in a quasiperiodic stacking of positive and negative refractive index materials, Phys. Lett. A 332 (2004) 461.
[66] F.F. de Medeiros, E.L. Albuquerque, M.S. Vasconcelos, Optical transmission spectra in quasiperiodic multilayered photonic structure, J. Phys. Condens. Matter 18 (2006) 8737.
[67] W.J. Hsueh, C.T. Chen, C.H. Chen, Omnidirectional band gap in Fibonacci photonic crystals with metamaterials using a band-edge formalism, Phys. Rev. A 78 (2008) 013836.
[68] C.A.A. Araujo, M.S. Vasconcelosc, P.W. Mauriz, E.L. Albuquerque, Omnidirectional band gaps in quasiperiodic photonic crystals in the THz region, Opt. Mater. 35 (2012) 18.
[69] Jose Jolly, Third-harmonic generation in resonant-metamaterial-based photonic-band-gap materials, Phys. Rev. A 88 (2013) 043849.
[70] Jose Jolly, Frequency selective bistable switching in metamaterial based photonic bandgap medium, Opt. Commun. 328 (2014) 116.
[71] D. Bria, B. Djafari-Rouhani, A. Akjouj, L. Dobrzynski, J.P. Vigneron, E.H. El Boudouti, A. Nougaoui, Band structure and omnidirectional photonic band gap in lamellar structures with left-handed materials, Phys. Rev. E 69 (2004) 066613.
[72] J.N. Fink, S. Winn, S. Fan, C. Chen, J. Michel, J.D. Joannopoulos, E.L. Thomas, A dielectric omnidirectional reflector, Science 282 (1998) 1679.
[73] J.P. Dowling, Mirror on the wall: you're omnidirectional after all?, Science 282 (1998) 1841.
[74] D.N. Chigrin, A.V. Lavrinenko, D.A. Yarotsky, S.V. Gaponenko, Observation of total omnidirectional reflection from a one-dimensional dielectric lattice, Appl. Phys. A, Mater. Sci. Process. 68 (1999) 25.
[75] D. Bria, B. Djafari-Rouhani, E.H. El Boudouti, A. Mir, A. Akjouj, A. Nougaoui, Omnidirectional optical mirror in a cladded-superlattice structure, J. Appl. Phys. 91 (2002) 2569.

[76] D. Bria, B. Djafari-Rouhani, A. Bousfia, E.H. El Boudouti, A. Nougaoui, Absolute acoustic band gap in coupled multilayer structures, Europhys. Lett. 55 (2001) 841.
[77] D. Bria, B. Djafari-Rouhani, Omnidirectional elastic band gap in finite lamellar structures, Phys. Rev. E 66 (2002) 056609.
[78] R. Ruppin, Surface polaritons of a left-handed medium, Phys. Lett. A 227 (2000) 61.
[79] R. Ruppin, Surface polaritons of a left-handed material slab, J. Phys. Condens. Matter 13 (2001) 1811.
[80] L. Dobrzynski, Interface response theory of electromagnetism in composite dielectric materials, Surf. Sci. 180 (1987) 505.
[81] L. Dobrzynski, Interface response theory of discrete composite systems, Surf. Sci. Rep. 6 (1986) 118.
[82] L. Dobrzynski, Interface response and scattering matrix theories for composite systems, Surf. Sci. 175 (1986) 1.
[83] L. Dobrzynski, Interface response theory of continuous composite materials, Surf. Sci. 180 (1987) 489.
[84] L. Dobrzynski, Interface response theory of continuous composite systems, Surf. Sci. Rep. 11 (1990) 139.
[85] L. Dobrzynski, A. Akjouj, E.H. El Boudouti, et al., Interface Response theory, in: Phononics, in: Interface Transmission Tutorial Book Series, Elsevier, 2017, p. 1.
[86] M.G. Cottam, A.A. Maradudin, in: Surface Excitations, in: Modern Problems in Condensed Matter Sciences, vol. 9, North-Holland, Amsterdam, 1986.
[87] M.L. Bah, A. Akjouj, L. Dobrzynski, Response functions in layered dielectric media, Surf. Sci. Rep. 16 (1992) 95.

Superluminal, negative delay times and selective transmission in isotropic–anisotropic layered media

22

El Houssaine El Boudouti[a], *Bahram Djafari-Rouhani*[b], *Abdellatif Akjouj*[b], *and Leonard Dobrzyński*[b]
[a]Laboratoire de Physique de la Matière et de Rayonnement, Département de Physique, Faculté des Sciences, Université Mohammed I, Oujda, Morocco, [b]Institute of Electronic, Microelectronics and Nanotechnology (IEMN), UMR CNRS 8520, Department of Physics, Faculty of Sciences and Technologies, Lille University, Villeneuve d'Ascq Cedex, France

22.1 Introduction

Photonic crystals (PCs) are periodic materials characterized by the modulation of the refractive index [1]. The main property of these structures is the existence of bandgaps where photons are prohibited from propagation for certain directions and/or polarizations in much the same way as electrons in periodic potentials [2]. These structures can be arranged along one-dimensional (1D), 2D, and 3D directions [2,3]. Several types of PCs have been proposed for various applications such as: optical insulators, waveguides, demultiplexers, microstructured fibers, and sensors [2]. In addition to these properties, it was proved that PCs can show superluminal phenomena [4] where the group velocity (v_g) of light can become greater than the speed of light in vacuum ($v_g > c$), infinite ($v_g = \pm\infty$), or even negative ($v_g < 0$). These behaviors can occur in anomalous dispersion regions of electrical [5,6], photonic [7–10], or even acoustic [11–13] bandgap materials. Recently, slow and fast light technologies have received a great deal of attention because of their potential applications in optical information processing, with the realization of temporal storage components for the synchronization of information packets [14], high speed optical switches [15], and sensing precision [16].

Such anomalous behavior of electromagnetic pulse has been predicted in nonperiodic linear dispersive media [17–21]. It was shown that this phenomenon is easily achieved by exploiting the dispersive properties of absorptive or active media [17,18]. Wang and Zhu [19] have demonstrated theoretically a large negative group delay for a reflected pulse near a resonance from a weakly absorbing dielectric slab. Furthermore, the same authors have predicted simultaneous superluminal reflected and transmitted pulses for a dielectric slab system doped with passive (absorptive) two-level atoms or with active (gain) three level atoms [20]. A similar study has been performed in a degenerate two-level atomic system by Jiang et al. [21]. Some other works have paid

Photonics. https://doi.org/10.1016/B978-0-12-819388-4.00033-2

special attention to the superluminal or even negative group delay through simple structures operating in the microwave domain such as Mach–Zehnder, multiple-beam interferometers [22–24], and double stubs grafted at one [25] or two [26] sites on a waveguide. The anomalous behavior has been also observed for pulses propagating through a Fabry–Perot cavity with silver mirror [27], erbium-doped optical fiber [28], as well as Fibonacci photonic loop structure [29].

In addition to the aperiodic photonic structures mentioned above, periodic systems have been proposed also to explore the possibility of superluminal or even negative group delays. Longhi [9] has shown superluminal pulse reflection in asymmetric 1D photonic bandgaps, and he first observed with his coworkers [8] a reflected optical pulse from a double Lorentzian fiber Bragg grating. It was demonstrated that the negative group velocity is due to the interference of two resonance modes of this structure [8]. Contrary to the anomalous behavior of the reflected pulses from linear and passive 1D periodic structures, negative group delays of the transmitted pulses have been reported for a coaxial photonic crystal [7] made of two coaxial cables with different impedances. This finding has stimulated a controversial debate (see [30,31]) since negative group velocities should not be possible in a periodic structure with or without loss as it was demonstrated by Poirier et al. [30] (see also [32]). However, several works [10,29,33,34] have clearly demonstrated that negative group delays may occur in coaxial photonic crystals based on asymmetric serial loops [10,29], anisotropic semiconductor metamaterial structures [33], as well as in 1D active dielectric photonic crystal [34]. Let us mention that the observation of negative delay times and therefore negative group velocities in a photonic crystal is due to the existence of absorption in these systems [7,10,31,33], as well as to the presence of amplification in 1D dielectric PC [34].

Owing to the importance of the control of the group delay in slow and fast light technologies, the modulation of this parameter in the above works varies from one system to another. For example, in Fabry–Perot cavities (with silver mirror) containing absorbing atoms, the pulse propagation can be controlled from subluminal to superluminal by adjusting atomic density [27] or controlling the slab's thickness in a dielectric slab system [20], or simply by changing solely the power of the backward coupling field in a degenerate two-level atomic system [21]. The crossover from slow to fast propagation is controlled by the system's characteristics (attenuation and/or length difference between adjacent arms) like in Mach–Zehnder and multiple-beam interferometers [22–24], whereas in optical fibers the slow/fast propagation can be achieved by changing the cross-gain modulation phase [28]. It is worth mentioning that the experimental works cited above have clearly shown the evidence of anomalous dispersion phenomenon in optical and lower frequency ranges like microwave [8], radio frequency [7,10,23,26,31,35], and even acoustic waveguides [11–13].

In this chapter we provide a theoretical evidence of superluminal and negative delay times of light transmission in linear and passive PC formed by alternating isotropic and anisotropic media [36]. This layered structure proposed here exhibits such phenomena without introducing any loss or gain in the system. Our goal is to show that this kind of structures may exhibit abnormal group delays for both P and S polarized transmitted waves. Tunneling and superluminal regimes appear at some frequencies around the transmission minima in the allowed bands and inside the photonic bandgaps (PBGs). We show that the possibility of the transition from subluminal to apparently superluminal and vice versa can be achieved by adjusting the orientations of the anisotropic layers of the structure. To investigate the properties of the structure when the group delay approaches zero or even becomes negative, we evaluate in addition to the transmission coefficient, the behavior of the phase and the delay time of the transmitted waves. Also, we highlight the impact of the orientation of the principal axes of the anisotropic layer with respect to the laboratory axes, on the tunneling and superluminal regimes. The effect of the number of the unit cells of the PC on the magnitude and the sign of the group delays for both *P* and *S* polarized transmitted waves is also investigated. The theoretical method adopted for the analysis of the 1D multilayered PC is the Green's function approach, which permits deriving the transmission and reflection coefficients, densities of states, as well as delay times of electromagnetic modes propagating through the structure. This approach enables obtaining directly the density of states which plays an essential role in determining many physical properties of a system. The theoretical predictions are illustrated with numerical simulations for a finite periodic structure consisting of alternating layers of antimony sulpho-iodide (SbSI) and air. Because of its interesting linear and nonlinear optical properties in paraelectric and ferroelectric phases, the SbSI material is considered suitable for applications as an optoelectronic device in visible and infrared regions [37]. The advantage of using the material selected with respect to an absorptive or an active media lies in the fact that the existence of negative delay times is an intrinsic property of the anisotropic layers of the structure without a need to dope the structure with an absorptive or an active media. Furthermore, the anisotropic structure exhibits abnormal group delays for both *P* and *S* polarized transmitted waves. This is a characteristic of anisotropic media which is without analogue in absorptive or active media as well as in pure isotropic layers.

Another subject of interest in anisotropic layered media concerns the effect of an anisotropic defect layer on the transmission properties in an isotropic PC. Indeed, PCs with a defect layer are more desired than the perfect ones for several applications. The physical properties of the localized or defect modes within the photonic bandgaps (PBGs) have been increasingly studied. According to such combination, several materials are used as the defect layer in the design of the photonic devices. Several works have investigated the optical and microwave properties of defect modes in 1D isotropic dielectric PCs. Liu et al. [38] have studied a 1D PC structure consisting of alternate layers of Ta_2O_5 and MgF_2 films with a defect layer in the visible region. They have shown that the number and the frequencies of the defect modes can be controlled in the visible range by adjusting the thickness of the defect layer. This structure can be useful in the design of blue–green color filters. The green and red color filters were studied by Xiang et al. [39].

Superconductor material and liquid crystal are among potential materials to serve as defect layers in tunable photonic devices, as their optical properties can be controlled by external excitation. Indeed, Wu and Gao [40] have studied the effect of the temperature on the defect modes in a 1D dielectric PC heterostructure with a superconducting defect. They have shown that this structure has very high temperature sensitivity. Dadoenkova et al. [41] have investigated PBG spectra of a 1D dielectric PC with a complex defect layer, consisting of ultrathin superconducting and dielectric sublayers. Some other research works [42–45] have investigated the effect of an electric field on the defect modes induced by a liquid crystal defect layer (anisotropic medium) or by an Ag/$LiNbO_3$/Ag sandwich structure into 1D PC [44,45]. Tang et al. [46] have explored the influence of the anisotropy of the single-negative material on photonic bandgaps and tunneling modes of 1D PCs containing dispersive anisotropic single-negative metamaterials. Penninck et al. [47] have studied the emission of light in uniaxially anisotropic thin film devices. This emission is performed through a dipole antenna inserted inside a thin isotropic film sandwiched between two anisotropic layers. The whole structure is placed between glass and air.

One-dimensional birefringent photonic crystal based on anisotropic porous silicon layers has been designed to control simultaneous enhancement of second and third harmonic generation [48]. The structure consists of 12 quarter-wavelength pairs of layers separated by a half-wavelength cavity spacer. Also, chiral PCs with defects have been shown of particular interest due to their rich optical properties. Various types of defects have been considered, such as thin isotropic [49–51] or anisotropic [52–54] defect layers inserted between two chiral PCs. The properties of the defect modes induced by anisotropic dielectric layer in 1D birefringent photonic crystal have been investigated theoretically by Ouchani et al. [55].

In all the above works, only the reflectance and/or the transmittance of defect modes have been explored. In this chapter, we are interested in some optical transmission properties of the defect modes, such as the transmission delay times and the behavior of total density of states (DOS) in the visible and infrared regions of a 1D PC constituted by a combination of a finite isotropic superlattice (SL) and a dielectric anisotropic defect layer. In particular, we analyze the delay times of the defect modes and compare their behavior with the total density of state (DOS). In particular, we show the possibility of the existence of transmission zeros induced by the anisotropic defect layer, giving rise to negative delay times for some wavelengths in the visible and infrared regions [56]. Let us recall that the existence of transmission zeros has been shown in 2D photonic crystals [57,58], however, the possibility of existence of such forbidden transmission regions in 1D lamellar systems has not been investigated before. This property is a characteristic of the anisotropic defect layer and is without analogue in their counterpart isotropic layer. It is worth mentioning that transmission zeros and the corresponding negative delay times have been demonstrated in quasi-one-dimensional circuits made of coaxial cables in the radio-frequency domain [10]. Also, we highlight the impact of the orientation of the principal axes of the defect layer with respect to the laboratory axes on the behavior of the optical modes in the forbidden bandgap of the system. The effect of the position of the cavity layer within the PC on the transmittance and the delay time of the defect modes will be also investigated. In general, the main theoretical method adopted for the analysis of the 1D multilayered PCs is the transfer matrix method [38,59,60]. In this chapter, we use the Green's function approach which enables deriving the transmission and reflection coefficients, the dispersion curves, the delay times, as well as the total and local DOS of electromagnetic modes propagating through the structure.

This chapter is organized as follows: in Section 22.2 we present the theoretical model used to derive the expressions of the reflection and transmission coefficients, the density of states, and the delay time when S or P polarized wave is incident upon the interface separating the substrate and the PC. Section 22.3 gives some numerical results of the variation of the transmittance and the delay time of the transmitted waves with respect to the frequency and their dependence on the azimuthal angle of the anisotropic layers, as well as on the number of unit cells in the PC. In Section 22.4, we give the numerical results of the transmittance and the delay time of the optical modes with respect to the wavelength and their dependence on the azimuthal angle and the position of the defect layer within the PC. The spatial behavior of optical modes along the z-axis of the PC has been investigated through the local density of states. Finally, a summary of this chapter is presented in Section 22.5.

22.2 Theoretical model

First, we consider a photonic crystal made of alternating layers of air (isotropic) and SbSI (anisotropic) media as depicted in Fig. 22.1. It is denoted by air/$(HL)^N H$/air, where N is the number of the HL bilayers (periods); H and L refer to high and low refractive indices of SbSI and air, respectively. All the interfaces of the layers are taken to be parallel to the (X, Y)-plane of a Cartesian (laboratory) coordinate system, and the Z-axis is along the normal to the interfaces (Fig. 22.1). The thicknesses of the layers H and L are denoted d_H and d_L, respectively. The materials constituting the layers are assumed to be homogeneous and nonmagnetic. The layers L and H are characterized by the dielectric constants ϵ_L and by the diagonal dielectric tensor, respectively, where ϵ_x, ϵ_y, and ϵ_z are the dielectric constants in the corresponding principal axes (x, y, z) of the crystal. For a given orientation of the crystal, its dielectric tensor with respect to the fixed (X, Y, Z) coordinate system is described by using the Euler angles [61,62] θ, ϕ, and ψ. For the sake of simplicity and without loss of generality, we investigate in this chapter the effect of only the azimuthal angle ϕ and omit the effect of the two other angles ($\theta = \psi = 0$). The axes of the birefringent layer H are shown in Fig. 22.1.

The propagation of electromagnetic waves in the periodic structure is performed within the framework of the interface response theory of continuous media [63]. The objective of this theory is to calculate the Green's function of a composite system containing a large number of interfaces that separate different homogeneous media. The knowledge of this Green's function enables us to obtain different physical properties of the system such as the reflection and transmission coefficients, the dispersion relations, the delay times, and the total and local densities of states of the optical modes. In this theory, the Green's function of a composite system can be written as [63]

$$\begin{aligned} g(D, D) = G(D, D) + G(D, M)\big\{[G(M, M)]^{-1}[g(M, M) - G(M, M)] \\ \times [G(M, M)]^{-1}\big\}G(M, D), \end{aligned}$$

where D and M are respectively the whole space and the space of the interfaces in the lamellar system; G is a block-diagonal matrix in which each block G_i cor-

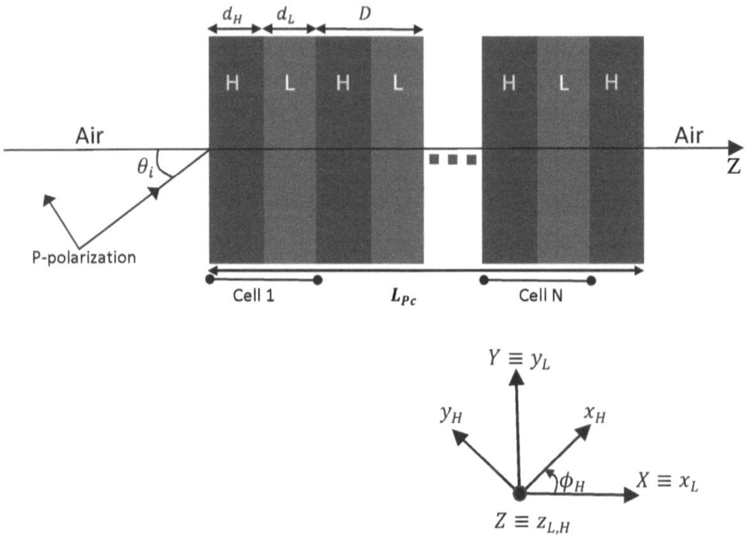

Figure 22.1 Schematic representation of a finite 1D photonic crystal formed by air (layer L) and anisotropic layer made of SbSI (layer H). The layers H and L are characterized by their thicknesses d_H and d_L; $D = d_H + d_L$ is the period of the structure. The crystal axes of the anisotropic layers are rotated with respect to the fixed ones by the azimuth angle ϕ_H around the Z-axis. The input and output isotropic media are air and the incoming light has P polarization; θ_i is the incidence angle.

responds to the bulk Green's function of the subsystem i. All the matrix elements $g(D, D)$ of the composite material can be obtained from the knowledge of the matrix elements $g(M, M)$ in the interface space M; $g(M, M)$ is calculated from its inverse $g^{-1}(M, M)$. The latter is formed using a linear superposition of the surface matrix elements $g_i^{-1}(M, M)$ of any independent film bounded by perfectly-free interfaces with appropriate boundary conditions. The matrix elements $g_i^{-1}(M, M)$ for an anisotropic medium are given in Appendix 22.A.

Within this theory, the reflected and transmitted waves $u(D)$, resulting from a uniform plane wave $U(D)$ incident upon a plane boundary between two different media, are given by [63]

$$u(D) = U(D) + G(D, M)\{[G(M, M)]^{-1}[g(M, M) - G(M, M)] \times [G(M, M)]^{-1}\}U(M).$$

Let us mention that the incident wave, generated in the substrate s, can have two different polarizations, namely S and P. In general, each wave entering the anisotropic medium generates two transmitted waves and two reflected waves with different polarizations. Let us call E_{iS} and E_{iP} the amplitudes of the S and P components of the incident field, respectively. Then, the amplitudes of the reflected and transmitted fields can be written as [55]:

$$\vec{E}_{RS}(z) = r_{SS}\vec{E}_{iS}(z) + r_{SP}\vec{E}_{iP}(z), \tag{22.1}$$

$$\vec{E}_{TS}(z) = t_{SS}\vec{E}_{iS}(z) + t_{SP}\vec{E}_{iP}(z), \tag{22.2}$$

$$\vec{E}_{RP}(z) = r_{PS}\vec{E}_{iS}(z) + r_{PP}\vec{E}_{iP}(z), \tag{22.3}$$

$$\vec{E}_{TP}(z) = t_{PS}\vec{E}_{iS}(z) + t_{PP}\vec{E}_{iP}(z). \tag{22.4}$$

The expressions of r_{ij} and t_{ij} in these equations ($i, j = S$ or P) for an anisotropic structure are given in Appendix 22.A. Therefore, the total reflection coefficients for the P and S modes are given respectively by

$$R_P = R_{PP} + R_{SP} = |r_{PP}|^2 + |r_{SP}|^2, \tag{22.5}$$

$$R_S = R_{SS} + R_{PS} = |r_{SS}|^2 + |r_{PS}|^2, \tag{22.6}$$

and the total transmission coefficients for the P and S modes are given respectively by

$$T_P = T_{PP} + T_{SP} = |t_{PP}|^2 + |t_{SP}|^2, \tag{22.7}$$

$$T_S = T_{SS} + T_{PS} = |t_{SS}|^2 + |t_{PS}|^2. \tag{22.8}$$

From the expressions of r_{ij} and t_{ij}, one can deduce also the phases of the transmission and reflection coefficients. The derivatives of these phases with respect to the pulsation ω yield to an important physical quantity. The latter, called the delay time, indicates the time needed by the wave to complete the transmission or reflection process through the structure. Four different delay times can be defined for the transmission coefficients, namely

$$\tau_{ss} = \frac{d\phi_{ss}}{d\omega}, \quad \tau_{ps} = \frac{d\phi_{ps}}{d\omega}, \quad \tau_{pp} = \frac{d\phi_{pp}}{d\omega}, \quad \text{and} \quad \tau_{sp} = \frac{d\phi_{sp}}{d\omega}, \tag{22.9}$$

where ϕ_{ss}, ϕ_{ps}, ϕ_{pp}, and ϕ_{sp} are respectively the phases of the transmission coefficients t_{ss}, t_{ps}, t_{pp}, and t_{sp}.

The delay time is used to describe the tunneling and superluminal propagation regime. Thus, regions of strong normal dispersion give rise to a very slow group velocity (subluminal or slow light regime) while regions of strong anomalous dispersion yield an apparent superluminal behavior in wave propagation (fast light regime). The abnormal group velocities was regarded as unphysical, but several works [10,22] have demonstrated that this superluminal behavior is in no way in contradiction with either causality or Einstein theory of special relativity [18], since the amount of transmitted energy is always smaller than that in vacuum.

As mentioned above, the Green's function enables us also to calculate the density of states (DOS), especially one can determine the variation of the density of states Δn between the structure $s_1/(HL)^N H/s_2$ and a reference system formed out of the same volumes of the bulk semiinfinite substrates s_1 and s_2 and the finite superlattice. This quantity is given by [64]:

$$\Delta n(\omega) = -\frac{1}{\pi}\frac{d}{d\omega}\text{Arg}\det\{\frac{g(M_0, M_0)}{[g_{s_1}(0,0)g_{s_2}(L,L)]^{1/2}}\}, \tag{22.10}$$

where $g(M_0, M_0)$ is the Green's function of the whole system at its both extremities $M_0 = \{0, L\}$, whereas $g_{s_1}(0,0)$ and $g_{s_2}(L, L)$ are the elements of the Green's functions at the surfaces 0 and L of the two substrates (air).

As mentioned above, for the numerical results and discussion, first we shall consider superluminal and negative delay times in isotropic–anisotropic PCs (Section 22.3) and then we shall investigate the effect of an anisotropic defect layer inserted inside a finite isotropic PC on the transmission properties in this system (Section 22.4).

22.3 Superluminal and negative delay times in isotropic–anisotropic photonic crystal

The PC air/$(HL)^N H$/air consists of two alternating anisotropic SbSI (layer H) and isotropic air (layer L) media. The number N of the HL bilayers (periods) of the PC is chosen equal to $N = 2$ and the thicknesses d_H and d_L of these layers are chosen in accordance with the quarter-wave-stack condition, i.e., $n_{z_H} d_H = n_L d_L$, where $n_L = 1$ is the refractive index of air and n_{z_H} is one of the principal optical indices of the anisotropic layer [65]: $n_{x_H} = 2.7$, $n_{y_H} = 3.2$, $n_{z_H} = 3.8$; $D = d_H + d_L$ is the period of the structure. Let us note that D is chosen equal to $D = 1$ μm and all numerical results are presented in the dimensionless frequency region $0.6 < \Omega < 1.8$ where $\Omega = \frac{\omega D}{2\pi c} = \frac{D}{\lambda}$. The corresponding wavelengths fall in the visible and near-infrared domains (0.5 μm $< \lambda <$ 1.6 μm). The orientation of the principal axes of the layer H is characterized by the azimuthal angle ϕ_H, as illustrated in Fig. 22.1. In most of our numerical results, we study the behavior of the optical modes exhibiting superluminal and negative group delay times when the incoming light has P polarization. After multiple reflection and transmission in the anisotropic layer, the incoming wave produces two reflected and two transmitted waves with S and P polarizations. For special cases when the principal axes of the anisotropic layer are parallel or perpendicular to the fixed axes, the S and P polarized waves remain uncoupled [55], whereas the transmitted and reflected S and P polarized waves are coupled for other orientations of the birefringent layer.

We display in Fig. 22.2 the variation of DOS ($\pi \Delta n$), the transmission coefficients (T_P), the phase (ϕ_P), and the delay time (τ_P) of the electromagnetic waves versus the dimensionless frequency Ω when a P polarized wave is launched normally ($\theta_i = 0°$) to the superlattice. The curves in the left panel of Fig. 22.2 correspond to the case where the principal axes of the anisotropic layers are parallel to the fixed axes of the coordinates system (X, Y, Z), i.e., $\phi_H = 0°$, whereas the curves in the right panel of Fig. 22.2 correspond to the case $\phi_H = 90°$, i.e., the principal axes of the birefringent layers are perpendicular to the laboratory axes. One can notice that for these two values of the azimuthal angle ϕ_H, the transmittance curves (Figs. 22.2(c)–(d)) exhibit the behavior of a PC with frequency regions where transmission is allowed and frequency regions where the transmission is forbidden. The transmission dips coincide

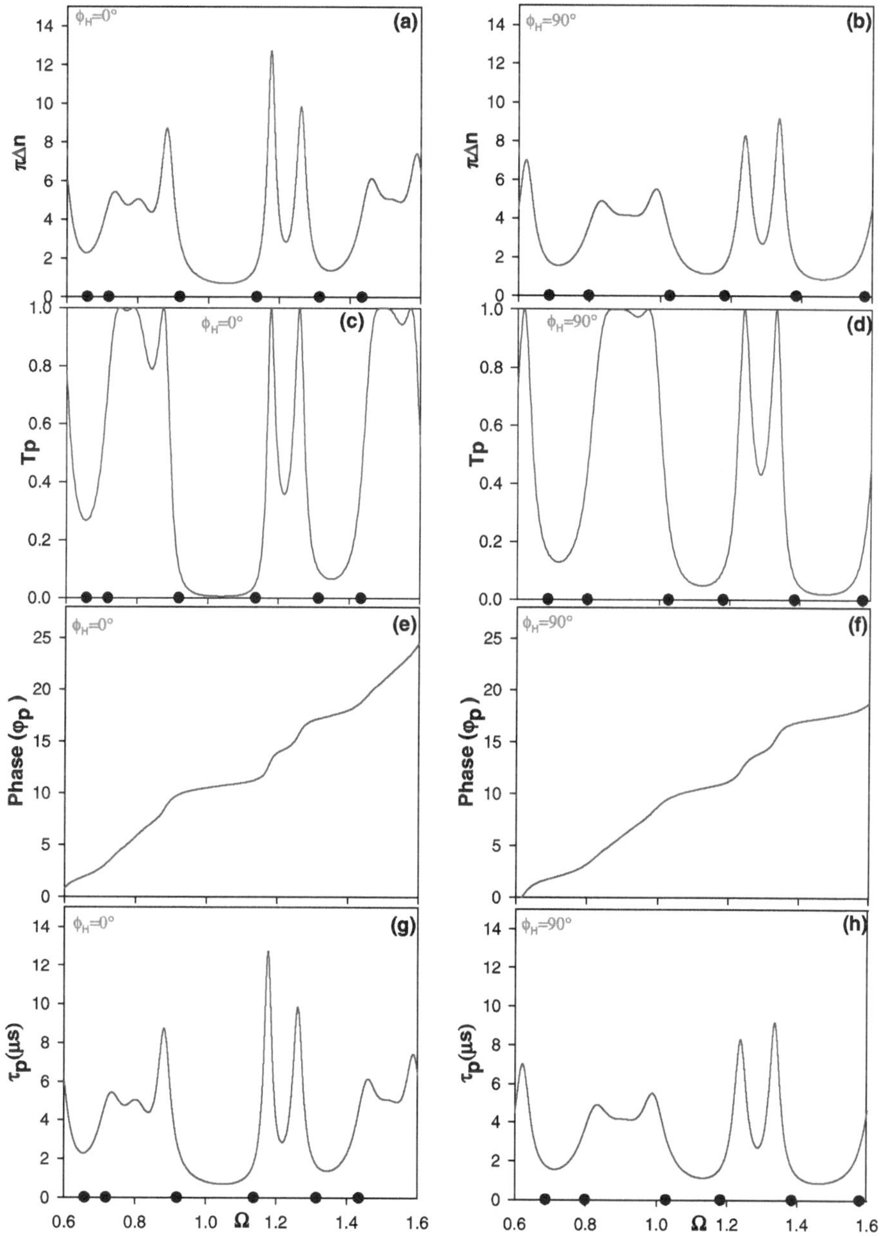

Figure 22.2 Variation of densities of states (a)–(b), transmission coefficients (c)–(d), phases (e)–(f) and delay times (g)–(h) versus the dimensionless frequency Ω for $\phi_H = 0°$ (left panel) and $\phi_H = 90°$ (right panel). The black dots on the Ω-axis represent the edges of the band gaps of the infinite PC structure. The incidence angle of the incoming P-polarized light is $\theta_i = 0°$ and the number of the bilayers in the structure is $N = 2$.

with the limits of the bandgaps of an infinite superlattice indicated by black circles on the frequency axis. These photonic bandgaps give rise to small positive slopes in the transmission phase function (Figs. 22.2(e)–(f)), and consequently to very small positive values of the delay time as it is shown in Figs. 22.2(g) and 22.2(h). In addition, the delay times exhibits the same behavior as the DOS ($\pi \Delta n$) for $\phi_H = 0°$ (Fig. 22.2(a)) and for $\phi_H = 90°$ (Fig. 22.2(b)). For these particular orientations of the principal axes of the layers, the density of states is directly proportional to the transmission delay time, i.e. $\tau_p = \pi \Delta n$, in accordance with the results demonstrated for isotropic PC [32]. Furthermore, it is shown that an anomalous dispersion occurs mostly within the photonic stopbands when principal axes of the anisotropic layers are parallel (Fig. 22.2(g)) or perpendicular (Fig. 22.2(h)) to the laboratory axes. In particular, the group delay converges to 0.69 μs inside the bandgap located between $\Omega = 0.918$ and $\Omega = 1.136$ when $\phi_H = 0°$ (Fig. 22.2(g)), whereas when $\phi_H = 90°$ (Fig. 22.2(h)) the delay time approaches 0.78 μs for $1.384 \leq \Omega \leq 1.578$. One can notice also in Figs. 22.2(g) and 22.2(h) that outside the bandgaps (i.e., inside the allowed bands), the light propagates with subluminal group velocity, i.e., lower than the speed of light in vacuum.

It was clearly pointed out in several works that superluminal behavior results exclusively as a wave interference effect in an anomalous dispersion region. As the magnitude of the interference increases with the number of unit cells N, the transmitted waves appear to traverse the photonic crystal in zero time. This is the apparent superluminal phenomenon observed empirically. It occurs well behind the light front of the original signal and is not an indication of true superluminal propagation [66]. It is worth mentioning that for these particular values of the azimuthal angle ϕ_H, anomalous behavior of the transmitted light occurs inside stopbands in the same way as for other photonic bandgap systems. For example, Haché et al. [35] have reported highly superluminal group velocity near $3c$ inside a photonic stopband located between 9 and 11 MHz for a coaxial PC, which is a scaled version of 1D PC made of 16 unit cells. Also, some of us have shown superluminal group velocities between $3c$ and $3.5c$ inside photonic bandgaps of periodic [10] and Fibonacci [29] photonic loop structures. However, in asymmetric 1D PC, Longhi et al. [9] have demonstrated this phenomenon only for the reflected pulse, while the transmitted pulse remains subluminal.

Our simple structure made of air/$(HL)^2H$/air shows anomalous dispersion and supports superluminal delay time. In the following, we demonstrate that this structure may also support negative group delay. Fig. 22.3 shows the evolution of the transmission coefficients and the delay times versus Ω but for an arbitrary orientation of the anisotropic layers, namely $\phi_H = 50°$. It is worth noticing here that the incident plane wave with P polarization generates two reflected and two transmitted plane waves containing both S and P polarized plane waves. Thus, we presented in Figs. 22.3(a), 22.3(c), 22.3(e) (left panel) and in Figs. 22.3(b), 22.3(d), 22.3(f) (right panel) the results associated to P and S polarized transmitted waves, respectively. Figs. 22.3(a) and 22.3(c) show that the slope of the transmission phase function changes sign for some frequencies in the allowed bands for P polarized transmitted wave. Similar results are obtained inside and outside the photonic bandgaps for S polarized transmitted wave (Figs. 22.3(b) and 22.3(d)). These frequency regions with negative slope of phases

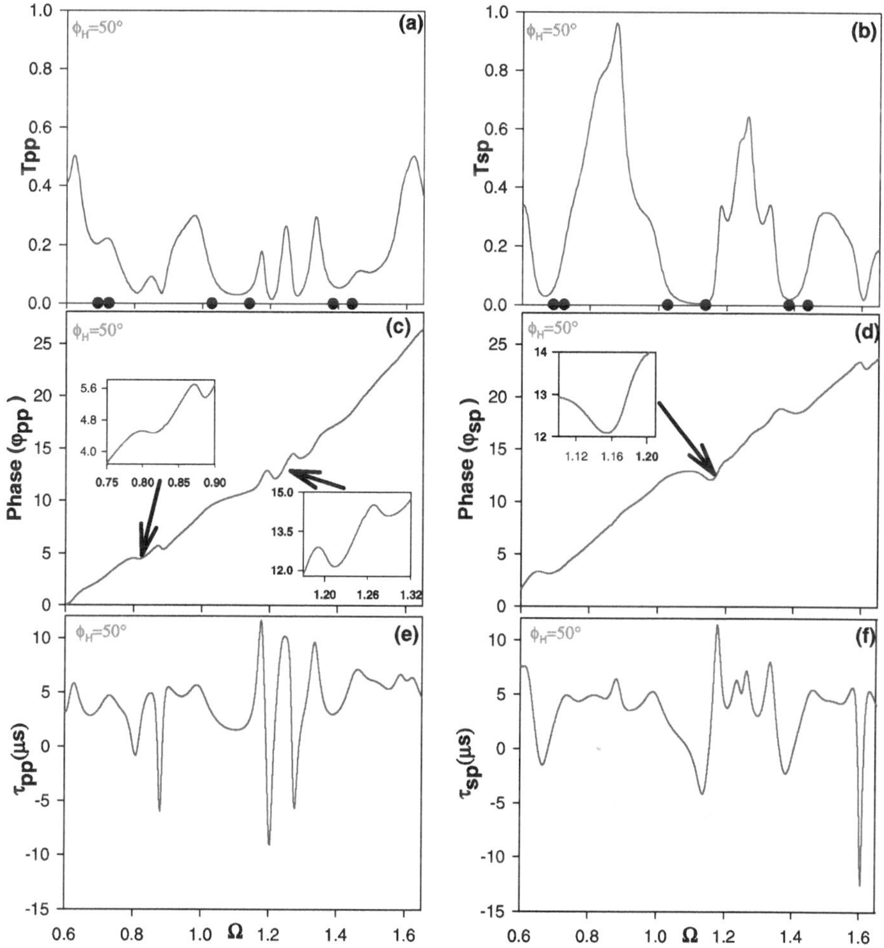

Figure 22.3 Evolution of transmission coefficients (a)–(b), phases (c)–(d), and delay times (e)–(f) as function of the dimensionless frequency Ω for the azimuthal angle $\phi_H = 50°$. The results in the left (a), (c), (e) and right (b), (d), (f) panels correspond to P and S polarized transmitted waves, respectively. The arrows in (c) and (d) indicate the abnormal frequency regions where the phase decreases as function of Ω. The black dots on the Ω-axis represent the edges of the bandgaps of the infinite PC structure. The incoming P-polarized light is launched normally to the structure.

(ϕ_{pp}, ϕ_{sp}) lead to negative delay times τ_{pp} and τ_{sp} as it is shown in Figs. 22.3(e) and 22.3(f), respectively. The delay time τ_{pp} exhibits negative peaks in a very narrow frequency interval around the transmission minima in the allowed bands (see Fig. 22.3(e)), whereas τ_{sp} presents negative peaks in the allowed and at the edges of photonic bandgaps (see Fig. 22.3(f)). Large negative delay times appear around $\Omega = 1.20$ and $\Omega = 1.60$ for P (Fig. 22.3(e)) and S (Fig. 22.3(f)) polarized transmitted waves, respectively. At these frequencies, the transmitted waves leave the PC before the incident wave has even entered the structure.

The frequency regions of abnormal dispersion are indicated by arrows in Figs. 22.3(c) and 22.3(d). The negative delay times are the result of the negative off-diagonal elements of the dielectric tensor (Eq. (22.A.2) in Appendix 22.A) of the anisotropic layers of the photonic crystal. Let us recall that negative group delays were achieved before in the frequency region where Mach–Zehnder interferometer [23], anisotropic semiconductor metamaterial, and isotropic negative-refractive-index metamaterial (NIM) structures [33] exhibit negative effective refractive index. Also, tunneling and superluminal regimes of transmitted pulses were detected in passive and linear aperiodic structures [22–24], as well as in an asymmetric loop structure made of coaxial cables in the radio-frequency domain [10,29]. In comparison with these structures, our system enables achieving large superluminal and negative group delays for both P and S polarized transmitted waves at different frequency intervals inside and outside the forbidden bandgaps. Also, it is worth mentioning that the observation of negative delay times and therefore negative group velocities in coaxial cable PCs is due to the existence of absorption in these systems [7,10,31,33] and to the presence of amplification in the 1D active dielectric PC [34], whereas the layered structure proposed here exhibits such phenomena without introducing any loss or gain in the system. Let us mention that the choice of the isotropic layers within the PC may affect considerably the existence of negative delay times. Indeed, if we replace air by another isotropic media such as SiO_2, AlAs, or GaAs, we have remarked the existence of negative group delays as in Figs. 22.3(e) and 22.3(f) but the negative peaks fall at different frequencies.

Let us now consider the effect of the birefringence of the SbSI layers constituting the PBG structure on the behavior of the tunneling and superluminal regimes when the incident P polarized wave is launched normally to the structure. We present in Fig. 22.4 the evolution of the delay times τ_{pp} as a function of Ω for different values of the azimuthal angle ϕ_H. It is clear that the birefringence of the anisotropic layers has a strong impact on the number and the magnitude of the peaks of the delay time spectra. The tunneling behavior of light is more pronounced when the absolute values of off-diagonal elements of the dielectric tensor of the SbSI layers are larger, i.e., around $\phi_H = 45°$ (Fig. 22.4(b)). However, this phenomenon decreases or even vanishes when the orientation of the principal axes of anisotropic layers remains almost parallel (Fig. 22.4(a)) or perpendicular (Fig. 22.4(d)) with respect to the laboratory axes. These results clearly show that by adjusting ϕ_H, we can perform a transition from subluminal to apparently superluminal, and vice versa. It is worth mentioning that the total attenuation of the Mach–Zehnder and multiple-beam interferometers drives the change in the group delay regime at the minima, which range from tunneling to superluminal and finally into normal regime as the attenuation increases [22–24]. Furthermore, the transition from tunneling to superluminal regime can also be achieved by changing slightly the length of any of the interferometer arms. Radovanovic et al. [33] have reported that the anisotropic materials offer the possibility of tuning the tunneling times by changing the doping profile or selecting different semiconductor materials combinations.

To widen the scope of our findings, we have also investigated the effect of the number of unit cells N on the tunneling and superluminal behavior of the P polarized

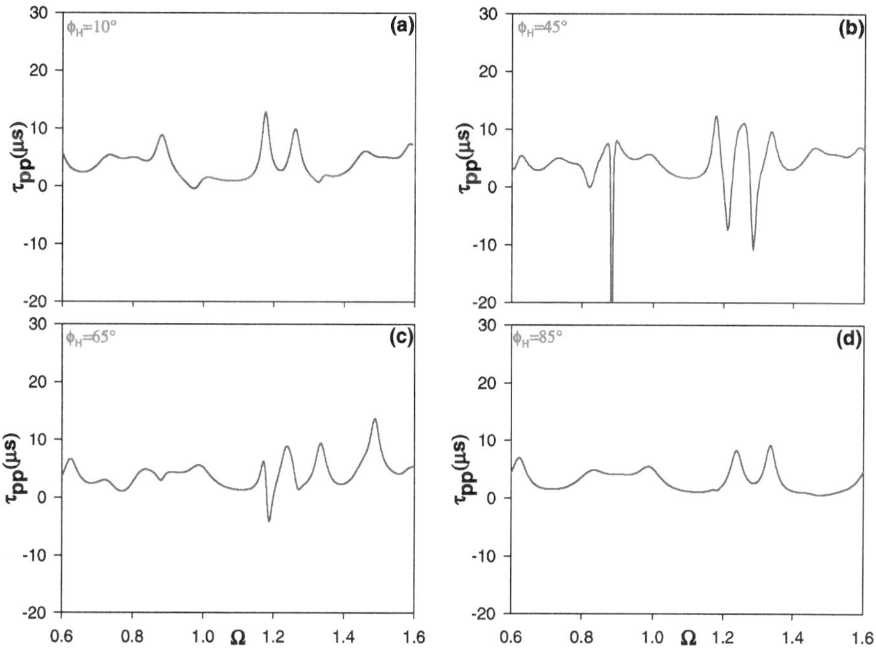

Figure 22.4 Variation of the delay time τ_{PP} versus the dimensionless frequency Ω for $\phi_H = 10°$ (a), $\phi_H = 45°$ (b), $\phi_H = 65°$ (c), and $\phi_H = 65°$ (d). The number of the periods of the structure is $N = 2$ and the incoming P-polarized light is launched normally to the PC structure.

transmitted wave as described in Fig. 22.5. These results clearly show that a simple structure air/$(HL)^1H$/air with one period ($N = 1$) is enough to produce tunneling and superluminal delay times (Fig. 22.5(a)). When N increases, other frequency intervals appear for which the light propagates through the structure in zero time and in negative delay times (Figs. 22.5(b), 22.5(c), and 22.5(d)). The abnormal (superluminal or negative) group delay is due to the wave interference effect and is in no way in contradiction with Einstein causality. Fig. 22.6 illustrates the evolution of the delay time τ_{sp} versus Ω for different values of the azimuthal angle ϕ_H where the P polarized wave is launched obliquely ($\theta_i = 45°$) to the structure. One can notice that τ_{sp} are slightly sensitive to the variation of the azimuthal angle when the incident light is launched normally to the PC structure. These findings at oblique incidence clearly show that the origin of negative delay times inside the PBGs, is due to destructive interference phenomenon in the superlattice. A large negative delay time is observed around $\Omega \simeq 1.31$; its position and magnitude vary slightly with the azimuthal angle of the layers ϕ_H. We can deduce from these numerical results that the birefringence of the anisotropic layers of the PBG structure has a significant impact on the tunneling and superluminal propagation of P polarized transmitted waves as compared to S waves. In order to see the effect of the size of the PC on the magnitude of the large negative time around $\Omega \simeq 1.31$ in Fig. 22.6, we have plotted in Fig. 22.7(b) the variation of the absolute value of the delay time τ_{sp} versus the number of periods N. Fig. 22.7(a) recalls the

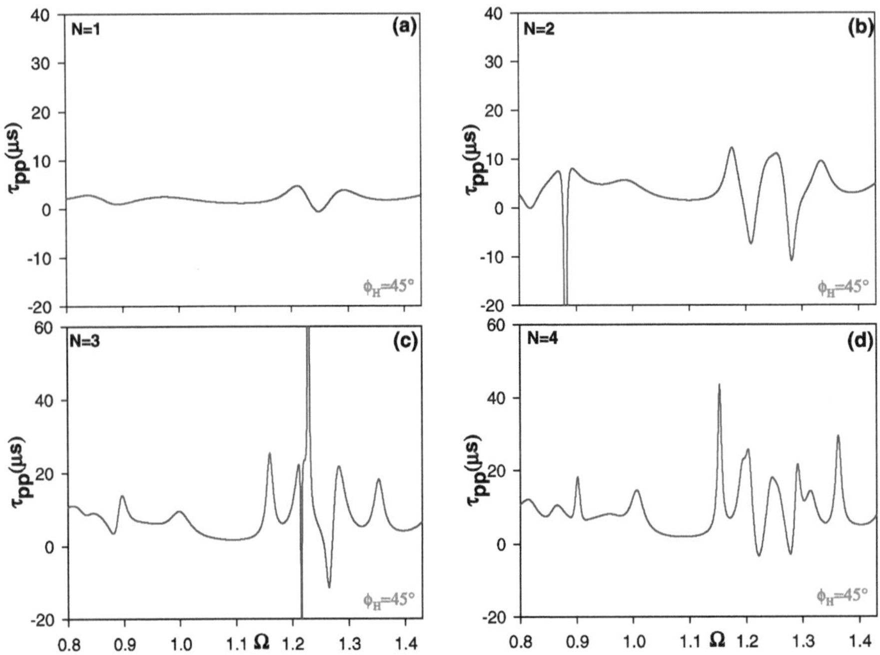

Figure 22.5 Evolution of the delay times τ_{PP} as function of the dimensionless frequency Ω for different number of unit cells: $N = 1$ (a), $N = 2$ (b), $N = 3$ (c), and $N = 4$ (d) when $\phi_H = 45°$. The incidence angle of the incoming light is $\theta_i = 0°$.

behavior of the delay time spectrum for $\phi_H = 45°$ and $N = 5$. The delay time noted A in Fig. 22.7(a) increases as a function of N and converges asymptotically to a finite value for $N \simeq 10$ (Fig. 22.7(b)). This result is similar to the Hartman effect in a quantum system [67]. The Hartman effect exists in the tunneling of electromagnetic waves, acoustic waves, and all other kinds of waves. Experimentally it has been observed using electromagnetic waves and sound waves. Winful [4,68] interprets that the group delay in tunneling is not a transit time but a lifetime. Since the group delay is proportional to stored energy, it saturates with barrier length as the stored energy saturates.

22.4 Selective transmission of an anisotropic cavity in 1D isotropic photonic crystal

In this section we shall consider the effect of an anisotropic cavity layer on the propagation and localization of optical modes in 1D isotropic PC. The perfect PC $s_1/(AB)^N/s_2$ consists of two alternating isotropic media: TiO_2 (layer A) and SiO_2 (layer B). The refractive indices of these materials are respectively $n_A = 2.35$ and $n_B = 1.46$. The thicknesses d_A and d_B of these layers are chosen in accordance

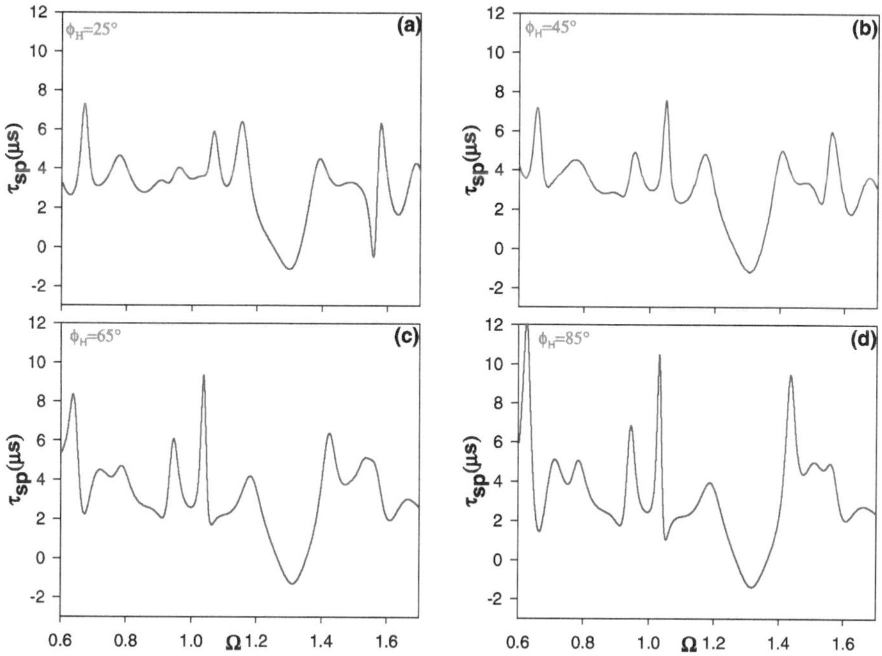

Figure 22.6 Same as in Fig. 22.4 but for S polarized transmitted wave and the incidence angle of the incoming light is $\theta_i = 45°$. The number of the periods of the structure is $N = 2$.

with the quarter-wave-stack condition, i.e., $n_A d_A = n_B d_B = \frac{\lambda_0}{4}$ where $\lambda_0 = 1550$ nm, which leads to $d_A = 164.89$ nm and $d_B = 265.41$ nm, respectively; $\lambda_0 = 1550$ nm is chosen for its use in optical communications. The number N of the AB bilayers (periods) of the PC is chosen equal to $N = 6$, for which the PBGs are well defined. Both substrates surrounding the PC are simply assumed to be air. The periodicity of the perfect SL is broken by inserting the defect layer C in the center of the structure as depicted in Fig. 22.8. The defect layer is made of an anisotropic dielectric material SbSI, with its principal optical indices [65] being $n_x^{(C)} = 2.7$, $n_y^{(C)} = 3.2$, $n_z^{(C)} = 3.8$. The orientation of the principal axes of the layer C is characterized by the azimuthal angle ϕ_C, as illustrated in Fig. 22.1, and the thickness of this layer is chosen equal to $d_C = 263.50$ nm, for which the defect modes are well defined in PBGs. In most of our numerical results, we study the behavior of the optical modes in a 1D PC with a defect layer when the incoming light has S or P polarization. After multiple reflection and transmission in the anisotropic layer C, the incoming wave produces two reflected and two transmitted waves with S and P polarizations. For special cases when the principal axes of the defect layer are parallel or perpendicular to the fixed axes, S and P polarized waves remain uncoupled [55]. Therefore, a single incident S or P polarized plane wave generates a single reflected R_S or R_P polarized plane wave and a single transmitted T_S or T_P polarized plane wave, respectively.

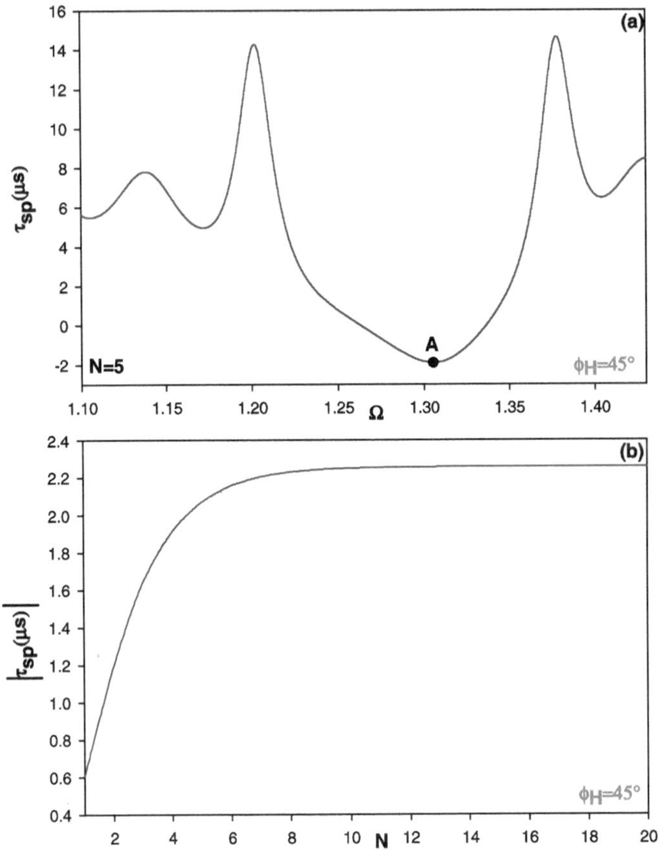

Figure 22.7 (a) Variation of the delay time τ_{SP} as function of the dimensionless frequency Ω for $\phi_H = 45°$ and $N = 5$. The incidence angle of the incoming light is $\theta_i = 45°$. (b) Variation of the modulus of the delay time as function of the number of periods N of PC at $\Omega \simeq 1.31$ (point **A** in (a)).

We display in Fig. 22.9 the transmittances T_P (Figs. 22.9(a) and 22.9(b)) and T_S (Figs. 22.9(c) and 22.9(d)) of the optical modes as a function of the wavelength λ for a perfect PC (solid lines) and a PC with a defect (dashed lines). In this illustration, we assume that the incident electromagnetic wave is launched normally ($\theta_i = 0°$) to the SL, and the principal axes of the defect layer C are parallel $\phi_C = 0°$ (Figs. 22.9(a) and 22.9(c)), or perpendicular $\phi_C = 90°$ (Figs. 22.9(b) and 22.9(d)) to the fixed axes of the coordinates system (X, Y, Z). It is shown that for these two values of the azimuthal angle ϕ_C, the defect modes appear inside the PBGs of the transmission spectrum. The localized modes in the electromagnetic forbidden bands of the structure arise due to the constructive interference of incident and reflected waves in the defect layer. From the analysis of Fig. 22.9, it can be observed that when the principal axes of anisotropic defect layer were parallel to the fixed axes of the coordinates system, the defect modes (dashed curves) are localized at $\lambda_{P_0}^{VIS} = 535.81$ nm, $\lambda_{P_0}^{IR} = 1611.47$ nm

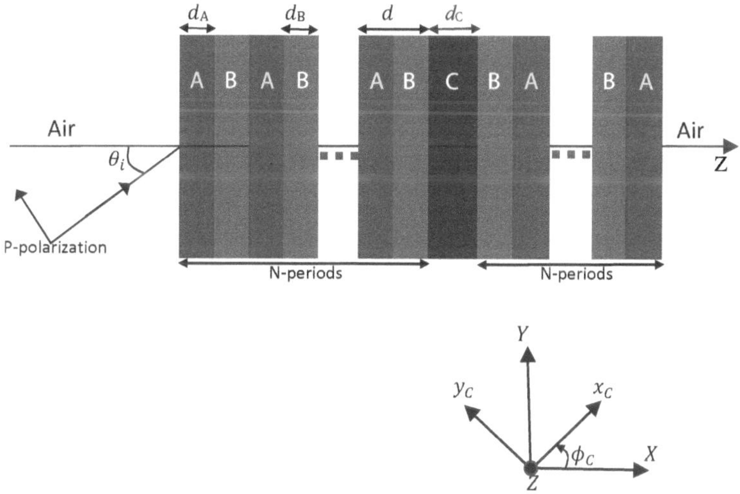

Figure 22.8 Schematic representation of a finite 1D superlattice composed of $2N$ cells with an anisotropic defect cavity SbSI (layer C) in the middle of the structure. The unit cell of the perfect structure is composed by TiO_2 (layer A) and SiO_2 (layer B). The layers A, B, and C are characterized by their thicknesses d_A, d_B, and d_C, respectively. The principal axes of the defect layer are oriented by the azimuthal angle ϕ_C with respect to the laboratory axes (X, Y, Z); $d = d_A + d_B$ is the period of the SL. The input and output isotropic media are air and the incident electromagnetic wave is launched with an angle θ_i with respect to the normal to the superlattice.

in the visible (VIS) and infrared (IR) regions, respectively, for transmission spectrum of P-polarized wave (Fig. 22.9(a)). Moreover, when the azimuthal angle of the layer C is equal to 90°, both modes shift to shorter wavelengths and appear respectively at $\lambda_{P_{90}}^{VIS} = 500.22$ nm and $\lambda_{P_{90}}^{IR} = 1497.83$ nm (Fig. 22.9(b)). The comparison of the results for S (Figs. 22.9(c) and 22.9(d)) and P (Figs. 22.9(a) and 22.9(b)) polarized incident waves, shows that the S-polarized defect modes for $\phi_C = 0°$ have the same behavior as the P-polarized defect modes when $\phi_C = 90°$, i.e., $\lambda_{S_0}^{VIS} = \lambda_{P_{90}}^{VIS}$ and $\lambda_{S_0}^{IR} = \lambda_{P_{90}}^{IR}$. Moreover, the same behavior of the localized modes in PBGs has been observed when the principal axes of defect layer are perpendicular to the fixed axes and the incoming light has S polarization, i.e., $\lambda_{S_{90}}^{VIS} = \lambda_{P_0}^{VIS}$ and $\lambda_{S_{90}}^{IR} = \lambda_{P_0}^{IR}$. These results are in good accordance with those presented in [69]. In addition, the transmission amplitudes of these peaks reach approximately unity for $\phi_C = 0°$ and $\phi_C = 90°$. The black dots on the λ-axis indicate the limits of the bandgaps of the infinite isotropic SL. These findings clearly show that the orientation of the principal axes of the anisotropic defect layer changes the properties of the defect modes within the VIS and IR regions bandgaps. In order to understand the spatial distribution of the S-polarized defect modes labeled 1 and 2 in Fig. 22.9(c) along the Z-axis of the SL, we have plotted in Fig. 22.10 the local density of states versus the reduced position Z/d for the azimuthal angle $\phi_C = 0°$ of the defect layer. This quantity reflects the spatial behavior of the square modulus of the electric field inside the structure. Figs. 22.10(a) and 22.10(b) show a strong localization of the defect modes 1 and 2

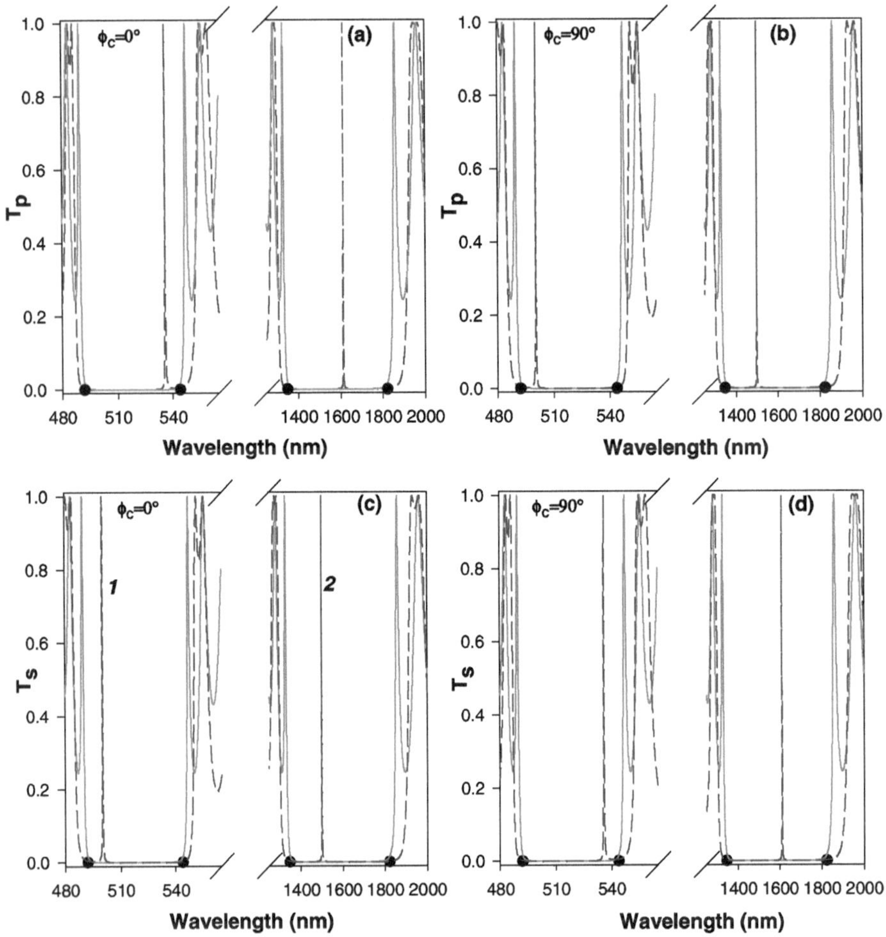

Figure 22.9 Variation of the transmittances T_P (a)–(b) and T_S (c)–(d) as a function of the wavelength λ for $\phi_C = 0°$ (a)–(c) and $\phi_C = 90°$ (b)–(d). The solid and dashed lines correspond to the perfect and defective structures, respectively. The black dots on the λ-axis represent the edges of the band gaps of the infinite isotropic SL. The incidence angle of the incoming light is $\theta_i = 0°$.

in the vicinity of the defect layer C and a decaying behavior on both sides of the interfaces of this layer. Similar behavior of the localized mode can be observed for other orientations of the principal axes of the defect layer and also for P-polarized waves.

In the following, we consider (Fig. 22.11) the effect of the incidence angle on the optical properties of the defect modes within the forbidden bandgaps in the VIS and IR regions for $\phi_C = 0°$ and $\phi_C = 90°$. Fig. 22.11 gives the dependence of the peak wavelengths on the angle of incidence, when the input light has P (Figs. 22.11(a) and 22.11(b)) or S (Figs. 22.11(c) and 22.11(d)) polarization. One can notice that when the incidence angle increases, the positions of P and S polarized defect modes

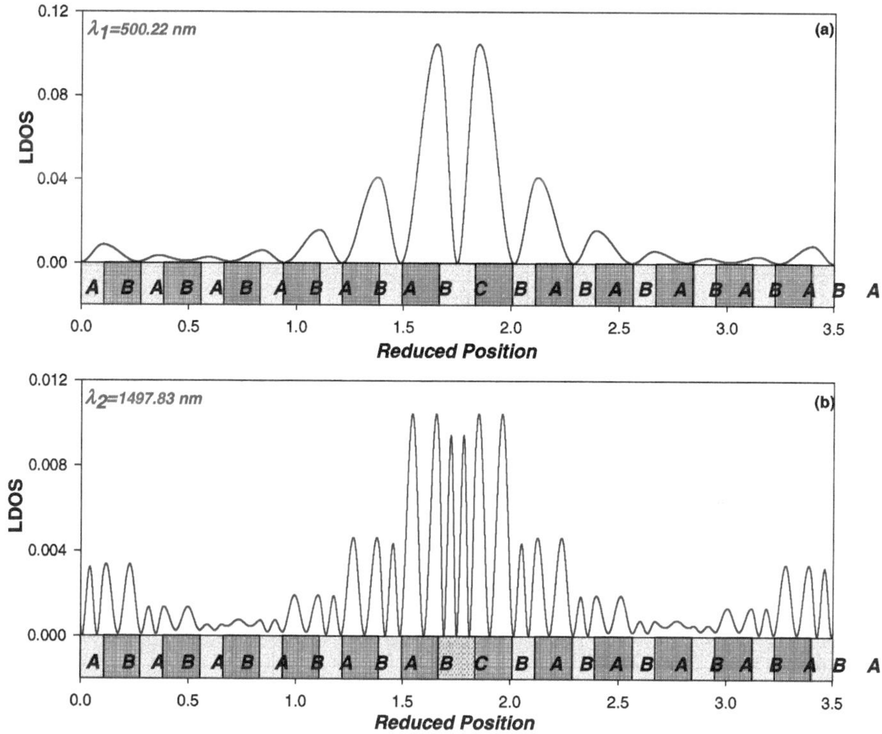

Figure 22.10 Spatial representation of the local density of states (in units of d/c) of the modes labeled 1 and 2 in Fig. 22.9(c) at $\lambda = 500.22$ nm (a) and $\lambda = 1497.83$ nm (b), respectively. The input S-polarized wave is launched normally to the SL and the azimuthal angle of the layer C is $\phi_C = 0°$.

shift towards the lower wavelengths inside the PBGs (hatched areas) as illustrated in Fig. 22.11. In addition, when the incidence angle is greater than 35°, the defect branch within the first gap for P polarization and $\phi_C = 0°$ (Fig. 22.11(a)) and for S polarization and $\phi_C = 90°$ (Fig. 22.11(d)) disappear at the higher edge of the forbidden gap and falls in the allowed band. Furthermore, the transmission amplitudes of these modes remain almost unity whatever the incidence angle for both cases ($\phi_C = 0°$ and $\phi_C = 90°$).

Till now, we have depicted only the properties of the defect modes in the transmission spectrum of the defective 1D PC when the orientation of the principal axes of defect layer is parallel or perpendicular with respect to the laboratory axes. Let us now consider the effect of other orientations of the axes of the layer C on the behavior of the defect mode when the incident electromagnetic wave is launched normally to the SL. We present in Figs. 22.12(a) and 22.12(b), the evolution of the transmission spectra T_{SS} and T_{PS} of the S-polarized wave as a function of the wavelength λ for the azimuthal angle $\phi_C = 60°$. It is clear that the birefringence of the defect layer has a strong impact on the number and the magnitudes of the defect modes inside the PBGs in the VIS and IR regions. For this azimuthal angle ($\phi_C = 60°$) of the layer C,

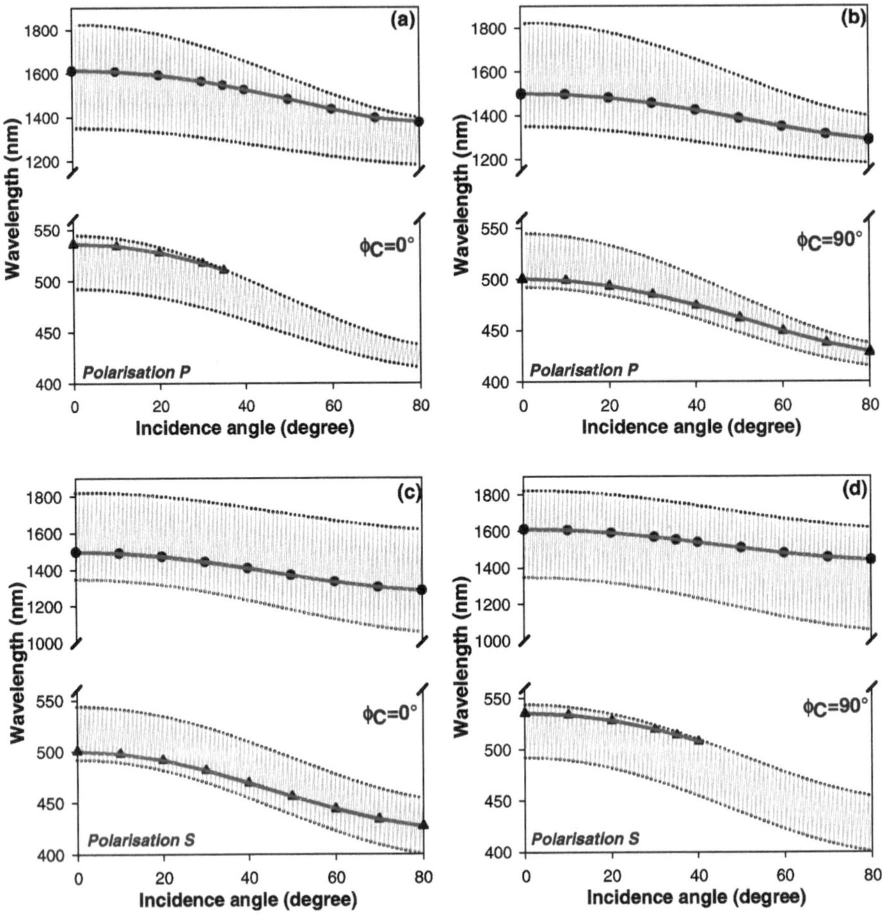

Figure 22.11 Dependence of the peak wavelengths of the transmitted P- and S-polarized waves for $\phi_C = 0°$ (a)–(c) and $\phi_C = 90°$ (b)–(d) versus the incidence angle θ_i. The black triangles and dots denote the defect modes in PBGs. The dotted curves represent the edges of the bandgaps (hatched areas) of the infinite isotropic SL.

two defect modes appear inside each PBGs as every incident plane wave with S or P polarization generates two reflected and two transmitted plane waves containing both S and P polarized plane waves. These two propagating waves are coupled, thus each mode is composed of two components of the transmission (T_{SS} and T_{PS}) spectrum for the same wavelength. These results correlated with the results in our previous work [55], when the PC is composed only by anisotropic media.

Moreover, the intensities of the defect modes depend strongly on the azimuthal angle of the anisotropic defect cavity as illustrated in Figs. 22.12(c) and 22.12(d), where we have plotted the dependence of peak heights T_{SS} (Fig. 22.12(c)) and T_{PS} (Fig. 22.12(d)) of the modes labeled 1 and 2 in Figs. 22.12(a) and 22.12(b) versus the azimuthal angle ϕ_C. The results of Fig. 22.12(c) show that by adjusting ϕ_C, the

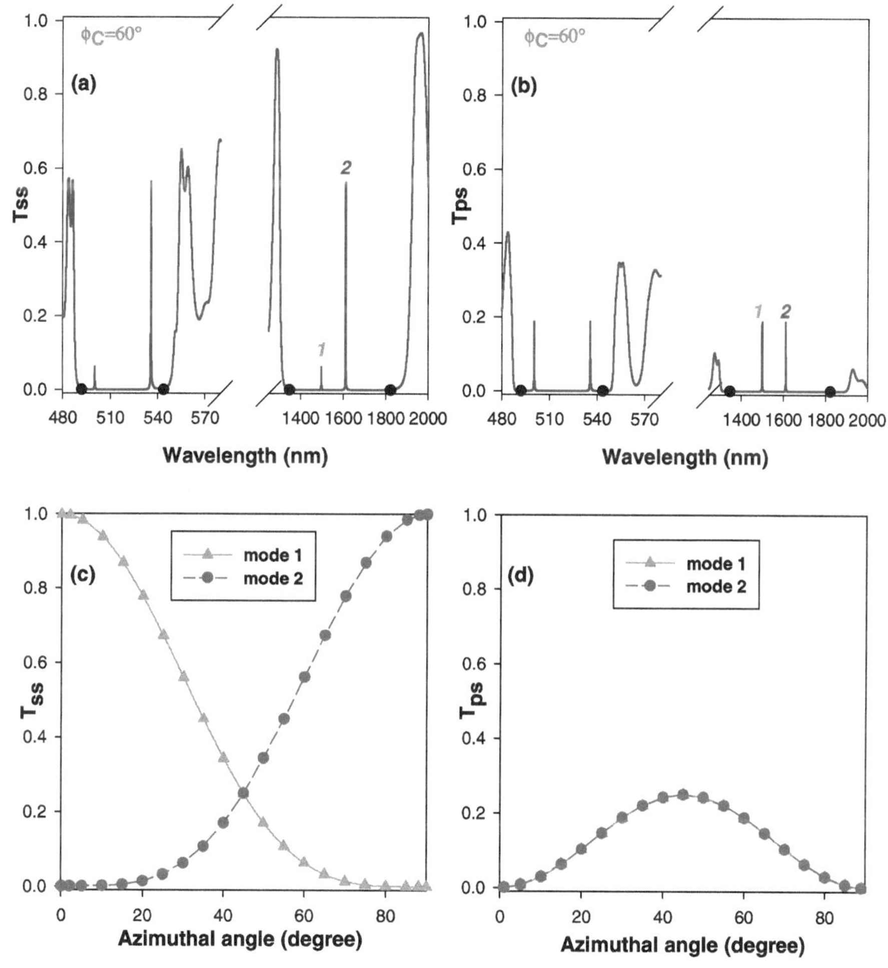

Figure 22.12 Variation of the transmissions T_{SS} (a) and T_{PS} (b) versus the wavelength λ for $\phi_C = 60°$; (c) and (d) represent the dependence of peak heights T_{SS} and T_{PS} of the defect modes labeled 1 and 2 in (a) and (b) on the azimuthal angle ϕ_C, respectively. The incoming S-polarized light is launched normally to the SL.

intensities of the defect modes exhibit different behaviors inside the PBGs of the structure. Indeed, by increasing ϕ_C from 0° to 90°, the transmission amplitudes of the mode 2 (indicated by the black dots) increase from 0 to 1, while the amplitudes of the mode 1 (indicated by black triangles) decrease from 1 to 0. However, the transmitted P-polarized modes (Fig. 22.12(d)) exhibit similar behavior. Their intensities gradually increase by rotating the azimuthal angle of the defect layer, reach a maximum value around $\phi_C = 45°$ and then decrease to vanish for $\phi_C = 90°$. Also, both transmitted P- and S-polarized defect modes have identical magnitudes (0.25) around $\phi_C = 45°$.

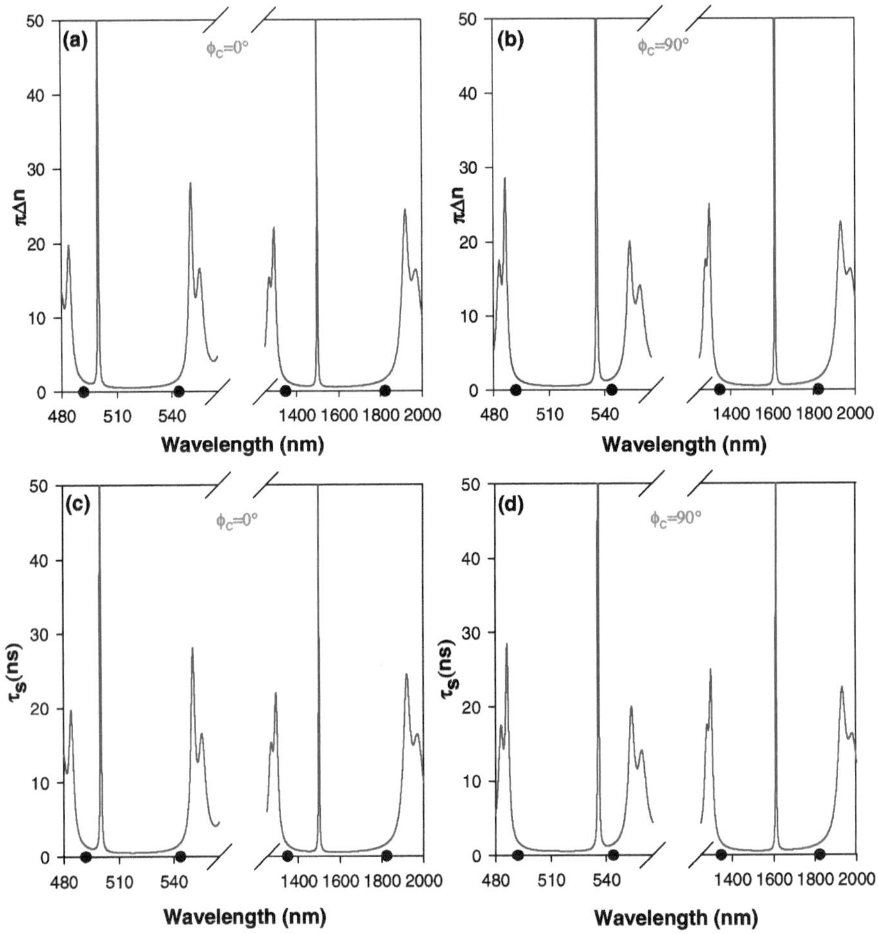

Figure 22.13 Variation of the density of states Δn (in units of d/c) (a)–(b) and the delay time τ_S (c)–(d) of S incident polarized waves versus the wavelength λ for $\phi_C = 0°$ (a)–(c) and $\phi_C = 90°$ (b)–(d). The black dots on the λ-axis represent the edges of the bandgaps of the infinite isotropic SL at normal incidence.

The Green's function method enables us to derive other physical properties of the composite system under study, through the analysis of the transmission delay time τ and the total density of states Δn. The delay time describes the time needed for photons to complete the transmission process and provides information about the interaction of an incident wave with the modes confined in the anisotropic defect layer. In the following, we study these two quantities and give a comparison between the total DOS and the delay time of the optical waves. We display in Fig. 22.13, the variation of DOS ($\pi \Delta n$) (Figs. 22.13(a) and 22.13(b)) and the delay time τ (Figs. 22.13(c) and 22.13(d)) as a function of the wavelength for S-polarized incident wave and when the principal axes of defect layer are parallel or perpendicular to the fixed ones. It is clear that the delay time of the transmitted peaks in the VIS and IR regions, is very sensitive to the azimuthal angle of the anisotropic defect layer. The frequencies of the defect modes obtained from the total DOS (Figs. 22.13(a) and 22.13(b)) coincide with those given by the delay time (Figs. 22.13(c) and 22.13(d)) and the transmission coefficient (see Figs. 22.10(c) and 22.10(d)). In addition, the delay time τ_s exhibits the same behavior as the DOS ($\pi \Delta n$) for $\phi_C = 0°$ (Figs. 22.13(c) and 22.13(a)) and for $\phi_C = 90°$ (Figs. 22.13(d) and 22.13(b)). For these particular orientations of the principal axes of the defect layer, the DOS is directly proportional to the transmission delay time, i.e., $\tau_s = \pi \Delta n(\omega)$, in accordance with the results demonstrated for isotropic PC [32]. These results remain also valid for P-polarized waves. However, for $\phi_C \neq 0°$ and $\phi_C \neq 90°$, different transmission delay times (τ_{ss}, τ_{ps}) and (τ_{pp}, τ_{sp}) for S and P transmitted polarized waves, respectively, show different behaviors in comparison with the DOS as it is illustrated in Fig. 22.14. One can notice that the delay times spectra for both S and P transmitted polarized waves show negative peaks for some wavelengths in the VIS and IR regions. We have checked that the corresponding transmission amplitudes reach almost zero as a consequence of a destructive interference between S and P waves inside the anisotropic defect layer. The negative peaks in τ_{pp} (Fig. 22.14(b)) and τ_{ss} (Fig. 22.14(c)) spectra are localized between the two defect modes in both VIS and IR regions, whereas those in τ_{sp} and τ_{ps} (Fig. 22.14(d)) appear at the vicinities of the PBG edges. It is worth mentioning that the negative delay times in 1D systems are generally observed by taking into account the absorption in these systems [10,70]. However, in the present system, the negative delta peaks are intrinsically broadened. Moreover, because of the symmetry of the structure, the delay time τ_{sp} presents the same behavior as τ_{ps}. Indeed, rendering the structure asymmetric, by moving for example the cavity layer C inside the PC, gives rise to $\tau_{ps} \neq \tau_{sp}$. Let us notice that the existence of transmission zeros in one of the two components of the transmission, may give rise also to Fano and electromagnetic induced transparency resonances [58] even though these properties have not been explored here and postponed to a future work.

To widen the scope of our findings, we have also investigated the effect of the position of the cavity layer within the PC on the optical properties of defect modes in the PBGs of the structure. We have plotted in Fig. 22.15 the transmittance (Figs. 22.15(a) and 22.15(b)) and the delay time (Figs. 22.15(c) and 22.15(d)) of the localized modes in the IR region for different positions of the cavity layer within the PC and for $\phi_C = 90°$. The PC is composed of 13 cells and the cavity layer can be inserted at different positions labeled n inside the PC ($0 \leq n \leq 12$). The transmission spectra are displayed for S-polarized waves. One can notice that the transmittance and the delay time of the defect modes depend strongly on the position of the cavity layer within the finite SL. The magnitudes of the filtered modes in the PBGs reach their maximum value when the defect cavity is inserted in the middle of the structure (i.e., $n = 6$) and decrease gradually when the defect layer is moved far from the middle of the PC, as shown in Figs. 22.15(b) and 22.15(d). The transmittance and the delay time almost vanish when the cavity layer is placed at both extremities of the structure. These results remain also valid for other orientations of the principal axes of the defect layer and for P-polarized waves.

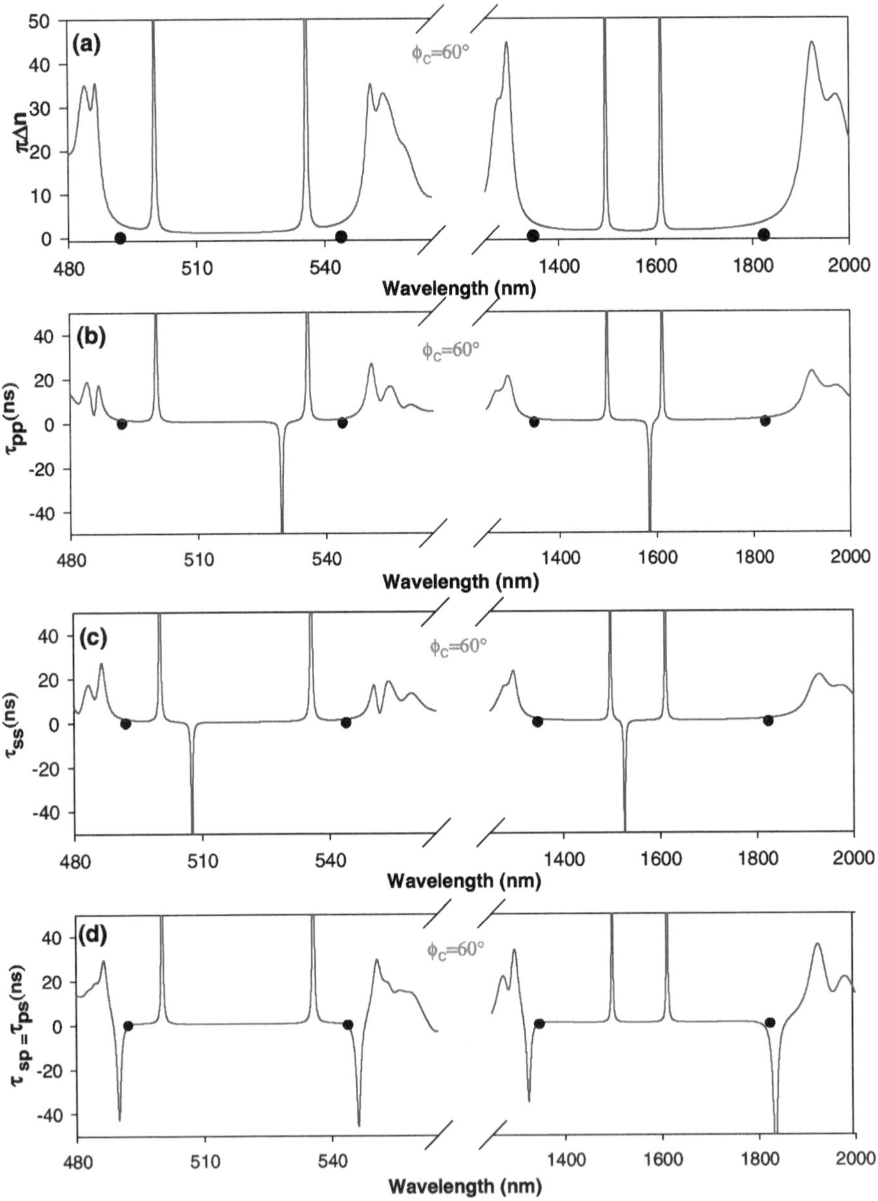

Figure 22.14 Variation of the density of states Δn (in units of d/c) (a) and different delay times: τ_{PP} (b), τ_{SS} (c), and $\tau_{PS} = \tau_{SP}$ (d), as a function of the wavelength λ for $\phi_C = 60°$. The black dots on the λ-axis represent the edges of bandgaps of the infinite isotropic SL and for normal incidence.

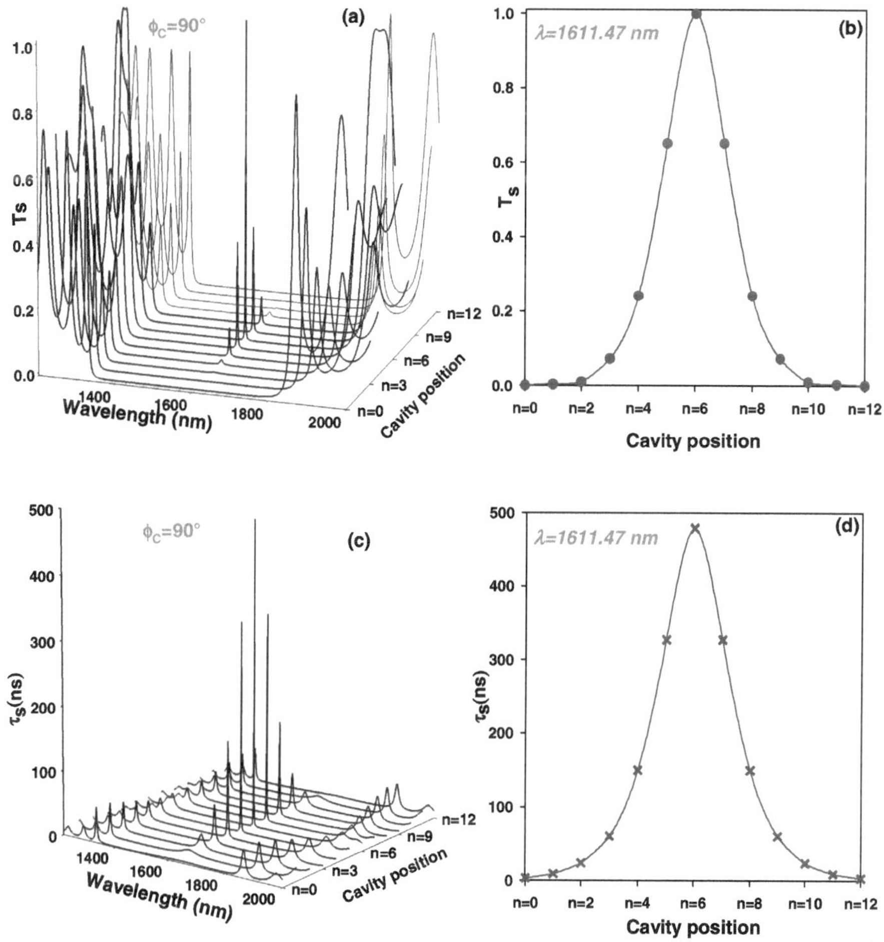

Figure 22.15 Variation of the transmittance T_S and the delay time (τ_S) of S polarized modes versus the wavelength for different positions of the cavity layer n ($0 \leq n \leq 12$) inside a finite PC composed of 13 cells. The curves are sketched in IR region and for $\phi_C = 90°$; (b) and (d) represent the dependence of the transmission amplitude and the delay time of the defect modes as a function of the defect cavity position n.

22.5 Summary and conclusions

In summary, we have investigated theoretically the existence of the tunneling and superluminal regimes in a linear, passive, and lossless periodic structure consisting of alternating isotropic and anisotropic media. By adjusting the orientations of the anisotropic layers, we obtained a rapid decrease of the phase versus the frequency at the transmission minima giving rise to anomalous effective dispersion. This behavior is a characteristic of the anisotropic layers and is without analogue in the case of isotropic–isotropic PCs [30,32]. The birefringence of the anisotropic layers drives a change in the sign and the magnitude of the delay times, which transit from subluminal to apparently superluminal regime, and vice versa. The transmission spectrum shows photonic bandgaps, which sustains superluminal delay times when the orientation of the principal axes of the SbSI layers is parallel or perpendicular to the laboratory axes. For other orientations, the light propagates through the structure in zero times and in negative delay times in a very narrow frequency interval around the transmission minima at the photonic bandgap edges and in the allowed bands. This study is performed by means of the Green's function approach, through an investigation of the DOS, transmission coefficients, and delay times for the transmitted waves through a finite PC structure. We have shown that the delay time reproduces exactly the total density of states for zero off-diagonal elements of the dielectric tensor of anisotropic layers. Furthermore, we have investigated that a simple structure composed of only a few periods may exhibit anomalous dispersion and may support tunneling and superluminal group delay for the transmitted waves with both P and S polarizations. Also, we have demonstrated that the isotropic–anisotropic photonic crystal exhibits the Hartman effect, in which the tunneling delay time of an electromagnetic wave through the photonic bandgap of the structure converges asymptotically to a finite value with increasing numbers of layers. This structure can be a good candidate for the implementation of photonic devices with different delay times.

Also, we have theoretically investigated the effect of an anisotropic defect cavity on the optical transmission properties in one-dimensional isotropic photonic crystals. This study is performed through an investigation of the transmission coefficients, delay times, and DOS for electromagnetic modes in the finite structure. The transmittance and delay times are discussed as functions of the azimuthal angle and the position of the anisotropic defect cavity within the PC. We have presented a comparison between the total DOS and different transmission delay times in such systems. We have shown that the delay time reproduces exactly the DOS when the principal axes of the defect layer are parallel or perpendicular to the laboratory axes. However, far from these directions, the birefringence of the cavity layer induces, in addition to mode conversion, transmission zeros in one of the two components of the transmission, for some wavelengths in the VIS and IR regions, giving rise to negative delta peaks in the delay times. These properties are without analogue in their counterpart isotropic defect layers. This structure can be of potential interest in the fields of photonics and optoelectronics as a high-precision optical filter.

Acknowledgments

The authors would like to thank Noama Ouchani, Abdelaziz El Moussaouy, Hassan Aynaou, and Youssef El Hassouani for their fruitful contribution in the original papers from which this chapter is inspired.

Appendix 22.A

1. The bulk Green's function elements of an infinite anisotropic medium are given by

$$
\begin{aligned}
G_{XX}(Z,Z') &= -\frac{C}{2\alpha_-\alpha_+}[\alpha_- A_- \exp(-\alpha_+|Z-Z'|) \\
&\quad - \alpha_+ A_+ \exp(-\alpha_-|Z-Z'|)], \\
G_{YX}(Z,Z') &= -\frac{B}{D}\frac{C}{2\alpha_-\alpha_+}[\alpha_- \exp(-\alpha_+|Z-Z'|) - \alpha_+ \exp(-\alpha_-|Z-Z'|)], \\
G_{ZX}(Z,Z') &= -\frac{iq_Y C}{2}\frac{\epsilon_{XY}}{\epsilon_{ZZ}}[\exp(-\alpha_+|Z-Z'|) \\
&\quad - \exp(-\alpha_-|Z-Z'|)]\mathrm{sgn}(Z-Z'), \\
G_{XY}(Z,Z') &= -\frac{B}{D}\frac{C}{2\alpha_-\alpha_+}[\alpha_- \exp(-\alpha_+|Z-Z'|) - \alpha_+ \exp(-\alpha_-|Z-Z'|)], \\
G_{YY}(Z,Z') &= -\frac{C}{2\alpha_-\alpha_+ D}[\alpha_- A_+ \exp(-\alpha_+|Z-Z'|) \\
&\quad - \alpha_+ A_- \exp(-\alpha_-|Z-Z'|)], \\
G_{ZY}(Z,Z') &= -\frac{C}{2}\frac{iq_Y}{q_0^2\epsilon_{ZZ}}[A_+ \exp(-\alpha_+|Z-Z'|) \\
&\quad - A_- \exp(-\alpha_-|Z-Z'|)]\mathrm{sgn}(Z-Z'), \\
G_{XZ}(Z,Z') &= -\frac{iq_Y C}{2}\frac{\epsilon_{XY}}{\epsilon_{ZZ}}[\exp(-\alpha_+|Z-Z'|) \\
&\quad - \exp(-\alpha_-|Z-Z'|)]\mathrm{sgn}(Z-Z'), \\
G_{YZ}(Z,Z') &= -\frac{C}{2}\frac{iq_Y}{q_0^2\epsilon_{ZZ}}[A_+ \exp(-\alpha_+|Z-Z'|) \\
&\quad - A_- \exp(-\alpha_-|Z-Z'|)]\mathrm{sgn}(Z-Z'), \\
G_{ZZ}(Z,Z') &= \frac{1}{q_Y^2 - q_0^2\epsilon_{ZZ}}[\frac{1}{D}\delta(Z-Z') \\
&\quad + \frac{C}{2}\frac{q_Y^2}{q_0^2\epsilon_{ZZ}}(\alpha_+ A_+ \exp(-\alpha_+|Z-Z'|) \\
&\quad - \alpha_- A_- \exp(-\alpha_-|Z-Z'|))],
\end{aligned} \tag{22.A.1}
$$

where $A_\pm = (q_Y^2 - q_0^2\epsilon_{XX} - \alpha_\pm^2)$, $B = q_0^2\epsilon_{XY}$, $C = (\alpha_+^2 - \alpha_-^2)^{-1}$, $D = q_0^2\epsilon_{ZZ}(q_Y^2 - q_0^2\epsilon_{ZZ})^{-1}$, and $\delta(Z-Z')$ is the Dirac delta function. Moreover, $q_0 = \frac{\omega}{c}$ is the vacuum wave vector, c is the velocity of light in vacuum, and ω is the frequency of the optical wave; ϵ is the dielectric tensor defined in the laboratory system (X, Y, Z), with $\theta = 0$, $\psi = 0$ and for an arbitrary value of the azimuthal angle ϕ, by

$$\epsilon = \begin{pmatrix} \epsilon_x \cos^2(\phi) + \epsilon_y \sin^2(\phi) & (\epsilon_x - \epsilon_y)\cos(\phi)\sin(\phi) & 0 \\ (\epsilon_x - \epsilon_y)\cos(\phi)\sin(\phi) & \epsilon_x \sin^2(\phi) + \epsilon_y \cos^2(\phi) & 0 \\ 0 & 0 & \epsilon_z \end{pmatrix}, \quad (22.A.2)$$

also $\alpha_\pm$ are defined as

$$-q_Z^2 = \alpha_\pm^2, \quad (22.A.3)$$

$$\alpha_\pm^2 = \frac{1}{2\epsilon_{ZZ}}\Big[\Big[q_Y^2(\epsilon_{ZZ} + \epsilon_{YY}) - q_0^2\epsilon_{ZZ}(\epsilon_{XX} + \epsilon_{YY})\Big] \pm \Big[\Big[k_Z^2\epsilon_{YY} - \epsilon_{ZZ}(q_Y^2 - q_0^2\epsilon_{XX})\Big]^2 - 4q_0^2k_Z^2\epsilon_{ZZ}\epsilon_{XY}^2\Big]^{\frac{1}{2}}\Big], \quad (22.A.4)$$

where $k_Z^2 = q_Y^2 - q_0^2\epsilon_{ZZ}$.

2. The surface response function of the finite layer of a homogeneous medium i, extending in the region $\frac{-d_0}{2} \leq Z \leq \frac{d_0}{2}$ with free surfaces is given by

$$[g_i(M, M)]^{-1} = \begin{pmatrix} A_i & B_i \\ B_i & A_i \end{pmatrix}, \quad (22.A.5)$$

where $M = \{\frac{-d_0}{2}, +\frac{d_0}{2}\}$ is the space of interfaces of the layer, whereas A_i and B_i are 2×2 matrices, whose elements have the forms

$$A_i = \begin{pmatrix} r_i & q_i \\ q_i & k_i \end{pmatrix} \quad \text{and} \quad B_i = \begin{pmatrix} h_i & f_i \\ f_i & e_i \end{pmatrix}, \quad (22.A.6)$$

where r_i, q_i, k_i, h_i, f_i, and e_i are defined by

$$\begin{aligned} r_i &= -C_i\left[\alpha_{i+}A_{i-}\coth\theta_{i+} - \alpha_{i-}A_{i+}\coth\theta_{i-}\right], \\ q_i &= C_iB_i\left[\alpha_{i+}\coth\theta_{i+} - \alpha_{i-}\coth\theta_{i-}\right], \\ k_i &= -C_iD_i\left[\alpha_{i+}A_{i+}\coth\theta_{i+} - \alpha_{i-}A_{i-}\coth\theta_{i-}\right], \\ h_i &= C_i\left[\alpha_{i+}A_{i-}(\sinh\theta_{i+})^{-1} - \alpha_{i-}A_{i+}(\sinh\theta_{i-})^{-1}\right], \\ f_i &= C_iB_i\left[\alpha_{i-}(\sinh\theta_{i-})^{-1} - \alpha_{i+}(\sinh\theta_{i+})^{-1}\right], \\ e_i &= -C_iD_i\left[\alpha_{i-}A_{i-}(\sinh\theta_{i-})^{-1} - \alpha_{i+}A_{i+}(\sinh\theta_{i+})^{-1}\right], \end{aligned} \quad (22.A.7)$$

with $A_{i\pm} = (q_Y^2 - q_0^2\epsilon_{XX}^{(i)} - \alpha_{i\pm}^2)$, $B_i = q_0^2\epsilon_{XY}^{(i)}$, $C_i = (\alpha_{i+}^2 - \alpha_{i-}^2)^{-1}$, $D_i = q_0^2\epsilon_{ZZ}^{(i)}(q_Y^2 - q_0^2\epsilon_{ZZ}^{(i)})^{-1}$, and $\theta_{i\pm} = \alpha_{i\pm}d_i$.

3. The expressions of the amplitude transmitted (t_{SS}, t_{PS}) and reflected coefficients (r_{SS}, r_{PS}), associated to a pure incident (S) wave are given by

$$
\begin{aligned}
r_{ss} &= -\left(2\alpha d_{11}+1\right), \\
r_{ps} &= -\left(2\alpha d_{21}\right)\frac{B_s}{B_p}, \\
t_{ss} &= -\left(2\alpha d_{13}\right), \\
t_{ps} &= -\left(2\alpha d_{23}\right)\frac{B_s}{B_p},
\end{aligned}
\tag{22.A.8}
$$

where $\alpha = \left(k_{\|}^2 - \frac{\omega^2\epsilon_s}{c^2}\right)^{\frac{1}{2}}$ and $k_{\|}$ is the wave vector parallel to the (X, Y) interfaces, ω is the frequency of the optical wave, c the speed of light in vacuum, and ϵ_s the permittivity of the substrate. Further, $B_s = 1$ is the amplitude of the S wave and B_p $\left(B_p = \frac{-i\alpha c}{\omega\sqrt{\epsilon_s}}\right)$ is the amplitude of the P input wave, while d_{11}, d_{21}, d_{13}, and d_{23} are the elements of the truncated matrix of the finite system inserted between the two substrates sketched in Fig. 22.1.

4. The expressions of the amplitudes of transmitted (t_{PP}, t_{SP}) and reflected (r_{PP}, r_{SP}) coefficients, when a pure TM (P) wave is incident, are given by

$$
\begin{aligned}
r_{pp} &= \left(\frac{2q_0^2\epsilon_s}{\alpha}d_{22}-1\right), \\
r_{sp} &= \left(\frac{2q_0^2\epsilon_s}{\alpha}d_{12}\right)\frac{B_p}{B_s}, \\
t_{sp} &= \left(\frac{2q_0^2\epsilon_s}{\alpha}d_{14}\right)\frac{B_p}{B_s}, \\
t_{pp} &= \left(\frac{2q_0^2\epsilon_s}{\alpha}d_{24}\right),
\end{aligned}
\tag{22.A.9}
$$

where d_{12}, d_{22}, d_{14}, and d_{24} are the elements of the truncated matrix of the finite system.

References

[1] J.D. Joannopoulos, S.G. Johnson, J.N. Winn, R.D. Meade, Photonic Crystals: Molding the Flow of Light, second ed., Princeton University Press, 2008.

[2] I.A. Sukhoivanov, I.V. Guryev, Photonic Crystals: Physics and Practical Modeling, Springer Series in Optical Sciences, 2009.

[3] N.W. Ashcroft, N. Mermin, Solid State Physics, Brooks Cole, 1976.

[4] H.G. Winful, Tunneling time, the Hartman effect, and superluminality: a proposed resolution of an old paradox, Phys. Rep. 436 (2006) 1–69.
[5] T. Nakanishi, K. Sugiyama, M. Kitano, Demonstration of negative group delays in a simple electronic circuit, Am. J. Phys. 70 (2002) 1117.
[6] J.N. Munday, R.H. Henderson, Superluminal time advance of a complex audio signal, Appl. Phys. Lett. 85 (2004) 503.
[7] J.N. Munday, W.M. Robertson, Negative group velocity pulse tunneling through a coaxial photonic crystal, Appl. Phys. Lett. 81 (2002) 2127.
[8] S. Longhi, M. Marano, P. Laporta, M. Belmonte, P. Crespi, Experimental observation of superluminal pulse reflection in a double-Lorentzian photonic band gap, Phys. Rev. E 65 (2002) 045602.
[9] S. Longhi, Superluminal pulse reflection in asymmetric one-dimensional photonic band gaps, Phys. Rev. E 64 (2001) 037601.
[10] E.H. El Boudouti, N. Fettouhi, A. Akjouj, B. Djafari-Rouhani, A. Mir, J.O. Vasseur, L. Dobrzynski, J. Zemmouri, J. Appl. Phys. 95 (2004) 1102.
[11] W.M. Robertson, J. Pappafotis, P. Flannigan, Sound beyond the speed of light: measurement of negative group velocity in an acoustic loop filter, Appl. Phys. Lett. 90 (2007) 014102.
[12] W.M. Robertson, J. Ash, J.M. McGaugh, Breaking the sound barrier: tunneling of acoustic waves through the forbidden transmission region of a one-dimensional acoustic band gap array, Am. J. Phys. 70 (2002) 689.
[13] E.H. El Boudouti, T. Mrabti, H. Al-Wahsh, B. Djafari-Rouhani, A. Akjouj, L. Dobrzynski, Transmission gaps and Fano resonances in an acoustic waveguide: analytical model, J. Phys. Condens. Matter 20 (2008) 255212.
[14] C. Liu, Z. Dutton, C.H. Behroozi, L.V. Hau, Observation of coherent optical information storage in an atomic medium using halted light pulses, Nature 409 (2001) 490.
[15] M. Bajcsy, S. Hofferberth, V. Balic, T. Peyronel, M. Hafezi, A.S. Zibrov, V. Vuletic, M.D. Lukin, Efficient all-optical switching using slow light within a hollow fiber, Phys. Rev. Lett. 102 (2009) 203902.
[16] M.S. Shahriar, G.S. Pati, R. Tripathi, V. Gopal, M. Messall, K. Salit, Ultrahigh enhancement in absolute and relative rotation sensing using fast and slow light, Phys. Rev. A 75 (2007) 053807.
[17] S. Chu, S. Wong, Linear pulse propagation in an absorbing medium, Phys. Rev. Lett. 48 (1982) 738.
[18] B. Macke, B. Segard, Propagation of light-pulses at a negative group-velocity, Eur. Phys. J. D 23 (2003) 125.
[19] L.G. Wang, S.Y. Zhu, Superluminal pulse reflection from a weakly absorbing dielectric slab, Opt. Lett. 31 (2006) 2223.
[20] L.G. Wang, H. Chen, S.Y. Zhu, Superluminal pulse reflection and transmission in a slab system doped with dispersive materials, Phys. Rev. E 70 (2004) 066602.
[21] Q. Jiang, Y. Zhang, D. Wang, S. Ahrens, J. Zhang, S. Zhu, Superluminal reflection and transmission of light pulses via resonant four-wave mixing in cesium vapor, Opt. Express 24 (2016) 24451.
[22] J. Arias, A. Sanchez-Merono, M.M. Sanchez-Lopez, I. Moreno, Slow and fast light in three-beam interferometers: theory and experiment, Phys. Rev. A 85 (2012) 033815.
[23] M.M. Sanchez-Lopez, A. Sanchez-Merono, J. Arias, J.A. Davis, I. Moreno, Observation of superluminal and negative group velocities in a Mach–Zehnder interferometer, Appl. Phys. Lett. 93 (2008) 074102.
[24] A. Sanchez-Merono, M.M. Sanchez-Lopez, J. Arias, Fast light in unbalanced low-loss Mach–Zehnder interferometers, Phys. Rev. A 89 (2014) 043828.

[25] A. Mouadili, E.H. El Boudouti, A. Soltani, A. Talbi, B. Djafari-Rouhani, A. Akjouj, K. Haddadi, Electromagnetically induced absorption in detuned stub waveguides: a simple analytical and experimental model, J. Phys. Condens. Matter 26 (2014) 505901.

[26] A. Mouadili, E.H. El Boudouti, A. Soltani, A. Talbi, A. Akjouj, B. Djafari-Rouhani, Theoretical and experimental evidence of Fano-like resonances in simple monomode photonic circuits, J. Appl. Phys. 113 (2013) 164101.

[27] V.S.C. Manga Rao, S. Dutta Gupta, G.S. Agarwal, Atomic absorbers for controlling pulse propagation in resonators, Opt. Lett. 29 (2004) 307.

[28] K. Qian, L. Zhan, L. Zhang, Z.Q. Zhu, J.S. Peng, Z.C. Gu, X. Hu, S.Y. Luo, Y.X. Xia, Group velocity manipulation in active fibers using mutually modulated cross-gain modulation: from ultraslow to superluminal propagation, Opt. Lett. 36 (2011) 2185–2187.

[29] H. Aynaou, E.H. El Boudouti, Y. El Hassouani, A. Akjouj, B. Djafari-Rouhani, J. Vasseur, A. Benomar, V.R. Velasco, Propagation and localization of electromagnetic waves in quasiperiodic serial loop structures, Phys. Rev. E 72 (2005) 056601.

[30] L. Poirier, R.I. Thompson, A. Haché, Impossibility of negative group velocities in a periodic layer structure with or without loss, Opt. Commun. 250 (2005) 258–265.

[31] J.N. Munday, W.M. Robertson, Observation of negative group delays within a coaxial photonic crystal using an impulse response method, Opt. Commun. 273 (2007) 32–36.

[32] M.L.H. Lahlaouti, A. Akjouj, B. Djafari-Rouhani, L. Dobrzynski, M. Hammouchi, E.H. El Boudouti, A. Nougaoui, B. Kharbouch, Theoretical analysis of the density of states and phase times: application to resonant electromagnetic modes in finite superlattices, Phys. Rev. B 63 (2001) 35312.

[33] J. Radovanovic, I. Ilic, P.P. Belicev, V. Milanovic, L. Hadzievski, Comparison of tunneling times in isotropic and anisotropic media, Appl. Phys. A 109 (2012) 997.

[34] Y.S. Dadoenkova, N.N. Dadoenkova, D.A. Korobko, I.O. Zolotovskii, D.I. Sementsov, I.L. Lyubchanskii, Tunnelling of frequency-modulated wavepackets in photonic crystals with amplification, J. Opt. 18 (2016) 015102.

[35] A. Haché, L. Poirier, Anomalous dispersion and superluminal group velocity in a coaxial photonic crystal: theory and experiment, Phys. Rev. E 65 (2002) 036608.

[36] N. Ouchani, A. El Moussaouy, H. Aynaou, Y. El Hassouani, E.H. El Boudouti, B. Djafari-Rouhani, Superluminal and negative delay times in isotropic–anisotropic one-dimensional photonic crystal, J. Appl. Phys. 122 (2017) 183106.

[37] H. Akkus, A.M. Mamedov, Linear and nonlinear optical properties of SbSI: first-principle calculation, Ferroelectrics 352 (2007) 148.

[38] J. Liu, P.D. Han, G.J. Qiao, J.F. Yang, Properties of PC filters in one-dimensional photonic crystals containing defects, J. Intense Pulsed Lasers Appl. Adv. Phys. 1 (2011) 69.

[39] Y.J. Xiang, X.Y. Dai, S.C. Wen, D.Y. Fan, Independently tunable omnidirectional multichannel filters based on the fractal multilayers containing negative-index materials, Opt. Lett. 33 (2008) 1255.

[40] J.-J. Wu, J.-X. Gao, Temperature-dependent optical properties of defect mode in dielectric photonic crystal heterostructure containing a superconducting layer, Mater. Chem. Phys. 171 (2016) 91.

[41] N.N. Dadoenkova, A.E. Zabolotin, I.L. Lyubchanskii, Y.P. Lee, Th. Rasing, One-dimensional photonic crystal with a complex defect containing an ultrathin superconducting sublayer, J. Appl. Phys. 108 (2010) 093117.

[42] C.-Y. Wu, Y.-H. Zou, I. Timofeev, Y.-T. Lin, V.Y. Zyryanov, J.-S. Hsu, W. Lee, Multichannel photonic devices based on tristable polymer-stabilized cholesteric textures, Opt. Express 19 (2011) 7349.

[43] H.T. Wang, I.V. Timofeev, K. Chang, V.Y. Zyryanov, W. Lee, Tunable narrow-bandpass filter based on an asymmetric photonic bandgap structure with a dual-mode liquid crystal, Opt. Express 22 (2014) 15097.
[44] K. Jamshidi-Ghaleh, B. Kazempour, Effect of incident angle and polarization on electrically-tunable defect mode in anisotropic photonic crystals, Appl. Opt. 55 (2016) 4350.
[45] W. Zhang, P. Han, A. Lan, Y. Li, X. Zhang, Defect modes tuning of one-dimensional photonic crystals with lithium niobate and silver material defect, Physica E 44 (2012) 813.
[46] T. Tang, W. Liu, X. Gao, X. He, J. Yang, Band gaps and nonlinear defect modes in one-dimensional photonic crystals with anisotropic single-negative metamaterials, Opt. Laser Technol. 43 (2011) 1016.
[47] L. Penninck, P. De Visschere, J. Beeckman, K. Neyts, Dipole radiation within one-dimensional anisotropic microcavities: a simulation method, Opt. Express 19 (2011) 18558.
[48] I.V. Soboleva, E.M. Murchikova, A.A. Fedyanin, O.A. Aktsipetrov, Second- and third-harmonic generation in birefringent photonic crystals and microcavities based on anisotropic porous silicon, Appl. Phys. Lett. 87 (2005) 241110.
[49] V.A. Belyakov, S.V. Semenov, Optical defect modes at an active defect layer in photonic liquid crystals, J. Exp. Theor. Phys. 118 (2014) 798.
[50] A.H. Gevorgyan, A. Kocharian, G.A. Vardanyan, Accumulation and transmission of the light energy in nonreciprocal multilayer systems, Opt. Commun. 259 (2006) 4551.
[51] Y.C. Yang, C.S. Kee, J.E. Kim, H.Y. Park, J.C. Lee, Y.J. Jeon, Photonic defect modes of cholesteric liquid crystals, Phys. Rev. E 60 (1999) 6852.
[52] A. Lakhtakia, V.C. Venugopal, M.W. McCall, Spectral holes in Bragg reflection from chiral sculptured thin films: circular polarization filters, Opt. Commun. 177 (2000) 57.
[53] A.H. Gevorgyan, M.Z. Harutyunyan, Chiral photonic crystals with an anisotropic defect layer, Phys. Rev. E 76 (2007) 031701.
[54] A.H. Gevorgyan, M.Z. Harutyunyan, Tuning of emission wavelength in chiral photonic crystals with an anisotropic defect layer, J. Mod. Opt. 56 (2009) 1163.
[55] N. Ouchani, D. Bria, B. Djafari-Rouhani, A. Nougaoui, Defect modes in one-dimensional anisotropic photonic crystal, J. Appl. Phys. 106 (2009) 113107.
[56] N. Ouchani, A. El Moussaouy, H. Aynaou, Y. El Hassouani, E.H. El Boudouti, B. Djafari-Rouhani, Optical transmission properties of an anisotropic defect cavity in one-dimensional photonic crystal, Phys. Lett. A 382 (2018) 231.
[57] K. Zhang, J. Fu, L.Y. Xiao, Q. Wu, L.W. Li, Total transmission and total reflection of electromagnetic waves by anisotropic epsilon-near-zero metamaterials embedded with dielectric defects, J. Appl. Phys. 113 (2013) 084908.
[58] W. Zhou, D. Zhao, Y.-C. Shuai, H. Yang, S. Chuwongin, A. Chadha, J.-H. Seo, K.X. Wang, V. Liu, Z. Ma, S. Fan, Progress in 2D photonic crystal Fano resonance photonics, Prog. Quantum Electron. 38 (2014) 1–74.
[59] M. Upadhyay, S.K. Awasthi, S.K. Srivastava, S.P. Ojha, Infrared omni-directional mirror based on one-dimensional birefringent-dielectric photonic crystal, Prog. Electromagn. Res. 25 (2012) 211.
[60] I. Abdulhalim, Omnidirectional reflection from anisotropic periodic dielectric stack, Opt. Commun. 174 (2000) 43.
[61] H. Goldstein, Classical Mechanics, Addison-Wesley, 1957, pp. 107–109.
[62] G.D. Landry, T.A. Maldonado, Complete method to determine transmission and reflection characteristics at a planar interface between arbitrarily oriented biaxial media, J. Opt. Soc. Am. A 12 (1995) 2048.

[63] L. Dobrzynski, Interface response theory of discrete composite systems, Surf. Sci. Rep. 180 (1987) 489;
L. Dobrzynski, Interface response theory of continuous composite systems, Surf. Sci. Rep. 11 (1990) 139.
[64] B. Djafari-Rouhani, L. Dobrzynski, Acoustic resonances of adsorbed wires and channels, J. Phys. Condens. Matter 5 (1993) 139.
[65] P. Yeh, Optical Waves in Layered Media, Wiley, 1988, pp. 201–253.
[66] A.D. Jackson, A. Lande, B. Lautrup, Apparent superluminal behavior in wave propagation, Phys. Rev. A 64 (2001) 044101.
[67] T.E. Hartman, Tunneling of a wave packet, J. Appl. Phys. 33 (1962) 3427.
[68] H.G. Winful, Energy storage in superluminal barrier tunneling: origin of the "Hartman effect", Opt. Express 10 (2002) 1491.
[69] V.I. Fesenko, I.A. Sukhoivanov, S.N. Shulǵa, J.A. Andrade Lucio, in: Vittorio M.N. Passaro (Ed.), Advances in Photonic Crystals, 2013, pp. 79–105.
[70] Y. El Hassouani, E.H. El Boudouti, B. Djafari-Rouhani, H. Aynaou, Sagittal acoustic waves in finite solid-fluid superlattices: band-gap structure, surface and confined modes, and omnidirectional reflection and selective transmission, Phys. Rev. B 78 (2008) 174306.

Multilayered structures based one dimensional photonic crystals for MEMS applications

23

Cécile Ghouila-Houri[a], *Abdelkrim Talbi*[a], *Abdellatif Akjouj*[a], *El Houssaine El Boudouti*[b], *Bahram Djafari-Rouhani*[a], *and Leonard Dobrzyński*[a]
[a]Univ. Lille, CNRS, Centrale Lille, Yncréa ISEN, Univ. Polytechnique Hauts-de-France, UMR 8520 IEMN, Lille, France, [b]LPMR, Department of Physics, Faculty of Sciences, University Mohammed I, Oujda, Morocco

23.1 Introduction

Bragg reflectors are dielectric materials based mirrors commonly used in waveguides like optical fibers. Compared to ordinary metallic mirrors, they present extremely low losses at optical and infrared frequencies. The structure of a Bragg mirror is formed from layers of alternating materials with high and low refractive indices. At each boundary layer, the optical wave is partially reflected. When the wavelength is close to four times the optical thickness (thickness depending on the refractive index) of the layer, the reflected waves interferences are constructive and the sequencing of layers acts as a high quality reflector. The range of wavelengths in which most of the incident intensity is reflected is called the photonic stopband or bandgap. In the limit, in which the reflector contains a very large number of layers, radiation in this range of wavelengths cannot propagate into the structure. Bragg reflectors are therefore multilayered based photonic crystals. They present several ways to be implemented as sensors and MEMS (micro-electro-mechanical systems) devices.

On the one hand, Bragg reflectors can be directly embedded inside an optic fiber, usually called fiber Bragg grating (FBG) [1], which can be functionalized for sensors applications, as described in [2] and [3], both dealing with hydrogen sensors. In both examples, the mirror characteristics are highly dependent on the hydrogen concentration and temperature. Another example of such utilization of Bragg reflectors inside an optical fiber is presented in [4] and deals with the realization of a Pirani vacuum sensor with optical-based detection. Pirani vacuum sensors are thermal sensors which exploit the pressure sensitivity of gas thermal conductivity. In [4], the heat source is around the FBG and the variation of temperature impacts the reflection peak of the FBG.

On the other hand, Bragg reflectors can be used to realize a photonic crystal cavity. The photonic cavity can indeed be elaborated by introducing a defect between two identical Bragg reflectors [5,6] or by associating two different Bragg mirrors [7,8]. The defect will introduce a cavity mode exploitable for sensors applications. This

Photonics. https://doi.org/10.1016/B978-0-12-819388-4.00034-4

chapter will present the temperature sensor described in [5]. In [6], the authors present an original MEMS accelerometer in which the detection of the acceleration is made by measuring the wavelength and intensity of the defect-induced resonance. Indeed, the MEMS accelerometer is based on a mass–spring design, but the mass is mechanically linked to the photonic crystal and its displacement impacts the defect resonance. In [8], the authors associated two different mirrors, a nonmagnetic one and a magnetic one. The resulting resonance peak is thereby used as a magnetic sensor.

In this chapter, we will focus on the modeling of multilayered photonic crystals based on Bragg mirrors, using analytical calculation and finite element method based on COMSOL Multiphysics software. Section 23.2 will focus on the mirror principle and parametric study, and will present an experimental MEMS fabrication. Section 23.3 will present the modeling and study of a one-dimensional photonic crystal cavity and its application as a numerical MEMS pressure sensor and an experimental MEMS temperature sensor.

23.2 Multilayered Bragg mirror: simulations and experiments

23.2.1 Theoretical aspects on dielectric mirrors

We consider a monochromatic electromagnetic wave of vacuum wavelength λ_0 and associated pulsation $\omega = 2\pi c/\lambda$, propagating through a dielectric medium composed of periodic bilayers. The two layers are respectively called medium 1, with a refractive index n_1, and medium 2, with a refractive index n_2. A schematic in Fig. 23.1 illustrates the configuration.

In each medium, the electric field of the electromagnetic wave is given by the real part of:

$$E_j(x) = E_{r,j}e^{-ik_jx} + E_{l,j}e^{+ik_jx} \tag{23.1}$$

where $j = 1, 2$, and the index r (respectively l) corresponds to a wave propagating to the right (respectively left). Two different phenomena are to be taken into account: first, the propagation in each medium and, second, the boundary conditions from one medium to the other.

First, the propagation of the wave in medium I until the next limit induces a phase shift

$$\begin{pmatrix} E_{l,j}^{(1)} \\ E_{r,j}^{(1)} \end{pmatrix} = A_j \begin{pmatrix} E_{l,j}^{\prime(1)} \\ E_{r,j}^{\prime(1)} \end{pmatrix}, \tag{23.2}$$

with A_j being the phase shift matrix

$$A_j = \begin{pmatrix} e^{-i\varphi_j} & 0 \\ 0 & e^{+i\varphi_j} \end{pmatrix}, \tag{23.3}$$

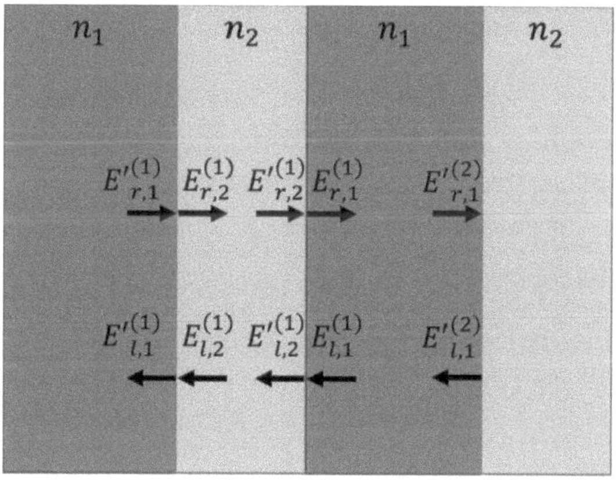

Figure 23.1 Schematic of a dielectric medium composed of periodic bilayers with representation of the electric fields going to the left and to the right before and after the passing through an interface. Index 1 or 2 represents the medium, index l or r represents the direction (left or right) and the number between parentheses corresponds to the pattern, i.e., (1) is the first biinterface pattern, (2) the second biinterface pattern, etc. (inspired from [9]).

where $\varphi_j = k_j l_j = 2\pi n_j l_j/\lambda_0$ is the phase shift of the wave in layer j of length l_j.

Second, the boundary condition states that the electric and magnetic fields are continuous. The conditions written in matrix form are then at the first medium 1–medium 2 interface:

$$\begin{pmatrix} E'^{(1)}_{l,1} \\ E'^{(1)}_{r,1} \end{pmatrix} = \frac{1}{t_{12}} \begin{pmatrix} 1 & r_{12} \\ r_{12} & 1 \end{pmatrix} \begin{pmatrix} E^{(1)}_{l,2} \\ E^{(1)}_{r,2} \end{pmatrix} = P_{12} \begin{pmatrix} E^{(1)}_{l,2} \\ E^{(1)}_{r,2} \end{pmatrix}, \tag{23.4}$$

where P_{12} is the passing matrix, and t_{12} and r_{12} are respectively the transmission and reflection coefficients of the wave, when considering that the incident wave comes from the left ($E^{(1)}_{l,2} = 0$). These two coefficients can be rewritten as functions of the refractive indexes of each medium:

$$t_{12} = \frac{2n_1}{n_1 + n_2}, \quad r_{12} = \frac{n_1 - n_2}{n_1 + n_2}. \tag{23.5}$$

We note that $t_{12} \neq t_{21}$ and $r_{12} = -r_{21}$. Moreover, the conservation of the flow gives $t_{12}t_{21} + r_{12}^2 = 1$.

The different obtained relations allow the definition of the propagation matrix S through the bilayer medium 1–medium 2. As shown in Fig. 23.1, the wave crosses the interface 1–2, then propagates in medium 2, crosses interface 2–1, and propagates in medium 1, leading to the relation

$$\begin{pmatrix} E'^{(1)}_{l,1} \\ E'^{(1)}_{r,1} \end{pmatrix} = S \begin{pmatrix} E'^{(2)}_{l,1} \\ E'^{(2)}_{r,1} \end{pmatrix} = P_{12} A_2 P_{21} A_1 \begin{pmatrix} E'^{(2)}_{l,1} \\ E'^{(2)}_{r,1} \end{pmatrix}. \tag{23.6}$$

Thus, $S = P_{12}A_2P_{21}A_1$ represents the propagation matrix of the wave thought the bilayer pattern.

We can rewrite S as a function of $r = r_{12} = -r_{21}$ and $t^2 = t_{12}t_{21}$:

$$S = \frac{1}{t^2}\begin{pmatrix} e^{-i\varphi_1}(e^{-i\varphi_2} - r^{2e^{+i\varphi_2}}) & 2i\sin\varphi_2 r e^{+i\varphi_1} \\ -2i\sin\varphi_2 r e^{-i\varphi_1} & e^{-i\varphi_1}(e^{+i\varphi_2} - r^{2e^{-i\varphi_2}}) \end{pmatrix}. \tag{23.7}$$

Considering then that the wave is propagating through N bilayers, the propagation matrix is equal to S^N. The transmission and reflection probabilities are given assuming that the wave comes from the left, that is to say, $E'^{(N)}_{l,1} = 0$. We can then deduce the coefficients r_N and t_N and probabilities R_N and T_N of reflection and transmission:

$$r_N = \frac{(S^N)_{12}}{(S^N)_{22}} \quad \text{and} \quad R_N = |r_N|^2, \tag{23.8}$$

$$t_N = \frac{1}{(S^N)_{22}} \quad \text{and} \quad T_N = |t_N|^2. \tag{23.9}$$

The phase shifts φ_1 and φ_2 introduced by the bilayer have an impact on R_N. Calculating it for different values of N, we can conclude that when N tends to infinity (infinite periodicity), the probability of reflection R_N tends to 1 when $\varphi_1 + \varphi_2 = \pi[2\pi]$. We consider here the quarter-wave case with $\varphi_1 = \varphi_2 = \pi/2[\pi]$ which can be rewritten as $n_1l_1 + n_2l_2 = \frac{p\lambda}{2}$, $p \in \mathbb{N}$.

There is then a Bragg reflection when the phase accumulated by the wave on the periodic structure is equal to π. With this condition on the phase, we find the more general form of the Bragg reflection: $2d\sin\theta = n\lambda_0$, considering here that $d = n_1l_1 + n_2l_2$ and that there is reflection on a pattern perpendicular to the wave propagation direction ($\theta = \pi/2$).

With this hypothesis and $\varphi_1 = \varphi_2 = \frac{\pi}{2}[\pi]$, layers 1 and 2 are quarter-wavelength layers, that is to say,

$$n_1l_1 = n_2l_2 = (2p+1)\frac{\lambda_0}{4}. \tag{23.10}$$

In this case, S can be simplified as

$$S = -\frac{1}{t^2}\begin{pmatrix} 1+r^2 & 2r \\ 2r & 1+r^2 \end{pmatrix}, \tag{23.11}$$

and S^N can be rewritten as

$$S^N = (-1)^N \frac{1}{t^{2N}}\begin{pmatrix} (1+r)^{2N} + (1-r)^{2N} & (1+r)^{2N} - (1-r)^{2N} \\ (1+r)^{2N} - (1-r)^{2N} & (1+r)^{2N} + (1-r)^{2N} \end{pmatrix}. \tag{23.12}$$

We can then deduce the probability for reflection

$$R_N = \left| \frac{(1+r)^{2N} - (1-r)^{2N}}{(1+r)^{2N} + (1-r)^{2N}} \right|^2 = \left| \frac{1-(n_2/n_1)^{2N}}{1+(n_2/n_1)^{2N}} \right|^2, \tag{23.13}$$

where we consider arbitrarily that $n_2 < n_1$.

In the case where the number of bilayers N becomes high, we see that the probability of reflection tends towards 1 forming a dielectric mirror, called a Bragg mirror.

23.2.2 Finite element method simulations of multilayered Bragg mirror

This chapter is devoted to the experimental investigation of multilayered photonic crystal for MEMS applications. Therefore, the numerical simulations will be based on experimental targets and possibilities. The equipment and materials available in the IEMN (Institute of Electronics Microelectronics and Nanotechnology) cleanroom facilities led us to realize an SiO_2/Si_3N_4 Bragg mirror. We targeted wavelength $\lambda_0 =$ 670 nm, because given the refractive index of silicon dioxide ($n = 1.45$) and silicon nitride ($n = 2.05$), the thicknesses to consider are respectively 116 nm for silicon dioxide and 82 nm for silicon nitride. Both materials can easily be deposited on a silicon wafer, these thicknesses are feasible in microfabrication, and making several bilayers of them does not create too many constraints in the mirror. Indeed, 6 bilayers lead to a total mirror thickness of less than 600 nm that is correct for avoiding stress in the film. The calculations will thereby have these as parameter's basis. The mirrors presented are all deposited on a silicon substrate ($n = 3.48$).

23.2.2.1 Method

In this section, we present finite element method simulations done on Bragg mirrors. We used COMSOL Multiphysics software. Two COMSOL physics modules can be used, namely the Ray Optic geometrical optics module and the Wave Optics module.

The two have their advantages and drawback as we will detail in this section. As shown in Fig. 23.2, for a given mirror configuration, the simulations using both modules give the same result.

In the Ray optics module, the geometry is defined as a 3D bilayers cylinder, as shown in Fig. 23.3(a). In the physics part of COMSOL Multiphysics, we defined one layer with the medium refractive index of air and the other with that of silicon. The physic principle resolves ray tracing equations:

$$\begin{cases} \dfrac{d\boldsymbol{q}}{dt} = \dfrac{\partial \omega}{\partial \boldsymbol{k}}, \\ \dfrac{d\boldsymbol{k}}{dt} = \dfrac{\partial \omega}{\partial \boldsymbol{q}}, \end{cases} \tag{23.14}$$

where $\boldsymbol{q}$ and $\boldsymbol{k}$ are respectively the ray position vector and the wavevector, and ω the angular frequency.

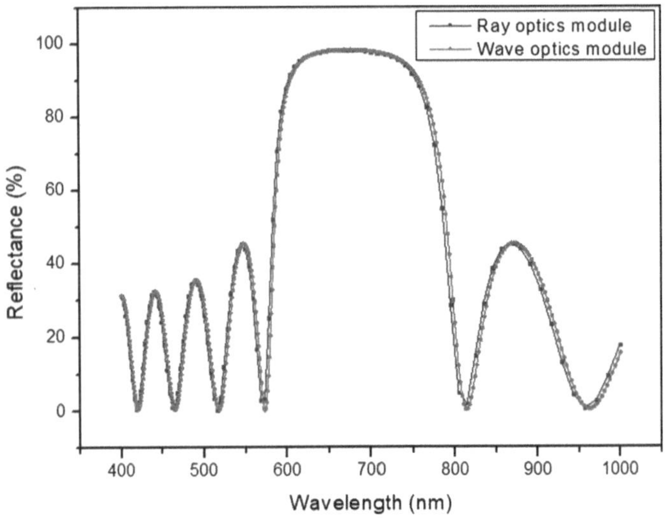

Figure 23.2 6 SiO_2/Si_3N_4 bilayer Bragg mirror reflectance calculation using two different COMSOL modules.

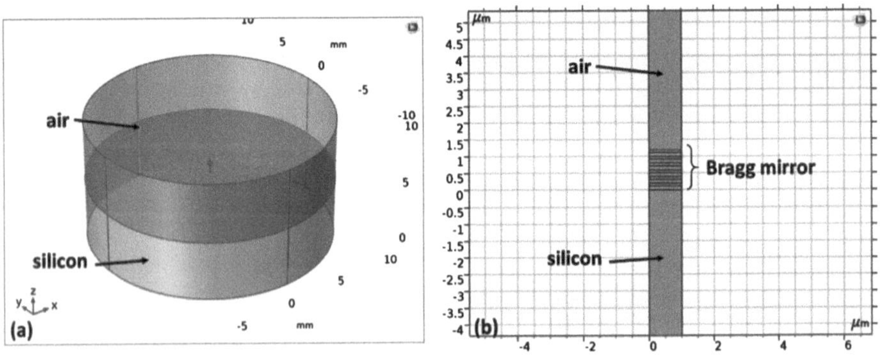

Figure 23.3 Model geometry (a) in Ray Optics module and (b) in Wave Optics module.

The mirror is modeled using "Material Discontinuity" feature on the Ray Optics model and does not appear in the geometry section of the model.

On the contrary, in the Wave Optics module, the mirror is defined in the geometry section. In Fig. 23.3(b), one can see the layer-by-layer defined Bragg mirror and the two media above (air) and below (silicon). For computation time issues, we chose a 2D configuration, whereas in Ray Optics we used a 3D configuration. This point will be explained later on. Periodic limit conditions allow to consider the structure as infinite in the horizontal direction.

The physics resolves electromagnetic wave propagation from Maxwell's equations:

$$\nabla \times (\nabla \times \boldsymbol{E}) - k_0^2 \epsilon_r \boldsymbol{E} = \boldsymbol{0} \tag{23.15}$$

where $\boldsymbol{E}$ is the electric field, ϵ_r the relative permittivity, and k_0 the wavevector. In this work, $\boldsymbol{E}$ is represented by a monochromatic progressive plane wave.

Fig. 23.2 demonstrated that both modules gave the same result, as regards the reflectance result of such SiO_2/Si_3N_4 Bragg mirror on silicon. However, they do not calculate the same quantities and do not resolve the same equations. Consequently, the two simulations have their own advantages and drawbacks. Indeed, the calculation time is very different when using one module or the other. To get the reflectance spectrum from Fig. 23.2, the Ray Optics module needed only 1 second whereas the same result took 2 hours, 7 minutes, and 20 seconds in Wave Optics module.

However, the Wave Optics module allows a better understanding of the physical principles at stake in this configuration. Indeed, we can not only measure the total reflectance but also the absorbance and transmittance, as shown in Fig. 23.4(a), following the relation $T + R + A = 1$. In this particular model where the refractive index is real, the absorbance (in red [light gray in print version] in Fig. 23.4(a)) $A = 0$. Moreover, the amplitude of the electric field can be visualized for the calculated frequencies and, for instance, at $f = 450$ THz, a frequency in the middle of the mirror bandwidth, Fig. 23.4(b) displays the total reflection of the electric field in the air and the rapid attenuation in the top of the mirror.

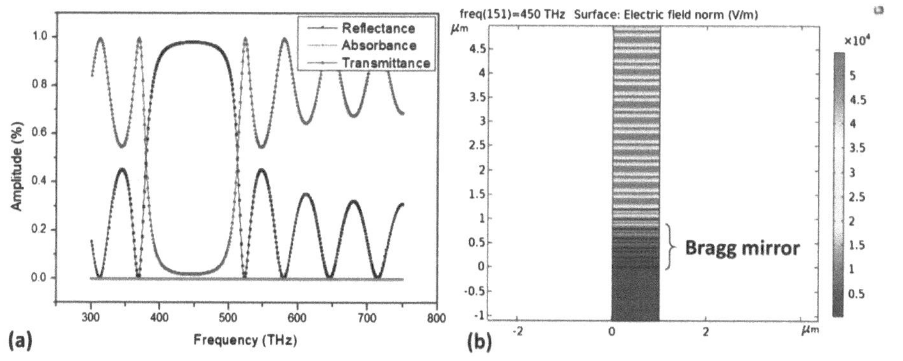

Figure 23.4 (a) Reflectance, absorbance, and transmittance of the Bragg mirror calculated in Wave Optics; (b) Visualization of the electric field amplitude at $f = 450$ THz, which is in the bandwidth of the mirror.

Therefore, in the following results, calculations were realized using the Ray Optics module when they focus on the spectra variation with different parameters (parametric studies). The utilization of the Wave Optics module was reserved for physics related to the discussion of better understanding of the physical phenomenon.

23.2.2.2 Parameters' impact

This section presents some parametric studies realized using the Ray Optics module for rapid computations. The Bragg mirror is highly dependent on the number of bilayers of low/high refractive indexes. As shown in Fig. 23.5, increasing the number of bilayers improves the mirror reflectance towards 100% but shortens the photonic stopband, which corresponds to the width between the two first zeros of reflection

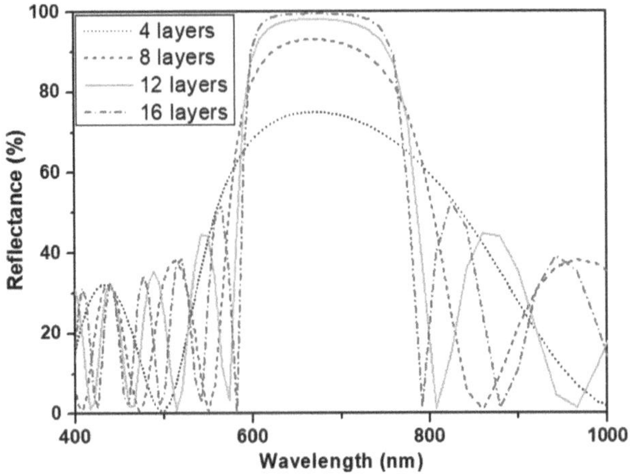

Figure 23.5 Impact of the number of layers on the Bragg mirror response.

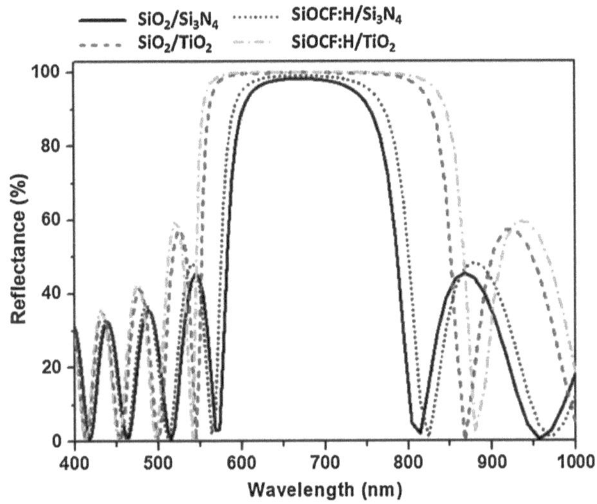

Figure 23.6 COMSOL Simulations of Bragg Mirror with different materials.

around the central wavelength. Considering practical aspects, a compromise choice needs to be done considering the following parameters: the photonic stopband width, the reached maximum reflectance, and the total thickness of the device which depends on the number of layers.

Evidently, the materials refractive indexes have also a great impact on the mirror behavior. Considering a 6 bilayer basis structure for the Bragg mirror, we performed the calculation presented in Fig. 23.6. One can see the impact of different refractive indexes materials, by replacing SiO_2 ($n = 1.45$) by SiOCF:H ($n = 1.38$) [10] as low-index materials and replacing Si_3N_4 ($n = 2.05$) by TiO_2 ($n = 2.56$) as high index materials. SiOCF:H (resp. TiO_2) has a lower (resp. higher) refractive index than SiO_2

(resp. Si_3N_4). By increasing the difference between the refractive indexes of the low and high index materials, one can widen the photonic stopband.

23.2.3 Experimental SiO_2/Si_3N_4 Bragg mirror

We processed a Bragg mirror based on silicon dioxide (SiO_2) and silicon nitride (Si_3N_4) in the IEMN cleanroom. We targeted a mirror with $\lambda_0 = 670$ nm. As mentioned before, given the refractive index of silicon dioxide ($n = 1.45$) and silicon nitride ($n = 2.05$), the targeted thicknesses were respectively 116 nm for silicon dioxide and 82 nm for silicon nitride.

The process was performed on a 3 in. P-doped silicon wafer. Using plasma enhanced chemical vapor deposition (PECVD), layers of SiO_2 and Si_3N_4 were successfully deposited alternatingly. We deposited 6 pairs of SiO_2/Si_3N_4 layers, beginning with the low index material, SiO_2, and finishing with the high index material, Si_3N_4.

Fig. 23.7(a) presents a cross-view of the processed mirror, realized using focused ion beam (FIB) technique.

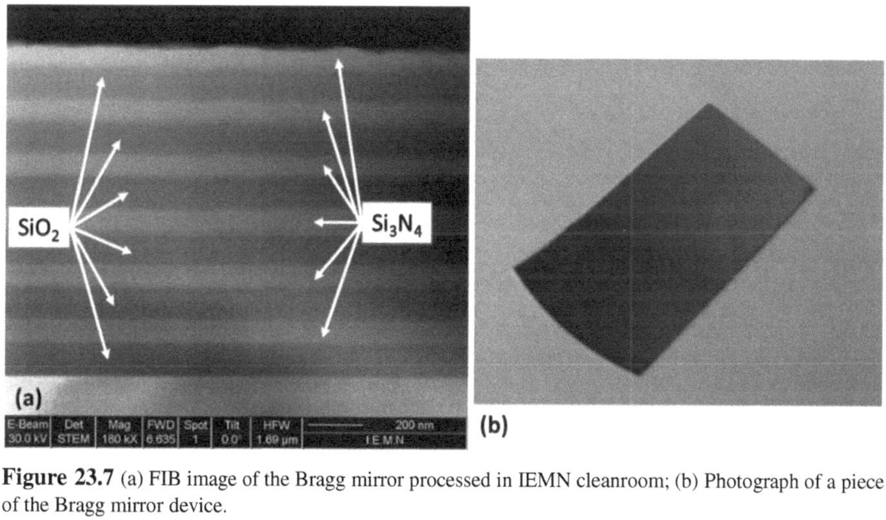

Figure 23.7 (a) FIB image of the Bragg mirror processed in IEMN cleanroom; (b) Photograph of a piece of the Bragg mirror device.

From bottom to top, the SiO_2/Si_3N_4 thicknesses are displayed in Table 23.1. Each layer was measured on the picture highlighting differences between what was targeted and what was effectively deposited.

We performed COMSOL simulation of the Bragg mirror using the measured thicknesses for SiO_2 and Si_3N_4 and compared it to the measured reflectance, as displayed in Fig. 23.8. The two results are close to each other even if the reflectance intensity is lower in the experimental results. The width in wavelength of the Bragg mirror is 220 nm, going from 555 to 775 nm.

Table 23.1 Thicknesses of bilayers constituting the Bragg mirror.

Bilayer from bottom to top	SiO_2 thickness	Si_3N_4 thickness
No. 1	122 nm	79 nm
No. 2	122 nm	76 nm
No. 3	117 nm	78 nm
No. 4	114 nm	74 nm
No. 5	114 nm	76 nm
No. 6	115 nm	76 nm

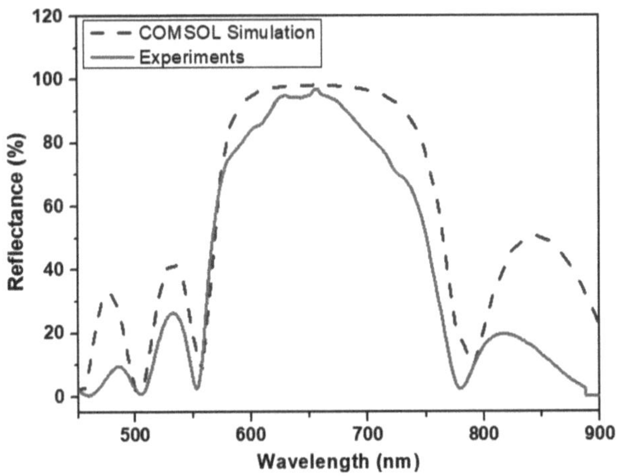

Figure 23.8 Measured reflectance of the realized Bragg mirror and comparison with COMSOL simulation.

23.3 1D photonic cavity crystals for MEMS applications

23.3.1 Theoretical principles

The concept of photonic crystals (PC) exploited in this work consists in introducing a defect inside a Bragg reflector lattice in order to create an optical cavity. In the basic design, we have therefore two mirrors separated by the defect. The defect breaks the periodicity of the crystal and excites one or several particular wavelengths. Consequently, given the defect characteristics, one or several peaks can appear inside the bandgap of the mirror. Indeed, for particular wavelengths, the energy is attenuated by the first mirror, then it excites the cavity modes and is highly amplified, and finally attenuated by the second mirror. Given the quality factor of the resonance, it results in a transmission peak inside the reflection band of the mirror.

In Fig. 23.9, we computed such a photonic crystal with two 6 bilayer Bragg reflectors and a PDMS layer between them. The 6 bilayer mirrors are the same as those presented in the previous part, made of SiO_2/Si_3N_4. PDMS stands for polydimethylsiloxane and is an elastomeric material with a refractive index of 1.41. Using the wave optics module of COMSOL, the amplitude of the electric field can be studied, as well

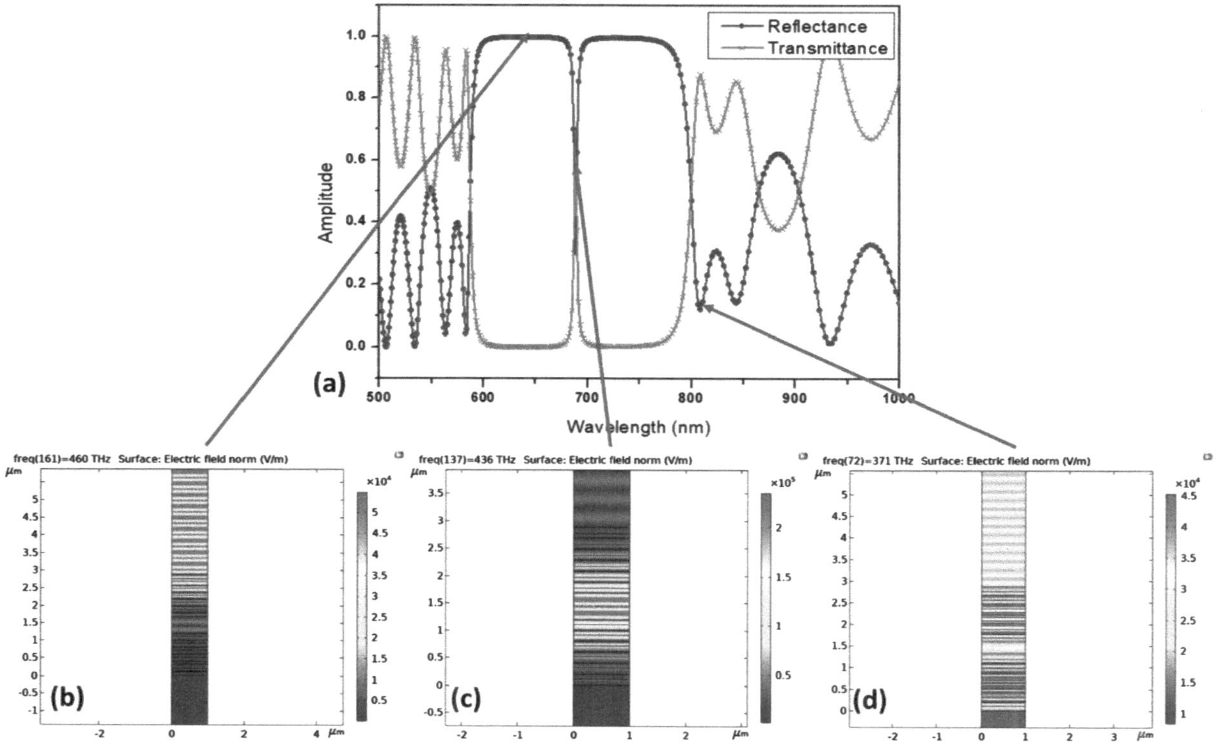

Figure 23.9 (a) Transmittance and reflectance of the 1D photonic crystal from 500 to 1000 nm; (b) electric field at 652 nm in the quasitotal reflection band of the mirror; (c) electric field at the resonance peak; and (d) electric field at a low reflection peak after the mirror bandwidth.

as the transmittance and reflectance spectra. In Fig. 23.9(a), we displayed the transmittance and reflectance from 500 nm to 1 μm. We considered again the real refractive index, therefore $R + T = 1$. The two mirrors allow a maximum reflection bandwidth going from about 600 nm to about 800 nm. The PDMS defect induces a transmission peak (reflection dip) inside the bandwidth, at 688 nm, with a quality factor of 36. Figs. 23.9(b)–(d) display the electric field for different wavelength. Fig. 23.9(b) corresponds to a wavelength of 652 nm which is in the total reflection band, and the total electric energy is in the air medium due to the reflection on the top mirror. Fig. 23.9(d) corresponds to the first transmission peak after the bandgap. The transmission is not equal to 1 (and thus the reflection is not 0) but the figure allows seeing the electric field energy transmitting through the mirrors and the cavity. Finally, Fig. 23.9(c) corresponds to the defect related transmission peak wavelength. It shows that the defect will concentrate the electromagnetic energy at one particular wavelength and allow transmission despite being in the reflection bandwidth.

23.3.2 Numerical 1D PC and parametric study

As presented above, introducing a defect in the photonic reflector impacts its reflection response with one or several transmission peaks appearing in the mirror bandwidth. Given the material and the mirror characteristics, the peaks can present a more or less high quality factor.

In this section, we will investigate how to tune or shift the resonance peak using the defect parameters (thickness and refractive index). Indeed, for designing MEMS sensors with such a principle, the basic idea is to tune the peak frequency by the effect of the targeted measurand. In the following simulations, we chose the PDMS for the defect material to form the cavity because, as an elastomeric component, its thickness and refractive index can be sensitive to external conditions, such as being deformed under pressure or having characteristic change with temperature.

The simulations can be realized using, once again, either the Ray Optics Module on COMSOL Multiphysics or the Wave Optics Module, as shown in Fig. 23.10. As mentioned before, we will prefer the Ray Optics Module for performing parametric studies as the calculation time is much shorter.

In simulations, we still consider Bragg reflectors composed of 6 bilayers of SiO_2/Si_3N_4 on each side of the elastomer defect, with the same thicknesses as before: nitride thickness of about 82 nm and oxide thickness of about 116 nm.

The elastomer thickness is first considered at 500 nm and its refractive index is 1.41 at room temperature. The thickness and the refractive index are the two parameters of the defect that have an impact on the electromagnetic wave propagation.

As shown in Fig. 23.11, increasing the defect thickness changes the wavelength of the transmission peaks and increases their number. However, the thicknesses considered in this simulation are too different to be of interest for applications, e.g., sensors.

On the contrary, in Fig. 23.12, the thickness is changed from 400 to 550 nm by step of 50 nm. In these configurations, there is only one transmission peak in the bandgap, except for 550 nm where a second peak appears at the beginning of the mirror bandwidth (in terms of wavelength). In Fig. 23.12, the shift in the transmission

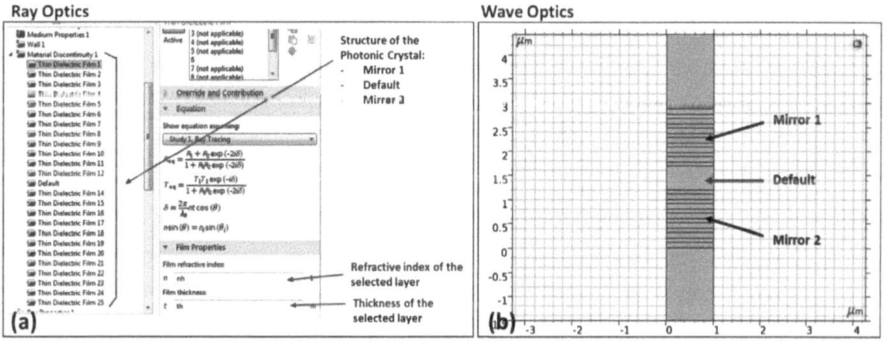

Figure 23.10 1D photonic crystal definition in (a) Ray Optics Module and in (b) Wave Optics Module.

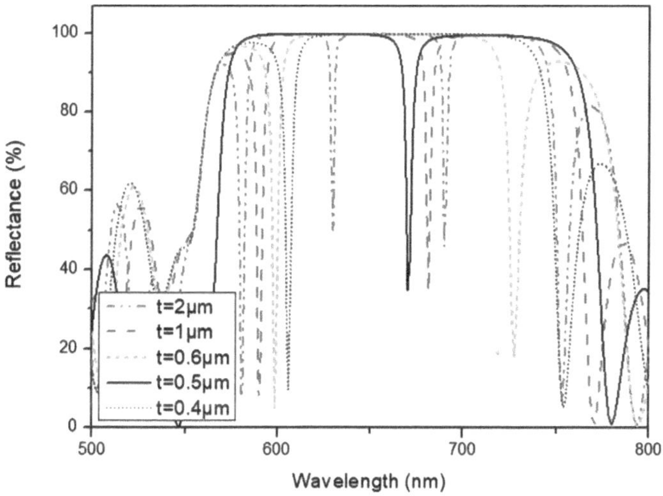

Figure 23.11 Reflectance of the 1D photonic crystal for various defect thicknesses.

peak is measurable when the elastomer thickness changes. One example of application of such a principle is the design of a pressure or strain sensor. When the pressure increases, the elastomer thickness is decreased, and consequently, the transmission peak wavelength shifts toward lower wavelengths.

The other parameter of interest is the refractive index of the defect. External conditions like temperature can change the value of the refractive index. Fig. 23.13 presents the resonance peak shifting depending on the defect refractive index.

23.3.3 Numerical 1D photonic crystal based MEMS pressure sensor

In this section, we will illustrate the previous parametric study for the design and study of an optical cavity based pressure sensor. Indeed, with the previously presented

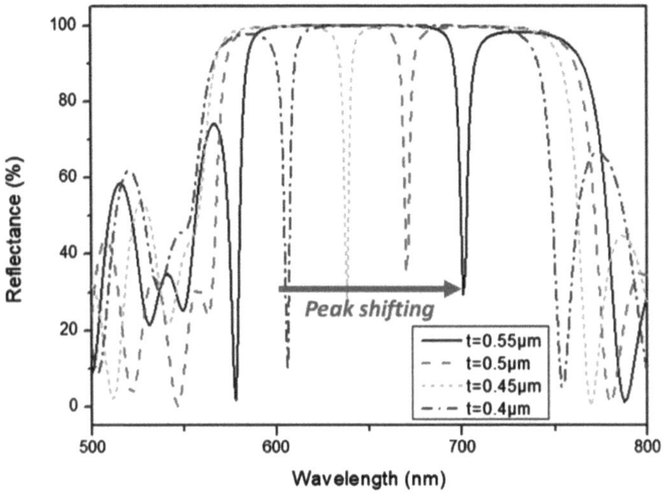

Figure 23.12 Reflectance of the 1D photonic crystal for various but close to defect thicknesses.

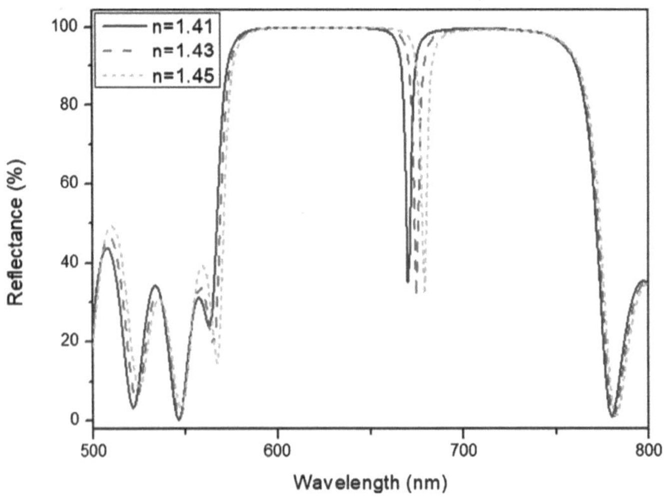

Figure 23.13 Reflectance spectrum for various refractive indexes.

structure with two Bragg reflectors and a PDMS based defect layer, and when pressure is applied on the device surface, the elastomer layer is deformed and its thickness decreases as a function of applied pressure. As a consequence, the resonance wavelength will shift, as described in Fig. 23.11.

For studying such a pressure sensor, the numerical simulation needs to be divided into two parts. As the resonance wavelength is directly dependent on the cavity size, the first study was a mechanical one and aimed at determining the evolution of the elas-

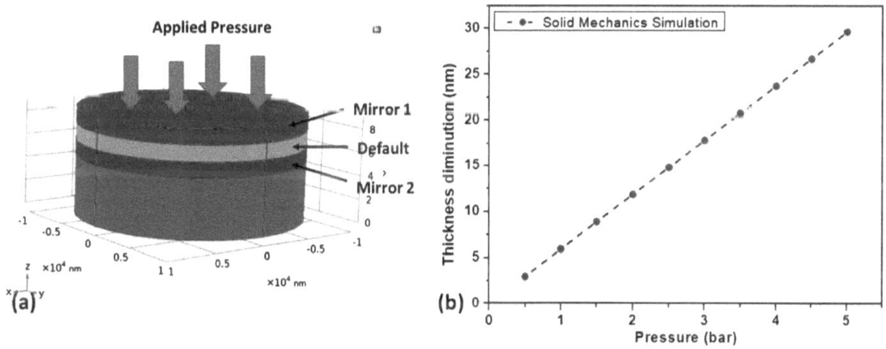

Figure 23.14 (a) Geometry of the structure on Solid Mechanics; (b) simulation results displaying the amplitude of the thickness diminution.

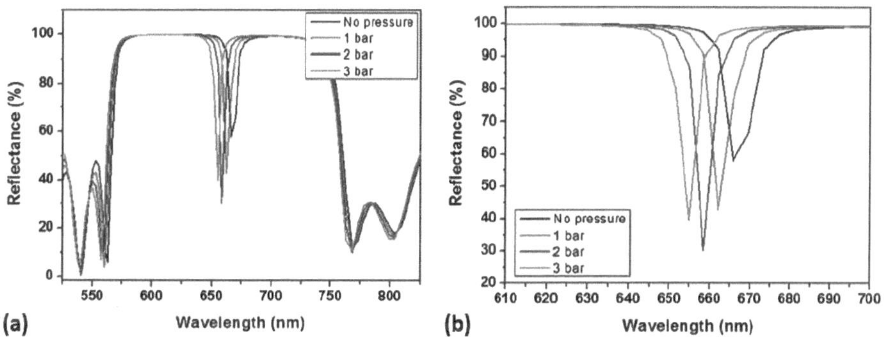

Figure 23.15 Simulation results on Ray Optics combining the results from Solid Mechanics: (a) reflectance versus wavelength for 0 to 3 bar of pressure, (b) zoom on the transmission peak resonance.

tomer thickness when pressure is applied. The mechanical behavior of the structure can be deduced from the structural mechanics equations. Then we can apply optical physics equations, using either the Ray Optics Module or the Wave Optics Module. In this particular example, the resolution of the sensor design is performed associating two different physical modules of COMSOL.

Indeed, we used the COMSOL Module Solid Mechanics for simulated this behavior. We considered a homogeneous pressure applied on the device, as shown by the schematic in Fig. 23.14(a). The obtained results show that the elastomer thickness varies by about 6 nm/bar (Fig. 23.14(b)).

The value was introduced into the photonic crystal model and we performed the optical simulations in COMSOL Ray Optics to determine the resonance shift under pressure. Figs. 23.15(a) and 23.15(b) illustrate the simulated reflectance device for pressure from 0 to 3 bar. The simulated sensitivity is around −3.7 nm/bar (minus sign of the shift to the left of the wavelength peak).

23.3.4 Experimental 1D photonic crystal based MEMS temperature sensor

Experiments were conducted to realize this photonic crystal structure. We deposited Bragg reflectors on silicon wafers, each composed of 6 bilayers of Si_3N_4/SiO_2 as presented in Fig. 23.7. The process is the same as used before, that is to say, PECVD deposition process in cleanroom.

On one wafer with mirror, we deposited Solaris Elastomer using spin coating. However, given the viscosity of the substance and despite several tests on deposition parameters, a thickness of 3.3 μm was the smallest we got, as shown by the FIB picture of Fig. 23.16(a). The second mirror was stuck on the top of the elastomer before the elastomer went dry. We then realized deep silicon etching to remove the top mirror from silicon, as shown by Fig. 23.16(b).

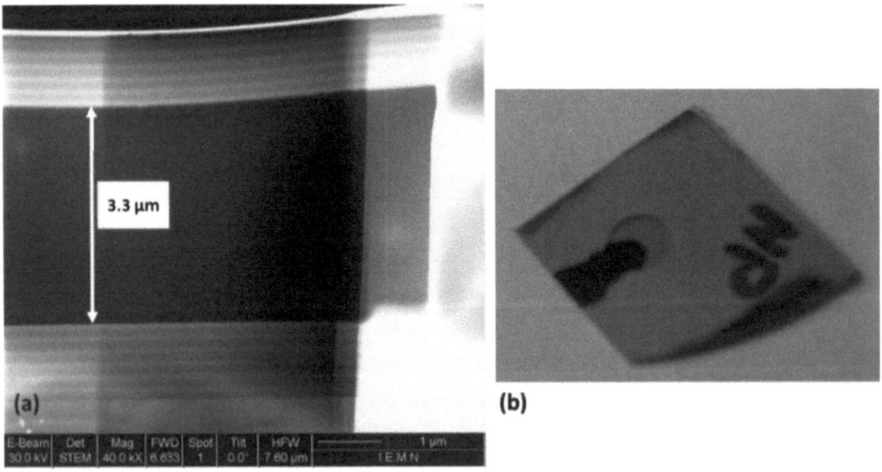

Figure 23.16 (a) FIB picture of the experimental photonic crystal; (b) photograph of the device.

The high thickness of the PDMS-based defect had an impact on the measured reflectance as shown in Fig. 23.17. Indeed, contrary to simulation results, the measured reflectance displays multiple transmission peaks due to the 3.3 μm thickness.

In this work, originated from the basic ideas presented before, we propose to exploit the change in the refractive index of the defect with the temperature. Fortunately, the Solaris Elastomer used in the experiments is indeed sensitive to temperature. Therefore, we measured the reflectance while changing the whole device temperature using a Peltier device as calibration system. Figs. 23.18(a) and 23.18(b) show experimental results under temperature variations. The resonance peaks' wavelengths increase with temperature.

Measurements under temperature variations demonstrated the shift of the resonance peaks' wavelengths. Considering the initial peak wavelength at 723 nm as shown in Fig. 23.18(b), we obtained an experimental sensitivity close to 3.5 nm/°C (Fig. 23.19).

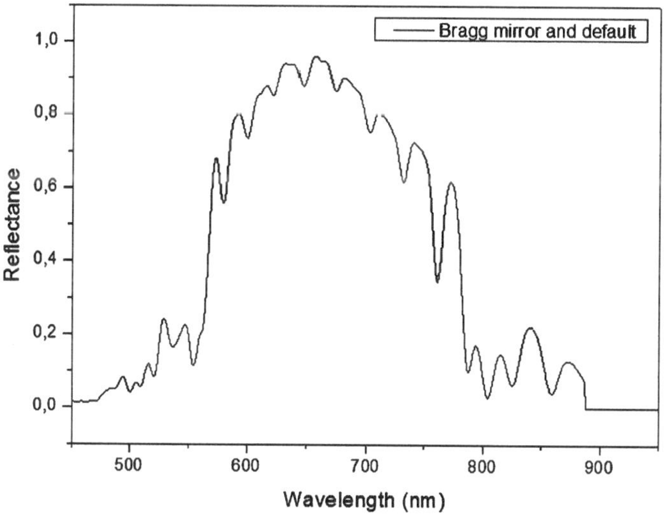

Figure 23.17 Measured reflectance of the fabricated photonic crystal.

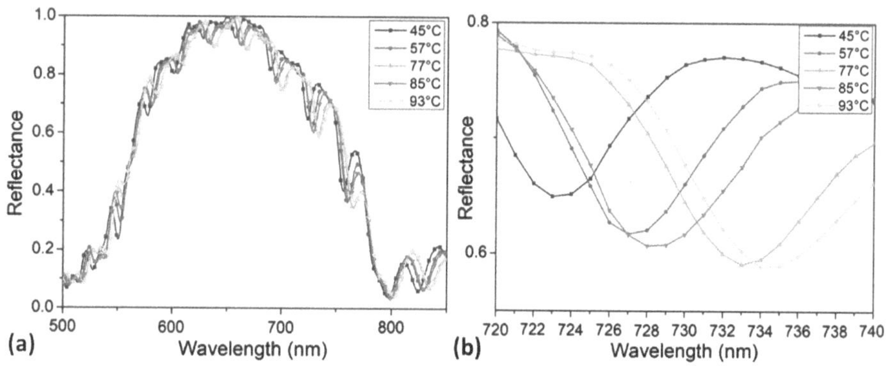

Figure 23.18 (a) Experimental device reflectance with temperature variation; (b) zoom on a wavelength resonance [5].

23.4 Conclusions

In conclusion, this chapter compared two different methods for calculating multilayered 1D photonic crystals structures using COMSOL Multiphysics and demonstrated the utilization of such structures for the design of micro-electro-mechanical systems. We first studied the multilayered Bragg reflectors that produce a transmission photonic stopband in a given bandwidth. Theoretical aspects were completed by simulation studies and parametric characterizations. An experimental SiO_2/Si_3N_4 Bragg mirror was realized using conventional micromachining techniques and characterized. Then we studied a 1D photonic crystal cavity formed by the introduction of a different layer between two reflectors. The defect induces one or several transmission peaks

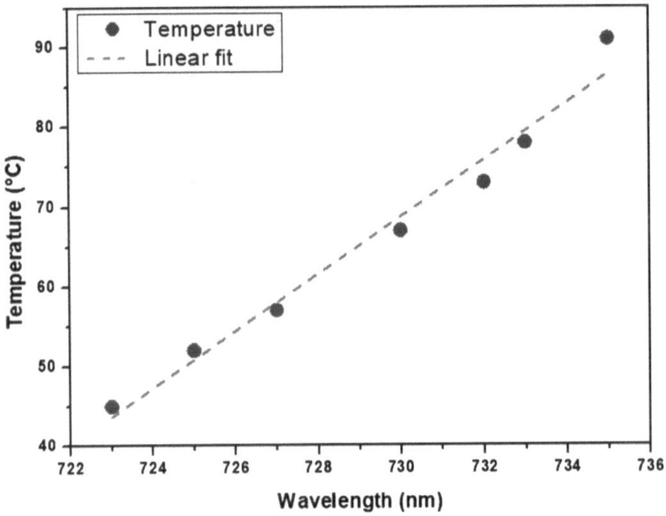

Figure 23.19 Wavelength shift with temperature for the initial peak at 723 nm, and linear fit.

inside the mirror bandwidth. Given the physical properties of the defect, principally its thickness and its refractive index, the transmission peaks can be shifted. Such a structure was studied for MEMS sensor applications as external conditions like pressure, strain, or temperature can affect the defect. A numerical pressure sensor was presented associating a COMSOL simulation on Solid Mechanics and another one on Optical Physics module. It numerically demonstrated a sensitivity of 3.7 nm/bar. Then, an experimental temperature sensor based on a PDMS defect between two SiO_2/Si_3N_4 Bragg mirrors illustrated this concept, and the peak shifting demonstrated a sensitivity to temperature of 3.5 nm/°C.

Acknowledgments

The authors would like to thank J.-C. Gerbedoen, R. Viard, A. Merlen, P. Pernod, for their fruitful contributions to the original paper from which this chapter is inspired.

References

[1] A. Othonos, Fiber Bragg gratings, Rev. Sci. Instrum. 68 (1997) 4309–4341.

[2] Y. Zhang, H. Peng, X. Qian, Y. Zhang, G. An, Y. Zhao, Recent advancements in optical fiber hydrogen sensors, Sens. Actuators B, Chem. 244 (2017) 393–416.

[3] A. Hosoki, M. Nishiyama, H. Igawa, A. Seki, Y. Choi, K. Watanabe, A surface plasmon resonance hydrogen sensor using $Au/Ta_2O_5/Pd$ multi-layers on hetero-core optical fiber structures, Sens. Actuators B, Chem. 185 (2013) 53–58.

[4] B. McMillen, C. Jewart, M. Buric, K.P. Chen, Y. Lin, W. Xu, Fiber Bragg grating vacuum sensors, Appl. Phys. Lett. 87 (2005) 234101.

[5] C. Ghouila-Houri, J.-C. Gerbedoen, R. Viard, A. Talbi, A. Merlen, P. Pernod, Joint International Laboratory LIA Lics/Lemac, Design and elaboration of 1D photonic crystal cavity based on highly flexible elastomer thin layer for sensors applications, Proc. Eng. 120 (2015) 744–747.

[6] A. Sheikhaleh, K. Abedi, K. Jafari, R. Gholamzadeh, Micro-optoelectromechanical systems accelerometer based on intensity modulation using a one-dimensional photonic crystal, Appl. Opt. 55 (2016) 8993–8999.

[7] T. Goto, A.V. Dorofeenko, A.M. Merzlikin, A.V. Baryshev, A.P. Vinogradov, M. Inoue, A.A. Lisyansky, A.B. Granovsky, Optical Tamm states in one-dimensional magnetophotonic structures, Phys. Rev. Lett. 101 (2008) 113902.

[8] T. Goto, A.V. Baryshev, M. Inoue, A.V. Dorofeenko, A.M. Merzlikin, A.P. Vinogradov, A.A. Lisyansky, A.B. Granovsky, Tailoring surfaces of one-dimensional magnetophotonic crystals: optical Tamm state and Faraday rotation, Phys. Rev. B 79 (2009) 125103.

[9] C. Fabre, Miroirs de Bragg pour ondes de matière et apport de la supersymétrie aux potentiels exponentiels, PhD thesis, Université Paul Sabatier - Toulouse III, 2012.

[10] S.G. Yoon, W.J. Park, H. Kim, S.W. Kim, D.H. Yoon, Characterization of low refractive index SiOCF: H films designed to enhance the efficiency of light emission, J. Electroceram. 16 (2006) 469–472.

Index

Kolibri 360 GmbH
Ferdinand-Jühlke-Straße 7
99095 Erfurt, Deutschland
produktsicherheit@kolibri360.de